I0063877

Mathematical Methods for Curves and Surfaces

Innovations in Applied Mathematics

An international series devoted to the latest research in modern areas of mathematics, with significant applications in engineering, medicine, and the sciences.

Mathematical Methods for Curves and Surfaces

Oslo 2000

Edited By

TOM LYCHE
Institutt for informatikk
University of Oslo
Oslo, Norway

LARRY L. SCHUMAKER
Department of Mathematics
Vanderbilt University
Nashville, Tennessee

Vanderbilt University Press
Nashville

First Edition 2001

05 04 03 02 01 5 4 3 2 1

Mathematical methods for curves and surfaces : Oslo 2000 / edited by Tom Lyche, Lar
L. Schumaker.
p. cm — (Innovations in applied mathematics)
ISBN 0-8265-1378-6 (alk. paper)
1. Curves, Algebraic—Data processing—Congresses. 2. Surfaces—Data processing-
Congresses. 3. Computer-aided design—Congresses I. Lyche, Tom. II. Schumaker,
Larry L., 1939- III. Series.

QA567 .M345 2001
516.3'5—dc21

200102618

CONTENTS

PREFACE

During the week of June 29–July 4, 2000, an international conference on *Mathematical Methods for Curves and Surfaces* was held in Oslo, Norway. The conference was attended by 126 mathematicians from 24 countries. There were 9 one hour survey talks, 12 talks in 3 minisymposia, and 65 contributed research talks. This volume contains papers based on the invited lectures, along with 39 full length refereed research papers. Unfortunately, there was not enough space to include all of the papers presented at the conference.

Topics discussed in this book include animation, beta continuity, bivariate splines, blossoming, Bézier curves, conic sections, data reduction, digital shapes, flow surfaces, implicit curves and surfaces, implicitization, mesh fairing, moving least squares, nurps, offsets, pedal curves and surfaces, radial basis functions, rational surfaces, refinable functions, ruled surfaces, scattered data fitting, subdivision, surface intersection, surface patches, tension methods, triangulated surfaces, and wavelets.

We would like to thank SINTEF, The Research Council of Norway, and the University of Oslo for their support of the conference. We would also like to thank our co-organizers Knut Mørken and Ewald Quak.

Nashville, March 25, 2001

CONTRIBUTORS and PARTICIPANTS

Numbers in parentheses indicate pages on which authors' contributions begin.

RON AGMON, *Cimatron Ltd, 11 Gush Etzion St., 54030 Givat Shmuel, Israel* [rona@cimatron.co.il]

SUNG JOON AHN (1), *Fraunhofer Institute IPA, Nobelstr. 12, 70569 Stuttgart, Germany* [sja@ipa.fhg.de]

GUDRUN ALBRECHT (15), *Zentrum Mathematik, TU Munich, Arcisstr. 21, 80290 Munich, Germany* [albrecht@ma.tum.de]

GIAMPIETRO ALLASIA (25), *Department of Mathematics, University of Turin, Via Carlo Alberto 10, 10123 Torino, Italy* [allasia@dm.unito.it]

BRIAN BARSKY (35), *University of California, Berkeley, Computer Science Division #1776, 94720-1776 Berkeley, USA* [barsky@cs.berkeley.edu]

HANS-ULRICH BECKER, *CoCreate Software GmbH, Posener Strasse 1, 71065 Sindelfingen, Germany* [hans-ulrich_becker@hp.com]

RENATA BESENGHI (25), *Department of Mathematics, University of Turin, Via Carlo Alberto 10, 10123 Torino, Italy* [besenghi@dm.unito.it]

HENNING BIERMANN, *New York University, 715 Broadway, 12th Floor, Room 1206, 10003 New York, NY, USA* [biermann@mrl.nyu.edu]

KARL-HEINZ BRAKHAGE, *Geometry and Applied Mathematics, RWTH Aachen, Templergraben 55, 52056 Aachen, Germany* [brakhage@igpm.rwth-aachen.de]

PEER-TIMO BREMER (45), *University of Hannover, Am Isenbrink 2A, 30926 Seelze, Germany* [tbremer@ucdavis.edu]

TROND BRENNA, *University of Oslo, Institute of Informatics, PO Box 1080, Blindern, 0316 Oslo, Norway* [trondbre@ifi.uio.no]

EMMANUELLE CALCOEN (347), *ESA, 55, rue Rabelais, BP 748, 49007 Angers, cedex, France* [e.calcoen@esa-angers.educagri.fr]

BJØRN CARLIN, *SINTEF Applied Mathematics, P. O. Box 124, Blindern, 0314 Oslo, Norway* [Bjorn.Carlin@math.sintef.no]

JESUS CARNICER (55), *Universidad de Zaragoza, Edificio de Matematicas, planta 1a, 50009 Zaragoza, Spain* [carnicer@posta.unizar.es]

PATRICK CHENIN, *Univ. Joseph Fourier, Grenoble, LMC-IMAG, BP 53, F-38041, Grenoble, cedex 9, France* [Patrick.Chenin@imag.fr]

ELAINE COHEN (183), *University of Utah, 50 S. Central Campus Dr. 3190 MEB, 84112 Salt Lake City, UT, USA* [cohen@cs.utah.edu]

ANDREW CRAMPTON (63), *School of Computing and Mathematics, University of Huddersfield, Huddersfield, UK* [a.crampton@hud.ac.uk]

ISABELLA CRAVERO, *Università di Torino, Dipartimento di Matematica, Via Carlo Alberto 10, 10123 Torino, Italy* [gravero@dm.unito.it]

MARC DANIEL, *ESIL, XAOlab, Campus de Luminy, case postale 925, 13288 Marseille, France* [Marc.Daniel@esil.univ-mrs.fr]

STEFANO DE MARCHI, *University of Udine, Via delle Scienze, 206, 33100 Udine, Italy* [demarchi@dimi.uniud.it]

ALESSANDRA DE ROSSI (25), *Dipartimento di Matematica, Università di Torino, Via Carlo Alberto 10, 10123 Torino, Italy* [derossi@dm.unito.it]

JOHAN DE VILLIERS, *University of Stellenbosch, Department of Mathematics, Private Bag X1, 7602 Matieland, Republic of South Africa* [jmdv@land.sun.ac.za]

TONY DEROSE (73), *Pixar Animation Studios, 1200 Park Ave., Emeryville, CA 94608, USA* [derose@pixar.com]

PAUL DIERCKX (527), *Katholieke Universiteit Leuven, Celestijnenlaan 200 A, 3001 Heverlee, Belgium* [paul.dierckx@cs.kuleuven.ac.be]

TOR DOKKEN (81,103), *SINTEF Applied Mathematics, P. O. Box 124, Blindern, 0314 Oslo, Norway* [Tor.Dokken@math.sintef.no]

SERGE DUBUC (113), *Département de mathématiques et de statistique, Université de Montréal, C.P. 6128, Succursale Centre-ville, Montréal (Québec), Canada H3C 3J7* [dubucs@dms.umontreal.ca]

NIRA DYN (123,135), *Tel Aviv University, Dept. of Mathematics, 69978 Tel Aviv, Israel* [niradyn@math.tau.ac.il]

MORTEN DÆHLEN, *Univ. of Oslo, Institute of Informatics, P. O. Box 1080, Blindern, 0316 Oslo, Norway* [mortend@ifi.uio.no]

SVEN EHRICH, *GSF National Research Center for Environment and Health, Institute of Biomathematics and Biometry, Ingolstädter Landstr. 1, D-85764 Neuherberg, Germany* [ehrich@gsf.de]

GERALD FARIN (545), *Arizona State University, Computer Science and Engineering, 85287-5406 Tempe, AZ, USA* [farin@asu.edu]

MORTEN FIMLAND, *Institute of Informatics, University of Oslo, Postbox 1080 Blindern, 0316 Oslo, Norway* [mortenj@ifi.uio.no]

SUSANNA FISHEL, *Mental Images GmbH & Co. KG, Fasanenstrasse 81, 10623 Berlin, Germany* [office@mental.com]

MICHAEL FLOATER (123), *SINTEF Applied Mathematics, P. O. Box 124, Blindern, 0314 Oslo, Norway* [mif@math.sintef.no]

MARIANO GASCA (55), *Universidad de Zaragoza, Edificio de Matematicas, planta 1a, 50009 Zaragoza, Spain* [gasca@posta.unizar.es]

GILI GOLANI, *Cimatron Ltd., Gush Etzyon 11, 54030 Givat Shmuel, Israel* [gili@cimatron.co.il]

RON GOLDMAN (35,325), *Computer Science Department, Rice University, Houston, TX 77251-1892, USA* [rng@cs.rice.edu]

TIMOTHY GOODMAN (147), *University of Dundee, Department of Mathematics, DD1 4HN Dundee, UK* [tgoodman@maths.dundee.ac.uk]

KARIN GOOSEN, *University of Stellenbosch, Department of Mathematics, Private Bag X1, 7602 Matieland, Republic of South Africa* [karin@goose.sun.ac.za]

ARDESHIR GOSHTASBY (163), *Wright State University, Computer Science and Engineering Dept., 45435 Dayton, OH, USA* [ardy@cs.wright.edu]

JENS GRAVESEN, *Department of Mathematics, Technical University of Denmark, Matematiktorvet, Building 303, DK-2800 Lyngby, Denmark* [J.Gravesen@mat.dtu.dk]

HANS HAGEN (173), *University of Kaiserslautern, Computer Science Department, 67653 Kaiserslautern, Germany* [hagen@informatik.uni-kl.de]

BERND HAMANN (45), *Center for Image Processing and Integrated Computing, Department of Computer Science, 1 Shields Avenue, University of California, Davis, CA 95616-8562, USA* [hamann@cs.ucdavis.edu]

YVON HALBWACHS, *SINTEF Applied Mathematics, Postboks 124, Blindern, N-0314 OSLO, Norway* [Yvon.Halbwachs@math.sintef.no]

DIANNE HANSFORD, *NURBS Depot, 4952 East Mockingbird Lane, 85253 Paradise Valley, AZ, USA* [nurbs@goodnet.com]

MASATAKE HIGASHI, *Toyota Technological Institute, 2-12-1, Hisakata, Tempaku-ku, 4 68-8511 Nagoya, Japan* [higashi@toyota-ti.ac.jp]

ØYVIND HJELLE, *SINTEF Applied Mathematics, P. O. Box 124, Blindern, 0314 Oslo, Norway* [Oyvind.Hjelle@math.sintef.no]

CHIH-CHENG HO (183), *Engineering Geometry Systems, 275 East South Temple, Suite 305, Salt Lake City, UT 84111, USA* [ho@cs.utah.edu]

KLAUS HÖLLIG (195), *Mathematisches Institut A, Universität Stuttgart, Pfaffenwaldring 57, 70569 Stuttgart, Germany* [hollig@mathematik.uni-stuttgart.de]

KAI HORMANN (135), *University of Erlangen-Nuremberg, IMMD IX, Am Weichselgarten 9, 91058 Erlangen, Germany* [hormann@informatik.uni-erlangen.de]

ARMIN ISKE (123,211), *Technische Universität München, Zentrum Mathematik, TU München, Arcisstrasse 21, 80290 München, Germany* [iske@ma.tum.de]

GASPER JAKLIC, *Institute for Mathematics, Physics and Mechanics, Jadranska 19, 1000 Ljubljana, Slovenia* [gasper.jaklic@fmf.uni-lj.si]

MICHAEL JOHNSON, *Kuwait University, P. O. Box 5969, 13060 Safat, Kuwait* [johnson@mcc.sci.kuniv.kw]

BERT JÜTTLER (223), *Johannes Kepler University Linz, Institute of Analysis and Computational Mathematics, Dept. of Applied Geometry, Altenberger Str. 69, 4040 Linz, Austria* [bert.juettler@jk.uni-linz.ac.at]

JOHANNES KAASA, *SimSurgery AS, Forskningsveien 1, 0314 Oslo, Norway* [jka@multimediacapital.no]

KĘSTUTIS KARČIAUSKAS (233), *Vilnius University, Naugarduko 24, LT2600 Vilnius, Lithuania* [kestutis.karciauskas@maf.vu.lt]

HERMANN KELLERMANN (103), *CoCreate, Posener Str. 1, 71065 Sindelfingen, Germany* [Hermann_Kellermann@hp.com]

MOHAMMED KHACHAN, *IRCOM-SIC, Boulevard 3 - Teleport 2 - BP 179, 86960 Futuroscope, France* [khachan@sic.sp2mi.univ-poitiers.fr]

JOERG KIEFER, *University of Kaiserslautern, Gottlieb Daimler Strasse, 67663 Kaiserslautern, Germany* [j_kiefer@informatik.uni-kl.de]

SUN-JEONG KIM (135), *Dept. of Applied Mathematics, School of Mathematical Sciences, Tel Aviv University, Tel Aviv 69978, Israel* [sjkim@math.tau.ac.il]

MARIUS KINTEL, *Systems in Motion, Nedre Vaskegang 6, 0186 Oslo, Norway* [kintel@sim.no]

LEIF KOBBELT (423,445,455), *Max-Planck-Institute for Computer Sciences, Im Stadtwald, 66123 Saarbruecken, Germany* [kobbelt@mpi-sb.mpg.de]

OLIVER KREYLOS (45), *Center for Image Processing and Integrated Computing, Department of Computer Science, 1 Shields Avenue, University of California, Davis, CA 95616-8562, USA* [kreylos@cs.ucdavis.edu]

BERNARD LACOLLE (347), *LMC-IMAG, BP 53, 38041 Grenoble, cedex 9, France* [Bernard.Lacolle@imag.fr]

PIERRE-JEAN LAURENT, *Univ. Joseph Fourier, Grenoble, LMC-IMAG, BP 53, cedex 9, 38041 Grenoble, France* [pjl@imag.fr]

ALAIN LE MEHAUTE, *University of Nantes, 2 rue de la Houssiniere, 44322, cedex 3 Nantes, France* [alm@math.univ-nantes.fr]

BYUNG GOOK LEE (243), *Dongseo University, Dept. of Applied Mathematics, 617-716 Pusan, South Korea* [lbg@dongseo.ac.kr]

DAVID LEVIN (135), *Tel Aviv University, Ramat Aviv, 69978 Tel Aviv, Israel* [levin@math.tau.ac.il]

CHARLES LOOP, *Microsoft Research, One Microsoft Way, 98052 Redmond, WA, USA* [cloop@microsoft.com]

HÉLIO LOPES (253), *Departamento de Matemática, Pontifícia Universidade Católica do Rio de Janeiro, Rua Marquês de São Vicente, 225, Gávea, Rio de Janeiro, RJ, Brazil, CEP:22.453-900* [lopes@mat.puc-rio.br]

ZHAOYING LU (263), *University of Bath, Computing Group, Dept. of Math. Sci., BA2 7AY Bath, UK* [mapzl@bath.ac.uk]

TOM LYCHE (243), *University of Oslo, Institute of Informatics, P. O. Box 1080, Blindern, 0316 Oslo, Norway* [tom@ifi.uio.no]

MARRYAT MA (273), *University of Waterloo, Computer Science Department, 200 University Avenue West, N2L 3G1 Waterloo, Ontario, Canada* [mma@cgl.uwaterloo.ca]

WEIYIN MA, *City University of Hong Kong, 83 Tat Chee Avenue, Kowloon, Hong Kong* [mewma@cityu.edu.hk]

ESMERALDA MAINAR (283), *Departamento de Matemática Aplicada, Universidad de Zaragoza, 50009 Zaragoza, Spain* [esme@posta.unizar.es]

STEPHEN MANN (273), *University of Waterloo, Computer Science Dept, 200 University Ave W, N2L 3G1 Waterloo, Canada* [smann@cgl.uwaterloo.ca]

CARLA MANNI (293), *Dipartimento di Matematica, Universita' degli Studi di Torino, Via Carlo Alberto, 10, 10123 Torino, Italy* [manni@dm.unito.it]

JOHN MASON (63), *University of Huddersfield, School of Comp and Maths, Queensgate, HD3 1DH Huddersfield, UK* [j.c.mason@hud.ac.uk]

Marie-Laurence Mazure, *Univ. Joseph Fourier, Grenoble LMC-IMAG, BP 53, 38041 Grenoble, cedex 9, France* [mazure@imag.fr]

Jean-Louis Merrien (113), *INSA de Rennes, 20 av des Buttes de Coesmes, CS 14315, 35043 Rennes, France* [Jean-Louis.Merrien@insa-rennes.fr]

Jon Anders Mikkelsen, *University of Oslo, SINTEF Applied Mathematics, P. O. Box 124, Blindern, 0314 Oslo, Norway* [jonmi@ifi.uio.no]

Rossana Morandi (315), *Dipartimento di Energetica, Via Lombroso 6/17, 50134 Florence, Italy* [morandi@de.unifi.it]

Geraldine Morin (325), *Computer Science Department, Rice University, Houston, TX 77251-1892, USA* [gege@rice.edu]

Knut Mørken (243), *University of Oslo, Institute of Informatics, P. O. Box 1080 Blindern, 0316 Oslo, Norway* [knutm@ifi.uio.no]

Manuela Neagu (347), *LMC-IMAG, Tour IRMA, BP 53, 38041 Grenoble, France* [manuela.neagu@imag.fr]

Marian Neamtu (355), *Department of Mathematics, Vanderbilt University, Nashville, TN 37240, USA* [neamtu@math.vanderbilt.edu]

Trygve Kastberg Nilssen, *University of Bergen, Johannes Bruns gate 12, 5008 Bergen, Norway* [trygvekn@mi.uib.no]

Günther Nürnberger (393), *Institut für Mathematik, Universität Mannheim, D-618131 Mannheim, Germany* [nuernberger@euklid.math.uni-mannheim.de]

Jens Olav Nygaard, *SINTEF Applied Mathematics, P. O. Box 124, Blindern, 0314 Oslo, Norway* [jnygaard@math.sintef.no]

Peeter Oja (405), *Tartu University, Faculty of Mathematics, Institute of Applied Mathematics, Liivi 2-206, 50409 Tartu, Estonia* [Peeter.Oja@ut.ee]

Derek Paddon (263), *University of Bath, Computing Group, Dept. of Math. Sci., BA2 7AY Bath, UK* [derek@bath.ac.uk]

Juan Manuel Peña (283), *Departamento de Matemática Aplicada, Universidad de Zaragoza, 50009 Zaragoza, Spain* [jmpena@posta.unizar.es]

Sinésio Pesco (253), *Departamento de Matemática, Pontifícia Universidade Católica do Rio de Janeiro, Rua Marquês de São Vicente, 225, Gávea, Rio de Janeiro, RJ, Brazil, CEP:22.453-900* [pesco@mat.puc-rio.br]

Martin Peternell (413), *Institut fuer Geometrie, Wiedner Hauptstrasse 8–10, A-1040 Wien, Austria* [martin@geometrie.tuwien.ac.at]

Jorg Peters, *Univ. of Florida CISE, 32611-6120 Gainesville, USA* [jorg@cise.ufl.edu]

Laura Pezza, *Univ. degli Studi di Roma, Via A. Scarpa 16, 00161 Roma, Italy* [pezza@dmmm.uniroma1.it]

Christine Potier, *Telecom Paris, 46 rue Barrault, 75634 Paris cedex 13, France* [Christine.Potier@infres.enst.fr]

Juergen Prestin, *University of Luebeck, Institute of Mathematics, Wallstr. 40, D-23560 Luebeck, Germany* [prestin@math.mu-luebeck.de]

Ewald Quak, *SINTEF Applied Mathematics, P. O. Box 124, Blindern, 0314 Oslo, Norway* [Ewald.Quak@math.sintef.no]

Atgeirr Rasmussen, *SINTEF Applied Mathematics, P. O. Box 124, Blindern, 0314 Oslo, Norway* [atgeirr@sintef.no]

Wolfgang Rauh (1), *Fraunhofer Institute IPA, Nobelstr. 12, D–70569 Stuttgart, Germany* [wor@ipa.fhg.de]

Ulrich Reif (195), *TU Darmstadt, Schlossgartenstr. 7, 64289 Darmstadt, Germany* [reif@mathematik.tu-darmstadt.de]

Martin Reimers, *SINTEF Applied Mathematics, P. O. Box 124 Blindern, 0314 Oslo, Norway* [Martin.Reimers@math.sintef.no]

Richard Riesenfeld, *University of Utah, 50 S. Central Campus Dr. 3190 MEB, 84112 Salt Lake City, UT, USA* [rfr@cs.utah.edu]

Christian Rössl (423), *Max-Planck-Institut f. Informatik, Im Stadtwald, 66123 Saarbruecken, Germany* [roessl@mpi-sb.mpg.de]

Ramon F. Sarraga, *General Motors Research & Development, 30500 Mound Road, P. O. Box 9055, 48090-9055 Warren, Michigan, USA* [sarraga@gmr.com]

Eric Saux, *Institut de Recherche de l'Ecole Navale, Ecole Navale, Lanvoc-Poulmic, B. P. 600, 29240 Brest-Naval, France* [saux@ecole-navale.fr]

Karl Scherer (433), *Inst. Angew. Math., Universität Bonn, Wegelerstr. 6, 53115 Bonn, Germany* [scherer@iam.uni-bonn.de]

Gerik Scheuermann (173), *Center for Image Processing and Integrated Computing, University of California, 2343 Academic Surge Bld., Davis, CA 95616, USA* [scheuer@ucdavis.edu]

Regine Schickentanz, *Darmstadt University of Technology, Schlossgartenstr. 7, D-64289 Darmstadt, Germany* [rs@mathematik.tu-darmstadt.de]

Robert Schneider (445), *Max-Planck Institute, Im Stadtwald, 66123 Saarbruecken, Germany* [schneider@mpi-sb.mpg.de]

LARRY L. SCHUMAKER (393), *Vanderbilt University, Dept. of Mathematics, 37240 Nashville, USA* [s@mars.cas.vanderbilt.edu]

BERND SCHWALD, *Universität Stuttgart, Math. Institut A, Postfach 80 11 40, Pfaffenwaldring 57, D-70511 Stuttgart, Germany* [schwald@gmx.net]

ULRICH SCHWANECKE (455), *Max-Planck-Institute for Computer Sciences, Im Stadtwald, 66123 Saarbrücken, Germany* [schwanecke@mpi-sb.mpg.de]

THOMAS SEDERBERG (467), *Brigham Young University, 3374 TMCB, 84097 Provo, UT, USA* [tom@cs.byu.edu]

HANS-PETER SEIDEL (423,445), *Max-Planck-Institute for Computer Sciences, Im Stadtwald, 66123 Saarbruecken, Germany* [hpseidel@mpi-sb.mpg.de]

ALESSANDRA SESTINI (315), *Dipartimento di Energetica, Via Lombroso 6/17, 50134 Florence, Italy* [sestini@de.unifi.it]

CHANG SHU, *National Research Council Canada, Montreal Road, Building M-50, K1A 0R6 Ottawa, Canada* [chang.shu@nrc.ca]

JO SIMOENS, *Dept. of Computer Science, K. U. Leuven, Celestijnenlaan 200A, B-3001 Leuven, Belgium* [Jo.Simoens@cs.kuleuven.ac.be]

VIBEKE SKYTT, *SINTEF Applied Mathematics, P. O. Box 124, Blindern, N-0314 Oslo, Norway* [Vibeke.Skytt@math.sintef.no]

NEIL STEWART, *Universite de Montreal, Dep't IRO, CP6128 Succ. Centre-Ville, H3C 3J7 Montreal, Canada* [stewart@iro.umontreal.ca]

HANS STRAUSS, *Universität Erlangen-Nuernberg, Angewandte Mathematik, Martensstr 3, 91058 Erlangen, Germany* [strauss@am.uni-erlangen.de]

KYRRE STRØM, *SINTEF Applied Mathematics, P.O. Box 124, Blindern, 0314 Oslo, Norway* [Kyrre.Strom@math.sintef.no]

ERICH SUTER (477), *SINTEF Applied Mathematics, P. O. Box 124, Blindern, 0314 Oslo, Norway* [erich.suter@math.sintef.no]

CHIEW LAN TAI, *Hong Kong University of Science and Technology, Clear Water Bay, Hong Kong, China* [taicl@ust.hk]

SALIM TALEB (487), *Lab. Macs, University of Valenciennes, 59313 Valenciennes, cedex 9, France* [salim.taleb@free.fr]

CHRISTIAN TARROU, *Metaphor AS, Soerkedalsvn. 90 A, 0482 Oslo, Norway* [tarrou@metaphor.no]

CATHRINE TEGNANDER (103), *SINTEF Applied Mathematics, P. O. Box 124, Blindern, 0314 Oslo, Norway* [Cathrine.Tegnander@math.sintef.no]

DAVID A. TURNER (63), *School of Computing and Mathematics, University of Huddersfield, Huddersfield, UK* [d.a.turner@hud.ac.uk]

KENJI UEDA (497), *Ricoh Company, Ltd. 1-1-17, Koishikawa, Bunkyo-ku, 112-0002 Tokyo, Japan* [ueda@src.ricoh.co.jp]

GEORG UMLAUF, *Gerwigstr. 4, 76131 Karlsruhe, Germany* [georg.umlauf@gmx.de]

MICHAEL UNGSTRUP, *Dept. Math, DTU, / Odense Steel Shipyard DTU Building 303, 2800 Kgs. Lyngby, Denmark* [M.Ungstrup@mat.dtu.dk]

LUIZ VELHO (507), *IMPA - Instituto de Matematica Pura e Aplicada, Estrada Dona Castorina, 110, 22460-320 Rio de Janeiro, Brazil* [lvelho@visgraf.impa.br]

MARSHALL WALKER, *York University, Atkinson College, 4700 Keele Street, M3J 1P3 Toronto, Canada* [walker@yorku.ca]

HANS-JÜRGEN WARNECKE (1), *Fraunhofer Society, Leonrodstr. 54, D–80636 Munich, Germany* [warnecke@zv.fhg.de]

OLA WEISTRAND, *Uppsala University, Box 256, 751 05 Uppsala, Sweden* [weistrand@math.uu.se]

HOLGER WENDLAND (517), *Universität Göttingen, Lotzestr. 16-18, 37083 Göttingen, Germany* [wendland@math.uni-goettingen.de]

CLAIRE WILLIS (263), *University of Bath, Computing Group, Dept. of Math. Sci., BA2 7AY Bath, UK* [cpw@bath.ac.uk]

JORIS WINDMOLDERS (527), *Katholieke Universiteit Leuven, Celestijnenlaan 200 A, 3001 Heverlee, Belgium* [joris@cs.kuleuven.ac.be]

PETER WINGREN, *Umeå University, Matematiska institutionen, S-90187 Umeå, Sweden* [Peter.Wingren@math.umu.se]

JOAB WINKLER (535), *The University of Sheffield, Department of Computer Science, 211 Portobello Street, S1 4DP Sheffield, UK* [j.winkler@dcs.shef.ac.uk]

JOACHIM WIPPER (195), *Mathematisches Institut A, Universität Stuttgart, Pfaffenwaldring 57, 70569 Stuttgart, Germany* [wipper@mathematik.uni-stuttgart.de]

FRANZ-ERICH WOLTER (45), *Lehrstuhl für graphische Datenverarbeitung, Institut für Informatik, Universität Hannover, Welfengarten 1, 30167 Hannover, Germany* [few@informatik.uni-hannover.de]

ZHIYONG XIE (545), *Arizona State University, Computer Science and Engineering, 85287-5406 Tempe, AZ, USA* [zxie@asu.edu]

EMIL ZAGAR, *Fakulty of Computer Science, Trzaska 25, 1000 Ljubljana, Slovenia* [emil@gollum.fri.uni-lj.si]

FRANK ZEILFELDER (393), *Institut für Mathematik, Universität Mannheim, D-618131 Mannheim, Germany* [zeilfeld@mpi-sb.mpg.de]

JIANMIN ZHENG (467), *Institute of Computer Images and Graphics, Department of Mathematics, Zhejiang University, Hangzhou 310027, People's Republic of China* [jm_zheng@sina.com]

ZVI ZIEGLER, *Technion, Technion City, 32000 Haifa, Israel* [ziegler@tx.technion.ac.il]

DENIS ZORIN, *New York University, 719 Broadway, 12th floor, 10003 New York, USA* [dzorin@mrl.nyu.edu]

Best-Fit of Implicit Surfaces and Plane Curves

Sung Joon Ahn, Wolfgang Rauh, and Hans-Jürgen Warnecke

Abstract. Dimensional model fitting to a set of given points is a relevant subject in various disciplines of science and engineering. In this paper, we present a universal, and very efficient, best-fit algorithm for implicit surfaces and plane curves, by which the square sum of the orthogonal error distances of the given points to the model feature will be minimized. The estimation parameters are grouped in form, position, and rotation parameters, and are simultaneously estimated. The form parameters determine the shape of the model feature, and the position/rotation parameters describe the rigid body motion of the model feature. The mathematical frame of the proposed algorithm is applicable to any kind of implicit surface and plane curve.

§1. Introduction

The least squares method (LSM) [5] is one of the best known, and most often applied, mathematical tools in various disciplines of science and engineering. With applications of LSM to dimensional model fitting, the natural, and best, choice of the error distance is the shortest distance between the given point and the model feature. This error definition is prescribed in coordinate metrology guidelines [4,8]. However, except for some simple model features, computing and minimizing the square sum of the shortest error distances are not simple tasks for general features. For applications, a fitting algorithm would be very advantageous, if estimation parameters are grouped and simultaneously estimated in terms of form, position, and rotation. Furthermore, it would be helpful, if the reliability (variance) of each parameter could be tested.

Algebraic fitting is a procedure whereby each model feature is implicitly described by $F(\mathbf{a},\mathbf{X}) = 0$ with parameters $\mathbf{a} = (a_1, \cdots, a_q)^{\mathrm{T}}$, and the error distances are defined with the deviations of functional values from the expected value (i.e. zero) at each given point. If $F(\mathbf{a},\mathbf{X}_i) \neq 0$, the given point $\mathbf{X}_i$ does not lie on the model feature (i.e. there is some error-of-fit). Most publications about LS fitting of implicit features have been concerned with the square sum of algebraic distances or their modifications [3,10,13]

$$\sigma_0^2 = \sum_i F(\mathbf{a}, \mathbf{X}_i)^2 \quad \text{or} \quad \sigma_0^2 = \sum_i [F(\mathbf{a}, \mathbf{X}_i)/\nabla F(\mathbf{a}, \mathbf{X}_i)]^2. \tag{1}$$

Mathematical Methods for Curves and Surfaces: Oslo 2000
Tom Lyche and Larry L. Schumaker (eds.), pp. 1–14.

ISBN 0-8265-1378-6

In spite of advantages in implementation and computing costs, the algebraic fitting has drawbacks in accuracy. The estimated parameters are not invariant to coordinate transformation (e.g. a simple parallel shift of the given points also causes changes in rotation parameters), and generally, we cannot find a physical interpretation of an algebraic error definition. Finally, resolving the algebraic parameters into the physical parameters (e.g. in terms of form, position, and rotation parameters) is not a simple task for general features [6].

In **geometric fitting**, also known as best fitting, the error distance is defined as the shortest error distance of a given point to the model feature. Sullivan et. al. [12] have presented a geometric fitting of algebraic features

$$F(\mathbf{a},\mathbf{X}) = \sum_{j}^{q} a_j X^{k_j} Y^{l_j} Z^{m_j} = 0, \tag{2}$$

minimizing the square sum of the geometric distances $d(\mathbf{a},\mathbf{X}_i)=\mathbf{X}_i-\mathbf{X}'_i$, where $\mathbf{X}_i$ is a given point, and $\mathbf{X}'_i$ is the nearest point on the model feature F from $\mathbf{X}_i$. In order to locate $\mathbf{X}'_i$, they have iteratively minimized the Lagrangian $L(\lambda,\mathbf{X})=\|\mathbf{X}_i-\mathbf{X}\|^2+\lambda F(\mathbf{a},\mathbf{X})$ individually for each $\mathbf{X}_i$. To obtain the partial derivatives $\partial d(\mathbf{a},\mathbf{X})/\partial \mathbf{a}$ (which are necessary for nonlinear iteration) they have utilized the orthogonal contacting equations in machine coordinate system XYZ

$$\mathbf{F}(\mathbf{a},\mathbf{X}_i,\mathbf{X}) = \begin{pmatrix} F \\ \nabla F \times (\mathbf{X}_i - \mathbf{X}) \end{pmatrix} = \mathbf{0}. \tag{3}$$

A weakness of Sullivan's algorithm, the same as in the case of algebraic fitting, is that physical parameters are combined into an algebraic parameter vector **a**.

We will now present a universal and very efficient best-fit algorithm for dimensional model features and describe algorithm implementations to implicit surfaces and plane curves. Our algorithm is a generalized extension of a best-fit algorithm for implicit plane curves [1]. Because no assumption is made about the mathematical handling of implicit features, the algorithm can be applied to any kind of implicit dimensional feature. The estimation parameters are grouped in three categories and will be simultaneously estimated.

First, the **form parameters** $\mathbf{a}_g$ describe the shape (e.g. radius r for circle/cylinder/sphere, three axis lengths a, b, c for ellipsoid) of the standard model feature f defined in **model coordinate system** xyz (Fig. 1)

$$f(\mathbf{a}_g,\mathbf{x}) = 0 \qquad \text{with} \qquad \mathbf{a}_g = (a_1,\cdots,a_l)^{\mathrm{T}}. \tag{4}$$

The form parameters are invariant to the rigid body motion of the model feature. The second and third parameters group, the **position parameter** $\mathbf{a}_p$, and the **rotation parameter** $\mathbf{a}_r$ respectively, describe the rigid body motion of the model feature in machine coordinate system XYZ

$$\mathbf{X} = \mathbf{R}^{-1}\mathbf{x} + \mathbf{X}_o \qquad \text{or} \qquad \mathbf{x} = \mathbf{R}(\mathbf{X} - \mathbf{X}_o), \quad \text{where} \tag{5}$$

$$\mathbf{R} = \mathbf{R}_\kappa \mathbf{R}_\varphi \mathbf{R}_\omega = (\mathbf{r}_1 \quad \mathbf{r}_2 \quad \mathbf{r}_3)^{\mathrm{T}}, \qquad \mathbf{R}^{-1} = \mathbf{R}^{\mathrm{T}}, \tag{6}$$

$$\mathbf{a}_p = \mathbf{X}_o = (X_o, Y_o, Z_o)^{\mathrm{T}}, \qquad \text{and} \qquad \mathbf{a}_r = (\omega,\varphi,\kappa)^{\mathrm{T}}.$$

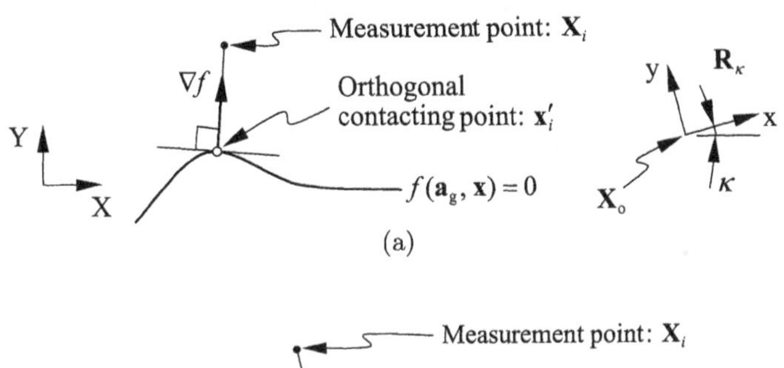

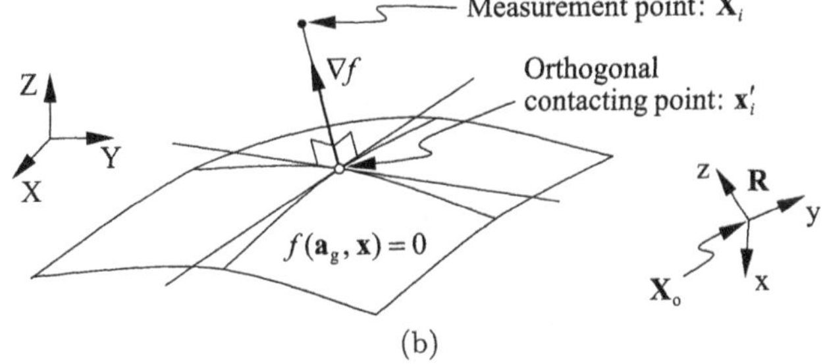

Fig. 1. Implicit features, and the orthogonal contacting point $\mathbf{x}'_i$ in frame xyz from the given point $\mathbf{X}_i$ in frame XYZ: (a) Plane curve; (b) Surface.

We characterize the form parameters as intrinsic parameters, and the position/rotation parameters as extrinsic parameters of the model feature according to their context. In this paper, we intend to simultaneously estimate all these parameters

$$\mathbf{a}^{\mathrm{T}} = (\mathbf{a}_{\mathrm{g}}^{\mathrm{T}},\ \mathbf{a}_{\mathrm{p}}^{\mathrm{T}},\ \mathbf{a}_{\mathrm{r}}^{\mathrm{T}}) = (a_1, \cdots, a_l, X_{\mathrm{o}}, Y_{\mathrm{o}}, Z_{\mathrm{o}}, \omega, \varphi, \kappa) = (a_1, \cdots, a_q).$$

For plane curve fitting, we simply ignore all terms concerning z, Z, ω, and φ (Fig. 1a). We may describe the model feature F in machine coordinate system XYZ, despite the fact that it has no interest for us from the viewpoint of our model fitting algorithm

$$F(\mathbf{a}, \mathbf{X}) = F(\mathbf{a}_{\mathrm{g}}, \mathbf{a}_{\mathrm{p}}, \mathbf{a}_{\mathrm{r}}, \mathbf{X}) = f(\mathbf{a}_{\mathrm{g}}, \mathbf{x}(\mathbf{a}_{\mathrm{p}}, \mathbf{a}_{\mathrm{r}}, \mathbf{X})) = 0.$$

The novelty of our algorithm, as will be described in Section 2, is that the necessary derivative values for nonlinear iteration are to be obtained from coordinate transformation equations (5), and from orthogonal contacting equations in the model coordinate system xyz (Fig. 1):

$$\mathbf{f}(\mathbf{a}_{\mathrm{g}}, \mathbf{x}_i, \mathbf{x}) = \begin{pmatrix} f \\ \nabla f \times (\mathbf{x}_i - \mathbf{x}) \end{pmatrix} = \mathbf{0} \qquad \text{with} \qquad \mathbf{x}_i = \mathbf{R}(\mathbf{X}_i - \mathbf{X}_{\mathrm{o}}). \quad (7)$$

For implementation of our algorithm to a new model feature, we need only provide the first and the second derivatives of the standard feature equation (4), which usually has only a few form parameters. Functional interpretations and treatments of the position/rotation parameters concerning coordinate transformation (5) are identical for all implicit model features.

§2. Best-Fit of Implicit Features

2.1 Nonlinear Least Squares Fitting

For solving nonlinear least squares problems, we have simply chosen Gauss-Newton iteration with initial parameters $\mathbf{a}_0$ and the step-size parameter λ

$$\mathbf{J}_{\mathbf{X}'_i,\mathbf{a}} = \begin{pmatrix} \frac{\partial X'_i}{\partial \mathbf{a}} \\ \frac{\partial Y'_i}{\partial \mathbf{a}} \\ \frac{\partial Z'_i}{\partial \mathbf{a}} \end{pmatrix}, \quad \begin{pmatrix} \mathbf{J}_{\mathbf{X}'_1,\mathbf{a}} \\ \vdots \\ \mathbf{J}_{\mathbf{X}'_m,\mathbf{a}} \end{pmatrix}_{\mathbf{a}_k} \Delta\mathbf{a} = \begin{pmatrix} \mathbf{X}_1 - \mathbf{X}'_1 \\ \vdots \\ \mathbf{X}_m - \mathbf{X}'_m \end{pmatrix}, \quad \mathbf{a}_{k+1} = \mathbf{a}_k + \lambda\Delta\mathbf{a}, \tag{8}$$

minimizing the performance index for m given points

$$\sigma_0^2 = \sum_{i=1}^{m} (\mathbf{X}'_i - \mathbf{X}_i)^{\mathrm{T}} (\mathbf{X}'_i - \mathbf{X}_i) = \sum_{i=1}^{m} \left[(X'_i - X_i)^2 + (Y'_i - Y_i)^2 + (Z'_i - Z_i)^2 \right].$$

For plane curve fitting, we ignore all terms concerning the coordinate Z in equations (8). To complete the linear system in equation (8), we must provide the nearest point $\mathbf{X}'_i$ on the surface/curve, and its Jacobian matrix $\mathbf{J}_{\mathbf{X}'_i,\mathbf{a}}$, corresponding to each measurement point $\mathbf{X}_i$. Additionally, if any constraint

$$f_{cj}(\mathbf{a}) - \mathrm{const}_j = 0, \quad j = 1, \cdots, n, \tag{9}$$

(e.g. on volume, position, rotation of the model feature) is to be applied, we append any of the following equations (10) with the large weighting values w_c to the linear system (8) of $p = \dim \times m$ equations

$$w_{cj} \left(\frac{\partial f_{cj}}{\partial \mathbf{a}} \right)_{\mathbf{a}_k} \Delta\mathbf{a} = -w_{cj} \left(f_{cj}(\mathbf{a}_k) - \mathrm{const}_j \right), \quad j = 1, \cdots, n, \tag{10}$$

$$\sigma_0^2 = \sum_{i=1}^{m} (\mathbf{X}'_i - \mathbf{X}_i)^{\mathrm{T}} (\mathbf{X}'_i - \mathbf{X}_i) + \sum_{j=1}^{n} \left[w_{cj} \left(f_{cj}(\mathbf{a}_k) - \mathrm{const}_j \right) \right]^2.$$

The Jacobian matrix in equation (8) will be decomposed by SVD [7,9]

$$\mathbf{J} = \mathbf{U}\mathbf{W}\mathbf{V}^{\mathrm{T}} \quad \text{with} \quad \mathbf{U}^{\mathrm{T}}\mathbf{U} = \mathbf{V}^{\mathrm{T}}\mathbf{V} = \mathbf{I}, \quad \mathbf{W} = [diag(w_1, \cdots, w_q)].$$

Then the linear system (8) can be solved for $\Delta\mathbf{a}$. After a successful termination of iteration (8), along with the performance index σ_0^2, the Jacobian matrix $\mathbf{J}$ provides useful information about the quality of parameter estimates:

- Covariance matrix: $\mathrm{Cov}(\hat{\mathbf{a}}) = (\mathbf{J}^{\mathrm{T}}\mathbf{J})^{-1} = \mathbf{V}\mathbf{W}^{-2}\mathbf{V}^{\mathrm{T}}$;
- Parameter covariance: $\mathrm{Cov}(\hat{a}_j, \hat{a}_k) = \sum_{i=1}^{q} \left(\frac{V_{ji} V_{ki}}{w_i^2} \right), \quad j, k = 1, \cdots, q$;
- Variance of parameters: $\sigma^2(\hat{a}_j) = \frac{\sigma_0^2}{p - q} \mathrm{Cov}(\hat{a}_j, \hat{a}_j), \; j = 1, \cdots, q$; (11)
- Correlation coefficients: $\rho(\hat{a}_j, \hat{a}_k) = \frac{\mathrm{Cov}(\hat{a}_j, \hat{a}_k)}{\sqrt{\mathrm{Cov}(\hat{a}_j, \hat{a}_j)\mathrm{Cov}(\hat{a}_k, \hat{a}_k)}}$. (12)

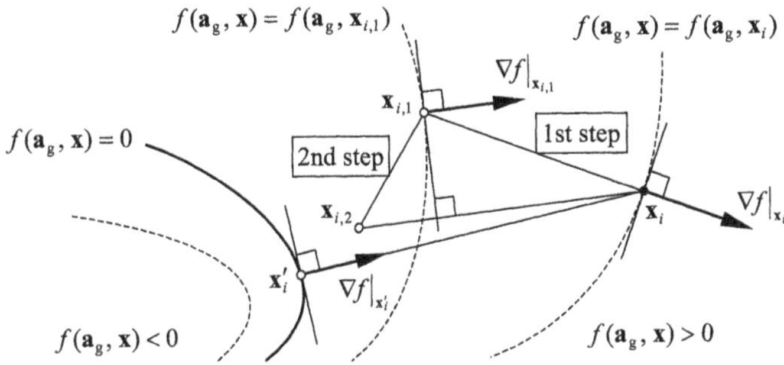

Fig. 2. Iterative search of the orthogonal contacting point $\mathbf{x}'_i$ on $f(\mathbf{a}_g,\mathbf{x}) = 0$ from the given point $\mathbf{x}_i$. The points $\mathbf{x}_{i,1}$ and $\mathbf{x}_{i,2}$ are the first and the second approximation of $\mathbf{x}'_i$, respectively.

2.2 Orthogonal Contacting Point

First, for each given point $\mathbf{x}_i$ in frame xyz,

$$\mathbf{x}_i = \mathbf{R}(\mathbf{X}_i - \mathbf{X}_o), \tag{13}$$

we determine the orthogonal contacting point $\mathbf{x}'_i$ on the standard model feature (4). Then, the orthogonal contacting point $\mathbf{X}'_i$ in frame XYZ to the given point $\mathbf{X}_i$ will be obtained through a backward transformation of $\mathbf{x}'_i$ into XYZ.

For some implicit features (e.g. circle, sphere, cylinder, cone and torus), the orthogonal contacting point $\mathbf{x}'_i$ can be determined in closed form. However, in general, we must iteratively find $\mathbf{x}'_i$ satisfying appropriate orthogonal contacting equations. The connecting line of $\mathbf{x}'_i$ with $\mathbf{x}_i$ must be parallel to the surface normal ∇f at $\mathbf{x}'_i$ (Fig. 1):

$$\nabla f \times (\mathbf{x}_i - \mathbf{x}) = \mathbf{0}, \quad \text{where} \quad \nabla f = (\partial f/\partial x,\ \partial f/\partial y,\ \partial f/\partial z)^{\mathrm{T}}. \tag{14}$$

Then, with equations (4) and (14), we build the orthogonal contacting equation (7), and solve it for $\mathbf{x}$ using a generalized Newton method with the initial point of $\mathbf{x}_0 = \mathbf{x}_i$ (Fig. 2):

$$\left.\frac{\partial \mathbf{f}}{\partial \mathbf{x}}\right|_k \Delta\mathbf{x} = -\mathbf{f}(\mathbf{x}_k), \quad \mathbf{x}_{k+1} = \mathbf{x}_k + \lambda\Delta\mathbf{x}. \tag{15}$$

How to compute the matrix $\partial\mathbf{f}/\partial\mathbf{x}$ in equation (15) will be shown in the next subsection. In the first iteration cycle with the initial point of $\mathbf{x}_0 = \mathbf{x}_i$, the linear system (15), and its solution (i.e. the first moving step), are equivalent to

$$\begin{pmatrix} \nabla f \cdot \Delta\mathbf{x} \\ \nabla f \times \Delta\mathbf{x} \end{pmatrix}_{\mathbf{x}_i} = \begin{pmatrix} -f \\ \mathbf{0} \end{pmatrix}_{\mathbf{x}_i}, \quad \text{and} \quad \Delta\mathbf{x} = \left.\frac{-f}{\|\nabla f\|^2}\nabla f\right|_{\mathbf{x}_i}. \tag{16}$$

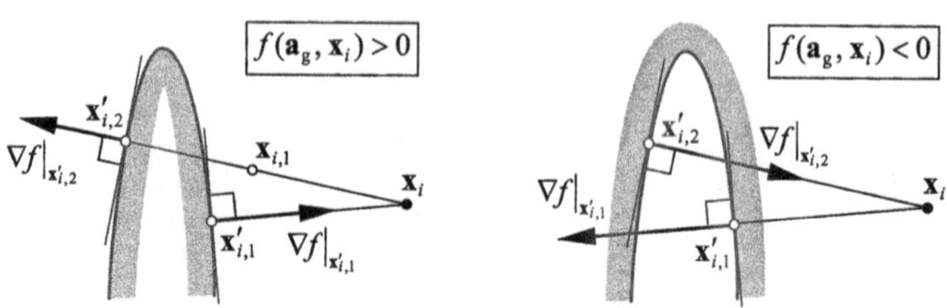

Fig. 3. Verification of the orthogonal contacting point. The point $\mathbf{x}'_{i,1}$ satisfies (17), while $\mathbf{x}'_{i,2}$ not.

It should be noted that the first moving step is the negative of the normalized algebraic error vector (compare with (1), which has been regarded in the literature [10,13] as a good approximation of the geometric error vector).

Once iteration (15) is successfully terminated, we verify the direction of the surface normal using another orthogonal contacting condition (Fig. 3):

$$\frac{\nabla f}{\|\nabla f\|} \cdot \frac{\mathbf{x}_i - \mathbf{x}'_i}{\|\mathbf{x}_i - \mathbf{x}'_i\|} - \operatorname{sign}(f(\mathbf{a}_g, \mathbf{x}_i)) = 0. \tag{17}$$

Rarely, if (17) could not be yet satisfied with a value other than zero (probably ± 2), then the connecting line of the two points $\mathbf{x}_i$ and $\mathbf{x}'_i$ pierced the model feature an odd number of times. In this case, we repeat iteration (15) with an adjusted starting point (e.g. $\mathbf{x}_{i,1}$ in Fig. 3) on the connecting line of the two points $\mathbf{x}_i$ and $\mathbf{x}'_i$.

2.3 Jacobian Matrix

We obtain the other necessary information, the Jacobian matrix for iteration (8), from derivatives of the coordinate transformation equation (5) relative to the parameters $\mathbf{a}$ as follows:

$$\begin{aligned} \mathbf{J}_{\mathbf{X}'_i,\mathbf{a}} &= \left.\frac{\partial \mathbf{X}}{\partial \mathbf{a}}\right|_{\mathbf{X}=\mathbf{X}'_i} = \left.\left(\mathbf{R}^{-1}\frac{\partial \mathbf{x}}{\partial \mathbf{a}} + \frac{\partial \mathbf{R}^{-1}}{\partial \mathbf{a}}[\mathbf{x}] + \frac{\partial \mathbf{X}_o}{\partial \mathbf{a}}\right)\right|_{\mathbf{x}=\mathbf{x}'_i} \\ &= \mathbf{R}^{-1}\left.\frac{\partial \mathbf{x}}{\partial \mathbf{a}}\right|_{\mathbf{x}=\mathbf{x}'_i} + \left(\mathbf{0} \;\middle|\; \mathbf{I} \;\middle|\; \frac{\partial \mathbf{R}^{-1}}{\partial \mathbf{a}_r}[\mathbf{x}'_i] \right), \quad \text{with} \end{aligned} \tag{18}$$

$$\mathbf{a}^T = (\mathbf{a}_g^T, \mathbf{a}_p^T, \mathbf{a}_r^T), \quad \frac{\partial \mathbf{R}}{\partial \mathbf{a}_r} = \left[\frac{\partial \mathbf{R}}{\partial \omega}\ \frac{\partial \mathbf{R}}{\partial \varphi}\ \frac{\partial \mathbf{R}}{\partial \kappa}\right], \quad \text{and} \quad [\mathbf{x}] = \begin{pmatrix} \mathbf{x} & & \mathbf{0} \\ & \ddots & \\ \mathbf{0} & & \mathbf{x} \end{pmatrix}.$$

The derivative matrix $\partial \mathbf{x}/\partial \mathbf{a}$ at $\mathbf{x}=\mathbf{x}'_i$ in (18) describes the variational behavior of the orthogonal contacting point $\mathbf{x}'_i$ in frame xyz relative to the differential changes of the parameters vector $\mathbf{a}$. Purposefully, we obtain $\partial \mathbf{x}/\partial \mathbf{a}$ from the orthogonal contacting equation (7). Because (7) has an implicit form, its derivatives lead to

$$\frac{\partial \mathbf{f}}{\partial \mathbf{x}}\frac{\partial \mathbf{x}}{\partial \mathbf{a}} + \frac{\partial \mathbf{f}}{\partial \mathbf{x}_i}\frac{\partial \mathbf{x}_i}{\partial \mathbf{a}} + \frac{\partial \mathbf{f}}{\partial \mathbf{a}} = \mathbf{0}, \quad \text{or} \quad \frac{\partial \mathbf{f}}{\partial \mathbf{x}}\frac{\partial \mathbf{x}}{\partial \mathbf{a}} = -\left(\frac{\partial \mathbf{f}}{\partial \mathbf{x}_i}\frac{\partial \mathbf{x}_i}{\partial \mathbf{a}} + \frac{\partial \mathbf{f}}{\partial \mathbf{a}}\right), \tag{19}$$

where, $\partial \mathbf{x}_i/\partial \mathbf{a}$ is, from (13),

$$\frac{\partial \mathbf{x}_i}{\partial \mathbf{a}} = \frac{\partial \mathbf{R}}{\partial \mathbf{a}}[\mathbf{X}_i - \mathbf{X}_\mathrm{o}] - \mathbf{R}\frac{\partial \mathbf{X}_\mathrm{o}}{\partial \mathbf{a}} = \left(\mathbf{0} \;\middle|\; -\mathbf{R} \;\middle|\; \frac{\partial \mathbf{R}}{\partial \mathbf{a}_\mathrm{r}}[\mathbf{X}_i - \mathbf{X}_\mathrm{o}]\right).$$

The other three matrices $\partial \mathbf{f}/\partial \mathbf{x}$, $\partial \mathbf{f}/\partial \mathbf{x}_i$, and $\partial \mathbf{f}/\partial \mathbf{a}$ in (15) and (19) are to be directly derived from (7). The elements of these three matrices are composed of simple linear combinations of components of the error vector $(\mathbf{x}_i - \mathbf{x})$ with elements of the following three vector/matrices ∇f, $\mathbf{H}$, and $\mathbf{G}$ (gradient, and Hessian matrix):

$$\nabla f = \left(\frac{\partial f}{\partial x}, \frac{\partial f}{\partial y}, \frac{\partial f}{\partial z}\right)^\mathrm{T}, \qquad \mathbf{H} = \frac{\partial}{\partial \mathbf{x}}\nabla f, \qquad \mathbf{G} = \frac{\partial}{\partial \mathbf{a}_\mathrm{g}}\begin{pmatrix} f \\ \nabla f \end{pmatrix}, \tag{20}$$

$$\frac{\partial \mathbf{f}}{\partial \mathbf{x}} = \begin{pmatrix} 0 & 0 & 0 \\ y_i - y & -(x_i - x) & 0 \\ -(z_i - z) & 0 & x_i - x \\ 0 & z_i - z & -(y_i - y) \end{pmatrix}\mathbf{H} + \begin{pmatrix} \frac{\partial f}{\partial x} & \frac{\partial f}{\partial y} & \frac{\partial f}{\partial z} \\ \frac{\partial f}{\partial y} & -\frac{\partial f}{\partial x} & 0 \\ -\frac{\partial f}{\partial z} & 0 & \frac{\partial f}{\partial x} \\ 0 & \frac{\partial f}{\partial z} & -\frac{\partial f}{\partial y} \end{pmatrix},$$

$$\frac{\partial \mathbf{f}}{\partial \mathbf{x}_i} = \begin{pmatrix} 0 & 0 & 0 \\ -\frac{\partial f}{\partial y} & \frac{\partial f}{\partial x} & 0 \\ \frac{\partial f}{\partial z} & 0 & -\frac{\partial f}{\partial x} \\ 0 & -\frac{\partial f}{\partial z} & \frac{\partial f}{\partial y} \end{pmatrix},$$

$$\frac{\partial \mathbf{f}}{\partial \mathbf{a}} = \begin{pmatrix} 1 & 0 & 0 & 0 \\ 0 & y_i - y & -(x_i - x) & 0 \\ 0 & -(z_i - z) & 0 & x_i - x \\ 0 & 0 & z_i - z & -(y_i - y) \end{pmatrix}\left(\ \mathbf{G} \mid \mathbf{0} \mid \mathbf{0}\ \right).$$

Now, (19) can be solved for $\partial \mathbf{x}/\partial \mathbf{a}$ at $\mathbf{x} = \mathbf{x}_i'$. For the sake of a common coding practice for surfaces and plane curves, we have interchanged, without loss of generality, the second and the fourth row of (7). Then, for plane curves, we consider only the upper-left block of the matrices in (20). We would like to stress that only the standard model feature equation (4), without involvement of the position/rotation parameters, is required in (20). The overall structure of our algorithm remains unchanged for all dimensional fitting problems of implicit features. All that is necessary for a new implicit feature is to derive ∇f, $\mathbf{H}$, and $\mathbf{G}$ of (20) from equation (4) of the new feature, and to supply a proper set of initial parameter values $\mathbf{a}_0$ for iteration (8). This fact makes possible the realization of a universal and very efficient best-fit algorithm for implicit surfaces and plane curves. An overall schematic information flow is shown in Fig. 4.

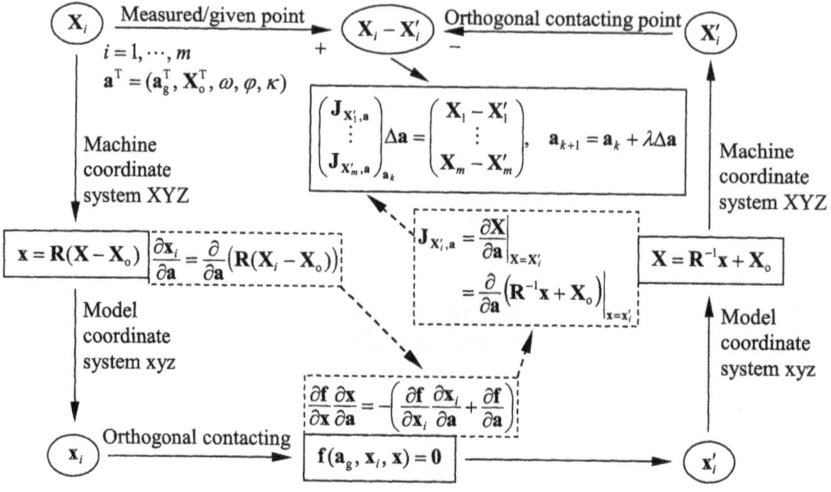

Fig. 4. Information flow for best-fit of implicit features.

§3. Best-Fit Examples

3.1 Cone Fitting

The standard model feature of a cone in frame xyz can be described as

$$f(\psi, \mathbf{x}) = x^2 + y^2 - (z \tan(\psi/2))^2 = 0,$$

where ψ, the only form parameter, is the vertex angle of a cone. The position $\mathbf{X}_o$, the origin of frame xyz, is defined at the vertex. The orientation is represented by the direction cosines of the z-axis, $\mathbf{r}_3$ in (6). However, from the viewpoint of coordinate metrology [8], a better parameterization of a cone is

$$f(\psi, r, \mathbf{x}) = x^2 + y^2 - (r - z \tan(\psi/2))^2 = 0,$$

with a constraint on the position and rotation parameters (see (9))

$$f_c(\mathbf{a}_p, \mathbf{a}_r) = (\mathbf{X}_o - \overline{\mathbf{X}}) \cdot \mathbf{r}_3(\omega, \varphi) = 0,$$

where the second form parameter r is the sectional radius of a cone at $z=0$, and $\overline{\mathbf{X}}$ is the gravitational center of the given points set.

As one of the test data sets used in an authorized certification [8] process of our algorithm, the 10 points in Table 1 representing only a quarter of a cone slice were prepared by the German authority PTB (Physikalisch-Technische Bundesanstalt). We have obtained the initial parameters set from a 3D-circle fitting [2] and a cylinder fitting successively, with $\psi=0$ (Fig. 5, Table 2). The

Tab. 1. 10 points representing a quarter of a cone slice.

X	734.8905	739.8980	736.4229	850.6449	850.6271
Y	−720.8340	−736.6202	−750.8837	−699.2051	−718.8401
Z	−735.4193	−731.4877	−731.9028	−645.0159	−645.7938
X	919.1539	921.6025	935.7441	931.8560	927.4975
Y	−711.7191	−727.6805	−766.1543	−781.5263	−809.7823
Z	−527.1499	−519.4751	−391.4352	−372.0004	−388.0452

Tab. 2. Result of the orthogonal distance fitting to the 10 points in Table 1.

Parameters $\hat{\mathbf{a}}$	σ_0	ψ	r	X_o
3D-Circle	38.848	−−	283.0367	694.5271
Cylinder	2.4655	−−	379.0909	561.5321
Cone	0.0357	1.4262	276.4373	706.7202
$\sigma(\hat{\mathbf{a}})$	−−	0.0026	0.3337	0.5105
Parameters $\hat{\mathbf{a}}$	Y_o	Z_o	ω	φ
3D-Circle	−889.7335	−498.1031	2.0534	0.5477
Cylinder	−702.1460	−398.2213	1.6090	−0.1578
Cone	−890.5186	−499.1046	2.0554	0.5877
$\sigma(\hat{\mathbf{a}})$	0.0186	0.2644	0.0012	0.0015

Tab. 3. Correlation coefficients of the estimated cone parameters.

$\rho(\hat{\mathbf{a}})$	ψ	r	X_o	Y_o	Z_o	ω	φ
ψ	1.00						
r	−0.99	1.00					
X_o	0.99	−1.00	1.00				
Y_o	−0.50	0.48	−0.50	1.00			
Z_o	−0.96	0.97	−0.95	0.48	1.00		
ω	0.95	−0.96	0.95	−0.50	−1.00	1.00	
φ	0.99	−0.99	1.00	−0.52	−0.94	0.94	1.00

cone fitting is terminated after 60 iteration cycles for $\|\Delta\mathbf{a}\|=2.3\times10^{-7}$ with the constraint weighting value of $w_c=1$. Table 3 shows strong parameter correlations. The convergence is somehow slow because the initial cone parameters (i.e. the cylinder parameters) already seem to provide a local minimum, and it is especially slow if the constraint weighting value w_c is large, and thus, locks the parameters from changing. If we perturb this quasi-equilibrium state, similarly to the random walking technique, through adding a small artifact error to the initial parameter values, then we could expect a faster convergence. With a nonzero initial vertex angle value of $\psi=\pi/10$, the iteration terminates after only 13 cycles for $\|\Delta\mathbf{a}\|=2.9\times10^{-6}$ (Fig. 6).

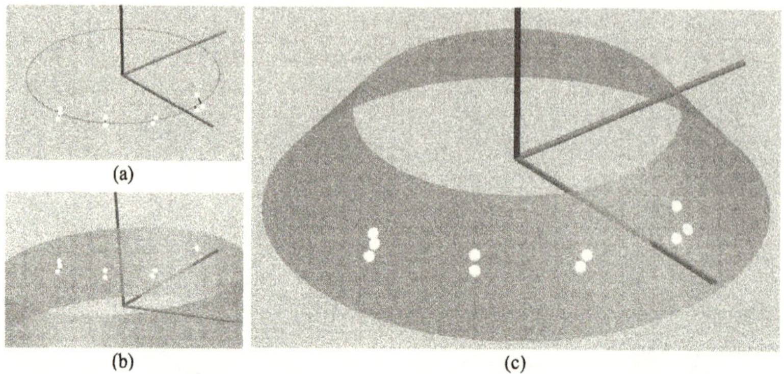

Fig. 5. Orthogonal distance fit to the 10 points in Table 1: (a) 3D-circle fit; (b) Circular cylinder fit; (c) Circular cone fit.

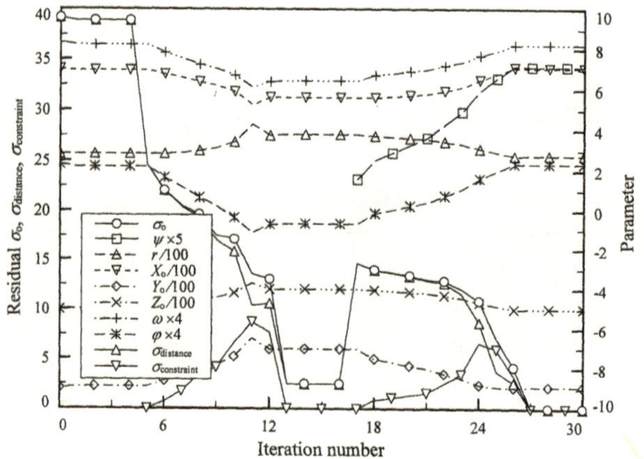

Fig. 6. Convergence of the cone fitting (Iteration number 0–4: 3D-circle, 5–16: cylinder, and 17–: cone fit with the initial value of $\psi = \pi/10$).

3.2 Superquadric Fitting

A superquadric is a generalization of an ellipsoid, and is described by

$$f(a, b, c, \varepsilon_1, \varepsilon_2, \mathbf{x}) = \left((x/a)^{2/\varepsilon_1} + (y/b)^{2/\varepsilon_1}\right)^{\varepsilon_1/\varepsilon_2} + (z/c)^{2/\varepsilon_2} - 1 = 0,$$

where a, b, c are the axis lengths, and exponents ε_1, ε_2 are the shape coefficients. In comparison with the algebraic algorithm [11], our algorithm can fit also extreme shapes of a superquadric (e.g. a box with $\varepsilon_{1,2} \ll 1$, or a star with

$\varepsilon_{1,2}>2$). After a series of experiments with numerous sets of data points, we have concluded that our algorithm safely converges within the parameter zone

$$0.002 \le (\varepsilon_1 \text{ and } \varepsilon_2) \le 10 \quad \text{and} \quad \varepsilon_1/\varepsilon_2 \le 500.$$

Otherwise, there is a danger of data overflow destroying the Jacobian matrices (15) and (18). Additionally, a sharp increase of computing cost is unavoidable.

We obtain the initial parameters set from a sphere fitting and an ellipsoid fitting successively, with $\varepsilon_1=\varepsilon_2=1$ (Fig. 7, Table 5). Superquadric fitting to the 30 points in Table 4 representing a box is terminated after 18 iteration cycles for $\|\Delta\mathbf{a}\|=7.9\times10^{-7}$ (Fig. 8). The intermediate ellipsoid fitting showed a slow convergence in its second half phase (iteration number 10 to 42 in Fig. 8), because of a relatively large variance of the major axis length a of the ellipsoid caused by the distribution of the given points (compare the standard deviation $\sigma(\hat{a})$ of the intermediate ellipsoid $\sigma(\hat{a})_{\text{Ellip.}}=6.8430$ with that of the superquadric $\sigma(\hat{a})_{\text{SQ}}=0.0507$).

§4. Summary

Dimensional model fitting finds its applications in various fields of science and engineering and is a relevant subject in coordinate metrology, computer aided geometric design, and computer/machine vision. In this paper, we have presented a new algorithm for best-fit of implicit surfaces and plane curves, which minimizes the square sum of orthogonal error distances between the model feature and given points. The new algorithm is universal, and very efficient, from the viewpoint of implementation and application to a new model feature. Our algorithm converges very well, and does not require a necessarily good initial parameter values set, which could also be internally supplied (e.g. gravitational center and RMS central distance of the given points set for sphere fitting, sphere parameters for ellipsoid fitting, and ellipsoid parameters for superquadric fitting, etc.). Memory space and computing time costs are proportional to the number of given points. The estimation parameters are grouped in form/position/rotation parameters, and are simultaneously estimated, thus providing useful algorithmic features for various applications. Additionally, the quality of parameter estimation (including the propriety of model selection) can be simply tested by using the covariance matrix. Together with other new best-fit algorithms [2] for parametric surfaces and curves, and for parametric curves on surface, we believe we have developed a complete set of algorithms for orthogonal distance fitting of dimensional model features. Our algorithm is certified by the German authority PTB [8], with a certification grade that the parameter estimation accuracy is higher than 0.1μm for length unit, and 0.1μrad for angle unit for all parameters of all tested model features.

Tab. 4. 30 points representing a box.

X	−4	1	4	20	−11	−26	−3	−7	6	11
Y	3	16	−11	−17	1	7	−13	−26	19	24
Z	13	29	10	22	4	−8	1	−15	9	19
X	3	15	18	−21	−4	−2	−14	20	4	6
Y	−18	9	−3	3	−14	19	14	−17	−20	20
Z	−25	21	22	−13	−22	11	1	15	−18	24
X	22	30	−8	−16	8	26	−22	−2	−3	7
Y	−8	−9	15	15	−13	−14	12	−22	−3	1
Z	4	12	−9	−18	−15	17	−13	−20	9	−5

Tab. 5. Result of the orthogonal distance fitting to the 30 points in Table 4.

Parameters $\hat{\mathbf{a}}$	σ_0	a	b	c	ε_1	ε_2
Sphere	33.8999	46.5113	−−	−−	−−	−−
Ellipsoid	14.4338	46.3303	25.5975	9.3304	−−	−−
Superquadric	0.9033	24.6719	20.4927	8.2460	0.0946	0.0374
$\sigma(\hat{\mathbf{a}})$	−−	0.0507	0.0503	0.0293	0.0074	0.0097
Parameters $\hat{\mathbf{a}}$	X_o	Y_o	Z_o	ω	φ	κ
Sphere	27.3890	18.2656	−20.8287	−−	−−	−−
Ellipsoid	1.6769	−1.2537	0.8719	0.7016	−0.7099	0.6925
Superquadric	1.9096	−1.0234	2.0191	0.6962	−0.6952	0.6960
$\sigma(\hat{\mathbf{a}})$	0.0368	0.0338	0.0380	0.0023	0.0015	0.0029

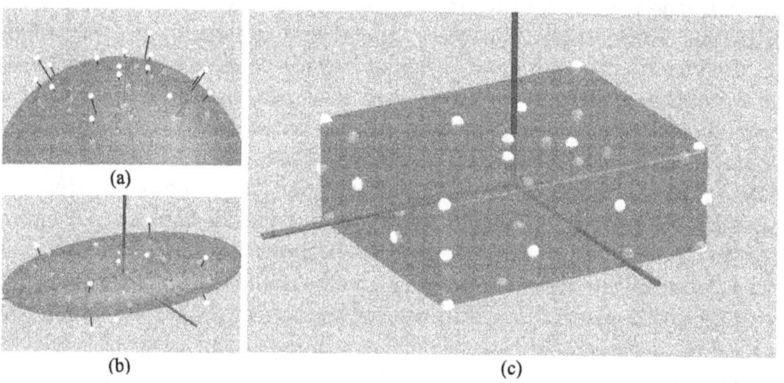

Fig. 7. Orthogonal distance fit to the 30 points in Table 4: (a) Sphere fit; (b) Ellipsoid fit; (c) Superquadric fit.

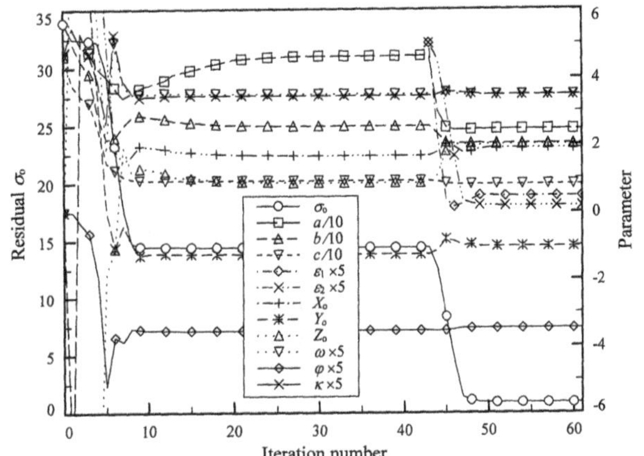

Fig. 8. Convergence of the superquadric fitting (Iteration number 0–42: ellipsoid, and 43–: superquadric fit).

References

1. Ahn, S. J. and W. Rauh, Geometric least squares fitting of circle and ellipse, Int. J. of Pattern Recognition and Artificial Intelligence **13** (1999), 987–996.

2. Ahn, S. J., Orthogonal distances fitting of surfaces and curves, Intern Research Report, Dept. 620, FhG-IPA, Stuttgart, Germany, Dec. 1999.

3. Bookstein, F. L., Fitting conic sections to scattered data, Comp. Graphics and Image Proc. **9** (1979), 56–71.

4. DIN 32880-1, Coordinate metrology; geometrical fundamental principles, terms and definitions, German Standard, Beuth Verlag, Berlin, 1986.

5. Gauss, C. F., *Theory of the motion of the heavenly bodies moving about the sun in conic sections* (*Theoria motus corporum coelestium in sectionibus conicis solem ambientum*), First published in 1809, Translation by C.H. Davis, Dover, New York, 1963.

6. Goldman, R. N., Two Approaches to a computer model for quadric surfaces, IEEE Computer Graphics & Applications **3** (1983), 21–24.

7. Golub, G. H. and C. Reinsch, Singular value decomposition and least squares solutions, Numer. Math. **14** (1970), 403–420.

8. ISO/DIS 10360-6, Geometrical product specification (GPS) - Acceptance test and reverification test for coordinate measuring machines (CMM) - Part 6: Estimation of errors in computing Gaussian associated features, Draft International Standard, ISO, Geneva, 1999.

9. Press, W. H., B. P. Flannery, S. A. Teukolsky, and W. T. Vetterling, *Numerical recipes in C: The art of scientific computing*, Cambridge University Press, Cambridge, UK, 1988.

10. Sampson, P. D., Fitting conic sections to "very scattered" data: An iterative refinement of the Bookstein algorithm, Comp. Graphics and Image Proc. **18** (1982), 97–108.

11. Solina, F. and R. Bajcsy, Recovery of parametric models from range images: The case for superquadrics with global deformations, IEEE Trans. Pattern Anal. and Machine Intelligence **12** (1990), 131–147.

12. Sullivan, S., L. Sandford, and J. Ponce, Using geometric distance fits for 3-D object modeling and recognition, IEEE Trans. Pattern Anal. and Machine Intelligence **16** (1994), 1183–1196.

13. Taubin, G., Estimation of planar curves, surfaces, and nonplanar space curves defined by implicit equations with applications to edge and range image segmentation, IEEE Trans. Pattern Anal. and Machine Intelligence **13** (1991), 1115–1138.

Sung Joon Ahn and Wolfgang Rauh
Fraunhofer Institute IPA
Nobelstr. 12, D–70569 Stuttgart, Germany
`{sja; wor}@ipa.fhg.de`
`http://www.ipa.fhg.de/english/600/Informationstechnik_e.php3`

Hans-Jürgen Warnecke
Fraunhofer Society
Leonrodstr. 54, D–80636 Munich, Germany
`warnecke@zv.fhg.de`
`http://www.fhg.de/english/contact/contact_executive.html`

Determination of Geometrical Invariants of Rationally Parametrized Conic Sections

Gudrun Albrecht

Abstract. We address the problem of finding geometrical invariants, such as foci, axes, center, and vertices of a non–degenerate conic section, uniquely from its rational quadratic parametrization, i.e., without knowing its implicit representation. The main idea, namely the determination of the conic's foci, is based on the projective definition of the foci of a non–degenerate conic section.

§1. Introduction

Conic sections, i.e., ellipses, hyperbolas and parabolas, have for a long time been very important curves in applications such as architecture, mechanical engineering (see, e.g., [8]), and the airplane industries (see, e.g., [1]). They have therefore been integrated in many modern CAD–systems; in Version 5 of Dassault Systèmes CAD–system CATIA, e.g., they may be constructed by means of their foci. On the other hand, due to their many advantages, the popularity of NURBS and rational Bézier representations of curves and surfaces has hugely increased in the past years. Therefore, also the *rational* Bézier representation of conic sections has been studied under different aspects, see, e.g., [2,5,6,7].

Due to these two main representations of conic sections in use, i.e., the one based on the geometrical invariants such as foci, center, axes, vertices on the one hand, and the rational quadratic Bézier form on the other hand, an easy transition between them has to be guaranteed. Thus, this paper deals with the problem of determining the above geometrical invariants of a non–degenerate conic section from its rational quadratic parametrization. This problem has already been solved by Lee [14] by means of methods from elementary geometry. But with a view towards generalization to quadric surfaces in three–dimensional space, this problem is now being reconsidered in [9] and the present paper. This paper presents an easy and unified geometrical approach for the determination of the conic's foci from which the remaining invariants such as center, axes, vertices are derived.

Mathematical Methods for Curves and Surfaces: Oslo 2000
Tom Lyche and Larry L. Schumaker (eds.), pp. 15–24.

ISBN 0-8265-1378-6

The organization of the paper is as follows. In §2 some basic knowledge from projective geometry regarding conic sections is stated, including the essential projective definition of a conic's foci. This uniform definition is then used in §3, where a method for determining the foci and the above remaining invariants of a non–degenerate conic section from its rational quadratic Bézier form is developed. This method is also illustrated for some examples. Finally, § 4 contains some concluding remarks including ideas for a generalization of the presented method to quadric surfaces in three–dimensional space.

§2. Theoretical Foundations

If given implicitly, conic sections are well–understood curves and their geometrical invariants such as foci, axes, center and vertices are easily obtained. In the Euclidean plane we distinguish three equivalence classes of non–degenerate conic sections, namely ellipses, hyperbolas and parabolas; for every such class there exists a representative normal form, which in homogeneous point coordinates (x_0, x_1, x_2) (x_0 being the homogenizing coordinate, i.e., $x = \frac{x_1}{x_0}, y = \frac{x_2}{x_0}$, where (x, y) are Cartesian coordinates) is shown in the following table, where $a_1 \geq a_2 > 0$. Table 1 also contains the foci of the respective conic section. In the case of the ellipse, the physical interpretation of the foci is the following. If we imagine a light source at one of the two foci, then the light rays which emanate from it are reflected by the ellipse such that they meet in the other focus. The same interpretation is valid for the parabola, if we imagine one of the two foci to lie at infinity, and in the case of the hyperbola the light rays emanated from one focus are reflected by the hyperbola such that they seem to have emanated from the other focus.

Conic section	Ellipse	Hyperbola	Parabola
Normal form in homogeneous coordinates	$\frac{x_1^2}{a_1^2} + \frac{x_2^2}{a_2^2} - x_0^2 = 0$	$\frac{x_1^2}{a_1^2} - \frac{x_2^2}{a_2^2} - x_0^2 = 0$	$\frac{x_1^2}{a_1^2} + 2x_0x_2 = 0$
Foci	$F_1(1, -\sqrt{a_1^2 - a_2^2}, 0)$, $F_2(1, \sqrt{a_1^2 - a_2^2}, 0)$	$F_1(1, -\sqrt{a_1^2 + a_2^2}, 0)$, $F_2(1, \sqrt{a_1^2 + a_2^2}, 0)$	$F(1, 0, -\frac{a_1^2}{2})$

Tab. 1. Conic normal forms and foci.

For the purpose of extracting the foci from the parametric form of a conic section, we are going to use a more uniform definition for the foci of a non–degenerate conic section, which we borrow from projective geometry, see, e.g., [3,11]. According to [13], this definition may be visualized with the geometry program Cinderella.

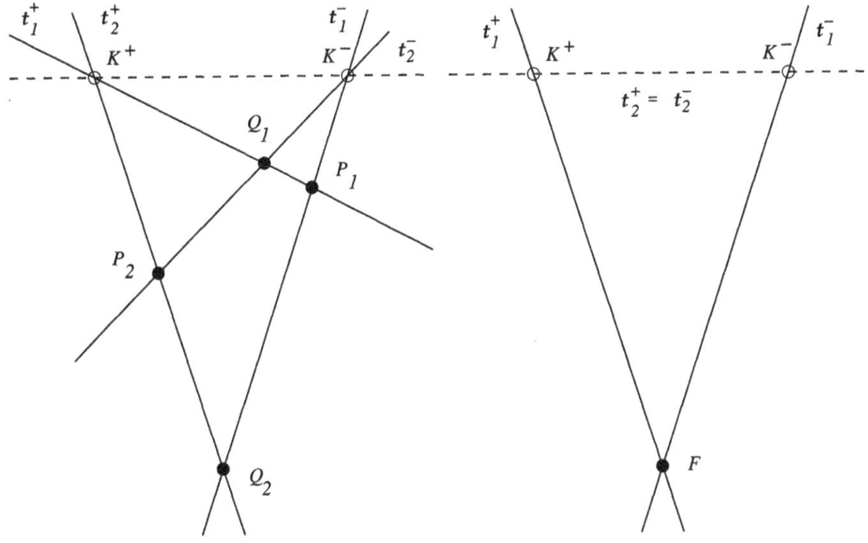

Fig. 1. Illustration of a conic's foci according to Definition 1 in the case of an ellipse or hyperbola (left) and parabola (right). The dashed line represents the line at infinity.

Definition 1. *Given a nondegenerate conic section c, let k be the algebraic curve of fourth order consisting of the four tangents of c through the circular points at infinity* $K^{\pm}(0,1,\pm i)$ $(i^2 = -1)$*. Then the foci of c are the singular points of the curve k excluding* $K^{\pm}$.

Let $\mathbf{t}_1^+, \mathbf{t}_2^+$ be the tangents of c through K^+, and $\mathbf{t}_1^-, \mathbf{t}_2^-$ the tangents of c through K^-. Then, as illustrated in Figure 1 in the case of an ellipse or a hyperbola, we obtain four foci, namely the four intersection points

$$P_1 := \mathbf{t}_1^+ \cap \mathbf{t}_1^- \,, \quad P_2 := \mathbf{t}_2^+ \cap \mathbf{t}_2^- \,, \quad Q_1 := \mathbf{t}_1^+ \cap \mathbf{t}_2^- \,, \quad Q_2 := \mathbf{t}_1^- \cap \mathbf{t}_2^+ . \qquad (1)$$

In the case of a parabola two of the above tangents, say $\mathbf{t}_2^+, \mathbf{t}_2^-$, coincide with the line at infinity. Thus we obtain one, necessarily real focus, namely

$$F := \mathbf{t}_1^+ \cap \mathbf{t}_1^- \,. \qquad (2)$$

We are going to look at the above definition in more detail for two reasons, firstly in order to verify that the so–defined real foci coincide with the "physically" defined foci from Table 1, and secondly in order to obtain an easy and practical way of determining the foci of an arbitrary non–degenerate conic section.

Motivated by the importance of certain tangents of the given conic in Definition 1 and according to the principle of duality in the projective plane, we are going to interpret the conic not as a set of points but as a set of lines, namely its tangents, see, e.g., [5].

Given a non–degenerate conic section

$$\mathbf{x}^T A\mathbf{x} = 0 \tag{3}$$

in homogeneous point coordinates $\mathbf{x} = (x_0, x_1, x_2)^T$ with regular and symmetric coefficient matrix A ($A = A^T \in \mathbb{R}^{3,3}$), its line representation reads

$$\mathbf{u}^T A^{-1}\mathbf{u} = 0, \tag{4}$$

where $\mathbf{u} = (u_0, u_1, u_2)^T$ are so–called homogeneous line coordinates in the projective plane. We also represent the pair of circular points at infinity $K^{\pm}(0, 1, \pm i)$ in homogeneous line coordinates obtaining the equation

$$u_1^2 + u_2^2 = 0.$$

We now consider a one–parameter system of conic sections in line coordinates

$$\mathbf{u}^T(A^{-1} - \lambda E)\mathbf{u} = 0, \tag{5}$$

where $\lambda \in \mathbb{R}$ and $E = diag(0, 1, 1)$. (5) is spanned by the conic (4) and the pair of circular points at infinity $K^{\pm}$. For every value of λ we obtain a "line conic" of the system (5). Due to the construction of (5), the four tangents of the original conic (4) through the circular points at infinity $K^{\pm}$ are common to all line conics of the system (5). Therefore, according to Definition 1 all the conics of the system (5) have the same foci, and the system (5) is called a system of *confocal* conic sections. For an illustration of confocal ellipses and hyperbolas, see, e.g., [10], p. 6.

As intersection points of the four common tangents of the conics of the system (5) through the points $K^{\pm}$, the foci might be interpreted as singular "line conics" of the system (5), and thus can be obtained from the condition

$$\det(A^{-1} - \lambda E) = 0, \tag{6}$$

which has to be solved for λ. From the condition (6) we can thus easily calculate the foci of an arbitrary non–degenerate conic section (3). Applying this to the conics in normal form from Table 1, we obtain the following foci. In the case of the ellipse ($A = diag(-1, \frac{1}{a_1^2}, \frac{1}{a_2^2})$) we obtain the two pairs of points $P_{1/2}(1, \pm\sqrt{a_1^2 - a_2^2}, 0)$, $Q_{1/2}(1, 0, \pm\sqrt{a_2^2 - a_1^2})$, in the case of the hyperbola ($A = diag(-1, \frac{1}{a_1^2}, -\frac{1}{a_2^2})$) we obtain $P_{1/2}(1, \pm\sqrt{a_1^2 + a_2^2}, 0)$, $Q_{1/2}(1, 0, \pm i\sqrt{a_1^2 + a_2^2})$, and in the case of the parabola ($A = \begin{pmatrix} 0 & 0 & 1 \\ 0 & \frac{1}{a_1^2} & 0 \\ 1 & 0 & 0 \end{pmatrix}$) we get $F(1, 0, -a_1^2/2)$ as the proper focus.

Thus, in the cases of the ellipse and hyperbola, one of the pairs (P_1, P_2), (Q_1, Q_2) is real and coincides with the foci known from real Euclidean geometry, and the other one is an imaginary (conjugate complex) pair of points. Supposing $a_1 \geq a_2 > 0$, the real foci are $P_{1/2}$, which coincide with the ones from Table 1. While the points $P_{1/2}$ lie on the major axis of the ellipse/hyperbola, $Q_{1/2}$ lie on the conic's minor axis in the complex plane.

Also in the case of the parabola, the calculated focus coincides with the one from Table 1.

§3. Geometrical Invariants of Rationally Parametrized Conic Sections

Due to the increasing practical importance of Bézier representations, and since every rationally parametrized curve of degree 2 may easily be converted to this format, we now consider a nondegenerate conic section c in rational quadratic Bézier standard form in the projective plane $\mathbb{P}^2$ (see [4]) as

$$c: \ \mathbf{X}(t) = \begin{pmatrix} w(t) \\ \mathbf{b}(t) \end{pmatrix} = \begin{pmatrix} 1 \\ \mathbf{b}_0 \end{pmatrix} B_0^2(t) + w_1 \begin{pmatrix} 1 \\ \mathbf{b}_1 \end{pmatrix} B_1^2(t) + \begin{pmatrix} 1 \\ \mathbf{b}_2 \end{pmatrix} B_2^2(t), \quad (7)$$

where $\mathbf{b}_i$ are the affine control points, $w_1 > 0$ the interior weight, and $B_i^2(t) := \binom{2}{i}(1-t)^{2-i}t^i$ the Bernstein polynomials $(i = 0, 1, 2)$. According to [4] we obtain an ellipse for $w_1 < 1$, a parabola for $w_1 = 1$, and a hyperbola for $w_1 > 1$.

We thus address the problem of determining the conic's foci from its parametric representation (7). From those foci and (7), the other geometrical characteristics such as center, axes, and vertices are easily obtained as described below. According to Definition 1, the foci are the intersection points of the tangents of c through the circular points at infinity $K^{\pm}$.

An arbitrary tangent $\mathbf{t}$ of c has the representation

$$\mathbf{t}: \ \mathbf{Y}(\mu_0, \mu_1; t) = \mu_0 \begin{pmatrix} w(t) \\ \mathbf{b}(t) \end{pmatrix} + \mu_1 \begin{pmatrix} 0 \\ \mathbf{v}(t) \end{pmatrix}, \quad (\mu_0, \mu_1) \in \mathbb{R}^2 \setminus (0,0), \quad (8)$$

where $\mathbf{v}(t) := \dot{\mathbf{b}}(t)w(t) - \mathbf{b}(t)\dot{w}(t)$ for an arbitrary but fixed parameter value $t \in \mathbb{R}$..

The condition for a tangent $\mathbf{t}$ to contain the circular point $K^{\pm}$ at infinity thus reads

$$\det \begin{pmatrix} w(t) & 0 & 0 \\ \mathbf{b}(t) & \mathbf{v}(t) & \begin{pmatrix} 1 \\ \pm i \end{pmatrix} \end{pmatrix} = 0, \quad (9)$$

or equivalently (since $w(t) \neq 0$)

$$\det \left(\mathbf{v}(t), \begin{pmatrix} 1 \\ \pm i \end{pmatrix} \right) = 0. \quad (10)$$

By introducing the quantities

$$\alpha_1^{\pm} := \det\left(\mathbf{b}_1 - \mathbf{b}_0, \begin{pmatrix} 1 \\ \pm i \end{pmatrix}\right), \quad \alpha_2^{\pm} := \det\left(\mathbf{b}_2 - \mathbf{b}_0, \begin{pmatrix} 1 \\ \pm i \end{pmatrix}\right), \tag{11}$$

we now solve (10) in order to obtain the parameter values of the tangent points on c.

In the case of an ellipse or a hyperbola, i.e., $w_1 \neq 1$, we obtain

$$t_{1/2}^{\pm} = \frac{1}{2(w_1 - 1)\alpha_2^{\pm}}\left(2w_1\alpha_1^{\pm} - \alpha_2^{\pm} \pm \sqrt{4w_1^2\alpha_1^{\pm}(\alpha_1^{\pm} - \alpha_2^{\pm}) + (\alpha_2^{\pm})^2}\right), \tag{12}$$

and in the case of a parabola, i.e., $w_1 = 1$, we get

$$t^{\pm} = \frac{\alpha_1^{\pm}}{2\alpha_1^{\pm} - \alpha_2^{\pm}}. \tag{13}$$

According to (12), (13) we now need to make a case distinction: ellipse/ hyperbola ($w_1 \neq 1$) versus parabola ($w_1 = 1$).

3.1. Ellipse and Hyperbola

By carrying out the intersection (1), we obtain the points

$$\begin{aligned} \mathbf{Z}(\rho, \tau) := {} & w(\tau)\det(\mathbf{v}(\rho), \mathbf{v}(\tau))\begin{pmatrix} w(\rho) \\ \mathbf{b}(\rho) \end{pmatrix} \\ & + \det\begin{pmatrix} w(\rho) & w(\tau) & 0 \\ \mathbf{b}(\rho) & \mathbf{b}(\tau) & \mathbf{v}(\tau) \end{pmatrix}\begin{pmatrix} 0 \\ \mathbf{v}(\rho) \end{pmatrix}, \end{aligned} \tag{14}$$

where $P_1 = \mathbf{Z}(t_1^+, t_1^-)$, $P_2 = \mathbf{Z}(t_2^+, t_2^-)$, $Q_1 = \mathbf{Z}(t_1^+, t_2^-)$, $Q_2 = \mathbf{Z}(t_2^+, t_1^-)$.

Remark 1. *It is easy to verify that* $K^+ = \mathbf{Z}(t_1^+, t_2^+)$, $K^- = \mathbf{Z}(t_1^-, t_2^-)$.

With the help of the computer algebra system MAPLE [15], (14) can be reduced to the form

$$\mathbf{Z}(\rho, \tau) = \begin{pmatrix} 1 \\ \mathbf{b}_0 \end{pmatrix}(-\rho\tau + \rho + \tau - 1) - w_1\begin{pmatrix} 1 \\ \mathbf{b}_1 \end{pmatrix}(-2\rho\tau + \rho + \tau) - \begin{pmatrix} 1 \\ \mathbf{b}_2 \end{pmatrix}\rho\tau. \tag{15}$$

From the calculated foci, the axes of the conic c are now obtained as $\lambda_1 P_1 + \lambda_2 P_2$ and $\mu_1 Q_1 + \mu_2 Q_2$, $((\lambda_1, \lambda_2), (\mu_1, \mu_2) \in \mathbb{R}^2 \setminus (0, 0))$, and the center of c is the intersection point $\lambda_1 P_1 + \lambda_2 P_2 \cap \mu_1 Q_1 + \mu_2 Q_2$, or affinely speaking the midpoint of the points P_1, P_2 or Q_1, Q_2. In the case of an ellipse, the vertices are the intersection points $\lambda_1 P_1 + \lambda_2 P_2 \cap c$ and $\mu_1 Q_1 + \mu_2 Q_2 \cap c$, and in the case of a hyperbola, the vertices are obtained as intersection points of the axis containing the *real* foci with c.

In the case of a circle all foci coincide yielding the circle's center; in this case there don't exist distinct axes and vertices. The radius of the circle may be obtained by intersecting the circle with any line through its center point and by calculating the distance between the center point and one of the two intersection points.

3.2. Parabola

By carrying out the intersection (2), we obtain the point

$$F := \det(\dot{\mathbf{b}}(t^+), \dot{\mathbf{b}}(t^-)) \begin{pmatrix} 1 \\ \mathbf{b}(t^+) \end{pmatrix} + \det(\mathbf{b}(t^-) - \mathbf{b}(t^+), \dot{\mathbf{b}}(t^-)) \begin{pmatrix} 0 \\ \dot{\mathbf{b}}(t^+) \end{pmatrix}, \quad (16)$$

which may be simplified to

$$\begin{aligned} F = {} & (\mathbf{b}_2 - \mathbf{b}_1)^T (\mathbf{b}_2 - \mathbf{b}_1) \begin{pmatrix} 1 \\ \mathbf{b}_0 \end{pmatrix} \\ & - 2\,(\mathbf{b}_1 - \mathbf{b}_0)^T (\mathbf{b}_2 - \mathbf{b}_1) \begin{pmatrix} 1 \\ \mathbf{b}_1 \end{pmatrix} + (\mathbf{b}_1 - \mathbf{b}_0)^T (\mathbf{b}_1 - \mathbf{b}_0) \begin{pmatrix} 1 \\ \mathbf{b}_2 \end{pmatrix}. \end{aligned} \quad (17)$$

Since in this case the axes of c are not yet determined by the sole focus F, we also calculate the vertex S of the parabola c. According to (7), the parabola c has the affine representation $\mathbf{b}(t)$, and its vertex is the point with maximum curvature

$$\kappa(t) = \frac{\det(\dot{\mathbf{b}}(t), \ddot{\mathbf{b}}(t))}{|\dot{\mathbf{b}}(t)|^3}. \quad (18)$$

We thus obtain the vertex for $\dot{\kappa}(t) = 0$, which results in the parameter value

$$t_S = \frac{-(\mathbf{b}_1 - \mathbf{b}_0)^T (\mathbf{b}_0 - 2\mathbf{b}_1 + \mathbf{b}_2)}{(\mathbf{b}_0 - 2\mathbf{b}_1 + \mathbf{b}_2)^T (\mathbf{b}_0 - 2\mathbf{b}_1 + \mathbf{b}_2)}, \quad (19)$$

and therefore $S = \begin{pmatrix} 1 \\ \mathbf{b}(t_S) \end{pmatrix}$. The axes of the parabola c thus are $\nu_1 S + \nu_2 F$, $((\nu_1, \nu_2) \in \mathbb{R}^2 \setminus (0,0))$, and the tangent $\mathbf{Y}(\mu_0, \mu_1; t_S)$ (see (8)) of c in S.

3.3. Examples

We now illustrate the above method for finding the geometrical characteristics, such as foci, axes, and center, of a non–degenerate conic section c in parameter form. According to (7), the input data are the affine control points $\mathbf{b}_0, \mathbf{b}_1, \mathbf{b}_2$ and the interior weight w_1. The method has been implemented as a MAPLE program.

Example 1:
Input: Hyperbola with the control structure

$$\mathbf{b}_0 = (0,0)^T\,, \quad \mathbf{b}_1 = (0,1)^T\,, \quad \mathbf{b}_2 = (1,1)^T\,, \quad w_1 = 2.$$

Output:

- Real foci: $P_{1/2}\left(1, -\frac{1}{6}\,(1 \mp \sqrt{7}), \frac{1}{6}\,(7 \mp \sqrt{7})\right)$
- Imaginary foci: $Q_{1/2}\left(1, -\frac{1}{6}\,(1 \pm i\sqrt{7}), \frac{1}{6}\,(7 \mp i\sqrt{7})\right)$

- Center: $\left(-\frac{1}{6}, \frac{7}{6}\right)$
- Major axis direction: $(-1, 1)$
- Minor axis direction: $(1, 1)$

Example 2:

Input: Ellipse with the control structure

$$\mathbf{b}_0 = (0,0)^T\ ,\ \ \mathbf{b}_1 = (0,1)^T\ ,\ \ \mathbf{b}_2 = (1,1)^T\ ,\ \ w_1 = 1/2.$$

Output:

- Real foci: $P_{1/2}\left(1, \frac{1}{3}\,(2 \pm \sqrt{2}), \frac{1}{3}\,(1 \pm \sqrt{2})\right)$
- Imaginary foci: $Q_{1/2}\left(1, \frac{1}{3}\,(2 \pm i\sqrt{2}), \frac{1}{3}\,(1 \mp i\sqrt{2})\right)$
- Center: $\left(\frac{2}{3}, \frac{1}{3}\right)$
- Major axis direction: $(1, 1)$
- Minor axis direction: $(1, -1)$

Example 3:

Input: Parabola with the control structure

$$\mathbf{b}_0 = (0,0)^T\ ,\ \ \mathbf{b}_1 = (0,1)^T\ ,\ \ \mathbf{b}_2 = (1,1)^T\ ,\ \ w_1 = 1.$$

Output:

- Focus: $F\left(1, \frac{1}{2}, \frac{1}{2}\right)$
- Vertex: $S\left(1, \frac{1}{4}, \frac{3}{4}\right)$
- Axis directions: $(1, 1)\ , (1, -1)$

Example 4:

Input: Ellipse with the control structure

$$\mathbf{b}_0 = (0,0)^T\ ,\ \ \mathbf{b}_1 = (0,1)^T\ ,\ \ \mathbf{b}_2 = (1,1)^T\ ,\ \ w_1 = \frac{\sqrt{2}}{2}.$$

Output:

- Real foci: $P_{1/2}(1, 1, 0)$
- Imaginary foci: $Q_{1/2}(1, 1, 0)$
- Center: $(1, 1, 0)$
- Conic type $=$ circle

§4. Conclusion

Based on a definition from classical projective geometry, we presented an easy and unified approach for determining the geometrical invariants of a non–degenerate conic section in rational quadratic parameter form. The main emphasis herein lies in the determination of the conic's foci.

The method presented here has several advantages over other methods serving the same purpose. Comparing it to the method of Lee [14], e.g., which uses *several different* results from elementary geometry, it has the advantage of being more uniform, since the derivation is mainly based upon only *one* simple definition. The method by Goldman and Wang [9] relies on the invariants of a conic's rational quadratic parametrization under rational linear reparametrizations. It results in long algebraic manipulations which have to be carried out separately for ellipses, hyperbolas and parabolas respectively. The results here therefore seem to have the advantage over this simultaneously developed algebraic approach of being much more compact, uniform and shorter.

Moreover, our method seems to be well–suited for a generalization to quadric surfaces, i.e., to accomplishing the task of determining the geometrical characteristics of quadrics such as focal curves, center, axes, and vertices. This is due to the immediate generalization of Definition 1 to the projective definition of a quadric's *focal conics*. These curves lie in the symmetry planes of the quadric, and play the analogous role for quadrics that the foci do for conics, see, e.g., [10]. According to their projective definition, they are singular curves of the quadric's so–called focal developable surface, see, e.g., [12]. An application of this classical theoretical knowledge to the interesting task of determining the geometrical characteristics of quadrics from their rational quadratic parametrization will be addressed in due course.

References

1. Ball, A. A., CONSURF: Part one: introduction of the conic lofting tile, CAD **6** (1974), 243–249.
2. Blanc, C. and C. Schlick, Accurate parametrization of conics by NURBS, IEEE Computer Graphics & Appl. Nov. 1996, 64–71.
3. Clebsch, A. and F. Lindemann, *Vorlesungen über Geometrie I*, B.G. Teubner, Leipzig, 1906.
4. Farin, G., *Curves and Surfaces for Computer Aided Geometric Design*, Academic Press Inc., 1990.
5. Farin, G., *NURB Curves and Surfaces: From Projective Geometry to Practical Use*, A.K. Peters, Wellesley Massachusetts, 1995.
6. Frey, W. H. and D. A. Field, Designing Bézier conic segments with monotone curvature, Comput. Aided Geom. Design **17** (2000), 457–483.
7. Fudos, I. and C. M. Hoffmann, Constraint–based parametric conics for CAD, CAD **28** (1996), 91–100.

8. Giering, O. and H. Seybold, *Konstruktive Ingenieurgeometrie*, Carl Hanser Verlag, München, 1987.
9. Goldman, R. and W. Wang, Extracting geometric characteristics of conic sections from quadratic parameterizations, preprint, 2000.
10. Hilbert, D. and S. Cohn–Vossen, *Geometry and the Imagination*, Chelsea Publ. Comp., New York, 1952.
11. Kommerell, K., *Vorlesungen über analytische Geometrie der Ebene*, K. F. Koehler Verlag, Leipzig, 1941.
12. Kommerell, K., *Vorlesungen über analytische Geometrie des Raumes*, K. F. Koehler Verlag, Leipzig, 1940.
13. Kortenkamp, U., private communication, (http://www.cinderella.de/), 2000.
14. Lee, E. T. Y., The rational Bézier representation for conics, in *Geometric Modeling: Algorithms and New Trends, G. Farin (ed.)*, SIAM, Piladelphia, 1985, 3–19.
15. Heal, K. M., M. Hansen, and K. Richard, *Maple V Learning Guide for Release 5*, Springer, Berlin, 1997.

Gudrun Albrecht
Zentrum Mathematik
Technische Universität München
D–80290 München, Germany
`albrecht@ma.tum.de`
`http://www-m10.mathematik.tu-muenchen.de/~albrecht/`

A Scattered Data Approximation Scheme for the Detection of Fault Lines

G. Allasia, R. Besenghi, and A. De Rossi

Abstract. We propose a method for the localization of unknown fault lines of a surface only known at scattered data. Our detection scheme is divided into three steps: first, a nearest-neighbor searching procedure is applied; second, all the nodes near a fault line are picked out and collected in a set; then, a polygonal curve approximating the fault line is obtained. To select the nodes near the faults we use a cardinal radial basis interpolation formula. Numerical results are given, which show the efficiency of our scheme in comparison with similar ones. The method has important applications in several fields, for example in the oil industry, where automatic algorithms are required for the detection of faults from geological scattered data. Furthermore, the output of our algorithm is the natural input of a method, that we proposed [4], for representing faulted surfaces by means of radial basis near-interpolants.

§1. Introduction

In this paper we consider the problem of the detection of fault lines of a surface only known at scattered data. Faults represent discontinuities in geological layers caused by severe movements of the earth's crust and their localization plays an important role in geological applications, for example in finding oil. The topic has already been studied by various authors with different approaches [5,6,7,8,10,11]. We propose a method for the detection of faults, assuming there is a quite large set of points, irregularly distributed in a region of the plane, and the corresponding function values. The method is strictly connected with another, that we proposed about constrained surface approximation [4], which requires the preliminary knowledge of fault lines. Therefore, the output of the detection algorithm can directly be used as input for representing faulted surfaces.

The proposed detection scheme works in three steps. First, a cell-based search method is used to find a point set neighboring to each point of the data set. Then a cardinal radial basis interpolant (CRBI), based on the nearest

Mathematical Methods for Curves and Surfaces: Oslo 2000
Tom Lyche and Larry L. Schumaker (eds.), pp. 25–34.

ISBN 0-8265-1378-6

neighbors, is used to determine the fault points, that is, the points near a fault line. More precisely, if the value of the local approximant at each point is quite different from the corresponding function value, then the point is classified as a fault point. Finally, having collected the fault points, each fault line is represented as a polygonal curve, either obtained by a weighted least squares method or by the connection of barycenters.

The paper is organized as follows. In Section 2 we briefly explain the interpolation formula. The detection scheme is detailed in Section 3. Finally, Section 4 contains some numerical results which show the efficacy of our method.

§2. Local Approximation by CRBIs

To identify the fault lines, we approximate the given surface near a fault by an interpolation method, which we briefly recall in this section.

Definition 2.1. *Suppose we are given a set of points $S_n = \{x_i,\ i = 1,\dots,n\}$, distinct and generally scattered, in a domain $D \subset \mathbb{R}^s$, $(s \geq 1)$, with associated real values $\{f_i,\ i = 1,\dots,n\}$, and a linear space $\Phi(D)$, spanned by continuous real basis functions*

$$g_i : D \to \mathbb{R}, \qquad i = 1,\dots,n,$$

such that

$$g_i(x) \geq 0, \quad \sum_{i=1}^{n} g_i(x) = 1, \quad g_i(x_j) = \delta_{ij},$$

where δ_{ij} is the Kronecker delta. We define the Cardinal Basis Interpolant *(CBI) by*

$$F(x) = \sum_{i=1}^{n} f_i\, g_i(x).$$

The points x_i are the nodes or centers and the f_i are the function values. The f_i can be generated by an interpolated function $f(x)$, i.e., $f_i \equiv f(x_i)$.

It is evident that each CBI is characterized by its cardinal basis and the family of possible bases is theoretically infinite. Our attention will be focused on a wide class of CBIs, specified by the following constructive procedure due to Cheney (see, e.g., [1]).

Definition 2.2. *Let α be a continuous real function such that*

$$\alpha(x,y) > 0, \quad \text{if } x \neq y; \qquad \alpha(x,x) = 0; \qquad \forall x,y \in D.$$

Define the functions g_i by the equations

$$g_i(x) = \frac{\prod_{k=1,k\neq i}^{n} \alpha(x,x_k)}{\sum_{j=1}^{n} \prod_{k=1,k\neq j}^{n} \alpha(x,x_k)},$$

and the interpolant F by

$$F(x) = \sum_{i=1}^{n} f_i \, g_i(x) = \sum_{i=1}^{n} f_i \, \frac{\prod_{k=1,k\neq i}^{n} \alpha(x, x_k)}{\sum_{j=1}^{n} \prod_{k=1,k\neq j}^{n} \alpha(x, x_k)} \,,$$

or equivalently by

$$F(x) = \sum_{i=1}^{n} f_i \, \frac{1/\alpha(x, x_i)}{\sum_{j=1}^{n} 1/\alpha(x, x_j)} \,, \qquad F(x_i) = f_i \,, \quad i = 1, \ldots, n \,.$$

Property 2.3. *The interpolant F satisfies the following properties*

$$\begin{aligned} &F(x) \geq 0 \quad \text{if} \quad f(x) \geq 0 \,, && F(x) \equiv f(x) \quad \text{if } f(x) \text{ is a constant} \,, \\ &\min_i f_i \leq F(x) \leq \max_i f_i \,, && F(x) = \min_{u \in \mathbf{R}} \sum_{i=1}^{n} (f_i - u)^2 \, \frac{1}{\alpha(x, x_i)} \,. \end{aligned} \tag{2.1}$$

Therefore F is a convex combination of the function values and a positive linear operator. In particular, properties in the second row of (2.1) are "shape preserving" attributes of F.

For most applications it is convenient to consider $\alpha(x, y)$ as a function of a distance $d(x, y)$ defined on $D \subset \mathbb{R}^s$, i.e., $\alpha(x, y) = \phi(d(x, y))$. If $d(x, y)$ is the Euclidean metric $\|x - y\|_2$, then the interpolant F becomes independent of Euclidean transformations of the data set. Such an interpolant is called a Cardinal Radial Basis Interpolant (CRBI). As a significant example of distance-weighted interpolation, we recall Shepard's formula (see, for instance, [1]):

$$S(x) = \sum_{i=1}^{n} f_i \, \frac{\|x - x_i\|_2^{-p}}{\sum_{j=1}^{n} \|x - x_j\|_2^{-p}} \,, \qquad S(x_i) = f_i \,, \quad i = 1, \ldots, n \,, \tag{2.2}$$

where $p > 0$. Shepard's formula is not only invariant under translations and rotations, but also scale invariant. The special case of Shepard's interpolation in $\mathbb{R}^2$ in which ϕ is the square of the Euclidean metric (i.e., $p = 2$) is particularly favourable, since the basis elements g_i are then analytic functions. CRBIs offer some advantages with regard to other interpolants, because important properties hold [1]:

a) a subdivision procedure allows us to partition the domain, and to find approximations on each subdomain and collect them conveniently;

b) a multistage procedure makes it possible to process a selected subset of the whole data set, and then to improve the obtained result by processing additional subsets;

c) a recurrence relation makes it easy to insert or remove nodes.

In all these cases the repetition of the whole computation is not needed.

§3. Detection of Fault Lines

Let $S_n = \{P_i,\ i = 1, \ldots, n\}$ be a set of distinct and scattered data points in a domain $D \subset \mathbb{R}^2$, and $\{f(P_i),\ i = 1, \ldots, n\}$ a set of values of an unknown function $f : D \to \mathbb{R}$, which is discontinuous across a set $\Gamma = \{\Gamma_j, j = 1, \ldots, m\}$ of fault lines

$$\Gamma_j = \{\gamma_j(t) : t \in [0, 1]\} \subset D\,,$$

where γ_j are unknown parametric continuous curves. We assume that the function f is smooth on $D \backslash \Gamma$, where $D \subset \mathbb{R}^2$ is a bounded, closed and simply connected domain which contains the convex hull of S_n. In the following the notation $F(P; f, X)$ will be used to denote any CRBI applied to the function f on a set X, which coincides with S_n or is a subset of S_n.

Our aim is to describe a detection algorithm which gives approximations of the unknown fault lines. The algorithm is divided into three steps:

1) Preprocessing the data set: a cell-based search method is applied to find the point set neighboring to each node of the data set S_n; namely, we make a subdivision in cells and identify the points nearest to a data point within each cell.

2) Classification of fault points: to determine the fault points, that is, the points near a fault line, a CRBI is used. The decision for the detection is based on the size of the difference between a given function value at a node and the corresponding value of the local approximant constructed on a neighborhood of the node. All nodes of S_n which can be classified as near a fault line are selected in this way and collected in a set $\mathcal{F}(f; S_n)$.

3) Representation of fault lines: polygonal curves or other curves which approximate the fault lines are obtained from $\mathcal{F}(f; S_n)$ or another set opportunely derived from it, by the discrete least squares approximation or by a method based on the connection of barycenters.

3.1. Preprocessing

First, it is necessary to determine the data point set $\mathcal{N}_{P_j}$ neighboring to the point $P_j \in S_n$, in order to localize the interpolation formula on this set (excluding P_j) and to evaluate the difference between the data function value $f(P_j)$ and the interpolant $F(P; f, \mathcal{N}_{P_j})$ at P_j; the number $|f(P_j) - F(P_j; f, \mathcal{N}_{P_j})|$ will give a measure of the smoothness of the function around P_j. This preprocessing phase is a classical nearest-neighbor searching procedure. Among other methods, we choose the cell-based search due to Bentley and Friedman [3]. This procedure, although less efficient and accurate than some triangle-based methods, is easily extendible to higher dimensions [9].

We start subdividing the domain into a prescribed number of cells (Bentley suggested at most $n/3$) and locating all nodes of S_n in cells. In this way we get a sequence $\{\mathcal{N}_{P_j},\ j = 1, \ldots, n\}$, where $\mathcal{N}_{P_j}$ is the nearest-neighbor set of P_j, excluding P_j. The number $N_{P_j} = \#\mathcal{N}_{P_j}$ of the nearest-neighbor points of P_j is considered fixed in each approximation procedure and independent by P_j (so in the sequel $N_{P_j} = N_P$ and $N_P \geq 3$). However, in general, N_P depends on the data set and the test function; as a consequence, it will be chosen in the experimental phase, analyzing the obtained results. After the determination of the nearest-neighbor set $\mathcal{N}_{P_j}$ of P_j, we use the barycentric form of Shepard's formula (2.2) calculated on $\mathcal{N}_{P_j}$.

3.2. Classification

The preprocessing step provides for each $P_j \in S_n$ the nearest-neighbor set $\mathcal{N}_{P_j}$, namely, the set of the N_P points of S_n neighboring to P_j, with $P_j \notin \mathcal{N}_{P_j}$, obtained by a nearest-neighbor procedure. Then we consider the local interpolation formula $F(P; f, \mathcal{N}_{P_j})$ applied on the set $\mathcal{N}_{P_j}$, that is,

$$F(P_i; f, \mathcal{N}_{P_j}) = f(P_i), \qquad \text{for} \quad \text{each} \quad P_i \in \mathcal{N}_{P_j}.$$

Supposing F gives a good approximation to f in D, we expect that the error

$$\sigma(P_j) = |f(P_j) - F(P_j; f, \mathcal{N}_{P_j})|$$

constitutes a measure of the smoothness of f around P_j. Hence to identify the points near a fault line, we set a suitable threshold value σ_0 and compare the values $\sigma(P_j)$ and σ_0. If $\sigma(P_j)$ is less than σ_0, then we regard f as smooth in a neighborhood of P_j. On the contrary, if $\sigma(P_j)$ assumes a value greater than σ_0, we conclude that there is a steep variation of the function f at P_j and, therefore, it will be marked as a fault point. When the classification procedure of the nodes is terminated, we have characterized the set of fault points

$$\mathcal{F}(f; S_n) = \{P_j \in S_n : |f(P_j) - F(P_j; f, \mathcal{N}_{P_j})| > \sigma_0\}$$

consisting of all the nodes which belong to the fault lines or, at least, are very close to them.

A crucial point is the best choice of the threshold value σ_0. It seems a difficult task, mainly when a finite number of function values is the only available information. In this case, we start computing the maximum deviation

$$S = \max\{|f_i - f_j| : i > j, \text{for all } i, j = 1, 2, \ldots, n\}.$$

S can be obtained by an appropriate sorting procedure, included in the preprocessing phase, which gives the set of all deviations and, at the same time, supplies some information on the variation of f. Then we set

$$\sigma_0 = \epsilon \, S,$$

where $0 < \epsilon < 1$. If the function f is constant or nearly constant in $D \setminus \Gamma$, the CRBI property of reproducing exactly constant functions allow us to consider a value of σ_0 close to zero. Unfortunately, such a favourable situation does not occur in general. On the other hand, if ϵ is close to one, then it can happen that no point of S_n is identified as a fault point. Therefore, it is clear that ϵ must assume an intermediate value such that the classification algorithm individualizes the fault points, but it leaves out the points at which the function has only a steep variation. So, after the automatic computation of S, we set in the algorithm a value of ϵ, suggested by the experience, and we start the detection procedure to find the fault points. The procedure is completed successfully if the function f is smooth on $D \setminus \Gamma$ as we supposed at the beginning of Section 3. On the contrary, if the function f shows strong variations, the procedure can be repeatedly applied using decreasing values of ϵ.

3.3. Representation

The third step of our scheme constructs the fault lines starting from the set $\mathcal{F}(f; S_n)$. Given a fault point $P_j \in \mathcal{F}(f; S_n)$ and the corresponding nearest-neighbor set $\mathcal{N}_{P_j}$, we define

$$D_{P_j} = \{P \in D : d(P, P_j) \le R_{P_j}\},$$

where

$$R_{P_j} = \max\{d(P_i, P_j) : P_i \in \mathcal{N}_{P_j}\},$$

and

$$\bar{\mathcal{N}}_{P_j} = \mathcal{N}_{P_j} \cup \{P_j\}.$$

Then we order the $N_P + 1$ points in $\bar{\mathcal{N}}_{P_j}$ so that

$$f(P_{j1}) \le f(P_{j2}) \le \cdots \le f(P_{jN_P+1}).$$

The expected jump δ_j of f in the subdomain D_{P_j} is evaluated by considering the index $l_j \in \{1, \ldots, N_P\}$ such that

$$l_j = \min\{1 \le l < N_P + 1 : \delta_j = \max_{1 \le l < N_P + 1} \Delta f(P_{jl})\},$$

where Δ indicates the forward difference.

The part $\Pi_j = D_{P_j} \cap \Gamma$ of a fault line separates the set

$$\Delta_j^L = \{Q \in \bar{\mathcal{N}}_{P_j} : f(Q) \le f(P_{jl_j})\}$$

of all nodes of $\bar{\mathcal{N}}_{P_j}$ with lower function values from the set

$$\Delta_j^H = \{Q \in \bar{\mathcal{N}}_{P_j} : f(Q) > f(P_{jl_j})\}$$

containing those with higher function values. If Δ_j^L or Δ_j^H is the empty set, then we enlarge N_P and repeat the process, so that $\bar{\mathcal{N}}_{P_j}$ contains points lying in the two parts of the subdomain separated by the fault line. Having determined Δ_j^L and Δ_j^H in this way, we calculate the barycenters B_j^L and B_j^H of Δ_j^L and Δ_j^H, respectively. Then we find $B_j = (B_j^L + B_j^H)/2$ and collect it in $\mathcal{B}(f; S_n)$, the set of the barycenters.

In order to obtain curves representing the fault lines, we can consider different approaches. If there is only one fault in D and the corresponding set of barycenters can be well approximated by a simple function of one coordinate variable, for example a polynomial (see Fig. 3), then a discrete least squares method can be advantageously applied.

On the other hand, when we deal with complex situations (several faults, intersections or bifurcations of faults), the least squares method does not work well. Therefore, we derived a different technique, briefly explained below, in order to obtain polygonal representations of the faults. First, we subdivide the domain D by a regular grid and consider the grid cells containing points of $\mathcal{B}(f; S_n)$. Then we calculate the barycenters of the points of $\mathcal{B}(f; S_n)$ lying in each grid cell and collect them in a set $\mathcal{BB}(f; S_n)$. After that, the points belonging to $\mathcal{BB}(f; S_n)$ are sorted by applying a nearest-neighbor searching procedure. More precisely, we begin from one of the points of $\mathcal{BB}(f; S_n)$ nearest to a side of the domain D and we find the nearest-neighbor point to it. Then, the search is repeated until all the points of $\mathcal{BB}(f; S_n)$ are ordered. Finally, the ordered points are connected by straight line segments, so as to obtain the polygonal curve $\mathcal{L}(f; S_n)$ (see Fig. 1 and Fig. 2).

§4. Numerical Results

Numerical results have been obtained by using Shepard's formula, but other choices of CRBIs could work as well. We have considered n randomly scattered points $P_i, (i = 1, 2, .., n)$, in the square $[0,1] \times [0,1] \subset \mathbb{R}^2$ and the corresponding function values f_i. The detection scheme has been successfully tested against several functions with different kinds of faults, varying the sample dimension n and the threshold value σ_0.

The cell-based search method and the CRBIs allow us to partition the domain, to process the data in different stages, and to insert or remove nodes. This is particularly important in surveying phenomena, such as geodetic or geophysical ones, whose data are distributed in regions with different characteristics. In fact, it is generally necessary to improve the approximation of $F(x; f, S_n)$ to the function $f(x)$ on those subsets on which the accuracy was in a first time unsatisfactory.

Our algorithm has been implemented on a supercomputer CRAY T3E. This very powerful MIMD machine highlights the features and the flexibility of the algorithm, allowing some parallel computation. However, the algorithm works well on any monoprocessor workstation.

We sketch the results yielded by the detection algorithm considering three significant examples among those analyzed. In these cases we have taken $\epsilon = 0.04$.

In the first example the test function (already studied in [4])

$$f_1(x,y) = \begin{cases} y - x + 1, & \text{if } 0 \le x < 0.5, \\ 0, & \text{if } 0.5 \le x < 0.6, \\ 0.3, & \text{if } x \ge 0.6. \end{cases}$$

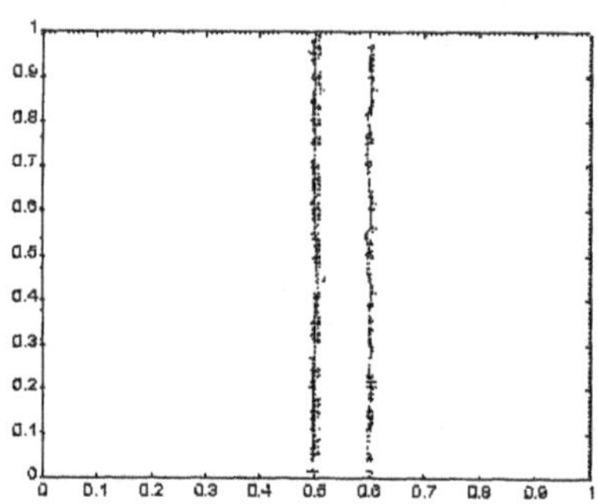

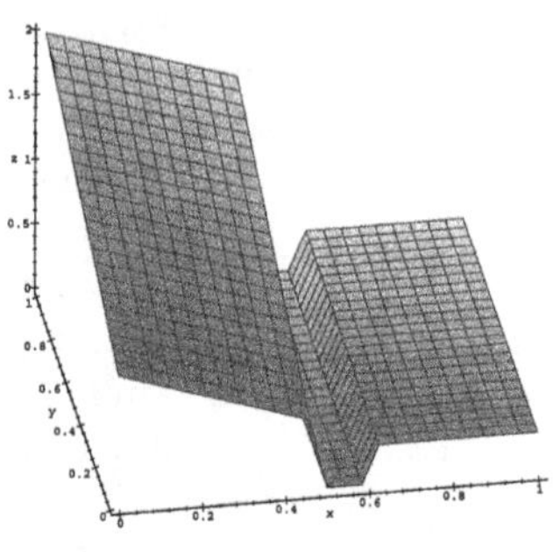

Fig. 1. Function $f_1(x, y)$.

has a sample of 10,000 points and the value of σ_0 is 0.077. Figure 1 shows the set of barycenter points and the reconstructed fault lines on a 15×15 grid (left), and the true surface (right).

In the second example the detection algorithm is applied to the function

$$f_2(x,y) = \begin{cases} 1 + 2\lfloor 3.5\|(x,y)^T\|_2 \rfloor, & \text{if } \|(x-0.5, y-0.5)^T\|_2 < 0.4, \\ 0, & \text{otherwise,} \end{cases}$$

studied in [7], where the symbol $\lfloor r \rfloor$ stands for the integer n satisfying $n \leq z < n+1$ for $z \in \mathbb{R}$. The sample includes 2,000 nodes and σ_0 is equal to 0.28.

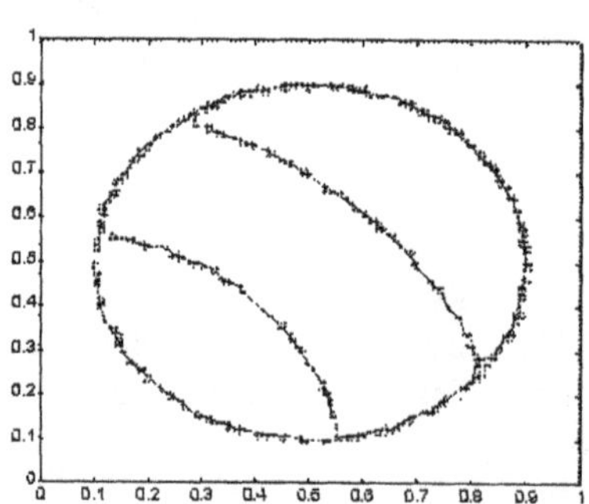

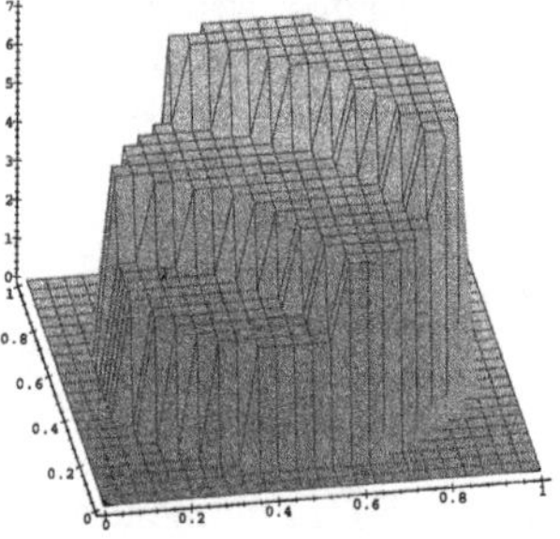

Fig. 2. Function $f_2(x, y)$.

Figure 2 shows the set of barycenter points and the reconstructed fault lines on a 23×23 grid (left) and the true surface (right). Three fault lines are represented in the picture: the circular curve which forms the boundary and the two arcs of curve in the interior.

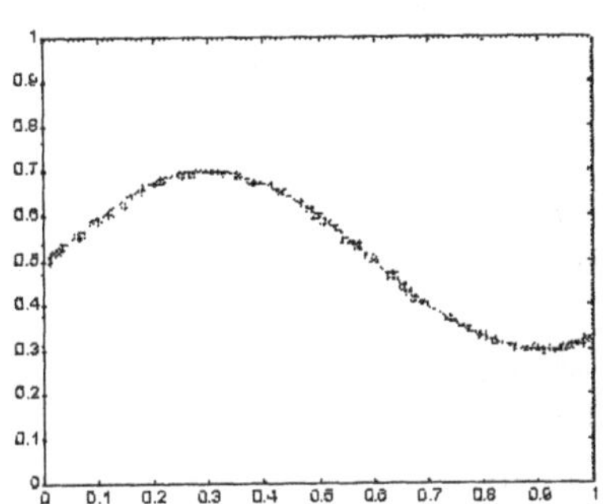

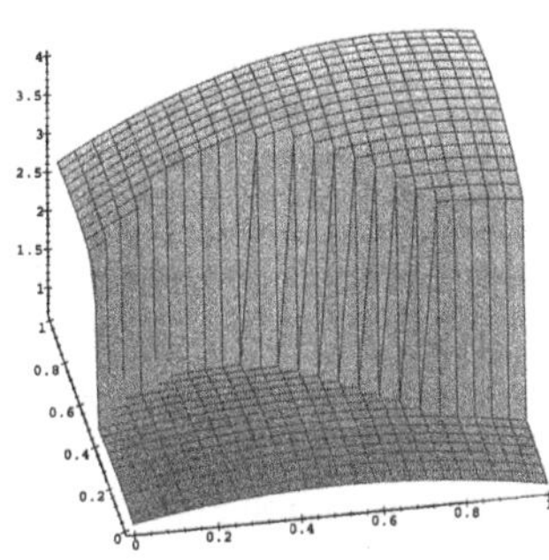

Fig. 3. Function $f_3(x, y)$.

For the first two examples we have used all the steps of the detection algorithm as described in Section 3, representing the fault lines by the polygonal curves.

The third test function [8]

$$f_3(x,y) = \begin{cases} (2-(x-0.8)^2)(2-(y-0.8)^2), & \text{if } y > 0.5 + 0.2\sin(5\pi x)/3, \\ (1-(x-0.5)^2)(1-(y-0.2)^2), & \text{otherwise}, \end{cases}$$

has a sample of 10,000 points and $\sigma_0 = 0.131$. Figure 3 shows the true surface on the right. On the left, we show the set of fault points computed by the first and second step of the detection algorithm, and the curve reconstructed using the least squares method.

§5. Concluding Remarks

In this paper we present an automatic algorithm for the detection of the unknown fault lines of a surface only known at scattered data. The only parameters involved are the number N_P of the nearest-neighbor points of a node and the threshold value σ_0. Both of them must be determined experimentally.

The multistage procedure and the recurrence relation for CRBIs allow the addition or reduction of data, if necessary, without repeating the whole computation. The entire procedure, from the detection and representation of faults to the rendering of the surface, could be considerably speeded up by using parallel computation [2].

Acknowledgments. This research was supported by the Italian Ministry of University, Scientific and Technological Research, and the University of Turin within the project "Advanced Numerical Methods for Scientific Computing". The authors would like to thank the referee and Larry L. Schumaker for suggesting improvements.

References

1. Allasia, G., A class of interpolating positive linear operators: theoretical and computational aspects, in *Approximation Theory, Wavelets and Applications*, S. P. Singh (ed.), Kluwer, Dordrecht, 1995, 1–36.
2. Allasia, G. and P. Giolito, Fast evaluation of cardinal radial basis intepolants, in *Surface Fitting and Multiresolution Methods*, A. Le Mèhautè, C. Rabut and L. L. Schumaker (eds.), Vanderbilt Univ. Press, Nashville, TN, 1997, 1–8.
3. Bentley, J. L. and J. H. Friedman, Data structures for range searching, Computing Surveys **11** (1979), 397–409.
4. Besenghi, R. and G. Allasia, Scattered data near-interpolation with applications to discontinuous surfaces, in *Curves and Surfaces Fitting*, A. Cohen, C. Rabut and L. L. Schumaker (eds.), Vanderbilt Univ. Press, Nashville, TN, 2000, 75–84.
5. Bozzini, M. and L. Lenarduzzi, Recovering a function with discontinuities from correlated data, in *Advanced Topics in Multivariate Approximation*, F. Fontanella, K. Jetter and P.-J. Laurent (eds.), World Scientific, Singapore, 1996, 1–16.
6. Fremming, N. P., Ø. Hjelle, and C. Tarrou, Surface modelling from scattered geological data, in *Numerical Methods and Software Tools in Industrial Mathematics*, M. Dæhlen and A. Tveito (eds.), Birkhäuser, Boston, 1997, 301–315.
7. Gutzmer, T. and A. Iske, Detection of discontinuities in scattered data approximation, Numerical Algorithms **16** (1997), 155–170.
8. Parra, M. C., M. C. Lopez de Silanes, and J. J. Torrens, Vertical fault detection from scattered data, J. Comput. Appl. Math. **73** (1996), 225–239.
9. Renka, R. J., Multivariate interpolation of large sets of scattered data, ACM Trans. Mathematical Software **14** (1988), 139–148.
10. Rossini, M., 2D-Discontinuity detection from scattered data, Computing **61** (1998), 215–234.
11. Springer, J., Modeling of geological surfaces using finite elements, in *Wavelets, Images and Surface Fitting*, P.-J. Laurent, A. Le Méhauté and L. L. Schumaker (eds.), A K Peters, Wellesley, Massachusetts, 1994, 467–474.

Giampietro Allasia, Renata Besenghi, Alessandra De Rossi
Department of Mathematics
University of Turin
Via Carlo Alberto 10
10123 Torino, Italy
allasia@dm.unito.it,besenghi@dm.unito.it,derossi@dm.unito.it

Beta-Continuity Revisited: Determining Bézier Control Vertices to Construct Geometrically Continuous Curves and Surfaces

Brian A. Barsky and Ronald N. Goldman

Abstract. In the previous Oslo conference held in June 1988, we observed that a sufficient condition for the affine combination of two geometrically continuous curves to be a geometrically continuous curve is that the two curves satisfy the Beta-constraints for the same β values [11]. This result was motivated by John Gregory's observation at the same conference (see page 361 of [15]) that an affine combination of two geometrically continuous curves does not necessarily yield a geometrically continuous curve. This anomaly causes havoc in various curve and surface constructions. For example, a ruled, lofted, or Boolean sum surface constructed from geometrically continuous curves need not be a geometrically continuous surface. Similarly, Catmull-Rom splines and rational curves constructed from geometrically continuous curves or geometrically continuous blending functions need not be geometrically continuous curves. Here, we examine geometrically continuous curves described in piecewise Bézier form that do not satisfy the Beta-constraints for the same β values. We know that we can reparametrize these Bézier curves so that they will satisfy the Beta-constraints for the same β values. From our previous work, it follows that after reparametrization, an affine combination of the two curves will necessarily be geometrically continuous. We show how to determine the locations of the Bézier control vertices of the same curves with a different parametrization so that any affine combination of the reparametrized curves will be geometrically continuous, thereby permitting classical smooth curve and surface constructions such as ruled and lofted surfaces and Catmull-Rom splines.

§1. Introduction

In the previous Oslo conference in June 1988, John Gregory [15] observed that an affine combination of two geometrically continuous [2,3,4,5,8,10,13] curves does not necessarily yield a geometrically continuous curve. Consequently, various curves and surfaces constructed from geometrically continuous curves or geometrically continuous blending functions might not be geometrically

Mathematical Methods for Curves and Surfaces: Oslo 2000
Tom Lyche and Larry L. Schumaker (eds.), pp. 35–44.

ISBN 0-8265-1378-6

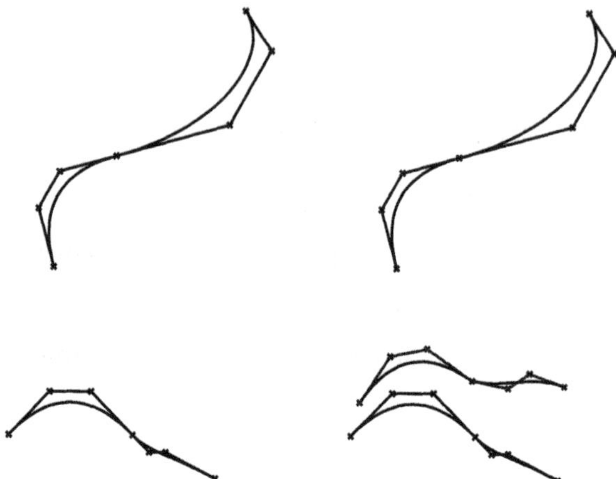

Fig. 1. Two 2-segment cubic Bézier curves with different β's (left) and their affine combination (right).

Fig. 2. Close-up of the affine combination of curves with different β's of Fig. 1.

continuous curves or surfaces. Such constructions include ruled surfaces, lofted surfaces, Boolean sum surfaces [14], Catmull-Rom splines [5,6,9], and rational curves.

We illustrate this anomaly with some examples. Figure 1 shows two piecewise Bézier curves where the value of the β_1 parameter at the joint in the top curve differs from that for the bottom curve (for a discussion of geometry continuity and β parameters, see the section entitled "Geometric Continuity, Beta-constraints, and Connection Matrices").

In Figure 1 (right), these curves are shown again, along with one of their affine combinations. Figure 2 shows a close-up of this affine combination, immediately revealing that it is not G^1 continuous, despite the fact that its two constituent curves are indeed G^1.

Figure 3 shows a ruled surface formed from two piecewise Bézier curves having different values of the β_1 parameter at the joint. An isoparametric curve is shown, again revealing a lack of G^1 continuity, despite the fact that the two rail curves are indeed G^1.

Figure 4 (left) depicts four boundary curves used to form the Boolean sum surface shown in Figure 4 (right). The boundary curves have $\beta_1 = \frac{1}{2}$, $\beta_2 = 0$; $\beta_1 = \frac{1}{3}$, $\beta_2 = \frac{1}{3}$; $\beta_1 = 2$, $\beta_2 = 2$; and $\beta_1 = 1$, $\beta_2 = \frac{1}{2}$. The uneven spacing of

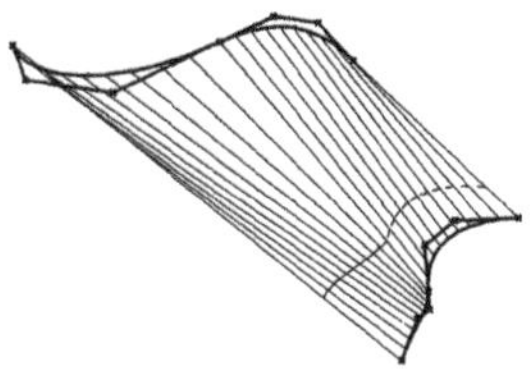

Fig. 3. Ruled surface with different β's.

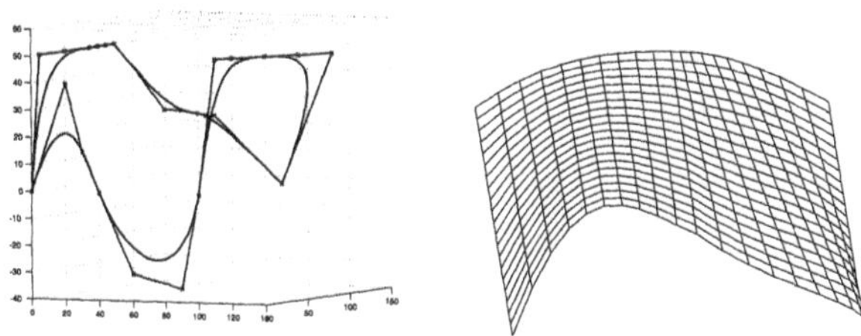

Fig. 4. Four boundary curves (left) and the Boolean sum surface, with different β's (right).

the isoparametric lines is indicative of the lack of geometric continuity.

At that same 1988 conference in Oslo, the present authors [11] observed that a sufficient condition for the affine combination of two geometrically continuous curves to be a geometrically continuous curve is that the two curves satisfy the Beta-constraints for the same values of all the β parameters. This result follows because functions that satisfy the Beta-constraints for the same values of all the β parameters form a linear space.

Consequently, this property enables the construction of various geometrically continuous curves and surfaces from geometrically continuous curves or geometrically continuous blending functions. These constructions include ruled surfaces, lofted surfaces, Boolean sum surfaces, Catmull-Rom splines, and rational curves.

Another pair of piecewise Bézier curves are shown in Figure 5 (left), but in this case the values of the β parameters at the joint are the same for both the top and bottom curves. An affine combination of these two curves is shown in Figure 5 (right), and it is evident that G^1 continuity has been maintained.

Figure 6 shows another ruled surface, where this time the surface is formed from two piecewise Bézier curves that have the same values of the β parameters at the joint. Here, again, the isoparametric curve illustrates that G^1 continuity has been maintained.

Fig. 5. Two piecewise Bézier curves with the same β's (left) and their affine combination (right).

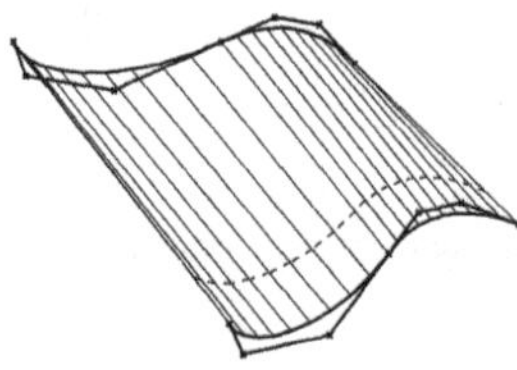

Fig. 6. Ruled surface with equal β's.

§2. Curves Described in Piecewise Bézier Form

In recognition of the return of this conference to Oslo, we now revisit this problem. Here, we examine geometrically continuous curves described in piecewise Bézier form that do not satisfy the Beta-constraints for the same values of the β parameters.

We can achieve geometric continuity by reparametrizing these curves so that they will satisfy the Beta-constraints for the same values of the β parameters. From our previous work [11], we know that if the two curves satisfy the Beta-constraints for the same values of all the β parameters, then an affine combination of these curves will be a geometrically continuous curve. Hence, it must be the case that after this reparametrization, an affine combination of the two curves will necessarily be geometrically continuous.

But there is a complication: If the reparametrization is not monotonic, surfaces constructed using the reparameterized curves will have undesirable features such as unwanted creases and folds. We shall show, however, that given certain mild assumptions, we can easily avoid such undesirable artifacts.

The reparameterization of these two curves that results in curves that will then satisfy the Beta-constraints for the same values of the β parameters will be achieved by determining new Bézier control vertices. Below, we will show how to determine the locations of these new control vertices that specify the reparametrized curves so that the two curves will satisfy the Beta-constraints for the same values of the β parameters.

An alternative approach for piecewise Bézier curves that are G^1 but not C^1 is given in [16]. Here we take an a very different approach that works not only for G^1, but for arbitrary G^n.

§3. Geometric Continuity, Beta-constraints, and Connection Matrices

Recall the definition of geometric continuity [2,3,4,8]. A piecewise regular curve $\mathbf{Q}(t)$ in $\mathbb{R}^d$, $d > 1$, is n^{th} order geometrically continuous, or G^n, at $\mathbf{Q}(t_0)$ if there exists a regular (nonzero derivative) reparametrization $t = h(u)$ such that $\mathbf{Q}(h(u))$ is n^{th} order parametrically continuous, or C^n, at $\mathbf{Q}(t_0)$.

It can be useful to consider the special cases of low order geometric continuity of G^1 and G^2. G^1 continuity is equivalent to the continuity of the unit tangent vector. G^2 continuity is equivalent to the continuity of the unit tangent and curvature vectors.

The Beta-constraints characterization of geometric continuity [2] provides a more concrete characterization of geometric continuity based on the parametrization at hand. This characterization uses the chain rule to express the jth lefthand derivative of the curve $\mathbf{Q}(t)$ at $t = t_0$ in terms of the first j righthand derivatives of the curve $\mathbf{Q}(t)$ at $t = t_0$. Specifically, a piecewise regular C^n parametrization $\mathbf{Q}(t)$ is n^{th} order geometrically continuous, or G^n, at $\mathbf{Q}(t_0)$ if and only if there exists real numbers $\beta_1, \ldots, \beta_n$ with $\beta_1 > 0$ such that

$$\mathbf{Q}^{(j)}(t_0^-) = \sum_{k=0}^{n} M_{jk}(\boldsymbol{\beta})\mathbf{Q}^{(k)}(t_0^+), \qquad j = 0, 1, \ldots, n$$

where the righthand side is the result of applying the chain rule to $\boldsymbol{Q}^{(s)}(h(u))$ with β_i substituted for $h^{(i)}(u_0)$, $i \geq 1$ and $\boldsymbol{\beta}$ denotes $\beta_1, \ldots, \beta_n$.

It follows that if $H(t) = t_0 + \beta_1(t - t_0) + \frac{\beta_2(t-t_0)^2}{2!} + \cdots + \frac{\beta_n(t-t_0)^n}{n!}$, then the curve

$$\begin{matrix} Q(t), & t < t_0, \\ Q(H(t)), & t > t_0 \end{matrix}$$

is C^n. Thus, the Beta-constraints indicate how to reparameterize a G^n curve so that the resulting curve is C^n.

Recall that the Beta-constraints can also be represented using a connection matrix [12]. A connection matrix $M(\boldsymbol{\beta}) = \{M_{jk}(\boldsymbol{\beta})\}$ is a lower triangular matrix. One new β parameter appears in each row of $M(\boldsymbol{\beta})$. The entries of a connection matrix are as follows:

$$M_{0k} = M_{k0} = \delta_{0k}$$
$$M_{j1}(\boldsymbol{\beta}) = \beta_j, \qquad j = 1, \ldots, n$$
$$M_{jj}(\boldsymbol{\beta}) = \beta_1^j, \qquad j = 1, \ldots, n$$

other entries of $M(\boldsymbol{\beta})$ are specific polynomials in $\beta_1, \ldots, \beta_n$.

Using this definition of a connection matrix enables the following characterization of geometric continuity. First, introduce the notation,

$$D_n(\boldsymbol{Q})(t) = (\boldsymbol{Q}(t), \boldsymbol{Q}^{(1)}(1)(t), \ldots, \boldsymbol{Q}^{(n)}(t))^t.$$

The Beta-constraints can then be written in matrix form as

$$D_n(\boldsymbol{Q})(t_0^-) = M(\boldsymbol{\beta}) \cdot D_n(\boldsymbol{Q})(t_0^+).$$

A β-continuous curve then satisfies this matrix equation for a specific set of values of the parameters $\boldsymbol{\beta} = (\beta_1, \ldots, \beta_n)$.

§4. Find Reparametrization Functions and Determine New Bézier Control Vertices

Now we must find reparametrization functions and Bézier control vertices. Given two geometrically continuous piecewise Bézier curves with connection matrices whose β values differ, we will find reparameterization functions so that the connection matrices have the same β values. Using these reparameterization functions, we will then find the new Bézier control vertices of the reparameterized Bézier curve.

First, we will reparametrize to construct connection matrices that have the same β values. Our objective is for two piecewise Bézier curves **P** and **Q** to have the same connection matrix (same β values) after we reparameterize both **P** and **Q**. Only one segment of each of these two curves needs to be reparametrized.

The easiest way to proceed is to reparametrize both curves so that their connection matrices are each the identity matrix; that is, so that the curves are both C^n. In the previous section, we saw that the β values indicate how to proceed. However, since **P** and **Q** are Bézier curves, we need to be careful not to change the domains of **P** and **Q**. Let

$$HP(t) = \hat{\beta}_1 t + \cdots + \frac{\hat{\beta}_n t^n}{n!}$$
$$HQ(t) = \tilde{\beta}_1 t + \cdots + \frac{\tilde{\beta}_n t^n}{n!}.$$

Assuming, as is most commonly the case, that the β's are nonnegative, then HP and HQ are monotonic on the interval (0,1). If $HP(1) < 1$, then we can simply add an additional term to HP so that

$$HP_{new}(t) = HP(t) + \frac{gt^{n+1}}{(n+1)!}, \qquad g > 0,$$

Fig. 7. Affine combination after reparametrization.

where the constant g is chosen so that $HP_{new}(1) = 1$. Notice that this new term does not affect the connection matrix of **P** because the connection matrix is only affected by the first n derivatives of **P** at $t = 0$. We do the same for HQ if $HQ(1) < 1$. On the other hand, if $HP(1) > 1$, then we can proceed in the following manner: Assume that $HP(1) > HQ(1)$. Choose any value of a so that $HP(a) < 1$, and set

$$HP_{new}(t) = HP(at)$$
$$HQ_{new}(t) = HQ(at).$$

Then **P** and **Q** still have the same connection matrices (no longer the identity) and we are back in the previous case. We conclude that if the original reparameterization functions are monotonic (for example, positive β's), then it is possible to introduce new polynomial reparameterizations so that after reparameterization:

(i) **P** and **Q** have the same connection matrix
(ii) the domains of **P** and **Q** are unchanged

With these reparameterization functions, we will determine the new Bézier control vertices of the reparameterized Bézier curve. To proceed, we convert the reparameterization functions from the Taylor basis to the Bernstein basis [1]. Then we compute the Bézier control vertices of the reparameterized segments of each curve, using either the algorithm given in [7] or equivalently the standard formula for blossoming the composite of two univariate polynomials **P** and **Q** of degrees m and n:

$$(p \circ q)(u_1, \ldots, u_{mn}) = \sum_{\tau} p(q(u_{\tau(1)}, \ldots, u_{\tau(n)}), \ldots, q(u_{\tau(mn-n+1)}, \ldots, u_{\tau(mn)}))$$

Thus, the overall algorithm is as follows:

1) Compute the appropriate polynomial reparameterization functions for both curves
2) Convert reparameterization functions from Taylor to Bernstein basis
3) Compute the Bézier control vertices of the reparameterized curves.

§5. Examples

We now provide some examples. Figure 7 shows an affine combination of the two piecewise Bézier curves having different β values that were shown in Figure 1. Recalling the affine combination that was shown in Figure 2, we show an affine combination of the reparameterized curves in Figure 7.

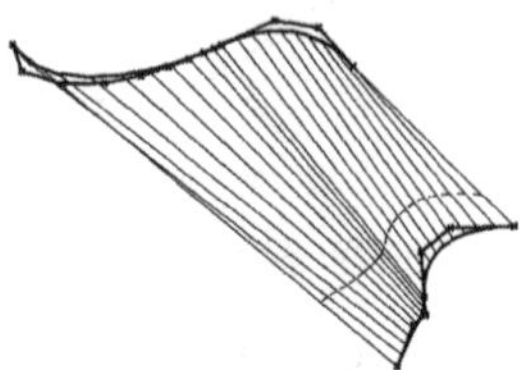

Fig. 8. Ruled surface after the rails are reparameterized.

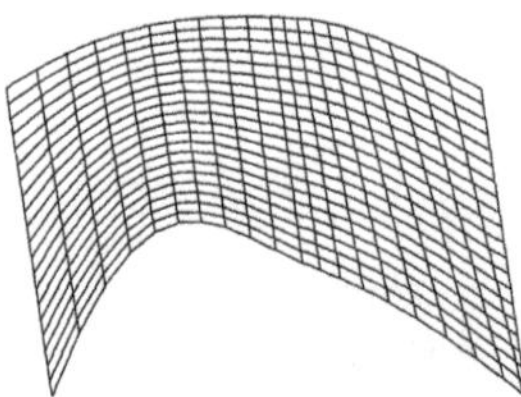

Fig. 9. Boolean sum surface after the rails are reparameterized.

Recall the ruled surface that was formed from the two piecewise Bézier curves having different values of the β_1 parameter at the joint that was shown in Figure 3. The reparameterized surface is shown in Figure 8.

An isoparametric curve is again displayed, and it shows G^1 smoothness. Finally, Figure 9 shows the Boolean sum surface that was shown in Figure 4 (right), after the rail curves have been reparameterized. Note how the isoparametric lines are now evenly spaced, indicative that geometric continuity has been maintained.

§6. Conclusion

We considered geometrically continuous curves described in piecewise Bézier form that satisfy the Beta-constraints for different values of β. We presented an algorithm to determine locations of the Bézier control vertices such that the two curves satisfy the Beta-constraints for the same values of β. Our algorithm permits classical smooth curve and surface constructions such as ruled surfaces, lofted surfaces, Boolean sum surfaces, Catmull-Rom splines, and rational curves.

Acknowledgments. We would like to thank Lillian Chu and Billy Chen, both of the Computer Science Division at the University of California, Berkeley, for their excellent assistance in the preparation of this paper.

References

1. Barry, Phillip J. and Ronald N. Goldman, Algorithms for progressive curves: extending B-spline and blossoming techniques to the monomial, power, and Newton dual bases, in *Knot Insertion and Deletion Algorithms for B-spline Curves and Surfaces*, Ronald N. Goldman and Tom Lyche (eds.), SIAM, 1993, 89–133.
2. Barsky, Brian A. and Tony D. DeRose, Geometric continuity of parametric curves, Technical Report No. UCB/CSD 84/205, Computer Science Division, Electrical Engineering and Computer Sciences Department, University of California, Berkeley, California, USA, October, 1984.
3. Barsky, Brian A. and Tony D. DeRose, Geometric continuity of parametric curves: three equivalent characterizations, IEEE Computer Graphics and Applications **9** (6) (1989), 60–68.
4. Barsky, Brian A. and Tony D. DeRose, Geometric continuity of parametric curves: constructions of geometrically continuous splines, IEEE Computer Graphics and Applications **10** (1) (1990), 60–68.
5. Bartels, Richard H., John C. Beatty, and Brian A. Barsky, *An Introduction to Splines for Use in Computer Graphics and Geometric Modeling*, Morgan Kaufmann Publishers, Inc., San Francisco, California, 1987.
6. Catmull, Edwin E. and Raphael J. Rom, A class of local interpolating splines, *Computer Aided Geometric Design*, Robert E. Barnhill and Richard F. Riesenfeld (eds.), Academic Press, New York, 1974, 317–326.
7. DeRose, Tony D., Composing Bézier simplexes, ACM Trans. on Graphics **7** (3) (1988), 198–221 (Figure 6 on page 208).
8. DeRose, Tony D. and Brian A. Barsky, An intuitive approach to geometric continuity for parametric curves and surfaces, *Proceedings of Graphics Interface '85*, Montreal, 343–351. Revised version published in *Computer-Generated Images – The State of the Art*, N. Magnenat-Thalmann and D. Thalmann (eds.), Springer-Verlag, 1985, 159–175.
9. DeRose, Tony D. and Brian A. Barsky, Geometric continuity, shape parameters, and geometric constructions for Catmull-Rom splines, ACM Trans. on Graphics **7** (1988), 1–41.
10. Dyn, Nira and Charles A. Micchelli, Piecewise polynomial spaces and geometric continuity of curves, RC 11390, IBM T.J. Watson Research Center, Yorktown Heights, New York, 1985.
11. Goldman, Ronald N. and Brian A. Barsky, On beta-continuous functions and their application to the construction of geometrically continuous curves and surfaces, *Mathematical Methods in Computer Aided Geometric Design*, T. Lyche and L. L. Schumaker (eds.), Academic Press, New York, 1989, 299–311.
12. Goldman, Ronald N. and Charles A. Micchelli, Algebraic aspects of geometric continuity, *Mathematical Methods in Computer Aided Geometric*

Design, T. Lyche and L. L. Schumaker (eds.), Academic Press, New York, 1989, 313–332.

13. Goodman, Tim N.T. and Keith Unsworth, Manipulating shape and producing geometric continuity in beta-spline curves, IEEE Computer Graphics and Applications **6** (1986), 50–56.
14. Gordon, William J., Spline-blended surface interpolation through curve networks, J. Math. Mech. **18** (1969), 931–952.
15. Gregory, John A., Geometry continuity, *Mathematical Methods in Computer Aided Geometric Design*, T. Lyche and L. L. Schumaker (eds.), Academic Press, New York, 1989, 353–371.
16. Hui, Kin-Chuen, Shape blending of curves and surfaces with geometric continuity, Computer-Aided Design **31** (13) (1999), 819–828.

Brian A. Barsky
University of California, Berkeley
EECS Computer Science Division
387 Soda Hall # 1776
Berkeley, CA 94720-1776
USA
barsky@cs.berkeley.edu
http://www.cs.berkeley.edu/~barsky

Ronald N. Goldman
Rice University
Department of Computer Science, MS 132
Duncan Hall
6100 Main Street
Houston, Texas 77005-1892
USA
rng@cs.rice.edu

Simplification of Closed Triangulated Surfaces Using Simulated Annealing

Peer-Timo Bremer, Bernd Hamann,
Oliver Kreylos, and Franz-Erich Wolter

Abstract. We describe a method to approximate a closed surface triangulation using simulated annealing. Our approach guarantees that all vertices and triangles in an approximating surface triangulation are within a user-defined distance of the original surface triangulation. We introduce the idea of atomic envelopes to guarantee error bounds that are independent of the surface geometry. Atomic envelopes also allow approximation distance to be different for different parts of the surface. We start with the original triangulation and perturb it randomly and improve an approximating triangulation by locally changing the triangulation, using a simulated annealing algorithm. Our algorithm is not restricted to using only original vertices; the algorithm considers every point inside the envelope triangulation as a possible position. The algorithm attempts to minimize the total number of vertices needed to approximate the original surface triangulation within the prescribed error bound.

§1. Introduction

Over the past two decades, data visualization has become increasingly important in several research areas, including medicine, fluid flow, and geographical data analysis. The speed of visualization algorithms has unfortunately not kept up with the speed of developing new technology producing high-resolution data. Every year, the quality of imaging and computational simulation technology — including laser scanners, digital cameras and radar systems — improves substantially. This results in such an increase in the amount of data that even state-of-the-art computers are stretched beyond their capacities. However, it has become apparent that for many applications, large parts of data sets are often not necessary for generating a good picture. The goal was and still is to reduce data sets in such a way that the pictures generated from a reduced data set are highly similar to those produced from the original one.

We are concerned with polygonal surfaces and their compression. Examples for polygonal surfaces are discretized height fields, parametric surfaces,

Mathematical Methods for Curves and Surfaces: Oslo 2000
Tom Lyche and Larry L. Schumaker (eds.), pp. 45–54.

ISBN 0-8265-1378-6

and manifold surfaces. We focus on triangulated two-dimensional (2D) manifolds with no boundaries. For an extensive overview of the field of polygonal surface simplification, we refer to Heckbert and Garland [6] and Rossignac [14].

We present a randomized algorithm that approximates triangulated, orientable 2D manifolds without boundaries, using a *min-#* approach, see §2. The algorithm preserves a specified error bound.

§2. Related Work

2.1 Two Approximation Types: Min-ϵ and Min-#

When approximating a polygonal surface using the min-ϵ approach, one has to determine, for a given number n, an approximation that consists of n vertices and minimizes the approximation error. Many of the common algorithms use min-ϵ optimization, and several references are given in [6,14]. Of special interest is Kreylos and Hamann [9], since they use a method closely related to the one presented here.

Using a min-# approximation approach, one tries to find an approximation with the minimal number of vertices that satisfies a tolerance condition [2]. This approach is relevant for scientific applications. For example, given the size of an object and the view-point distance, one can compute the error tolerance related to one pixel on the screen. Approximating the object within this tolerance results in a picture where each data point is no more than one pixel away from its original location. Computing min-# approximations can be very complicated and expensive. The error metric one wants to minimize is the number of vertices, faces or edges. Additionally, one has to stay inside an error bound. Our algorithm ensures that no point of the approximating surface deviates more than ϵ from the original surface. It is important to notice that this condition is stronger than to require that only the vertices be inside an error bound. It requires us to consider an *offset* around the original surface, and the approximation surface must stay inside this offset. Such an approach was first proposed by Cohen et al. [2], and was called simplification envelope. A simplification envelope is a linearized and, in some respects, simplified version of the exact offset.

2.2 Simplification Envelopes

The simplification envelope of a triangulated surface is constructed in the following way: For each vertex, one computes its normal $\vec{n}$ as a combination of the normals of the surrounding triangles, normalized to length ϵ; one defines two offset vertices, the $(+\epsilon)$-offset and the $(-\epsilon)$-offset vertices, by adding/subtracting $\vec{n}$ to/from the original vertex. This defines a so-called fundamental prism.

This approximation of the offset is close to the exact one as long as the original surface has low curvature. Our approach uses this type of envelope, but it provides the option to use better approximations.

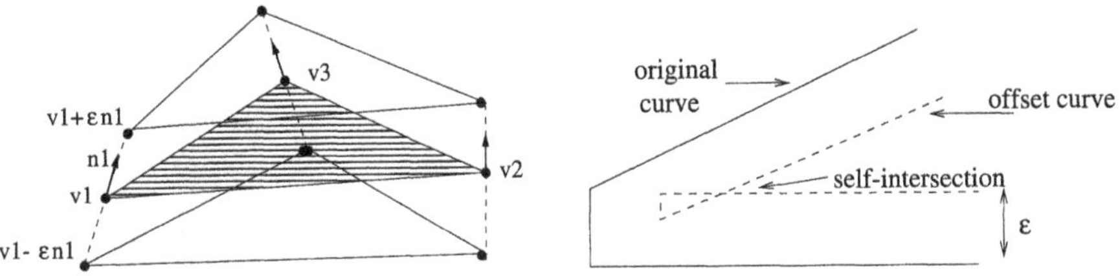

Fig. 1. (a) Fundamental prism; (b) self-intersecting offset curves.

The second problem is caused by self-intersections, see Fig. 1b. Cohen et al. [2] require a simplification envelope that does not self-intersect. They use the global ϵ-value whenever possible and decrease it in areas of possible self-intersections. Our approach is not impacted by self-intersections, and can handle every ϵ-value at any given vertex.

§3. Atomic Envelopes

To satisfy an a priori error bound, we define atomic envelopes. For each triangle, we construct an atomic envelope so that the simplification envelope equals the union of atomic envelopes. Our implementation uses fundamental prisms as atomic envelopes but different constructions are possible when higher accuracy is desired.

During simplification, we have to decide whether a triangle lies inside the simplification envelope. To answer this query we first find all atomic envelopes that might intersect the triangle. We are only interested in the top and bottom triangles that intersect the triangle in question. We use a bounding box test incorporating an R*-Tree [8] to speed up calculations. Especially for smaller error bounds, this results in roughly the same set of triangles one would get using triangulated offset surfaces to describe the simplification envelope. We intersect all resulting triangles with the triangle being tested. At each resulting intersection point, the triangle might leave the simplification envelope. It leaves the envelope if and only if the exit point is not covered by another atomic envelope. To test this, we use the fact that fundamental prisms are pentaeder Bézier volumina [10, 11] and solve the resulting non-linear system of equations.

Cohen et al. [2] define the side faces of a fundamental prism as bilinear patches, defined by the four corner points. Since we deal with closed triangulated surfaces, we do not have to consider the side patches. A triangle cannot leave the envelope through a side patch of a prism: The fundamental prisms of two neighboring triangles always share a common side patch, since the side patches correspond to the edges of the triangulated surface; thus a triangle leaving a prism through a side patch immediately enters another prism.

§4. Simplification

We simplify the given surface using a *simulated annealing* algorithm, also called *Metropolis algorithm* [12]. Simulated annealing models the state transition from fluid to crystalline state of metals. From the algorithmic viewpoint, this process is an optimization process with extremely high dimension. To apply simulated annealing to a general optimization problem, one needs to formulate the given problem as a cooling process. For our application, we interpret the configuration of a polygonal surface as the configuration of metal molecules. Our internal energy is represented by a target function, and the random heat movement of molecules is represented by random changes in the configuration. In general, we change a configuration randomly, see Section 4.2., accepting only changes that do not violate the error bound. We compute the new target function, see Section 4.1, and either accept or reject the change, following the rules of simulated annealing. We first applied this approach to the 2D case (simplification of closed polygons) and the good results encouraged us to extend it into 3D.

4.1. The Target Function

The target function describes the quality of an approximation. We experimented with different error norms, but the results were poor. In general, an error norm describes the difference between the original and the approximating surface. However, it is not the goal to minimize this difference but the number of vertices of the approximation. Furthermore, the target function should not only prefer configurations that consist of few vertices, but also configurations that lead to vertex removals. We use the sum of the square roots of the angles between triangle normals as the target function. This function is highly related to the number of vertices. Fewer vertices lead to fewer edges and to fewer angles to be added. This strategy prefers planar surfaces, since a large number of small angles has a higher target function value than a smaller number of large angles. This leads to near-planar platelets of triangles, where we can delete vertices. One could argue that, for planar surfaces, this target function is independent of the vertex number since all angles of neighboring triangles ("dihedral angles") are zero. However, in practice a mathematically planar surface cannot be represented by "truly co-planar" triangles due to numerical errors. Furthermore, this target function discourages self-intersections of the surface because self-intersections can only happen in regions of high curvature meaning high angles between triangles. It is also easy to compute and can be recomputed locally after local changes.

4.2. Configuration Changes

To change a configuration, we use the method of Kreylos and Hamann [9], adapted to our problem. In general, we approximate by decimation, like most of the published algorithms [3,4,5,13,14,15,16]. We use three different operations: edge flip, vertex removal, and vertex movement. The edge flip only changes the triangulation of two neighboring triangles. To move a vertex, we randomly choose a new position inside a small sphere around the original one. This enables us, like Hoppe et al. [7], to use more than just original points. However, they do not preserve a global error bound. We then move the vertex to the new position. To check geometric validity of the triangulation, we project all involved triangles onto the plane defined by the vertex normal of the changing vertex. In the case of non-convex platelets, this can lead to degenerate triangles or triangles with wrong orientations. We resolve these conflicts by flipping the appropriate edges. Nevertheless, it is possible that the projection method cannot detect all self-intersections. However, we are not aware of an approach for determining efficiently a plane for projection that is optimal. Our target function punishes self-intersections and therefore our method works well in practice. We store the sequence of the flipped edges in a stack, which allows us to undo the movement with minimal computational effort when the move is ultimately rejected by the simulated annealing scheme. To remove a vertex we collapse the shortest edge emanating from it. Therefore, we do not have to re-triangulate holes. (We have implemented the edge collapse operation by moving one vertex of the edge onto the other, using the standard vertex movement.)

There is a problem with vertex movement: The algorithm requires non-self-intersecting platelets. Yet it is possible to construct platelets where the projection onto the vertex-normal plane does self-intersect. However, these cases are highly unlikely to occur in real data sets and can be neglected for our purposes.

§5. Improvements and Future Research

The main drawback of our algorithm is its lack of computational efficiency. There are two reasons for this: First, tests involving the simplification envelope are expensive. However, considering our goal to satisfy an a priori error bound, this cost cannot be avoided. As mentioned before, even if triangulated representations of the offset surfaces were known, the complexity of the tests would not change significantly. Second, simulated annealing is expensive. On the other hand, a major advantage of simulated annealing is the fact that the random movements provide a mean to use any point inside the envelope as a possible vertex position. However, the large number of necessary movements and the tests involved result in a poor performance. Future work will be done to replace the simulated annealing approach with a more efficient alternative.

The same basic algorithm can also be used with more complicated atomic envelopes. An example is the construction shown in Fig. 2. This construction not only uses the vertex normal, but also the normal of the triangle to create the atomic envelope.

Compared to fundamental prisms, we add three bilinear patches to each atomic envelope, which can all have possible exit points. Furthermore, to test whether an exit point is covered by an atomic envelope is a consequently slower operation. The new atomic envelope embeds the the old one and one or two additional volumes above the top and below the bottom triangles. The

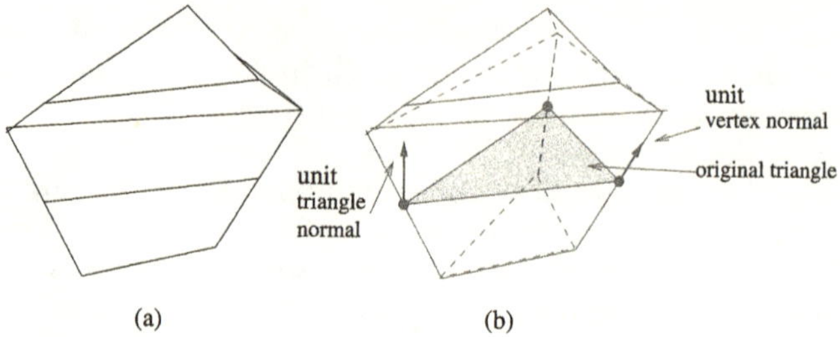

Fig. 2. Different atomic envelopes (a) solid view; (b) transparent view.

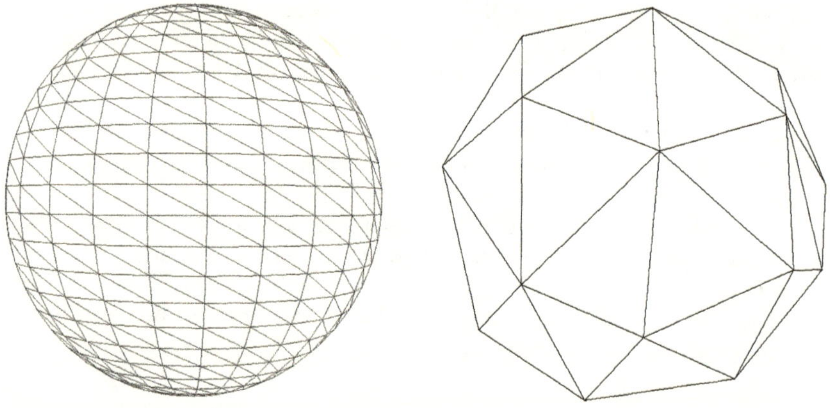

Fig. 3. Sphere data set: (a) original; (b) simplified within 3% error bound.

additional volumes can also be represented as pentaeder Bézier volumina. However, this new atomic envelope approximates the exact non-linear offset much better, especially in regions of high curvature.

§6. Results

All simplifications were performed on an SGI Octane with an R10000 processor, running at 250MHz and using 128MBytes of main memory. The two analytical data sets of the sphere and torus were the results of triangulating a parametric representation. The drill bit is available on the web pages of the Department of Computer Science at Stanford University [17]; it is a reconstruction of a laser-range scan. The cave data set was obtained by a range scan using a laser positioned in the center of the cave. For more results and pictures, we refer to [1].

Our absolute error bound, is the given percentage of the average side length of the bounding box of the original model. All times listed in Table 1 are in minutes. In Figures 9 and 10 the true shape is difficult to show, because from all meaningful viewing directions the points scanned from this cave are nearly co-planar.

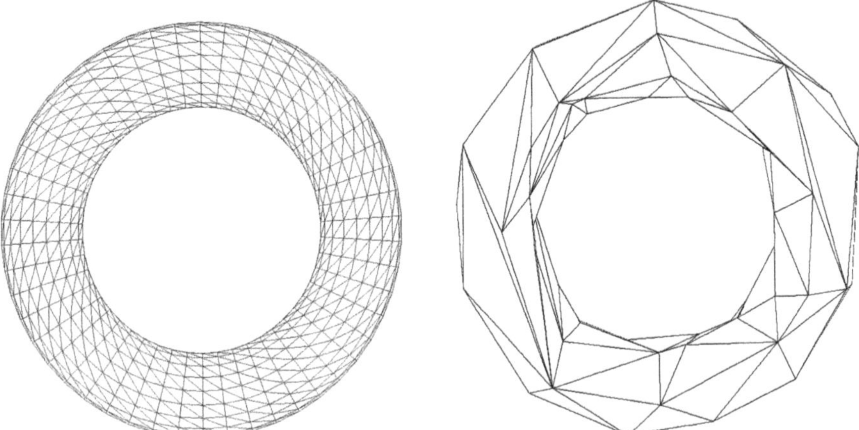

Fig. 4. Torus data set; (a) original; (b) simplified within 3% error bound.

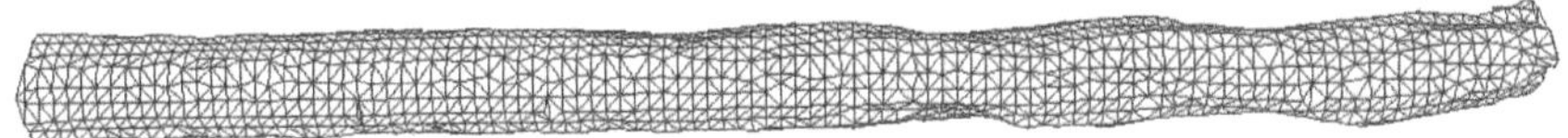

Fig. 5. Original drill bit data set.

Fig. 6. Drill bit data set simplified within 0.5% error bound.

Fig. 7. Original drill bit data set, flatshaded.

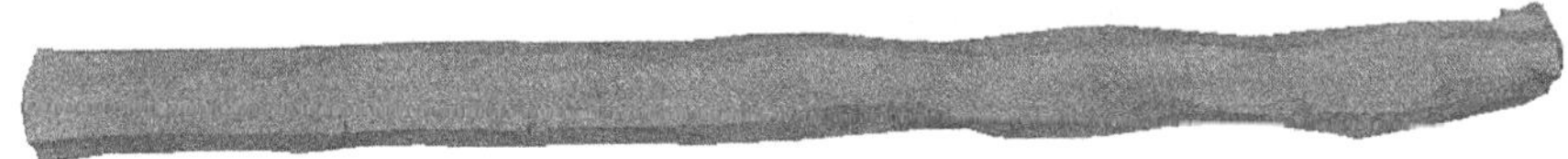

Fig. 8. Drill bit data set simplified within 0.5% error bound, flatshaded.

§7. Conclusions

We have presented an algorithm to simplify closed 2-manifold triangulations. The algorithm constructs a simplified triangulation within an a priori error bound. The concept of atomic envelopes allows us to use any error bound for any surface triangulation. The approach can easily be modified to be ap-

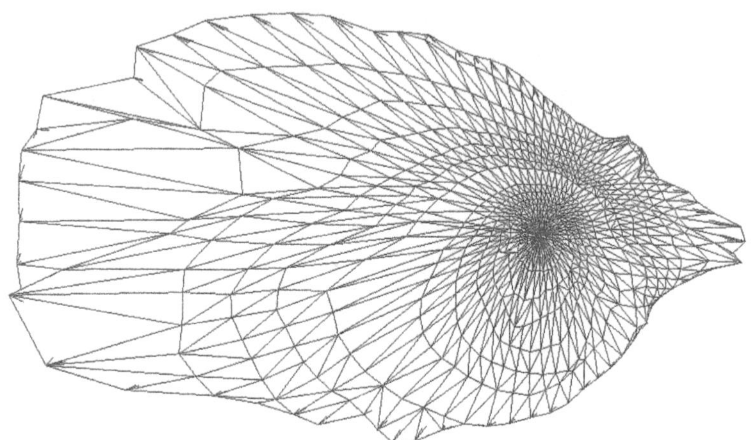

Fig. 9. Original cave data set.

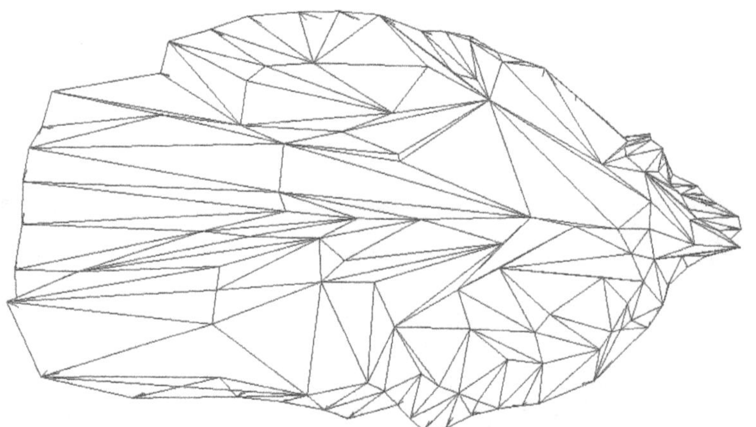

Fig. 10. Cave data set simplified within 1% error bound.

	Sphere		Torus		Drill Bit		Cave	
ϵ	Vertices	Time	Vertices	Time	Vertices	Time	Vertices	Time
0	602	0	1000	0	1946	0	1581	0
1/8	540	1	-	-	1589	2	849	1
1/4	358	1	843	1	981	2	611	2
1/2	200	1	493	1	430	2	374	2
1	101	1	231	1	144	3	198	3
3	29	2	72	2	18	9	81	7
5	18	2	39	2	-	-	-	-

Tab. 1. Results.

plicable to triangulated surfaces with boundaries or to non-manifold surfaces. Our approach can be extended to consider different error bounds in different regions, which would allow adaptive simplifications.

Acknowledgments.

This work was supported by the National Science Foundation under contracts ACI 9624034 (CAREER Award), through the Large Scientific and Software Data Set Visualization (LSSDSV) program under contract ACI 9982251, and through the National Partnership for Advanced Computational Infrastructure (NPACI); the Office of Naval Research under contract N00014-97-1-0222; the Army Research Office under contract ARO 36598-MA-RIP; the NASA Ames Research Center through an NRA award under contract NAG2-1216; the Lawrence Livermore National Laboratory under ASCI ASAP Level-2 Memorandum Agreement B347878 and under Memorandum Agreement B503159; the Lawrence Berkeley National Laboratory; the Los Alamos National Laboratory; and the North Atlantic Treaty Organization (NATO) under contract CRG.971628. We also acknowledge the support of ALSTOM Schilling Robotics and SGI. We thank the members of the Visualization and Graphics Research Group of the Center for Image Processing and Integrated Computing (CIPIC) at the University of California, Davis and the members of the Welfen Laboratory at the University of Hannover, Germany.

References

1. Bremer, P. T., Boundary simplification of a triangulated body, Diplomarbeit, Fachbereich Mathematik und Informatik, Universität Hannover, 10. Apr. 2000.
2. Cohen, J., A. Varshney, D. Manocha, G. Turk, H. Weber, P. Agarwal, F. Brooks and W. Wright, Simplification envelopes, Proceedings SIGGRAPH (1996), 119–128.
3. Gourdon, A., Simplification of irregular surface meshes in 3D medical images, Computer Vision, Virtual Reality and Robotics in Medicine, CVRMed 1995, Apr. 1995, 413–419.
4. Guéziec, A., Surface simplification with variable tolerance, Second Annual Intl. Symp. on Medical Robotics and Computer Assisted Surgery, MRCAS 1995, Nov. 1995, 132–139.
5. Hamann, B., A data reduction scheme for triangulated surfaces, Comput. Aided Geom. Design **11** (1994), 197–214.
6. Heckbert, P. S. and M. Garland, Survey of polygonal surface simplification algorithms, Multiresolution Surface Modeling Course, SIGGRAPH 1997.
7. Hoppe, H., T. DeRose, T. Duchamp, J. McDonald and W. Stuetzle, Mesh optimization, Proceedings SIGGRAPH, Aug. 1993, 19–26.

8. Klopp, S., Implementation von R*-Bäumen als benutzerdefinierte Indexstruktur in Oracle 8i, Studienarbeit, Fachbereich Mathematik und Informatik, Universität Hannover, 1999.
9. Kreylos, O. and B. Hamann, On simulated annealing and the construction of linear spline approximations for scattered data, Proceedings of the Joint EUROGRAPHICS-IEEE TVCG Symposium on Visualization, Vienna, Austria, May 1999, 189–198.
10. Lasser, D., Bernstein-Bézier-Darstellung trivarianter Splines, Dissertation, Fachbereich Mathematik, Technische Hochschule Darmstadt, Germany, 1987.
11. Lasser, D., Bernstein-Bézier representation of volumes, Comput. Aided Geom. Design **2** (1985), 145–150.
12. Metropolis, N., A. Rosenbluth, M. Rosenbluth, A. Teller, and E. Teller, Equations of state calculations by fast computing machines, Journal of Chemical Physics **21** (1953), 1087–1092.
13. Ronfard, J. and J. Rossignac, Full-Range approximation of triangulated polyhedra, Proceedings EUROGRAPIHCS, Computer Graphics Forum **15**, Aug. 1996.
14. Rossignac, J., Interactive exploration of distributed 3D databases over the internet, Proceedings CGI, Hannover, Germany, June 1998, 324–335.
15. Schroeder, W. J., J. A. Zarge, and W. E. Lorensen, Decimation of triangle meshes, Proceedings SIGGRAPH, Computer Graphics **26** (1992), 65–70.
16. Soucy, M. and D. Laurendau, Multiresolution surface modeling based on hierarchical triangulation, Computer Vision and Image Understanding **63** (1996), 1–14.
17. Web pages of the Department of Computer Science at Stanford University, http://www.graph- ics.stanford.edu/data/3Dscanrep/

Peer-Timo Bremer, Bernd Hamann, Oliver Kreylos
Center for Image Processing and Integrated Computing
Department of Computer Science
1 Shields Avenue
University of California
Davis, CA 95616-8562
USA
`tbremer@ucdavis.edu`, {`hamann`,`kreylos`}`@cs.ucdavis.edu`

Franz-Erich Wolter
Lehrstuhl für graphische Datenverarbeitung
Institut für Informatik
Universität Hannover
Welfengarten 1
30167 Hannover
Germany
`few@informatik.uni-hannover.de`

Planar Configurations with Simple Lagrange Interpolation Formulae

J. M. Carnicer and M. Gasca

Abstract. The geometric condition (GC) for multivariate interpolation is equivalent to the existence of a Lagrange formula whose terms are products of linear factors. In 1982, Gasca and Maeztu conjectured that any set of $(n+2)(n+1)/2$ points in the plane satisfying the GC condition must contain $n+1$ collinear points. The conjecture has only been proved for degrees $n \leq 4$. In this paper we classify some configurations of points in the plane satisfying the GC condition.

§1. Introduction and Auxiliary Results

Let $\Pi_n(\mathbb{R}^k)$ be the space of all polynomials in k variables of degree less than or equal to n, whose dimension is $\binom{n+k}{k}$. For any finite set $X \subseteq \mathbb{R}^k$ we may pose the

Lagrange interpolation problem. *Given $X \subseteq \mathbb{R}^k$ and $f \in \mathbb{R}^X$, find $p \in \Pi_n(\mathbb{R}^k)$ such that*

$$p(x) = f(x), \quad \forall x \in X. \tag{1.1}$$

Every polynomial p of degree not greater than n can be written in the form $p(x) = \sum_{|\alpha| \leq n} c_\alpha x^\alpha$, and the interpolation conditions give rise to the system of $|X|$ equations and $\binom{n+k}{k}$ unknowns

$$\sum_{|\alpha| \leq n} c_\alpha x^\alpha = f(x), \quad x \in X. \tag{1.2}$$

An interesting problem in multivariate interpolation is to infer the existence and uniqueness of the solution of the Lagrange interpolation problem from the distribution of the points in X. This leads to the following

Mathematical Methods for Curves and Surfaces: Oslo 2000
Tom Lyche and Larry L. Schumaker (eds.), pp. 55–62.

ISBN 0-8265-1378-6

Definition 1.1. We say that a set $X \subseteq \mathbb{R}^k$ is unisolvent in $\Pi_n(\mathbb{R}^k)$ if the Lagrange interpolation problem for X has a unique solution for any $f \in \mathbb{R}^X$.

Equation (1.2) confirms that a necessary condition for a set X to be unisolvent in $\Pi_n(\mathbb{R}^k)$ is that $|X| = \binom{n+k}{k}$. If $|X| = \binom{n+k}{k}$, the linear system (1.2) has the same number of equations and unknowns. Then any set X of $\binom{n+k}{k}$ points, X is unisolvent in $\Pi_n(\mathbb{R}^k)$ if and only if there exists no $p \in \Pi_n(\mathbb{R}^k)$ vanishing at all the points of X. This condition can be geometrically expressed by saying that not all points of X lie on the same algebraic hypersurface of degree less than or equal to n. The question of easily recognizing and generating unisolvent sets for posing Lagrange interpolation problems can be analyzed from several points of view. The Newton approach consists of finding a basis of functions of $\Pi_n(\mathbb{R}^k)$ vanishing on bigger and bigger subsets of X (see [3]). The Lagrange approach analyzes the existence and construction of certain functions called Lagrange polynomials.

Definition 1.2. For a set $X \subseteq \mathbb{R}^k$, we say that $l \in \Pi_n(\mathbb{R}^k)$ is a Lagrange polynomial associated to $x \in X$ if $l(x) = 1$ and $l(y) = 0$, for all $y \in X \setminus \{x\}$.

In view of this definition we can deduce the following proposition as a direct consequence of well-known results of Linear Algebra.

Proposition 1.3. *Let $X \subseteq \mathbb{R}^k$. Then the following properties are equivalent:*
(i) *X is unisolvent in $\Pi_n(\mathbb{R}^k)$.*
(ii) *For each $x \in X$ there exists a unique Lagrange polynomial $l_x \in \Pi_n(\mathbb{R}^k)$.*
(iii) *$|X| = \binom{n+k}{k}$ and there exists a Lagrange polynomial $l_x \in \Pi_n(\mathbb{R}^k)$ for all $x \in X$.*

Furthermore, the solution p of the Lagrange interpolation problem (1.1) can be expressed by the Lagrange formula

$$p = \sum_{x \in X} f(x) l_x. \tag{1.3}$$

The following properties of Lagrange polynomials will be useful throughout this paper:

Proposition 1.4. *Let $X \subseteq \mathbb{R}^k$ be unisolvent in $\Pi_n(\mathbb{R}^k)$, and let l_x be the Lagrange polynomial associated with $x \in X$. Then*
(i) *$\deg l_x = n$.*
(ii) *The factorization of l_x into irreducibles cannot have multiple factors.*
(iii) *For any polynomial g with $\deg g = r$, one has $|\{x \in X \mid g(x) = 0\}| \leq \binom{n+k}{k} - \binom{n-r+k}{k}$.*

Proof: (i) Let h be a polynomial of degree 1 vanishing on $x \in X$. If $\deg l_x < n$, then hl_x is a polynomial of degree less than or equal to n vanishing on X, contradicting the fact that X is unisolvent. (ii) If the factorization of l_x into irreducibles has repeated factors, then removing all the repeated factors, we would be able to construct a Lagrange polynomial of degree less than n which

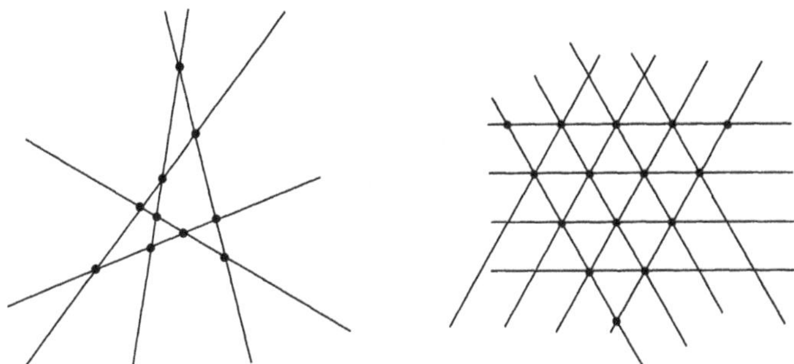

Fig. 1. A natural lattice (left) and a principal lattice (right).

is impossible by (i). (iii) Let $Y := \{x \in X \mid g(x) \neq 0\}$, and assume that $|Y| \leq \binom{n-r+k}{k}$. Then there exists a polynomial $f \in \Pi_{n-r}(\mathbb{R}^k)$ vanishing on Y. Now $fg \in \Pi_n(\mathbb{R}^k)$ vanishes on X, contradicting the unisolvence of X. □

In the case of polynomials of 1 variable, the Lagrange polynomials have a simple expression as a product of linear factors: $l_x(\xi) = \Pi_{y \in X \setminus \{x\}} \frac{\xi - y}{x - y}$. This formula does not have a simple extension to several variables, unless the points of X are structured in a special way.

Definition 1.5. Let $X \subseteq \mathbb{R}^k$ with $|X| = \binom{n+k}{k}$. The set X satisfies the geometric condition (GC_n) if for all $x \in X$, there exist affine functions h_i^x, $i = 1, \ldots, r$, $r \leq n$, such that the union of all hyperplanes $h_i^x = 0$ contains all points of $X \setminus \{x\}$, but not the point x. We say that $\{h_i^x = 0 \mid i = 1, \ldots, r\}$ is the set of hyperplanes associated with the point x. The set of all hyperplanes associated with some point $x \in X$ is denoted by Γ_X.

The GC condition, introduced by Chung and Yao [2], is equivalent to the existence of Lagrange polynomials which are a product of linear factors: $l_x = \Pi_{i=1}^n h_i^x$ (we may assume h_i^x normalized to have $h_i^x(x) = 1$). Therefore, if X satisfies the GC_n condition, then X must be unisolvent in $\Pi_n(\mathbb{R}^k)$. By Proposition 1.3, the set of hyperplanes associated with a point must be unique, and by Proposition 1.4 (i), (ii), it must have exactly n elements.

An interesting question is how to construct sets of points X satisfying the GC condition. Some important examples have been given in [2], such as natural lattices and principal lattices. Natural lattices are the set of intersection points of $n+2$ lines which are in general position, that is, no two of them are parallel and no three of them are concurrent. Principal lattices can be described as the intersection points of three families of $n+1$ parallel lines such that each point is the intersection of three lines, one of each family.

A generalization of principal lattices (also satisfying the GC condition) was provided in [4]. A pencil of lines is a set of lines intersecting at one point (the center of the pencil) or parallel lines (the center is at the infinity line). A 3-pencil lattice of order n is defined as a set of $\binom{n+2}{2}$ points generated by three pencils of $n+1$ lines each, in such a form that every point is the intersection

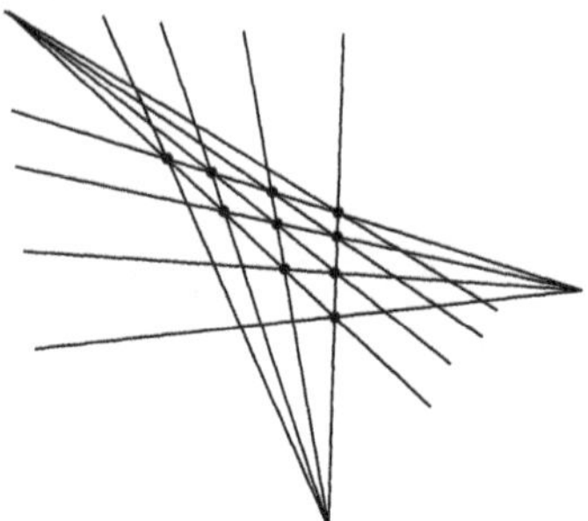

Fig. 2. A 3-pencil lattice.

of exactly one line of each pencil.

The distributions of points satisfying the GC condition have not been completely described even in the two-dimensional case. The combinatorics of the GC condition are so difficult to study that it still has not been possible to solve the conjecture on the GC_n condition on the plane made in [3]:

Conjecture 1.6. *Let $X \subseteq \mathbb{R}^2$ satisfy the GC_n condition. Then, there exists a line in Γ_X containing $n+1$ points of X.*

Based on the work of Bush [1], we know that the conjecture has been verified for degrees less than or equal to 4. The purpose of this paper is to offer a classification of some configurations of points satisfying the GC condition in the plane. This analysis could be a starting point for dealing with more complicated cases.

§2. Natural Lattices and Default

Let us summarize some properties of the GC configurations.

Proposition 2.1. *Let $X \subseteq \mathbb{R}^2$ satisfy the GC_n condition. Then*

(i) *Each line in Γ_X has at least 2 points of X.*

(ii) *For each point $x \in X$, there exist at least two lines in Γ_X containing x, associated with different points $y, z \in X$ ($n \geq 1$).*

(iii) *Γ_X contains at least $n+2$ lines.*

(iv) *A set of r lines cannot contain more than $r(2n+3-r)/2$ points of X. In particular, no line contains more than $n+1$ points of X.*

(v) *A line containing $n+1$ points of X must be in Γ_X, and it is associated with every point not lying on it ($n \geq 1$).*

(vi) *Two lines, each containing $n+1$ points of X, cannot be parallel, and meet at a point $x \in X$.*

(vii) *Three lines, each containing $n+1$ points of X, cannot be concurrent.*

(viii) *There are at most $n+2$ lines containing $n+1$ points of X.*

Proof: (i) Let $H \equiv h_j^y = 0$ be a line in Γ_X associated with $y \in X$. Assume that $\{x\} = H \cap X$ and let g be an affine function such that $g = 0$ is the line passing through x and y. Then $g\Pi_{i\neq j} h_i^y$ is a polynomial of degree n vanishing on X, contradicting the fact that X is unisolvent. (ii) Take any $y \neq x \in X$. Clearly, there exist H in the set of lines associated with y such that x lies in H. By (i), $H \cap X$ must contain a second point z. There must exist a line associated with z passing through x, which will be different from H. (iii) If we take $x \in X$, then there are n lines in Γ_X associated with x. By (ii), there are also 2 lines in Γ_X passing through x. (iv) follows directly from Proposition 1.4 (iii). (v) Let K be a line with $|K \cap X| = n+1$. Since $n \geq 1$, there must be at least one point $x \in X \backslash K$. Let $H_1, \dots, H_n$ be the lines associated with x, then $K \cap X = \bigcup_{i=1}^{n}(K \cap H_i)$. Since $|K \cap X| = n+1$, at least one of the sets $K \cap H_i$ has more than one point, and so $K = H_i$ is associated with x. (vi) If two lines containing $n+1$ points are parallel or they meet at a point not in the set X, then this set of two lines contains $2n+2$ points of X, a contradiction with (iv). (vii) If there were three concurrent lines, each of them with $n+1$ points of X, then this set of three lines would contain $3n+1$ points of X, which contradicts (iv). (viii) Let m be the number of lines with $n+1$ points of X. These lines cannot be either parallel or concurrent, and X must contain all pairs of intersections of lines. Therefore $\binom{m}{2} \leq |X| = \binom{n+2}{2}$, and $m \leq n+2$. □

In the sequel, the $\binom{n+2}{2}$ points of a set X satisfying the GC_n condition will be denoted by x_{ij}:

$$X = \{x_{ij} \in \mathbb{R}^2 | i < j \in \{1, \dots, n+2\}\}. \tag{2.1}$$

As we have seen in Proposition 2.1, an important subset of Γ_X is the set of lines $K_1, \dots, K_m$ containing $n+1$ points of X. From Proposition 2.1 (vi-viii) we deduce that these lines are in *general position* and their number allows us to establish a classification of sets satisfyng the GC_n condition.

Definition 2.2. Let $X \subseteq \mathbb{R}^2$ be a set satisfying the GC_n condition. We say that X has default d or that X is a d-lattice if the number of lines in Γ_X with $n+1$ points is just $n+2-d$.

Let $m = n+2-d$ and let $K_1, \dots, K_m$ be the lines with $n+1$ points of a set X with default d. From Proposition 2.1, all the intersection points of these lines are points of X. Then we can assume without loss of generality in formula (2.1) that

$$x_{ij} \in K_l \Longleftrightarrow l \in \{i, j\}, \tag{2.2}$$

which means that

$$x_{ij} = K_i \cap K_j,\ i < j \leq m;\ x_{ij} \in K_i,\ i \leq m < j;\ x_{ij} \notin \bigcup_{r=1}^{m} K_r,\ m < i < j.$$

By Proposition 2.1 (viii), $m \leq n+2$. Conjecture 1.6 means that $m \geq 1$. In fact, this number is at least 3 in all known examples. Principal lattices and 3-pencil lattices have exactly three lines with $n+1$ points of X. In other words, Conjecture 1.6 means that the default d of a set X satisfies $d \leq n+1$. We even conjecture that it is less than or equal to $n-1$. In the rest of the paper we completely describe sets with default 0,1 and 2.

Proposition 2.3. *Let X be a set satisfying the GC_n condition. Then the following properties are equivalent:*

(i) *X is a natural lattice.*

(ii) *X is a 0-lattice.*

(iii) *The lines associated with each $x \in X$ are the set of all lines of Γ_X not containing the point x.*

(iv) $|\Gamma_X| = n + 2$.

Proof: (i) $\Longrightarrow$ (iv) and (iii) follows from Proposition 2.1 (iii,v). (iii) $\Longrightarrow$ (iv): Let $n+2+k$ be the number of lines of Γ_X. By Proposition 2.1 (iii), $k \geq 0$. From (iii) we see that for each $x \in X$, there exist exactly n lines in Γ_X not vanishing at x and $k+2$ vanishing on it. Taking into account the intersections of the lines in Γ_X, we obtain that $\binom{k+2}{2}\binom{n+2}{2} \leq \binom{n+2+k}{2}$ which means that $k \leq 0$, and so, $k = 0$. (iv) $\Longrightarrow$ (ii): Let $\Gamma_X = \{K_1, \ldots, K_{n+2}\}$ and denote $r_i := |K_i \cap X|$. By Proposition 2.1 (iv), $r_i \leq n+1$, for all i and by Proposition 2.1 (ii), $r_1 + \cdots + r_{n+2} \geq 2|X| = (n+2)(n+1)$. So $r_i = n+1$ for all i. (ii) $\Longrightarrow$ (i): By Proposition 2.1 (vi)–(vii), the $n+2$ lines with $n+1$ points are in general position and X is formed by the $\binom{n+2}{2}$ intersection points. □

Now we describe all 1-lattices.

Proposition 2.4. *A set X given by (2.1) with $n > 1$ is a 1-lattice if and only if the following properties simultaneously hold:*

(i) *There exist lines $K_1, \ldots, K_{n+1}$ in general position such that (2.2) holds, that is,*

$$x_{ij} = K_i \cap K_j,\ i < j \in \{1, \ldots, n+1\}; \quad x_{i,n+2} \in K_i,\ i < n+2.$$

(ii) *Not all points $x_{i,n+2}$, $i = 1, \ldots, n+1$, lie on the same line.*

Proof: Assume that a set (2.1) satisfies (2.2). Clearly, each K_i has $n+1$ points. Let K_{ij} be the line containing $x_{i,n+2}$ and $x_{j,n+2}$. Then we have that the set of lines associated with x_{ij}, $i < j < n+2$, consists of the line K_{ij} and all the lines K_r, $r \neq i, j$. The set of lines associated with $x_{i,n+2}$ is K_r, $r \neq i$. Therefore the GC_n condition holds, and Γ_X consists of the lines K_i and K_{ij}. Since not all points $x_{i,n+2}$, $i = 1, \ldots, n+1$ lie on the same line, no line K_{ij} can contain $n+1$ points and then X is a 1-lattice. For the converse, if $K_1, \ldots, K_{n+1}$ are the lines with $n+1$ points, then the set X must contain all intersections $K_i \cap K_j$, and each line K_i must have an additional point. So, X satisfies (2.1)-(2.2). Since $K_1, \ldots, K_{n+1}$ must be the lines with $n+1$ points, not all points $x_{i,n+2}$ may lie on the same line. □

Let us observe that 1-lattices with $n = 1$ do not exist because a set satisfying GC_1 is trivially a natural lattice. In Figure 3, we show a lattice with default 1, and satisfying GC_3.

Now we provide a complete description of all 2-lattices.

Proposition 2.5. *A set X given by (2.1) with $n > 2$ is a 2-lattice if and only if the following properties simultaneously hold:*

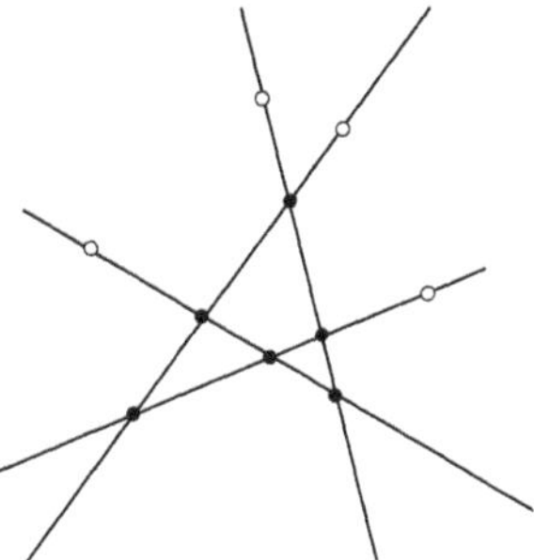

Fig. 3. A lattice with default 1.

(i) *There exist lines* $K_1, \ldots, K_n$ *in general position such that (2.2) holds.*
(ii) *There exist lines* L_1, L_2, L_3 *such that* $x_{n+1,n+2} = L_1 \cap L_2 \cap L_3$ *and* $\{x_{i,n+1}, x_{i,n+2}\} \subseteq K_i \cap (L_1 \cup L_2 \cup L_3)$ *for all* $i < n+1$.
(iii) *No line* L_r, $r = 1, 2, 3$, *contains* $n+1$ *points of* X.

Furthermore, each of the lines L_r *must have at least 3 points of* X.
Proof: Let us show first that a set (2.1) with (i), (ii), (iii) satisfies the GC_n condition. For $x_{n+1,n+2}$ the associated lines are $K_1, \ldots, K_n$. From (ii), there exist indices $j, k \in \{1, 2, 3\}$ such that

$$x_{i,n+1} = K_i \cap L_j, \quad x_{i,n+2} = K_i \cap L_k.$$

Then the set of lines associated with $x_{i,n+1}$ consists of L_k and K_r, $r \neq i$. Analogously, the set of lines associated with $x_{i,n+2}$ consists of L_j and K_r, $r \neq i$. Finally, given $i < j \leq n$, the three lines L_1, L_2, L_3 contain the four points $x_{i,n+1}, x_{i,n+2}, x_{j,n+1}, x_{j,n+2}$, and therefore there exists k such that L_k contains two of them. Let H be the line connecting the other two. So K_r, $r \neq i, j$, L_k and H are the lines associated with x_{ij}. We have thus checked the GC_n condition. On the other hand, X cannot be a natural or a 1-lattice. Indeed, if there exists another line with $n+1$ points, it would contain the point $x_{n+1,n+2}$ and one point of $(L_1 \cup L_2 \cup L_3) \cap K_i$ for each i. Since $n > 2$, this line must be one of the L_1, L_2 or L_3, contradicting (iii).

Conversely, let X be a 2-lattice. There exist lines $K_1, \ldots, K_n$ with $n+1$ points and X contains all the points $x_{ij} = K_i \cap K_j$, $i < j \leq n$. Each line K_i contains two additional points $x_{i,n+1}, x_{i,n+2}$, $i \leq n$. The set X must still have a point $x_{n+1,n+2}$ not belonging to any of the lines $K_1, \ldots, K_n$. So, (2.2) holds for $m = n$. By Proposition 2.1 (v), all lines K_r, $r \neq i, j$, are associated with x_{ij}, $i < j \leq n$. The set of lines associated with x_{ij} must contain two more lines with the five points $\{x_{i,n+1}, x_{i,n+2}, x_{j,n+1}, x_{j,n+2}, x_{n+1,n+2}\}$. Therefore three of these points lie on the same line, say H_{ij}: one is $x_{n+1,n+2}$, the second one is in $\{x_{i,n+1}, x_{i,n+2}\}$ and the third one in $\{x_{j,n+1}, x_{j,n+2}\}$. Since the default is 2, H_{ij} cannot contain $n+1$ points of X. So, for each i, j, there exists $k \neq i, j$ such that the lines H_{ij}, H_{ik}, H_{jk} are different, they are concurrent at $x_{n+1,n+2}$ and $x_{r,n+1}, x_{r,n+2} \in (H_{ij} \cup H_{ik} \cup H_{jk}) \cap K_r$, $r = i, j, k$. Let us define

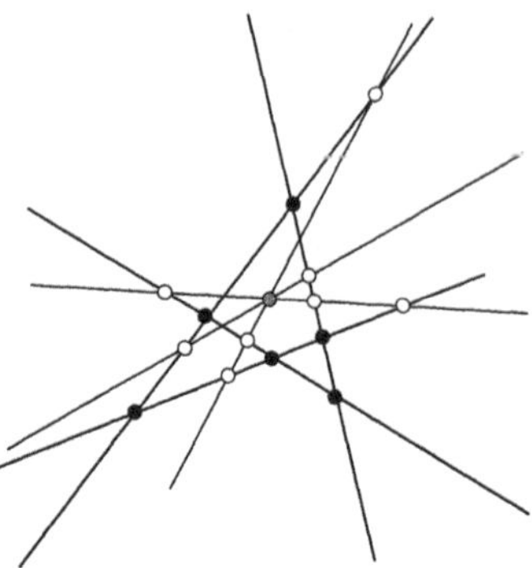

Fig. 4. A lattice with default 2.

$L_1 := H_{ij}, L_2 := H_{ik}, L_3 := H_{jk}$. Now, for $r \neq i, j, k$ one has $H_{ir} \in \{L_1, L_2\}$, $H_{jr} \in \{L_1, L_3\}$, $H_{kr} \in \{L_2, L_3\}$. Two of these lines are different. That means that $x_{r,n+1}, x_{r,n+2}$ lie in two of the three lines L_1, L_2, L_3, and we see that (ii) holds. Since Γ_X has only n lines with $n+1$ points, (iii) follows immediately. Finally, we have also shown that L_1 contains at least three points: one is $x_{n+1,n+2}$, a second one in K_i and a third one in K_j. On the other hand L_1 does not intersect K_k, and so it has at most $n-1$ points of $X \setminus \{x_{n+1,n+2}\}$. Analogously L_2, L_3 also contain at least three and at most n points of X. □

For $n = 2$ there are no GC_2 set X with default 2. Indeed, from Proposition 2.1, it is very easy to deduce that there exist at least 3 lines with 3 points. Figure 4 shows a lattice with default 2, satisfying GC_4.

Acknowledgments. This research was partially supported by the Spanish Research Grant DGES PB96–0730.

References

1. Busch, J. R., A note on Lagrange interpolation in $\mathbb{R}^2$, Revista de la Unión Matemática Argentina **36** (1990), 33–38.
2. Chung, K. C. and T. H. Yao, On lattices admitting unique lagrange interpolation, SIAM J. Numer. Anal. **14** (1977), 735–743.
3. Gasca, M. and J. I. Maeztu, On Lagrange and Hermite interpolation in $\mathbb{R}^n$, Numer. Math. **39** (1982), 1–14.
4. Lee, S. L. and G. M. Phillips, Construction of lattices for Lagrange interpolation in projective space, Constr. Approx. **7** (1991), 283–297.

J. M. Carnicer and M. Gasca
Universidad de Zaragoza
Edificio de Matemáticas, Planta 1a
50009 Zaragoza, Spain
`carnicer@posta.unizar.es` and `gasca@posta.unizar.es`

Approximating Semi-Structured Data with Different Errors Using Support Vector Machine Regression

A. Crampton, J. C. Mason, and D. A. Turner

Abstract. In this paper we present a new algorithm for approximating data that lie on a set of curved paths by using support vector machine (SVM) regression. The algorithm is designed to exploit the structure inherent in the abscissae, which is done by fitting a tensor-product surface to the data. This allows substantial computational savings to be made. In addition, it is assumed that the measurements made along each curved path may be corrupted by different levels or types of noise. SVM regression is ideally suited to cope with this type of problem, and we show how the algorithm can be modified to overcome any difficulties caused by different types of noise in the data.

§1. Introduction

Support vector machines (SVMs) are a new tool for the approximation of multivariate data, and we present here a new algorithm for using SVMs to approximate data collected on a number of curved paths in the xy-plane. This type of data set occurs, for example, in oceanographic data collection where several ships may gather measurements of the sea-bed level. Typically, a scattered data approach is used to approximate curved data, despite the structure inherent in the abscissae. Furthermore, some curves of data are likely to be corrupted by a different level or type of noise, since different sources have been used to gather each curve of data. However, this fact tends to be ignored in existing approaches.

In this paper, we describe a new algorithm for curved surface fitting which takes into account both the structure and noise of the data. Broadly speaking, the algorithm consists of two distinct stages. In the first stage, an SVM approximant to each curve of data is constructed, so that the original data set can be replaced by a new data set which lies on a family of parallel lines. SVMs are ideally suited to perform this task when there are differing

Mathematical Methods for Curves and Surfaces: Oslo 2000
Tom Lyche and Larry L. Schumaker (eds.), pp. 63–72.

ISBN 0-8265-1378-6

errors in the data, since an important feature of SVMs is that they can cope with a wide variety of types of noise. In the second stage, the new data on lines are approximated by a tensor-product surface. The structure of the new data set is exploited to allow substantial computational savings.

§2. The Secant Lines Method

In this section we show how the curved data can be replaced by a set of 'approximated' data that lie on parallel, or secant, lines. Initially, this is done by using SVM regression to approximate both the abscissae and the ordinate values for each path. This is a key feature of the algorithm, since by selecting an appropriate loss function in the approximation process we can remove the noise in the data. The approach also has the benefit of being computationally cheap, since we are only forming univariate approximants. Each curve of data is assumed to be corrupted by a different level or type of noise, and we discuss how SVM regression can be used to overcome this difficulty. Once the set of approximants have been constructed, a new set of data is obtained by finding the intersections between the approximants and a set of parallel lines.

2.1 Support Vector Machine Regression

SVM regression is not especially well-known, and therefore we give a brief overview of it here. In particular, we consider SVM regression for the case where we wish to approximate a set of curved data. We begin by discussing the most frequently used type of SVM regression, in which the so-called epsilon-insensitive loss function is minimized. We then consider SVM regression for general loss functions.

Suppose that we are given ordinate values collected on a set of m_l curved paths. We assume that each path is an explicit curve $y_l(x)$ and that $m^{(l)}$ observations are made on each path. We also assume that the kth ordinate value collected on the path $y_l(x)$ is given by $f_{k,l} \in \mathbb{R}$. We construct in turn local approximations $\{\hat{y}_l(x)\}$ to each curved path and local approximations $\{\hat{f}_l(x)\}$ to the ordinate values. These approximations are obtained after considering the error distribution associated with each source, selecting an appropriate loss function, and applying SVM regression. Thus, in the first step of the algorithm we remove the need for further error consideration. This involves approximating a set of discrete data $\{(\boldsymbol{x}_i, y_i)\}_{i=1}^n$ by a function of the linear form

$$f(\boldsymbol{x}) = \sum_{i=1}^{n} (\alpha_i^* - \alpha_i) K(\boldsymbol{x}_i, \boldsymbol{x}) + b, \tag{1}$$

where the parameters $\{(\alpha_i, \alpha_i^*)\}_{i=1}^n$ and b are determined as part of the approximation process. Note that we have written equation (1) in terms of $\boldsymbol{x} \in \mathbb{R}^d$, despite the fact that here we are only interested in the case $x \in \mathbb{R}$. We have done this so as to maintain consistency with standard discussions on SVM regression. For more details about the properties of the linear form (1), see Vapnik [7], for example. The function $K(\boldsymbol{x}_i, \boldsymbol{x})$ is known as a kernel function

Kernel Function	Type of Approximant
$K(\boldsymbol{x},\boldsymbol{y}) = \exp(-\|\boldsymbol{x}-\boldsymbol{y}\|^2)$	Gaussian
$K(\boldsymbol{x},\boldsymbol{y}) = (\|\boldsymbol{x}-\boldsymbol{y}\|^2+\rho^2)^{\frac{1}{2}}$	Multiquadric
$K(\boldsymbol{x},\boldsymbol{y}) = (\|\boldsymbol{x}-\boldsymbol{y}\|^2+\rho^2)^{-\frac{1}{2}}$	Inverse Multiquadric
$K(\boldsymbol{x},\boldsymbol{y}) = \tanh(\boldsymbol{x}^{\mathrm{T}}\boldsymbol{y}-\theta)$	Multi-layer Perceptron
$K(\boldsymbol{x},\boldsymbol{y}) = (1+\boldsymbol{x}^{\mathrm{T}}\boldsymbol{y})^d$	Polynomial of degree d
$K(x,y) = B_{2n+1}(x-y)$	B-spline

Tab. 1. Some examples of admissible kernel functions.

and determines the specific form of the approximant. Some examples of typical kernel functions are shown in Tab. 1. Note that in the case of B-splines, the kernel function is defined in terms of scalar rather than vector values. As a result, in SVM regression B-spline surfaces are constructed as tensor-products of univariate B-splines.

SVM regression proceeds by minimizing a functional of the form

$$H(f) = \frac{1}{n}\sum_{i=1}^{n} V(y_i, f(\boldsymbol{x}_i)) + \sigma\|f\|_K^2, \tag{2}$$

where $V(\cdot,\cdot)$ is a *loss function*, $\|\cdot\|_K$ is a norm in a reproducing kernel Hilbert space, and σ is a fixed regularization parameter. In SVM regression, the loss function is most commonly chosen to be Vapnik's epsilon-insensitive loss function

$$V(y_i, f(\boldsymbol{x}_i)) = |y_i - f(\boldsymbol{x}_i)|_\epsilon, \tag{3}$$

where

$$|x|_\epsilon = \begin{cases} 0, & \text{if } |x| \le \epsilon, \\ |x| - \epsilon, & \text{otherwise.} \end{cases}$$

This loss function is particularly suitable when the data are known to contain a mixture of uniform and arbitrary noise. In the left-hand picture of Fig. 1, we illustrate the use of the epsilon-insensitive loss function in approximating three curved paths of data. For each curve, the data are represented as crosses, the approximant is shown as the solid line, and the dashed lines represent what is know as the epsilon tube, which lies a distance ϵ above and below the approximant. Ordinate values which then lie outside this tube are traditionally termed support vectors. Because the hypothesis space from which the function f is chosen can be very large, the regularization term is added so as to maintain a well defined approximation problem. We discuss other types of loss functions in the next section.

By substituting (3) into (2), the epsilon-insensitive SVM regression problem can be written as

$$\min_{\boldsymbol{\alpha},b} \frac{1}{n}\sum_{i=1}^{n} |y_i - f(\boldsymbol{x}_i)|_\epsilon + \sigma\|f\|_K^2, \tag{4}$$

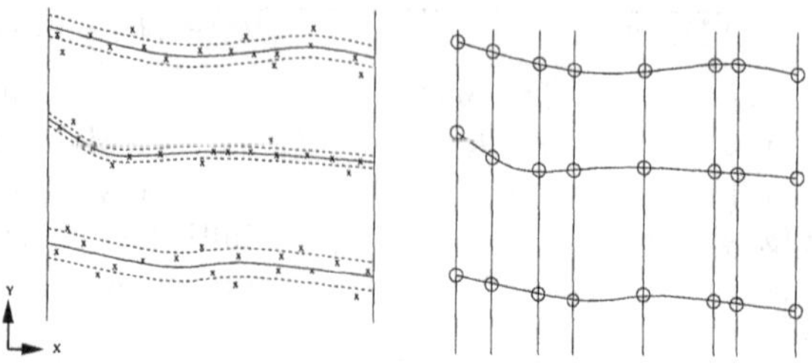

Fig. 1. The SVM approximants to three curved paths of data, and the new 'approximated' data set on parallel lines.

where $\boldsymbol{\alpha} = \{(\alpha_i, \alpha_i^*)\}_{i=1}^m$. It is difficult to solve explicitly problem (4). Instead, it is possible to use Lagrange multiplier techniques to reformulate problem (4) as a quadratic programming problem, specifically

$$\max_{\boldsymbol{\alpha}} \sum_{i=1}^{n} y_i(\alpha_i^* - \alpha_i) - \sum_{i=1}^{n} \epsilon(\alpha_i^* + \alpha_i) - \frac{1}{2}\sum_{i=1}^{n}\sum_{j=1}^{n}(\alpha_i^* - \alpha_i)(\alpha_j^* - \alpha_j)K(\boldsymbol{x}_i, \boldsymbol{x}_j), \tag{5}$$

subject to

$$\sum_{i=1}^{n}(\alpha_i^* - \alpha_i) = 0,$$
$$0 \leq \alpha_i, \alpha_i^* \leq C/n, \quad \text{for } i = 1\ldots, n,$$

where C is some specified constant. Smola and Schölkopf [6] discuss in detail techniques for solving problem (5).

2.2 Obtaining the Data on Parallel Lines

We are now in a position to consider how to approximate each curve of abscissae and ordinate values. Specifically, for each curve of data we construct an SVM approximation of the form (1) by solving the quadratic programming problem (5), to produce a set of m_l local approximants $\{\hat{y}_l(x)\}$ to the abscissae values and $\{\hat{f}_l(x)\}$ to the ordinate values. We then generate a set of secant lines $X = \{X_I\}_{I=1}^p$, say, such that the secant lines are distributed over the range $[x_{min}, x_{max}]$ so as to represent the global spread of the data. Appropriate values for the set of $\{X_I\}_{I=1}^p$ can be obtained by using a cluster analysis technique [4], for example. We are now able to form a new set of data which lie at the intersections of the approximants $\{\hat{y}_l(x)\}$ and the secant lines $X = \{X_I\}_{I=1}^p$ by using the localised SVM approximations. Hence we have determined

$$\hat{f}_l(X_I, \hat{y}_l(X_I)), \qquad \text{for } l = 1, \ldots, m_l, \quad I = 1, \ldots, p.$$

This process is illustrated in Fig. 1, where the problem of approximating a set of curved data is transformed into one of approximating data on a family of parallel lines.

2.3 SVM Regression for General Loss Functions

We commented earlier in this paper that if each curved path of data is collected by a different source, then it is likely that data on different paths will be corrupted by different types or levels of noise. In this case, the approximant to each curved path should be obtained by minimizing a different functional $H(f)$, where each particular functional is chosen to reflect the noise in each path. We show here that this can be done without substantial revision to the approach described in the previous section.

It may be the case that the distribution of the error attributed to each source remains the same (i.e., there is mixed uniform and arbitrary noise) but that the level of error is different. To include this consideration into the current model, we replace ϵ in expression (4) with ϵ_l, where the value of ϵ_l changes for each curve of data. This enables the band width of the tolerance to change with each curve, and in fact this process was used in the example illustrated in Fig. 1.

More generally, it may be that the type of noise is different for each curve, and in some of these cases the epsilon insensitive loss function is an inappropriate measure of error. Tab. 2 gives some examples of error distributions which are likely to occur in practice, along with the loss functions which should be applied when these types of error are known.

Noise type	Loss function	Distribution
Arbitrary	Laplacian	$V(x) = \lvert x\rvert$
Gaussian	Least squares	$V(x) = x^2/2$
Mixed Gaussian and arbitrary noise	Huber M-estimator	$V(x) = \begin{cases} x, & \text{if } \lvert x\rvert \le \gamma, \\ \gamma\lvert x\rvert - \gamma^2, & \text{otherwise.} \end{cases}$
Mixed uniform and arbitrary noise	Epsilon-insensitive	$V(x) = \lvert x\rvert_\epsilon$

Tab. 2. Error distributions and appropriate loss functions.

The loss functions shown in Tab. 2 can be written in the generalised form

$$V\big(y_i, f(\boldsymbol{x}_i)\big) = \begin{cases} 0, & \text{if } |y_i - f(\boldsymbol{x}_i)| \le \epsilon, \\ \tilde{V}\big(|y_i - f(\boldsymbol{x}_i)| - \epsilon\big), & \text{otherwise.} \end{cases} \tag{6}$$

This form has been chosen because of its similarity with the epsilon-insensitive loss function. In particular, if we replace the epsilon-insensitive loss function with the general loss function (6) in the SVM regression problem, then problem (5) becomes the more general problem

$$\max_{\boldsymbol{\alpha},\xi} \sum_{i=1}^{n} y_i(\alpha_i^* - \alpha_i) - \sum_{i=1}^{n} \epsilon(\alpha_i^* + \alpha_i) + \frac{C}{m}\sum_{i=1}^{n}\big(T(\xi_i) - T(\xi_i^*)\big)$$
$$- \frac{1}{2}\sum_{i=1}^{n}\sum_{j=1}^{n}(\alpha_i^* - \alpha_i)(\alpha_j^* - \alpha_j)K(\boldsymbol{x}_i, \boldsymbol{x}_j), \tag{7}$$

subject to

$$\sum_{i=1}^{n}(\alpha_i^* - \alpha_i) = 0,$$
$$0 \le \alpha_i \le (C/n)\partial\tilde{V}/\partial\xi_i, \quad \text{for } i = 1\ldots, n,$$
$$0 \le \alpha_i^* \le (C/n)\partial\tilde{V}/\partial\xi_i^*, \quad \text{for } i = 1\ldots, n,$$

where

$$T(\xi) = \tilde{V}(\xi) - \xi\frac{\partial\tilde{V}}{\partial\xi},$$

and $\boldsymbol{\xi} = \{(\xi_i, \xi_i^*)\}_{i=1}^n$. Problem (7) is closely related to the quadratic programming problem (5). In fact, it is solved by using generalizations of the algorithms proposed for solving problem (5), an issue which is discussed in detail by Smola and Schölkopf [6]. Consequently, the approach of Section 2.1 for approximating the curved data can be adapted to take into account different types of error in each curve without significant alteration. The main task is to adopt a loss function that corresponds to the noise in each particular curve, and then problem (7) can be solved as appropriate.

§3. Approximating Data on Lines

Once we have developed the new set of data on lines, we are in a position to construct a tensor-product surface interpolant or approximant, which corresponds to the second stage of our algorithm. The computation of an interpolant is reasonable, since we have removed the noise in the data in the previous stage of the algorithm. However, we may wish to compute an approximant in order to save computational time. The technique we use is based on that of Clenshaw and Hayes [2], who used Chebyshev polynomials to approximate data on lines bounded in the rectangular region $[-1, 1]^2$. The tensor-product method is not limited to Chebyshev polynomials, and in fact it can be applied to tensor-product surfaces defined in terms of a variety of basis functions. In this paper, we consider the cases of Gaussian radial basis functions (RBFs), which produce a least squares type approximant, and B-splines, which produce an interpolant to the data. Unlike the Chebyshev case, when using these bases it is necessary to ensure that certain conditions are enforced. Below, we discuss in detail the implementation process for the Gaussian RBF approximation and give a brief summary for constructing the B-spline interpolant. For more details on the tensor-product method, see Crampton and Mason [3] for Gaussians, and Anderson et al. for B-splines [1].

3.1 Gaussian Tensor-Product Approximation

The motivation for using separable Gaussian RBFs to approximate data on lines was borne out of the analysis for approximating data specified on finite rectangular grids. We develop the analysis for data on lines by first considering the finite grid. In radial basis function approximation, we are given a set of observed values $f_1, f_2, \ldots, f_m \in \mathbb{R}$ associated with the data abscissae $\boldsymbol{x}_1, \boldsymbol{x}_2, \ldots, \boldsymbol{x}_m \in \mathbb{R}^d$. A radial basis function approximation $S : \mathbb{R}^d \to \mathbb{R}$

takes the form

$$S(\boldsymbol{x}_i) \approx f_i, \quad \text{for } i = 1, 2, \ldots, m,$$

where

$$S(\boldsymbol{x}_i) = \sum_{j=1}^{n} b_j \phi(\|\boldsymbol{x}_i - \boldsymbol{\lambda}_j\|), \quad \text{for } i = 1, 2, \ldots, m,$$

and $\boldsymbol{\lambda}_1, \boldsymbol{\lambda}_2, \ldots, \boldsymbol{\lambda}_n \in \mathbb{R}^d$, for $n \leq m$, are a set of centres located at n distinct points within the abscissae domain. The basis function ϕ is usually chosen from the standard classes of radial functions. The Gaussian basis function is one of the standard functions used in RBF approximation and takes the form

$$\phi(\boldsymbol{x}) = e^{-\rho^2(\|\boldsymbol{x}-\boldsymbol{\lambda}\|^2)}. \tag{8}$$

By factorizing (8) in terms of the separable components of the abscissae we obtain

$$\begin{aligned}
\phi(\boldsymbol{x}; \boldsymbol{\lambda}) &= e^{-\rho^2(\|\boldsymbol{x}-\boldsymbol{\lambda}\|^2)} \\
&= e^{-\rho^2[(x_1-\lambda_1)^2+(x_2-\lambda_2)^2+\cdots+(x_d-\lambda_d)^2]} \\
&= \phi(x_1; \lambda_1)\phi(x_2; \lambda_2) \ldots \phi(x_d; \lambda_d),
\end{aligned}$$

thus enabling (8) to be expressed as a product of d one dimensional Gaussians. However, the factorization shows that associated with each component of the abscissae is the corresponding component of the centre. Therefore, in order to utilise this form for a tensor-product approximation, it will be a necessary condition for the centres to lie at the intersections of a rectangular mesh. Enforcing this condition, we express the abscissae and centres in the form

$$\begin{aligned}
\boldsymbol{x}_{k,l} &= (x_1^{(k)}, x_2^{(l)}), \quad \text{for } k = 1, \ldots, m_k, \quad l = 1, \ldots, m_l, \\
\boldsymbol{\lambda}_{i,j} &= (\lambda_1^{(i)}, \lambda_2^{(j)}), \quad \text{for } i = 1, \ldots, p, \quad j = 1, \ldots, q,
\end{aligned}$$

where m_k and m_l are the number of different values for the two components of the abscissae and p and q are the corresponding values for the centres.

We construct the approximant by summing over all $p \times q$ centres, and let the set $\{\phi_j\}_{j=1}^{p\times q}$ represent the Gaussian basis function in its separable form. Hence the approximating form is

$$f(\boldsymbol{x}) = \sum_{i=1}^{p} \left(\sum_{j=1}^{q} c_{i,j} \phi(\|x_2^{(l)} - \lambda_2^{(j)}\|) \right) \phi(\|x_1^{(k)} - \lambda_1^{(i)}\|). \tag{9}$$

Substituting the abscissae values into this expression, we obtain for each value of l the over-determined linear system

$$f(\boldsymbol{x}_{k,l}) = \sum_{i=1}^{p} b_i^{(l)} \phi(\|x_1^{(k)} - \lambda_1^{(i)}\|), \quad \text{for } k = 1, 2, \ldots, m_k. \tag{10}$$

For each value of i, we also have the over-determined system

$$b_i^{(l)} = \sum_{j=1}^{q} c_{i,j}\phi(\|x_2^{(l)} - \lambda_2^{(j)}\|), \quad \text{for } l = 1, 2, \ldots, m_l. \tag{11}$$

We can represent the system (10) in matrix notation as

$$\boldsymbol{f}^{(l)} = A_{x_1}\boldsymbol{b}^{(l)}, \quad \text{for } l = 1, 2, \ldots, m_l,$$

where A_{x_1} is a matrix whose (k,i)th element is $\phi(\|x_1^{(k)} - \lambda_1^{(i)}\|)$, $\boldsymbol{b}^{(l)}$ is a vector of the elements $b_i^{(l)}$, for $i = 1, 2, \ldots, p$, and $\boldsymbol{f}^{(l)}$ is a vector of the elements $f(\boldsymbol{x}_{k,l})$, for $k = 1, 2, \ldots, m_k$. We note that the matrix A_{x_1} is independent of l, and so it is only necessary to factorize A_{x_1} once. Similarly if we express (11) in the form

$$\boldsymbol{b}_i = A_{x_2}\boldsymbol{c}^{(i)}, \quad \text{for } i = 1, 2, \ldots, p,$$

we can see that the matrix A_{x_2} is independent of i, and so again only one factorization is required. The solutions of the linear systems (10) and (11) using QR factorization can be achieved in $\mathcal{O}(m_l p(m_k + q))$ operations, which is a great saving over the $\mathcal{O}(m_k m_l p^2 q^2)$ operations required to solve the linear system (9) without exploiting any structure.

To extend this process to data on parallel lines, we can assume that the x_1 coordinate is still fixed but that the x_2 coordinate is free to lie in any position on each line. We can then rewrite the data abscissae as $\{\boldsymbol{x}_{k,l} = (x_1^{(k)}, x_2^{(k,l)})\}$, where the x_2 values now vary with both k and l. The validity of the equations (10) and (11) still holds but with $x_2^{(l)}$ in equation (11) replaced by $x_2^{(k,l)}$. The solution of system (11) now requires a different matrix for each value of i and will in consequence no longer produce the true least squares solution of (9), but rather a close approximation to it. See Clenshaw and Hayes [2] for an analogous discussion about bivariate polynomials and the error associated with this technique.

3.2 Tensor-Product B-splines

To construct the tensor-product B-spline interpolant at the abscissae

$$\boldsymbol{x}_{k,l} = (x_1^{(k)}, x_2^{(k,l)}), \quad \text{for } k = 1, \ldots, m_k, \quad l = 1, \ldots, m_l,$$

we represent the interpolant as

$$f(\boldsymbol{x}) = \sum_{k=1}^{m_k}\sum_{l=1}^{m_l} c_{k,l} N_{n_1 k}(\lambda_1; x_1) N_{n_2 l}(\lambda_2; x_2), \tag{12}$$

where $N_{n_1 k}(\lambda_1; x_1)$ denotes the kth normalized B-spline of order n_1 (degree $n_1 - 1$) in x_1. The B-spline $N_{n_2 l}(\lambda_2; x_2)$ is defined analogously to the B-spline $N_{n_1 k}(\lambda_1; x_1)$ and the (k,l)th B-spline coefficient is represented by $c_{k,l}$.

The B-spline tensor-product method relies on the construction of a common knot set from which a suitable subset is selected for each line to produce a satisfactory interpolant. The spline surface is constructed in two separate stages. In the first stage, a subset of knots is chosen for each line from the common knot set. A spline interpolant to the data in x_1 is then constructed. A technique known as knot insertion [5] is used to re-express the spline such that it becomes defined over the full knot set; the underlying spline curve defined over the subset remains unchanged. The second stage constructs a spline interpolant to the coefficients in the orthogonal direction x_2 producing the required coefficients $c_{k,l}$ of the original tensor-product spline (12). The solution of the interpolation problem using tensor-product B-splines can be achieved in $\mathcal{O}(m_k m_l n_1^2 + m_l n_2 q)$ operations where q is the number of coefficients in (any one of) the univariate x_1-splines. In practice we would expect the operations count to be dominated by the first term $\mathcal{O}(m_k m_l n_1^2)$. In contrast, standard techniques for constructing B-spline tensor-product interpolants exploit only the fact that an observation matrix corresponding to equation (12) has band structure, and these routines typically require $\mathcal{O}(m_k^3 m_l n_2^2)$ operations.

§5. Numerical Example

In this section, we illustrate our algorithm by constructing an approximation to a set of data $\{f_{k,l}\}$ nominally lying on the function $F(x,y) = \sin(x) + \cos(y)$. The data $\{f_{k,l}\}$ were obtained by randomly sampling $n = 60$ points on each of a family of $m_l = 12$ exponential curves

$$y = \frac{r+1}{2} e^{((x+1)/10)} + \left(\frac{x+1}{20}\right) e^{((r+1)/2-1)} + \frac{r-1}{2},$$

where $r = -\pi, -\pi + h, -\pi + 2h, \ldots, \pi$, and $h = 2\pi/(n-1)$, and then introducing uniform noise. Also, a small number of the sampled data were replaced by outlying data, so that these observations were unreliable. The curved data were approximated by using the technique of Section 2 for a constant value of $\epsilon = 0.0125$ for each curve and regularization parameter $\sigma = 0.0004$, and this allowed us to obtain a new set of data lying on a family of parallel lines. These data are shown in the left-hand side of Fig. 2. We then fitted a tensor-product Gaussian surface to the data on lines by using the approach described in Section 3.1, and the resulting approximant is shown in the right-hand side of Fig. 2. The approximation produced a maximum absolute error between the function $F(x,y)$ and the approximant of approximately 2.26×10^{-3}.

§6. Conclusions

We have demonstrated how the application of support vector machine regression to data specified on curved paths provides a generalised approach for approximating data contaminated by a number of different types and level of error. The need to select a different approximation process when these errors vary between sources has been eliminated by introducing appropriate types of loss function into the SVM model.

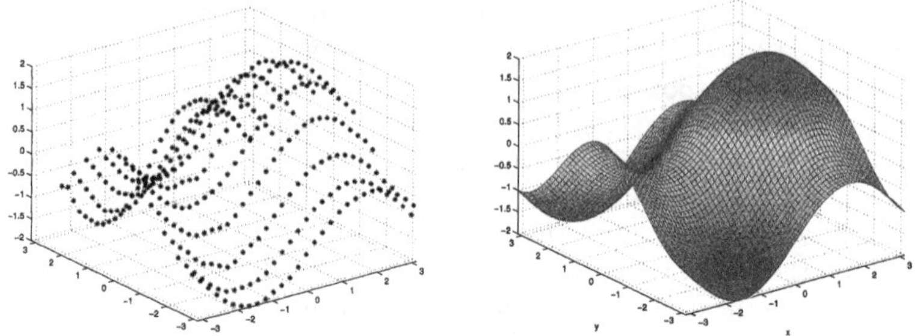

Fig. 2. A set of data on a family of parallel lines and a Gaussian tensor-product surface approximant to the function $\sin(x) + \cos(y)$.

Many of the existing algorithms used for solving the quadratic programming problem require a full Hessian matrix to be specified. If a scattered data approach is taken for the example problem of Section 5, this would mean that we need to store a Hessian matrix containing $2,073,600$ elements. However, the algorithm detailed in this paper requires only $14,400$ elements to be stored at any one time. Clearly, for large data sets the storage requirements are substantially reduced by using this algorithm.

References

1. Anderson, I. J., M. G. Cox, and J. C. Mason, Tensor-product spline interpolation to data on or near a family of lines, Numerical Algorithms **5** (1993), 193–204.
2. Clenshaw, C. W. and J. G. Hayes, Curve and surface fitting, Journal of the Institute of Mathematics and its Applications **1** (1965), 164–183.
3. Crampton, A. and J. C. Mason, Surface approximation of curved data using separable radial basis functions, to appear in *Advanced Mathematical and Computational Tools in Metrology V*, 2001.
4. Everitt, B. S., *Cluster Analysis*, Edward Arnold, UK, 3rd edition, 1998.
5. Farin, G., *Curves and Surfaces for Computer Aided Geometric Design*, Academic Press, UK, 3rd edition, 1993.
6. Smola, A. J. and B. Schölkopf, A tutorial on support vector regression, Technical Report NC2-TR-1998-030, Neural and Computational Learning 2, 1998.
7. Vapnik, V. N., *Statistical Learning Theory*, Wiley, US, 1998.

A. Crampton, J. C. Mason and D. A. Turner
School of Computing and Mathematics
University of Huddersfield
Huddersfield, UK
a.crampton, j.c.mason, d.a.turner@hud.ac.uk

Subdivision Surfaces in Feature Films

Tony D. DeRose

Abstract. The creation of believable and endearing digital characters in feature films presents a number of technical challenges, including the modeling, animation and rendering of complex shapes such as heads, hands, and clothing. Traditionally, these shapes have been modeled with NURBS surfaces despite the severe topological restrictions that NURBS impose. In order to move beyond these restrictions, we have introduced subdivision surfaces into our production environment. Subdivision surfaces are not new, but their use in high-end CG production has, until recently, been limited. Here we survey a series of developments that were required in order for subdivision surfaces to meet the demands of high-end production.

§1. Introduction

The most common way to model complex smooth surfaces such as those encountered in human character animation is by using a patchwork of trimmed NURBS. Trimmed NURBS are used primarily because they are readily available in existing commercial systems such as Alias-Wavefront and SoftImage. They do, however, suffer from at least two difficulties:

1) Trimming is expensive and prone to numerical error.
2) It is difficult to maintain smoothness, or even approximate smoothness, at the seams of the patchwork as the model is animated. As a case in point, considerable manual effort was required to hide the seams in the face and hands of Woody, a principal character in *Toy Story*.

Subdivision surfaces have the potential to overcome both of these problems: they do not require trimming, and smoothness of the model is automatically guaranteed, even as the model animates.

The use of subdivision in animation systems is not new, but for a variety of reasons (several of which we address in this paper), their use has not been widespread. In the mid 1980s for instance, Symbolics was possibly the first to use subdivision in their animation system as a means of creating detailed

Mathematical Methods for Curves and Surfaces: Oslo 2000
Tom Lyche and Larry L. Schumaker (eds.), pp. 73–79.

ISBN 0-8265-1378-6

Fig. 1. Geri.

polyhedra. The LightWave 3D modeling and animation system from NewTek also uses subdivision in a similar fashion.

This paper surveys two developments, the concise modeling of fillets and blends, and the texture mapping of subdivision surfaces that we made when we added Catmull-Clark [1] subdivision surfaces to our animation and rendering systems, Marionette and RenderMan [5], respectively. For a more complete description of these developments see DeRose *et.al.* [2]. The resulting extensions were initially used in the creation of Geri (Figure 1), a human character in our Academy Award winning short film *Geri's Game*. Subdivision surfaces have subsequently become the method of choice for representing virtually all surfaces that undergo deformation during animation. Examples include many of the faces in Disney and Pixar's film, *A Bug's Life*, as well as the faces, hands, and bodies of most characters in *Toy Story 2*.

§2. Background

When considering which of the many available subdivision surfaces to add first to our animation and rendering systems, we quickly decided upon Catmull-Clark surfaces for two main reasons:

1) Catmull-Clark surfaces reduce to uniform cubic B-splines in regular rectangular regions of the control mesh. This was helpful because our animators and model builders were used to using B-splines, and because the internals of our renderer were optimized for B-splines.
2) Quadrilaterally based subdivision schemes such as Catmull-Clark are better at capturing many of the symmetries that our models exhibit. For example, cylinder-like structures such as arms, legs, and fingers are better characterized by regular quadrilateral grids than by regular triangular ones.

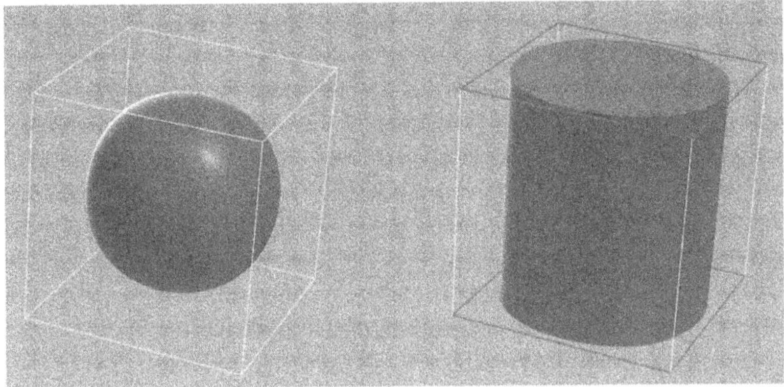

Fig. 2. Catmull-Clark surfaces: standard (left), and infinitely sharp (right).

Fig. 3. Geri's hand is a piecewise smooth Catmull-Clark surface.

A simple example of a Catmull-Clark surface is shown in Figure 2. The light lines indicate the edges of the initial control mesh; the grey surface is the limiting surface generated through Catmull-Clark subdivision.

Following Hoppe *et. al.* [3] it is possible to modify the subdivision rules to create piecewise smooth surfaces containing infinitely sharp features such as creases and corners. This is illustrated in Figure 2 (right), and Figure 3 which shows a close-up shot of Geri's hand. Infinitely sharp creases were used to separate the skin of the hand from the finger nails. Sharp creases can be modeled by marking a subset of the edges of the control mesh as sharp, and then using specially designed rules in the neighborhood of sharp edges. These sharp edges are indicated by dark lines in Figure 2 (right).

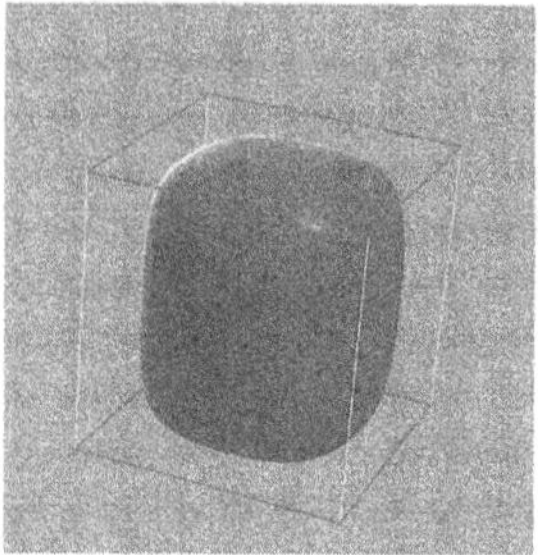
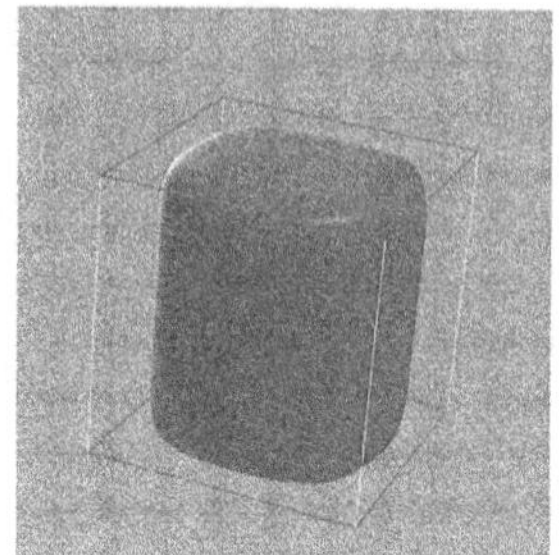
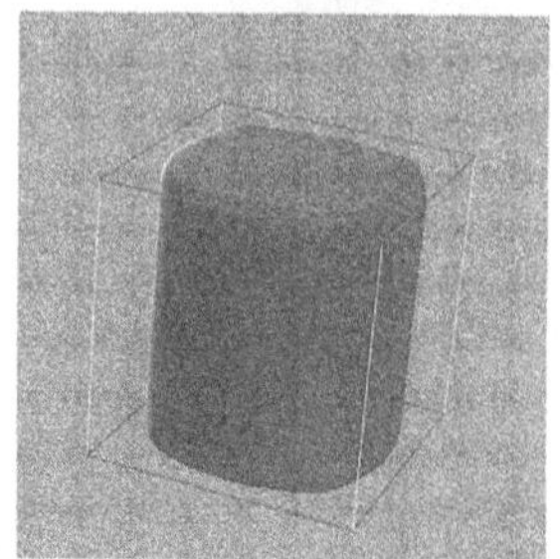

Fig. 4. Semi-sharp Catmull-Clark surfaces.

§3. Semi-sharp Creases: Modeling Fillets and Blends

As mentioned in Section 2 and shown in Figure 4, infinitely sharp creases are very convenient for representing piecewise-smooth surfaces. However, real-world surfaces are never infinitely sharp. The corner of a tabletop, for instance, is smooth when viewed sufficiently closely. For animation purposes it is often desirable to capture such tightly curved shapes.

To this end, we have developed a generalization of the Catmull-Clark scheme to admit semi-sharp creases – that is, creases of controllable sharpness, a simple example of which is shown in Figure 4.

One approach to achieve semi-sharp creases is to develop subdivision rules whose weights are parametrized by the sharpness of the crease. This approach is difficult because it can be quite hard to discover rules that lead to the desired smoothness properties of the limit surfaces. One of the roadblocks is that subdivision rules around a crease break a symmetry possessed by the smooth rules: typical smooth rules (such as the Catmull-Clark rules) are invariant under cyclic reindexing, meaning that discrete Fourier transforms can be used to prove properties for vertices of arbitrary valence (cf. Zorin [6]). In the absence of this invariance, each valence must currently be considered separately, as was done by Schweitzer [4]. Another difficulty is that such an approach is likely to lead to a zoo of rules depending on the number and configuration of creases through a vertex. For instance, a vertex with two semi-sharp creases passing through it would use a different set of rules than a vertex with just one crease through it.

Our approach is to use a very simple process we call hybrid subdivision. The general idea is to use one set of rules for a finite but arbitrary number of subdivision steps, followed by another set of rules that are applied to the limit. Smoothness therefore depends only on the second set of rules. Hybrid subdivision can be used to obtain semi-sharp creases by using infinitely sharp rules during the first few subdivision steps, followed by use of the smooth rules for subsequent subdivision steps. Intuitively, this leads to surfaces that are sharp at coarse scales, but smooth at finer scales.

The surface shown in Figure 4 (left) results if only the first subdivision step is done using the sharp rules and all subsequent steps are done using

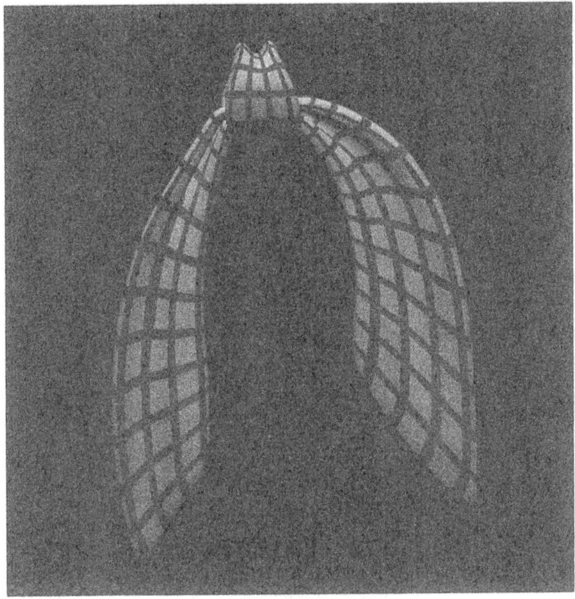

Fig. 5. A procedurally textured subdivision surface.

the smooth rules. The surface in Figure 4 (middle) results from two steps of sharp subdivision, and the surface in Figure 4 (right) results from three steps of sharp subdivision.

The hybrid subdivision technique can be extended to capture surfaces of non-integer sharpnesses, as well as to surfaces where the sharpness varies continuously along the sharp feature (see [2] for details). Such continuously varying sharpness surfaces were used in *Toy Story 2* to model the face of the Buzz Lightyear character.

§4. Texture Mapping

The lack of global parameterization may at first seem disastrous when it comes to texturing subdivision surfaces. However, in overcoming this deficiency, we have found that all the techniques we normally use to texture NURBS can also be applied to subdivision surfaces. We include the following in our toolset: solid texturing, 3D (i.e., projected) paint, texture mapping, and procedural texturing.

In our solution, we assign scalar values to each of the vertices of the control mesh, and subdivide them along with the geometric coordinates of the mesh vertices. Propagated in this way, these scalar fields can be shown to vary smoothly over the limit surface (see [2]), thus providing the necessarily smooth coordinates with which to perform parametric texturing. An example is shown in Figure 5.

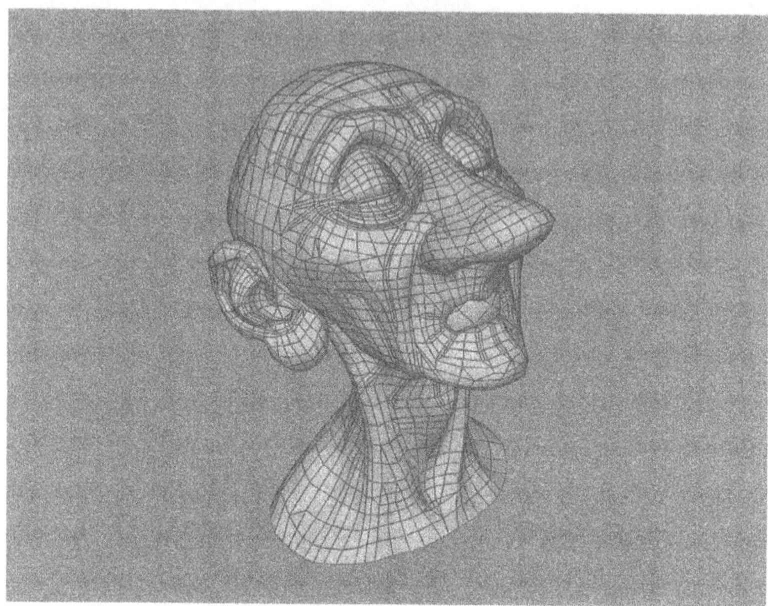

Fig. 6. The control mesh used for Geri's head.

§5. Summary

Our experience using subdivision surfaces in production has been extremely positive. The use of subdivision surfaces allows our model builders to arrange control points in a way that is natural to capture geometric features of the model (see Figure 6) without concern for maintaining a regular gridded structure as required by NURBS models. This freedom has two principal consequences. First, it dramatically reduces the time needed to plan and build an initial model. Second, and perhaps more importantly, it allows the initial model to be refined locally. Local refinement is not possible with a NURBS surface, since an entire control point row, or column, or both must be added to preserve the tensor product structure. Additionally, extreme care must be taken either to hide the seams between NURBS patches, or to constrain control points near the seam to create at least the illusion of smoothness.

References

1 Catmull, E. and J. Clark. Recursively generated B-spline surfaces on arbitrary topological meshes. Computer Aided Design **10** (6) (1978), 350–355.

2 DeRose, T., M. Kass, and T. Truong. Subdivision surfaces in character animation. Computer Graphics **32** (3) (1998), 85–94.

3 Hoppe, H., T. DeRose, T. Duchamp, M. Halstead, H. Jin, J. McDonald, J. Schweitzer, and W. Stuetzle. Piecewise smooth surface reconstruction. Computer Graphics **28** (3) (1994), 295–302.

4 Schweitzer, Jean E., Analysis and application of subdivision surfaces. PhD thesis, Department of Computer Science and Engineering, University of Washington, 1996.
5 Upstill, Steve, *The RenderMan Companion.* Addison-Wesley, 1990.
6 Zorin, Denis, Stationary subdivision and multiresolution surface representations, PhD thesis, Caltech, Pasadena, 1997.

Tony D. DeRose
Pixar Animation Studios
1200 Park Ave.
Emeryville, CA 94608
derose@pixar.com

Approximate Implicitization

Tor Dokken

Abstract. A method for the approximation of piecewise rational parametric manifolds of a chosen degree with an algebraic hypersurface is presented. This method has good convergence rate, uses singular value decomposition to find a set of alternative approximations and enables the selection of an approximation to be based on required behavior of the gradient of the resulting approximating algebraic surface.

§1. Introduction

In CAD-systems, both implicit and parametric descriptions of curves and surfaces are used for representing conic sections and quadric surfaces. The duality between the parametric and implicit representations simplifies the computational complexity. However, for rational parametric curves and surfaces of degree higher than two, the relationship between the parametric and the algebraic representations is more complex than for conic sections and quadric surfaces.

While all rational parametric 2D curves and 3D rational parametric surfaces have an algebraic representation, there do not exist rational parametric representations for all algebraic curves or surfaces of degree higher than two. The parametric and implicit representations of a 2D curve have the same polynomial degree. However, the algebraic representation of a tensor product surface of degrees (n_1, n_2) has the algebraic degree $2n_1n_2$. Thus a bicubic surface (degrees $(3, 3)$) has algebraic degree 18. Polynomials of degree 18 are not computationally attractive.

Methods for finding the exact algebraic representation of rational parametric curves and surfaces are called implicitization. A number of establish methods for exact implicitization exists such at resultants and Groebner basis [3,4,7]. For traditional resultant-based methods, base points, which are points where both numerator and denominator of a parametric curve or surface vanish simultaneously, make the implicitization fail. The methods described in [5,16] are a major step in the direction of more applicable exact implicitization methods. Instead of base points being a problem, these methods use the

Mathematical Methods for Curves and Surfaces: Oslo 2000
Tom Lyche and Larry L. Schumaker (eds.), pp. 81–102.

ISBN 0-8265-1378-6

existence of base points to reduce the degree of the algebraic curve or surface found.

As the use of floating point arithmetic is much more efficient than using exact arithmetic on digital computers, most implementations of industrial geometric algorithms use floating point arithmetic. The geometric objects and the corresponding implicit representations are thus inflicted with rounding errors. Thus, knowing how well a given algebraic curve or surface approximates the corresponding parametric curve or surface in a region of interest is more important than finding the exact representation. In addition the exact representation can have singular or near singular points in the region of computational interest. These singularities or near singularities complicate the use of exact algebraic methods if arithmetic with finite precision is used. In many computation problems, the use of the implicit representation is simplified if singular or near singular points are well separated from the region of computational interest. This paper describes an approach to finding approximate implicit representations of rational piecewise polynomial curves and surfaces.

Definition: Approximate Implicitization. *Let l and g be integers with $1 \leq g < l$, and let $\mathbf{p}(\mathbf{s})$, $\mathbf{s} \in \Omega \subset \mathbb{R}^g$ be a manifold of dimension g in $\mathbb{R}^l$. The nontrivial algebraic hypersurface $q(\mathbf{x}) = 0$, $q \in P_m(\mathbb{R}^l)$, is an approximate implicitization of $\mathbf{p}(\mathbf{s})$ within the tolerance $\epsilon \geq 0$ if we can find a continuous function $\mathbf{g}(\mathbf{s})$ describing the direction for error measurement and an error function $\eta(\mathbf{s})$ such that*

$$q(\mathbf{p}(\mathbf{s}) + \eta(\mathbf{s})\mathbf{g}(\mathbf{s})) = 0,$$

where

$$\|\mathbf{g}(\mathbf{s})\|_2 = 1, \quad |\eta(\mathbf{s})| \leq \epsilon, \qquad \mathbf{s} \in \Omega.$$

In the definition we use the concept of a manifold, as the approach is not limited to curves in 2D or surfaces in 3D. Note that if $\epsilon \equiv 0$ in the definition, then we have exact implicitization.

Approximate implicitization is illustrated in Section 2 for approximate implicitization of 2D rational parametric curves. Then the notation used for presenting the general theory is given in Section 3. Properties of the combination of an algebraic hypersurface $q(\mathbf{x}) = 0$ of total degree m and a parametric represented manifold $\mathbf{p}(\mathbf{s})$ are the topics in Section 4. The combination $q(\mathbf{p}(\mathbf{s}))$ is expressed as a matrix vector product

$$q(\mathbf{p}(\mathbf{s})) = (\mathbf{D}\mathbf{b})^T \alpha(\mathbf{s}),$$

where $\mathbf{D}$ is a matrix, $\mathbf{b}$ contains the coefficients of q and $\alpha(\mathbf{s})$ contains the basis functions related to the coordinate functions of $\mathbf{p}(\mathbf{s})$. This means that if $\mathbf{b}$ is in the null space of $\mathbf{D}$, then $q(\mathbf{p}(\mathbf{s})) = 0$. Further, assuming that the basis is a partition of unity, we have $\|\alpha(\mathbf{s})\|_2 \leq 1$, and thus

$$|q(\mathbf{p}(\mathbf{s}))| \leq \|\mathbf{D}\mathbf{b}\|_2 .$$

The matrix $\mathbf{D}$ is then shown to have desirable numeric properties if the coefficients of $\mathbf{p}(\mathbf{s})$ are contained in a simplex S, and this simplex is used for the description of q in barycentric coordinates. In Section 5 it is shown that singular value decomposition can be used for finding an algebraic approximation of a parametric represented manifold. When $\sigma_1 \geq 0$ is the smallest singular value of $\mathbf{D}$, we show that $\min_{\|\mathbf{b}\|_2=1} \max_{\mathbf{s}\in\Omega} \mid q(\mathbf{p}(\mathbf{s})) \mid \leq \sigma_1$. In Section 6 the convergence rate of the algebraic approximation is established. These rates are higher than what is normal in approximation theory due to the many degrees of freedom in algebraic hypersurfaces. We add constraints to the algebraic approximation in Section 7 to control the behavior. Measurement of error of approximate implicitization is addressed in Section 8. Successful use of approximate algebraic surfaces depends on well-behaved gradients. How to choose approximations from the approximate null-space of $\mathbf{D}$ is addressed in Section 9. Approximate implicitization used for simultaneous approximation of multiple manifolds is the topic of Section 10.

§2. Approximate Implicitization of 2D Curves

The section is divided into 3 parts:

- The theory of approximate implicitization of 2D Bernstein basis represented curves,
- Numerical examples of implicitization,
- A sketch of the algorithm for approximate implicitization.

2.1. Theory for Approximate Implicitization of 2D Curves

We will illustrate the principles of approximate implicitization by approximating a Bezier curve $\mathbf{p}(s)$ of degree n with an algebraic curve $q(x, y, h) = 0$ of degree m. The descriptions of $\mathbf{p}$ and q are

$$\mathbf{p}(s) = (p_x(s), p_y(s), p_h(s)) = \sum_{i=0}^{n} (x_i, y_i, h_i) \binom{n}{i} (1-s)^{n-i} s^i$$

$$q(x, y, h) = \sum_{0 \leq j_1 + j_2 \leq m} b_{j_1,j_2} x^{j_1} y^{j_2} h^{m-j_1-j_2}.$$

The description of the curve $\mathbf{p}(s)$ in $\mathbb{R}^2$ is $\frac{(p_x(s), p_y(s))}{p_h(s)}$. The number of terms in $q(x, y, h)$ is $\tilde{M} = \binom{m+2}{2}$. By inserting $\mathbf{p}(s)$ into q we get

$$q(\mathbf{p}(s)) = \sum_{0 \leq j_1 + j_2 \leq m} b_{j_1,j_2} \left(\sum_{i=0}^{n} x_i \binom{n}{i} (1-s)^{n-i} s^i \right)^{j_1}$$
$$\left(\sum_{i=0}^{n} y_i \binom{n}{i} (1-s)^{n-i} s^i \right)^{j_2} \left(\sum_{i=0}^{n} h_i \binom{n}{i} (1-s)^{n-i} s^i \right)^{m-j_1-j_2}$$

$q(\mathbf{p}(s))$ is a linear combination of $\tilde{M}$ polynomials of degree mn in the variable s with coefficients $\{d_{i,j_1,j_2}\}_{i=0,\dots,mn}$, $0 \le j_1 + j_2 \le m$. The polynomial order of $q(\mathbf{p}(s))$ is $\tilde{N} = nm + 1$. Let

$$\alpha_i(s) = \binom{mn}{i-1}(1-s)^{nm-i+1}s^{i-1}, i = 1, \dots, \tilde{N}.$$

Thus,

$$\begin{aligned} q(\mathbf{p}(s)) &= \sum_{0 \le j_1 + j_2 \le m} b_{j_1,j_2} \sum_{i=0}^{mn} d_{i,j_1,j_2} \binom{mn}{i} (1-s)^{nm-i} s^i \\ &= \sum_{0 \le j_1 + j_2 \le m} b_{j_1,j_2} \sum_{i=1}^{\tilde{N}} d_{i,j_1,j_2} \alpha_i(s). \end{aligned} \tag{2.1}$$

The coefficients d_{i,j_1,j_2} can be calculated in a numerically stable way by making repeated products of the coordinate functions of $\mathbf{p}(s)$, represented in terms of Bernstein bases. Thus, for example, $(x(s)y(s)h(s))$ can be expressed in a Bernstein basis by first multiplying $x(s)$ and $y(s)$, and then multiplying this with $h(s)$. See the end of this section for an example algorithm for multiplication of Bernstein basis represented functions.

We introduce a lexicographical ordering of the coefficients $\{b_{j_1,j_2}\}$ in (2.1) to organize the coefficients in a vector. We choose the lexicographical ordering

$$\nu(j_1, j_2) = \frac{1}{2}(j_1 + j_2 + 2)(j_1 + j_2 + 1) - j_1.$$

Then we define

$$\begin{aligned} \tilde{b}_{\nu(j_1,j_2)} &= b_{j_1,j_2}, & 0 \le j_1 + j_2 \le m \\ \tilde{d}_{i+1,\nu(j_1,j_2)} &= d_{i,j_1,j_2}, & 0 \le j_1 + j_2 \le m, \quad 0 \le i \le mn. \end{aligned}$$

We can now rewrite (2.1) as

$$q(\mathbf{p}(s)) = \sum_{j=1}^{\tilde{M}} \tilde{b}_j \sum_{i=1}^{\tilde{N}} \tilde{d}_{i,j} \alpha_i(s) = (\mathbf{D}\mathbf{b})^T \alpha(s), \tag{2.2}$$

where

$$\mathbf{b} = \left(\tilde{b}_j\right)_{j=1}^{\tilde{M}}, \qquad \mathbf{D} = \left(\tilde{d}_{i,j}\right)_{i,j=1,1}^{\tilde{N},\tilde{M}}, \qquad \alpha(s) = (\alpha_i(s))_{i=1}^{\tilde{N}}.$$

Since $\alpha(s)$ is a Bernstein basis we know that $\sum_{i=1}^{\tilde{N}} \alpha_i(s) = 1$, and that $0 \le \alpha_i(s) \le 1$ for $s \in [0, 1]$. Thus

$$\|\alpha(s)\|_2^2 = \sum_{i=1}^{\tilde{N}} (\alpha_i(s))^2 \le \sum_{i=1}^{\tilde{N}} \alpha_i(s) = 1.$$

Applying this to (2.2) we can eliminate the variable s

$$| q(\mathbf{p}(s)) |=| (\mathbf{Db})^T \alpha(s) |\leq \|\mathbf{Db}\|_2 \|\alpha(s)\|_2 \leq \|\mathbf{Db}\|_2 . \qquad (2.3)$$

This expression can be used in different ways:

- A vector $\mathbf{b} \neq \mathbf{0}$ that makes $\mathbf{Db} \equiv \mathbf{0}$ represents an exact implicit representation of $\mathbf{p}(s)$. Thus we have reformulated the implicitization problem to finding the null space of a matrix. For rational parameterized curves the algebraic degree is the same as the parametric degree. Thus if $m = n$ we know that a vector $\mathbf{b}$ exists such that $\mathbf{Db} \equiv \mathbf{0}$.
- Vectors that make $\frac{\|\mathbf{Db}\|_2}{\|\mathbf{b}\|_2} \approx 0$ are candidates for approximate implicit representation for $\mathbf{p}(s)$. In Section 5 we will show that singular value decomposition of $\mathbf{D}$ can be used for finding candidate approximations. The vectors $\mathbf{b}$ belonging to the smallest singular values are the best candidates. The accuracy of the approximation, also addressed in Section 8, can be checked by
$$\frac{\tilde{q}(\tilde{\mathbf{p}}(s))}{\nabla\tilde{q}(\tilde{\mathbf{p}}(s) + \theta\tilde{\mathbf{n}}(s))}.$$
Here $\tilde{q}(x, y) = q(x, y, 1)$, $\tilde{\mathbf{p}}(s) = \frac{(p_x(s), p_y(s))}{p_h(s)}$, $\tilde{\mathbf{n}}(s)$ is the unit normal of $\tilde{\mathbf{p}}(s)$ and $|\theta| \leq \delta$. The role of δ is to define a region around $\tilde{\mathbf{p}}(s)$ for which to check the approximation. As the gradient of the implicit approximation can vanish on or close to the curve, we will require that the gradient is nonvanishing in the region defined by $\tilde{\mathbf{p}}(s)+\theta\tilde{\mathbf{n}}(s)$, $| \theta | \leq \delta$ and $s \in [0, 1]$.

In Section 4, Theorem 4.3, we show that using Bernstein polynomials over a triangle containing $\mathbf{p}(s)$, for describing the algebraic curve, makes the rows of $\mathbf{D}$ a partition of unity. When using power basis for representing the algebraic curve, scaling the problem to be contained in the box $[-1, 1]\times[-1, 1]$, gives almost the same effect. Thus $\mathbf{D}$ has an attractive structure with respect to numerical algorithms.

To minimize numerical errors in the approximation, it is important to build the matrix $\mathbf{D}$ using stable numerical algorithms. By using the following algorithm for multiplying Bernstein basis represented functions the propagation of rounding error in $\mathbf{D}$ is controlled.

Example. *Multiplication of two Bernstein bases represented functions.* Let

$$p(s) = \sum_{i=1}^{n_1} p_i \binom{n_1}{i} (1-s)^{n_1-i} s^i,$$

$$q(s) = \sum_{j=1}^{n_2} q_j \binom{n_2}{j} (1-s)^{n_2-j} s^j.$$

Then the product of $p(s)$ and $q(s)$ can be expressed as

$$p(s)q(s) = \sum_{i=0}^{n_1+n_2} g_i \binom{n_1+n_2}{i} (1-s)^{n_1+n_2-i} s^i,$$

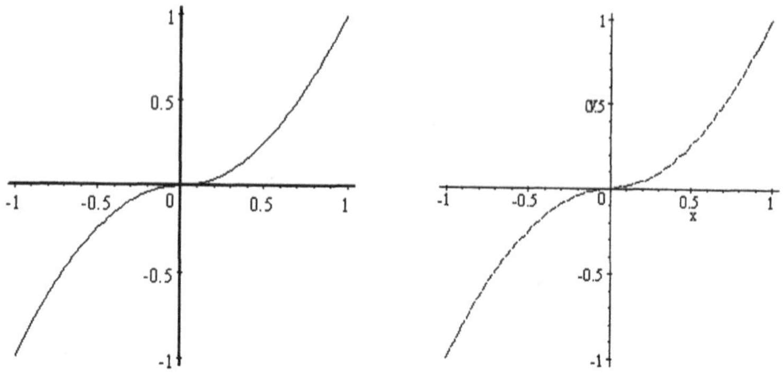

Fig. 1. Original parabola pieces (left). Degree 3 algebraic approximation (right).

where

$$g_i = \sum_{j=\max(0,i-n_2)}^{\min(i,n_1)} p_j q_{i-j} \frac{\binom{n_1}{j}\binom{n_2}{i-j}}{\binom{n_1+n_2}{i}}.$$

2.2. Numerical Example of Approximate Implicitization

Let a piecewise polynomial curve consist of two parabola pieces

$$\begin{aligned} \mathbf{p}_1(s_1) &= (-1,-1)\,(1-s_1)^2 + (-\tfrac{1}{2},0)\,2(1-s_1)s_1 + (0,0)\,s_1^2, \qquad s_1 \in [0,1] \\ \mathbf{p}_2(s_2) &= (0,0)\,(1-s_2)^2 + (\tfrac{1}{2},0)\,2(1-s_2)s_2 + (1,1)\,s_2^2, \qquad s_2 \in [0,1]. \end{aligned}$$

These parabola pieces are shown on the left in Figure 1. We now want to find an algebraic approximation $q(x,y) = 0$ of degree 3 to $\mathbf{p}_1(s_1)$ and $\mathbf{p}_2(s_2)$:

$$q(x,y) = b_1x^3 + b_2x^2y + b_3xy^2 + b_4y^3 + b_5x^2 + b_6xy + b_7y^2 + b_8x + b_9y + b_{10}.$$

Using the notation from the introduction, we have to find $q(\mathbf{p}_1(s_1))$ and $q(\mathbf{p}_2(s_2))$, the combination of the unknown algebraic curve q and the two curves $\mathbf{p}_1(s_1)$ and $\mathbf{p}_2(s_2)$. Both $\mathbf{p}_1(s_1)$ and $\mathbf{p}_2(s_2)$ are polynomials of degree 2 in s_1 and s_2, respectively. Thus, $q(\mathbf{p}_1(s_1))$ and $q(\mathbf{p}_2(s_2))$ will be polynomials of degree 6 in s_1 and s_2, respectively. We will express these polynomials in the Bernstein basis. We denote these degree 6 Bernstein bases by $\alpha_1(s_1)$ and $\alpha_2(s_2)$, giving

$$\begin{aligned} q(\mathbf{p}_1(s_1)) &= (\mathbf{D}_{\mathbf{p}_1}\mathbf{b})^T\alpha_1(s_1) \\ q(\mathbf{p}_2(s_2)) &= (\mathbf{D}_{\mathbf{p}_2}\mathbf{b})^T\alpha_2(s_2). \end{aligned}$$

The coefficients of the algebraic curve are located in the vector $\mathbf{b}$, and the

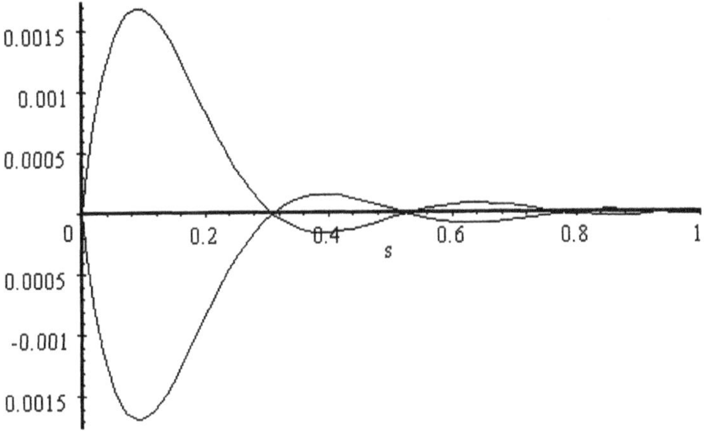

Fig. 2. Error for degree 3 implicit approximation of two curve pieces.

matrices are

$$\mathbf{D}_{\mathbf{p}_1} = \begin{pmatrix} -1 & -1 & -1 & -1 & 1 & 1 & 1 & -1 & -1 & 1 \\ -\frac{3}{6} & -\frac{2}{6} & -\frac{1}{6} & 0 & \frac{4}{6} & \frac{3}{6} & \frac{2}{6} & -\frac{5}{6} & -\frac{4}{6} & 1 \\ -\frac{3}{15} & -\frac{1}{15} & 0 & 0 & \frac{6}{15} & \frac{3}{15} & \frac{1}{15} & -\frac{10}{15} & -\frac{6}{15} & 1 \\ -\frac{1}{20} & 0 & 0 & 0 & \frac{4}{20} & \frac{1}{20} & 0 & -\frac{10}{20} & -\frac{4}{20} & 1 \\ 0 & 0 & 0 & 0 & \frac{1}{15} & 0 & 0 & -\frac{5}{15} & -\frac{1}{15} & 1 \\ 0 & 0 & 0 & 0 & 0 & 0 & 0 & -\frac{1}{6} & 0 & 1 \\ 0 & 0 & 0 & 0 & 0 & 0 & 0 & 0 & 0 & 1 \end{pmatrix}$$

$$\mathbf{D}_{\mathbf{p}_2} = \begin{pmatrix} 0 & 0 & 0 & 0 & 0 & 0 & 0 & 0 & 0 & 1 \\ 0 & 0 & 0 & 0 & 0 & 0 & 0 & \frac{1}{6} & 0 & 1 \\ 0 & 0 & 0 & 0 & \frac{1}{15} & 0 & 0 & \frac{5}{15} & \frac{1}{15} & 1 \\ \frac{1}{20} & 0 & 0 & 0 & \frac{4}{20} & \frac{1}{20} & 0 & \frac{10}{20} & \frac{4}{20} & 1 \\ \frac{3}{15} & \frac{1}{15} & 0 & 0 & \frac{6}{15} & \frac{3}{15} & \frac{1}{15} & \frac{10}{15} & \frac{6}{15} & 1 \\ \frac{3}{6} & \frac{2}{6} & \frac{1}{6} & 0 & \frac{4}{6} & \frac{3}{6} & \frac{2}{6} & \frac{5}{6} & \frac{4}{6} & 1 \\ 1 & 1 & 1 & 1 & 1 & 1 & 1 & 1 & 1 & 1 \end{pmatrix}.$$

Combining the two matrices, we get a matrix $\mathbf{D}$ for simultaneous approximation of both curves.

$$\mathbf{D} = \begin{pmatrix} \mathbf{D}_{\mathbf{p}_1} \\ \mathbf{D}_{\mathbf{p}_2} \end{pmatrix}.$$

The singular values of this matrix are 4.252, 3.919, 1.972, 1.307, 0.3836, 0.3760, 0.1076, 0.05044, 0.0294, 0.007937. The vector corresponding to the smallest singular value represents the algebraic curve plotted on the right in Figure 1.

This curve has the description

$$\begin{pmatrix} x^3 & x^2y & xy^2 & y^3 & x^2 & xy & y^2 & x & y & 1 \end{pmatrix} \begin{pmatrix} -.4602815334 \\ .6983314313 \\ -.5087222523 \\ .1433800204 \\ 0 \\ 0 \\ 0 \\ -.0170228764 \\ .1443197267 \\ 0 \end{pmatrix} = 0.$$

The error of this approximation is shown in Figure 2. Note that the largest error is close to the point where the two parabolas meet.

If we build the matrices $\mathbf{D}_{\mathbf{p}_1}$ and $\mathbf{D}_{\mathbf{p}_1}$ using the power basis, then the singular values resulting are $\sqrt{6}+\sqrt{2}$, $\sqrt{2}$ (eight times) and $\sqrt{6}-\sqrt{2}$. These values are all greater than 1. Thus the smallest singular value will not have a coefficient vector giving a good approximation. In the following sections we will show the importance of using a basis that is the partition of unity such as the Bernstein basis.

2.3. Algorithm for Approximate Implicitization

The notation employed in this subsection is related to 2D curves. However, approximate implicitization of 3D surfaces and hypersurfaces in higher dimensions has the same structure.

1) Translate the curve $\mathbf{p}(s)$ to the box $[-1,1] \times [-1,1]$ around the origin. (Alternatively if the algebraic curve is represented by barycentric coordinates: Find a triangle containing the curve $\mathbf{p}(s)$, and describe the curve in the barycentric coordinates.) Select the algebraic degree for the approximation.
2) Build the matrix $\mathbf{D}$.
3) Perform singular value decomposition of $\mathbf{D}$. In Section 5 we show that
$$\min_{\|\mathbf{b}\|_2=1} \max_{s\in[0,1]} |\, q(\mathbf{p}(s)) \,| \leq \sigma_1,$$
where σ_1 is the smallest singular value of $\mathbf{D}$. Thus the coefficient vectors corresponding to the smallest singular values of $\mathbf{D}$ are candidates for algebraic approximations to $\mathbf{p}(s)$.
4) If $\mathbf{D}$ has at least one singular value equal to 0, then an exact implicit representation of $\mathbf{p}(s)$ is found.
5) If all singular values of $\mathbf{D}$ are different from 0, then check if the algebraic curves corresponding to the smaller singular value have nonvanishing gradient in a sufficient large area. Then estimate the approximation error for the candidate approximations.
6) If the approximation is outside the required tolerance, then increase the algebraic degree and continue from Step 2. If the error is smaller than the required tolerance, a solution is found.

§3. Notation

We will use barycentric coordinates over a simplex S in $\mathbb{R}^l$ for describing an algebraic hypersurface $q(\mathbf{x}) = 0$ of total degree m. The use of this approach was first proposed in [11]. Assume that $(\beta_1(\mathbf{x}), \ldots, \beta_{l+1}(\mathbf{x}))$ are the barycentric coordinates of the point $\mathbf{x}$ with respect to the simplex S. In this notation the algebraic hypersurface is

$$q(\mathbf{x}) = \sum_{\mathbf{i}\in I(m)} b_{\mathbf{i}} \frac{m!}{\mathbf{i}!} \beta^{\mathbf{i}}(\mathbf{x}),$$

with

$$\mathbf{i}! = i_1! \ldots i_{l+1}!, \qquad \text{and} \qquad \beta^{\mathbf{i}}(\mathbf{x}) = \prod_{j=1}^{l+1} (\beta_j(\mathbf{x}))^{i_j}.$$

We have introduced the index set $I(m)$ to simplify the notation. The number of elements in $I(m)$ is denoted by $\tilde{M}$. For an algebraic hypersurface in $\mathbb{R}^l$, the index set is constructed as follows

$$I(m) = \{(i_1, \cdots, i_l, i_{l+1}) \in \mathbb{N}_0^{l+1} \mid i_1 + \ldots + i_l + i_{l+1} = m\}.$$

The parametric manifolds of degree $\mathbf{n}$ approximated are denoted $\mathbf{p} : \Omega \to \mathbb{R}^l$, with $\Omega \subset \mathbb{R}^g$ a compact set. We assume that the manifolds have a regular parameterization

$$\mathbf{p}(\mathbf{s}) = \sum_{\mathbf{i}\in\mathcal{I}_\mathbf{n}} \frac{(c_\mathbf{i}^1, \ldots, c_\mathbf{i}^l)}{c_\mathbf{i}^h} \alpha_{\mathbf{i},\mathbf{n}}(\mathbf{s}), \ \mathbf{s} \in \Omega \text{ with}$$

$$\alpha_{\mathbf{i},\mathbf{n}}(\mathbf{s}) = \frac{c_\mathbf{i}^h B_{\mathbf{i},\mathbf{n}}(\mathbf{s})}{\sum_{\mathbf{j}\in\mathcal{I}_\mathbf{n}} c_\mathbf{j}^h B_{\mathbf{j},\mathbf{n}}(\mathbf{s})}, \ \mathbf{i} \in \mathcal{I}_\mathbf{n}, \ c_\mathbf{i}^h > 0, \ \mathbf{i} \in \mathcal{I}_\mathbf{n}.$$

We have introduced the index set $\mathcal{I}_\mathbf{n}$ to simplify the notation. The number of elements in $\mathcal{I}_\mathbf{n}$ is denoted by $\tilde{N}_\mathbf{n}$. The basis functions $B_{\mathbf{i},\mathbf{n}}(\mathbf{s}), \mathbf{i} \in \mathcal{I}_\mathbf{n}$, are assumed to be piecewise polynomials and to form a partition of unity. Thus, when collecting the basis function in a vector $\alpha_\mathbf{n}(\mathbf{s})$, we will have that $\|\alpha_\mathbf{n}(\mathbf{s})\|_2 \leq 1$ for $\mathbf{s} \in \Omega$.

To illustrate the indices contained in $\mathcal{I}_\mathbf{n}$, we look at a rational B-spline represented surface in $\mathbb{R}^3$ with positive weights of polynomial orders (k_1, k_2) and with $N_1 \times N_2$ basis functions:

$$\mathbf{p}(s,t) = \frac{\sum_{i_1=1}^{N_1}\sum_{i_2=1}^{N_2} (x_{i_1,i_2}, y_{i_1,i_2}, z_{i_1,i_2})\, B_{i_1,k_1}(s) B_{i_2,k_2}(t)}{\sum_{i_1=1}^{N_1}\sum_{i_2=1}^{N_2} h_{i_1,i_2} B_{i_1,k_1}(s) B_{i_2,k_2}(t)}$$

$$= \sum_{\mathbf{i}\in\mathcal{I}_\mathbf{n}} \frac{(x_\mathbf{i}, y_\mathbf{i}, z_\mathbf{i})}{h_\mathbf{i}} \alpha_{\mathbf{i},\mathbf{n}}(\mathbf{s}).$$

Thus, we have the following definitions of $\mathbf{n}$, $\mathcal{I}_{\mathbf{n}}$, $\mathbf{s}$ and $\alpha_{\mathbf{i},\mathbf{n}}(\mathbf{s})$:

$$\begin{aligned}
\mathbf{n} &= (k_1 - 1, k_2 - 1)\\
\mathcal{I}_{\mathbf{n}} &= \{(i_1, i_2) \in \mathbb{N}^2 \mid 1 \le i_1 \le N_1 \wedge 1 \le i_2 \le N_2\}\\
\mathbf{s} &= (s, t)\\
\alpha_{\mathbf{i},\mathbf{n}}(\mathbf{s}) &= \frac{h_{i_1,i_2} B_{i_1,k_1}(s) B_{i_2,k_2}(t)}{\sum_{j_1=1}^{N_1} \sum_{j_2=1}^{N_2} h_{j_1,j_2} B_{j_1,k_1}(s) B_{j_2,k_2}(t)}, \quad \mathbf{i} = (i_1, i_2) \in \mathcal{I}_{\mathbf{n}}.
\end{aligned}$$

§4. Combining Parametric Manifolds and Algebraic Hypersurfaces

The results in this section are useful for keeping the relative rounding errors in $q(\mathbf{p}(\mathbf{s}))$ at a minimal level. To do this we use a barycentric basis description of the algebraic hypersurfaces. This is, however, not necessary in the implementation of the method, as the power basis yields good results if the hypersurface (curve/surface) approximated is translated into a box around the origin with coordinates in the interval $[-1, 1]$. However, to use only one notation for algebraic hypersurfaces we use the barycentric representation.

The first step is Theorem 4.1 reformulating $q(\mathbf{p}(\mathbf{s}))$ to an expression using rational basis functions. Based on this, Corollary 4.1 reformulates the sum to an expression of vectors and matrices

$$q(\mathbf{p}(\mathbf{s})) = (\mathbf{D}\mathbf{b})^T \alpha_{m\mathbf{n}}(\mathbf{s}).$$

Theorem 4.2 then follows by bounding $|q(\mathbf{p}(\mathbf{s}))|$ by $\|\mathbf{D}\mathbf{b}\|_2$, and using thc the assumption that the basis $\alpha_{m\mathbf{n}}(\mathbf{s})$ is a partition of unity.

The remainder of the section is devoted to analyzing the numerical properties of the matrix $\mathbf{D}$. To simplify the presentation, we introduce vectors, matrices and lexicographical ordering of coefficients and basis functions. The coefficients of the algebraic hypersurface are isolated in a vector

$$\mathbf{b} = \left(\tilde{b}_j\right)_{j=1}^{\tilde{M}}.$$

The entries in the vector $\mathbf{b}$ are related to the coefficients, $b_{\mathbf{j}}$, $\mathbf{j} \in I(m)$, of the algebraic hypersurface $q(\mathbf{x}) = 0$ by a lexicographical ordering ν

$$\tilde{b}_{\nu(\mathbf{j})} = b_{\mathbf{j}},\ \mathbf{j} \in I(m).$$

Using the inverse ν^{-1} of the lexicographical ordering, we have

$$b_{\nu^{-1}(j)} = \tilde{b}_j,\ j = 1, \ldots, \tilde{M}.$$

We call the basis functions in the variables used for the parametrization of the manifold an α-basis, and collect them in a vector $\alpha_{m\mathbf{n}}(\mathbf{s})$,

$$\alpha_{m\mathbf{n}}(\mathbf{s}) = \left(\tilde{\alpha}_{i,m\mathbf{n}}(\mathbf{s})\right)_{i=1}^{\tilde{N}_{m\mathbf{n}}}.$$

The entries in the vector $\alpha_{m\mathbf{n}}(\mathbf{s})$ are related to the rational basis functions, $\alpha_{\mathbf{i},m\mathbf{n}}(\mathbf{s})$, $\mathbf{i} \in \mathcal{I}_{m\mathbf{n}}$, by a lexicographical ordering γ

$$\tilde{\alpha}_{\gamma(\mathbf{i}),m\mathbf{n}}(\mathbf{s}) = \alpha_{\mathbf{i},m\mathbf{n}}(\mathbf{s}), \quad \mathbf{i} \in \mathcal{I}_{m\mathbf{n}}.$$

Using the inverse γ^{-1} of the lexicographical ordering γ, we have

$$\tilde{\alpha}_{i,m\mathbf{n}}(\mathbf{s}) = \alpha_{\gamma^{-1}(i),m\mathbf{n}}(\mathbf{s}), \quad i = 1, \ldots, \tilde{N}_{m\mathbf{n}}.$$

Expressions depending on the coefficients of the manifold are inserted in the matrix $\mathbf{D}$. The entries in $\mathbf{D}$ are products of powers of the coefficients of the coordinate functions of $\mathbf{p}(\mathbf{s})$:

$$\mathbf{D} = \left(\tilde{d}_{i,j}\right)_{i,j=1,1}^{\tilde{N}_{m\mathbf{n}},\tilde{M}} \tag{4.1}$$

We also use the lexicographical ordering ν and γ and their inverse for the ordering of the entries of the $\mathbf{D}$ matrix. In the proof a transformation to the α-basis plays a central role. Thus, the first step is to represent $q(\mathbf{p}(\mathbf{s}))$ in this basis.

Theorem 4.1. *Let* $\mathbf{p}(\mathbf{s})$ *and* $q(\mathbf{x}) = 0$ *be as described in Section 3. Then*

$$q(\mathbf{p}(\mathbf{s})) = \sum_{\mathbf{i} \in \mathcal{I}_{m\mathbf{n}}} \left(\sum_{\mathbf{j} \in I(m)} d_{\mathbf{i}}^{\mathbf{j}} b_{\mathbf{j}} \right) \alpha_{\mathbf{i},m\mathbf{n}}(\mathbf{s}), \tag{4.2}$$

where $d_{\mathbf{i}}^{\mathbf{j}}$ *satisfy*

$$\frac{m!}{\mathbf{j}!} \beta^{\mathbf{j}}(\mathbf{p}(\mathbf{s})) = \sum_{\mathbf{i} \in \mathcal{I}_{m\mathbf{n}}} d_{\mathbf{i}}^{\mathbf{j}} \alpha_{\mathbf{i},m\mathbf{n}}(\mathbf{s}),\ \mathbf{j} \in I(m),$$

with

$$\beta^{\mathbf{j}}(\mathbf{x}) = \beta_1^{j_1}(\mathbf{x}), \ldots, \beta_{l+1}^{j_{l+1}}(\mathbf{x}).$$

Proof: Remembering that $\beta^{\mathbf{j}}(\mathbf{p}(\mathbf{s}))$ is a product of m of the coordinate functions of $\mathbf{p}(\mathbf{s})$, we have that the polynomial degree of $\beta^{\mathbf{j}}(\mathbf{p}(\mathbf{s}))$ is $m\mathbf{n}$. Thus, $\frac{m!}{\mathbf{j}!}\beta^{\mathbf{j}}(\mathbf{p}(\mathbf{s}))$ can be expressed in the α-basis of degree $m\mathbf{n}$. We denote these coefficients by $d_{\mathbf{i}}^{\mathbf{j}}$, and get

$$\frac{m!}{\mathbf{j}!} \beta^{\mathbf{j}}(\mathbf{p}(\mathbf{s})) = \sum_{\mathbf{i} \in \mathcal{I}_{m\mathbf{n}}} d_{\mathbf{i}}^{\mathbf{j}} \alpha_{\mathbf{i},m\mathbf{n}}(\mathbf{s}), \qquad \mathbf{j} \in I(m),$$

giving

$$q(\mathbf{p}(\mathbf{s})) = \sum_{\mathbf{j} \in I(m)} b_{\mathbf{j}} \sum_{\mathbf{i} \in \mathcal{I}_{m\mathbf{n}}} d_{\mathbf{i}}^{\mathbf{j}} \alpha_{\mathbf{i},m\mathbf{n}}(\mathbf{s}).$$

Now rearranging we get (4.2). □

Corollary 4.1. *Let* $\mathbf{p}(\mathbf{s})$ *and* $q(\mathbf{x})$ *be defined as in Theorem 4.1. Then*

$$q(\mathbf{p}(\mathbf{s})) = (\mathbf{Db})^T \alpha_{mn}(\mathbf{s}).$$

Proof: Rearranging (4.2) in Theorem 4.1 and remembering that $\tilde{d}_{i,j} = d^{\nu^{-1}(j)}_{\gamma^{-1}(i)}$, $\tilde{b}_j = b_{\nu^{-1}(j)}$ and $\tilde{\alpha}_{i,mn}(\mathbf{s}) = \alpha_{\gamma^{-1}(i),mn}(\mathbf{s})$, we have

$$\begin{aligned} q(\mathbf{p}(\mathbf{s})) &= \sum_{i=1}^{\tilde{N}_{mn}} \left(\sum_{j=1}^{\tilde{M}} d^{\nu^{-1}(j)}_{\gamma^{-1}(i)} b_{\nu^{-1}(j)} \right) \alpha_{\gamma^{-1}(i),mn}(\mathbf{s}) \\ &= \sum_{i=1}^{\tilde{N}_{mn}} \left(\sum_{j=1}^{\tilde{M}} \tilde{d}_{i,j}\, \tilde{b}_j \right) \tilde{\alpha}_{i,mn}(\mathbf{s}) = (\mathbf{Db})^T \alpha_{mn}(\mathbf{s}). \quad \square \end{aligned}$$

In the following theorem we assume that the basis functions are a partition of unity, thus limiting the vector $\alpha_{mn}(\mathbf{s})$ containing the variable(s) $\mathbf{s}$ used in the parametrization of the manifold.

Theorem 4.2. *Let the algebraic hypersurface and the manifold be as in Corollary 4.1 and assume that the basis* $\alpha_{mn}(\mathbf{s})$ *is a partition of unity. Let* $\mathbf{s} \in \Omega$, *e.g. in the support of the basis functions of the manifold. Then*

$$|q(\mathbf{p}(\mathbf{s}))| \le \|\mathbf{Db}\|_2 .$$

Proof: For any $\mathbf{s} \in \Omega$ we have by definition $\sum_{i=1}^{\tilde{N}_{mn}} \alpha_{i,mn}(\mathbf{s}) = 1$ and $0 \le \alpha_{i,mn}(\mathbf{s}) \le 1$, $i = 1, \ldots, \tilde{N}_{mn}$. Thus, $\|\alpha_{mn}(\mathbf{s})\|_2 \le 1$, giving

$$|q(\mathbf{p}(\mathbf{s}))| = \left|(\mathbf{Db})^T \alpha_{mn}(\mathbf{s})\right| \le \|\mathbf{Db}\|_2 \,\|\alpha_{mn}(\mathbf{s})\|_2 \le \|\mathbf{Db}\|_2 \quad \square$$

If a barycentric coordinate system containing the manifold to be approximated is used for representing the manifold and the algebraic hypersurface, then the matrix $\mathbf{D}$ has desirable numerical properties. The use of a Bernstein basis for representing the algebraic hypersurface is crucial when proving these properties:

- All entries are nonnegative.
- The sum of all entries in a row is one.
- The square of the Frobenius norm of $\mathbf{D}$ is limited by the number of rows.

Theorem 4.3. *Let* $\mathbf{D}$ *be defined as in (4.1). Then*

$$\sum_{j=1}^{\tilde{M}} \tilde{d}_{i,j} = 1, \qquad i = 1, \ldots, \tilde{N}_{mn}. \tag{4.3}$$

If the coefficients of the manifold are contained in the simplex defining the barycentric coordinate system in which $q(\mathbf{x})$ *is described, then*

$$\tilde{d}_{i,j} \ge 0, \quad i = 1, \ldots, \tilde{N}_{mn}, \quad j = 1, \ldots, \tilde{M}, \tag{4.4}$$

and

$$\|\mathbf{D}\|_F^2 = \sum_{i=1}^{\tilde{N}_{mn}} \sum_{j=1}^{\tilde{M}} \left(\tilde{d}_{i,j}\right)^2 = \sum_{j=1}^{\tilde{M}} \sigma_j^2 \leq \tilde{N}_{\mathbf{mn}}. \tag{4.5}$$

Here $\| \ \|_F$ *is the Frobenius norm, and* σ_i, $i = 1, \ldots, \tilde{M}$ *are the singular values of* $\mathbf{D}$.

Proof: Choose $\mathbf{b} = (1, \ldots, 1)^T$. Then $q(\mathbf{x}) = 1$ giving

$$q(\mathbf{x}) = \sum_{i=1}^{\tilde{N}_{mn}} \left(\sum_{j=1}^{\tilde{M}} \tilde{d}_{i,j} \right) \tilde{\alpha}_{i,mn}(\mathbf{s}) = 1.$$

As $\sum_{i=1}^{\tilde{N}_{mn}} \tilde{\alpha}_{i,mn}(\mathbf{s}) = 1$, and $\tilde{\alpha}_{i,mn}(\mathbf{s})$ are linearly independent, we have

$$\sum_{j=1}^{\tilde{M}} \tilde{d}_{i,j} = 1.$$

Thus, (4.3) is proved. Since the coefficients of the manifold are contained in the simplex defining the barycentric coordinate system, the coefficients of the barycentric representation of the manifold are all nonnegative. In addition, we have assumed that the weights of the manifold are positive. Thus, all the elements in $\mathbf{D}$ are greater than or equal to zero, as all functions involved in the multiplication process making $\mathbf{D}$ have nonnegative coefficients. This proves (4.4).

Since $0 \leq \tilde{d}_{i,j} \leq 1$, we have $0 \leq (\tilde{d}_{i,j})^2 \leq \tilde{d}_{i,j} \leq 1$, resulting in

$$\|\mathbf{D}\|_F^2 = \sum_{i=1}^{\tilde{N}_{mn}} \sum_{j=1}^{\tilde{M}} \left(\tilde{d}_{i,j}\right)^2 \leq \sum_{i=1}^{\tilde{N}_{mn}} \sum_{j=1}^{\tilde{M}} \tilde{d}_{i,j} = \sum_{i=1}^{\tilde{N}_{mn}} 1 = \tilde{N}_{mn}$$

proving (4.5). The relation between the Frobenius norm and the singular values can be found in standard books on linear algebra, e.g. in [18]. □

In CAGD applications, we are most often interested in planar curves (1-manifolds in $\mathbb{R}^2$), space curves (1-manifolds in $\mathbb{R}^3$) or surfaces (2-manifolds in $\mathbb{R}^3$). The results in Theorem 4.2 are independent of the dimension of the space in which the manifold lies. It can also be shown that when proper algorithms are used for building the matrix $\mathbf{D}$ using floating point arithmetic the relative rounding errors of the entries in $\mathbf{D}$ are limited by $m\epsilon_{\max}$ where $\epsilon_{\max}$ is the maximal relative rounding error of the coefficients of the manifold $\mathbf{p}(\mathbf{s})$ being approximated. For details, see [6].

§5. Implicitization by Singular Values

How can the results from Section 4 be used when approximating a manifold in $\mathbb{R}^l$? In Theorem 4.2 we established a relationship $|q(\mathbf{p}(\mathbf{s}))| \leq \|\mathbf{D}\mathbf{b}\|_2$. We now relate this expression to the singular values of $\mathbf{D}$. The result is used in the algorithm for approximate implicitization described in Section 2.

Theorem 5.1. *Let* $\mathbf{p}(\mathbf{s})$ *and* $q(\mathbf{x}) = 0$ *be as described in Section 3, and let* $\sigma_1 \geq 0$ *be the smallest singular value of* $\mathbf{D}$. *Then*

$$\min_{\|\mathbf{b}\|_2=1} \max_{\mathbf{s}\in\Omega} | q(\mathbf{p}(\mathbf{s})) | \leq \sigma_1.$$

Proof: Since $|q(\mathbf{p}(\mathbf{s}))| \leq \|\mathbf{Db}\|_2$, we have $|q(\mathbf{p}(\mathbf{s}))|^2 \leq \mathbf{b}^T\mathbf{D}^T\mathbf{Db}$, and thus $\max_{\mathbf{s}\in\Omega} |q(\mathbf{p}(\mathbf{s}))|^2 \leq \mathbf{b}^T\mathbf{D}^T\mathbf{Db}$. Now restricting $\|\mathbf{b}\|_2 = 1$, and taking the minimum we get

$$\min_{\|\mathbf{b}\|_2=1} \max_{\mathbf{s}\in\Omega} | q(\mathbf{p}(\mathbf{s})) |^2 \leq \min_{\|\mathbf{b}\|_2\neq 0} \frac{\mathbf{b}^T\mathbf{D}^T\mathbf{Db}}{\mathbf{b}^T\mathbf{b}} = \lambda_{\min}.$$

Where $\lambda_{\min} \geq 0$ is the smallest eigenvalue of $\mathbf{D}^T\mathbf{D}$. The last equal sign is based on $\mathbf{D}^T\mathbf{D}$ being a Hermitian matrix, and the fact that for a Hermitian matrix the Raleigh quotient $\frac{\mathbf{x}^T\mathbf{D}^T\mathbf{Dx}}{\mathbf{x}^T\mathbf{x}}$ satisfies $\lambda_{\min} \leq \frac{\mathbf{x}^T\mathbf{D}^T\mathbf{Dx}}{\mathbf{x}^T\mathbf{x}} \leq \lambda_{\max}$. Here $\lambda_{\min}$ is the smallest and $\lambda_{\max}$ the largest eigenvalue of $\mathbf{D}^T\mathbf{D}$. The smallest singular value of $\mathbf{D}$ is $\sigma_1 = \sqrt{\lambda_{\min}}$. □

It is important to take account of the number of singular values in $\mathbf{D}$ that are identically zero or very small. If the number exceeds 1, then it is possible that we have used a higher algebraic degree than necessary in the approximation. This also is an indication that the parametric description of the manifold contains base points.

§6. Convergence Rate of Approximate Implicitization

For approximate implicitization to be useful we need to establish results on how the approximation error behaves, as the part of the manifold being approximated is reduced. How the actual error is measured is addressed in Section 8. The convergence rates established are best for low dimensional manifolds. The results show that approximate implicitization of 2D curves has convergence rate

$$O(h^{\frac{(m+1)(m+2)}{2}-1}).$$

Thus approximate implicitization with cubic algebraic curves has convergence rate $O(h^9)$. Approximation of 3D surfaces has convergence rate

$$O(h^{\left\lfloor \frac{1}{6}\sqrt{(9+12m^3+72m^2+132m)}-\frac{1}{2}\right\rfloor}).$$

Thus approximate implicitization with cubic algebraic surface has convergence rate $O(h^5)$. However, the convergence rate can be misleading if the gradient of the curve or surface found, is vanishing or near vanishing close to the manifold being approximated.

The proof of the theorem is based on the power expansion around a point $\mathbf{v} \in \Omega$, where Ω is the parameter domain of the manifold $\mathbf{p}$ approximated. The value of h is used to define a closed hyperbox in Ω around $\mathbf{v}$ in the following way:

$$\Omega(\mathbf{v},h) = \Omega \cap \{\mathbf{s} \in \mathbb{R}^g \mid \|\mathbf{v}-\mathbf{s}\|_\infty \le h\}.$$

It should be noted that the matrix $\mathbf{D}$ now depends on the value of $\mathbf{v}$ and h. Remember that m is the degree of the algebraic surface, g is the dimension of the manifold, and l the dimension of the space containing the manifold approximated.

Theorem 6.1. *Let $\mathbf{p}(\mathbf{s})$ and $q(\mathbf{x}) = 0$ be as described in Section 3, and assume that $\mathbf{p}(\mathbf{s})$, is C^{n+1}-continuous. Then for $\mathbf{v} \in \Omega$ and $h > 0$,*

$$\min_{\|\mathbf{b}\|_2=1} \max_{\mathbf{s}\in\Omega(\mathbf{v},h)} |q(\mathbf{p}(\mathbf{s})| \le \min_{\|\mathbf{b}\|_2=1} \|\mathbf{Db}\|_2 \le O(h^{n+1}),$$

where n satisfies

$$\binom{n+g+1}{g} \ge \binom{m+l}{l} > \binom{n+g}{g}.$$

Proof: We use the description of the algebraic hypersurface in Section 3 and let $r_{\mathbf{i}}(\mathbf{s}) = \frac{m!}{\mathbf{i}!}\beta^{\mathbf{i}}(\mathbf{p}(\mathbf{s}))$. Then $q(\mathbf{p}(\mathbf{s})) = \sum_{\mathbf{i}\in I(m)} b_{\mathbf{i}}\, r_{\mathbf{i}}(\mathbf{s})$. Now assumimg that $\mathbf{s} \in \Omega(\mathbf{v},h)$, then with $\mathbf{s} = \mathbf{v}+\delta$, we have that $\|\delta\|_\infty \le h$. Taylor expansion of $r_{\mathbf{i}}(\mathbf{s})$ around $\mathbf{v}$ with a polynomial of total degree n gives $r_{\mathbf{i}}(\mathbf{v}+\delta) = t_{\mathbf{i}}(\delta) + e_{\mathbf{i}}(\delta)$, where $t_{\mathbf{i}}(\delta)$ has total degree n, and

$$e_{\mathbf{i}}(\delta) = O\left((\|\delta\|_\infty)^{n+1}\right) \le O(h^{n+1}).$$

The number of terms in a polynomial of total degree n in g variables is $\binom{n+g}{g}$. The number of coefficients in the algebraic hypersurface is $\binom{m+l}{l}$. If the number of coefficients in the algebraic hypersurface is greater than number of coefficients in the polynomial of total degree n, then the polynomials $t_{\mathbf{i}}(\delta)$ are linearly dependent. I.e. if

$$\binom{m+l}{l} > \binom{n+g}{g}$$

we can find coefficients $\mathbf{b}'$ with $\|\mathbf{b}'\|_2 = 1$, such that $\sum_{\mathbf{i}\in I(m)} b'_{\mathbf{i}}\, t_{\mathbf{i}}(\delta) = 0$. We are interested in the largest natural number n with this property, thus we state

$$\binom{n+1+g}{g} \ge \binom{m+l}{l} > \binom{n+g}{g}.$$

Using the above Taylor expansion, the associated error functions, and that $\delta = \mathbf{s}-\mathbf{v}$, we get

$$q(\mathbf{p}(\mathbf{s})) = (\mathbf{Tb})^T\alpha(\mathbf{s}) + (\mathbf{Eb})^T\alpha_{\mathbf{mn}}(\mathbf{s}),$$

where

$$(\mathbf{Tb})^T \alpha(\mathbf{s}) = \sum_{\mathbf{i} \in I(m)} b_{\mathbf{i}}\, t_{\mathbf{i}}(\mathbf{s} - \mathbf{v}),$$

and

$$(\mathbf{Eb})^T \alpha(\mathbf{s}) = \sum_{\mathbf{i} \in I(m)} b_{\mathbf{i}}\, e_{\mathbf{i}}(\mathbf{s} - \mathbf{v}).$$

Recalling that $e_{\mathbf{i}}(\mathbf{s} - \mathbf{v}) = e_{\mathbf{i}}(\delta) \leq O(h^{n+1})$, we have

$$(\mathbf{Eb})^T \alpha(\mathbf{s}) \leq O(h^{n+1}).$$

All entries in $\alpha(\mathbf{s})$ are linearly independent, and thus all entries in $\mathbf{Eb}'$ must be $O(h^{n+1})$, giving $\|\mathbf{Eb}'\|_2 \leq O(h^{n+1})$. Now $\mathbf{Tb}' = \mathbf{0}$, and we get

$$\min_{\|\mathbf{b}\|_2=1} \max_{\mathbf{s} \in \Omega(\mathbf{v},h)} |q(\mathbf{p}(\mathbf{s}))| \leq \min_{\|\mathbf{b}\|_2=1} \|\mathbf{Db}\|_2 \leq \frac{\|\mathbf{Db}'\|_2}{\|\mathbf{b}'\|_2}$$

$$= \frac{\|(\mathbf{T}+\mathbf{E})\mathbf{b}'\|_2}{\|\mathbf{b}'\|_2} = \frac{\|\mathbf{Eb}'\|_2}{\|\mathbf{b}'\|_2} \leq O(h^{n+1}). \quad \square$$

Example. *Approximate implicitization of a 2D curve.* For 2D curves, $l = 2$ and $g = 1$. Thus,

$$\binom{n+2}{1} \geq \binom{m+2}{2} > \binom{n+1}{1}.$$

Rewriting, we get

$$n + 2 \geq \frac{(m+1)(m+2)}{2} > n + 1,$$

giving

$$n = \frac{(m+1)(m+2)}{2} - 2,$$

and convergence rate

$$O(h^{\frac{(m+1)(m+2)}{2} - 1}).$$

Example. *Approximate implicitization of a 3D surface.* For 3D surfaces, $l = 3$ and $g = 2$. Thus,

$$\binom{n+3}{2} \geq \binom{m+3}{3} > \binom{n+2}{2},$$

resulting in

$$\binom{m+3}{3} - 1 = \binom{n+2}{2},$$

with solution

$$n = \left\lfloor \frac{1}{6}\sqrt{(9 + 12m^3 + 72m^2 + 132m)} - \frac{3}{2} \right\rfloor.$$

The convergence rate is thus

$$O(h^{\left\lfloor \frac{1}{6}\sqrt{(9+12m^3+72m^2+132m)} - \frac{1}{2} \right\rfloor}).$$

§7. Constraining the Approximation

Two different types of constraints can be added to the approximate implicitization: interpolating constraints and controlling constraints. For a more detailed discussion on constraints, see [6].

- Interpolating constraints for 2D curves are typically interpolation of points and tangents on the manifold being approximated. For surfaces, the interpolation can be located at points or along curves. Interpolating constraints do not reduce the convergence rate of approximate implicitization.
- Constraints controlling the behavior of the approximate implicitization reduce the convergence rate. One example of such a constraint is to enforce that the algebraic surface is a monoid manifold as shown in [17]. Assume that the algebraic hypersurface is of total degree m. By adding a singular point of multiplicity $(m-1)$, the manifold will be a monoid. Thus, any straight line going through the point of multiplicity $(m-1)$ can only intersect the algebraic surface at one additional point.

Corollary 7.1. *Let* $\mathbf{p}(\mathbf{s})$ *and* $q(\mathbf{x}) = 0$ *be as described in Theorem 6.1, and assume that* k *linear non-interpolating constraints are added to* q*. Then the convergence rate is* $O(h^{n+1})$*, where* n *satisfies*

$$\binom{n+g+1}{g} \geq \binom{m+l}{l} - k > \binom{n+g}{g}.$$

Corollary 7.2. *Let* $\mathbf{p}(\mathbf{s})$ *and* $q(\mathbf{x}) = 0$ *be as described in Corollary 7.1 and constrain* q *to be monoid, e.g. by adding* $\binom{m-2+l}{l}$ *non-interpolating constraints. Then the convergence rate is* $O(h^{n+1})$*, where* n *satisfies*

$$\binom{n+g+1}{g} \geq \binom{m+l}{l} - \binom{m-2+l}{l} > \binom{n+g}{g}.$$

Example. *Approximate implicitization of 2D monoid curve.* For 2D curves, $l = 2$ and $g = 1$. Thus,

$$\binom{n+2}{1} \geq \binom{m+2}{2} - \binom{m}{2} > \binom{n+1}{1}.$$

Rewriting, we get $n+2 \geq 2m+1 > n+1$, giving $n = 2m-1$, and convergence rate $O(h^{2m})$ for approximation with 2D monoids.

Example. *Approximate implicitization of 3D monoid surface.* For 3D surfaces, $l = 3$ and $g = 2$. Thus,

$$\binom{n+3}{2} \geq \binom{m+3}{3} - \binom{m+1}{3} > \binom{n+2}{2},$$

resulting in

$$\binom{m+3}{3} - \binom{m+1}{3} - 1 = \binom{n+2}{2},$$

with solution

$$n = -\frac{3}{2} + \frac{1}{2}\sqrt{(1+8m^2+16m)}.$$

The convergence rate is thus

$$O(h^{\lfloor -\frac{1}{2}+\frac{1}{2}\sqrt{(1+8m^2+16m)}\rfloor}).$$

Thus, cubic approximate implicitization with monoid surfaces gives convergence $O(h^5)$. This seems to be the same as for interpolation with no constraints given in the previous section. However, looking at the unconstrained approximation before the convergence rate is truncated, we get $\frac{3}{2}\sqrt{17} - \frac{1}{2} \approx 5.6847$ in the non-monoid case.

§8. Accuracy of Approximate Implicitization

In many cases we want an approximation to be within a certain tolerance $\epsilon > 0$ of the object approximated. When we measure the distance from a manifold $\mathbf{p}(\mathbf{s})$ to a manifold $\mathbf{q}(\mathbf{t})$, a natural choice for distance measure is

$$d(\mathbf{p}, \mathbf{q}) = \max_{\mathbf{s}} \inf_{\mathbf{t}} \|\mathbf{p}(\mathbf{s}) - \mathbf{q}(\mathbf{t})\|_2 .$$

In this definition inf is used since the manifolds are not required to be closed. The distance measure can be interpreted as follows:

- For all points on $\mathbf{p}$ find the closest point on $\mathbf{q}$,
- Use the maximum distance found as the distance measure.

To use the distance function $d(\mathbf{p}, \mathbf{q})$, we require a parametric description of both manifolds. In the case we shall discuss, this is not the case. We have an algebraic hypersurface $q(\mathbf{x}) = 0$ and a parametric represented manifold $\mathbf{p}(\mathbf{s})$. To measure the error, we now introduce a direction for error measurement $\mathbf{g}(\mathbf{s})$ satisfying $\|\mathbf{g}(\mathbf{s})\|_2 = 1$. To be a useful direction for error measurement, we assume that there exists an error function $\rho(\mathbf{s}) \in C^0$ such that

$$q(\mathbf{p}(\mathbf{s}) - \rho(\mathbf{s})\mathbf{g}(\mathbf{s})) = 0.$$

Not all choices for direction for error measurement will satisfy this requirement. To simplify the notion, we now let $\mathbf{p} = \mathbf{p}(\mathbf{s})$, $\mathbf{g} = \mathbf{g}(\mathbf{s})$ and $\rho = \rho(\mathbf{s})$. The above equation can then be expressed $q(\mathbf{p} - \rho\mathbf{g}) = 0$. Taylor expansion with respect to ρ now gives $q(\mathbf{p}) - \nabla q(\mathbf{p} - \theta\mathbf{g}) \cdot \rho\mathbf{g} = 0$, with $\theta\rho \geq 0$ and $|\theta| \leq |\rho|$. Rearranging, we get $\rho\,(\nabla q(\mathbf{p} - \theta\mathbf{g}) \cdot \mathbf{g}) = q(\mathbf{p})$. Now setting $\rho_{\max} = \max_{\mathbf{s}\in\Omega} |\rho(\mathbf{s})|$, and assuming that $\nabla q(\mathbf{p}-\theta\mathbf{g})\cdot\mathbf{g} \neq 0$, $\mathbf{s} \in \Omega$, $|\theta| \leq \rho_{\max}$ we get, by reintroducing $\mathbf{p}(\mathbf{s})$, $\mathbf{g}(\mathbf{s})$ and $\rho(\mathbf{s})$

$$\rho(\mathbf{s}) = \frac{q(\mathbf{p}(\mathbf{s}))}{\nabla q(\mathbf{p}(\mathbf{s}) - \theta\mathbf{g}(\mathbf{s})) \cdot \mathbf{g}(\mathbf{s})}$$

with $|\theta| \leq |\rho(\mathbf{s})|$. We see that the relative orientation of the direction for error measurement and the gradient of q play a central role in estimating the error. Thus, an optimal solution is to find a direction for error measurement that gives the maximum value of the denominator in the expression above. This requirement can be formulated as finding $\mathbf{g}(\mathbf{s})$ with $\|\mathbf{g}(\mathbf{s})\|_2 = 1$ that satisfies

$$\begin{aligned}&\min_{\mathbf{s}\in\Omega,|\theta|\leq\rho_{\max}} |\nabla q(\mathbf{p}(\mathbf{s}) - \theta\mathbf{g}(\mathbf{s}))\cdot\mathbf{g}(\mathbf{s})| \\ &\quad = \max_{\|\tilde{\mathbf{g}}(\mathbf{s})\|_2=1}\ \min_{\mathbf{s}\in\Omega,|\theta|\leq\rho_{\max}} |\nabla q(\mathbf{p}(\mathbf{s}) - \theta\tilde{\mathbf{g}}(\mathbf{s}))\cdot\tilde{\mathbf{g}}(\mathbf{s})|\end{aligned}.$$

However, to find such a maximum is not simple. Thus, for practical purposes we want to make a simpler choice of direction for error measurement. We see that if the gradient gets small, the error grows. We also see that if the gradient and direction for error measurement are near normal, the error will grow. Thus, to make the direction for error measurement dependent on the gradient direction seems natural. At the same time for curves in $\mathbb{R}^2$ or surfaces in $\mathbb{R}^3$, it is natural to base the direction for error measurement on the normal.

§9. Selecting an Approximate Implicitization

If the algebraic degree is sufficiently high, we can find the exact algebraic representation of any curve and surface that is represented by a rational polynomial parameterization, see e.g. [1]. However, when we deal with piecewise polynomials, there are in general internal discontinuities that prohibit a representation by one algebraic hypersurface. Generally, we use algebraic representations of low degree for efficiency and stability reasons. Thus, approximation methods that give a "good" enough solution is of great interest. By "good" solutions we mean hypersurfaces that do not have singularities in the area of interest and are within a given tolerance.

The approach employed to find a "good" approximation described in this section has the following steps:

- Select an approximate null-space for the matrix $\mathbf{D}$. This can be done by either performing singular value decomposition of $\mathbf{D}$ and selecting a number of coefficient vectors belonging to the smallest singular values, or using a direct search for an approximate null-space, and then describing this by orthonormal coefficient vectors.
- By using a property function $\omega(\mathbf{b})$ on the different orthonormal coefficient vectors, assign property values to the coefficient vectors.
- Find a coefficient vector $\mathbf{b}'$ with $\|\mathbf{b}'\| = 1$ that is a linear combination of the orthogonal vectors spanning the approximate null space and with a maximal value of ω, the property function.

The error estimate $\rho(\mathbf{s}) = \frac{q(\mathbf{p}(\mathbf{s}))}{\nabla q\mathbf{p}(\mathbf{s})-\theta\mathbf{g}(\mathbf{s}))\cdot\mathbf{g}(\mathbf{s})}$ is the basis of this approach. Here $\mathbf{g}(\mathbf{s})$ is the direction for error measurement. By maximizing the value of

$$\min_{(\mathbf{s},\theta)\in\Omega\times[-\epsilon,\epsilon]} |\nabla q(\mathbf{p}(\mathbf{s}) - \theta\mathbf{g}(\mathbf{s}))\cdot\mathbf{g}(\mathbf{s})|$$

for some chosen $\epsilon > 0$, we tend to reduce the total error. However, the error problem is easier facilitated by maximizing the integral of the expression above in the region in question. To measure the behavior of the denominator in this expression, the following integral can be used

$$\int_{\Omega}\int_{-\epsilon}^{\epsilon}(\nabla q(\mathbf{p}(\mathbf{s})-\theta\mathbf{g}(\mathbf{s})))\cdot\mathbf{g}(\mathbf{s})d\theta ds.$$

This integral has some attractive properties:

- If $(\nabla q(\mathbf{p}(\mathbf{s})-\theta\mathbf{g}(\mathbf{s})))\cdot\mathbf{g}(\mathbf{s})d\theta ds$ changes sign in the region of interest, the sign change tends to reduce the absolute value of the integral. As ridges and sinks introduce a near vanishing gradient, maximizing the integral tends to minimize the existence of near vanishing gradients.
- If $(\nabla q(\mathbf{p}(\mathbf{s})-\theta\mathbf{g}(\mathbf{s})))\cdot\mathbf{g}(\mathbf{s})d\theta ds$ has no sign changes, then this tends to increase the absolute value of the integral.
- If $q(\mathbf{p}(\mathbf{s})-\theta\mathbf{g}(\mathbf{s}))$ and $\mathbf{g}(\mathbf{s})$ are nearly parallel, the contribution to the total value of the integral is significant. Thus, the maximization favors solutions where the gradient of q and the direction for error measurement are parallel.

To evaluate these integrals exactly (if possible) is resource consuming. Thus, it is practical to employ numerical integration when the property functions are defined.

§10. Approximate Implicitization of a Number of Manifolds

There are a set of applications for a simultaneous approximation of more than one manifold. In some cases it is desirable to control the shape of the algebraic approximation in a region larger than covered by one manifold. In other cases, it may be necessary to model an algebraic surface approximating or interpolating a set of curves. Alternatively, the goal can be to make a rough approximation of a set of manifolds to simplify a complex structure of manifolds. For special purposes it can be desirable to combine a surface with additional problem dependent curves or surfaces to direct the surface to behave in a specific way.

The dimension of the manifolds approximated is not fixed in the following theorem. The manifolds can touch in a smooth way, or they can be separated by some distance. Some of the manifolds can be the target of the approximation. Other manifolds can be used to control the behavior of the approximation in a given region. The following theorem shows that we can combine the **D** matrices of the different manifolds to build a matrix for approximating a number of manifolds.

Theorem 8.1. *Let $\mathbf{p}_i(\mathbf{s}_i)$, $i = 1,\ldots,r$ be manifolds in $\mathbb{R}^l$ of respectively dimension g_i, $i = 1,\ldots,r$ with a regular parameterization $\mathbf{s}_i \in \Omega_i$, as described in Section 3. The combination of these manifolds with an algebraic* *hypersurface q of total degree m satisfies*

$$\sum_{i=1}^{r} (q(\mathbf{p}_i(\mathbf{s}_i)))^2 \leq \left\| \mathbf{b}^{\mathrm{T}} \left(\mathbf{D}^{\mathrm{T}}{}_1, \ldots, \mathbf{D}^{\mathrm{T}}{}_r \right) \right\|_2^2,$$

where $q(\mathbf{p}_i(\mathbf{s}_i)) = (\mathbf{D}_i\mathbf{b})^{\mathrm{T}} \alpha_i(\mathbf{s}_i),\ i = 1, \ldots, r.$

Proof:

$$q(\mathbf{p}_i(s))^2 = \sum_{i=1}^{r} ((\mathbf{D}_i\mathbf{b})^{\mathrm{T}} \alpha_i(\mathbf{s}_i))^2 \leq \sum_{i=1}^{r} \|\mathbf{D}_i\mathbf{b}\|_2^2 = \sum_{i=1}^{r} \mathbf{b}^{\mathrm{T}}\mathbf{D}_i^{\mathrm{T}}\mathbf{D}_i\mathbf{b}. \quad \square$$

The result is in correspondence with approximation of a piecewise represented manifold when we break a NURBS represented manifold into rational Bernstein basis represented manifolds by introducing the Bernstein knot vector. However, the result is more general, because the manifolds can be of different dimension with no correspondence of the parametrizations.

Acknowledgments. The preparation of the paper was partly funded by the European Commission through the project IST-1999-29010 - GAIA.

References

1. Bajaj, C., The Emergence of Algebraic Curves and Surfaces in Geometric Design, in *Directions in Geometric Computing*, Information Geometers, 1993,1–28.
2. Chionh, E.W. and R. N. Goldmann, Using multivariate resultants to find the implicit equation of a rational surface, The Visual Computer **8** (1992), 171–180.
3. Cox, D., J. Little, and O'Shea, D., *Ideals, Varieties and Algorithms*, Springer-Verlag, New York, 1992 and 1997.
4. Cox, D., J. Little, and O'Shea, D., *Using Algebraic Geometry*, Springer-Verlag, New York, 1998.
5. Cox D., R. Goldman, and M. Zhang, On the validity of implicitization by moving quadrics for rational surfaces with no base points, J. Symbolic Computation **11** (1999).
6. Dokken, T., Aspects of Intersection Algorithms and Approximation, Thesis for the doctor philosophiae degree, University of Oslo, Norway, 1997.
7. Hoffmann, C. M., Implicit Curves and Surfaces in CAGD, Comp. Graphics and Applics. , **13** (1993), 79–88.
8. Press, Flannery, Teukolsky and Vetterling, *Numerical Recipes in C*, Cambridge University Press, Cambridge, 1988, 60–71.
9. Sederberg, T. W., Implicit and Parametric Curves and surfaces for Computer Aided Geometric Design, Ph.D. Thesis, Purdue University, 1983.

10. Sederberg, T. W., D. C. Anderson and R. N. Goldman, Implicit representation of parametric curves and surfaces, Computer Vision, Graphics and Image Processing **28** (1984), 72–84.
11. Sederberg, T. W., Planar piecewise algebraic curves, Comput. Aided Geom. Design **1** (1984), 241-255.
12. Sederberg, T. W., An algorithm for algebraic curve intersection, Computer-Aided Design **21** (1989), 547–554.
13. Sederberg, T. W., J. Zhao and A. K. Zundel, Approximate parametrization of algebraic curves, in *Theory and Practice of Geometric Modeling*, W. Strasser and H.-P. Seidel (eds), Springer, Berlin, 1989, 33–54.
14. Sederberg, T. W., Techniques for cubic algebraic Surfaces, IEEE Comp. Graphics and Applics., July 1990, 14–25.
15. Sederberg, T. W. and F. Chen, Implicitization using moving curves and surfaces, in *Computer Graphics Annual Conference Series (1995)*, 301–308.
16. Sederberg, T. W., R. N. Goldman, and H. Du, Implicitization rational curves by the method of moving algebraic curves, J. Symbolic Computation **23** (1997), 153–175.
17. Sederberg, T.W., J. Zheng, K. Klimaszewski and T. Dokken, Approximate implicitization using monoid curves and surfaces, Graphical Models and Image Processing **61**(1999), 177–198.
18. Stewart, G. W., *Introduction to Matrix Computations*, Academic Press, New York, 1973.

Tor Dokken
SINTEF Applied Mathematics
P.O. Box 124 Blindern
0314 Oslo
Norway
`tor.dokken@math.sintef.no`
`http://www.math.sintef.no/Geom/`

An Approach to Weak Approximate Implicitization

Tor Dokken, Hermann K. Kellermann, and Cathrine Tegnander

Abstract. We present an approach for finding approximate implicitizations to a 2D-curve (or 3D-surface) using a weak formulation (L_2-norm). The corresponding numerical algorithm is presented, and the results are illustrated by numerical examples.

§1. Introduction

In CAGD, geometric objects are often represented parametrically. Due to the importance of stable numerical evaluation, the choice of parametrizations can be crucial. To avoid poorly behaved parameterization, it can thus be useful to define the curve or the surface by its implicit expression whenever this exists. In classical algebraic geometry there are well known methods [6] for finding an implicit equation given a polynomial parameterization. Several methods for performing exact implicitization have been developed [5]. These methods are often based on the classical determinant principle like Sylvester's and Cayley's methods [2], building up large matrices whose size depend on the degree of the implicit polynomial. Another possibility is to approximate the parameterized object by an implicit one [4,7]. For an overview of these methods, consult the article [4] on Approximate Implicitization in this book. From a numerical point of view, it is convenient to deal with implicit polynomials of low degree (i.e. degree < 5). In cases where the exact implicitization exists, it is often of high degree [8], and it is thus beneficial to approximate using lower degree polynomials.

Intersection between geometrical objects is an important issue within CAGD. Given a parameterized object, one can use efficient methods for intersecting this with a hypersurface (plane curve or space surface) defined implicitly. One important function used in computing intersections is a "point test". That is, given a point, check whether this is "over" or "under" the hypersurface. Efficient and accurate algorithms based on approximate implicitization are expected to improve the performance of e.g. ray-tracers (used for

Mathematical Methods for Curves and Surfaces: Oslo 2000
Tom Lyche and Larry L. Schumaker (eds.), pp. 103–112.

ISBN 0-8265-1378-6

visualization). When approximate implicitization is used for solving a specific task (like intersection of two parametric curves/surfaces), it might improve the performance to use an approximation constructed to handle that specific application (like keeping the approximated normals close to the originals, or keeping the total approximation close to the original in the plane/space). For other useful applications of implicit surfaces, see [1].

In this paper we introduce an approximate implicitization that is weakly defined, i.e., an implicit approximation is constructed with respect to the $L_2(\mathbb{R})$-norm. Thus, we construct an implicit object that approximates the parameterized one on a bounded parameter interval. We can thereby choose to approximate a small part of the parameterized object. Since we are interested in low degree implicitization (for efficient numerical computation), we will also approximate in cases where an exact solution exists.

In the following, we consider plane curves (in the real plane), but the results as well as the methods are also valid for 3D surfaces. The notation used is presented in Section 2. In Section 3 we present the principle for approximate implicitization. Then in Section 4 we present approximating implicit expressions based on a weak 1st order approximation of the original problem. In Section 5, we consider some numerical solutions based on the weak approximate implicitization. Finally, in Section 6 we make a discussion and conclude our study.

§2. Notation

The parametric curve to be approximated is denoted $\mathbf{p}(t)$, $t \in [a,b] \subset \mathbb{R}$. The algebraic curve of degree d that is used for approximating $\mathbf{p}(t)$ is denoted $q(x,y) = 0$, where

$$q(x,y) = \sum_{0 \le i+j \le d} a_{i,j} x^i y^j .$$

To later address the coefficient of q in a simple way, we organize the coefficients $a_{i,j}$, $0 \le i+j \le d$, in a vector $\mathbf{a}$. Note that the index set of $\mathbf{a}$ consists of multi-indices (i,j). Later (in subsection 5.1) we will see that the coefficient $a_{0,0}$ plays a special role. In order to make this more transparent, we structure $\mathbf{a}$ as

$$\mathbf{a} = (a_{0,0}, \bar{\mathbf{a}}),$$

where $\bar{\mathbf{a}}$ contains all $a_{i,j}$ with $i+j>0$.

If $\mathbf{p}(t)$ is a part of an algebraic curve in $\mathbb{R}^2$, the condition for exact implicitization is given by

$$q(\mathbf{p}(t)) = 0, \qquad \forall t \in [a,b] \subset \mathbb{R}.$$

§3. The Principle of Approximate Implicitization

Our goal is to construct a polynomial $q(x,y)$ that approximates an arbitrary parametrization $\mathbf{p}(t)$, i.e. the function $q(x,y)$ fulfills $q(\mathbf{p}(t)) \approx 0$. However, the approximation is not well-defined by saying that $q(\mathbf{p}(t))$ is small. We must add some requirements:

- It must be possible to measure the error of the approximation. To do this, we introduce a direction for error measurement $\mathbf{g}(t)$, with the same parametrization as the curve approximated. This direction for error measurement must in particular satisfy

$$q(\mathbf{p}(t) - e(t)\mathbf{g}(t)) = 0, \tag{1}$$

  where $e\,(t)$ denotes the size of the error and $\|\mathbf{g}(t)\|_2 = 1$. This ensures that moving from the parametric curve in the direction $\mathbf{g}(t)$, we meet a branch of the algebraic approximation. If the curve $\mathbf{p}(t)$ has nonvanishing normal, then it is natural to use the normal direction as the direction for error measurement. If the normal vanishes, then the direction of the gradient $\frac{\nabla q(\mathbf{p}(t))}{\|\nabla q(\mathbf{p}(t))\|_2}$ of the approximation along $\mathbf{p}(t)$ is an alternative, provided it is nonvanishing.
- We have to ensure that the direction of error measurement $\mathbf{g}(t)$ and the gradient of the algebraic approximation are not orthogonal to each other close to the parametric curve. This can be stated as

$$\mathbf{g}(t) \cdot \nabla q(\mathbf{p}(t) - \theta\mathbf{g}(t)) \neq 0, \ |\theta| < \delta.$$

  Here $\delta > 0$ with $\delta \in \mathbb{R}$ is used for defining an area around the curve approximated where we want the approximation to be well behaved. This ensures that neither the gradient nor the direction for error measurement vanishes in this area.

With the above conditions satisfied we can Taylor expand (1) with respect to $e(t)$

$$q(\mathbf{p}(t)) - e(t)\mathbf{g}(t) \cdot \nabla q(\mathbf{p}(t) - \theta(t)\mathbf{g}(t)) = 0. \tag{2}$$

Here $|\theta(t)| < |e(t)|$ with $\theta(t)e(t) \geq 0$. This equality can be applied in different ways. In [3] the error $e(t)$ is limited in the following way:

$$e(t) = \frac{q(\mathbf{p}(t))}{\mathbf{g}(t) \cdot \nabla q(\mathbf{p}(t) - \theta(t)\mathbf{g}(t))}, \tag{3}$$

$$|e(t)| \leq \frac{|q(\mathbf{p}(t))|}{\min_{|\theta|<|e(t)|} |\mathbf{g}(t) \cdot \nabla q(\mathbf{p}(t) - \theta\mathbf{g}(t))|}. \tag{4}$$

§4. Weak Approximate Implicitization

Instead of using the pointwise error in (3), we want to use an error estimate that is a quotient of the average of $q(\mathbf{p}(t))$ along the curve and an average of $\nabla q(\mathbf{p}(t))$. The average is taken for $t \in [t_1, t_2]$, that is we consider a bounded part of the curve. We thus define

$$\epsilon^2 = \frac{\int_{t_1}^{t_2} q(\mathbf{p}(t))^2 dt}{\int_{t_1}^{t_2} \|\nabla q(\mathbf{p}(t)\|_2^2\, dt}. \tag{5}$$

Doing this we have made the following assumptions:

- We set the direction for error measurement $\mathbf{g}(t) = \frac{\nabla q(\mathbf{p}(t)}{\|\nabla q(\mathbf{p}(t)\|_2}$, and thus assume that $\|\nabla q(\mathbf{p}(t)\|_2 \neq 0$.
- We assume that this choice satisfies $\mathbf{g}(t) \cdot \nabla q(\mathbf{p}(t) - \theta \mathbf{g}(t)) \neq 0$, $|\theta| < \delta$.
- We assume that the gradient of q is not varying too much in the region of interest, and thus that the error does not vary too much.
- By construction, the polynomial q is linear in the coefficients $\mathbf{a}$, so $q(\mathbf{p}(t))^2$ as well as $\|\nabla q(\mathbf{p}(t)\|_2^2$ are quadratic expressions with respect to the coefficients. Thus, the integrals in (5) can be expressed as

$$\mathbf{a}^T\mathbf{D}\mathbf{a} = \int_{t_1}^{t_2} q(\mathbf{p}(t))^2 dt,$$

$$\mathbf{a}^T\mathbf{E}\mathbf{a} = \int_{t_1}^{t_2} \nabla q(\mathbf{p}(t))^2 dt.$$

Using this, (5) can be expressed as

$$\epsilon^2 = \frac{\mathbf{a}^T\mathbf{D}\mathbf{a}}{\mathbf{a}^T\mathbf{E}\mathbf{a}}. \tag{6}$$

Rearranging, we get

$$\mathbf{a}^T\mathbf{D}\mathbf{a} = \epsilon^2 \mathbf{a}^T\mathbf{E}\mathbf{a}.$$

However, the approximation is better the larger $\mathbf{a}^T\mathbf{E}\mathbf{a}$ is compared to $\mathbf{a}^T\mathbf{D}\mathbf{a}$. We thus replace $=$ with $\leq$.

$$\mathbf{a}^T\mathbf{D}\mathbf{a} \leq \epsilon^2 \mathbf{a}^T\mathbf{E}\mathbf{a}.$$

$$0 \leq \epsilon^2 \mathbf{a}^T\mathbf{E}\mathbf{a} - \mathbf{a}^T\mathbf{D}\mathbf{a}. \tag{7}$$

By the definition of the weak error (5), we want the smallest possible value of $|\epsilon|$ satisfying (7). It is important to note that it is the quotient between the integrals we want to minimize, as the quotient is related to the error of the approximation. The rearrangment (7) has converted the problem of minimizing (5) to an expression containing matrices and vectors. The minimum of the absolute value of the right hand side of (7) satisfies

$$\mathbf{D}\mathbf{a} = \epsilon^2 \mathbf{E}\mathbf{a}, \tag{8}$$

which is a generalized eigenvalue problem. Looking in more detail at (8) two uses are possible:

- If $\mathbf{D}$ is singular, then a vector $\mathbf{a} \neq \mathbf{0}$ exists such that $\mathbf{D}\mathbf{a} = \mathbf{0}$, and thus an exact implicitization is found. However, if this exact implicitization has singularities in the region of interest, then an approximate implicitization satisfying (7) can be better suited for the later use.
- If $\mathbf{D}$ is nonsingular then the algebraic degree is not sufficient to perform exact implicitization. Thus (8) will give an approximate implicitization. Further, if we use numeric integration for the calculation of the matrices $\mathbf{D}$ and $\mathbf{E}$, then the method can be employed for approximate implicitization of procedural defines curves and surface.

§5. Example of Numerical Solutions

The approximation problem above was formulated as a generalized eigenvalue problem in equation (8). In the first subsection below, we will discuss the solution of this with respect to geometric interpretation. Then in the following subsection we will give some numerical examples.

To perform exact implicitization, exact arithmetic has to be used. As soon as we resort to floating point arithmetic, rounding errors are introduced. Thus, results are inflicted with small rounding errors. By proper choice of numerical methods, these errors can be controlled. In numerical integration the degree of precision reflects the degree of polynomials that are exactly integrated when using exact arithmetic. In the examples we have not stressed this issue, and used the trapezoidal rule that has degree of precision 1. If the curve to be approximated $\mathbf{p}(s)$ has polynomial degree n, then a quadrature formula with degree of precision mn will minimize the errors in $\mathbf{D}$ and $\mathbf{E}$.

5.1 Generalized Eigenvalue Problems

Equation (8) can be solved numerically as a generalized eigenvalue problem. The best solution will be to define $\mathbf{a}$ as the eigenvector corresponding to the smallest positive eigenvalue of the system (8).

Whenever dealing with power basis, the term $a_{0,0}$ will always be zero in ∇q since it defines the constant. This shows easily that the matrix $\mathbf{E}$ is always singular, so that a standard generalized eigenvalue method cannot be used. However, as long as the matrix $\mathbf{D}$ is invertible, we can turn the problem around, and solve the generalized eigenvalue problem

$$\lambda \mathbf{D}\mathbf{a} = \mathbf{E}\mathbf{a},$$

where $\epsilon = 1/\sqrt{\lambda}$. In this case, the best solution will be the eigenvector corresponding to the maximum (positive) eigenvalue. In the case when $\mathbf{D}$ is nonsingular, this problem can be converted to an ordinary eigenvalue problem. However, we cannot guarantee that $\mathbf{D}$ is nonsingular. In the case $\mathbf{D}$ is singular, the vectors $\mathbf{a}$ corresponding to the 0 singular values of $\mathbf{D}$ make the integral $\int_{t_1}^{t_2} q(\mathbf{p}(t))^2 dt \equiv 0$, thus we have an exact solution of the implicitization problem. Two alternatives then exist:

- Lower the degree of the algebraic curve to make $\mathbf{D}$ nonsingular.
- Transform the problem to a reduced generalized eigenvalue problem.

In the last case, in order to avoid singular matrices, we reduce the size of the matrices. Using the previous convention about the vector $\mathbf{a}$ and how this controls the indices of $\mathbf{E}$ (see Section 2, $a_{0,0}$ comes first in $\mathbf{a}$) we obtain after a short calculation

$$\mathbf{E} = \begin{pmatrix} 0 & [0] \\ [0]^T & \mathbf{E}_1 \end{pmatrix},$$

where $\mathbf{E}_1$ is an $(n-1)\text{x}(n-1)$ matrix. Therefore $\mathbf{E}$ is singular, and the established solvers cannot be used for $\mathbf{D}\mathbf{a} = \epsilon^2\mathbf{E}\mathbf{a}$.

In order to overcome this problem, we write $\mathbf{D}$ in the form

$$\mathbf{D} = \begin{pmatrix} d_{0,0} & \mathbf{D}_0 \\ \mathbf{D}_0{}^T & \mathbf{D}_1 \end{pmatrix},$$

were $\mathbf{D}_1$ is an $(n-1)\text{x}(n-1)$ matrix. The generalized eigenvalue problem $\mathbf{Da} = \epsilon^2\mathbf{Ea}$ can then be written as

$$a_{0,0} = -\frac{\mathbf{D}_0}{d_{0,0}}\bar{\mathbf{a}}$$

$$\left(\mathbf{D}_1 - \frac{\mathbf{D}_0^T\mathbf{D}_0}{d_{0,0}}\right)\bar{\mathbf{a}} = \epsilon^2\mathbf{E}_1\bar{\mathbf{a}}. \tag{9}$$

This means that $a_{0,0}$ can be calculated from $\bar{\mathbf{a}}$, and $\bar{\mathbf{a}}$ is a solution of the reduced generalized eigenvalue problem. The best solution is the eigenvector corresponding to the minimal (positive) eigenvalue $\lambda = \epsilon^2$. In case $\mathbf{E}_1$ is singular, we can repeat this procedure.

The reduced matrix $\mathbf{E}_1$ only seems to be singular when the problem is underdetermined, i.e. when one tries to approximate the parametric expression with an implicit one whose degree is higher than the minimal degree exact implicitization. We will show through examples in the next subsections that this last formulation handles exact implicitization as well as some underdetermined problems.

5.2 Numerical Examples for the Weak Implicitization

In the following set of experiments, we illustrate the efficiency of the weak approach. Exact solutions exist for all cases, but we will illustrate what happens when we approximate with lower degree polynomials. The different parametric curves are chosen in order to behave differently from a geometric point of view (with/without inflection point and singularities). The approximate implicitizations are defined with respect to the complete monomial basis of the used total degrees.

The results are collected in Tables 1–3. The calculations of errors are based on a sample-set of 100 points $\{t_i\}_{i=1}^{100}$ on the original parametric curve. This is also the case for the numerical integration which defines the matrices. We have found this low number of points sufficient since the curves under consideration are relatively simple. A solver from LaPack was used to solve the generalized eigenvalue problems. The tables contain the ϵ-error defined in (5). In addition, we calculate minimum and maximum geometric error (without absolute value) and the minimum and maximum length of the gradients from sampled points. Finally, a total discrete L_2-error is calculated. The errors are defined as follows:

$$\text{min error} = \min_{i=1,\ldots,100} \frac{q\left(\mathbf{p}(t_i)\right)}{|\nabla F\left(\mathbf{p}(t_i)\right)|},$$

$$\text{max error} = \max_{i=1,\ldots,100} \frac{q\left(\mathbf{p}(t_i)\right)}{\left|\nabla F\left(\mathbf{p}(t_i)\right)\right|},$$

$$\text{total error} = \frac{1}{2}\sum_{i=1}^{99}\left(\left(\frac{q\left(\mathbf{p}(t_i)\right)}{\nabla q\left(\mathbf{p}(t_i)\right)}\right)^2 + \left(\frac{q\left(\mathbf{p}(t_{i+1})\right)}{\nabla q\left(\mathbf{p}(t_{i+1})\right)}\right)^2\right)(t_{i+1} - t_i).$$

Example I. *Inflection point.* We consider the curve (t, t^3) with one inflection point on the interval $(-1, 1)$. In the experiments we have chosen linear, quadratic and cubic (close to exact) implicitizations. The results are collected in Table 1.

Degree	Weak error	Min. error	Max. error	Tot. error
1	0.13	-0.29	0.29	0.034
2	0.049	-3.27	0.069	0.57
3	7.0×10^{-9}	2.0×10^{-14}	2.0×10^{-14}	7.0×10^{-9}

Tab. 1. Error in example "Inflection point".

Example II. *Local parabolic behavior in a general cubic curve.* Now we consider a more general cubic curve, given by $(0.2t, -t^3 + 4t^2 + 1)$. In this case we consider the interval $(-1, 1)$, where the curve has a local parabolic behavior. This is done to illustrate the local behavior of the method (and also to show an example where the power basis is not well balanced). We repeat the experimental procedure, and obtain Table 2.

Degree	Weak error	Min. error	Max. error	Tot. error
1	0.11	-0.24	0.11	0.025
2	2.0×10^{-4}	-2.0×10^{-3}	4.0×10^{-4}	5.0×10^{-7}
3	1.76×10^{-9}	9.7×10^{-12}	9.5×10^{-11}	7.0×10^{-22}

Tab. 2. Error in example local "Parabolic behavior".

Example III. *A node.* In the 3rd example we consider a singularity, a node, parameterized by $(t^5 - 10t^3, t^4 - t^3 + t)$ in the interval $(-4, 4)$. Again we have constructed a quite nasty parametrization from a numerical point of view, and the results are collected in Table 3.

Degree	Weak error	Min. error	Max. error	Tot. error
3	0.311	-2.79	21.6	53.44
4	2.0×10^{-3}	-0.055	0.99	0.37
5	5.0×10^{-6}	-0.13	0.07	4.0×10^{-3}

Tab. 3. Error in example "A node".

5.3 Weak Approximate Implicitization for Surfaces

We have conducted similar numerical experiments for surfaces, testing out the numerical solutions. The preliminary results correspond very well with what we have found for curves, and we have made the same observations.

§6. Discussion and Conclusion

We have presented a weak approximate implicitization. We implemented a numerical solution of the approximation, and we have illustrated this with some examples. However, we stress that more investigation needs to be done. Our observations for numerical experiments for the weak approximate implicitization can so far be summarized as follows:

- The weak approximate implicitization is satisfactory in the examples we have tested. It handles in particular "near exact" cases as well as some underdetermined cases (i.e. cases where one tries to "approximate" by an polynomial of higher degree than the exact solution.) In such cases, the matrix **E** will be singular or near singular. These cases can be sorted out by singular value decomposition of the **E** matrix. As we have used floating point arithmetic, it is not surprising that we only obtain near exact solutions, and not exact solutions although these exist in our example. The lack of exact solutions is also due to the choice of numerical approximation (we have e.g. approximated the integration with few points as well as using a generalized eigenvalue method that does not consider the choice of basis). We might say that this method is in particular interesting for higher order implicitizations or complicated parameterizations that should be approximated locally. If we allow the special case $\epsilon = 0$, exact solutions will occur, but we will have to use other numerical solvers.
- The weak implicitization approximates the curves locally only over the given interval.
- The weak formulation is less dependent on the choice of basis than the method of Dokken in [3,4]. In this method the use of a Bernstein basis for the representation of the parametric curve is crucial. However, we should expect that also in this weak approach the use of a Bernstein basis over a simplex for representing the algebraic curve will improve the accuracy of the approximation.

For a more complete study, there are obvious points that must be dealt with:

- The numerical approaches do not take particular care of the length of the gradient, nor do they generally control the direction of the gradient (i.e. the normal direction).
- The choice of numerical method has to be improved (to reduce the building and solving of matrices).
- The choice of basis functions needs study (although we have seen satisfactory behavior for the power basis).

We end with general conclusions about approximate implicitization:

- Approximate implicitization has the potential of being a fast and useful tool in many applications (such as e.g. ray-tracing, detection of intersections or separation of different geometric objects).
- The choice of algorithm (like ordinary approximate implicitization, weak approximate implicitization or variation of these) depends on the applications.
- The choice of error measure (e.g. geometric error measured pointwise in sup-norm or globally in L_2-norm) will also depend on the applications.

Acknowledgments. The preparation of the paper is partly funded by the European Commission through the project IST-1999-29010 - GAIA.

References

1. Bloomenthal, J., *Introduction to Implicit Surfaces*, Morgan Kaufmann Publishers Inc., 1997.
2. Cayley, A., Note sur la méthode d'élimination de Bezout, J. Reine Angew. Math. **53** (1857), 366–367.
3. Dokken, T., Aspects of intersection algorithms and approximation, thesis for the doctor philosophiae degree, University of Oslo, Norway, 1997.
4. Dokken, T., Approximate Implicitization, in *Mathematical Methods in CAGD: Oslo 2000*, T. Lyche and L. L. Schumaker (eds.), Vanderbilt University Press, Nashville, 2001, 81-102.
5. Goldman, R. N., and Sederberg T. W., Some applications of resultants to problems in computational geometry, The Visual Computer, (1985), 101–107.
6. Salmon, G., *Analytic Geometry of Three Dimensions*, Cambridge, 1915.
7. Sederberg, T. W., J. Zheng, K. Klimaszewski, and T. Dokken, Approximate Implicitization Using Monoid Curves and Surfaces, Graphical Models and Image Processing **61**(1999), 177–198.
8. Walker, R. J., *Algebraic Curves*, Dover publications Inc, 1950.

Tor Dokken
SINTEF Applied Mathematics
P.O. Box 124 Blindern
0314 Oslo
Norway
`tor.dokken@math.sintef.no`
`http://www.math.sintef.no/Geom`

Hermann Kellermann
CoCreate Software
Posener Strasse 1
71065 Sindelfingen
Germany
Hermann_Kellermann@cocreate.com
http://www.cocreate.com

Cathrine Tegnander
SINTEF Applied Mathematics
P.O. Box 124 Blindern
0314 Oslo
Norway
Cathrine.Tegnander@math.sintef.no
http://www.math.sintef.no/Geom

A 4-Point Hermite Subdivision Scheme

Serge Dubuc and Jean-Louis Merrien

Abstract. A subdivision scheme based on 4 points with Hermite data (function and first derivatives) on $\mathbb{Z}$ is studied. For a large region in the parameter space, the scheme is C^1 convergent or at least is convergent in the space of Schwartz distributions. The Fourier transform of any interpolating function can be computed through products of matrices of order 2. The main tools for proving these results are the Paley-Wiener-Schwartz theorem on the characterization of the Fourier transforms of distributions with compact support, and a theorem of Heil-Colella about the convergence of some products of matrices.

§1. Introduction

Hermite interpolatory subdivision schemes have been introduced by Merrien [7]. He, Dyn and Levin [3,4] studied the convergence of these schemes to regular functions. In this paper, we would like to consider the most general case of a 4-point Hermite interpolatory symmetrical scheme using function and first derivatives values. Such a scheme will be called HS41. We will study the conditions giving a C^1 interpolant (Section 2), and weaker conditions allowing convergence in the space of Schwartz distributions. The last task will be done by a computation of Fourier transforms (Section 3). This harmonic analysis provides an additional tool to study the scheme, and allows extension to functions which are not necessarily of class C^1. To get the convergence in distributions, we will use a result proved by Heil and Colella [6] on the convergence of some infinite products of matrices arising in matrix refinement equations.

§2. The Hermite Subdivision Scheme HS41

We assume that the function f and its first derivative p are known on $\mathbb{Z}$. Precisely, we have two sequences $\{y_k, y'_k\}_{k\in\mathbb{Z}}$, and we suppose $f(k) = y_k$ and $p(k) = y'_k$. We build f and p on $D_n = \mathbb{Z}/2^n$ by induction. At step n, if

Mathematical Methods for Curves and Surfaces: Oslo 2000
Tom Lyche and Larry L. Schumaker (eds.), pp. 113–122.

ISBN 0-8265-1378-6

$\frac{i-1}{2^n}, \frac{i}{2^n}, \frac{i+1}{2^n}, \frac{i+2}{2^n}$ are four successive points of D_n, we compute f and p at $x = \frac{2i+1}{2^{n+1}}$ by the formulae:

$$\begin{aligned} f(\frac{2i+1}{2^{n+1}}) &= a_1[f(\frac{i+1}{2^n}) + f(\frac{i}{2^n})] + b_1[f(\frac{i+2}{2^n}) + f(\frac{i-1}{2^n})] \\ &\quad + \frac{c_1}{2^n}[p(\frac{i+1}{2^n}) - p(\frac{i}{2^n})] + \frac{d_1}{2^n}[p(\frac{i+2}{2^n}) - p(\frac{i-1}{2^n})], \\ p(\frac{2i+1}{2^{n+1}}) &= 2^n a_2[f(\frac{i+1}{2^n}) - f(\frac{i}{2^n})] + 2^n b_2[f(\frac{i+2}{2^n}) - f(\frac{i-1}{2^n})] \\ &\quad + c_2[p(\frac{i+1}{2^n}) + p(\frac{i}{2^n})] + d_2[p(\frac{i+2}{2^n}) + p(\frac{i-1}{2^n})]. \end{aligned} \tag{1}$$

Hence f and p are defined on D_{n+1}. The construction depends on eight parameters. When we reiterate the process, we define f and p on the set of dyadic numbers $D_\infty = \bigcup D_n$ which is dense in $\mathbb{R}$.

For $b_1 = d_1 = b_2 = d_2 = 0$, we recover 2-point subdivision schemes. We call them HS21. They have been studied by Merrien [7] to get a C^1 function f with $f' = p$. Recently, Dubuc and Merrien [2] gave a new study of the convergence of these schemes in the space of Schwartz distributions. We want a generalization of these results for new 4-point schemes.

Now, suppose that f and p built on D_∞ can be extended to $\mathbb{R}$ in smooth enough functions with $f' = p$. As n tends to ∞, a Taylor expansion of $f(\frac{1}{2^n})$ and $p(\frac{1}{2^n})$ at the origin gives necessary conditions on the parameters. This is to be connected to the reproducibility of polynomials. Like Dyn and Levin in [3], we will say that the scheme is C^r if, for any data $\{y_k, y'_k\}_{k\in\mathbb{Z}}$, there exist two functions, $f \in C^r(\mathbb{R})$ and $p \in C^{r-1}(\mathbb{R})$ such that $f' = p$ and $f(k) = y_k, p(k) = y'_k$. Dyn and Levin have proved that if the scheme is C^r, then it reproduces polynomials of degree less or equal to r. More precisely, if $\{y_k, y'_k\}_{k\in\mathbb{Z}}$ are two sequences such that there exists a polynomial P of degree less than or equal to r with $P(k) = y_k, P'(k) = y'_k$, then $f = P$ and $p = P'$ on D_∞.

Proposition 1. *Assume that the scheme is C^r. Then the following conditions are necessary:*

for $r = 0$, $a_1 + b_1 = 1/2$, *for* $r = 1$, $a_2 + 3b_2 + 2c_2 + 2d_2 = 1$,
for $r = 2$, $2a_1 - c_1 - 3d_1 = 9/8$, *for* $r = 3$, $6b_2 + c_2 + 13d_2 = -1/4$,
for $r = 4$, $-c_1 + 3d_1 = 9/64$, *for* $r = 5$, $-c_2 + 9d_2 = 9/32$,
for $r = 6$, $a_1 = 243/512, b_1 = 13/512, c_1 = -81/512, d_1 = -3/512$,
for $r = 7$, $a_2 = 405/256, b_2 = 5/256, c_2 = -81/256, d_2 = -1/256$.

2.1. Convergence and Smoothness Analyses of HS41

We now study when the scheme is C^1. The necessary conditions given above imply that $a_1 + b_1 = 1/2$ and $a_2 + 3b_2 + 2c_2 + 2d_2 = 1$. Set

$$U_n^i = \begin{pmatrix} p(\frac{i+1}{2^n}) - p(\frac{i}{2^n}) \\ 2^n[f(\frac{i+1}{2^n}) - f(\frac{i}{2^n})] - \frac{p(\frac{i+1}{2^n})+p(\frac{i}{2^n})}{2} \end{pmatrix}.$$

With the help of a computer algebra system, we immediately get

Proposition 2.

$$U_{n+1}^{2i} = A_1 U_n^i + B_1 U_n^{i+1} + C_1 U_n^{i-1},$$
$$U_{n+1}^{2i+1} = A_{-1} U_n^i + B_{-1} U_n^{i+1} + C_{-1} U_n^{i-1},$$

with

$$A_j = \begin{pmatrix} \frac{1}{2} & j(a_2+b_2) \\ j(-2a_1+\frac{5}{4}+2c_1+2d_1) & 1-\frac{a_2+b_2}{2} \end{pmatrix},$$

$$B_j = \begin{pmatrix} j(\frac{b_2}{2}+d_2) & jb_2 \\ j(-a_1+\frac{1}{2}+2d_1)-\frac{b_2}{4}-\frac{d_2}{2} & j(-2a_1+1)-\frac{b_2}{2} \end{pmatrix}, \qquad j=\pm 1,$$

$$C_j = \begin{pmatrix} -j(\frac{b_2}{2}+d_2) & jb_2 \\ j(-a_1+\frac{1}{2}+2d_1)+\frac{b_2}{4}+\frac{d_2}{2} & j(2a_1-1)-\frac{b_2}{2} \end{pmatrix}.$$

Theorem 3. *If there exists a matrix norm* $\|\cdot\|$ *on* $\mathbb{R}^{2\times 2}$ *such that* $\|A_j\| + \|B_j\| + \|C_j\| < 1, j = \pm 1$*, then the scheme is* C^1.

Proof: The functions p and f are built on D_∞. We want to extend them to $\mathbb{R}$ and prove that $f' = p$. We will do it on $[0,1]$, and the extension to $\mathbb{R}$ will be obvious.

Set $\kappa = \max(\|A_j\| + \|B_j\| + \|C_j\|, j = \pm 1)$. Then for all $n \in \mathbb{N}$ and for all $i \in \{0, \ldots, 2^n - 1\}$, $\|U_n^i\| \leq \gamma_1 \kappa^n$, where the constant γ_1 depends on the initial data on $[-3,3] \cap \mathbb{Z}$. Let γ_2 be a real number such that for any vector $V \in \mathbb{R}^2, \|V\|_\infty \leq \gamma_2 \|V\|$.

Firstly, for $n \in \mathbb{N}$, let p_n be the continuous piecewise linear function on $[0,1]$ defined by $p_n(i2^{-n}) = p(i2^{-n})$, $i \in \{0, \ldots, 2^n\}$. Then we have

$$\begin{aligned}\|p_{n+1} - p_n\|_\infty &= \sup_{i\in\mathbb{Z}} |p((2i+1)2^{-n-1}) - \frac{1}{2}(p((i+1)2^{-n}) + p(i2^{-n}))| \\ &\leq \frac{1}{2}(\|U_{n+1}^{2i+1}\|_\infty + \|U_{n+1}^{2i}\|_\infty) \\ &\leq \frac{1}{2}\gamma_2(\|U_{n+1}^{2i+1}\| + \|U_{n+1}^{2i}\|) \leq \gamma_1\gamma_2\kappa^{n+1}.\end{aligned}$$

Hence we can deduce that p_n is a Cauchy sequence in $C([0,1])$. Therefore it has a continuous limit that we still call p since it is an extension.

Secondly, let us set $\varphi(x) = f(0) + \int_0^x p(t)dt$. φ is in $C^1(\mathbb{R})$ with $\varphi' = p$. Let us prove that $\varphi = f$.

Given $x \in D_\infty \cap [0,1]$ and $\epsilon > 0$, there exists n such that $x \in D_n$; since $x \in D_{n'}$, for all $n' \geq n$, we can choose n as large as we need. Since p is uniformly continuous on $[0,1]$, for n large enough and $i \in \{0, \ldots, 2^n - 1\}$, we have

$$\forall t \in [i2^{-n}, (i+1)2^{-n}], \quad |p(t) - (p(i2^{-n}) + p((i+1)2^{-n}))/2| \leq \epsilon.$$

Similarly we suppose n large enough to ensure $\gamma_1\gamma_2\kappa^n \leq \epsilon$.
Writing $x = k2^{-n}, k \in \{0, \ldots, 2^n\}$, we have

$$\begin{aligned}\varphi(x) - f(x) &= f(0) + \int_0^x p(t)dt - f(x) \\ &= \sum_{i=0}^{k-1}\{\int_{i2^{-n}}^{(i+1)2^{-n}} p(t)dt - [f((i+1)2^{-n}) - f(i2^{-n})]\} \\ &= \sum_{i=0}^{k-1}\int_{i2^{-n}}^{(i+1)2^{-n}} [p(t) - \frac{p((i+1)2^{-n}) + p(i2^{-n})}{2}]dt \\ &+ \sum_{i=0}^{k-1}\{\frac{p((i+1)2^{-n}) + p(i2^{-n})}{2^{n+1}} - [f((i+1)2^{-n}) - f(i2^{-n})]\}.\end{aligned}$$

For all $i \in \{0, \ldots, 2^n\}$, $\|U_n^i\|_\infty \leq \gamma_2\|U_n^i\| \leq \gamma_1\gamma_2\kappa^n$, and with the hypotheses on n, we can deduce that $| \varphi(x) - f(x) | \leq \sum_{i=0}^{k-1} 2^{-n}\epsilon + 2^{-n}\sum_{i=0}^{k-1} \epsilon \leq 2\epsilon$. As ϵ can be chosen arbitrarily small, we obtain $\varphi = f$ on $D_\infty \cap [0, 1]$. Therefore φ is a continuous extension of f on $[0, 1]$, $f \in C^1([0, 1])$ and $f' = p$.
□

Theorem 3 gives a sufficient condition for the operator $U_n \to U_{n+1}$ to be contractive. A weaker condition for the C^1-convergence of the scheme can be obtained when the operator $U_n \to U_{n+m}$ is a contraction for an integer $m > 1$.

2.2. Examples

Before giving examples, we introduce a norm on $\mathbb{R}^2$ by $\|X\|_\theta = |x_1| + \theta|x_2|$ for $X = (x_1, x_2)^T$ where θ is a real positive number. Then it is easy to prove that for $M = (m_{ij}) \in \mathbb{R}^{2\times 2}$, $\|M\|_\theta = \max_{\|X\|_\theta=1}(\|MX\|_\theta) = \max(|m_{11}| + \theta|m_{21}|, \frac{|m_{12}|}{\theta} + |m_{22}|)$. In some cases, it is convenient to know that we can find a $\theta > 0$ to get $\|M\|_\theta < 1$ if and only if $|m_{11}| < 1, |m_{22}| < 1$ and $|m_{12}| \cdot |m_{21}| < (1 - |m_{11}|)(1 - |m_{22}|)$.

Example 4. For HS21, we have $b_1 = d_1 = b_2 = d_2 = 0$. Adding the conditions $a_1 + b_1 = 1/2$ and $a_2 + 3b_2 + 2c_2 + 2d_2 = 1$, so that $a_1 = 1/2$ and $c_2 = (1 - a_2)/2$, we have

$$A_j = \begin{pmatrix} \frac{1}{2} & ja_2 \\ j(\frac{1}{4} + 2c_1) & 1 - \frac{a_2}{2} \end{pmatrix}, \qquad B_j = \begin{pmatrix} 0 & 0 \\ 0 & 0 \end{pmatrix} = C_j, \qquad j = \pm 1.$$

The scheme is C^1 if there exists a matrix norm $\|\cdot\|$ such that $\|A_j\| < 1$. For example, sufficient conditions are $0 < a_2 < 4$ and $|a_2| \cdot |1 + 8c_1| < 2 - |2 - a_2|$.

Example 5. Assume that the necessary conditions to get a C^7 interpolant are satisfied, i.e., $a_1 = 243/512, b_1 = 13/512, c_1 = -81/512, d_1 = -3/512$, $a_2 = 405/256, b_2 = 5/256, c_2 = -81/256, d_2 = -1/256$. Then, for $j = \pm 1$,

$$A_j = \begin{pmatrix} \frac{1}{2} & j\frac{205}{128} \\ -j\frac{7}{256} & \frac{51}{256} \end{pmatrix}, \qquad B_j = \begin{pmatrix} j\frac{3}{512} & j\frac{5}{256} \\ j\frac{7}{512} - \frac{3}{1024} & j\frac{13}{256} - \frac{5}{512} \end{pmatrix},$$

$$C_j = \begin{pmatrix} -j\frac{3}{512} & j\frac{5}{256} \\ j\frac{7}{512} + \frac{3}{1024} & -j\frac{13}{256} - \frac{5}{512} \end{pmatrix}.$$

For $\theta = 4$, a numerical evaluation gives $\|A_j\|_\theta + \|B_j\|_\theta + \|C_j\|_\theta = 459/512, j = \pm 1$. Therefore the scheme is C^1.

Example 6. Let a_1 and a_2 be two real numbers. Set $b_1 = 1/2 - a_1$, $c_1 = a_1/2 - 3/8, d_1 = a_1/2 - 1/4, b_2 = 2 - a_2, d_2 = -b_2/2, c_2 = (1 - a_2 - 3b_2 - 2d_2)/2$ then the matrices A_j, B_j, C_j can be written

$$A_j = \begin{pmatrix} \frac{1}{2} & j2 \\ 0 & 0 \end{pmatrix}, \quad B_j = \begin{pmatrix} 0 & j(2 - a_2) \\ 0 & j(1 - 2a_1) - \frac{2-a_2}{2} \end{pmatrix},$$

$$C_j = \begin{pmatrix} 0 & j(2 - a_2) \\ 0 & j(2a_1 - 1) - \frac{2-a_2}{2} \end{pmatrix}.$$

If the condition $|-a_2 + 4a_1| + |4 - a_2 - 4a_1| < 1$ is satisfied, then the scheme is C^1. To get the result, we can use the norm $\|\cdot\|_\theta$ with a θ large enough.

§3. Convergence in a Distributional Sense of HS41

We consider convergence in a distributional sense of the Hermite scheme HS41. Such convergence has already been shown by Derfel, Dyn and Levine [5] in the context of non-Hermite subdivision schemes. There are two basic solutions of our recursive system (1): the first one is the pair (f_0, p_0) with data $f_0(k) = \delta_{k,0}, p_0(k) = 0, k \in \mathbb{Z}$, and the second one is the pair (f_1, p_1) with data $f_1(k) = 0, p_1(k) = \delta_{k,0}, k \in \mathbb{Z}$. These two pairs are important because we can express all the solutions (f, p) of (1) with linear combinations of their translates. For all $n \in \mathbb{N}, j \in \mathbb{Z}$,

$$\begin{aligned} f(j/2^n) &= \sum_{k=-\infty}^{\infty} [f(k)f_0(j/2^n - k) + p(k)f_1(j/2^n - k)], \\ p(j/2^n) &= \sum_{k=-\infty}^{\infty} [f(k)p_0(j/2^n - k) + p(k)p_1(j/2^n - k)]. \end{aligned} \tag{2}$$

Notice that all these sums are finite since the supports of $f_i, p_i, i = 0, 1$ are contained in $[-3, 3]$.

Now, applying relation (2) successively to the pair of functions $(f(x) = f_0(x/2), p(x) = p_0(x/2)/2)$ and then to the pair $(f_1(x/2), p_1(x/2)/2)$ and evaluating the functions f_0, p_0, f_1, p_1 at the half-integers by using (1), we obtain a system of functional equations for f_0, p_0, f_1, p_1 which can be written in a matrix equation as

$$\varphi(x/2) = \sum_{k\in\mathbb{Z}} M_k \varphi(x-k) \begin{pmatrix} 1 & 0 \\ 0 & 2 \end{pmatrix}, \tag{3}$$

where $\varphi(x) = \begin{pmatrix} f_0(x) & p_0(x) \\ f_1(x) & p_1(x) \end{pmatrix}$ and

$$M_0 = \begin{pmatrix} 1 & 0 \\ 0 & 1/2 \end{pmatrix}, \quad M_1 = \begin{pmatrix} a_1 & -a_2/2 \\ -c_1 & c_2/2 \end{pmatrix}, \quad M_{-1} = \begin{pmatrix} a_1 & a_2/2 \\ c_1 & c_2/2 \end{pmatrix},$$

$$M_3 = \begin{pmatrix} b_1 & -b_2/2 \\ -d_1 & d_2/2 \end{pmatrix}, \quad M_{-3} = \begin{pmatrix} b_1 & b_2/2 \\ d_1 & d_2/2 \end{pmatrix}, \quad M_k = \begin{pmatrix} 0 & 0 \\ 0 & 0 \end{pmatrix},$$

otherwise.

3.1. Fourier Transform of HS41

Let us begin with a computation without proper justification. We will suppose that the system (3) of functional equations is valid not only when x is a dyadic number, but also when x is an arbitrary real number. We must suppose that f_0, p_0, f_1, p_1 have been extended by continuity on $\mathbb{R}$. Now, we compute the Fourier transform $\hat{f}$ of a function f by $\hat{f}(\xi) = \int_{-\infty}^{+\infty} f(x)e^{-i\xi x}dx$. Using this Fourier operator on (3), we get

$$\hat{\varphi}(\xi) = A(\xi/2)\hat{\varphi}(\xi/2) \begin{pmatrix} 1 & 0 \\ 0 & 2 \end{pmatrix}, \tag{4}$$

where $A(\xi)$ is given by

$$\frac{1}{2}\sum_{k\in\mathbb{Z}} M_k e^{-ik\xi} = \begin{pmatrix} \frac{1}{2} + a_1\cos\xi + b_1\cos 3\xi & \frac{i}{2}(a_2\sin\xi + b_2\sin 3\xi) \\ i(c_1\sin\xi + d_1\sin 3\xi) & \frac{1}{4} + \frac{1}{2}(c_2\cos\xi + d_2\cos 3\xi) \end{pmatrix}.$$

To study the matrix equation in (4), we now look at the matrix product

$$P_n(\xi) = A(\xi/2)A(\xi/4)\cdots A(\xi/2^n). \tag{5}$$

More precisely, we look for conditions on the parameters to get the convergence of the sequence of matrices $P_n(\xi)$. These conditions should be independent of the real or complex value ξ. The study of this sequence for complex values of ξ is motivated by a generalization of Paley-Wiener theorem proposed by Schwartz [8].

Theorem 7. [Schwartz] *Let F be a continuous function on the real axis which is the Fourier transform of a tempered distribution T. Then the support of T is contained in $[-C, C]$ if and only if F may be extended on the complex plane to an analytic entire function of exponential type $\leq C$.*

We recall that an entire function $F(z)$ is of exponential type $\leq C$ if $\limsup_{|z|\to\infty} \frac{\log|F(z)|}{|z|} \leq C$. To study the convergence of the matrix products $P_n(\xi)$, we will also use Proposition 5.2 of Heil and Colella [6].

Proposition 8. *If* $\lim[A(0)]^n$ *exists and is not trivial, then the sequence* $P_n(\xi)$ *converges uniformly on compact sets of* $\mathbb{C}$ *to a continuous matrix-valued function* $P(\xi)$ *whose restriction to the real line has at most polynomial growth at infinity.*

We are now ready for the main result.

Theorem 9. *For* $a_1 + b_1 = 1/2$ *and* $-\frac{5}{2} < c_2 + d_2 \leq \frac{3}{2}$, *for any complex number* ξ, *the sequence of matrices* $P_n(\xi)$ *defined in (5) converges, and the convergence is uniform whenever* ξ *lies in the disk* $|\xi| \leq R$. *As functions of* ξ, *the four components of the limit matrix* $P(\xi)$ *are entire functions of exponential type* ≤ 3.

Proof: $A(0) = \begin{pmatrix} 1 & 0 \\ 0 & \frac{1}{4} + \frac{c_2+d_2}{2} \end{pmatrix}$, so that with the hypotheses and the previous proposition, the sequence $P_n(\xi)$ converges on each compact set. As the sequence $P_n(z)$ converges to a matrix $P(z)$, and as the moduli of the components of the matrices $P_n(z)$ are uniformly bounded whenever $|z| \leq R$, the Lebesgue dominated convergence theorem and Cauchy formula give us the proof that all the components of the matrix $P(z)$ are analytic in z.

Finally, let us verify that each element of the matrix $P(z)$ is an entire function of exponential type. Firstly, there exists a real positive number C which depends on the parameters such that for all $z \in \mathbb{C}$, $||A(z)||_\infty \leq Ce^{3|z|}$. Secondly, we know that there exists a real positive number M (depending on the parameters again) such that for all $z \in \mathbb{C}, |z| \leq 1, ||P(z)||_\infty \leq M$.

Let z be a complex number such that $2^n \leq |z| \leq 2^{n+1}$.
As $P(z) = A(z/2)A(z/4)\dots A(z/2^n)P(z/2^{n+1})$, we obtain the bound

$$||P(z)||_\infty \leq ||P(z/2^{n+1})||_\infty \prod_{k=1}^{n} ||A(z/2^k)||_\infty \leq M \prod_{k-1}^{n} [Ce^{3|z|/2^k}] \leq MC^n e^{3|z|},$$

and thus $\limsup_{|z|\to\infty} \frac{\log ||P(z)||_\infty}{|z|} \leq 3$. Then the functions composing the matrix $P(z)$ are entire functions of exponential type ≤ 3. □

The Schwartz version of the Paley-Wiener theorem implies the following corollary.

Corollary 10. *Set* $a_1+b_1 = 1/2$ *and* $-\frac{5}{2} < c_2+d_2 \leq \frac{3}{2}$. *Then each component function of the limit matrix* $P(z) = \lim P_n(z)$ *is the Fourier transform of a distribution whose support is contained in the interval* $[-3, 3]$.

3.2. Schwartz Distributions Associated with the Scheme

We will connect the computation of Fourier transforms of the previous subsection with the limit matrix $P(\xi)$. This link will come from a sequence of Schwartz matrix distributions. We set

$$T^{(n)} = \frac{1}{2^n} \sum_m \begin{pmatrix} f_0(m/2^n) & p_0(m/2^n) \\ f_1(m/2^n) & p_1(m/2^n) \end{pmatrix} \delta_{m/2^n},$$

where δ_h is the Dirac distribution at point h defined by $\delta_h(\phi) = \phi(h)$. Notice that these sums are finite and that the distributions are compactly supported since the supports of $f_i, p_i, i = 0, 1$ are in $[-3, 3]$.

Now, let us evaluate the Fourier transform of this matrix distribution: $\hat{T}^{(n)}(\xi) = T^{(n)}(e^{-i\xi x})$. Using the system of equations (3), we verify that a simple induction links the sequence of Fourier transforms through the matrix $A(\xi)$. Indeed,

$$\hat{T}^{(n+1)}(\xi) = \frac{1}{2^{n+1}} \sum_m \begin{pmatrix} f_0(m/2^n) & p_0(m/2^n) \\ f_1(m/2^{n+1}) & p_1(m/2^{n+1}) \end{pmatrix} e^{-i\xi m/2^{n+1}}.$$

We substitute $f_0(m/2^{n+1}), p_0(m/2^{n+1}), f_1(m/2^{n+1}), p_1(m/2^{n+1})$ by the right member of the first, second, third, fourth equation of system (3), respectively, with $x = m/2^n$ in the last equation to obtain the recursion

$$T^{(n+1)}(\xi) = A(\xi/2)T^{(n)}(\xi/2) \begin{pmatrix} 1 & 0 \\ 0 & 2 \end{pmatrix}. \tag{6}$$

Since $\hat{T}^{(0)}(\xi) = I$, the identity matrix, we get

$$T^{(n)}(\xi) = P_n(\xi) \begin{pmatrix} 1 & 0 \\ 0 & 2^n \end{pmatrix}.$$

The sequence of column vectors $\begin{pmatrix} \hat{T}_{11}^{(n)}(\xi) \\ \hat{T}_{21}^{(n)}(\xi) \end{pmatrix}$ converges to the first column of the matrix $P(\xi)$. Now, Schwartz has noticed that the Fourier transform in the space of tempered distributions is a linear continuous transformation and that its inverse is equal to its conjugate ([8] p. 107 of Vol. 2). Therefore the sequences $T_{11}^{(n)}, T_{21}^{(n)}$ converge to the distributions T_0, T_1, respectively, which are the components of the inverse Fourier transform applied to the first column of the matrix $P(\xi)$.

Theorem 11. *For $a_1 + b_1 = 1/2$ and $-\frac{5}{2} < c_2 + d_2 \leq \frac{3}{2}$, the sequences $T_0^{(n)}, T_1^{(n)}$ converge to the distributions T_0, T_1, respectively, which are the components of the inverse Fourier transform applied to the first column of the matrix $P(\xi)$.*

Now, we can prove that the subdivision scheme is always convergent in the space of distributions $\mathcal{D}'(\mathbb{R})$ whenever $a_1 + b_1 = 1/2$ and $-\frac{5}{2} < c_2 + d_2 \leq \frac{3}{2}$. In the following, we use Schwartz notation for the translation operator τ_h, where h is a real number. Given a function ϕ in C_0^∞ and a distribution T, then $\tau_h\phi(x) = \phi(x - h)$ and $\tau_h T(\phi) = T(\tau_h\phi)$.

Theorem 12. *Assume that $a_1 + b_1 = 1/2$ and $-\frac{5}{2} < c_2 + d_2 \leq \frac{3}{2}$. If we build the pair (f, p) by the subdivision scheme (1) from the data $\{y_k, y_k'\}_{k\in\mathbb{Z}}$,* *then the sequence of distributions* $F_n = \frac{1}{2^n}\sum_{m=-\infty}^{\infty} f(m/2^n)\delta_{m/2^n}$ *converges to the distribution* $F = \sum_{k=-\infty}^{\infty}[y_k\tau_{-k}T_0 + y'_k\tau_{-k}T_1]$.

Proof: Let ϕ be a function in C^∞ with support in $[-N, N]$. Then $F_n(\phi) = \frac{1}{2^n}\sum_{m=-N2^n}^{N2^n} f(m/2^n)\phi(m/2^n)$. We use relation (2) to get

$$\begin{aligned} 2^n F_n(\phi) &= \sum_{m=-N2^n}^{N2^n}\sum_{k=-N-1}^{N+1}[y_k f_0(m/2^n - k) + y'_k f_1(m/2^n - k)]\phi(m/2^n) \\ &= \sum_{k=-N-1}^{N+1}\sum_{m=-N2^n}^{N2^n}[y_k f_0(m/2^n) + y'_k f_1(m/2^n)]\phi(m/2^n + k) \\ &= 2^n\sum_{k=-N-1}^{N+1}[y_k T_0^{(n)} + y'_k T_1^{(n)}](\tau_{-k}\phi). \end{aligned}$$

As n tends to infinity, the limit of the sequence $F_n(\phi)$ is

$$\sum_{k=-\infty}^{\infty}[y_k\tau_{-k}T_0 + y'_k\tau_{-k}T_1](\phi). \quad \square$$

Theorem 13. *For* $a_1+b_1 = 1/2$ *and* $-\frac{3}{2} < c_2+d_2 \leq \frac{1}{2}$, *both sequences of distributions* $T_{12}^{(n)}, T_{22}^{(n)}$ *converge. They converge respectively to the derivatives* T'_0, T'_1 *of the distributions* T_0, T_1 *if and only if* $a_2 + 3b_2 + 2c_2 + 2d_2 = 1$.

Proof: Using the relations (5) and (6), we have

$$\begin{aligned} \begin{pmatrix} \hat{T}_{12}^{(n)}(\xi) \\ \hat{T}_{22}^{(n)}(\xi) \end{pmatrix} &= 2^n P_{n-1}(\xi)A(\xi/2^n)\begin{pmatrix} 0 \\ 1 \end{pmatrix} \\ &= 2^n P_{n-1}(\xi)\begin{pmatrix} \frac{i}{2}[a_2\sin(\xi/2^n) + b_2\sin(3\xi/2^n)] \\ \frac{1}{4} + \frac{c_2}{2}\cos(\xi/2^n) + \frac{d_2}{2}\cos(3\xi/2^n) \end{pmatrix} \\ &= \frac{i}{2}(a_2 + 3b_2)\xi\begin{pmatrix} \hat{T}_{11}^{(n-1)}(\xi) \\ \hat{T}_{21}^{(n-1)}(\xi) \end{pmatrix} + (\frac{1}{2} + c_2 + d_2)\begin{pmatrix} \hat{T}_{12}^{(n-1)}(\xi) \\ \hat{T}_{22}^{(n-1)}(\xi) \end{pmatrix} \\ &\quad + O(2^{-n}) \\ &= \frac{i}{2}(a_2 + 3b_2)\xi\sum_{k=0}^{n-1}(\frac{1}{2} + c_2 + d_2)^k\begin{pmatrix} \hat{T}_{11}^{(n-1-k)}(\xi) \\ \hat{T}_{21}^{(n-1-k)}(\xi) \end{pmatrix} \\ &\quad + O(n[\max(\frac{1}{2}, \frac{1}{2} + c_2 + d_2)]^n). \end{aligned}$$

If $|\frac{1}{2}+c_2+d_2| < 1$ which is one of the hypotheses, then the right member of the last vector equation tends to a column vector whose components are respectively $i\xi\frac{(a_2+3b_2)/2}{1/2-c_2-d_2}\hat{T}_0(\xi)$ and $i\xi\frac{(a_2+3b_2)/2}{1/2-c_2-d_2}\hat{T}_1(\xi)$. They are the two respective limits of the sequences $\hat{T}_{12}^{(n)}(\xi), \hat{T}_{12}^{(n)}(\xi)$. These limits are $i\xi\hat{T}_0\xi, i\xi\hat{T}_1(\xi)$ if and only if $\frac{(a_2+3b_2)/2}{1/2-c_2-d_2} = 1$. Using the inverse Fourier transform on each sequence, it is clear that $T_{12}^{(n)}, T_{22}^{(n)}$ converge to the distributions T'_0, T'_1, respectively. $\square$

Let us conclude with a theorem whose proof is similar to the proof of Theorem 12.

Theorem 14. *Suppose that* $a_1 + b_1 = 1/2$, $-\frac{3}{2} \le c_2 + d_2 \le \frac{1}{2}$, and $a_2 + 3b_2 + 2c_2 + 2d_2 = 1$. *If we build the pair* (f, p) *by the subdivision scheme (1) from the data* $\{y_k, y'_k\}_{k\in\mathbb{Z}}$, *then the sequence of distributions*

$$G_n = \frac{1}{2^n} \sum_{m=-\infty}^{\infty} p(m/2^n)\delta_{m/2^n}$$

converges to the distribution $G = \sum_{k=-\infty}^{\infty}[y_k\tau_{-k}T'_0 + y'_k\tau_{-k}T'_1]$.

References

1. Cohen A., N. Dyn, and D. Levin, Stability and inter-dependance of matrix subdivision, in *Advanced Topics in Multivariate Approximation*, F. Fontanella, K. Jetter, and P.-J. Laurent (eds.) World Scientific Publications, 1996, 1–13.
2. Dubuc S. and J.-L. Merrien, Fourier transform of Hermite interpolatory subdivision schemes, submitted.
3. Dyn N. and D. Levin, Analysis of Hermite-type subdivision schemes, in *Approximation Theory VIII,* Vol 2: *Wavelets and Multilevel Approximation,* C. K. Chui and L.L. Schumaker (eds.). World Scientific, Singapore, 1995, 117–124.
4. Dyn N. and D. Levin, Analysis of Hermite-interpolatory subdivision schemes, in *Spline Functions and the Theory of Wavelets*, S. Dubuc and G. Deslauriers (eds), A.M.S., Providence R. I., 1999, 105–113.
5. Derfel G., N. Dyn, and D. Levin, Generalized functional equations and subdivision schemes, J. Approx. Theory **80** (1995), 272–297.
6. Heil C. and D. Colella, Matrix refinement equations: existence and uniqueness. J. Fourier Anal. Appl. **2** (1996), 363–377.
7. Merrien J.-L., A family of Hermite interpolants by bissection algorithms. Numer. Algorithms **2** (1992), 187–200.
8. Schwartz L., *Théorie des Distributions*, Tomes 1 and 2. Hermann, Paris, 1956.

Serge Dubuc,
Département de mathématiques et de statistique,
Université de Montréal, C.P. 6128
Succursale Centre-ville,
Montréal (Québec), Canada H3C 3J7,
dubucs@dms.umontreal.ca

Jean-Louis Merrien,
INSA de Rennes,
20 av. des Buttes de Coësmes, CS 14315,
35043 Rennes cedex, France,
Jean-Louis.Merrien@insa-rennes.fr

Univariate Adaptive Thinning

Nira Dyn, Michael S. Floater, and Armin Iske

Abstract. In this paper we approximate large sets of univariate data by piecewise linear functions which interpolate subsets of the data, using adaptive thinning strategies. Rather than minimize the global error at each removal (AT0), we propose a much cheaper thinning strategy (AT1) which only minimizes errors locally. Interestingly, the two strategies are equivalent in all our numerical tests and we prove this to be true for convex data. We also compare with non-adaptive thinning strategies.

§1. Introduction

In applications such as visualization, it is often desirable to generate a hierarchy of coarser and coarser representations of a given discrete data set. Though we are primarily interested in hierarchies of scattered data sets, and in particular piecewise linear approximations over triangulations in the plane [1], we focus in this paper on univariate data sets and propose several adaptive thinning strategies. Thinning algorithms generate hierarchies of subsets by removing points from the given data set one by one, in such a way that the 'least' significant point is removed at each step, according to some desirable criterion. Our criterion here will primarily be the minimization of approximation error, so our thinning algorithms are adaptive. This is in contrast, for example, to the thinning strategies of [2,3], where the criterion was to generate subsets of well distributed points, independent of the height values.

Thinning algorithms for piecewise linear approximation to univariate data have appeared before in the literature as decimation algorithms, as in Heckbert and Garland [4], and as knot removal for linear splines, as in Lyche [6].

In this paper, we design, test, and compare four methods for anticipating the error incurred by the removal of a point from the current subset. Our algorithms choose the point to be removed as the one of minimal anticipated error. Our main conclusion is that the algorithm AT1, which is based on making a local error estimate, but taking account of all previously removed points, is the best algorithm from the point of view of our numerical results

Mathematical Methods for Curves and Surfaces: Oslo 2000
Tom Lyche and Larry L. Schumaker (eds.), pp. 123–134.

ISBN 0-8265-1378-6

and theoretical analysis. In fact our theoretical analysis shows that its computational complexity is $O(N \log N)$, with N the number of points in the data set, provided one uses a heap to store the anticipated errors. Moreover, we prove that for data sampled from a convex function, AT1 minimizes the global approximation error at every step. These latter two results extend to piecewise linear functions over triangulations for scattered data in the plane; see [1].

§2. Adaptive Thinning

Suppose $[a, b]$ is a real interval and that $X = (x_1, \ldots, x_N)$ is a given sequence of points in $[a, b]$ such that

$$a = x_1 < x_2 < \cdots < x_N = b.$$

Suppose further that some unknown function $f : [a, b] \to \mathbb{R}$ is sampled at these points, giving the values $f(x_1), \ldots, f(x_N)$.

For each n, $1 < n < N$, we are interested in finding a subset $Y = (y_1, \ldots, y_n)$ of X, such that

$$a = x_1 = y_1 < y_2 < \cdots < y_n = x_N = b, \tag{1}$$

and such that the piecewise linear interpolant $L(f, Y)$ to the data

$$\{(y, f(y)) : y \in Y\}$$

is close to the given data $\{(x, f(x)) : x \in X\}$, in the sense that the error

$$E(Y; X; f) = \max_{x \in X} |L(f, Y)(x) - f(x)| \tag{2}$$

is small relative to the errors corresponding to other subsets of X of cardinality n. To guarantee that $E(Y; X; f)$ is well defined, we refrain from removing the points x_1, x_N, so that $L(f, Y)$ is defined on $[a, b]$.

Ideally, for any given n, $1 < n < N$, we would like to find a subset Y of X of cardinality n for which the error in (2) is minimal. However, it is clearly impractical to search amongst all possible subsets, and this motivates the more pragmatic approach of thinning.

The idea of thinning is to remove points from X one by one in order to reach a subset Y of a certain size. In general we want to remove a point of 'least' significance. Our criterion for removing a point from the current subset is to minimize its *anticipated error*, which is an estimate of the error incurred by the removal of the point with respect to some error measure. Thus the thinning algorithm is a greedy algorithm, choosing the current step to do the optimal step in the current situation.

We define our thinning algorithm by saying that a point y_i in Y, $1 < i < n$, is removable if

$$e(y_i) = \min_{j=2,3,\ldots,n-1} e(y_j), \tag{3}$$

where $e(\cdot)$ is our chosen anticipated error.

Thinning Algorithm

1) Set $X_N = X$.
2) For $i = N, N-1, \ldots, 3$:

   locate a removable point x in X_i and set $X_{i-1} = X_i \setminus x$.

The result of the thinning algorithm is a hierarchical sequence of subsets of X,

$$\{a, b\} = X_2 \subset X_3 \subset \cdots \subset X_N = X,$$

with $|X_i| = i$.

Now we consider various anticipated error measures $e(\cdot)$, each of which defines a removable point in (3), and results in a different algorithm. The algorithm is termed Adaptive Thinning whenever the anticipated error depends on some of the function values $\{f(x) : x \in X\}$.

Algorithm AT0

In this algorithm the anticipated error of a point y_i is the maximum of the errors incurred by the removal of y_i at all the points of X,

$$e_0(y_i; Y) = E(Y \setminus y_i; X; f) = \max_{x \in X} |L(f, Y \setminus y_i)(x) - f(x)|. \qquad (4)$$

Indeed, $e_0(y_i)$ is the actual error incurred by the removal of y_i, measured in the sup-norm over X.

Algorithm AT1

A less expensive to compute measure of anticipated error is

$$e_1(y_i; Y) = e_{[y_{i-1}, y_{i+1}]}, \qquad (5)$$

where for any interval I whose endpoints I_B belong to X, e_I is defined as

$$e_I = \max_{x \in I \cap X} |L(f, I_B)(x) - f(x)|.$$

Note that here we consider only the error incurred by the removal of a point at those points of X which belong to the current interval of the removed point.

Algorithm AT2

In this algorithm the anticipated error is similar to the one in AT1, but does not depend on the points that are already removed. This anticipated error is simpler to compute than the anticipated error of AT1,

$$e_2(y_i; Y) = |L(f, \{y_{i-1}, y_{i+1}\})(y_i) - f(y_i)|. \qquad (6)$$

Algorithm NAT

Here the removal of a point depends only on the density of the points Y, and is independent of f. Thus it is a Non-Adaptive Thinning algorithm. In fact $e(\cdot)$ is such that the removal of points results in approximately equidistributed sets of points X_i, for intermediate values of i,

$$e_3(y_i; Y) = (y_i - y_{i-1})(y_{i+1} - y_i). \tag{7}$$

When the set Y is fixed we use also the notation $e_j(y) = e_j(y; Y)$ for $j = 0, 1, 2, 3$.

§3. Theoretical Aspects

In order to better understand univariate thinning, and in particular why AT1 is almost as good as AT0, we study the antipicated error used in AT1.

Notice that for any $i \in \{2, 3, \ldots, n-1\}$, and $x \in [y_{i-1}, y_{i+1}] \cap X$, the error in the linear interpolation at y_{i-1}, y_{i+1} is given by

$$L(f, \{y_{i-1}, y_{i+1}\})(x) - f(x) = (x - y_{i-1})(y_{i+1} - x) f[y_{i-1}, x, y_{i+1}],$$

where $f[a, b, c]$ denotes the usual second order divided difference of the function f at the abscissae a, b, c. It follows that

$$e_1(y_i) = \max_{x \in [y_{i-1}, y_{i+1}] \cap X} (x - y_{i-1})(y_{i+1} - x) \big| f[y_{i-1}, x, y_{i+1}] \big|.$$

For f a quadratic polynomial, this identity leads us to a relationship between adaptive and non-adaptive thinning.

Proposition 3.1. *If f is a quadratic polynomial, then the adaptive univariate thinning algorithms AT1 and AT2 are non-adaptive. A point y_i in Y is removable if and only if*

$$\max_{x \in [y_{i-1}, y_{i+1}] \cap X} (x - y_{i-1})(y_{i+1} - x) = \min_{1 < j < n} \max_{x \in [y_{j-1}, y_{j+1}] \cap X} (x - y_{j-1})(y_{j+1} - x) \tag{8}$$

in AT1, and

$$(y_i - y_{i-1})(y_{i+1} - y_i) = \min_{1 < j < n} (y_j - y_{j-1})(y_{j+1} - y_j) \tag{9}$$

in AT2.

Proof: Since for any $a, b, c \in \mathbb{R}$, the divided difference $f[a, b, c]$ is a constant, the anticipated errors (5) and (6) reduce, after a scaling, to

$$\max_{x \in [y_{i-1}, y_{i+1}] \cap X} (x - y_{i-1})(y_{i+1} - x),$$

and $(y_i - y_{i-1})(y_{i+1} - y_i)$ respectively. □

Note that AT2 reduces to NAT, in case f is a quadratic polynomial. The criteria (8) and (9) clearly favour well distributed subsets of data: a data point y_i is likely to be removed if it is close to its neighbours. To see this in the case of (8), we replace the discrete set X by the whole interval $[a, b]$, and the dicrete antipicated error $e_1(y_i)$ becomes

$$e_a(y_i) = \max_{x \in [y_{i-1}, y_{i+1}]} |L(f, \{y_{i-1}, y_{i+1}\})(x) - f(x)|.$$

Thus, if f is a quadratic polynomial, then

$$e_a(y_i) = \frac{1}{8}(y_{i+1} - y_{i-1})^2 |f''|,$$

and y_i is removable if and only if

$$y_{i+1} - y_{i-1} = \min_{1<j<n} (y_{j+1} - y_{j-1}).$$

This is the removal criterion of the non-adaptive Thinning Algorithm 3 of [3].

Next we give an explanation of why minimizing the anticipated error $e_1(\cdot)$ instead of the actual error $e_0(\cdot)$ in the thinning algorithm AT1 results in a good algorithm. We do it by considering convex functions f. First we establish a lemma.

Lemma 3.2. *Suppose f is convex, and let Y be any subset of X of the form (1). Then for any $i \in \{2, \ldots, n-1\}$,*

$$e_0(y_i) = \max\{e_1(y_i), E(Y; X; f)\}. \tag{10}$$

Proof: Due to the convexity of f, we have

$$e_1(y_i) = e_{[y_{i-1}, y_{i+1}]} \geq \max\{e_{[y_{i-1}, y_i]}, e_{[y_i, y_{i+1}]}\},$$

and since

$$e_0(y_i) = E(Y \setminus y_i; X; f) = \max\{e_1(y_i), \max_{\substack{k=1,\ldots,n-1 \\ k \neq i-1, i}} e_{[y_k, y_{k+1}]}\}, \tag{11}$$

we find

$$e_0(y_i) = \max\{e_1(y_i), \max_{k=1,\ldots,n-1} e_{[y_k, y_{k+1}]}\} = \max\{e_1(y_i), E(Y; X; f)\}. \quad \square$$

Proposition 3.3. *Suppose f is convex and let Y be any subset of X of the form (1). Then for any $i, j \in \{2, \ldots, n-1\}$,*

$$e_1(y_i) \leq e_1(y_j) \quad \Longrightarrow \quad e_0(y_i) \leq e_0(y_j). \tag{12}$$

Proof: From (10), we have

$$e_0(y_i) = \max\{e_1(y_i), E(Y; X; f)\} \leq \max\{e_1(y_j), E(Y; X; f)\} = e_0(y_j). \quad \square$$

Thus for convex data, the thinning algorithm AT1 performs as AT0. We show in the next example that there are arbitrary subsets of non-convex data for which (12) does not hold.

Example 3.4. *((12) does not hold for non-convex data.) Let*

$$X = (x_1, \ldots, x_7) = (1, 2, 3, 4, 5, 6, 7),$$

and let the non-convex function f be the piecewise linear interpolant over X satisfying $f(x_i) = f_i$, where

$$(f_1, \ldots, f_7) = (0, 0, -1, 1, 0, 0, 0).$$

Consider the subset $Y = X \setminus x_4$. Then $e_1(x_6) = 0$ and $e_1(x_3) = 1$, while $e_0(x_6) = 3/2$ and $e_0(x_3) = 1$.

Note however, that if the thinning algorithm AT1 were applied to this example, Y would not be the subset generated by the first removal as the first point to be removed would be x_6. We have not been able to construct an example of a data set X and a non-convex function f where AT0 acts differently from AT1 at any stage of the thinning algorithm. In the absence of such an example, and since typical data sets are locally convex or concave in large regions, i.e. there are relatively few inflection points, we arrive at the conclusion, supported by our numerical experiments, that AT1 is a good, computationally inexpensive thinning algorithm.

§4. Algorithmic Aspects

In this section, we discuss details concerning our implementation of the four thinning algorithms AT0, AT1, AT2, and NAT. Moreover, we shall compute their asymptotic complexity.

We first discuss AT2 and NAT. The interior points of the current set Y are stored in a heap, according to the sizes of their anticipated errors; $e_2(.)$ for AT2 and $e_3(.)$ for NAT. A heap is a binary tree which can be used for the implementation of a priority queue. Each point y in the heap bears its anticipated error as its significance value. Due to the heap condition, the significance of a node is smaller than the significances of its two children. Therefore, the root of the heap contains a removable point. It is well-known [7] that each insertion, removal, or update of one node in the heap costs $O(\log n)$ operations, where n is the number of nodes in the heap. In consequence, building the initial heap costs $O(N \log N)$ operations.

Now suppose Y is of the form (1), of size n. The number of points already removed is $N - n$. We perform Step 2) of the thinning algorithm of Section 2 as follows.

1) Pop the root y_i from the heap.
2) Compute $e(y_{i-1}; Y \setminus y_i)$ and $e(y_{i+1}; Y \setminus y_i)$ and update the heap.
3) Let $Y = Y \setminus y_i$.

As regards the number of operations, Steps 1) and 2) both require $O(\log n)$ operations, while Step 3) requires $O(1)$ operations. Therefore, summing the costs of Steps 1) to 3) for all n, we find that the total cost of the thinning algorithms AT2 and NAT is $O(N \log N)$.

Next we describe Step 2) of the thinning algorithm in Section 2 for AT1, which is somewhat more complicated than the previous ones. For this algorithm, we store $\{e_1(y) : y \in Y\}$ in a heap where the root contains the removable point for AT1.

1) Pop the root y_i from the heap.
2) Attach y_i and the previously removed points attached to the intervals $[y_{i-1}, y_i]$ and $[y_i, y_{i+1}]$ to the new interval $[y_{i-1}, y_{i+1}]$ (generated after the removal of y_i).
3) Compute $e_1(y_{i-1}; Y \setminus y_i)$ and $e_1(y_{i+1}; Y \setminus y_i)$ and update the e_1-heap.
4) Let $Y = Y \setminus y_i$.

Thus, during the adaptive thinning algorithm AT1, each of the points already removed is attached to an interval corresponding to the current subset. These attachments facilitate the computation of the anticipated error of the neighbouring points of y_i in Y, whose anticipated errors in $Y \setminus y_i$ differ from their anticipated errors in Y.

As regards the number of operations, Steps 1) and 4) are as in the previous algorithm. Step 2) requires $O((N-n)/n)$ operations under the additional assumption that the number of points attached to an interval is of the order of $(N-n)/n$. The computation of the anticipated error in step (3) is also $O((N-n)/n)$. So altogether the total cost is $O(N \log N)$, just as for AT2 and NAT, though with a higher constant.

The algorithm for AT0 is a variant of the algorithm for AT1, but is more complicated. Yet it can be organized so that the total cost remains $O(N \log N)$. We now employ two heaps, the first of which is the e_1-heap we used for AT1. The second heap, which we call the I-heap, consists of the values $\{e_{[y_i, y_{i+1}]}; i = 1, 2, \ldots, n-1\}$ so that the root of the heap points to the *maximal* element. Using the identity (11), it can easily be shown that there is always a removable point amongst the three points y_i, the root of the e_1-heap, and y_j and y_{j+1}, where $[y_j, y_{j+1}]$ is the root of the I-heap. Thus, it is only necessary to compute $e_0(.)$ at these three points and take the minimum, and using the two heaps, this can be achieved in just $O(\log n)$ operations. The update of the e_1-heap after the removal requires $O(\log n)$ operations as in the algorithm for AT1, and the update of the I-heap also requires just $O(\log n)$ operations. Thus the thinning algorithm AT0 requires $O(N \log N)$ operations, but with a larger constant than for AT1.

§5. Numerical Examples

We have implemented the four thinning algorithms AT0, AT1, AT2, and NAT corresponding to the error measures (4), (5), (6), (7) in Section 2. In this section we compare the performance of these algorithms in terms of their approximation error and computational costs. For the purpose of illustration,

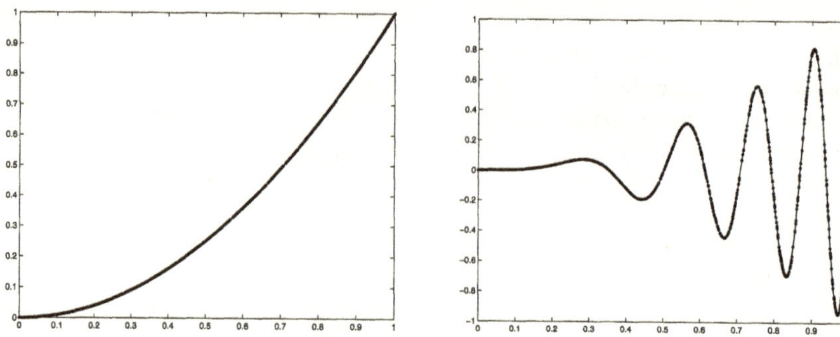

Fig. 1. Data sets sampled from f_1 (left) and f_2 (right).

we use the two test functions $f_1(x) = x^2$ and $f_2(x) = x^2 \sin(25x^2)$, which we sampled at a set X of 1000 randomly chosen points in the unit interval $(0, 1)$, together with the two boundary points $0, 1$, cf. Figure 1.

We have computed all subsets $X_n \subset X$ for $n = N, N-1, \ldots, 2$, where $N = 1002$, output by the four thinning algorithms. For each test case, we have recorded both the resulting max error $E_\infty(X_n) \equiv E(X_n; X; f)$ and ℓ_2 error

$$E_2^2(X_n) \equiv E_2^2(X_n; X; f) = \sum_{x \in X} |L(f, X_n)(x) - f(x)|^2.$$

For the test cases involving the quadratic function f_1, we observe that the four subsets X_n obtained by the four thinning algorithms have nearly equal approximation errors, $E_\infty(X_n)$ and $E_2(X_n)$. For $n = 22$, the resulting values are displayed in Table 1 which also shows the required computational costs in CPU time. Not surprisingly, the thinning method NAT is the fastest, followed by AT2 and AT1, whereas AT0 is the slowest. Table 1 also shows the *mesh ratio*,

$$\rho(\{y_1, \cdots, y_n\}) = \min_{0<j<n} |y_{j+1} - y_j| / \max_{0<j<n} |y_{j+1} - y_j|,$$

for each subset. From the values $\rho(X_{22})$ for the four subsets, we conclude that these subsets are well distributed in $[0, 1]$. Figure 2 shows the subset X_{22} selected by the methods AT0, AT1 (left) and AT2, NAT (right).

Method	$E_\infty(X_{22})$	$E_2(X_{22})$	$\rho(X_{22})$	CPU
AT0	0.0013880	0.0166962	0.4029	0.70
AT1	0.0013880	0.0166962	0.4029	0.49
AT2	0.0010135	0.0164082	0.4081	0.27
NAT	0.0010135	0.0164082	0.4081	0.26

Tab. 1. Thinning to 22 points with f_1.

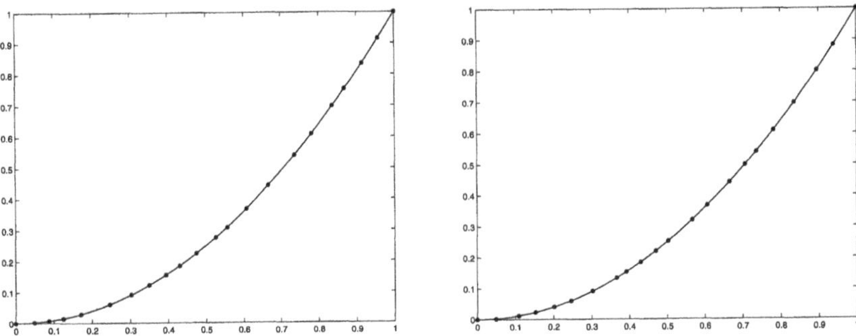

Fig. 2. The subsets X_{22} output by AT0, AT1 (left) and AT2, NAT (right).

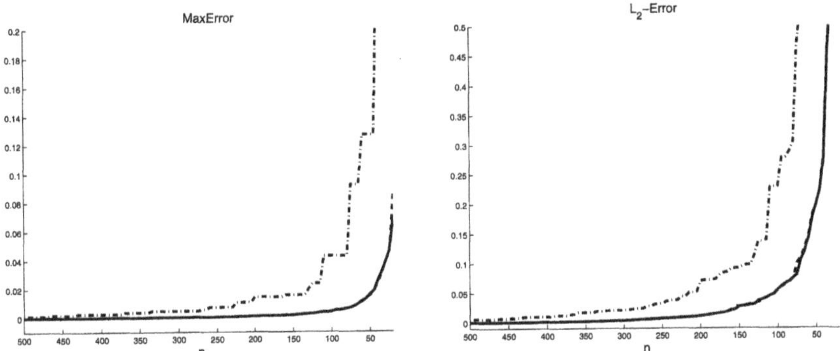

Fig. 3. Error of AT0,AT1 (solid), AT2 (dashed) and NAT (dash-dotted).

As expected from the theoretical results of Section 3, we see in Table 1 that AT0 is identical with AT1 since f_1 is convex. Also, AT2 and NAT are identical since f_1 is a quadratic polynomial. Thus all four algorithms are non-adaptive, and generate well distributed subsets.

Now let us turn to the test case involving the oscillating function f_2. In contrast to the results for f_1, we find that the three adaptive thinning methods AT0, AT1, and AT2 are, especially for the selection of small subsets, clearly superior to NAT in terms of approximation error. This is confirmed by Figure 3 showing the four graphs of $E_\infty(X_n; X; f_2)$ and $E_2(X_n; X; f_2)$, for $n = 500, 499, \ldots, 22$.

Observe from Figure 3 that the approximation behaviour of the three methods AT0, AT1, and AT2 is quite similar. In fact, we found that for any n the two subsets X_n output by AT0 and AT1 coincide. Taking a closer look at the approximation errors of AT1 and AT2, we see from Figure 4 that for very large numbers of removed points, AT1 is superior to AT2, in terms of the error $E_\infty(X_n; X; f_2)$. The trade-off is that AT1 typically required about 60% more CPU time than AT2.

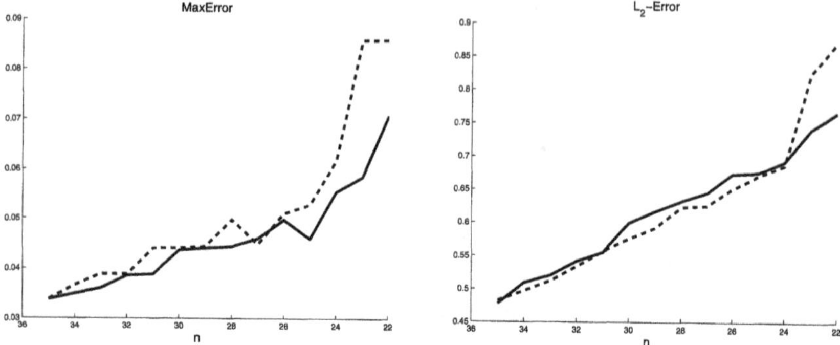

Fig. 4. Error of AT1 (solid), and AT2 (dashed) for $n = 35, 34, \ldots, 22$.

Finally, we wish to demonstrate the utility of adaptive thinning for a class of approximation methods other than piecewise linear interpolation. For subsets Y generated by the thinning algorithms, we computed the least squares approximation $s^*_{\phi,Y} \in S_{\phi,Y}$ satisfying

$$\eta^2_{\phi,2}(Y) = \sum_{x \in X} |s^*_{\phi,Y}(x) - f(x)|^2 = \min_{s \in S_{\phi,Y}} \sum_{x \in X} |s(x) - f(x)|^2,$$

where $S_{\phi,Y} = \text{span}\,\{\phi(|\cdot - y|) : y \in Y\}$ denotes the linear space of all linear combinations of Y-translates of the multiquadrics $\phi(r) = \sqrt{c^2 + r^2}$, $c > 0$ (see [5] for more details). This gives two additional criteria $\eta_{\phi,2}(Y)$ and

$$\eta_{\phi,\infty}(Y) = \max_{x \in X} |s^*_{\phi,Y}(x) - f(x)|,$$

for judging the *quality* of a subset Y of X. In order to show one concrete example, we let $c = 0.2$, $n = 22$. Table 2 reflects the numerical results, and Figure 5 show the two subsets $Y = X_{22}$ selected by the method AT1 (left) and NAT (right) along with the graphs of their corresponding least squares approximations $s^*_{\phi,Y}$.

Method	$E_\infty(X_{22})$	$E_2(X_{22})$	$\eta_{\phi,\infty}(X_{22})$	$\eta_{\phi,2}(X_{22})$
AT1	0.07088	0.76592	0.0196591	0.091160
NAT	0.84983	5.76083	0.0497937	0.567353

Tab. 2. Thinning to 22 points with f_2.

Table 2 indicates that small subsets of X output by an adaptive thinning algorithm can serve as good sets of centres for approximating the data $(x, f(x)) : x \in X$, by a sum of translates of ϕ to the chosen centres, and that these approximations are superior to the piecewise linear interpolants on these subsets.

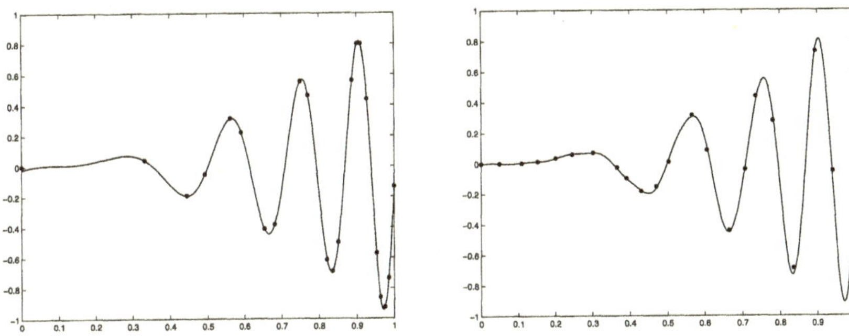

Fig. 5. Least squares approximation, AT1 (left), NAT (right).

Acknowledgments. The authors were partly supported by the European Union within the project MINGLE (Multiresolution in Geometric Modelling), contract no. HPRN-CT-1999-00117.

References

1. Dyn, N., M. S. Floater, and A. Iske, Adaptive thinning of bivariate scattered data, preprint.
2. Floater, M. S. and A. Iske, Multistep Scattered Data Interpolation using Compactly Supported Radial Basis Functions, J. Comp. Appl. Math. **73** (1996), 65–78.
3. Floater, M. S. and A. Iske, Thinning algorithms for scattered data interpolation, BIT **38** (1998), 705–720.
4. Heckbert, P. S. and M. Garland, Survey of Surface Simplification Algorithms, Technical Report, Computer Science Dept., Carnegie Mellon University, 1997.
5. Iske, A., Reconstruction of smooth signals from irregular samples by using radial basis function approximation, in *Proceedings of the 1999 International Workshop on Sampling Theory and Applications*, Y. Lyubarskii (ed.), The Norwegian University of Science and Technology, Trondheim, 1999, 82–87.
6. Lyche, T., Knot removal for spline curves and surfaces, in *Approximation Theory VII*, E. W. Cheney, C. K. Chui, and L. L. Schumaker, (eds.), Academic Press, Boston, 1993, 207–227.
7. R. Sedgewick, *Algorithms*, Addison-Wesley, Reading, MA, 1983.

Nira Dyn
School of Mathematical Sciences
Sackler Faculty of Exact Sciences
Tel-Aviv University
Tel Aviv 69978, Israel
niradyn@math.tau.ac.il

Michael S. Floater
SINTEF
Post Box 124, Blindern
N-0314 Oslo, Norway
`mif@math.sintef.no`

Armin Iske
Technische Universität München
Zentrum Mathematik
D-80290 München, Germany
`iske@ma.tum.de`

Optimizing 3D Triangulations Using Discrete Curvature Analysis

Nira Dyn, Kai Hormann, Sun-Jeong Kim, and David Levin

Abstract. A tool for constructing a "good" 3D triangulation of a given set of vertices in 3D is developed and studied. The constructed triangulation is "optimal" in the sense that it locally minimizes a cost function which measures a certain discrete curvature over the resulting triangle mesh. The algorithm for obtaining the optimal triangulation is that of swapping edges sequentially, such that the cost function is reduced maximally by each swap. In this paper three easy-to-compute cost functions are derived using a simple model for defining discrete curvatures of triangle meshes. The results obtained by the different cost functions are compared. Operating on data sampled from simple 3D models, we compare the approximation error of the resulting optimal triangle meshes to the sampled model in various norms. The conclusion is that all three cost functions lead to similar results, and none of them can be said to be superior to the others. The triangle meshes generated by our algorithm, when serving as initial triangle meshes for the butterfly subdivision scheme, are found to improve significantly the limit butterfly-surfaces compared to arbitrary initial triangulations of the given sets of vertices. Based upon this observation, we believe that any algorithm operating on triangle meshes such as subdivision, finite element solution of PDE, or mesh simplification, can obtain better results if applied to a "good" triangle mesh with small discrete curvatures. Thus our algorithm can serve for modelling surfaces from sampled data as well as for initialization of other triangle mesh based algorithms.

§1. Introduction

Triangle meshes are commonly used for representing 3D surfaces. Given a set of vertices sampled from a smooth surface, the triangle mesh with these vertices serves as a representation (approximation) of the sampled surface. This representation depends on the choice of the connections among the vertices.

In this paper we investigate good choices of triangulations for a fixed set of vertices under the assumption that the sampled surface is smooth. Our point of view is that the given discrete set of points represents the surface in the sense that its most prominent features (creases, curvatures, etc.) can be

Mathematical Methods for Curves and Surfaces: Oslo 2000
Tom Lyche and Larry L. Schumaker (eds.), pp. 135–146.

ISBN 0-8265-1378-6

extracted from the data and that the representation as a triangle mesh should not add features not present in the data.

This leads us to choices of optimal triangulations relative to cost functions which measure different kinds of discrete curvatures. Starting with an arbitrary initial triangulation of the given vertices, the main algorithm we present swaps edges in a greedy way so as to maximally reduce the cost function at each step and terminates at a local minimum of the cost function. Such an algorithm was also used for optimizing 2D triangulations (approximation of bivariate functions), *e.g.* for deriving **data dependent triangulations** [3,6,8,9].

Alboul and van Damme, in a series of papers [1,2,11], consider a cost function which is a discrete measure of the L_1-norm of the Gaussian curvature over the triangle mesh. Although this cost function requires heavy computations, it has a very important property. As proved in [2], for data sampled from a convex surface (convex data), swapping with this cost function leads to its unique global minimum which corresponds to the unique convex triangulation. For two of the cost functions we introduce here, the convex triangulation also seems to be the global minimum while their computation is simpler. The theoretical investigation of this observation is beyond the scope of this paper and we leave it to future work. Unfortunately, swapping edges with the greedy swapping algorithm does not always lead to the convex triangulation because the cost functions may have local minima.

We have made many numerical experiments with our cost functions and conclude that we have at hand a tool which improves significantly the visual appearance of a triangle mesh for a fixed set of vertices and enhances feature lines ("sharp edges"). For complex models, our different cost functions yield very similar results. For simple models, we also tested the approximation quality of the triangle meshes generated by the swapping algorithm with our cost functions and found all of them to reduce the approximation error significantly (see the torus example in Sect. 4).

We also realize that our algorithm can serve as a preprocessor for other algorithms operating on 3D triangle meshes (such as subdivision, finite element, simplification, etc.) by providing a better starting point for these algorithms, which results in better performance.

The outline of the paper is as follows: Section 2 presents a calculation of several discrete measures of curvature. In Section 3 we discuss the swapping algorithm and the various cost functions. Section 4 includes experiments with our different methods of optimizing 3D triangulations. An appendix with explicit formulas needed for the computation of our cost functions is the last section of the paper.

§2. Discrete Curvature Computation

From a theoretical point of view, triangle meshes do not have any curvature at all, since all faces are flat and the curvature is not properly defined along edges and at vertices because the surface is not C^2-differentiable there. But thinking of a triangle mesh as a piecewise linear approximation of an unknown

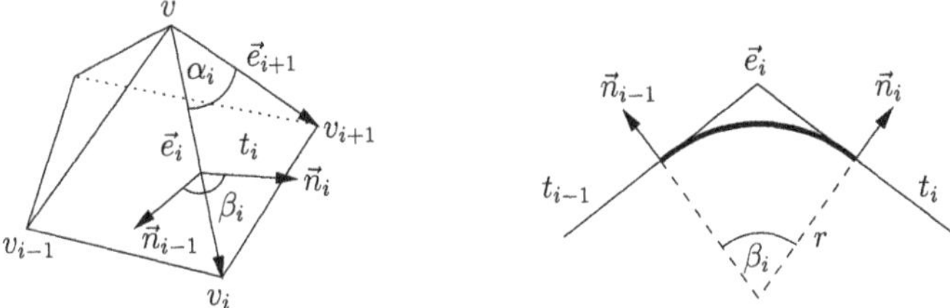

Fig. 1. A vertex v and the related variables for this local configuration (left). The blending cylinder along $\vec{e}_i$ between triangles t_{i-1} and t_i, seen from the side (right).

smooth surface, one can try to estimate the curvatures of that unknown surface using only the information that is given by the triangle mesh itself. We are particularly interested in computing the Gaussian curvature K and the absolute mean curvature $|H|$ at the vertices of the triangle mesh, since we base the cost functions to be minimized in the mesh optimization process on these values. But let us first fix the notation before explaining how to derive K and $|H|$ from the given data.

Notation. A triangle mesh M consists of a set of vertices $V = \{v_i\}_i \subset \mathbb{R}^3$, which are connected by a set of edges $E = \{e_j = (v_{j_1}, v_{j_2})\}_j$ and a set of triangles $T = \{t_k = \Delta(v_{k_1}, v_{k_2}, v_{k_3})\}_k$. Let $v \in V$ be a vertex of a triangle mesh M and let $v_1, \ldots, v_n$ be the ordered neighboring vertices of v (cf. Fig. 1). We define the edges $\vec{e}_i = v_i - v$ and the angles between two successive edges $\alpha_i = \angle(\vec{e}_i, \vec{e}_{i+1})$. The triangle between $\vec{e}_i$ and $\vec{e}_{i+1}$ is named $t_i = \Delta(v, v_i, v_{i+1})$, the corresponding face normal $\vec{n}_i = \frac{\vec{e}_i \times \vec{e}_{i+1}}{\|\vec{e}_i \times \vec{e}_{i+1}\|}$. The dihedral angle at an edge $\vec{e}_i$ is the angle between the normals of the adjacent triangles, $\beta_i = \angle(\vec{n}_{i-1}, \vec{n}_i)$. Note that in these definitions we identify the index 0 with n and the index $n+1$ with 1.

Now we can define the integral Gaussian curvature $\bar{K} = \bar{K}_v$ and the integral absolute mean curvature $|\bar{H}| = |\bar{H}_v|$ with respect to the area $S = S_v$ attributed to v by

$$\bar{K} = \int_S K = 2\pi - \sum_{i=1}^{n} \alpha_i \qquad \text{and} \qquad |\bar{H}| = \int_S |H| = \frac{1}{4} \sum_{i=1}^{n} \|\vec{e}_i\| \, |\beta_i|. \tag{1}$$

These formulas are also used by other authors [1,2,4,11], and can be understood in the following way. Suppose we replace each edge by a small cylinder of radius r that joins the adjacent faces tangentially (cf. Fig. 1) and blend these cylinders smoothly at the vertices in a C^2 manner. Now the triangle mesh is approximated by a smooth surface, K and $|H|$ are integrable functions on it and we can apply well-known theorems from differential geometry [5]. A straightforward computation [1,11] finally results in formulas (1) that depend neither on the choice of r nor on the specific blending method at the vertices.

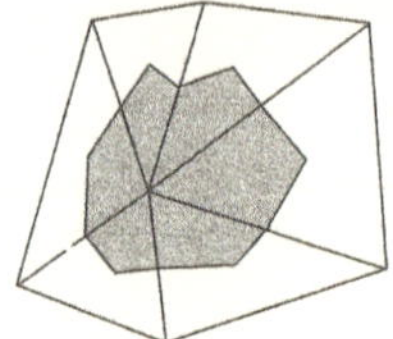 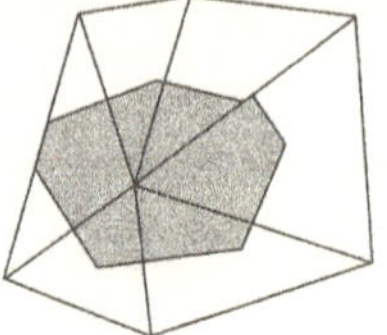

Fig. 2. Barycentric area S^B (left) and Voronoi area S^V (right) around a vertex.

To derive the curvatures at the vertex v from these integral values, we assume the curvatures to be uniformly distributed around the vertex, and simply normalize by the area:

$$K = \frac{\bar{K}}{S} \qquad \text{and} \qquad |H| = \frac{|\bar{H}|}{S}.$$

Of course there are different ways of defining the area S_v attributed to a vertex v, which result in different curvature values. We restrict ourselves to those methods for which the areas around all vertices sum up to the area of the triangle mesh M, *i.e.*, $\sum_{v \in V} S_v = M$, since this enables us to write an integral over M as the sum of integrals over the single area patches, *e.g.* $\int_M K = \sum_{v \in V} \int_{S_v} K$. The area that is most commonly used in the literature is the barycentric area S^B which is one third of the area of the triangles adjacent to v, and can be constructed by connecting the edge midpoints with the barycenters of the adjacent triangles (cf. Fig. 2). However, inspired by [12], we decided to use the Voronoi area S^V instead, which sums up the areas of v's local Voronoi cells restricted to the triangles adjacent to v, according to the Euclidean distance to the vertices of the triangle mesh (cf. Fig. 2). The explicit formulas for computing the Voronoi area are given in the appendix (Sec. 5).

Besides the Gaussian and the absolute mean curvature, we are also interested in the sum of the absolute principle curvatures $|\kappa_1|$ and $|\kappa_2|$. From the relations $K = \kappa_1 \kappa_2$ and $H = \frac{1}{2}(\kappa_1 + \kappa_2)$, we get $\kappa_{1,2} = H \pm \sqrt{H^2 - K}$. Moreover, we can get the sum of the absolute principle curvatures without knowing H but only $|H|$:

$$|\kappa_1| + |\kappa_2| = \begin{cases} 2\,|H|, & \text{if } K \geq 0, \\ 2\,\sqrt{|H|^2 - K}, & \text{otherwise.} \end{cases}$$

Note that $|\kappa_1| + |\kappa_2|$ is always a real number, even if $|H|^2 = H^2 < K$, which corresponds to complex principle curvature values. Of course, this cannot happen for smooth surfaces, but since we are dealing with discrete curvatures, it can occur for some vertices.

§3. Mesh Optimization

Over the past few years the problem of fairing (or smoothing) meshes has received a lot of attention. The need for these methods ranges from technical applications, where the noise that is due to measurement errors has to be removed from measured data, to entertainment applications, that require triangulated 3D models with a pleasing visual appearance. The usual approach in mesh fairing is to move the vertices of the mesh such that a certain energy functional is minimized [4,7,10]. However, these methods cannot be applied whenever the position of the original data points must not be changed, *e.g.* in numerical simulations or surface interpolation. The only parameter that is left to change is the triangulation of the data points. While the optimization of 2D triangulations has been studied thoroughly in the early nineties [3,6,8,9], little is known so far for the 3D case [1,2,11].

One of the energy functionals that has often been used for the fairing of meshes as well as for the fairing of continuous surfaces is the thin plate energy

$$F_{\mathrm{TP}} = \int 4a\, H^2 + 2\,(1-a-b)\, K,$$

with certain parameters $a, b \in \mathbb{R}$. However, for closed surfaces or surfaces with a fixed boundary this expression can be simplified to $F_{\mathrm{TP}} = 4a \int H^2$, because the theorem of Gauss-Bonnet states that $\int K$ is constant in these cases. Note that this also holds for the discrete version of the integral Gaussian curvature:

$$\int_M K = \sum_v \int_{S_v} K = \sum_v \left(2\pi - \sum_{i=1}^{n_v} \alpha_{v_i} \right) = (\#V)\, 2\pi - (\#T)\, \pi = 2\pi\, \chi(M),$$

where $\chi(M)$ is the Euler characteristic of M. Since we are not going to swap the boundary edges of meshes with boundary, we can assume constant integral Gaussian curvature in our setting.

Minimizing F_{TP} equals the minimization of H in the L_2-norm, $\|H\|_2 = \left(\int H^2\right)^{1/2}$. Likewise, the minimization of the integral absolute mean curvature relates to the L_1-norm of H, $\|H\|_1 = \int |H|$. Besides these two energy functionals we have also used the L_1-norm of the principle curvatures, $\|\kappa\|_1 = \int |\kappa_1| + |\kappa_2|$ as an optimization criterion. Note that the minimum of $\|\kappa\|_2$ equals the minimum of $\|H\|_2$, since $\kappa_1^2 + \kappa_2^2 = 4H^2 - 2K$. Using (1), these three energy functionals are given by

$$F_1 = \|H\|_2^2 = \int_M |H|^2 = \sum_{v \in V} \frac{1}{S_v} |\bar{H}_v|^2,$$

$$F_2 = \|H\|_1 = \int_M |H| = \sum_{v \in V} |\bar{H}_v|,$$

$$F_3 = \|\kappa\|_1 = \int_M |\kappa_1| + |\kappa_2| = \sum_{v \in V} \begin{cases} 2\,|\bar{H}_v|, & \text{if } \bar{K}_v \geq 0, \\ 2\,\sqrt{|\bar{H}_v|^2 - S_v \bar{K}_v}, & \text{otherwise.} \end{cases}$$

Choosing one of these three energy functionals as a cost function F and starting with an initial triangle mesh M, we perform a local swapping algorithm that decreases the cost function in each step. The key ingredients of this algorithm are the determination of a swap value s_j for each edge e_j and the use of a priority queue P. The swap value is the difference between the value of the cost function before and after swapping the corresponding edge and indicates the reduction of the cost function caused by this edge swap. Note that the swapping operation is not defined for boundary edges, and should be forbidden for edges that connect to a vertex of valence three, since it would result in two identical edges and two triangles glued together (cf. Fig. 3). We avoid swapping those edges by simply setting their swap value to $-\infty$. The priority queue P is a permutation on the set of integers $\{1, \ldots, \#E\}$, such that $s_{P(i)} \geq s_{P(j)}$ for all $1 \leq i < j \leq \#E$. The main advantage of using such a priority queue is the low complexity in building and updating it. Furthermore, $P(1)$ is always the index of the edge with the largest swap value and testing $s_{P(1)} > 0$ tells whether the cost function can be further reduced by swapping one of the edges or not. When an edge swap is actually carried out, the swap values of the edges in the 2-neighborhood of this edge change (cf. Fig. 3) and the priority queue has to be updated. Defining the 1-neighborhood $N(I)$ of a set of indices $I \subset \{1, \ldots, \#E\}$ by

$$N(I) = \{j : \exists i \in I \text{ and } t \in T : e_i \text{ and } e_j \text{ are edges of } t\},$$

the index set of the 2-neighborhood of an edge e_i is given by $N^2(\{i\}) = N(N(\{i\}))$. With these definitions, the local swapping algorithm can be stated in the following way.

The local swapping algorithm.

```
swap_value (j)
    if e_j is swappable
        return (F_before − F_after)
    else
        return (−∞)
initialization
    for j = 1,...,#E do
        s_j := swap_value (j)
    set_up (P)
optimization
    while s_P(1) > 0 do
        swap e_P(1)
        for j ∈ N²({P(1)}) do
            s_j := swap_value (j)
            update (P)
```

As the number of all possible triangulations of the given data is finite and the cost function is decreased by each swap, this algorithm is guaranteed to terminate in a finite number of steps. Unfortunately, it is generally impossible to determine whether the algorithm reaches a global minimum or not. For the L_1-norm of the Gaussian curvature, $\|K\|_1$, Alboul and van Damme could show

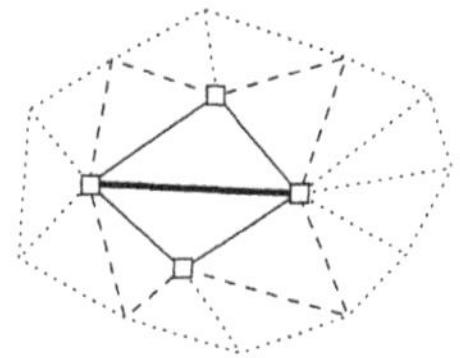
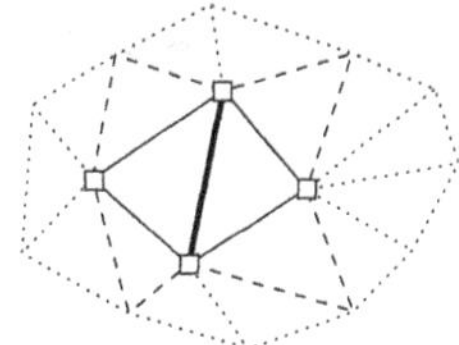
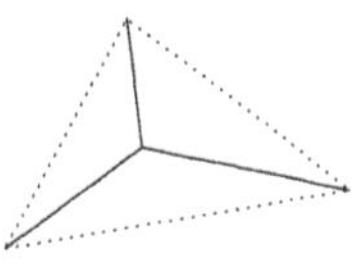

Fig. 3. Swapping an edge (left), the computation of its swap value involves only the discrete curvatures at the vertices marked by squares. After the edge swap, the swap values of the swapped edge (thick line), the edges in the 1-neighborhood (normal lines) and the edges in the 2-neighborhood (dashed lines) have to be updated. Swapping edges that connect to vertices of valence three is forbidden (right).

that in case of convex data this optimization strategy always converges to the global minimum, which is the convex triangulation [2]. We have also tested the three cost functions on convex data and observed that the convex triangulation seems to be the global optimum for F_2 and F_3. Proving this property is a subject for future investigations. Nevertheless, all three cost functions may have local minima at which the local swapping algorithm might get trapped, *i.e.*, it stops at a triangulation of the convex data that still contains concave edges.

We have also tested a slight modification of the local swapping algorithm that is guaranteed to find the same or a better local minimum. This **look-ahead strategy** determines the swap value s_j in a different way [13]. Whenever $s_j \leq 0$, it is tested whether swapping one of the four edges e_k, $k \in N(\{j\})$ in the 1-neighborhood can decrease the cost function, *i.e.*, the combined swap value $s_{j,k} = s_j + s_k$ is positive for some k. If such a **double swap** can be found, the largest $s_{j,k}$ is used as a sorting criterion for the priority queue instead of s_j and both edges e_j and e_k are swapped when e_j reaches the head of the queue, *i.e.*, $P(1) = j$.

§4. Examples

We have tested our optimization method with the various cost functions on many different data sets, ranging from small sets sampled from simple objects (*e.g.* sphere, cylinder, or torus) to complex models with several thousand vertices. The first class of data sets we investigated were convex sets of vertices, since there are many reasons (*e.g.* shape preservation or tightness [11]) to judge the convex triangulation of the vertices to be the best among all possible triangulations. In almost all cases we have tested, minimizing any of the three cost functions led to the convex triangulation, regardless of the initial triangulation. But there exist configurations for which the local as well as the look-ahead swapping algorithm terminate at a non-convex triangulation (cf. Fig. 4).

Furthermore we have tested the approximation quality of the triangle meshes generated by the swapping algorithm with our three cost functions for simple objects like the torus in Fig. 5. The initial triangle mesh consists of

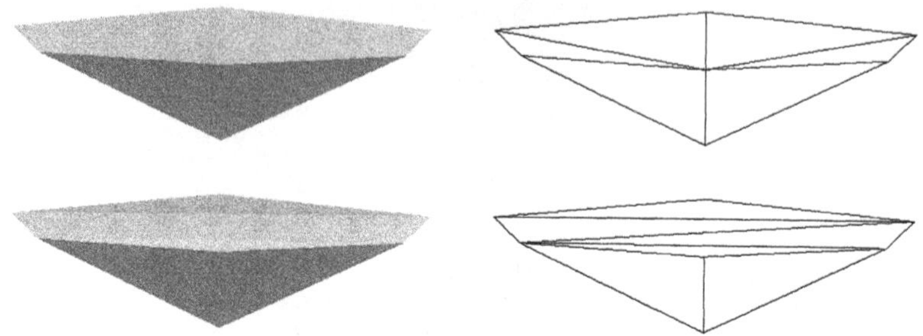

Fig. 4. A convex data set with 7 vertices $v_1 = (-12, 0, 0)$, $v_2 = (-10, -2, -\frac{1}{10})$, $v_3 = (0, -\frac{5}{2}, -\frac{1}{2})$, $v_4 = (10, -2, -\frac{1}{5})$, $v_5 = (12, 0, 0)$, $v_6 = (0, \frac{5}{2}, -\frac{1}{2})$, $v_7 = (0, 0, -10)$, for which the 10 triangles $\triangle(v_1, v_2, v_3)$, $\triangle(v_3, v_4, v_5)$, $\triangle(v_1, v_3, v_6)$, $\triangle(v_3, v_5, v_6)$, $\triangle(v_7, v_2, v_1)$, $\triangle(v_7, v_3, v_2)$, $\triangle(v_7, v_4, v_3)$, $\triangle(v_7, v_5, v_4)$, $\triangle(v_7, v_6, v_5)$, $\triangle(v_7, v_1, v_6)$ form a non-convex triangulation (top) that corresponds to a local minimum for any of the cost functions F_1, F_2, F_3. The convex triangulation of the data is shown on the bottom.

$m = 12$ by $n = 6$ data points

$$P_{ij} = \begin{pmatrix} \cos\sigma_i(R + r\cos\varphi_{ij}) \\ \sin\sigma_i(R + r\cos\varphi_{ij}) \\ r\sin\varphi_{ij} \end{pmatrix}, \quad i = 0, \ldots, m-1, \quad j = 0, \ldots, n-1,$$

$$\text{with } \sigma_i = \frac{i}{m}2\pi, \text{ and } \varphi_{ij} = \begin{cases} \frac{j}{n}2\pi, & \text{if } i \text{ is even,} \\ \frac{j+0.5}{n}2\pi, & \text{otherwise,} \end{cases}$$

sampled from the surface of a torus with radii $R = 5$, $r = 2$ and $2mn$ triangles $\triangle(P_{ij}, P_{i+1,j}, P_{i,j+1})$, $\triangle(P_{i,j+1}, P_{i+1,j}, P_{i+1,j+1})$, $i = 0, \ldots, m-1$, $j = 0, \ldots, n-1$.

Fig. 5 shows the initial triangle mesh as well as the results of the optimization process using the different cost functions, their approximation errors to the torus are listed in Tab. 1. By testing many choices of the parameters (m, n, R, r), we observed that the minimization of any of the three cost functions reduces the approximation error significantly but none of the results can be said to be superior to the others.

	L_1	L_2	L_∞
initial	0.2342743546	0.2824122092	0.7939588898
F_1	0.1581226238	0.1887044119	0.4019238949
F_2	0.1640362102	0.1931971435	0.4019238949
F_3	0.1660948138	0.1945930432	0.3892151477

Tab. 1. Approximation error of the initial and the optimized meshes.

One important aspect of our optimization process is, that it can be used as a preprocessor for other algorithms operating on triangle meshes. As an

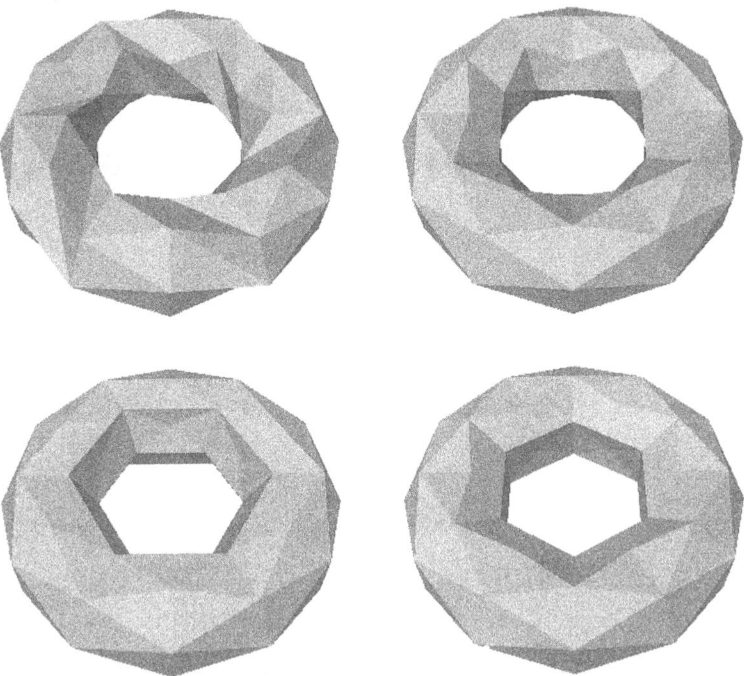

Fig. 5. Optimizing a triangle mesh sampled from a torus (top left) using different cost functions: F_1 (top right), F_2 (bottom left), and F_3 (bottom right).

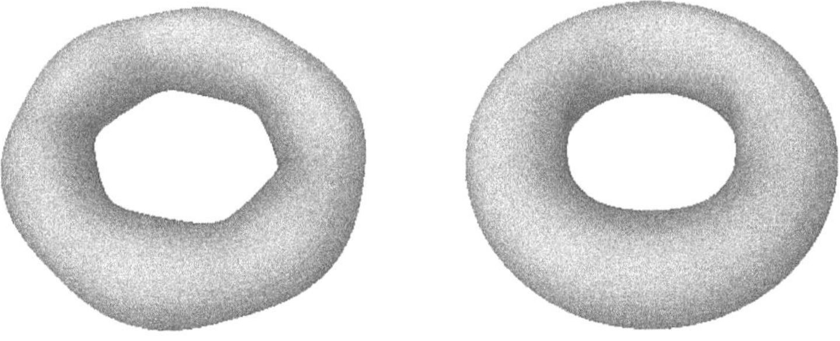

Fig. 6. The initial (left) and the optimized mesh (right) after three butterfly-subdivision steps.

example, Fig. 6 shows the result of applying three butterfly-subdivision steps to the initial triangle mesh and to the one obtained by minimizing F_1.

We have also tested our method with more complex data sets, like the Spock head in Fig. 7 with 16,386 vertices, the technical data set with 4,100 vetices shown in Fig. 8, or the data set of a tank cap with 3,374 vertices in

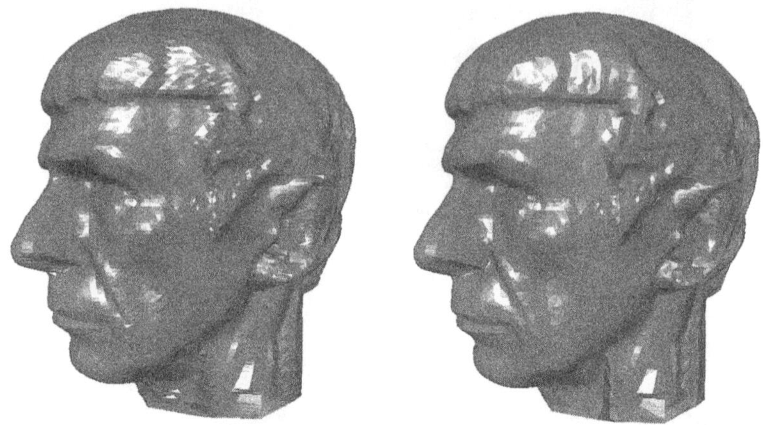

Fig. 7. Mr. Spock's head: initial (left) and optimized triangle mesh (right).

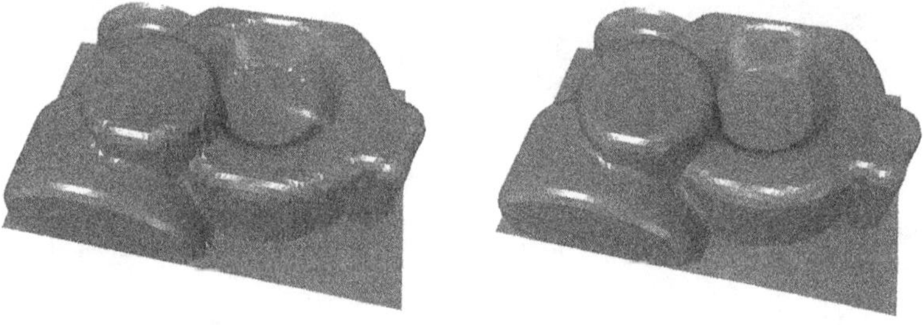

Fig. 8. A technical data set: initial (left) and optimized triangle mesh (right).

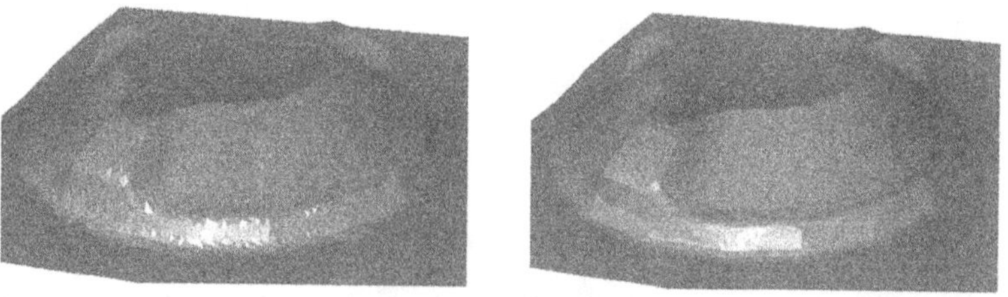

Fig. 9. Data set of a tank cap: initial (left) and optimized triangle mesh.

Fig. 9. All figures show the initial triangle mesh and the one obtained by minimizing F_2; the results of minimizing the other two cost functions are very similar. Note how the optimized meshes look much smoother and how the feature lines are enhanced in Fig. 8 and Fig. 9.

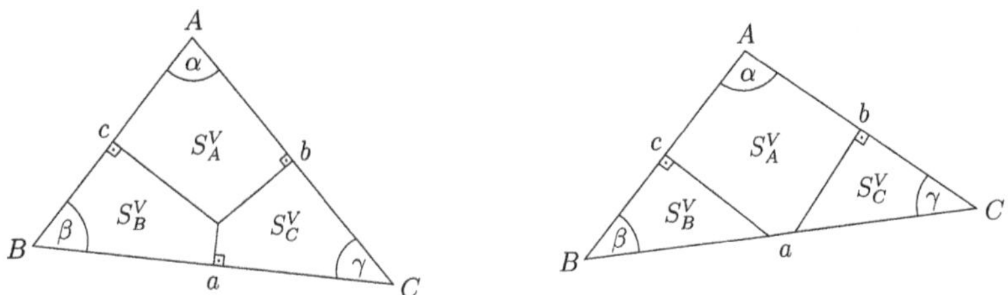

Fig. 10. Areas of the local Voronoi cells restricted to a triangle for a non-obtuse (left) and an obtuse (right) triangle.

As a concluding remark we would like to stress again that all three cost functions behaved very similar within the scope of our investigations except for convex data, and it is hard to tell which one performs best. But since the computation of the absolute mean curvature is the simplest and does not involve the calculation of the area around the vertices, and since with F_2, in the convex data case the optimal triangulation seems to be the convex one, we favor the use of F_2.

§5. Appendix

To determine the areas of the local Voronoi cells restricted to a triangle, we have to distinguish between obtuse and non-obtuse triangles (cf. Fig. 10). In the latter case they are given by

$$S_A^V = \frac{1}{8}\,(b^2 \cot(\beta) + c^2 \cot(\gamma)),$$

and likewise for S_B^V and S_C^V. For obtuse triangles,

$$S_B^V = \frac{1}{8}\,c^2 \tan(\beta), \qquad S_C^V = \frac{1}{8}\,b^2 \tan(\gamma), \qquad S_A^V = S_\Delta - S_B^V - S_C^V.$$

Acknowledgments. This work was supported in part by the European Union research project "Multiresolution in Geometric Modeling (MINGLE)" under grant HPRN-CT-1999-00117, by the Deutsche Forschungsgemeinschaft through the Sonderforschungsbereich 603 "Modellbasierte Analyse und Visualisierung komplexer Sensordaten" and by The Israel Science Foundation – Center of Excellence Program.

References

1. Alboul, L. and R. van Damme, Polyhedral metrics in surface reconstruction, in *The Mathematics of Surfaces VI*, G. Mullineux (ed.), Clarendon Press, Oxford, 1996, 171–200.

2. Alboul, L. and R. van Damme, Polyhedral metrics in surface reconstruction: tight triangulations, in *The Mathematics of Surfaces VII*, T. Goodman and R. Martin (eds.), Clarendon Press, Oxford, 1997, 309–336.
3. Baszenski, G. and L. L. Schumaker, Use of simulated annealing to construct triangular facet surfaces, in *Curves and Surfaces*, P.-J. Laurent, A. Le Méhauté, and L. L. Schumaker (eds.), Academic Press, New York, 1991, 27–32.
4. Desbrun, M., M. Meyer, P. Schröder, and A. H. Barr, Implicit fairing of irregular meshes using diffusion and curvature flow, Computer Graphics (SIGGRAPH '99 Proceedings) **33** (1999), 317–324.
5. do Carmo, M. P., *Differential Geometry of Curves and Surfaces*, Prentice-Hall Inc., New Jersey, 1976.
6. Dyn, N., D. Levin, and S. Rippa, Data dependent triangulations for piecewise linear interpolation, IMA J. Numer. Anal. **10** (1990), 137–154.
7. Kobbelt, L., Discrete fairing, in *The Mathematics of Surfaces VII*, T. Goodman and R. Martin (eds.), Clarendon Press, Oxford, 1997, 101–131.
8. Quak, E. and L. L. Schumaker, Cubic spline interpolation using data dependent triangulations, Comput. Aided Geom. Design **7** (1990), 293–301.
9. Schumaker, L. L., Computing optimal triangulations using simulated annealing, Comput. Aided Geom. Design **10** (1993), 329–345.
10. Taubin, G., A signal processing approach to fair surface design, Computer Graphics (SIGGRAPH '95 Proceedings) **29** (1995), 351–358.
11. van Damme, R. and L. Alboul, Tight triangulations, in *Mathematical Methods for Curves and Surfaces*, M. Dæhlen, T. Lyche, and L. L. Schumaker (eds.), Vanderbilt University Press, Nashville, 1995, 517–526.
12. Warren, J., private communication, 1999.
13. Yu, X., B. Morse, and T. Sederberg, Image reconstruction using data-dependent triangulations, preprint.

Nira Dyn, Sun-Jeong Kim and David Levin
Dept. of Applied Mathematics
School of Mathematical Sciences
Tel Aviv University
Tel Aviv 69978, Israel
`{niradyn,sjkim,levin}@math.tau.ac.il`

Kai Hormann
Computer Graphics Group
University of Erlangen-Nürnberg
Am Weichselgarten 9
91058 Erlangen, Germany
`hormann@informatik.uni-erlangen.de`

Refinable Spline Functions and Hermite Interpolation

Tim N. T. Goodman

Abstract. We consider r-vectors of spline functions with compact support and stable integer translates which satisfy a refinement equation with a finite number of terms. It is seen that for $r = 1$ or 2, the choice of such vectors is very limited. However for $r \geq 3$ there is sufficient flexibility, even for simple uniform knots, to allow the satisfaction of further useful properties. Mention is made of biorthogonality and orthogonality properties. Moreover it is shown that for each $r \geq 3$ such vectors of splines can be constructed which are fundamental for r-Hermite interpolation and have arbitrarily high smoothness.

§1. Introduction

Take $r \geq 1$ and let $\phi_1, \dots, \phi_r$ be real-valued functions on $\mathbb{R}$ and write $\phi = (\phi_1, \dots, \phi_r)^T$. We say that ϕ is **refinable** if it satisfies the **refinement equation**

$$\phi(x) = \sum_{j=-\infty}^{\infty} A_j \phi(2x - j), \quad x \in \mathbb{R}, \tag{1}$$

where $A_j, j \in \mathbb{Z}$, are $r \times r$ matrices.

Much work has been done in studying conditions on the matrices (A_j) that will ensure that (1) has a unique solution ϕ satisfying certain properties, e.g. smoothness, polynomial reproduction or orthogonality. The solution ϕ is then defined implicitly from the refinement equation. We here take the different approach of studying conditions on spline functions, i.e. piecewise polynomials, which ensure that they satisfy a refinement equation of form (1). In this case we know explicitly the smoothness and polynomial reproduction of ϕ, but the values of the matrices (A_j) will be defined implicitly from (1) and will not directly concern us.

We shall keep matters simple. Hence we assume that only a finite number of the matrices (A_j) in (1) are non-zero. Moreover we suppose that $\phi_1, \dots, \phi_r$

Mathematical Methods for Curves and Surfaces: Oslo 2000
Tom Lyche and Larry L. Schumaker (eds.), pp. 147–161.

ISBN 0-8265-1378-6

have compact support and their integer translates form a Riesz basis, i.e. for some constants $A, B > 0$, when

$$f = \sum_{i=1}^{r} \sum_{j=-\infty}^{\infty} a_{ij}\phi_i(. - j), \text{ for any } a_{ij} \in \mathbb{R}, \quad i = 1, \ldots, r, \quad j \in \mathbb{Z},$$

we have

$$A \sum_{i=1}^{r} \sum_{j=-\infty}^{\infty} a_{ij}^2 \leq \int_{-\infty}^{\infty} f^2 \leq B \sum_{i=1}^{r} \sum_{j=-\infty}^{\infty} a_{ij}^2. \tag{2}$$

For simplicity we shall say ϕ is stable when the above condition (2) holds. In the refinement equation (1) it would be possible to replace the dilation factor 2 by any integer $m \geq 2$, but for simplicity we consider only $m = 2$.

In Section 2 we shall review some general results, which show in particular that for $r = 1$ and 2 there is little flexibility in the choice of refinable spline functions as above. However, for $r \geq 3$ there is in contrast a great deal of flexibility, and this can be used to construct refinable spline functions with properties which are useful for the two main applications of refinable functions. The first of these concerns wavelets, for which it is usual to require some orthogonality properties of the refinable functions. At the end of Section 2 we recall briefly the construction for $r = 3$ of refinable spline functions which are biorthogonal to B-splines with simple uniform knots. The B-splines here have arbitrary degree $n \geq 1$ and simple knots at $\frac{1}{3}\mathbb{Z}$, while their dual functions have the same degree and simple knots in $\frac{1}{4}\mathbb{Z}$. We also mention the construction of orthogonal refinable spline functions, again for $r = 3$, arbitrary degree $n \geq 1$ and simple knots in $\frac{1}{4}\mathbb{Z}$, the details of which will appear in a later paper.

The other main application of refinable functions concerns subdivision schemes. In Section 3, which is all new work, we construct refinable spline functions ϕ which are fundamental for r-Hermite interpolation, i.e. we have

$$\phi_i^{(j-1)}(k) = \delta_{ij}\delta_{ko}, \quad i, j = 1, \ldots, r, \quad k \in \mathbb{Z}. \tag{3}$$

We show that for each $r \geq 3$, it is possible to construct such fundamental refinable splines for arbitrary degree $n \geq r+1$ and simple uniform knots, thus giving smoothness C^{n-1}. These give rise to Hermite interpolatory subdivision schemes, i.e. schemes in which we are given the values of a function and its derivatives up to order $r - 1$ at the integers and the subdivision successively fills in the values of the function and its derivatives up to order $r - 1$ at the dyadic points $2^{-j}k, k \in \mathbb{Z}, j = 1, 2, \ldots$, of the unique refinable function which interpolates the data. As far as we know, these are the only interpolatory subdivision schemes which, for $r \geq 2$, are known to give arbitrarily high smoothness of the interpolating functions.

§2. Some Generalities

For the case $r = 1$ of a single refinable spline function, it was shown by Lawton, Lee and Shen that the choice is very restricted indeed.

Theorem 1. [10] *A function ϕ is a stable refinable spline function (of compact support satisfying a refinement equation with a finite number of terms) if and only if ϕ is a constant multiple of a B-spline with simple knots at the integers.*

Before analysing the case $r = 2$, we consider some general results. We say a vector $\phi = (\phi_1, \ldots, \phi_r)^T$ **generates** a space $\mathcal{S}$ if $\mathcal{S}$ comprises all finite linear combinations of integer translates of elements of ϕ. Thus $\psi_1, \ldots, \psi_r$ lie in $\mathcal{S}$ if and only if for $\psi = (\psi_1, \ldots, \psi_r)^T$,

$$\psi = \sum_{i=1}^{r} \sum_{k=-\infty}^{\infty} A_k \phi(. - k), \tag{4}$$

for a finite collection (A_k) of $r \times r$ matrices. Equation (4) can be expressed more neatly by introducing the Fourier transform

$$\hat{f}(u) := \int_{-\infty}^{\infty} e^{-iux} f(x) dx, \quad u \in \mathbb{R}.$$

Then taking Fourier transforms of (4) gives

$$\hat{\psi}(u) = A(e^{-iu}) \hat{\phi}(u), \quad u \in \mathbb{R}, \tag{5}$$

where A denotes the $r \times r$ matrix of Laurent polynomials

$$A(z) := \sum_{k=-\infty}^{\infty} A_k z^k, \quad z = e^{-iu}, \quad u \in \mathbb{R}. \tag{6}$$

We shall say a $r \times r$ matrix A of Laurent polynomials, as in (6), is **invertible** if it has an inverse A^{-1} which is also a matrix of Laurent polynomials. It is easily seen that A is invertible if and only if its determinant $\det A$ is a non-trivial monomial.

The Fourier transform also gives an elegant condition for the stability of ϕ. It is shown by Jia and Micchelli [9] that $\phi = (\phi_1, \ldots, \phi_r)^T$ is stable if and only if for each u in $\mathbb{R}$, there are integers $k_1, \ldots, k_r$ with

$$\det[\hat{\phi}_i(u + 2\pi k_j)]_{i,j=1}^{r} \neq 0. \tag{7}$$

From the above discussion we can easily deduce the following.

Lemma 2. *Suppose ϕ is stable and generates $\mathcal{S}$. Then ψ is stable and also generates $\mathcal{S}$ if and only if*

$$\hat{\psi}(u) = A(e^{-iu}) \hat{\phi}(u), \quad u \in \mathbb{R},$$

for an invertible matrix A.

We shall call ϕ and ψ **equivalent** if they satisfy the conditions of Lemma 2. Clearly if ϕ and ψ are equivalent, then they have the same number of components.

Now suppose that ϕ is refinable and generates $\mathcal{S}$. Then any element of $\mathcal{S}$ is a finite linear combination of integer translates of the elements of ϕ, and hence by (1), is a finite linear combination of $\{\phi_i(2.-j) : i = 1, \ldots, r, j \in \mathbb{Z}\}$. Thus $f(\frac{\cdot}{2})$ is a finite linear combination of integer translates of the elements of ϕ, and hence lies in $\mathcal{S}$. Conversely, suppose that ϕ generates $\mathcal{S}$ and

$$f \in \mathcal{S} \Rightarrow f(\frac{\cdot}{2}) \in \mathcal{S}. \tag{8}$$

Since the components of ϕ lie in $\mathcal{S}$, we know that the components of $\phi(\frac{\cdot}{2})$ lie in $\mathcal{S}$. Since ϕ generates $\mathcal{S}$, we know that the components of $\phi(\frac{\cdot}{2})$ are finite linear combinations of integer translates of the elements of ϕ, and so ϕ satisfies an equation of form (1), i.e. ϕ is refinable. Refinability is thus a property of a space $\mathcal{S}$ rather than one particular generator ϕ, and it makes sense to call a space $\mathcal{S}$ refinable when (8) holds. We have seen that if ϕ generates $\mathcal{S}$, then ϕ is refinable if and only if $\mathcal{S}$ is refinable.

Next we shall consider refinable spaces of spline functions. Although our applications in Section 3 and at the end of this Section generally require only splines with simple knots, we shall give here a very general definition of spline spaces which allow discontinuities in non-consecutive derivatives. Take $n \geq 1$ and let S_n be a finite subset of $[0, 1)$. For $k = 0, 1, \ldots, n-1$, we let S_k be a subset of S_n. We then define $\mathcal{S}$ to be the space of all piecewise polynomials f of degree n with compact support in $\mathbb{R}$ whose only discontinuities of derivatives are in $f^{(k)}$ at $S_k + \mathbb{Z}$.

For example, suppose that for some μ, $1 \leq \mu \leq n+1$,

$$S_k = S_n, \quad n - \mu + 1 \leq k \leq n,$$

$$S_k = \emptyset, \quad 0 \leq k \leq n - \mu.$$

Then $\mathcal{S}$ comprises all spline functions f of degree n with compact support and discontinuities in $f^{(n-\mu+1)}, \ldots, f^{(n)}$ at $S_n + \mathbb{Z}$, i.e. knots of multiplicity μ at $S_n + \mathbb{Z}$.

Note that for $f \in \mathcal{S}$, f has a discontinuity in a derivative at α if and only if $f(\frac{\cdot}{2})$ has a similar discontinuity at 2α. Thus from (8), $\mathcal{S}$ is refinable if and only if

$$\alpha \in S_k \Rightarrow 2\alpha (\mathrm{mod}\mathbb{Z}) \in S_k, \quad k = 0, \ldots, n. \tag{9}$$

For a detailed analysis of refinable spline spaces, see [5], from which the next result derives.

Theorem 3. *Let $\mathcal{S}$ be a refinable spline space as above with a total of r knots (mod$\mathbb{Z}$), i.e. $\sum_{k=0}^{n} |S_k| = r$. Let $\phi_1, \ldots, \phi_r$ be elements of $\mathcal{S}$ whose integer translates are linearly independent. Then $(\phi_1, \ldots, \phi_r)$ generates $\mathcal{S}$.*

We note that linear independence is stronger than stability and we shall see later that, in general, the condition of linear independence in Theorem 3 cannot be relaxed to stability. We see from Theorem 3 and Lemma 2 that if $\phi = (\phi_1, \ldots, \phi_r)^T$ and $\psi = (\psi_1, \ldots, \psi_r)^T$ both lie in $\mathcal{S}$ and have linearly independent integer translates, then ϕ and ψ are equivalent. In [5] a canonical choice is given for a vector $\phi = (\phi_1, \ldots, \phi_r)^T$ generating $\mathcal{S}$, which reduces to the usual B-splines when the discontinuities in derivatives are consecutive.

In general for $\mathcal{S}$ as in Theorem 3, there may be refinable vectors $\phi = (\phi_1, \ldots, \phi_s)$ in $\mathcal{S}$ which are stable and for which either $s < r$ or the integer translates are linearly dependent. In this case, ϕ will not generate $\mathcal{S}$. However the following result is a consequence of results in [6].

Theorem 4. *Let $\mathcal{S}$ be a refinable spline space for which the elements of $\mathcal{S}$ all have odd denominator. If $\phi_1, \ldots, \phi_s$ lie in $\mathcal{S}$ and $\phi = (\phi_1, \ldots, \phi_s)$ is refinable and stable, then $s = r$ and the integer translates of ϕ are linearly independent.*

So from Theorems 3 and 4 we see that a refinable stable vector of splines in a space with knots having odd denominators is unique up to equivalence,and thus can be considered as the canonical choice in [5].

We shall now analyse all pairs of stable refinable spline functions $\phi = (\phi_1, \phi_2)$. It follows from the work of [6] that ϕ lies in a space $\mathcal{S}$ with 2 knots (mod$\mathbb{Z}$), and it is easily seen from (9) that there are only three cases, as below.

Case 1. *For some $0 \le k < n$, $S_k = S_n = \{0\}$.*

Here the only discontinuities are in $f^{(n)}(j)$ and $f^{(k)}(j)$ for $j \in \mathbb{Z}$. Since 0 can be taken to have odd denominator, ϕ is equivalent to the canonical refinable pair in [5]. In particular, for $k = n-1$, ϕ comprises B-splines with double knots at the integers.

Case 2. $S_n = \{\frac{1}{3}, \frac{2}{3}\}$.

Here ϕ is equivalent to B-splines with simple knots at $\frac{1}{3} + \mathbb{Z}$ and $\frac{2}{3} + \mathbb{Z}$.

Case 3. $S_n = \{0, \frac{1}{2}\}$.

It is shown in [6] that if the integer translates of ϕ are linearly independent, then ϕ is equivalent to B-splines with simple knots at $\frac{1}{2}\mathbb{Z}$. To be more precise, let N denote a B-spline of degree n with simple knots at $0, 1, \ldots, n+1$. Then ϕ is equivalent to $(N(2.), N(2.-1))^T$. Since

$$\hat{N}(u) = \left(\frac{1-e^{-iu}}{iu}\right)^{n+1}, \quad u \in \mathbb{R},$$

we have for $z = e^{-\frac{iu}{2}}$,

$$\hat{\phi}(u) = A(z^2)\begin{bmatrix}1\\ z\end{bmatrix}\left(\frac{1-z}{iu}\right)^{n+1}, \quad u \in \mathbb{R},$$

for an invertible matrix A of Laurent polynomials. This can be rewritten as

$$\hat{\phi}(u) = \begin{bmatrix}a(z)\\ b(z)\end{bmatrix}\left(\frac{1-z}{iu}\right)^{n+1}, \quad u \in \mathbb{R}, \tag{10}$$

where a,b are Laurent polynomials with

$$A(z^2) = \frac{1}{2}\begin{bmatrix} a(z)+a(-z) & z^{-1}(a(z)-a(-z)) \\ b(z)+b(-z) & z^{-1}(b(z)-b(-z)) \end{bmatrix}.$$

Writing

$$M(z) := \begin{bmatrix} a(z) & a(-z) \\ b(z) & b(-z) \end{bmatrix},$$

the condition that A is invertible is equivalent to M being invertible, i.e.

$$\det M(z) = cz^{2l+1}, \text{ some } l \in \mathbb{Z} \text{ and } c \neq 0.$$

If we merely assume that ϕ is stable, then the situation is more complex. It is shown in [6] that ϕ satisfies (10), where a and b satisfy the weaker conditions that $\det M(z)$ is non-zero for $|z| = 1$ and divides both components of

$$M(z)\begin{bmatrix} (1-z)^{n+1} \\ -(1+z)^{n+1} \end{bmatrix}.$$

Thus, even if we allow linear dependencies, the choice of ϕ for $r = 1$ or 2 is very limited. The situation is radically different for $r \geq 3$, provided that we have some knots without odd denominators, i.e. in $\frac{1}{2}\mathbb{Z}$. We illustrate the resulting flexibility with a simple case for $r = 3$.

As before we let N denote a B-spline of degree $n \geq 1$ with simple knots at $0, 1, \ldots, n+1$. Let $\phi_1 = N(2.)$, $\phi_2 = N(2.-1)$. We let $\mathcal{S}$ denote the space of splines of degree n with simple knots in $\frac{1}{4}\mathbb{Z}$, i.e. $S_n = \{0, \frac{1}{4}, \frac{1}{2}, \frac{3}{4}\}$ and $S_k = \emptyset$ for $k < n$. Then we take $\phi = (\phi_1, \phi_2, \phi_3)$, where ϕ_3 is an arbitrary element of $\mathcal{S}$ which has a discontinuity in $\phi_3^{(n)}$ at some point not in $\frac{1}{2}\mathbb{Z}$. Now $\{\phi_i(2.-j) : i = 1, 2, j \in \mathbb{Z}\}$ spans $\mathcal{S}$ and hence ϕ is refinable. Here ϕ generates a proper subspace of $\mathcal{S}$. Thus we have a large choice of refinable vectors of splines and in general their integer translates will be linearly independent.

In [4] it was shown that for any $n \geq 1$, $\psi = (\psi_1, \psi_2, \psi_3)$ could be chosen equivalent to ϕ as above so that ψ is biorthogonal to B-splines of degree n with simple knots in $\frac{1}{3}\mathbb{Z}$. To be precise, let N be a B-spline of degree n with simple knots at $0, 1, \ldots, n+1$. Let $\eta_j = N(3.-j), j = 1, 2, 3$. Then

$$\int_{-\infty}^{\infty} \psi_i \eta_j(.-k) = \delta_{ij}\delta_{ko}, \quad i, j = 1, 2, 3, \quad k \in \mathbb{Z}.$$

In a forthcoming paper it will be shown that for $n \geq 1$ we can construct ψ equivalent to ϕ as above so that ψ is orthogonal, i.e.

$$\int_{-\infty}^{\infty} \psi_i \psi_j(.-k) = \delta_{ij}\delta_{ko}.$$

This is proved, as in Daubechies' famous construction [1], by using the Riesz Lemma to take a square root of a positive solution of a Bezout's equation. That the solution is positive is guaranteed by a lemma of Micchelli and the author in [7]. The proof also depends on a very recent result on the factorisation of matrices of Laurent polynomials due to Hardin [8]. We note that in [2] it has already been shown that the orthogonal refinable spline functions with compact support of arbitrary smoothness can be constructed. However the nice general theory needs individual consideration for explicit cases and the cases considered are limited, e.g. C^0 linear (which coincides with ours), C^1 cubic for $r = 6$ and C^2 of degree 11 for $r = 10$, [2,3]. For our case, for arbitrary n, we have C^{n-1} splines of degree n and $r = 3$.

In the next section we extend the construction of ϕ above to general $r \geq 3$ to allow construction of refinable spline functions of arbitrary degree which are fundamental for r-Hermite interpolation.

§3. Interpolation

Recall that $\phi = (\phi_1, \dots, \phi_r)$ is fundamental for r-Hermite interpolation if equation (3) holds. Thus a function of form

$$s = \sum_{i=1}^{r} \sum_{k=-\infty}^{\infty} a_{ik}\phi_i(\cdot - k)$$

satisfies the interpolation conditions

$$s^{(j-1)}(k) = a_{jk}, \quad j = 1, \dots, r, \quad k \in \mathbb{Z}.$$

Clearly if ϕ is fundamental, then the integer translates of $\phi_1, \dots, \phi_r$ are linearly independent. We are here assuming that the interpolation conditions are on consecutive derivatives. Similarly we shall consider spline spaces $\mathcal{S}$ for which the discontinuities are on consecutive derivatives. In our previous notation this means that if $t \in S_k$ for some k, $0 \leq k \leq n$, then $t \in S_j$ for $k \leq j \leq n$. However, it is simpler to use the customary terminology that t is a knot of multiplicity μ if for f in $\mathcal{S}$, discontinuities are allowed only in $f^{(j)}$, $n - \mu + 1 \leq j \leq n$.

Our next result shows that in the circumstances of Theorem 3, there is little choice for fundamental functions and, in particular, they must have smoothness at most C^{r-1}. The result holds even without the assumption of refinability.

Theorem 5. *Let $\mathcal{S}$ denote a space of spline functions (of compact support) of degree $n \geq 1$ which has a total of r knots $(\mathrm{mod}\,\mathbb{Z})$, and suppose that 0 is a knot of multiplicity μ, $0 \leq \mu \leq r$. Then $\mathcal{S}$ contains a vector $\phi = (\phi_1, \dots, \phi_r)$ of fundamental functions for r-Hermite interpolation if and only if $n = r + \mu - 1$. In this case ϕ is unique and has support in $[-1, 1]$.*

Proof: Suppose that $\mathcal{S}$ does contain such fundamental functions ϕ. We may assume that they all have support in $[-M, M]$ for an integer $M \geq 1$. Let $\mathcal{J} = \{f \in \mathcal{S} : f = 0 \text{ outside } [-M, M]\}$ and let t_j, $j = 1, \dots, 2Mr + \mu$, denote in increasing order the knots of $\mathcal{S}$ in $[-M, M]$, counted with multiplicity. Then

$\mathcal{J}$ has a basis $N_1, \ldots, N_d$, where $d = 2Mr+\mu-n-1$ and N_i denotes a B-spline with knots $t_i, \ldots, t_{i+n+1}$.

Let x_i, $i = 1, \ldots, (2M-1)r-1$, denote in increasing order the integers $-M+1, \ldots, M-1$, where 0 has multiplicity $r-1$ and the other integers each have multiplicity r. We shall suppose that $n \geq r+\mu$ and reach a contradiction. Now N_1 has support $[t_1, t_{n+2}]$ and since $t_{n+2} \geq t_{r+\mu+2} > -M+1$, we see that $-M+1$ lies in the interior of the support of N_1. Indeed $-M+1$ lies in the interior of the support of N_i for $i = 1, \ldots, r$. The support of N_r if $[t_r, t_{r+n+1}]$ and since $t_{r+n+1} \geq t_{2r+\mu+1} > -M+2$, the interior of the support of N_r also contains $-M+2$. More generally, $-M+j+1$ lies in the interior of the support of N_{jr+i} for $j \geq 1$, $0 \leq i \leq r$, $jr+i \leq d$. Since $d = 2Mr+\mu-n-1 \leq (2M-1)r-1$, we see that x_i lies in the interior of the support of N_i for $i = 1, \ldots, d$. Now the fundamental function ϕ_r satisfies

$$\phi_r^{(j-1)}(k) = 0, \quad j = 1, \ldots, r, \quad k \neq 0,$$

$$\phi_r^{(j-1)}(0) = 0, \quad j = 1, \ldots, r-1,$$

and hence vanishes at $x_1, \ldots, x_d$ (with multiplicity). Since ϕ_r lies in $\mathcal{J}$, the Schoenberg-Whitney Theorem [11] implies that $\phi_r = 0$, which contradicts $\phi_r^{(r-1)}(0) = 1$. Thus, we have $n \leq r+\mu-1$.

Now suppose that $\mu = 0$. Then, from above, $n \leq r-1$. If $n \leq r-2$, then $f^{(r-1)} = 0$ for all f in $\mathcal{J}$ which contradicts $\phi_r^{(r-1)}(0) = 1$. Hence $n = r-1$.

Next suppose that $\mu \geq 1$. We assume $n \leq r+\mu-2$ and reach a contradiction. For r-Hermite interpolation to make sense, ϕ must be C^{r-1} at $\mathbb{Z}$ and hence $C^{n-\mu+1}$ at $\mathbb{Z}$. Now for any f in $\mathcal{S}$, let

$$g = \sum_{i=1}^{r} \sum_{k=-\infty}^{\infty} f^{(i-1)}(k)\phi_i(. - k).$$

Then $g-f$ has compact support and satisfies $(g-f)^{(j-1)}(k) = 0$, $j = 1, \ldots, r$, $k \in \mathbb{Z}$, and it follows from the Schoenberg-Whitney Theorem that $g-f = 0$. Thus f is $C^{n-\mu+1}$ at $\mathbb{Z}$, which contradicts $\mathcal{S}$ comprising spline functions with knots of multiplicity μ at $\mathbb{Z}$. Thus $n = r+\mu-1$.

Thus we have shown that if $\mathcal{S}$ contains fundamental functions ϕ, then $n = r+\mu-1$. Now suppose $n = r+\mu-1$. Take any integer $M \geq 1$ and define $\mathcal{J}$ as before. We have seen that $\mathcal{J}$ has a basis of B-splines $N_1, \ldots, N_d$, where $d = (2M-1)r$. Let $x_1, \ldots, x_d$ denote in increasing order the integers $-M+1, \ldots, M-1$, each with multiplicity r. Then for $i = 1, \ldots, d$, x_i lies in the interior of the support of N_i. So by the Schoenberg-Whitney Theorem, $\mathcal{J}$ contains unique fundamental functions ϕ for r-Hermite interpolation. Since this holds in particular for $M = 1$, ϕ has support in $[-1, 1]$. □

Corollary 6. *In the situation of Theorem 5, ϕ has smoothness at most C^{r-1}, and smoothness C^{r-1} is attained if and only if the maximal multiplicity of the knots of $\mathcal{S}$ occurs at $\mathbb{Z}$.*

Proof: If $\mu = 0$, then $n = r - 1$ and so the smoothness of ϕ is at most C^{r-2}. If $\mu \geq 1$, then $n - \mu = r - 1$ and so the smoothness of ϕ is C^{r-1} at $\mathbb{Z}$. For ϕ to be C^{r-1} everywhere, the multiplicity of all knots must be at most μ. □

Now suppose that ϕ, as in Theorem 5, is a vector of fundamental functions for r-Hermite interpolation which is also refinable. First take $r = 1$, i.e. Lagrange interpolation. Then ϕ lies in a space $\mathcal{S}$ with only one knot (mod $\mathbb{Z}$), i.e. simple knots at the integers. So $\mu = 1$ and from Theorem 5, $n = 1$. The function $\phi = \phi_1$, is the linear B-spline with knots $-1, 0, 1$, i.e. the 'hat function' which is even, has support on $[-1, 1]$ and satisfies

$$\phi(x) = 1 - x, \quad 0 \leq x \leq 1.$$

Next take $r = 2$. As we saw at the end of Section 2, there are three cases. As we are assuming discontinuities only of consecutive derivatives, Case 1 comprises double knots at the integers. So $\mu = 2$ and from Theorem 5, $n = 3$. The functions ϕ_1, ϕ_2 have support on $[-1, 1]$, are respectively even and odd, and satisfy

$$\phi_1(x) = (2x + 1)(x - 1)^2, \quad 0 \leq x \leq 1,$$

$$\phi_2(x) = x(x - 1)^2, \quad 0 \leq x \leq 1.$$

For Case 2 we have simple knots at $\frac{1}{3} + \mathbb{Z}$ and $\frac{2}{3} + \mathbb{Z}$. So $\mu = 0$ and by Theorem 5, $n = 1$. The functions ϕ_1, ϕ_2 have support on $[-\frac{2}{3}, \frac{2}{3}]$, are respectively even and odd, and satisfy

$$\phi_1(x) = \begin{cases} 1, & 0 \leq x \leq \frac{1}{3}, \\ 2 - 3x, & \frac{1}{3} < x \leq \frac{2}{3}, \end{cases}$$

$$\phi_2(x) = \begin{cases} x, 0 \leq x \leq \frac{1}{3}, \\ \frac{2}{3} - x, & \frac{1}{3} < x \leq \frac{2}{3}. \end{cases}$$

Finally in Case 3 there are simple knots at $\frac{1}{2}\mathbb{Z}$. So $\mu = 1$ and Theorem 5 gives $n = 2$. The functions ϕ_1, ϕ_2 have support on $[-1, 1]$, are respectively even and odd, and satisfy

$$\phi_1(x) = \begin{cases} 1 - 2x^2, & 0 \leq x \leq \frac{1}{2}, \\ 2(x - 1)^2, & \frac{1}{2} < x \leq 1, \end{cases}$$

$$\phi_2(x) = \begin{cases} x(1 - \frac{3}{2}x), & 0 \leq x \leq \frac{1}{2}, \\ \frac{1}{2}(x - 1)^2, & \frac{1}{2} < x \leq 1. \end{cases}$$

So the choice of fundamental refinable functions for r-Hermite interpolation is very limited for $r = 1$ and $r = 2$, and in particular the smoothness is respectively C^0 and either C^0 or C^1. The situation for $r \geq 3$ is radically different: for any such r there are fundamental refinable functions of arbitrarily high smoothness.

Theorem 7. *Take integers $r \geq 3$, $n \geq r+1$, and let $\mathcal{S}$ denote the space of spline functions of degree n with simple knots at $\frac{1}{2r-2}\mathbb{Z}$. Let $n-2 = (r-2)p-q, p, q \in \mathbb{Z}, 0 \leq q \leq r-3$, and put $L = n-r+p-\frac{q}{2r-2}$. Then for any integer J with $J+1 \leq 0 \leq J+L-1$, $\mathcal{S}$ contains a unique vector of fundamental functions $\phi = (\phi_1, \ldots, \phi_r)$ for r-Hermite interpolation with support $[J, J+L]$ such that the space generated by ϕ includes all spline functions (of compact support) of degree n with simple knots at $\frac{1}{r-1}\mathbb{Z}$.*

Before proving Theorem 7 we make some remarks. Since ϕ as above generates splines of degree n with simple knots at $\frac{1}{r-1}\mathbb{Z}$, we see that the span of $\{\phi_i(2.-j) : i = 1, \ldots, r, j \in \mathbb{Z}\}$ includes $\phi_1, \ldots, \phi_r$, and hence ϕ is refinable.

The definition of the support length L of ϕ is somewhat obscure, and so we make some further comments. It is easily seen that

$$L \geq n - r + \frac{n-2}{r-2} \geq 1 + \frac{r-1}{r-2} > 2.$$

When $n = r+1$, then $p = 2$ and $q = r-3$ and so $L = 3 - \frac{r-3}{2r-2}$, which approaches $\frac{5}{2}$ for large r. When $r = 3$, then $p = n-2$ and $q = 0$, and so $L = 2n-5$. For any fixed r, $\lim_{n\to\infty} \frac{L}{n} = \frac{r-1}{r-2}$.

Proof of Theorem 7. *For any integer $l \geq 0$, define*

$$V_l := \{f \in \mathcal{S} : \text{supp} f \subset [0, \frac{l}{2r-2}], \quad f^{(j)}(k) = 0, \quad k \in \mathbb{Z}, \quad j = 0, \ldots, r-1\}.$$

From the Schoenberg-Whitney Theorem we can deduce the following:

$$\dim V_{j+n+(p-2)r} = j, \quad j = 1, \ldots, q,$$

$$\dim V_{j+n+(p-1)r} = j, \quad j = q+1, \ldots, r-2.$$

Thus we may define functions $\psi_1, \ldots, \psi_{r-2}$ such that

$$\text{span}\{\psi_1, \ldots, \psi_j\} = \begin{cases} V_{j+n+(p-2)r}, & j = 1, \ldots, q, \\ V_{j+n+(p-1)r}, & j = q+1, \ldots, r-2. \end{cases}$$

We shall extend the definition of ψ_j, $j \in \mathbb{Z}$, by

$$\psi_{j+r-2} = \psi_j(.-1), \quad j \in \mathbb{Z}. \tag{11}$$

Since $\text{span}\{\psi_1, \ldots, \psi_{r-2}\} = V_{p(2r-2)-q} = V_{(2r-2)(L-n+r)}$, we have for $d = (r-2)(n-r+1)$,

$$\text{span}\{\psi_1, \ldots, \psi_d\} = V_{(2r-2)L}.$$

So defining

$$V := \{f \in \mathcal{S} : \text{supp} f \subset [J, J+L], \quad f^{(j)}(k) = 0, \quad k \in \mathbb{Z}, \quad j = 0, \ldots, r-1\},$$

we see that

$$V = \mathrm{span}\{\psi_{1+J(2r-2)}, \ldots, \psi_{d+J(2r-2)}\}.$$

By the Schoenberg-Whitney Theorem, we can choose fundamental functions in $\mathcal{S}$ with support in $[J, J+L]$, and for any choice $\eta = (\eta_1, \ldots, \eta_r)$ the general form of such functions is

$$\phi_i = \eta_i + \sum_{j=1}^{d} M_{ij}\psi_{j+J(2r-2)}, \quad i = 1, \ldots, r, \tag{12}$$

where M is any constant $r \times d$ matrix.

For $j \in \mathbb{Z}$, let N_j denotes the B-spline of degree n with simple knots at $\frac{j+i}{r-1}, i = -1, 0, \ldots, n$. We shall show that there is a unique choice of the matrix M in (12) such that the space generated by ϕ contains $N_1, \ldots, N_{r-1}$. Since $N_i(.-1) = N_{i+r-1}, i \in \mathbb{Z}$, the space generated by ϕ will then include all spline functions of degree n with simple knots at $\frac{1}{r-1}\mathbb{Z}$, and hence Theorem 7 will be proved. Let $n = (r-1)s + t, \quad s, t \in \mathbb{Z}, \quad 0 \le t \le r-2$. Note that

$$0 \le \frac{j-1}{r-1} < 1, \quad \frac{j+n}{r-1} > s + \frac{t+1}{r-1} > s, \quad j = 1, \ldots, r-1,$$

$$\frac{j+n}{r-1} \le s+1, \quad j = 1, \ldots, r-t-1,$$

$$s+1 < \frac{j+n}{r-1} < s+2, \quad j = r-t, \ldots, r-1.$$

For $j = 1, \ldots, r-1$, we let f_j denote the linear combination of $\phi_i(.-k), i = 1, \ldots, r, k \in \mathbb{Z}$, which is the r-Hermite interpolant of N_j. To be precise, if

$$\lambda(j,k,l) = N_j^{(k-1)}(l), \quad j = 1, \ldots, r-1, \quad k = 1, \ldots, r, \quad l \in \mathbb{Z}, \tag{13}$$

then

$$f_j = \sum_{k=1}^{r}\sum_{l=1}^{s_j} \lambda(j,k,l)\phi_k(.-l), \quad j = 1, \ldots, r-1, \tag{14}$$

where

$$s_j = \begin{cases} s, & j = 1, \ldots, r-t-1, \\ s+1, & j = r-t, \ldots, r-1. \end{cases}$$

We note that

$$\mathrm{supp}(f_j - N_j) \subset [J+1, J+s_j+L], \quad j = 1, \ldots, r-1.$$

So recalling (11) we can write

$$f_j - N_j = \sum_{i=1}^{m_j} a(j,i)\psi_{i+J(2r-2)+r-2}, \quad j = 1, \ldots, r-1, \tag{15}$$

for some constants $a(j,i)$, where $m_j = d + (s_j - 1)(r-2)$. The equations

$$f_j - N_j = 0, \quad j = 1, \ldots, r-1,$$

are equivalent to

$$a(j,i) = 0, \quad i = 1, \ldots, m_j, \quad j = 1, \ldots, r-1,$$

which give a total of $(r-1)[d + (s-1)(r-2)] + t(r-2) = (r-1)d + (r-2)(n-r+1) = rd$ equations in the rd entries of the matrix M.

From (11),(12),(14) and (15) the corresponding homogeneous equations are

$$\sum_{k=1}^{r} \sum_{l=1}^{s_j} \lambda(j,k,l) \sum_{m=1}^{d} M_{km} \psi_{m+J(2r-2)+l(r-2)} = 0, \quad j = 1, \ldots, r-1,$$

or

$$\sum_{m=1}^{m_j} \psi_{m+J(2r-2)+r-2} \sum_{l=1}^{s_j} \sum_{k=1}^{r} \lambda(j,k,l) M_{k,m-(l-1)(r-2)} = 0, \quad j = 1, \ldots, r-1,$$

which is equivalent to

$$\sum_{l=1}^{s_j} \sum_{k=1}^{r} \lambda(j,k,l) M_{k,m-(l-1)(r-2)} = 0, \quad m = 1, \ldots, m_j, \quad j = 1, \ldots, r-1, \tag{16}$$

where we have put $M_{k\nu} = 0$ unless $1 \le \nu \le d$. Putting

$$m - (l-1)(r-2) = \alpha(r-2) + i, \quad i = 1, \ldots, r-2,$$

gives $m = i + \beta(r-2)$, where $\beta = \alpha + l - 1$. Hence (16) splits into $r-2$ systems of equations: for $i = 1, \ldots, r-2$,

$$\sum_{\alpha=0}^{n-r} \sum_{k=1}^{r} \lambda(j,k,\beta-\alpha+1) M_{k,\alpha(r-2)+i} = 0,$$
$$\beta = 0, \ldots, n-r+s_j-1, \quad j = 1, \ldots, r-1.$$

For $i = 1, \ldots, r-2$, the matrix of this system is square of order $r(n-r+1)$ which is independent of i, given by

$$A_{(j,\beta),(k,\alpha)} = \lambda(j,k,\beta-\alpha+1) = N_j^{(k-1)}(\beta-\alpha+1),$$

on recalling (13). Now this is the collocation matrix for interpolating $f^{(k-1)}(l)$, $k = 1, \ldots, r, l = 0, -1, \ldots, -n+r$, by the B-splines $N_j, j = 0, -1, \ldots, 1-r(n-r+1)$. By the Schoenberg-Whitney Theorem, this is non-singular. Thus our system of equations for the entries of M has a unique solution and the proof is complete. □

It is easily seen that L as in Theorem 7 is an even integer if and only if r is even and $n = (r-2)p+2$, where p is even. In this case for $J = -\frac{1}{2}L$, ϕ is symmetric, i.e. ϕ_j is even or odd as j is odd or even respectively. In all other cases ϕ is not symmetric. However, we can define a symmetric vector $\tilde{\phi}$ of fundamental functions by

$$\tilde{\phi} = \frac{1}{2}(\phi_j + (-1)^{j-1}\phi_j(-.)), \quad j = 1, \ldots, r.$$

Now let f by any spline function of degree n with simple knots in $\frac{1}{r-1}\mathbb{Z}$. We know that

$$f = \sum_{j=1}^{r} \sum_{k=-\infty}^{\infty} f^{(j-1)}(k)\phi_j(.-k).$$

Also for $x \in \mathbb{R}$,

$$\begin{aligned} f(-x) &= \sum_{j=1}^{r} \sum_{k=-\infty}^{\infty} (-1)^{j-1} f^{(j-1)}(-k)\phi_j(x-k) \\ &= \sum_{j=1}^{r} \sum_{k=-\infty}^{\infty} (-1)^{j-1} f^{(j-1)}(k)\phi_j(x+k) \end{aligned}$$

and so

$$f(x) = \sum_{j-1}^{r} \sum_{k=-\infty}^{\infty} (-1)^{j-1} f^{(j-1)}(k)\phi_j(-x+k).$$

Thus, for $x \in \mathbb{R}$,

$$\begin{aligned} &\sum_{j=1}^{r} \sum_{k=-\infty}^{\infty} f^{(j-1)}(k)\tilde{\phi}_j(x-k) \\ &= \frac{1}{2}\sum_{j=1}^{r} \sum_{k=-\infty}^{\infty} f^{(j-1)}(k)\phi_j(x-k) + \frac{1}{2}\sum_{j=1}^{r} \sum_{k=-\infty}^{\infty} (-1)^{j-1} f^{(j-1)}(k)\phi_j(-x+k) \\ &= \frac{1}{2}f(x) + \frac{1}{2}f(x) = f(x). \end{aligned}$$

Thus, the space generated by $\tilde{\phi}$ includes all spline functions of degree n with simple knots at $\frac{1}{r-1}\mathbb{Z}$ and it follows, as before, that $\tilde{\phi}$ is refinable.

Finally, we note that any refinable vector ϕ gives rise to a vector subdivision scheme. Let

$$s = \sum_{i=1}^{r} \sum_{k=-\infty}^{\infty} a_{ik}\phi_i(.-k) \tag{17}$$

for constants (a_{ik}). If ϕ satisfies the refinement equation (1), then it can be seen that

$$s = \sum_{i=1}^{r} \sum_{k=-\infty}^{\infty} b_{ik}\phi_i(2.-k), \tag{18}$$

where putting $a_k = (a_{1k}, \ldots, a_{rk})^T$, $b_k = (b_{1k}, \ldots, b_{rk})^T$,

$$b_k = \sum_{l=-\infty}^{\infty} A_{k-2l}^T a_l, \quad k \in \mathbb{Z}. \tag{19}$$

Now if ϕ is fundamental, we see from (3), (17), (18) that for $i = 1, \ldots, r$, $k \in \mathbb{Z}$,

$$a_{ik} = s^{(i-1)}(k) = 2^{i-1} b_{i,2k}, \tag{20}$$

$$s^{(i-1)}(k + \frac{1}{2}) = 2^{i-1} b_{i,2k+1}. \tag{21}$$

Thus from (19) and (20),

$$(A_{2k})_{ij} = \delta_{ko}\delta_{ij} 2^{1-i}, \quad i, j = 1, \ldots, r, \quad k \in \mathbb{Z}.$$

Putting $a_{ik} = \delta_{io}\delta_{kj}$ gives, by (17), $s = \phi_j$ and substituting in (19), (21) gives

$$(A_{2k+1})_{ij} = 2^{1-j}\phi_i^{(j-1)}(k + \frac{1}{2}), \quad i, j = 1, \ldots, r, \quad k \in \mathbb{Z}.$$

In particular for the situation of Theorem 7, we see that (19) for odd k has $n - r + p$ non-zero terms in the summation. The subdivision relation (19) for odd k gives, by (21), the values $s^{(i-1)}(k + \frac{1}{2}), k \in \mathbb{Z}, i = 1, \ldots, r$, in terms of $a_{ik} = s^{(i-1)}(k), k \in \mathbb{Z}, i = 1, \ldots, r$. Repeating the subdivision process will successively give $s^{(i-1)}(2^{-j}k), i = 1, \ldots, r, k \in \mathbb{Z}, j = 0, 1, \ldots$.

For the situation of Theorem 7, the subdivision process converges to the C^{n-1} function s and reproduces all spline functions of degree n with simple knots at $\frac{1}{r-1}\mathbb{Z}$, and hence in particular all polynomials of degree n.

Acknowledgments. Much of this work was undertaken during a visit to Singapore funded by the Wavelets Strategic Research programme. Our thanks go to Prof. S. L. Lee for arranging the visit and to him and Dr. Z. W. Shen for helpful discussions.

References

1. Daubechies, I., Orthonormal bases of compactly supported wavelets, Comm. Pure and Applied Math. **41** (1988), 909–996.
2. Donovan, G. C., J. S. Geronimo, and D. P. Hardin, Intertwining multiresolution analyses and the construction of piecewise polynomial wavelets, SIAM J. Math. Anal. **27** (1996), 1791–1815.

3. Donovan, G. C., J. S. Geronimo, and D. P. Hardin, Orthogonal polynomials and the construction of piecewise smooth wavelets, SIAM J. Math. Anal. **30** (1999), 1029–1056.

4. Goodman, T. N. T., Biorthogonal refinable spline functions, in *Curve and Surface Fitting: Saint-Malo 1999*, Albert Cohen, Christophe Rabut, and Larry L. Schumaker (eds.), Vanderbilt University Press, Nashville, 2000, 219–226.

5. Goodman, T. N. T. and S. L. Lee, Refinable vectors of spline functions, in *Mathematical Methods for Curves and Surfaces II*, M. Dæhlen, T. Lyche, and L. L. Schumaker (eds.), Vanderbilt University Press, Nashville, 1998, 213–220.

6. Goodman, T. N. T. and S. L. Lee, Properties of stable refinable splines, preprint.

7. Goodman, T. N. T and C. A. Micchelli, Orthonormal cardinal functions, in *Wavelets: Theory, Algorithms and Applications*, C. K. Chui et al. (eds.) Academic Press. San Diego, 1994, 53–88.

8. Hardin, D. P., private communication.

9. Jia, R. Q. and C. A. Micchelli, On linear independence of integer translates of a finite number of functions, Proc. Edinburgh Math. Soc. **36** (1992), 69–85.

10. Lawton, W., S. L. Lee, and Z. W. Shen, Characterization of compactly supported refinable splines, Adv. Comp. Math. **3** (1995), 137–145.

11. Schoenberg, I. J. and A. Whitney, On Pólya frequency functions III: The positivity of translation determinants with an application to the interpolation problem by spline curves, Trans. Amer. Math. Soc. **74** (1953), 246–259.

Tim N. T. Goodman
Dept. of Mathematics
The University
Dundee DD1 4HN
Scotland, UK
`tgoodman@maths.dundee.ac.uk`

Approximating Digital Shapes by Parametric Surfaces

A. Ardeshir Goshtasby

Abstract. A method for parametrizing discrete points sampled from a smooth shape and fitting a rational Gaussian surface to the points with a required accuracy is presented. With the proposed method, a very complex shape can be represented by a single surface, and the surface can be rendered at a desired level of detail by adjusting a smoothness parameter.

§1. Introduction

Given a set of scattered points in 3D, $\{\mathbf{p}_i = (x_i, y_i, z_i) : i = 1, \ldots, n\}$, we would like to determine a parametric surface $\mathbf{P}(u, v)$ that approximates the points with a required error tolerance:

$$\max_i ||\mathbf{P}(u_i, v_i) - \mathbf{p}_i|| < \varepsilon \tag{1}$$

ε is the error tolerance and (u_i, v_i) are the parameters at $\mathbf{p}_i$.

We will consider a special case where the given points are voxels covering an object in a discretized 3D space. We will call such a data set a **digital shape.** Digital shapes are typically obtained by segmenting tomographic images obtained by industrial or medical scanners. We assume that the given shape is closed and contains no holes.

The objective in this approximation is threefold. First, we want to approximate the points with a parametric surface where the number of control points in the surface is much smaller than the number of points in the shape, and maximum distance between the shape points and the surface is within a required tolerance. Second, we want to have the ability to revise the obtained surface so that a reconstructed shape can be edited. Therefore, the surface formulation used should lend itself to easy editing. Third, the obtained surface should smooth noise among the given points, with the degree of smoothing adjustable by the user.

In the following sections, first, an algorithm to parametrize a set of points by mapping them to a sphere is described. Then, surface fitting using rational Gaussian (RaG) surfaces is discussed. Finally, examples of the proposed surface-fitting method using medical data are presented.

Mathematical Methods for Curves and Surfaces: Oslo 2000
Tom Lyche and Larry L. Schumaker (eds.), pp. 163–172.

ISBN 0-8265-1378-6

§2. Definitions

In this section, terminologies used in the paper are defined.

Digital shape: A set of points covering a smooth shape in a discretized 3D space.

Point: A point in a digital shape. Shape points will be denoted by $\mathbf{p}$'s.

Adjacent points: Two points, $\mathbf{p}_i$ and $\mathbf{p}_j$, are considered adjacent if $1 \leq ||\mathbf{p}_i - \mathbf{p}_j|| \leq \sqrt{3}$. Two adjacent points are also called **neighbors**.

Connected points: Two points are said to be connected if a path can be formed between them by connecting adjacent points.

Path: A path between points $\mathbf{p}_i$ and $\mathbf{p}_j$ is a connected set of points starting from $\mathbf{p}_i$ and ending at $\mathbf{p}_j$ where no point is repeated and every point has exactly two neighbors except for $\mathbf{p}_i$ and $\mathbf{p}_j$, which may have only one neighbor. A path is also called a **contour**.

Triangular mesh approximation of a shape: A triangular mesh that interpolates some points and approximates the rest in a shape. The mesh vertices will be denoted by $\mathbf{P}$'s. Note that $\mathbf{P}$'s are a subset of the $\mathbf{p}$'s.

Edge contour: A contour in a digital shape that is delimited by the end points of a mesh edge and lies in the plane passing through the edge and bisecting the angle between the two triangular faces that share the edge. Note that due to the digital nature of shape points, some points in an edge contour may not fall exactly in the bisecting plane, however, they will be closer to the plane than points in any other path connecting the edge end points.

Distance of point $\mathbf{p}_i$ to edge $\mathbf{P}_j\mathbf{P}_k$: Assuming the plane passing through the point and normal to the edge intersects the edge at $\mathbf{P}_l$, if $\mathbf{P}_l$ is between $\mathbf{P}_j$ and $\mathbf{P}_k$, the distance will be $||\mathbf{p}_i - \mathbf{P}_l||$. Otherwise, if $\mathbf{P}_l$ is closer to $\mathbf{P}_j$ than to $\mathbf{P}_k$, the distance will be $||\mathbf{p}_i - \mathbf{P}_j||$, and if $\mathbf{P}_l$ is closer to $\mathbf{P}_k$ than to $\mathbf{P}_j$, the distance will be $||\mathbf{p}_i - \mathbf{P}_k||$.

Distance of an edge to an edge contour: Assuming $\mathbf{p}_i$ is a contour point with distance d_i to the associating edge, we will take the maximum distance from points on the contour to the edge as the distance of the contour to the edge. That is, $D_e = \max_i\{d_i\}$.

Distance of point $\mathbf{p}_i$ to triangular face $\mathbf{P}_j\mathbf{P}_k\mathbf{P}_l$: Assuming the line passing through $\mathbf{p}_i$ and normal to the triangle intersects the triangle at $\mathbf{P}_m$, if $\mathbf{P}_m$ is inside the triangle, the distance will be $||\mathbf{p}_i - \mathbf{P}_m||$. If $\mathbf{P}_m$ is outside the triangle and assuming $\mathbf{P}_n$ is the point on a triangle edge closest to $\mathbf{P}_m$, the distance will be $||\mathbf{p}_i - \mathbf{P}_n||$.

Triangular patch: A connected set of points in a digital shape delimited by three contours whose end points are the vertices of a triangle.

Distance of a triangular patch to the associating triangle: Assuming point $\mathbf{p}_i$ belongs to the triangular patch and the distance between $\mathbf{p}_i$ and the triangle is d_i, the distance to be determined is the maximum of such distances when all points in the patch are tested. That is, $D_t = \max_i\{d_i\}$.

Major axis of a digital shape: This is the axis defined by the largest eigenvector of the inertia matrix [2] of points defining the shape.

Subdividing edge $\mathbf{P}_j\mathbf{P}_k$: Replacing the edge with edges $\mathbf{P}_j\mathbf{P}_i$ and $\mathbf{P}_i\mathbf{P}_k$, where $\mathbf{P}_i$ is the farthest point on the associating edge contour to the edge and distance of $\mathbf{P}_i$ to the edge is larger than the given tolerance ε.

Subdividing a triangle: A triangle may be subdivided in four different ways:

1) If distances between all three edges of the triangle and the corresponding contours are larger than the specified tolerance, then by subdividing each edge into two and connecting the obtained points to each other and to the vertices of the triangle, four smaller triangles are obtained.
2) If distances between two of the edges and corresponding contours are larger then the specified tolerance, then two of the edges are subdivided. By connecting the newly obtained points to each other and to the vertices of the triangle, three new smaller triangles are obtained.
3) If the distance between only one of the edges and the corresponding contour is larger than the required tolerance, then only one of the edges is subdivided into two. By connecting the newly obtained point to the opposing triangle vertex, two smaller triangles are obtained.
4) If distances between all three edges and the corresponding contours do not reach the required tolerance, then the distance between the patch and the triangle is determined, and if it is larger than the required tolerance, the point in the patch farthest from the triangle is connected to the vertices of the triangle to produce three smaller triangles.

In this way, a triangle is subdivided into 2, 3, or 4 smaller triangles. Subdivision will take place in the order specified above. That is, only if subdivision by case 1 is not possible subdivision by case 2 will be considered, and subdivision by case 3 will be considered only when subdivision by case 2 is not possible, and so on.

Parametrizing points in an edge contour: By knowing the parameters at the end points of an edge, parameters along the edge are determined by linear interpolation. Parameters at a contour point are then set equal to the parameters at the edge point closest to it. At a coarse resolution, multiple contour points may map to the same edge point, thus producing the same parameters. However, as the subdivision proceeds, contour points will more likely map to unique edge points, producing unique parameters.

Parametrizing points in a triangular patch: By knowing parameters at the vertices of a triangle, parameters of points in the triangle and along its edges can be determined from the barycentric coordinates [5, pp. 289–291]. Parameters of a point in a triangular patch are set equal to the parameters of the point colsest to it in the associating triangle. Initially, depending on the complexity of a shape, some points in a patch may receive the same parameters. However, as the subdivision proceeds, the probability of such cases decreases, producing unique parameters for points in a patch.

§3. The Subdivision Algorithm

Using the above definitions, we now describe an algorithm that approximates a digital shape by a triangular mesh with a required accuracy. We start by approximating the shape with an octahedron. Then, we subdivide the triangular faces into smaller triangles until error in the approximation reaches a required tolerance.

Algorithm 1: Subdivision of a digital shape to a triangular mesh

1) **Initialization:** Determine the major axis of the shape and approximate the shape with an octahedron whose major axis lies on the shape's major axis and whose vertices lie on the shape. Then, enter the triangular faces of the obtained octahedron into a list.
2) **Main Step:** Remove a triangle from the list. If the distance between the triangle and the corresponding patch is larger than the required tolerance, subdivide it and enter the newly obtained triangles into the list.
3) **Stopping Criterion:** If the list is empty, stop. Otherwise, go to the Main Step.

The reason for orienting the octahedron so that its major axis lies on the major axis of the shape is to maximize overlap between the shape and the octahedron and, thereby, minimize the distance between the approximating mesh and the shape. When subdivision is complete, the maximum distance between the shape and its approximating mesh is guaranteed to be smaller than the required tolerance. Some of the properties of this approximation are:

1) A unique subdivision is obtained independent of the orientation or position of a shape. This is achieved by aligning the major axis of the octahedron with the major axis of the shape.
2) The process avoids subdivision into triangles with acute angles or long edges. This is achieved by subdividing the edges of the mesh first.
3) Compression rate depends on the complexity of the shape. During subdivision, large triangles are generated at smooth areas and small triangles are created at detailed areas. The process automatically adjusts triangle sizes to reproduce local details in a shape.

§4. Parametrizing the Shape Points

To parametrize the mesh vertices, first parameters at the octahedral vertices approximating the shape are determined. This is achieved by fitting an octahedron to a sphere, establishing correspondence between vertices in the shape approximation and in the sphere approximation, and assigning parameters of mesh vertices in the sphere approximation to parameters of mesh vertices in the shape approximation. As a triangle in the shape approximation is subdivided, the corresponding triangle in the sphere approximation is subdivided also and, again, parameters at newly obtained mesh vertices in the sphere are assigned to corresponding mesh vertices in the shape.

When an edge contour in the shape approximation is divided into two, in the sphere approximation, the corresponding arc is divided into two in such a way that the proportion of the lengths of newly obtained arcs are the same as the proportion of the lengths of contour segments in the shape. At any stage of the process, by knowing parameters of mesh vertices in the sphere, parameters of corresponding mesh vertices in the shape are known. Therefore, when the subdivision ends, spherical parameters of all mesh vertices approximating a shape will be known. Algorithm 1, therefore, provides the means to determine the parameters at vertices of the triangular mesh approximating a shape. By knowing parameters at three vertices of a triangle, parameters at points in the triangle can be determined using the barycentric coordinates, and by knowing the parameters of points in a triangle, parameters of points in the associating triangular patch can be determined from the correspondence between the two. In this manner, the spherical parameters of all points in a digital shape can be determined.

§5. Approximating a Digital Shape by a Rational Gaussian Surface

Knowing the coordinates and the parameters of points in a digital shape, we can find a smooth parametric surface that approximates the shape. We will use the mesh vertices as the control points of the parametric surface. Since the vertices are irregularly spaced, we will need a surface formulation that does not require a regular grid of control points. A rational Gaussian (RaG) surface [3,4] can have irregularly spaced control points and, therefore, will be used in this approximation. Assuming vertices of the triangular mesh obtained by Algorithm 1 are $\{\mathbf{P}_i : i = 1, \ldots, N\}$ and the parameters associated with them are $\{(u_i, v_i) : i = 1, \ldots, N\}$, a RaG surface that approximates the vertices can be written as [3,4]

$$\mathbf{P}(u, v) = \sum_{i=1}^{N} \mathbf{P}_i g_i(u, v), \tag{2}$$

where $g_i(u, v)$ is the ith basis function of the surface defined by

$$g_i(u, v) = \frac{G_{\sigma_i}(u - u_i, v - v_i)}{\sum_{j=1}^{N} G_{\sigma_j}(u - u_j, v - v_j)}, \tag{3}$$

and $G_{\sigma_i}(u - u_i, v - v_i) = \exp\{[(u - u_i)^2 + (v - v_i)^2]/2\sigma_i^2\}$.

The standard deviations of Gaussians will be set in such a way to reproduce local shape details. The sizes of triangles obtained in Algorithm 1 contain information about local details in a shape. We will set the standard deviations of Gaussians proportional to the perimeters of the triangles.

If the surface is required to interpolate the vertices, we let $\mathbf{P}(u_i, v_i) = \mathbf{P}_i$ and compute the control points of the surface, $\{\mathbf{V}_i : i = 1, \ldots, N\}$, from three systems of N linear equations:

$$\mathbf{P}_i = \sum_{i=1}^{N} \mathbf{V}_i g_i(u, v), \qquad i = 1, \ldots, N. \tag{4}$$

Because of the nature of the rational Gaussian bases, the obtained matrix of coefficients will be diagonally dominant. For very large standard deviations, however, the system will become unstable because it may not be possible to fit a surface with a desired smoothness to fit to points in a very detailed area.

In Algorithm 1, the decision to subdivide a triangle was based initially on distances between edges of the triangle and the associating edge contours, and then on the distance between the triangle and the associating patch. In order to fit a RaG surface to a shape, we redefine the error criteria in Algorithm 1 as follows:

1) Instead of determining the distance between an edge and its corresponding edge contour, we will determine the distance between an edge contour and the approximating/interpolating surface. Since parameters of all points in an edge contour are known, for each point $\mathbf{p}_i$ in an edge contour, we can determine the corresponding point $\mathbf{P}(u_i, v_i)$ in the approximating/interpolating RaG surface. We then let the maximum distance between corresponding points in the contour and the surface be the distance between an edge contour and the RaG surface: $D_E = \max_i ||\mathbf{P}(u_i, v_i) - \mathbf{p}_i||$.
2) Instead of determining the distance between a triangle and its associating patch, we will determine the distance between the patch and the approximating/interpolating RaG surface. This is possible because, by knowing parameters (u_i, v_i) at each point $\mathbf{p}_i$ in the patch, we can determine the corresponding point $\mathbf{P}(u_i, v_i)$ in the surface. We will then define the distance between a triangular patch and its approximating/interpolating RaG surface to be the maximum distance between corresponding points in the patch and the surface: $D_T = \max_i ||\mathbf{P}(u_i, v_i) - \mathbf{p}_i||$.

Replacing error measures D_e and D_t in Algorithm 1 with error measures D_E and D_T, respectively, we will obtain an algorithm that fits a RaG surface to a digital shape, ensuring that distance between the given shape and the approximating/interpolating surface is within the required tolerance. We will call this new algorithm, **Algorithm 2**.

Note that σ's in the RaG formulation are in the same units as u's and v's. As the subdivision progresses, the sizes of triangles in the sphere subdivision become smaller. Roughly, the perimeter of a triangle reduces to half its size in each subdivision. Therefore, the σ to be assigned at a particular subdivision level will depend on the depth of the subdivision.

Suppose at level 0 an octahedron is fitted to the shape and another octahedron is fitted to a sphere. If we were to stop the approximation at level 0, we would set all σ's to 1 to obtain a smooth approximation to the shape. As the σ's are reduced, the obtained surface will resemble the approximating triangular mesh, and as the σ's are increased the surface approaches a sphere. In the case of interpolation, for very large values of σ's it may not be possible to obtain a surface that would pass through the points. Typically, proper values for the σ's at level 0 are between 0.2 and 2.

At level 1, perimeters of triangles in the sphere are roughly half the perimeters of triangles at level 0. Therefore, σ's at level 1 should be half the σ's at level 0. Analogously, σ's at level n should be 2^{-n} of σ's at level 0. Denoting the σ assigned to vertices at the ith level by s_i, we will have $s_i = 2^{-i}s_0$. In this manner, σ at all vertices will depend on a single parameter s_0. By adjusting this single parameter, the overall smoothness of the reconstructed surface can be controlled. At one extreme when s_0 is close to zero, the surface approaches the approximating triangular mesh. At the other extreme, when s_0 is very large, the shape will approach a sphere, and if an interpolating surface is required, beyond a certain point it may not be possible to obtain a surface that would fit the mesh vertices.

Note that since σ's are associated with the control points of a RaG surface, and the control points are the vertices of the approximating triangular mesh, whenever an edge is divided into two, σ's associated with the mesh vertices corresponding to the end points of the edge are also divided by two. A vertex may be shared by many triangles obtained at different levels. Assigning a σ to the vertex that is proportional to the triangle at the highest level will enable reproduction of details differently at different sides of a point. This can be explained by examining the spatial frequency characteristics of Gaussians.

From the signal processing point of view, as the standard deviation of Gaussians in the spatial domain increases (decreases), the Fourier transform of the Gaussians, which are also Gaussians, will become narrower (wider) in the frequency domain. This means, in areas where smaller σ's are used, the obtained surface can reproduce high and low spatial frequencies in the surface, and in areas where only large σ's are used, high spatial frequencies are not reproduced, thus creating a smooth surface. Narrower Gaussians enable reproduction of both low and high spatial frequencies. For any value of s_0, relative details obtained in different areas in a reconstructed surface will depend on relative details of local areas in the original shape.

Also note that although some triangles at level n could have the same size as some triangles at levels $m < n$ in the xyz space, in the uv space the sizes of triangles at level n are smaller than those at level m. By appropriately reducing the σ's at higher levels, we are in effect preserving information about the sizes of triangles at different levels. When a small portion of a sphere is mapped to a large portion of a shape, the σ's assigned to different subdivision levels enable proper reproduction of details in the shape.

§6. Examples

The digital shape depicted in Fig. 1a shows a femoral stem obtained by segmenting a volumetric X-ray Computer Tomography (CT) image. There were holes at the top and bottom of the original bone. The holes were covered with planar patches to obtain a closed shape. The process of subdividing the closed shape into triangles using Algorithm 1 with error tolerance of 0.5 units is shown in Figs. 1b–1e. It is assumed that the length of each side of elements (voxels) in the shape is 1 unit. Figure 1f shows the rendered femoral stem using the obtained triangular mesh. The triangulation process has placed larger triangles in smoother areas and smaller triangles in more detailed areas.

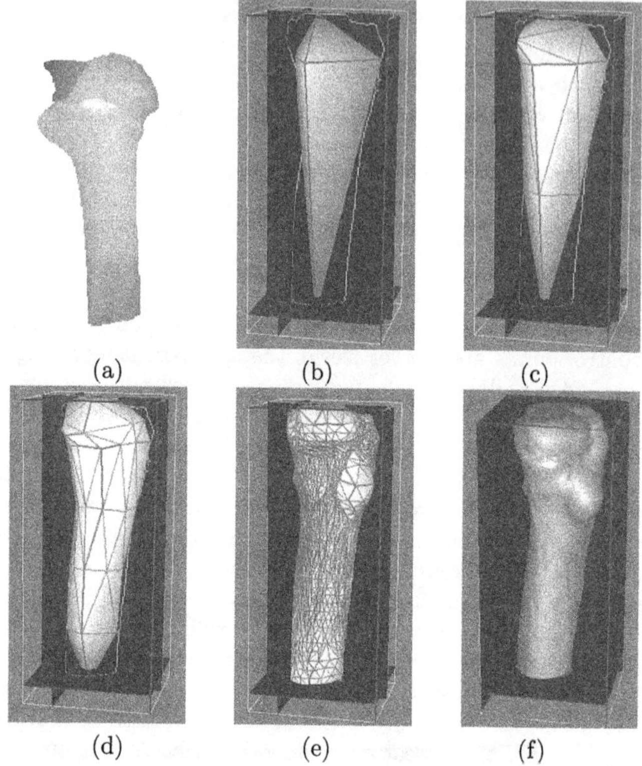

Fig. 1. Approximation of a digital shape by a triangular mesh.

Since subdivision in the shape and in the sphere are performed in parallel, at any stage of the process, parameters at the vertices of the triangular mesh are known from the parameters of corresponding points in the sphere. Using Algorithm 2 with the same number of control points as the number of vertices obtained in the mesh approximation, we obtain Figs. 2a–2c when s_0 is set to 0.2, 0.5, and 1, respectively.

A second set of examples is shown in Figs. 3a–3c. The data set used in this experiment was obtained by segmenting a Magnetic Resonance (MR) image of a person's head. The segmentation has extracted the skin of the head. Using Algorithm 2 with error tolerance equal to 0.5, we obtain the surface shown in Fig. 3a with $s_0 = 0.2$. Increasing s_0 to 0.5 we obtain the surface shown in Fig. 3b. Increasing s_0 further to 1, we obtain the surface shown in Fig. 3c.

Figures 2 and 3 show two examples of the proposed surface-fitting method. The number of control points obtained in an approximation is not a mere function of the error tolerance; it is also a function of the smoothness of the required surface. Preliminary results show that to achieve a high compression rate, when a large error tolerance is given a large s_0 should be used, and when

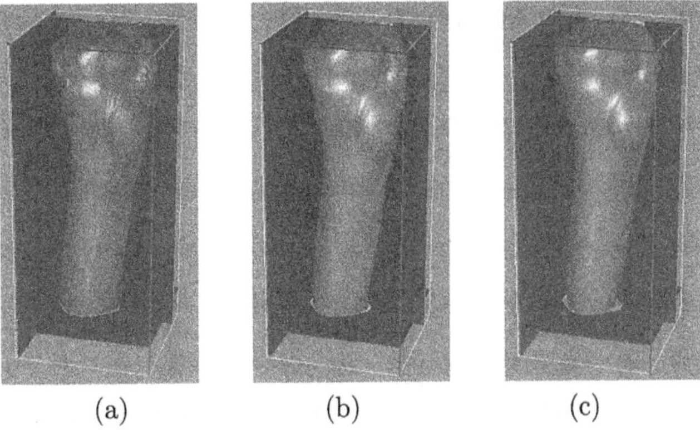

(a) (b) (c)

Fig. 2. Approximating a digital femoral stem by RaG surfaces.

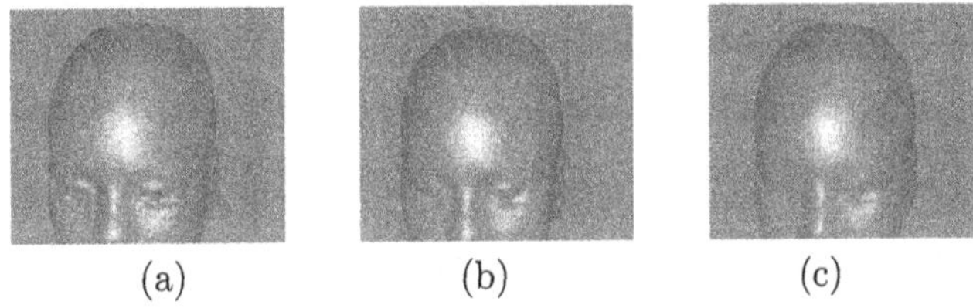

(a) (b) (c)

Fig. 3. Approximating a digital head by RaG surfaces.

a small error tolerance is given a small s_0 should be used.

§7. Concluding Remarks

An algorithm to subdivide a digital shape into a triangular mesh, parametrize the mesh vertices as well as the shape points, and fit a rational Gaussian surface to the points was presented. Attempts to parametrize mesh vertices have been made before. Lee *et al.* [6] simplified a mesh to a base mesh, assigned parameters to the vertices of the base mesh, and determined parameters at the original mesh vertices through conformal mapping of the base mesh to the original mesh. Rogers and Fog [7] developed a nonlinear optimization method for determining the parameters of a mesh to be approximated by B-spline patches. Brechbühler *et al.* [1] developed an optimization method for mapping vertices of a simple polyhedron into a sphere and thereby parametrizing the polyhedral vertices.

Once the shape points or the mesh vertices are parametrized, a single RaG surface can be fitted to the points to reconstruct the shape. A RaG surface enables editing of a shape by moving its control points just like a NURBS surface. This representation is especially useful when a noisy data set is given and there is a need to smooth noise in the data. The RaG formulation has a smoothness parameter that can be varied to obtain surfaces at different levels of detail.

Acknowledgments. The author would like to thank the National Science Foundation for partially funding this work under grant IIS-9906340. The assistance of Marcel Jakowski in preparation of images presented in this paper is also greatly appreciated. Images used in this work were provided by the Kettering Medical Center, Kettering, OH.

References

1. Brechbühler, Ch., G. Gerig, and O. Kübler, Parametrization of closed surfaces for 3D shape description, Computer Vision and Image Understanding **61**:2 (1995), 154–170.
2. Galvez, J.M. and M. Canton, Normalization and shape recognition of three-dimensional objects by 3D moments, Pattern Recognition **26**:5 (1993), 667–682.
3. Goshtasby, A., Design and recovery of 2D and 3D shapes using rational Gaussian curves and surfaces, Int'l J. Computer Vision **10**:3 (1993), 233–256.
4. Goshtasby, A.,Geometric modeling using rational Gaussian curves and surfaces, Computer-Aided Design **27**:5 (1995), 363–375.
5. Hoschek, J. and D. Lasser, *Computer Aided Geometric Design*, A. K. Peters, 1989.
6. Lee, A. W. F., W. Sweldens, P. Schröder, L. Cowsar, and D. Dobkin, MAPS: Multiresolution adaptive parametrization of surfaces, Computer Graphics Proceedings (1998), 95–104.
7. Rogers, D. F. and N. G. Fog, Constrained B-spline curve and surface fitting, Comput. Aided Geom. Design **21** (1989), 641–648.

A. Ardeshir Goshtasby
Department of Computer Science and Engineering
Wright State University
Dayton, OH 45435, USA
`ardy@cs.wright.edu`
`http://www.cs.wright.edu/~agoshtas/`

Clifford Algebra and Flows

Hans Hagen and Gerik Scheuermann

Abstract. Clifford algebra is an extension of the usual vector space description for geometric spaces. It combines Grassmann's inner and outer products into a single associative product. The resulting algebra allows a description of projections and rotations without matrices and has other nice properties. With this new model for geometric spaces, an analysis can be introduced that uses the additional properties compared to conventional vector spaces. After an introduction to the subject, some advantages of this model with respect to planar vector fields are discussed. For three-dimensional vector fields, we derive flow surfaces in piecewise linear vector fields. It is assumed that the flow surfaces are defined by linear splines. It can be shown that these stream surfaces can be calculated with the same precision as stream lines. The computation of stream surfaces enables a better understanding of tridimensional fluid flows than stream lines, since they contain information about rotation and divergence.

§1. Introduction

Vector fields and their associated flows have enjoyed increasing interest among different scientific communities in recent years. Mechanical and aeronautical engineers study fluid flows, usually air or water, around car, ship or aircraft bodies and inside turbines. Chemists try to understand the changing geometry of the border between reacting fluids, for example in combustion. Physicists have a strong interest in the field lines of magnetic, electric and gravitational fields.

The mathematical description of these phanomena uses vector fields for the local tangential movement. The integral curves of the vector field lead to the actual movement. Most engineers, scientists and applied mathematicians take the usual vector algebra to express their considerations. Nevertheless, some people have rediscovered Clifford algebra and calculus for this purpose over the last decades [1,2,3,4]. Clifford algebra extends the usual vector space description for geometric spaces by combining Grassmann's inner and outer products into a single associative product. This results in an algebra allowing a description of projections and rotations by elements of the algebra along

Mathematical Methods for Curves and Surfaces: Oslo 2000
Tom Lyche and Larry L. Schumaker (eds.), pp. 173–182.

ISBN 0-8265-1378-6

with other nice properties. The algebra can be modeled by matrices, but this is only useful for general considerations, and is rather distracting in practical calculations. With a new language for linear algebra at hand, a new language for analysis, Clifford analysis, can be developed. The additional operation of Clifford algebra allows here a unification of different differential operators. In the case of vector fields, we obtain a union of curl and divergence. Also, the formulation of curl by a plane segment instead of a normal vector generalizes nicely to higher dimensions.

For some time now, we have experienced an increasing interest in topological and related properties of vector fields. We recall a simplification of topological questions about planar vector fields by considerations based on Clifford analysis. A planar vector can be seen as a unit vector multiplied by a spinor or complex number. Therefore, a vector field, can be seen as an unit vector multiplied by a complex function. For polynomial vector fields, we derive a strong relation between algebraic properties of the polynomial, and the topological properties of the vector field. The roots of the polynomial are the critical points of the vector field. If the polynomial splits into factors, the critical points of the "factor fields" are the critical points of the whole field and their Poincaré-indices sum up. In three dimensions, we discuss some ideas on the calculation of stream surfaces in piecewise linear vector fields.

§2. Clifford Algebra

Clifford algebra is a way to extend the usual description of geometry by a multiplication of vectors. We give a basic introduction in the two-dimensional case but it can be done in any dimension. We start with a vector $v \in \mathbb{R}^2$. Together with the Euclidean standard basis $\{e_1, e_2\}$ it can be written as

$$v = v_1 e_1 + v_2 e_2.$$

The standard description as a column vector gives

$$v = v_1 \begin{pmatrix} 1 \\ 0 \end{pmatrix} + v_2 \begin{pmatrix} 0 \\ 1 \end{pmatrix} = \begin{pmatrix} v_1 \\ v_2 \end{pmatrix}.$$

If we use square matrices instead, we could take

$$v = v_1 \begin{pmatrix} 0 & 1 \\ 1 & 0 \end{pmatrix} + v_2 \begin{pmatrix} 1 & 0 \\ 0 & -1 \end{pmatrix} = \begin{pmatrix} v_2 & v_1 \\ v_1 & -v_2 \end{pmatrix}.$$

This looks a little bit strange, but it allows a matrix multiplication of vectors:

$$\begin{aligned} vw &= \begin{pmatrix} v_2 & v_1 \\ v_1 & -v_2 \end{pmatrix} \begin{pmatrix} w_2 & w_1 \\ w_1 & -w_2 \end{pmatrix} \\ &= \begin{pmatrix} v_1 w_1 + v_2 w_2 & v_2 w_1 - v_1 w_2 \\ v_1 w_2 - v_2 w_1 & v_1 w_1 + v_2 w_2 \end{pmatrix}. \end{aligned}$$

With a suitable choice of the remaining two basis vectors of the square matrices, we get

$$vw = (v_1 w_1 + v_2 w_2) \begin{pmatrix} 1 & 0 \\ 0 & 1 \end{pmatrix} + (v_1 w_2 - v_2 w_1) \begin{pmatrix} 0 & -1 \\ 1 & 0 \end{pmatrix}.$$

With the terms

$$1 := \begin{pmatrix} 1 & 0 \\ 0 & 1 \end{pmatrix}, \qquad i := \begin{pmatrix} 0 & -1 \\ 1 & 0 \end{pmatrix},$$

we end up with a four-dimensional algebra G with the following rules for the multiplication:

$$\begin{aligned} 1e_j &= e_j, \qquad j = 1, 2, \\ e_j 1 &= e_j, \qquad j = 1, 2, \\ 1^2 &= 1, \\ e_j^2 &= 1, \qquad j = 1, 2, \\ i^2 &= -1, \\ e_1 e_2 &= -e_2 e_1 = i. \end{aligned}$$

The following projections are useful for computations:

$$\begin{aligned} < \cdot >_0 \colon G^2 &\to \mathbb{R} \subset G^2 \\ a1 + be_1 + ce_2 + di &\mapsto a1 \\ < \cdot >_1 \colon G^2 &\to \mathbb{R}^2 \subset G^2 \\ a1 + be_1 + ce_2 + di &\mapsto be_1 + ce_2 \\ < \cdot >_0 \colon G^2 &\to \mathbb{R}i \subset G^2 \\ a1 + be_1 + ce_2 + di &\mapsto di. \end{aligned}$$

We get

$$vw = (v \bullet w) + (v \wedge w)$$

for two vectors $v, w \in \mathbb{R}^2 \subset G^2$, where $\bullet$ is the usual scalar product and $\wedge$ the outer product of Grassmann. Now, we have a unification of these two products into an associative multiplication.

We can find constructions like this one for every dimension n by using 2^n-dimensional subalgebras of a complex matrix algebra $Mat(m, \mathbb{C})$ [4]. These algebras are models for a Clifford algebra describing n-dimensional Euclidean space. More details can be found in the literature [3,4,5]. In our 2D-case we have another important fact, because we can interpret the elements

$$a1 + bi \in G^2$$

of our algebra as complex numbers.

§3. Clifford Analysis

After extending the structure of the linear algebra, one may also change the analysis. This leads to a differential operator that does not depend on the coordinates as we demonstrate below. Our maps will be multivector fields

$$\begin{aligned} A : \mathbb{R}^2 &\to G^2 \\ r &\mapsto A(r) . \end{aligned}$$

A Clifford vector field is just a multivector field with values in $\mathbb{R}^2 \subset G^2$:

$$\begin{aligned} v : \mathbb{R}^2 &\to \mathbb{R}^2 \subset G^2 \\ xe_1 + ye_2 &\mapsto v_1(x,y)e_1 + v_2(x,y)e_2. \end{aligned}$$

The directional derivative of A in direction $b \in \mathbb{R}^2$ is defined by

$$A_b(r) = \lim_{\epsilon \to 0} \frac{1}{\epsilon}[A(r + \epsilon b) - A(r)].$$

This allows the definition of the vector derivative of A at $r \in \mathbb{R}^2$,

$$\begin{aligned} \partial A(r) : \mathbb{R}^2 &\to G^2 \\ r &\mapsto \partial A(r) = \sum_{k=1}^{2} g^k A_{g_k}(r). \end{aligned}$$

This is independent of the basis $\{g_1, g_2\}$ of $\mathbb{R}^2$. The vectors

$$g^1 = \frac{i}{\gamma} g_2 \qquad g^2 = -\frac{i}{\gamma} g_1$$

with

$$g_1 \wedge g_2 = \gamma i$$

are called reciprocal vectors.

For a vector field $v : \mathbb{R}^2 \to \mathbb{R}^2$, we get in Euclidean coordinates,

$$\begin{aligned} \partial v &= \sum_{j=1}^{2} e_j v_{e_j} = \sum_{j=1}^{2} e_j (\frac{\partial v_1}{\partial e_j} e_1 + \frac{\partial v_2}{\partial e_j} e_2) \\ &= (\frac{\partial v_1}{\partial e_1} + \frac{\partial v_2}{\partial e_2})1 + (\frac{\partial v_2}{\partial e_1} - \frac{\partial v_1}{\partial e_2})i \\ \partial v &= (\text{div} v)1 + (\text{curl} v)i. \end{aligned}$$

This differential operator integrates divergence and rotation.

§4. Planar Vector Fields and Topology

For our analysis of vector fields, it is necessary to look at $v : \mathbb{R}^2 \to \mathbb{R}^2 \subset G^2$ in suitable coordinates. Let $z = x + iy$, $\bar{z} = x - iy$ be complex numbers in the algebra. This means

$$x = \frac{1}{2}(z + \bar{z}) \quad y = \frac{1}{2i}(z - \bar{z}).$$

We get

$$\begin{aligned} v(r) &= v_1(x, y)e_1 + v_2(x, y)e_2 \\ &= [v_1(\frac{1}{2}(z + \bar{z}), \frac{1}{2i}(z - \bar{z})) - iv_2(\frac{1}{2}(z + \bar{z}), \frac{1}{2i}(z - \bar{z}))]e_1 \\ &= E(z)e_1, \end{aligned}$$

where

$$\begin{aligned} E &: \mathbb{C} \to \mathbb{C} \subset G^2 \\ z &\mapsto v_1(\frac{1}{2}(z + \bar{z}), \frac{1}{2i}(z - \bar{z})) - iv_2(\frac{1}{2}(z + \bar{z}), \frac{1}{2i}(z - \bar{z})) \end{aligned}$$

is a complex-valued function of a complex variable. The idea is to analyze E instead of v and get topological results directly from the formulas in some interesting cases. Let us first assume that E and v are linear.

Theorem 1. *Let*

$$v(r) = (az + b\bar{z} + c)e_1$$

be a linear vector field. For $|a| \neq |b|$, it has a unique zero at $z_0 e_1 \in \mathbb{R}^2$. For $|a| > |b|$, v has one saddle point with index -1. For $|a| < |b|$, it has one critical point with index $+1$. The special types in this case can be obtained from the following list :

(1) *$Re(b) = 0 \Leftrightarrow$ circle at z_0.*
(2) *$Re(b) \neq 0$, $|a| > |Im(b)| \Leftrightarrow$ node at z_0.*
(3) *$Re(b) \neq 0$, $|a| < |Im(b)| \Leftrightarrow$ spiral at z_0.*
(4) *$Re(b) \neq 0$, $|a| = |Im(b)| \Leftrightarrow$ focus at z_0.*

In cases (2) – (4) one has a sink for $Re(b) < 0$ and a source for $Re(b) > 0$. For $|a| = |b|$, one gets a whole line of zeros.

We included this simple theorem to show that this description gives topological information directly. Let us look now at the general polynomial case.

Theorem 2. *Let $v : \mathbb{R}^2 \to \mathbb{R}^2 \subset G^2$ be an arbitrary polynomial vector field with isolated critical points. Let $E : \mathbb{C} \to \mathbb{C}$ be the polynomial so that $v(r) = E(z)e_1$. Let $F_k : \mathbb{C} \to \mathbb{C}$, $k = 1, \ldots, n$ be the irreducible components of E, so that $E(z) = \prod_{k=1}^{n} F_k(z)$. Then the vector fields $w_k : \mathbb{R}^2 \to \mathbb{R}^2$,*

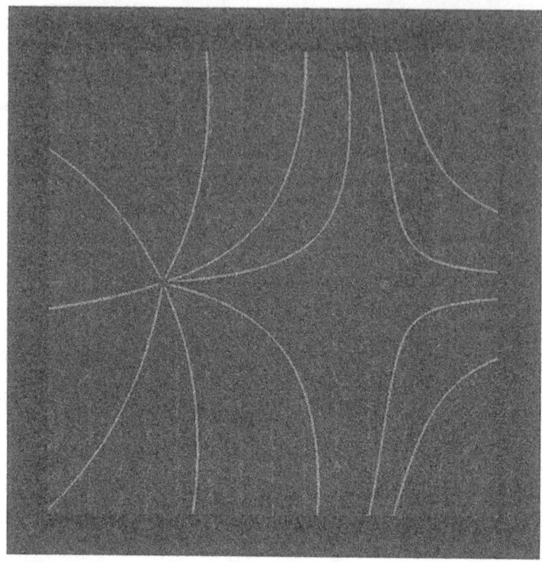

Fig. 1. Two critical points in an analytic vector field.

$w_k(r) = F_k(r)e_1$, have only isolated zeros $z_1, \ldots, z_m$. These are the zeros of v, and for the Poincaré-indices we have

$$\text{ind}_{z_j} v = \sum_{k=1}^{n} \text{ind}_{z_j} w_k.$$

For illustration, we include two small examples. In Figure 1, we show the vector field

$$v : \mathbb{R}^2 \to \mathbb{R}^2$$
$$ze_1 \to (z-2)\bar{z}e_1.$$

In Figure 2, we show the vector field

$$v : \mathbb{R}^2 \to \mathbb{R}^2$$
$$ze_1 \to (\bar{z} + \frac{1}{4} - \frac{1}{2}i)(z - \frac{1}{2}i)(z - \frac{1}{4})e_1.$$

More examples can be found in [6].

§5. Piecewise Linear Vector Fields in 3-Space and Flow Surfaces

The definitions and constructions of the previous two sections can be extended to three dimensions. Since the notations do not change significantly, we omit this here. We study piecewise linear vector fields over a tetrahedrization $\mathcal{T}$ of

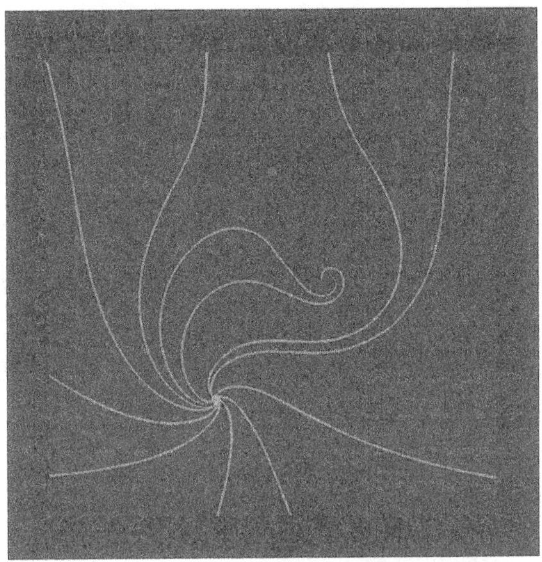

Fig. 2. Three critical points in an analytic vector field.

a domain $D \subset \mathbb{R}^3$,

$$\begin{aligned} v : D &\to \mathbb{R}^3 \\ v|_T : T &\to \mathbb{R}^3 \\ x = \sum_{i=0}^{3} \beta_i p_i &\mapsto v = \sum_{i=0}^{3} \beta_i v_i, \end{aligned}$$

with $\sum_{i=0}^{3} \beta_i = 1$. p_i are the vertices of the tetrahedron T, and v_i given vector values. A stream line through a point $a \in D$ is defined by

$$\begin{aligned} c_a : \mathbb{R} \supset I_a &\to \mathbb{R}^3 \\ \tau &\mapsto c_a(\tau) \\ \partial c_a(\tau) &= v(c_a(\tau)) \\ c_a(0) &= a. \end{aligned}$$

On a tetrahedron T, we have an exact solution for the stream lines,

$$\begin{aligned} c_a : \mathbb{R} \supset I_a &\to T \\ \tau &\mapsto \exp(A\tau)a, \end{aligned}$$

where A is a matrix describing our linear vector field $v : T \to \mathbb{R}^3$, $v(x) = Ax + b$. Detailed formulas for the calculation of $\exp(A\tau)$ have been published by Nielson [7]. A stream surface through a curve

$$\begin{aligned} b : [0, 1] &\to D \\ \sigma &\mapsto b(\sigma) \end{aligned}$$

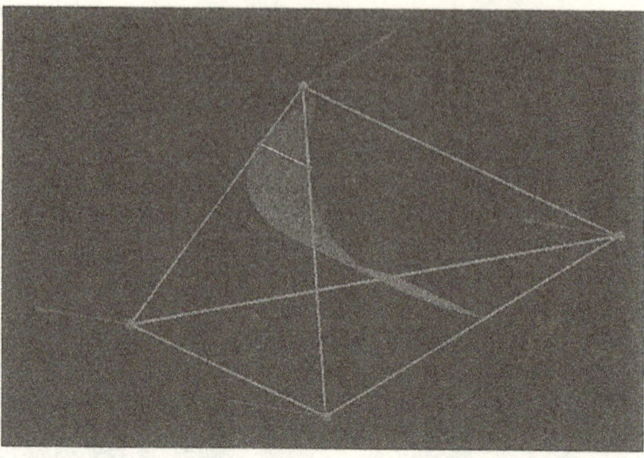

Fig. 3. A stream surface inside a tetrahedron defined by a line segment.

Fig. 4. A stream surface in a domain divided into tetrahedra.

is defined by

$$\begin{aligned} S_b : [0,1] \times \mathbb{R} &\to D \\ (\sigma, \tau) &\mapsto S(\sigma, \tau) \\ \partial_\tau S(\sigma, \tau) &= v(S(\sigma, \tau)) \\ S(\sigma, 0) &= b(\sigma). \end{aligned}$$

(We assume that each streamline on the surface stays at the boundary of D forever to allow the simple notation $[0,1] \times \mathbb{R}$ for the domain of the stream surface.) On a tetrahedron T, there is an exact solution for a stream surface

$$\begin{aligned} S_b : [0,1] \times \mathbb{R} &\to T \\ (\sigma, \tau) &\mapsto (1-\sigma)\exp(A\tau)b(0) + \sigma \exp(A\tau)b(1). \end{aligned}$$

This allows a stream surface calculation with arbitrary precision $\epsilon > 0$ in each tetrahedron, and finally a quite precise stream surface calculation in the whole domain. The calculation in one tetrahedron is illustrated in Figure 3. In Figure 4, a complete stream surface is shown.

§6. Conclusion

We have introduced Clifford algebra and analysis as an alternative to conventional vector spaces. The application to planar vector fields has given relations between algebraic description and topological properties. In three dimensions, we have discussed some ideas for the precise calculation of stream surfaces which look promising. So far, we are not using the full power of Clifford analysis here, so there is still a lot of work to do.

References

1. Brackx, F., R. Delanghe, and F. Sommen, *Clifford Analysis*, Research Notes in Mathematics 76. Pitman, London, 1982.
2. Hestenes, D., and G. Sobczyk, *Clifford Algebra to Geometric Calculus*, D. Reidel Publishing Company, Dordrecht, 1984.
3. Hestenes, D., *New Foundations for Classical Mechanics*, Kluwer Academic Publishers, Dordrecht, 1986.
4. Snygg, J., *Clifford Algebra : A Computational Tool for Physicists*, Oxford University Press, Oxford, 1997.
5. Gilbert, J. E., and M. A. M. Murray, *Clifford Algebras and Dirac Operators in Harmonic Analysis*, Cambridge University Press, Cambridge, 1991.
6. Scheuermann, G., Topological vector field visualization with Clifford algebra, PhD thesis, University of Kaiserslautern, Kaiserslautern, Germany, 1999.
7. Nielson, G. M., Tools for computing tangent curves for linearly varying vector fields over tetrahedral domains, IEEE Transactions on Visualization and Computer Graphics **5**(4) (1999), 360–372.

Hans Hagen
Computer Science Department
University of Kaiserslautern
PO Box 3049
D-67653 Kaiserslautern, Germany
hagen@informatik.uni-kl.de

Gerik Scheuermann
Center for Image Processing and Integrated Computing
University of California
2343 Academic Surge Bld.
Davis, CA 95616, USA
scheuer@ucdavis.edu

Surface Self-Intersection

Chih-Cheng Ho and Elaine Cohen

Abstract. To unambiguously represent a solid volume, it is necessary to identify and trim away extraneous and distracting parts caused by self-intersecting regions of the boundary surface. We define self-intersection as a global intrinsic property of the geometry and introduce a necessary condition for surface self-intersection, that can be computed from the normal and tangent bounding cones of the surface. Therefore, a surface that fails the condition cannot have any self-intersection. Using this property, we develop a divide-and-conquer algorithm to find the self-intersection curves of surfaces. We also introduce a method to locate a miter point which is the open end point of a vanishing self-intersection curve. Miter points can cause slow or false convergence using existing numerical intersection methods.

§1. Introduction

A boundary representation used in solid modeling may not represent a solid volume unambiguously because of boundary surface self-intersections. Surface self-intersection can inadvertently occur as a result of common geometric modeling operations, such as sweeping, interpolation and offsetting (see Fig. 1, from left to right). Surface self-intersections must be detected, located, and trimmed away to obtain a valid solid model.

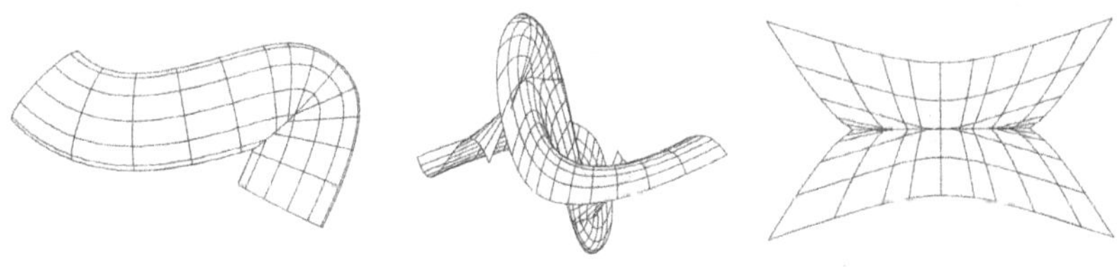

Fig. 1. Examples of self-intersecting surfaces.

Mathematical Methods for Curves and Surfaces: Oslo 2000
Tom Lyche and Larry L. Schumaker (eds.), pp. 183–194.

ISBN 0-8265-1378-6

Existing methods apply surface/surface intersection techniques directly on the same surface [4,6]. This approach either subdivides the whole surface and finds intersections between subsurface pairs, or needs to solve $\sigma(s,t) - \sigma(u,v) = 0$ with four independent variables s, t, u, and v. One problem of this approach is the difficulty of distinguishing between a real self-intersection point and a trivial root with $s = u$ and $t = v$ which should be ignored, especially when the self-intersection point can have nearly identical parametric values. This problem can be avoided by taking the quotient of the Euclidean difference to the parametric difference of the surface points. A formulation of such a function for triangular Bézier patches is presented in [1]. Roots of the quotient function represent critical points where some directional derivative is null, as well as, points of self-intersection.

In this paper, we define self-intersection as a *global intrinsic* property of the surface, independent of its parametrization. We start with the intersection problem of two surfaces, i.e., detecting and locating the intersection curves between two surfaces. Problems that may arise if an intersection algorithm is to be applied directly on the same surface, or surfaces sharing a common boundary are discussed. Then, we introduce a *necessary* condition of surface self-intersection. Using this condition, a divide-and-conquer surface self-intersection algorithm is presented, which subdivides the surface only on regions where a self-intersection may occur. We also identify a common but special type of surface point, called miter point, which represents the *open* end point of a self-intersection curve that vanishes in the middle of the surface. A special method is presented to find the exact location of a miter point. Finally, we present a new method for finding surface self-instersections.

§2. Intersection of Surfaces

The intersection points of surfaces $\sigma_1(u,v)$ and $\sigma_2(s,t)$ are the points where $\sigma_1(u,v) = \sigma_2(s,t)$. Common surface intersection finding techniques include divide-and-conquer methods [8,13], intersection curve tracing [2,3], and hybrid marching method [5]. We use the divide-and-conquer method as an example to demonstrate the problems of applying surface/surface intersection technique to the self-intersection problem.

In computing the intersection of two surfaces using a divide-and-conquer algorithm, a test of whether the two surfaces can intersect is first performed. If the test fails, there is no intersection between them. Otherwise, the two surfaces are subdivided, and the same procedure is applied on each of the subsurface pairs. The subdivision ends when the surfaces are simple enough to compute or approximate the intersection curve directly. A bounding box test is used to prevent unnecessary subdivision [13]. When two surfaces intersect, their bounding boxes must intersect too. This test prunes the subdivision tree and improves the efficiency of the divide-and-conquer algorithm.

In the case of surface self-intersection, the bounding box test will always be positive when the two surfaces are identical or are neighbors from previous subdivision. This in turn results in a full subdivision tree and an exhaus-

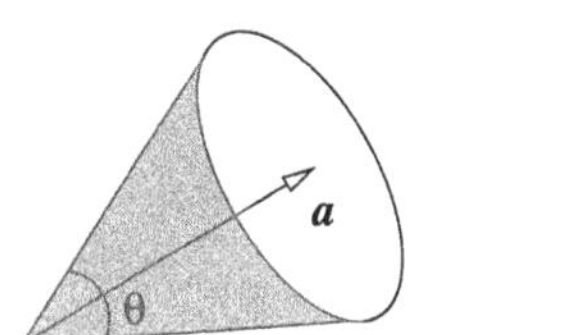

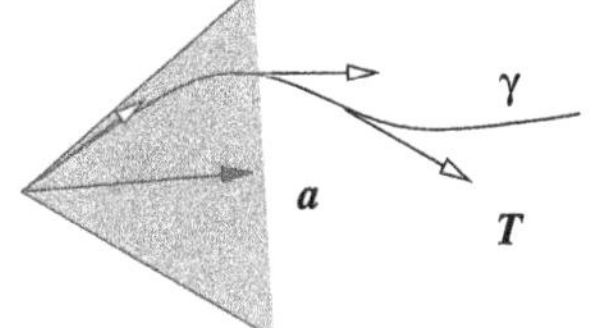

Fig. 2. Direction cone and tangent bounding code.

tive search for intersections among the subsurfaces. To apply the divide-and-conquer method to surface self-intersection, a different test is needed. This new test must detect surfaces that cannot self-intersect and neighboring surfaces that cannot intersect except at the common boundary. It is preferred that the computation is as simple and efficient as the bounding box method.

§3. Curve Self-Intersection

In this section, we derive a *necessary* condition of curve self-intersection. Let $\gamma(s)$ be an arc length parametrized curve. Then, $\gamma(s)$ is self-intersecting if and only if $\gamma(s_1) = \gamma(s_2)$ and $s_1 \neq s_2$. For parametric curves, we want to ensure that a self-intersecting curve is still self-intersecting after reparametrization by its arc length. We assume the curve is piecewise smooth $(C^{(1)})$.

The smallest tangent bounding cone of a curve is the smallest direction cone that contains all unit tangent vectors of the curve (see Fig. 2). For curves with tangent discontinuities, the tangent bounding cone should also include the spanning angle between the two end tangents at any tangent discontinuity. For NURBS curves, the tangent bounding cone can be approximated by bounding the first derivative which also has a NURBS representation. A more efficient but less accurate approximation is to bound the direction vectors of the control polygon.

Proposition 1. *The smallest tangent bounding cone of a self-intersecting curve spans at least π.*

Proof: From the definition, a self-intersecting curve must form at least one non-degenerate loop between the intersection point. From differential geometry, we know that the tangent spherical image of a closed curve does not lie in any open hemisphere, i.e., a closed curve rotates at least 2π [10]. Since the angle of intersection is at most π, the smallest tangent bounding cone of any self-intersecting curve must span at least π. □

Proposition 1 provides the key to a test for a self-intersecting curve that can be answered by "*no*" or "*maybe*." To test whether two connected curves can intersect at places other than the junction point, we concatenate the two curves together, then test whether the concatenated curve can self-intersect. The bounding cone of the concatenated curve can be approximated by combining the bounding cones of the two subcurves. The combined tangent bounding cone should also include the possible tangent discontinuity at the junction point.

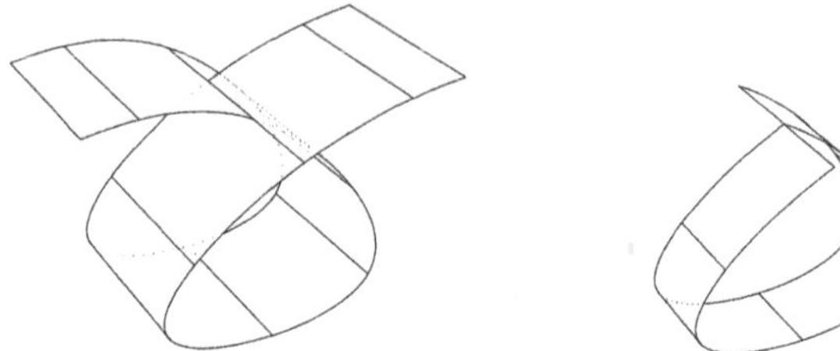

Fig. 3. Minimal self-intersecting surface.

§4. Surface Self-Intersection

An immediate surface analogy to the curve tangent bounding cone condition would be to assert that the minimum bounding cone of surface normal vectors must span at least π. This condition appears to hold for examples shown in Fig. 1. However, in general, the normal bounding cone test alone is insufficient since planar surfaces can also self-intersect.

Let $\sigma(u, v)$ be a continuous surface. Then, σ is self-intersecting if and only if there exist $\sigma(u_1, v_1) = \sigma(u_2, v_2)$, and every curve on the surface through $\sigma(u_1, v_1)$ and $\sigma(u_2, v_2)$ is also self-intersecting. In this section, a necessary condition for planar self-intersecting surfaces is first presented. Then, any planar bijective (one-to-one and onto) projection of a 3D self-intersecting surface should also self-intersect and satisfy the condition.

Definition 2. *A* minimal self-intersecting surface *σ is a self-intersecting surface such that every nontrivial subdivision of σ results in subsurfaces that are not self-intersecting. A nontrivial subdivision means that the subdivision does not occur along a boundary in the subdivision direction.*

The self-intersection of a minimal self-intersecting surface occurs only at the boundaries. Moreover, the self-intersection point is located either at the two diagonal corners of the surface or at the end points of an isoline. In the latter case, the minimal self-intersecting surface is degenerate and its image is a single isoline curve [7].

Fig. 3 shows two self-intersecting surfaces. On the left, the surface can be subdivided further into smaller self-intersecting subsurfaces, and is not a minimal self-intersecting surface. On the right is one of the subsurfaces that is a minimal self-intersecting surface. It has one self-intersection point. Any nontrivial subdivision of the surface in either parametric directions will separate the two touching corners.

In the following, we use three bounding cones of the surface $\sigma(u, v)$ to derive the surface self-intersection condition. The normal bounding cone $\langle \boldsymbol{a}_n, \theta_n \rangle$ bounds all surface normal vectors. The isoparametric tangent bounding cones $\langle \boldsymbol{a}_u, \theta_u \rangle$, and $\langle \boldsymbol{a}_v, \theta_v \rangle$ bound all tangent vectors in the corresponding isoparametric direction [9]. Vector $\boldsymbol{a}$ is the axis of the cone and angle θ is the spanning

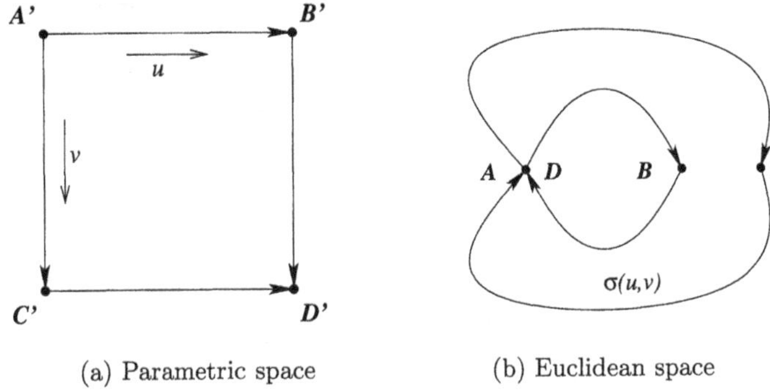

(a) Parametric space (b) Euclidean space

Fig. 4. Planar minimal self-intersecting surface.

of the cone. We start with planar case where the spanning angle of the minimum normal bounding cone is either 0 or π.

Proposition 3. *A self-intersecting planar surface satisfies at least one of the following conditions:*

1) *the spanning angle of the smallest normal bounding cone is π, or*
2) *at least one of the two smallest isoparametric tangent bounding cones has a span of at least π.*

Proof: Since the smallest normal bounding cones of a surface are always greater than or equal to those of a subsurface, all we need to show is that the conditions hold for any planar minimal self-intersecting surface. There are only two types of minimal self-intersecting surfaces. In the case of a self-intersecting isoline, condition 2 holds because of Proposition 1. That leaves only the diagonal corners case.

Consider a planar minimal self-intersecting surface $\sigma(u, v)$ with four corner points $\boldsymbol{A}$, $\boldsymbol{B}$, $\boldsymbol{C}$, and $\boldsymbol{D}$ and the corresponding parametric points $\boldsymbol{A}'$, $\boldsymbol{B}'$, $\boldsymbol{C}'$, and $\boldsymbol{D}'$ (see Fig. 4). It is necessary to show that when condition 1 is false, i.e., when surface normals do not flip, one of the smallest isoparametric tangent bounding cones must span at least π, i.e., condition 2 is true.

Assume that $\boldsymbol{A}$ and $\boldsymbol{D}$ are the two intersecting corners ($\boldsymbol{B}$ and $\boldsymbol{C}$ intersecting is the same as reversing the surface orientation, which does not change spanning angles of the normal bounding cones). In Fig. 4(b), the boundary curves $\boldsymbol{AB}$ and $\boldsymbol{BD}$ form a simple closed loop from $\boldsymbol{A}$ through $\boldsymbol{B}$ to $\boldsymbol{D}$, and have $\pm 2\pi$ total rotation angle. The boundary curves $\boldsymbol{AC}$ and $\boldsymbol{CD}$ also form a closed loop from $\boldsymbol{A}$ through $\boldsymbol{C}$ to $\boldsymbol{D}$, and have $\pm 2\pi$ total rotation angle. Since σ intersects itself only at $\boldsymbol{A}$ and $\boldsymbol{D}$, the two loops do not cross each other in any other places. They can either be on the same side of the self-intersection points, i.e., one inside the other like one shown in Fig. 4, or on the opposite sides of the points as two disjoint regions. Additionally, the two loops can be in the same orientation or in opposite orientation. There are only four possible configurations on a plane geometry, with the other four being mirrors.

When condition 1 does not hold, all surface normal vectors point in the same direction. Two of the four possible configurations have surface normals at $\boldsymbol{A}$ and $\boldsymbol{D}$ pointing in the opposite directions, and they can be ruled out quickly. Among the remaining two cases, the one with two disjoint regions and opposite loop orientations, like the shape of 8, is also impossible. In such a configuration, isolines in same parametric direction will cross each other, which contradicts the assumption that the surface self-intersects only at $\boldsymbol{A}$ and $\boldsymbol{D}$.

The only valid configuration has both loops in the same orientation, one inside the other and touching at points $\boldsymbol{A}$ and $\boldsymbol{D}$ (see Fig. 4). Since $\boldsymbol{A} = \boldsymbol{D}$, the four boundary edges form a double loop $\boldsymbol{ABD}$-$\boldsymbol{ACD}$ whose total rotation angle is $\pm 4\pi$. The double loop can be broken at points $\boldsymbol{B}$ and $\boldsymbol{C}$ into $\boldsymbol{BD}$-$\boldsymbol{AC}$ and $\boldsymbol{CD}$-$\boldsymbol{AB}$, each representing the two boundary isolines of the same parametric direction. Since the jump angles at $\boldsymbol{B}$ and $\boldsymbol{C}$ are at most π, sum of the spanning angles of $\boldsymbol{BD}$-$\boldsymbol{AC}$ and $\boldsymbol{CD}$-$\boldsymbol{AB}$ is at least $4\pi - 2\pi = 2\pi$. That means one of them has to span at least π, and therefore, the smallest isoparametric tangent bounding cone at the corresponding direction also spans at least π. □

An orthographic projection is a projection that maps points in a direction perpendicular to the projection plane. To extend the planar self-intersection condition to 3D surfaces, we want to project the 3D self-intersecting surface onto a plane. The projected planar surface must self-intersect and must satisfy the conditions given above.

Lemma 4. *An orthographic projection of a direction cone $\langle \boldsymbol{a}, \theta \rangle$ in the normal direction $\boldsymbol{n}$ has a projected spanning angle less than π if and only if*

$$|\boldsymbol{n} \cdot \boldsymbol{a}| < \cos \frac{\theta}{2}. \tag{1}$$

What Lemma 4 says is that for the projected cone to span less than π, the projection vector must come from outside of the direction cone. Obviously, this can be true only if the direction cone spans less than π too.

Proposition 5. *A self-intersecting nonplanar surface $\sigma(u, v)$ satisfies at least one of the following conditions:*

$$\theta_n \geq \pi, \tag{2}$$

$$|\boldsymbol{a}_n \cdot \boldsymbol{a}_u| \geq \cos \frac{\theta_u}{2} \tag{3}$$

$$|\boldsymbol{a}_n \cdot \boldsymbol{a}_v| \geq \cos \frac{\theta_v}{2}. \tag{4}$$

Proof: If (2) is not true, an orthographic projection of σ in direction $\boldsymbol{a}_n$ is a bijective mapping. The projected surface is a self-intersecting planar surface whose normal vectors point in one direction. If (3) and (4) are also false, from Lemma 4, the projected isoparametric tangent bounding cones span less than π. This contradicts Proposition 3 and is impossible. □

Proposition 5 provides an easy test for surface self-intersection. Like the bounding box test in surface/surface intersection, a surface that fails this test cannot have any self-intersection. The isoparametric tangent bounding cones of surface $\sigma(u,v)$ can be computed from the partial derivatives σ_u and σ_v. The surface normal bounding cone can be approximated using normals of the control mesh. A tighter bound could be obtained by the cross produce of σ_u and σ_v, which can be computed symbolically as a NURBS surface. A fast but nonoptimal method to bound any cross product between vectors from the two tangent bounding cones is given in [12]. This creates a conservative normal cone whose axis is mutually perpendicular to the axes of the isoparametric tangent bounding cones. With such a loosely bounded normal cone, (2) dominates the other two conditions.

§5. Algorithms and Results

We generalize the subdivision surface/surface intersection algorithm for surface self-intersection. The `CanIntersect` algorithm listed below detects, in general, whether there can be any intersection between two surfaces, or, specifically, whether there exists any self-intersection if the two surfaces are the same.

```
Algorithm CanIntersect( Srf1, Srf2 )
Begin
  If Srf1 and Srf2 are the same surface,
    return CanSelfInter( Srf1 ).
  If Srf1 and Srf2 are neighbors,
    return CanNeighborsInter( Srf1, Srf2 ).
  Otherwise, return CanBBoxInter( Srf1, Srf2 ).
End
```

The `CanSelfInter` function is a direct translation of the equations given by Proposition 5. Function `CanNeighborsInter` detects whether there may be any intersection between two adjacent surfaces at places other than the common boundary. This can be performed by testing the self-intersection conditions on the merged bounding cones of the two neighbors, i.e., by applying `CanSelfInter` on the concatenated surface. Function `CanBBoxInter` is the bounding box test for regular surface/surface intersection.

The rest of the self-intersection algorithm is straightforward. Details of the intersection computation can be found in the references. Note that degenerate self-intersection requires special handling. A divide-by-half subdivision method may not terminate if a degenerate self-intersection exists, and will need to be terminated early when the subsurface is too small. Properties and techniques to find this type of self-intersection are discussed in the next section.

Fig. 5 shows results of the adaptive subdivision surface self-intersection algorithm applied on examples from Fig. 1. On the top are the Euclidean space self-intersection curves and subdivision lines. On the bottom are the corresponding curves in the parametric space. The dotted lines indicate subdivision locations and directions. A surface is subdivided whenever an intersection is possible, with itself or with another surface, until the surface is

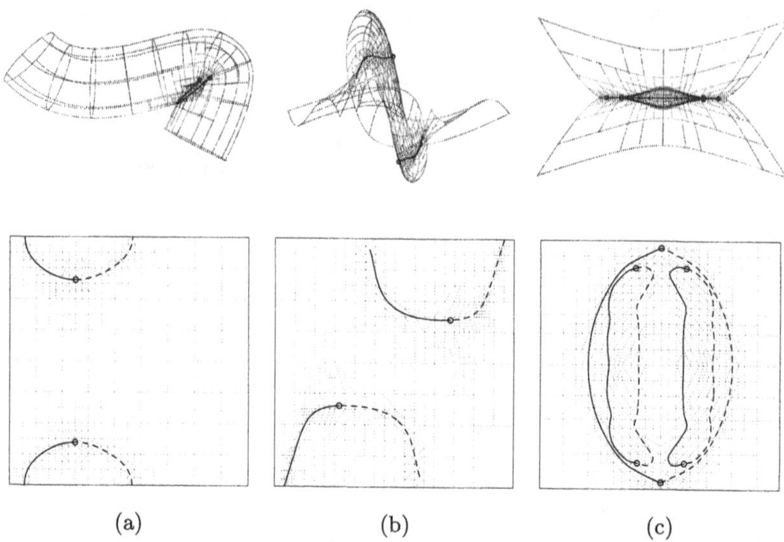

Fig. 5. Adaptive subdivision for surface self-intersection.

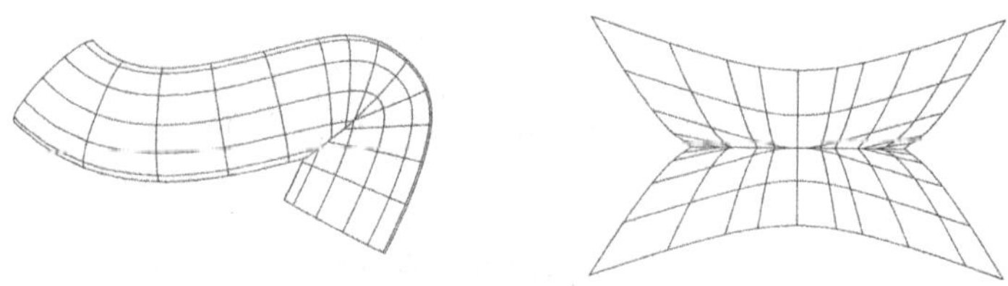

Fig. 6. Trimmed self-intersecting surfaces.

flat or small enough to within a coarse initial tolerance. The two parametric self-intersection curve pair μ_1 and μ_2 are shown in solid and dashed lines, respectively.

In Fig. 5(a), the surface has a seam along its top and bottom edges. A topological adjacency has been declared in the surface model. In addition to subdivision lines, *a priori* adjacency information can be used as part of neighboring surfaces detection. Fig. 5(b) shows a self-intersecting surface whose isolines do not self-intersect. Thus, it would be insufficient to design a self-intersection algorithm based on self-intersecting isolines. Fig. 5(c) shows three self-intersection curves found in an offset surface. Two of them are from the imperfect cusps of the offset approximation. Using Boolean set operations, e.g., union and intersection, surface self-intersections can be trimmed away. Fig. 6 shows the trimmed self-intersecting surfaces.

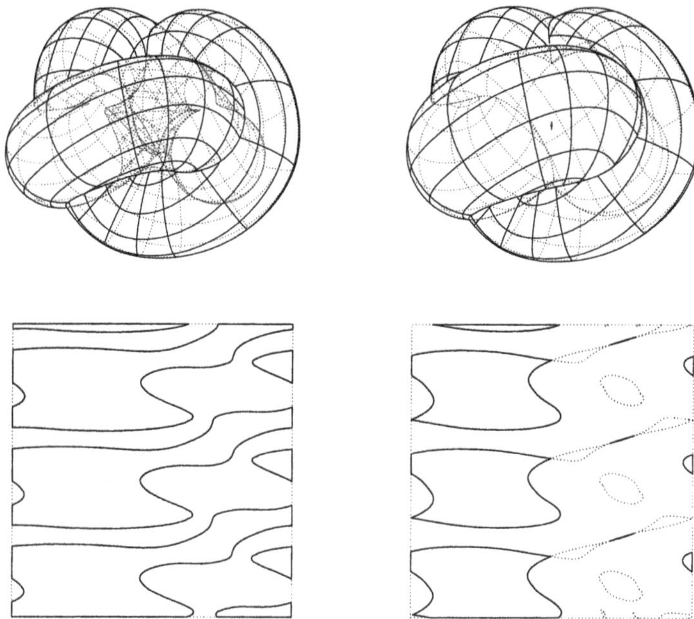

Fig. 7. Self-intersection trimming of sweep surfaces.

More complex examples of surface self-intersection trimming are shown in Fig. 7. The figure on the top shows two sweep surfaces of circlar cross-sections, with self-intersections trimmed away. The surface on the left has a smaller cross-section. This leaves a hollow vertical passage through the center. The surface on the right has a larger cross-section radius which blocks the passage and leaves a small void at the center. The corresponding parametric trimming loops are shown below the surfaces. The dashed lines represent the parametric space self-intersection curves that get trimmed away by the Boolean operations. Details of the surface self-intersection trimming can be found in [7].

§6. Miter Point

Note that some of the self-intersection curves shown in Fig. 5 have circles marked on their end points. From the corresponding parametric space drawings, these are the junction points where parametric self-intersection curve pairs join together, i.e., $\mu_1 = \mu_2$. We call them miter points.

Definition 6. *A* miter point *is an open end point of a surface self-intersection curve where $u_1 = u_2$ and $v_1 = v_2$.*

Fig. 8 shows a self-intersecting surface and its miter point. On the left, the Euclidean self-intersection curve vanishes in the middle of the surface. The two corresponding parametric space self-intersection curves are shown

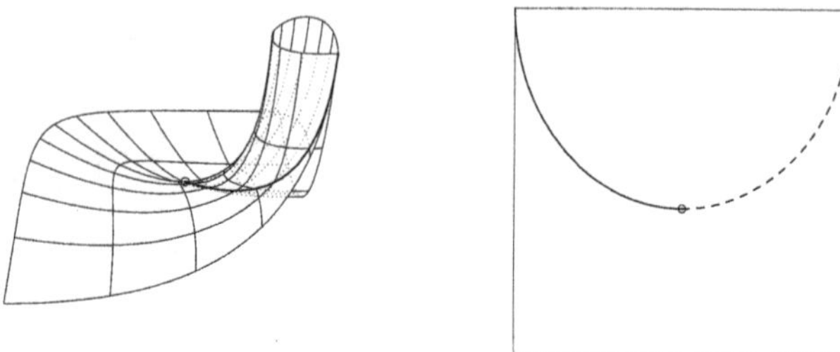

Fig. 8. Miter point of surface self-intersection.

on the right, μ_1 in solid lines and μ_2 in dashed lines. The two curves join together at the point marked by a circle. This is the parameter value of the miter point. Strictly speaking, miter points are not self-intersection points. A self-intersection curve with miter points is a curve whose domain is defined on an open, or semi-open interval.

Miter points occur frequently in sweep surfaces along an elbow bend or a collapsed offset surface. However, pinpointing the exact location of a miter point is not easy. Approaching from one side of a miter point, the surface is not self-intersecting. Approaching from the other side of a miter point, the self-intersection curve vanishes when $\mu_1 \to \mu_2$. Existing numerical methods for surfaces intersection require distinct surface normals at intersection points for intersection computation [3,11]. If marching is used along a self-intersection curve toward its miter point, any overshoot will converge to a single surface point on the other side of the miter point. No easy way to distinguish between them is known. To closely approximate a miter point, the step size must be very small. This gives slow convergence, at best.

Since the miter point is an open end point of the self-intersection curve, any small neighborhood of the miter point is also self-intersecting. This means that unless the subdivision occurs right across the miter point and in a direction to separate the two parametric self-intersection curves, the subdivided surface containing the miter point will always pass the self-intersection bounding cone test. To prevent an endless calculation of unbounded depth, a divide-and-conquer subdivision algorithm should terminate when the surface is small enough.

The following special condition is established to numerically improve the accuracy of a miter point. We assume the self-intersecting surface is C^1 near the miter point.

Proposition 7. *At the miter point of a C^1 self-intersecting surface $\sigma(u, v)$,*

$$\sigma_u \times \sigma_v = 0. \tag{5}$$

Proof: Since any arbitrarily small neighborhood of a miter point is self-intersecting and satisfies the conditions given in Proposition 5, there is no

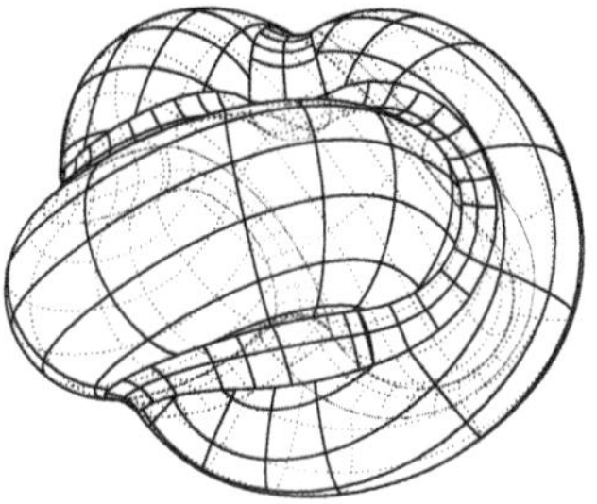

Fig. 9. Self-filleting operation.

unique surface normal defined around a miter point, i.e., the surface is not regular at miter points. For a C^1 surface, the cross product of its partial derivatives is null at a point that is not regular. □

The condition given above is only a necessary condition for miter points. To verify that a point satisfying the condition is actually a miter point, a self-intersection curve must be traced to end at the miter point. When the subdivision of a self-intersecting surface is terminated early due to the afore mentioned reason, we find the potential miter point using Proposition 7. Then, we trace the self-intersection curve from the miter point toward the self-intersection point at the subdivision boundary.

§8. Conclusions

This paper presents a theoretical and pratical solution to the surface self-intersection problem. The surface self-intersection conditions and algorithms presented in this paper apply not only to a single surface but can also be used on models consisting of multiple surfaces. The general self-intersection algorithm can also be used to trim away self-intersections of offset surfaces which is important in applications such as NC machining and solid modeling. Fig. 9 shows an example of self-filleting operation on the self-intersecting sweep surface shown in Fig. 7. The self-filleting operation consists of three different operations. First, offset the surface by the fillet radius. Then, trim away any self-intersections. Finally, offset the trimmed offset surface back.

Acknowledgments. This work was supported in part by the NSF Science and Technology Center for Computer Graphics and Scientific Visualization (ASC-89-20219). All opinions, findings, conclusions or recommendations expressed in this document are those of the authors and do not necessarily reflect the views of the sponsoring agencies.

References

1. Andersson, L., T. Peters, and N. Stewart, Selfintersection of composite curves and surfaces, Comput. Aided Geom. Design **15** (1998), 507–527.

2. Bajaj, C., C. Hoffmann, R. Lynch, and J. Hopcroft, Tracing surface intersections, Comput. Aided Geom. Design **5** (1988), 285–307.
3. Barnhill, R., G. Farin, M. Jordan, and B. Piper, Surface/surface intersection, Comput. Aided Geom. Design **4** (1987), 3–16.
4. Barnhill, R., T. Frost, and S. Kersey, Self-intersections and offset surfaces, in *Geometry Processing for Design and Manufacturing*, R. Barnhill, (ed), Society for Industrial and Applied Mathematics, Philadelphia, 1992, 35–44.
5. Barnhill, R. and S. Kersey, A marching method for parametric surface/surface intersection, Comput. Aided Geom. Design **7** (1990), 257–280.
6. Elber, G., Free form surface analysis using a hybrid of symbolic and numeric computation, dissertation, Univ. of Utah, Salt Lake City, 1992.
7. Ho, C.-C., Feature-based process planning and automatic numerical control part programming, dissertation, Univ. of Utah, Salt Lake City, 1997.
8. Houghton, E., R. Emnett, J. Factor, and C. Sabharwal, Implementation of a divide-and-conquer method for intersection of parametric surfaces, Comput. Aided Geom. Design **2** (1985), 173–183.
9. Kim, D., P. Papalambros, and T. Woo, Tangent, normal, and visibility cones on Bézier surfaces, Comput. Aided Geom. Design **12** (1995), 305–320.
10. Millman, R. and G. Parker, *Elements of Differential Geometry*, Prentice-Hall, Inc.,, Englewood Cliffs, New Jersey, 1977.
11. Müllenheim, G., On determining start points for a surface/surface intersection algorithm, Comput. Aided Geom. Design **8** (1991), 401–408.
12. Sederberg, T. and R. Meyers, Loop detection in surface patch intersections, Comput. Aided Geom. Design **5** (1988), 161–171.
13. Thomas, S., Modeling volumes bounded by B-spline surfaces, dissertation, Univ. of Utah, Salt Lake City, 1984.

Chih-Cheng Ho
Engineering Geometry Systems
275 East South Temple, Suite 305
Salt Lake City, UT 84111, USA
`ho@cs.utah.edu`

Elaine Cohen
University of Utah
50 South Central Campus Drive, 3190 MEB
Salt Lake City, UT 84112, USA
`cohen@cs.utah.edu`

Error Estimates for the web-Spline Method

Klaus Höllig, Ulrich Reif, and Joachim Wipper

Abstract. The web-spline method is a new finite element technique for solving elliptic boundary value problems on bounded domains in $\mathbb{R}^m$. It is based on **w**eighted **e**xtended **B**-splines and provides high accuracy approximations with relatively low dimensional subspaces. Further, it does not require any non-trivial grid generation process. In this paper, we review the method, prove stability, and present a Jackson-type error estimate showing that web-splines have optimal approximation order.

§1. Introduction

Tensor-product B-splines are a powerful tool for the approximation of multivariate functions providing high accuracy with a relatively low number of coefficients. Nevertheless, they are rarely used as finite elements for solving elliptic boundary value problems. Supposedly, this is due to two reasons:

- Dirichlet boundary conditions are not immediately compatible with approximations on regular grids. Two approaches devised by Babuška, the Lagrange multiplier method [2] and the penalty method [1], are in principle capable of solving this problem, but reveal some severe drawbacks in applications.
- When restricted to a bounded domain, the B-spline basis loses its uniform stability with respect to the knot spacing, and thus one of the most important prerequisites for approximation purposes. This phenomenon is caused by the fact that certain B-splines can have very small support inside the domain.

In [11,12] we propose **w**eighted **e**xtended **B**-splines (web-splines) as a natural generalisation of standard B-splines to overcome these difficulties. As will be described in the next section, we employ a weight function w to ensure that all web-splines vanish on the boundary. The stability problem is overcome as follows: We distinguish between inner B-splines and outer B-splines, where outer B-splines are characterised by a very small support inside the domain. Now, the inner B-splines are extended by coupling them with outer B-splines

Mathematical Methods for Curves and Surfaces: Oslo 2000
Tom Lyche and Larry L. Schumaker (eds.), pp. 195–209.

ISBN 0-8265-1378-6

in such a way that both the locality of the support and the approximation power are retained.

The idea of using weight functions to satisfy Dirichlet boundary conditions goes back to Kantorovich [14]. In the meantime, a comprehensive theory of this approach has been developed by Rvachev and others, see e.g. [16,19]. Besides polynomials and trigonometric functions, also B-splines were suggested as shape functions [17,20], but the hitherto unsolved stability problem remained as a serious drawback of the method. An extension process to stabilise wavelet bases, which reveals some parallels to our approach, was suggested by Oswald [15].

The paper is organised as follows: After introducing some basic notation in the next section, we review the construction of web-splines and some of their properties in Section 3. In Sections 4 and 5, we investigate stability and approximation power of web-splines, and finally, in Section 6, we discuss their application as finite elements

§2. Notation

We denote by b the cardinal tensor-product B-spline of order n with support $[0, n]^m$, see Figure 1 (left). The scaled translates

$$b_k^h(x) = h^{-m/2}\, b(x/h - k), \quad k \in \mathbb{Z}^m \tag{1}$$

are $(n-2)$-times continuously differentiable and polynomials of coordinate order n on each of the grid cells

$$Q_\ell^h = h\big([0,1)^m + \ell\big), \quad \ell \in \mathbb{Z}^m,$$

of width h.

Since the non-zero polynomial segments of b are linearly independent, for any $\ell \in \{0, \dots, n-1\}^m$ there exists a function λ_ℓ with support in $[\frac{1}{4}, \frac{3}{4}]^m + \ell$ such that

$$\int b(\cdot - k)\lambda_\ell = \delta_{k,0}.$$

Two out of the three different functions $\lambda_0, \lambda_1, \lambda_2$ for the quadratic case are shown in Figure 1. Hence, for arbitrary $\ell(i) \in \{0, \dots, n-1\}$, the functions

$$\lambda_{\ell(i),i}^h(x) := h^{-m/2}\, \lambda_{\ell(i)}(x/h - i), \quad \ell \in \mathbb{Z}^m, \tag{2}$$

form a family of dual functionals in the L_2-sense for the B-splines $b_i^h, i \in \mathbb{Z}^m$. Because of the normalization factor, the L_2-norms of the B-splines and the dual functionals are bounded independent of h and i,

$$\|b_i^h\| \sim \|\lambda_{\ell(i),i}^h\| \sim 1. \tag{3}$$

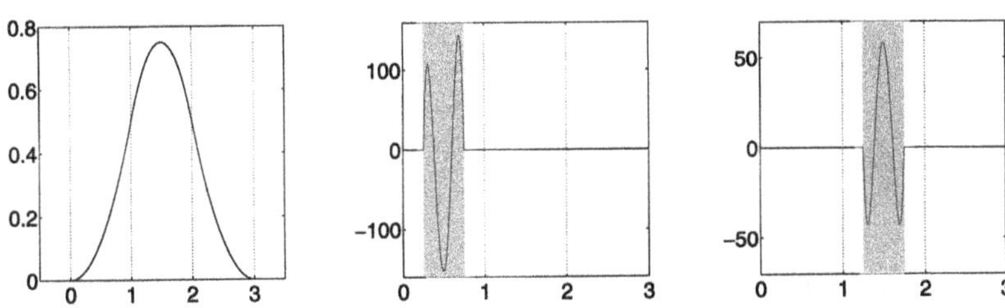

Fig. 1. Univariate quadratic cardinal B-spline (left), dual functional λ_0 (middle), and dual functional λ_1 (right).

Let $\Omega \subset \mathbb{R}^m$ be a bounded domain. Then the Ω-relevant B-splines have indices in

$$K := \{k \in \mathbb{Z}^m : \operatorname{supp} b_k \cap \Omega \neq \emptyset\}.$$

A spline s on Ω is a linear combination of Ω-relevant B-splines,

$$s = \sum_{k \in K} a_k b_k^h, \quad a_k \in \mathbb{R}.$$

The sequence of coefficients is denoted by $A = \{a_k\}_{k \in K}$. As with all vectors here, it will be measured in the Euclidean norm

$$\|A\| := \left(\sum_{k \in K} a_k^2 \right)^{1/2}.$$

The Sobolev space $H^p(\Omega), p \in \mathbb{N}_0$, on Ω is endowed with the norm

$$\|f\|_p := \left(\sum_{|\alpha| \leq p} \int_\Omega |D^\alpha f|^2 \right)^{1/2},$$

where $\alpha \in \mathbb{N}_0^m$ denotes a multi-index. If the norm is restricted to a subdomain $\Omega' \subset \Omega$, we write $\|f\|_{r,\Omega'}$. The inner-product in $L_2(\Omega) = H^0(\Omega)$ is denoted by $\langle f, g \rangle := \int_\Omega fg$. The space $H_0^1(\Omega)$ consists of all functions in $H^1(\Omega)$ which vanish on the boundary $\partial\Omega$ of Ω. Further, we write

$$f \preceq g,$$

if $f \leq cg$ with a constant c which does not depend on the grid width h, indices, or arguments of f, g. The symbols $\succeq$ and $\sim$ are defined analogously.

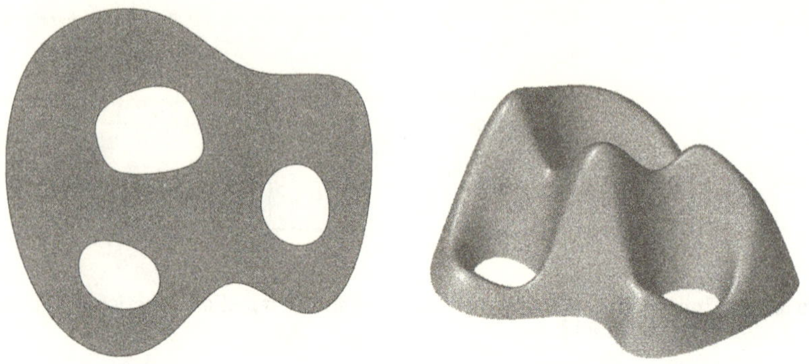

Fig. 2. Domain Ω (left) and solution u to $-\Delta u = 1$ (right).

§3. The web-Spline Basis

In this section, we recall the construction of web-splines as introduced in [11], and present a result on their stability in Sobolev spaces.

The construction of web-splines is geared to solving second order elliptic problems with Dirichlet boundary conditions. As a generic example, we may consider Poisson's equation

$$\begin{aligned} -\Delta u &= f \quad \text{in} \quad \Omega, \\ u &= 0 \quad \text{on} \quad \partial\Omega, \end{aligned} \tag{4}$$

on a smoothly bounded domain $\Omega \subset \mathbb{R}^m$, see Figure 2.

For $f \in H^{p-2}(\Omega)$, $p \geq 1$, there exists a weak solution $u \in H_0^1(\Omega) \cap H^p(\Omega)$ which will be approximated by a linear combination of shape functions B_i,

$$u \approx u^h = \sum_{i \in I} a_i B_i^h.$$

Standard tensor-product splines are not well suited since interpolation of the boundary value can in general only be achieved if the spline vanishes on all grid cells intersecting $\partial\Omega$. Also approximation techniques, e.g. following Babuška [1,2], provide no easy solution. Instead, we advocate the following natural approach. Let $w \sim \operatorname{dist}(\cdot, \partial\Omega)$ be a smooth weight function which is equivalent to the boundary distance, i.e., there exist constants $r, R > 0$ such that

$$rw(x) \leq \operatorname{dist}(x, \partial\Omega) \leq Rw(x), \quad x \in \Omega.$$

Then, the weighted B-splines

$$B_k^h := wb_k^h, \quad k \in K,$$

are a set of shape functions guaranteeing interpolation of the boundary value. The drawback of this approach, which was already considered in [17,20], is

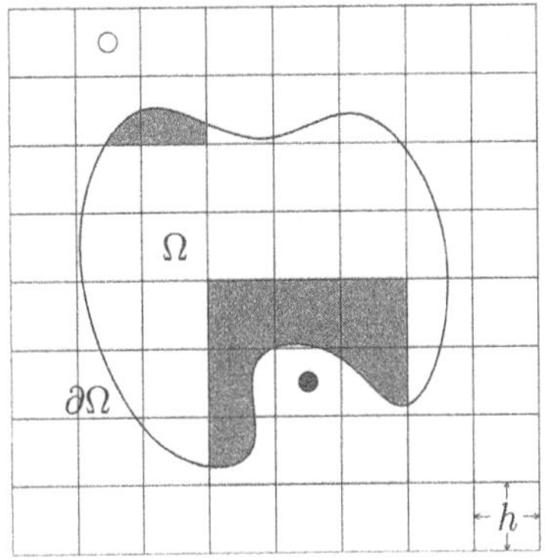

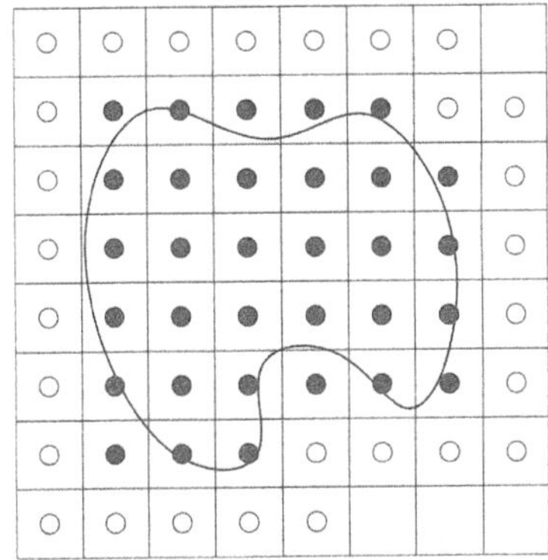

Fig. 3. Support of B-splines (left) and arrangement of outer (∘) and inner (•) B-splines (right).

that the basis $\{B_k^h, k \in K\}$ is not uniformly stable with respect to the knot spacing. More precisely, the constant

$$M^h := h^r \sup_A \frac{\|A\|}{\|\sum_k a_k B_k^h\|_r}$$

is unbounded as $h \to 0$. This implies for instance that L_2-Gramian or Galerkin matrices can be extremely badly conditioned.

This problem is due to those B-splines which have very small support in Ω, and can be resolved as follows: We divide the set of Ω-relevant B-splines into two classes. The inner B-splines b_i have at least one grid cell of their support inside the domain. contained in Ω. That is, there exists an $\ell(i) \in \{0, \ldots, n-1\}^m$ such that $Q^h_{\ell(i)+i} \subset \operatorname{supp} b_i \cap \Omega$. The set off all indices of inner B-splines is denoted by I. The complementary class of outer B-splines b_j has indices in $J := K \backslash I$. For the quadratic case, this classification is illustrated in Figure 3, where the markers are located at the centers of the corresponding B-splines.

Then, we extend the inner B-splines by linear combinations of outer B-splines,

$$\tilde{B}_i^h := b_i^h + \sum_{j \in J} e_{i,j} b_j^h.$$

The coefficients $e_{i,j}$ are chosen such that the size of the support of $\tilde{B}_i^h$ is $\preceq h$ and that all polynomials of order n remain in the span of the functions $\tilde{B}_i^h$. Thus, two basic properties of standard B-splines, which account for their approximation power, are retained. A convenient construction for the coefficients is based on Marsden's identity, saying that a control mesh which is polynomial in the index corresponds to a uniform spline which is a polynomial of equal order. For fixed $j \in J$, let

$$I(j) = \{i \in I : \alpha_\mu \le i_\mu < \alpha_\mu + n\}$$

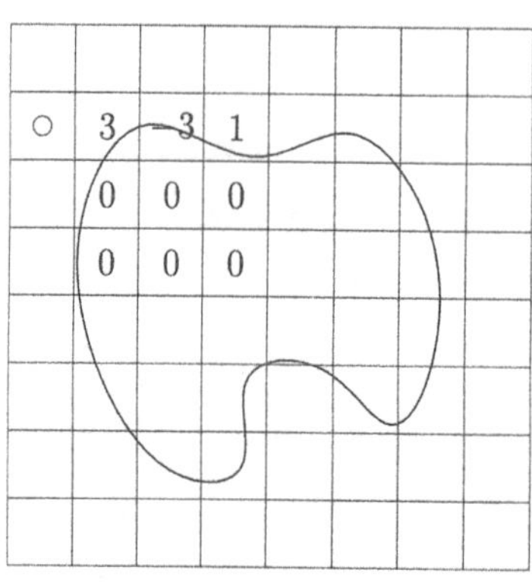

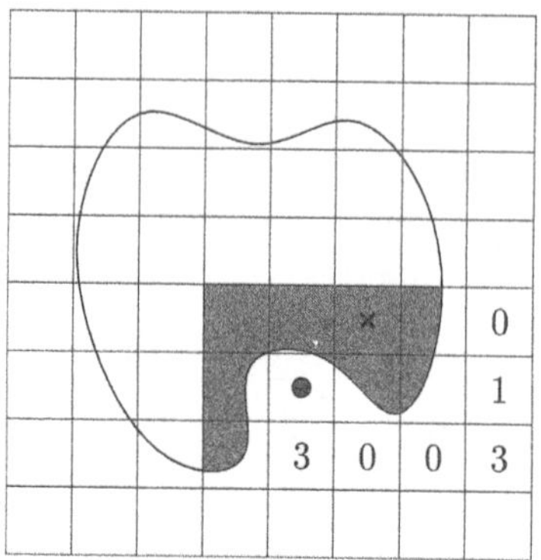

Fig. 4. Coefficients $e_{i,j}$ (left) and support of a web-spline (right).

be the nearest $(n \times n)$-array of indices in I, and $l_{j,i}$ the Lagrange polynomial with

$$l_{j,i}(k) = \delta_{i,k}, \quad i, k \in I(j).$$

Then, the choice

$$e_{i,j} := \begin{cases} l_{j,i}(j) & \text{for } i \in I(j), \\ 0 & \text{otherwise,} \end{cases} \tag{5}$$

satisfies all requirements. In particular, $e_{i,j} = 0$ for $|i-j| \succ 1$. The relevant coefficients $e_{i,j}, i \in I(j)$, can be expressed explicitly by the formula

$$e_{i,j} = \prod_{\mu=1}^{m} \prod_{\substack{\ell=\alpha_\mu \\ \ell \neq i_\mu}}^{\alpha_\mu+n-1} \frac{j_\mu - \ell}{i_\mu - \ell}.$$

A more detailed derivation can be found in [11]. Now, the web-splines are obtained from the functions $\tilde{B}_i^h$ by multiplication with the weight function w and appropriate scaling.

Definition 1. *For coefficients $e_{i,j}$ according to (5), the extended B-splines (eb-splines) $\tilde{B}_i^h$ and the weighted extended B-splines (web-splines) B_i^h are defined by*

$$\tilde{B}_i^h := b_i + \sum_{j \in J} e_{i,j} b_j, \quad B_i^h := \frac{w}{w(x_i)} \tilde{B}_i^h, \quad i \in I,$$

where x_i is the center of the grid cell $Q^h_{\ell(i)+i}$. The corresponding spline spaces are called eb-space *and* web-space*, respectively, and are denoted by*

$$\tilde{\mathcal{B}}^h := \operatorname{span}\{\tilde{B}_i^h : i \in I\}, \quad \mathcal{B}^h := \operatorname{span}\{B_i^h : i \in I\}.$$

Figure 4 illustrates the construction for the quadratic case. On the left-hand side, the values of the coefficients $e_{i,j}$ with $i \in I(j)$ are given for an

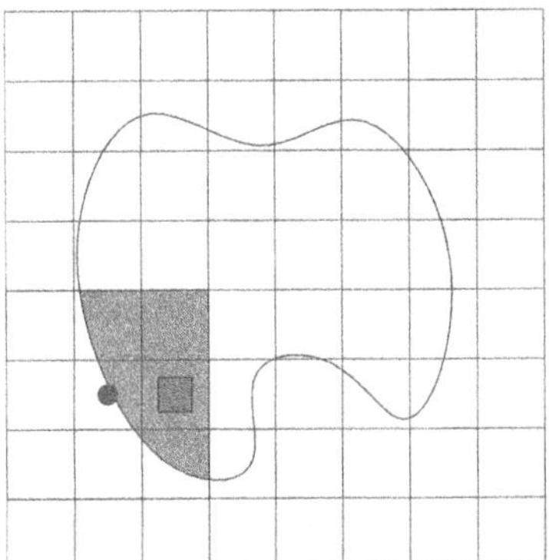

Fig. 5. Support of the dual functional Λ_k^h.

outer B-spline b_j^h. The zeros in the second and third row indicate that this outer B-spline is actually involved only in 3 web-splines B_i^h. This is because $j_1 = i_1$ for the top row of the lattice points $i \in I(j)$, causing the Lagrange polynomials associated with the other lattice points in $I(j)$ to vanish at j. On the right-hand side, the support of a web-spline B_i^h is shown together with the coefficients $e_{i,j}$ of those adjoined outer B-splines b_j^h with $i \in I(j)$. Again, we note some zeros due to the evaluation of the Lagrange polynomials. In particular, the outer B-spline corresponding to the uppermost coefficient does not contribute to the support of B_i^h. The point x_i is marked by a cross. Again, markers and coefficients are assigned to the center of the support of the corresponding B-spline.

§4. Stability

In this section, we discuss some stability properties of web-splines. As usual, stability is established by means of dual functionals. For their construction, we note that by definition, all inner B-splines $b_k^h, k \in I$, possess a dual functional $\lambda_{\ell(k),k}^h$ according to (2). For the B-spline marked in Figure 5, the support of the dual functional for the choice $\ell(i) = [2, 1]$ is highlighted.

Lemma 2. *The functions*

$$\tilde{\Lambda}_k^h := \lambda_{\ell(k)+k}^h, \quad \Lambda_k^h := \frac{w(x_k)}{w} \tilde{\Lambda}_k^h, \quad k \in I$$

are dual to eb-splines and web-splines, respectively, i.e.,

$$\langle \tilde{B}_i^h, \tilde{\Lambda}_k^h \rangle = \langle B_i^h, \Lambda_k^h \rangle = \delta_{i,k}, \quad i, k \in I.$$

Proof: Since the supports of the outer B-splines $b_j, j \in J$, and the dual functionals $\lambda_{\ell(k),k}^h, k \in I$ are disjoint, we have

$$\langle b_i^h, \lambda_{\ell(k),k}^h \rangle = \delta_{i,k}, \quad i \in K, \ k \in I.$$

This implies the bi-orthogonality relation for eb-splines, and with this, the result for web-splines follows immediately. □

An immediate consequence of the existence of a set of dual functions is the fact that eb-splines and web-splines are linearly independent. Hence, they form a basis of the spaces $\tilde{\mathcal{B}}^h$ and $\mathcal{B}^h$, respectively.

Lemma 3. *For $0 < n$, eb-splines, web-splines, and their dual functions are bounded according to*

$$\|\tilde{B}_i\|_\ell \preceq h^{-\ell}, \quad \|B_i\|_\ell \preceq h^{-\ell}, \quad \|\tilde{\Lambda}_k\|_0 \preceq 1, \quad \|\Lambda_k\|_0 \preceq 1, \quad i,k \in I.$$

Proof: Scaling according to (1) implies $\|D^\beta b_i^h\|_0 \preceq h^{-|\beta|}$. Since the number and magnitude of the non-zero coefficients $e_{i,j}$, $j \in J$, is bounded independent of i and h, the result for eb-splines follows. To verify the estimate for web-splines, we consider Leibniz' formula for the derivatives of a product,

$$D^\alpha B_i^h = \frac{w}{w(x_i)} D^\alpha \tilde{B}_i^h + \frac{1}{w(x_i)} \sum_{\beta<\alpha} \binom{\alpha}{\beta} D^{\alpha-\beta} w \, D^\beta \tilde{B}_i^h.$$

We note that $w/w(x_i)$ and $D^{\alpha-\beta}w$ are bounded, while $1/w(x_i) \preceq h^{-1}$. Hence,

$$\left\|D^\alpha B_i^h\right\|_0 \preceq h^{-|\alpha|} + \sum_{\beta<\alpha} h^{-|\beta|-1} \preceq h^{-|\alpha|},$$

as required. The estimates for the dual functions follow from (3) and the boundedness of $w(x_k)/w$ on the support of $\tilde{\Lambda}_k^h$, which, by construction, has distance $\succeq h$ to the boundary. □

The following stability result is crucial for approximation purposes in general, and for finite element applications in particular.

Theorem 4. *The web-basis is L_2-stable, i.e.,*

$$\left\|\sum_{i\in I} a_i B_i^h\right\|_0 \sim \|A\|. \tag{6}$$

More generally, in terms of Sobolev norms,

$$\|A\| \preceq \left\|\sum_{i\in I} a_i B_i^h\right\|_\ell \preceq h^{-\ell}\|A\|. \tag{7}$$

The same statements are true for eb-splines.

Proof: The lower bound on the norm of the spline $f = \sum_i a_i B_i^h$ is based on the uniform boundedness of the dual functionals,

$$\sum_{i\in I} a_i^2 = \sum_{i\in I} \langle f, \Lambda_i^h\rangle^2 \leq \sum_{i\in I} \|f\|_{0,\,\mathrm{supp}\,\Lambda_i^h}^2 \|\Lambda_i^h\|_0^2 \preceq \|f\|_0^2 \leq \|f\|_\ell^2.$$

The upper bound relies on the locality and boundedness of the web-basis. The number of indices in $I(x) := \{i \in I : B_i^h(x) \neq 0\}$ is bounded independent of x and h. Hence, by Lemma 3,

$$\begin{aligned}\|D^\alpha f\|_0^2 &= \int_\Omega \Big(\sum_{i\in I(x)} a_i D^\alpha B_i^h(x)\Big)^2 dx \preceq \int_\Omega \sum_{i\in I(x)} a_i^2 \left(D^\alpha B_i^h(x)\right)^2 dx \\ &= \sum_{i\in I} \int_\Omega a_i^2 \left(D^\alpha B_i^h(x)\right)^2 dx \le \|A\|^2 \max_{i\in I} \|D^\alpha B_i^h\|_0^2 \preceq h^{-|\alpha|},\end{aligned}$$

proving our claim. The proof for eb-splines is analogous. □

§5. Approximation Power

In this section, we prove a Jackson inequality which demonstrates the full approximation power of web-splines. We omit the discussion of an analogous result for eb-splines, which can be obtained using standard arguments. The situation for web-splines is more subtle since we approximate functions u, which vanish on $\partial\Omega$, by linear combinations of weighted B-splines. Hence, the error bounds will depend on the regularity of the quotient $v := u/w$. In analysing the smoothness of v, we start with a simple observation.

Lemma 5. *For any subdomain $\Omega' \subset \Omega$ with distance δ to the boundary and $k \ge 1$,*

$$\|v\|_{k,\Omega'} \preceq \delta^{-1} \left(\|u\|_{k,\Omega'} + \|v\|_{k-1,\Omega'}\right).$$

Proof: This lemma is a direct consequence of Leibniz' formula for the derivatives of the product $u = wv$. Rearranging the terms appropriately, we have

$$wD^\alpha v = D^\alpha u - \sum_{\beta<\alpha} \binom{\alpha}{\beta} D^{\alpha-\beta} w \, D^\beta v.$$

Dividing by w and recalling that w is smooth and equivalent to the distance, the desired inequality follows. □

Lemma 5 indicates a loss of regularity of the quotient v near the boundary. This can be made more precise with the aid of the following generalisation of Hardy's inequality.

Lemma 6. *For $u \in H_0^1(\Omega) \cap H^k(\Omega)$ and $v := u/w$,*

$$\|v\|_{k-1} \preceq \|u\|_k. \tag{8}$$

Proof: The proof is based on the following univariate estimate: If $p(0) = 0$ and $q(t) := p(t)/t$, then

$$\left|q^{(k-1)}\right| \preceq \left|p^{(k)}\right|, \tag{9}$$

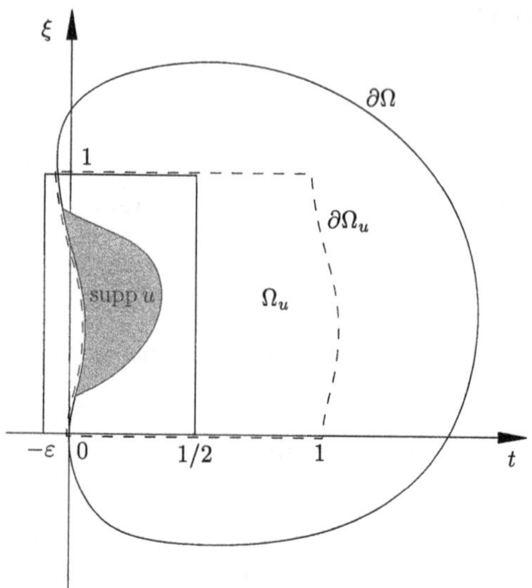

Fig. 6. Estimate near the boundary.

where $|\cdot|$ denotes the L_2-norm on $[0,1]$. This estimate follows from the identity

$$q(t) = \frac{1}{t}\int_0^t p'(\tau)\,d\tau = \int_0^1 p'(t\tau)\,d\tau.$$

Differentiating $(k-1)$-times and taking norms, we obtain

$$\left|q^{(k-1)}\right| \le \int_0^1 \left|\tau^{k-1}p^{(k)}(\cdot\tau)\right| d\tau = \int_0^1 \tau^{k-1}\left(\int_0^1 \left|p^{(k)}(t\tau)\right|^2 dt\right)^{1/2} d\tau.$$

Changing variables in the inner integral, the right hand side can be bounded by $\int_0^1 \tau^{k-3/2}\left|p^{(k)}\right| d\tau$, which proves (9). This estimate is readily generalised to several variables. By a standard partition of unity technique, a local proof is sufficient. Hence, we may assume that u has small support near the boundary (the interior estimate is trivial) as depicted in Figure 6. With an appropriate choice of the coordinate system, the support of u is contained in $[0,1]^{m-1} \times [-\varepsilon, 1/2]$. Moreover, the relevant portion of the domain Ω can be represented as a strip

$$\Omega_u := \{x = (\xi, t) :\ 0 \le \xi_i \le 1,\ \psi(\xi) \le t \le 1 + \psi(\xi)\}$$

with

$$\psi(0) = 0, \quad \nabla\psi(0) = 0, \quad \|\nabla\psi(\xi)\| \le \varepsilon$$

and $\varepsilon \ll 1$.

We first conclude from (9) that the lemma is valid for the cube $Q_* = [0,1]^m$ if the boundary is flat ($\psi \equiv 0$) and $w(x) = t$. We simply observe that tangential derivatives do not affect the estimate.

For a curved boundary, we map the strip to the cube Q_* with the diffeomorphism

$$x = (\xi, t) \leftrightarrow (\xi, s) = y, \quad s = t - \psi(\xi).$$

Denoting by $\tilde{u}, \tilde{v}$ functions with respect to the transformed variables, the lemma follows from the inequalities

$$\|v\|_{k-1} = \|v\|_{k-1,\Omega_u} \preceq \|\tilde{v}\|_{k-1,Q_*} \preceq \|\tilde{u}/s\|_{k-1,Q_*} \preceq \|\tilde{u}\|_{k,Q_*} \preceq \|u\|_{k,\Omega_u}.$$

The first and last inequalities hold because Sobolev norms remain bounded after a smooth change of variables. For the second inequality we note that

$$\tilde{v}(y) = \frac{u(\xi, s+\psi(\xi))}{w(\xi, s+\psi(\xi))} = \frac{s}{w(\xi, s+\psi(\xi))} \frac{\tilde{u}(y)}{s},$$

where the first quotient on the right hand side is smooth. □

We are now prepared to prove our main result.

Theorem 7. *The quasi-interpolant $\mathcal{P}^h u := \sum_{i \in I} \langle \Lambda_i^h, u \rangle B_i^h$ approximates functions $u \in H_0^1(\Omega) \cap H^p(\Omega)$ with optimal order, i.e.*

$$\|u - \mathcal{P}^h u\|_\ell \preceq h^{k-\ell} \|u\|_k \tag{10}$$

for $0 \le \ell < k \le \min(n, p)$.

Proof: With the aid of the regularity results in the preceding section, the proof of Theorem 7 is standard. Because of the local support of the basis functions, we can prove the estimate separately for each grid cell Q. To this end, we denote by $I(Q)$ the set of all indices i, for which B_i^h is non-zero on Q, and by $\tilde{Q}$ the union of the supports of these B_i^h intersected with Ω. Clearly, the number of lattice points in $I(Q)$ is $\preceq 1$, and the diameter of $\tilde{Q}$ is $\preceq h$.

Let p be a polynomial of order n approximating v with

$$\|v - p\|_{\nu,\tilde{Q}} \preceq h^{\mu-\nu} \|v\|_{\mu,\tilde{Q}} \tag{11}$$

for $0 \le \nu < \mu \le \min(n, p)$. Using that $\mathcal{P}^h$ reproduces the weighted polynomial wp, which is a function in $\mathcal{B}^h$, we write

$$u - \mathcal{P}^h u = w(v-p) - \sum_{i \in I} a_i B_i^h, \quad a_i = \langle \Lambda_i^h, w(v-p) \rangle, \tag{12}$$

and estimate each term separately.

For the first term, the simple estimate

$$\|w(v-p)\|_{\ell,\tilde{Q}} \preceq \|v-p\|_{\ell,\tilde{Q}}$$

does not yield the optimal approximation order for functions with minimal regularity. We have to exploit the fact that w is small near the boundary, where the highest derivatives of v become large. To this end, we note that

$$\|w(v-p)\|_{\ell,Q\cap\Omega} \preceq \|w(v-p)\|_{\ell,\tilde{Q}} \preceq \ell\|v-p\|_{\ell-1,\tilde{Q}} + \left(\max_{\tilde{Q}} w\right) \|v-p\|_{\ell,\tilde{Q}},$$

and claim that both summands on the right can be bounded by

$$h^{k-\ell}\left(\|u\|_{k,\tilde{Q}} + \|v\|_{k-1,\tilde{Q}}\right). \tag{13}$$

For the first summand, this follows from (11) with $\nu = \ell - 1, \mu = k - 1$. For the second summand, we have to consider two cases. If $\delta = \operatorname{dist}(\tilde{Q}, \partial\Omega) \geq h$, we apply (11) with $\nu = \ell, \mu = k$ and Lemma 5, noting that

$$\max w \preceq \delta + h \preceq \delta.$$

If $\delta \leq h$, the bound follows from (11) with $\nu = \ell, \mu = k-1$ since we gain an additional factor h by estimating $\max w$.

For the second term in (12), we obtain

$$\left\| \sum_{i\in I(Q)} a_i B_i^h \right\|_{\ell,Q\cap\Omega} \preceq h^{-\ell} \left(\sum_{i\in I(Q)} |a_i|^2 \right)^{1/2},$$

where we have omitted the summands for which B_i^h has no support in Q. Since the number of indices in $I(Q)$ is $\preceq 1$, we can replace the 2-norm on the right hand side by a maximum. By the boundedness of the dual functionals Λ_i according to Lemma 3, each of the coefficients satisfies

$$|a_i| \preceq \|w(v-p)\|_{0,\tilde{Q}}.$$

Combining this with the bound (13) for $\|w(v-p)\|_{0,\tilde{Q}}$, ($\ell = 0$), we have shown that

$$\|u - \mathcal{P}_h^h u\|^2_{\ell,Q\cap\Omega} \preceq h^{2(k-\ell)}\left(\|u\|^2_{k,\tilde{Q}} + \|v\|^2_{k-1,\tilde{Q}}\right).$$

Summing this inequality over all grid cells Q and applying Lemma 6 to bound $\|v\|_{k-1}$ completes the proof. □

§6. Finite Elements

In this section, we outline the application of web-splines to finite element problems using Poisson's equation with Dirichlet boundary conditions (4) as a model problem. Further details and proofs can be found in [11,12].

The variational form of (4) is

$$\langle \nabla u, \nabla v\rangle = \langle f, v\rangle \quad \text{for all} \quad v \in H_0^1(\Omega).$$

Approximating the solution $u = u_*$ by a web-spline

$$u_* \approx u^h = \sum_{i \in I} a_i^h B_i^h \tag{14}$$

and restricting the space of test functions to $\mathcal{B}^h$, we obtain the Galerkin system

$$G^h A_*^h = F^h, \quad g_{k,i}^h = \langle \nabla B_k^h, \nabla B_i^h \rangle, \quad f_k^h = \langle f, B_k^h \rangle, \quad i, k \in I \tag{15}$$

for the coefficients A_*^h of u_*^h. The approximation order derived in the preceding section remains valid for the Galerkin approximation.

Theorem 8. *For a solution u_* of (14) in $H_0^1(\Omega) \cap H^p(\Omega)$ and $0 \leq \ell < k \leq \min(n, p)$, the approximation u^h defined by the Galerkin system (15) satisfies*

$$\|u_* - u^h\|_\ell \preceq h^{k-\ell} \|u\|_k.$$

The proof is standard and involves Céa's Lemma and Nitsche's duality argument. Thus, for smooth solutions, accurate approximations can be constructed with a relatively low number of coefficients.

Estimating the Rayleigh quotient $A^t G^h A / A^t A$ by means of Theorem 4, it can be shown that the growth of the condition number of the Galerkin matrix is moderate as $h \to 0$.

Theorem 9. *The condition number of the Galerkin matrix G^h is bounded by* $\operatorname{cond} G^h \preceq h^{-2}$.

It is worth noting that a similar statement is not true for weighted (non-extended) B-splines. First numerical examples show that the Galerkin system can usually be solved efficiently using standard iterative solvers like an SSOR-preconditioned conjugate gradient algorithm [10]. However, for very small h, the number of iteration might grow unacceptably. This problem can be overcome by multigrid techniques as described for instance in [3,6,7,21]. The required algorithms for hierarchical refinement can be derived from the well known subdivision techniques for B-splines [9].

For example, using Richardson's method as smoothing iteration and the quasi-interpolant $\mathcal{P}^h : \mathcal{B}^{2h} \to \mathcal{B}^h$ for the grid transfer as building blocks for the multigrid algorithm, our main result in [12] is the following.

Theorem 10. *The convergence rate of the standard multigrid W-cycle is bounded independent of the grid width h, if the number of smoothing steps is sufficiently large.*

This theorem guarantees that the total complexity of solving elliptic boundary value problems via the web-method can be bounded by a constant times the number of unknowns.

§7. Conclusion

Our results obtained so far show that web-splines are promising new finite elements. In particular,

- the web-basis is stable,
- the web-basis has full approximation power,
- the condition number of Galerkin matrices grows as h^{-2},
- no mesh generation is required,
- multigrid techniques yield linear complexity.

Future research will focus on boundary value problems with non-smooth boundary and on adaptive refinement strategies.

Acknowledgments. We are grateful to Carl de Boor for valuable comments and for introducing us to R-functions and RFM-methods.

References

1. Babuška, I., The finite element method with penalty, Math. Comp. **27**-122 (1973), 221–228.
2. Babuška, I., The finite element method with Lagrangian multipliers, Numer. Math. **20** (1973), 179–192.
3. Bank, R. E. and T. Dupont, An optimal order process for solving finite element equations, Math. Comp. **36** (1981), 35–51.
4. de Boor, C., *A Practical Guide to Splines*, Springer, 1978.
5. Braess, D., *Finite Elemente*, Springer, 1992.
6. Braess, D., M. Dryja, and W. Hackbusch, A multigrid method for nonconforming fe-discretizations with application to non-matching grids, Computing **63**-1 (1999), 1–25.
7. Brandt, A., Multi-level adaptive solutions to boundary value problems, Math. Comp. **31** (1977), 333–390.
8. Ciarlet, P., *The Finite Element Method for Elliptic Problems*, North-Holland, 1978.
9. Cohen, E., T. Lyche, and R. Riesenfeld, Discrete B-splines and subdivision techniques in computer-aided geometric design and computer graphics, Comp. Graphics and Image Proc. **14** (1980), 87–111.
10. Höllig, K., *Grundlagen der Numerik*, MathText, 1998.
11. Höllig, K., U. Reif, and J. Wipper, Weighted extended B-spline approximation of Dirichlet problems, Preprint 2000-8, University of Stuttgart.
12. Höllig, K., U. Reif, and J. Wipper, Multigrid methods with weighted extended B-splines, Preprint 2000-10, University of Stuttgart.
13. Höllig, K., U. Reif, and J. Wipper, Weighted B-spline approximation of Neumann problems, in preparation.

14. Kantorowitsch, L. W. and W. I. Krylow, *Näherungsmethoden der Höheren Analysis*, VEB Deutscher Verlag der Wissenschaften, Berlin, 1956.
15. Oswald, P., Multilevel solvers for elliptic problems on domains, in *Multiscale Wavelet Methods for PDEs*, W. Dahmen, A. Kurdila, P. Oswald (eds.), Academic Press, New York, 1997, 3–58.
16. Rvachev, V. L. and T. I. Sheiko, R-functions in boundary value problems in mechanics, Appl. Mech. Rev. **48**-4 (1995), 151–188.
17. Rvachev, V. L., T. I. Sheiko, V. Shapiro, and I. Tsukanov, On completeness of RFM solution structures, Comp. Mech. **25** (2000), 305–316.
18. Schoenberg, I. J., Contributions to the problem of approximation of equidistant data by analytic functions, Quart. Appl. Math. **4** (1946), 45–99, and 112–141.
19. Shapiro, V., Theory of R-functions and applications: a primer, Technical Report CPA88-3 (1988), Cornell Programmable Automation, Sibley School of Mechanical Engineering, Ithaca, NY.
20. Shapiro, V. and I. Tsukanov, Meshfree simulation of deforming domains, Computer-Aided Design **31** (1999), 459–471.
21. Strang, G. and G. J. Fix, *An Analysis of the Finite Element Method*, Prentice-Hall, Englewood Cliffs, NJ, 1973.

Klaus Höllig
Mathematisches Institut A
Universität Stuttgart
Pfaffenwaldring 57
70569 Stuttgart, Germany
hollig@mathematik.uni-stuttgart.de

Ulrich Reif
Fachbereich Mathematik, AG 3
TU Darmstadt
Schlossgartenstr. 7
64289 Darmstadt, Germany
reif@mathematik.tu-darmstadt.de

Joachim Wipper
Mathematisches Institut A
Universität Stuttgart
Pfaffenwaldring 57
70569 Stuttgart, Germany
wipper@mathematik.uni-stuttgart.de

Hierarchical Scattered Data Filtering for Multilevel Interpolation Schemes

Armin Iske

Abstract. Multilevel scattered data interpolation requires decomposing the given data into a hierarchy of nested subsets. This paper concerns the efficient construction of such hierarchies. To this end, a recursive filter scheme for scattered data is proposed which generates hierarchies of locally optimal nested subsets. The scheme is a composition of greedy thinning, a recursive point removal strategy, and exchange, a local optimization procedure. The utility of the filter scheme for multilevel interpolation using radial basis functions is shown by numerical examples.

§1. Introduction

Scattered data approximation requires recovering a function $f : \mathbb{R}^d \to \mathbb{R}$, $d \geq 1$, from a given data vector $D_Z(f) = (f(z_1), \ldots, f(z_N))^T \in \mathbb{R}^N$ of function values sampled from f at a finite set $Z = \{z_1, \ldots, z_N\} \subset \mathbb{R}^d$ of locations. Especially when N is extremely large, and the points in Z are unevenly distributed, multilevel interpolation schemes are appropriate techniques. One such scheme was introduced in [4], where compactly supported radial basis functions were used. For a recent survey on radial basis functions, we recommend [1]. The starting point in [4] is a decomposition of the given data into a hierarchy

$$X_L \subset X_{L-1} \subset \cdots \subset X_1 \subset X_0 = Z \tag{1}$$

of $L+1$ nested subsets. As confirmed in Section 4, the performance of the multilevel interpolation scheme [4] heavily depends on the choice of the hierarchy (1). Due to available error estimates for radial basis function interpolation, we wish to keep for each level index $1 \leq j \leq L$ the covering radius

$$r(X_j, X_{j-1}) = \max_{y \in X_{j-1}} d_{X_j}(y)$$

of X_j on X_{j-1} small. Throughout this paper, we use the notation

$$d_X(y) = \min_{x \in X} \|y - x\|$$

Mathematical Methods for Curves and Surfaces: Oslo 2000
Tom Lyche and Larry L. Schumaker (eds.), pp. 211–221.

ISBN 0-8265-1378-6

for the Euclidean distance between a point $y \in \mathbb{R}^d$ and a non-empty set $X \subset \mathbb{R}^d$. This paper concerns the *efficient* construction of hierarchies (1) whose covering radii $r(X_j, X_{j-1})$, $1 \le j \le L$, are small. This shall be accomplished by recursively applying TTX-filters, a family of operators on scattered data sets which output *locally optimal* subsets. The resulting TTX-filtering is a composition of greedy thinning, a recursive point removal scheme [4,5], and exchange, a local optimization procedure. Details are provided in Section 3. The multilevel scattered data interpolation scheme [4] is briefly discussed in Section 2. Finally, the performance of TTX-filtering and its utility for multilevel interpolation is illustrated by the numerical results in Section 4.

§2. Multilevel Interpolation

Let us briefly discuss how to use the hierarchy (1) for multilevel interpolation. According to the scheme proposed in [4], also subject of the papers [3,9], a sequence $s_L, \ldots, s_0$ of approximations to f is recursively computed as follows. Let $s_{L+1} \equiv 0$. For $j = L, \ldots, 0$, compute an interpolant $\Delta s_j : \mathbb{R}^d \to \mathbb{R}$ to the residual $D_{X_j}(f - s_{j+1})$ on X_j, then let $s_j = s_{j+1} + \Delta s_j$. Altogether, the following $L+1$ interpolation problems are to be solved one after the other:

$$\begin{aligned} D_{X_L}(f) &= D_{X_L}(\Delta s_L) \\ D_{X_{L-1}}(f - s_L) &= D_{X_{L-1}}(\Delta s_{L-1}) \\ &\cdots \\ D_{X_1}(f - s_2) &= D_{X_1}(\Delta s_1) \\ D_{X_0}(f - s_1) &= D_{X_0}(\Delta s_0). \end{aligned} \tag{2}$$

Note that each function s_j matches f at the subset X_j, i.e. $D_{X_j}(s_j) = D_{X_j}(f)$, $0 \le j \le L$. The basic idea behind this multilevel scheme is to capture a global trend of the function f by its coarse approximation s_L, before finer details of f are gradually added by using the sequence of functions Δs_j. Thus, the interpolants $s_L, \ldots, s_0$ provide a sequence of representations for f at $L+1$ different resolutions.

In [4], it was suggested to use compactly supported radial basis functions. For a fixed positive definite radial function $\phi : [0, \infty) \to \mathbb{R}$ with support $\text{supp}(\phi) = [0,1]$, let $\phi_\rho(\cdot) = \phi(\cdot/\rho)$ for $\rho > 0$, so that $\text{supp}(\phi_\rho) = [0, \rho]$. Each of the interpolants Δs_j, $0 \le j < L$, in (2) is then supposed to be of the form

$$\Delta s_j = \sum_{x \in X_j} c_x \phi_{\rho_j}(\| \cdot - x \|),$$

where the support radii ρ_j are monotonically decreasing: $\rho_{L-1} > \cdots > \rho_0 > 0$. Further details can be found in [4]. For the initial step $j = L$, however, we recommend to use the globally supported thin plate splines, $\phi(r) = r^2 \log(r)$, according to whose interpolation scheme the function Δs_L in (2) is of the form

$$\Delta s_L = \sum_{x \in X_L} c_x \| \cdot - x \|^2 \log(\| \cdot - x \|) + p,$$

where $p : \mathbb{R}^d \to \mathbb{R}$ denotes a linear polynomial. In contrast to interpolation by compactly supported radial basis functions, the thin plate spline method reproduces linear polynomials. This carries over to the above multilevel interpolation scheme. Indeed, if f is linear, then $s_L \equiv f$. In this case, every subsequent residual $f - s_j$ vanishes identically, which implies $\Delta s_j \equiv 0$ and thus $s_j \equiv f$ for $0 \leq j \leq L$.

We remark that for the stability of the multilevel interpolation, we require the separation distance $q(X) = \min_{x \in X} d_{X \setminus x}(x)$ in each subset $X = X_j$, $1 \leq j \leq L$, to be not too small. This is because the condition number of a collocation matrix arising in (2) blows up, whenever the separation distance of the corresponding point set is too small [10]. We shall come back to this important point in the following section.

Finally, note that the multilevel scheme (2) would allow us to replace *interpolation* with *least squares approximation* in each of its steps. This makes especially sense in situations, where there is liable to be noise in the data. However, as confirmed in our numerical examples, this leads to significantly higher computational costs (see the details in [8] concerning the required computations for least squares approximation by using radial basis functions).

§3. Hierarchical Scattered Data Filtering

In this section, we shall introduce thinning and exchange, our two main ingredients for scattered data filtering. But let us first fix some notations.

Definition 1. *Let $\mathcal{X}_Z = \{X : X \subset Z, X \neq \emptyset\}$ denote the power set of the $2^N - 1$ non-empty subsets of Z. We say that an operator $F : \mathcal{X}_Z \to \mathcal{X}_Z$ is a* filter *on Z, iff it satisfies $F(X) \subset X$ and $|F(X)| < |X|$ for all $X \in \mathcal{X}_Z$. A filter F is said to be a* thinning operator, *iff $|F(X)| = |X| - 1$ for all $X \in \mathcal{X}_Z$.*

The idea for the construction of the hierarchy (1) is to recursively apply a sequence $F_1, \ldots, F_L$ of TTX-filters, so that $X_j = F_j(X_{j-1})$, $1 \leq j \leq L$. The form of these particular filters shall be specified at the end of this section. For this purpose, it is sufficient to restrict ourselves to the special case of one sublevel, i.e. $L = 1$ in (1). Therefore, in this section we discuss how to extract merely *one* subset $X \equiv X_1$ from Z, where $|X| \ll |Z|$.

3.1. Thinning

Recall that we wish to select one subset $X \in \mathcal{X}_Z$, $|X| \ll |Z|$, whose covering radius $r(X, Z)$ on Z is small. This shall be accomplished by recursive point removals from Z. To this end, let us fix one thinning operator T, and let T^n denote its n-fold composition. Then, by

$$T^n(Z) \subset T^{n-1}(Z) \subset \cdots \subset T(Z) \subset T^0(Z) = Z$$

we obtain a nested sequence of subsets, like in (1), but with a much finer granulation. Now, the idea is to select one particular subset $X = T^k(Z) \in \mathcal{X}_Z$ by picking one suitable breakpoint $k \gg 0$ at run time (i.e. during the removal).

Ideally, we wish to select one subset X^* of size $|Z| - k$ which is optimal by minimizing its covering radius on Z among all subsets $X \in \mathcal{X}_Z$ of equal size, i.e.

$$r_k^*(Z) = r(X^*, Z) = \min_{\substack{X \in \mathcal{X}_Z \\ |X|=|Z|-k}} r(X, Z).$$

The problem of finding for a fixed $\alpha \geq 1$ an algorithm which outputs for any given finite set Z and any k, $1 \leq k < |Z|$, a subset $X \in \mathcal{X}_Z$ of size $|X| = |Z| - k$ satisfying

$$r(X, Z) \leq \alpha \cdot r_k^*(Z) \tag{3}$$

is one variant of the k-center problem (for more details see [6], Section 9.4.1). However, for any $\alpha < \alpha^* = \sqrt{2 + \sqrt{3}}$ there is no α-approximation algorithm for the k-center problem, unless P=NP (cf. [11], Section 4). There are a few approximation algorithms available for $\alpha = 2$, the first one dating back to [7], but for $\alpha^* \leq \alpha < 2$ no such algorithm is known.

Note that the universal constant α in (3) depends neither on Z nor on X. The following proposition provides one dynamic bound, of a similar form to (3), which allows us to pick several subsets $X \in \mathcal{X}_Z$ at run time whose covering radii $r(X, Z)$ approach the optimal value up to a factor $\alpha \equiv \alpha(X, Z) \ll \alpha^*$. To this end, assume without loss of generality that the points in $Z = \{z_1, \ldots, z_N\}$ are ordered such that their *significances* $\sigma_k = d_{Z \setminus z_k}(z_k)$ are monotonically increasing, i.e. $\sigma_k \leq \sigma_{k+1}$, $1 \leq k < N$.

Proposition 2. *Let $X \in \mathcal{X}_Z$ be of size $|X| = |Z| - k$. Then,*

$$\sigma_k \leq r_k^*(Z) \leq r(X, Z) \leq \alpha(X, Z) \cdot r_k^*(Z), \tag{4}$$

where $\alpha(X, Z) = r(X, Z)/\sigma_k$.

Proof: We shall prove the inequality $r(X, Z) \geq \sigma_k$ for all $X \in \mathcal{X}_Z$ of size $|X| = |Z| - k$. This implies $r_k^*(Z) \geq \sigma_k$, and moreover

$$r(X, Z) = \alpha(X, Z) \cdot \sigma_k \leq \alpha(X, Z) \cdot r_k^*(Z).$$

To this end, let $Y = Z \setminus X$. Since $d_X(y) = d_{Z \setminus Y}(y) \geq d_{Z \setminus y}(y)$ for all $y \in Y$, using $|Y| = k$ we conclude

$$r(X, Z) = \max_{z \in Z} d_X(z) = \max_{y \in Y} d_X(y) \geq \max_{y \in Y} d_{Z \setminus y}(y) \geq \sigma_k,$$

which completes our proof. □

We shall record the quality indices $\alpha(X, Z)$ during the point removal at run time. The above bound (4) allows us to control for any current subset $X \in \mathcal{X}_Z$, $|X| = |Z| - k$, the relative error by

$$\left| \frac{r(X, Z) - r_k^*(Z)}{r_k^*(Z)} \right| \leq \alpha(X, Z) - 1,$$

and this provides a useful criterion for the breakpoint selection.

NP-complete optimization problems, like the k-center problem, can effectively be tackled by using greedy approximation algorithms, algorithms of polynomial complexity which make an optimal decision at each single step. This motivates the usage of greedy thinning operators.

Definition 3. *A thinning operator $T_* : \mathcal{X}_Z \to \mathcal{X}_Z$ is said to be a* greedy thinning operator *iff T_* minimizes for any $X \in \mathcal{X}_Z$ the covering radius $r(T(X), X)$ among all thinning operators T.*

The following observation provides a useful characterization of this operator class.

Proposition 4. *A thinning operator T_* is greedy iff for any $X \in \mathcal{X}_Z$, the action of T_* on X removes one point $x^* \in X$ from X which satisfies*

$$\sigma_X(x^*) = \min_{x \in X} \sigma_X(x), \tag{5}$$

where $\sigma_X(x) = d_{X \setminus x}(x)$ denotes the significance of $x \in X$ in X.

Proof: Let T be any thinning operator whose action on $X \in \mathcal{X}_Z$ is given by $T(X) = X \setminus x_T$ for a specific $x_T \in X$. Then,

$$r(T(X), X) = \max_{x \in X} d_{T(X)}(x) = d_{X \setminus x_T}(x_T) = \sigma_X(x_T).$$

Hence, T minimizes $r(T(X), X)$ iff x_T minimizes $\sigma_X(x)$ among all $x \in X$. □

The above proposition yields an additional argument in favour of using greedy thinning operators: Recall that we wish to have the separation distances $q(X)$ not too small. Since $q(X) = \sigma_X(x^*)$, where x^* in (5) is a least significant point in X, we see that a greedy thinning operator serves to eliminate closest point pairs.

3.2. Exchange

In order to balance the short-sightedness of greedy thinning, we offer one additional useful ingredient for the construction of suitable filters. The idea is, for any given $X \in \mathcal{X}_Z$, to exchange a point pair between X and the complementary set $Y = Z \setminus X$, whenever the exchange pays off in terms of the reduction of the covering radius $r(X, Z)$. This leads us to the following

Definition 5. *We say that a point pair $(x, y) \in X \times Y$ is* exchangeable *iff it satisfies*

$$r(X, Z) > r((X \setminus x) \cup y, Z).$$

A subset $X \in \mathcal{X}_Z$ is said to be locally optimal *in Z iff there is no exchangeable point pair $(x, y) \in X \times Y$. By $\mathcal{X}_Z^*$ we shall denote the set of all subsets from $\mathcal{X}_Z$ which are locally optimal in Z.*

We shall interpret each exchange as a non-trivial operation of an exchange operator $E : \mathcal{X}_Z \to \mathcal{X}_Z$ on a subset $X \in \mathcal{X}_Z$:

$$E(X) = \begin{cases} X, & \text{iff } X \text{ is locally optimal,} \\ (X \setminus x) \cup y, & \text{for an exchangeable pair } (x, y) \in X \times Y. \end{cases}$$

Let E^n be the n-fold composition of E, and E^0 the identity on $\mathcal{X}_Z$. Note that for every $X \in \mathcal{X}_Z \setminus \mathcal{X}_Z^*$, there is one unique positive index $m = m(X) < \infty$, such that

$$E^{m+1}(X) = E^m(X) \neq E^{m-1}(X).$$

In other words, the exchange process terminates after finitely many steps, in which case the subset $E^\infty(X) = E^m(X)$ is locally optimal. Indeed, this is true since Z was assumed to be finite, and in each exchange the (non-negative) covering radius is strictly reduced:

$$0 \leq r(E^m(X), Z) < r(E^{m-1}(X), Z) < \ldots < r(E(X), Z) < r(X, Z).$$

Therefore, the operator $E^\infty : \mathcal{X}_Z \to \mathcal{X}_Z^*$ is a projector onto the locally optimal subsets of Z, where we let $m(X) = 0$ for $X \in \mathcal{X}_Z^*$.

The remainder of this subsection is devoted to the characterization of exchangeable point pairs. The following theorem provides one sufficient criterion which is useful for the efficient localization of such pairs.

Theorem 6. *Let $y^* \in Y$ satisfy $d_X(y^*) > d_X(y)$ for all $y \in Y \setminus y^*$. Then, for $x \in X$ the pair $(x, y^*) \in X \times Y$ is exchangeable if it satisfies the following two conditions.*

1) $r(X, Z) > \sigma_X(x)$;
2) $r(X, Z) > d_{X \setminus x}(y)$ *for all* $y \in (Y \setminus y^*) \cap V_X(x)$.

In the above condition 2), $V_X(x) = \{z \in \mathbb{R}^d : d_X(z) = \|z - x\|\}$ stands for the Voronoi tile of $x \in X$ w.r.t. X.

Proof: Note that $r(X, Z) = d_X(y^*)$, which together with 2) implies

$$r(X, Z) > d_{X \setminus x}(y) \qquad \text{for all } y \in Y \setminus y^*. \tag{6}$$

Indeed, condition 2) covers (6) for all $y \in (Y \setminus y^*) \cap V_X(x)$, and by using

$$r(X, Z) > d_X(y) = d_{X \setminus x}(y) \qquad \text{for all } y \in Y \setminus y^* \text{ with } y \notin V_X(x),$$

we see that (6) holds. Now, (6) in combination with condition 1) implies

$$r(X, Z) > \max\left(\max_{y \in Y \setminus y^*} d_{X \setminus x}(y), d_{X \setminus x}(x)\right),$$

and therefore

$$r(X, Z) > \max_{y \in (Y \setminus y^*) \cup x} d_{X \setminus x}(y) \geq \max_{y \in (Y \setminus y^*) \cup x} d_{(X \setminus x) \cup y^*}(y) = r((X \setminus x) \cup y^*, Z)$$

which completes our proof. □

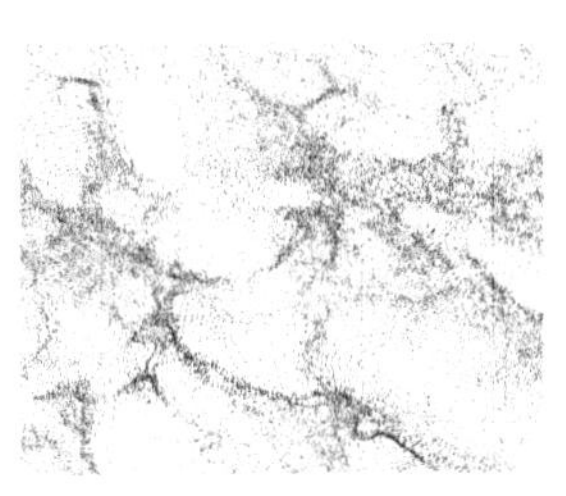

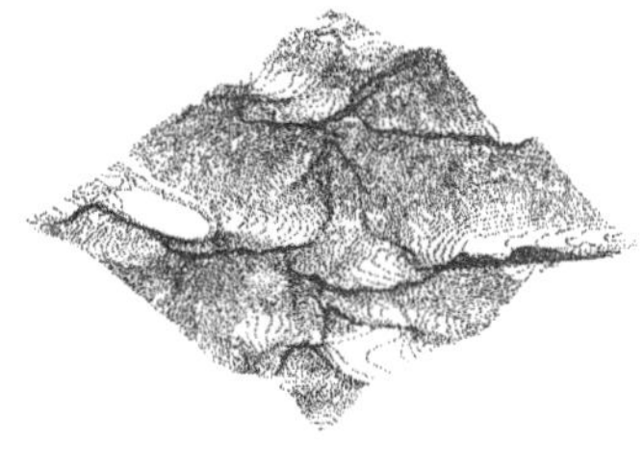

Fig. 1 and 2. The data set Hurrungane (23092 points), 2D view and 3D view.

3.3. Scattered Data Filtering

Let T_* be a fixed greedy thinning operator and E a fixed exchange operator. For the purpose of selecting one subset $X \in \mathcal{X}_Z$ with a small covering radius $r(X, Z)$, we shall use TTX-filters, filters of the form $F_n = E^\infty \circ T_*^n$. Mnemonically, the acronym TTX stands for "**T**hinning, ..., **T**hinning, **EX**change". These filters generate a (not necessarily nested) sequence $\{F_n(Z)\}_n$ of locally optimal subsets of decreasing size $|F_n(Z)| = |Z| - n$. In our numerical examples, we have used their quality indices $\alpha(F_n(Z), Z)$ for picking one subset $X = F_n(Z) \in \mathcal{X}_Z^*$ whose covering radius $r(X, Z)$ is close to the optimum $r_n^*(Z)$. We finally remark that an efficient implementation of F_n can be accomplished in two dimensions by using Delaunay triangulations and heaps, as suggested in [5] for greedy thinning.

§4. Numerical Results

We have implemented TTX-filtering for bivariate data, $d = 2$, by using Delaunay triangulations and heaps, with the exchange being based on the sufficient criterion in Theorem 6. In our numerical examples, we considered using a data set called *Hurrungane*, a sample of height values $\{f(z)\}_{z \in Z}$ of a mountain area at $|Z| = 23092$ distinct locations in the domain $\Omega = [437000, 442000] \times [6812000, 6817000]$ with minimum height $\min_{z \in Z} f(z) = 1100$ and maximum height $\max_{z \in Z} f(z) = 2400$. The data set is displayed in the Figures 1,2.

4.1. Non-hierarchical Filtering

Let us briefly illustrate how the performance of the TTX-filtering scheme compares with approximation algorithms for the k-center problem. For this purpose, we have computed the sequence $\{F_n(Z)\}_n$ of subsets. Figure 3 shows the graph of their covering radii $r(F_n(Z), Z)$, $n \in [300, 20000]$, along with that of the initial significances σ_n. Figure 4 displays the graph of the quality indices $\alpha(F_n(Z), Z)$. We find $\alpha(F_n(Z), Z) < \alpha^*$ for $n = 300, \ldots, 18599$, and

$$\min_{500 \leq n \leq 20000} \alpha(F_n(Z), Z) = \alpha(F_{2976}(Z), Z) = 1.1851.$$

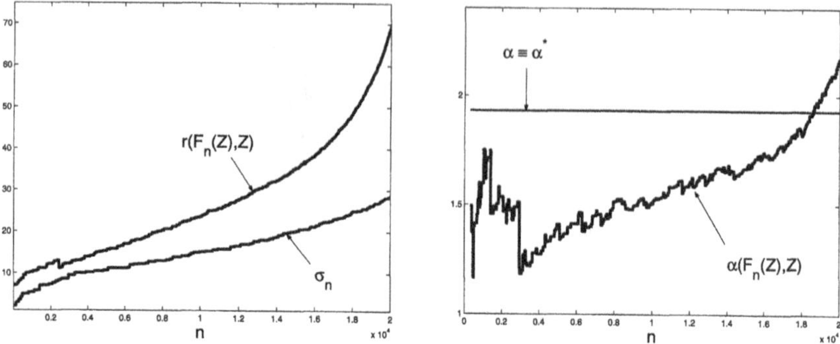

Fig. 3 and 4. The graphs σ_n, $r(F_n(Z), Z)$ and $\alpha(F_n(Z), Z)$, $n \in [300, 20000]$.

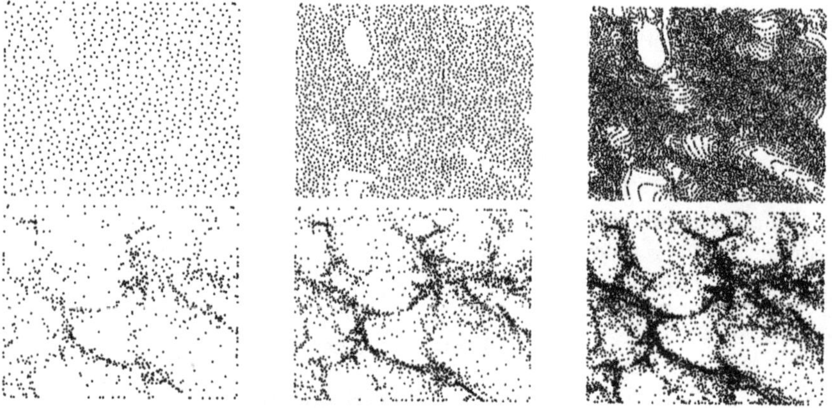

Fig. 5. The sets X_3^F, X_2^F, X_1^F (first row) and X_3^A, X_2^A, X_1^A (second row).

4.2. Hierarchical Filtering and Multilevel Interpolation

We have decomposed the given superset Z into four nested subsets, i.e. $L = 3$ in (1), by recursively using TTX-filters. The three subsets $X_3^F \subset X_2^F \subset X_1^F$ of sizes $|X_3^F| = 855$, $|X_2^F| = 2566$ and $|X_1^F| = 7697$ are displayed in the Figure 5, first row. Observe that the points of these subsets are uniformly distributed in the domain Ω.

In contrast, the second row of Figure 5 displays a nested sequence $X_3^A \subset X_2^A \subset X_1^A$ of non-uniformly distributed subsets of the same sizes, i.e. $|X_3^A| = 855$, $|X_2^A| = 2566$ and $|X_1^A| = 7697$. This sequence has been generated by using adaptive thinning [2], a recent method for selecting subsets from bivariate scattered data whose piecewise linear interpolants over their Delaunay triangulations are close to the original data. This particular thinning method prefers to remove data points in flat regions of the surface, while it keeps data points where the surface's curvature is high. Adaptive thinning is essentially

j	$r_j(F)$	$r_j(A)$	$\alpha_j(F)$	$\alpha_j(A)$	$\delta_j(F)$	$\delta_j(A)$	u(F)	u(A)
1	34.41	252.53	1.67	12.26	185.24	6.14	19.80	195.91
2	65.49	372.83	1.53	13.12	356.66	31.42	3.68	52.94
3	117.44	463.01	1.47	9.64	424.71	69.26	0.78	15.00

Tab. 1. Performance of TTX-filtering (F) and adaptive thinning (A).

j	$\eta_\infty(F)$	$\eta_\infty(A)$	$\eta_2(F)$	$\eta_2(A)$	u(F)	u(A)
3	169.63	541.51	18.44	35.87	4.81	4.84
2	126.33	342.39	10.70	18.91	1.76	30.47
1	83.81	222.05	7.97	12.60	10.83	247.78

Tab. 2. Multilevel interpolation: approximation quality and computational costs.

a data-dependent method, i.e. it depends on the function values of f at Z, whereas TTX-filtering is data-independent, because it merely depends on the data set Z.

The computational costs, in CPU units time, required for the subset selection at each level are, along with the subsets' covering radii $r_j = r(X_j, X_{j-1})$, and the quality indices $\alpha_j = \alpha(X_j, X_{j-1})$, $j = 1,2,3$, documented in Table 1, where 'F' stands for TTX-filtering, and 'A' for adaptive thinning. In addition, Table 1 shows the values $\delta_j = \max_{z\in Z} |L(f, X_j)(z) - f(z)|$ of the maximal deviations between f and $L(f, X_j)$ at Z, where $L(f, X_j)$ is the piecewise linear interpolant over the Delaunay triangulation of X_j satisfying $D_{X_j}(L(f, X_j)) = D_{X_j}(f)$.

Note that the covering radii $r_j(F)$ and quality indices $\alpha_j(F)$ of the subsets X_j^F are much smaller than the corresponding values $r_j(A)$ and $\alpha_j(A)$, even at significantly lower computational costs. On the other hand, however, the piecewise linear interpolants $L(f, X_j^A)$ are compared with $L(f, X_j^F)$ much closer to the given function values of f at Z, i.e. $\delta_j(A) \ll \delta_j(F)$, $j = 1,2,3$. The latter is the goal of adaptive thinning in [2], and therefore we refrain from expanding further details here.

Now let us turn to multilevel interpolation. Often has it been suggested to work with a data-dependent method, like adaptive thinning, instead of a data-independent one, like TTX-filtering, when it comes to the construction of the hierarchy (1). If the piecewise linear interpolant $L(f, X_j)$ was close to the function f, then the interpolant s_j in (2) would be close to f, i.e. $f \approx L(f, X_j) \approx s_j$, so the expectation.

The comparison between the two different sequences X_1^F, X_2^F, X_3^F and X_1^A, X_2^A, X_3^A shows, however, that the opposite can be true. For the purpose of illustration, we considered using the compactly supported radial basis function $\phi(r) = (1-r)_+^4(4r+1)$ [12], and we let $\rho_2 = 327.45, \rho_1 = 177.05$ for the support radii at the levels $j = 2,1$ in (2). Table 2 reflects our numerical results, where

$$\eta_\infty = \max_{x\in X_{j-1}} |f(x) - s_j(x)|, \qquad \eta_2^2 = \frac{1}{|X_{j-1}|} \sum_{x\in X_{j-1}} |f(x) - s_j(x)|^2.$$

Already at the initial level $L = 3$, the thin plate spline interpolant $s_3(F)$ of f at X_3^F yields much smaller residual norms $\eta_\infty(F), \eta_2(F)$ than its counterpart $s_3(A)$. Moreover, the poor approximation quality of $s_3(A)$ cannot be recovered by the subsequent interpolants $s_2(A), s_1(A)$, and the computational costs u(A) blow up in the remaining steps of (2). Observe that the method 'A' looses dramatically against the method 'F' at both ends, approximation quality and computational costs. In fact, we haven't encountered any numerical example where this is different. In summary, the choice of the hierarchy (1) can significantly affect the performance of the multilevel interpolation scheme (2), where special attention is to be paid to the approximation quality of s_L.

Finally, coming back to one comment in Section 2, we remark that working with least squares approximation rather than interpolation does dramatically increase the computational costs. Indeed, if we e.g. replace $s_3(F)$ by the corresponding least squares approximation, this would reduce the least squares error by $\eta_2 = 15.31$, but at $u = 53.28$ CPU units time.

Acknowledgments. The author was partly supported by the European Union within the project MINGLE (Multiresolution in Geometric Modelling), contract no. HPRN-CT-1999-00117.

References

1. Buhmann, M. D., Radial basis functions, Acta Numerica, (2000), 1–38.
2. Dyn, N., M. S. Floater, and A. Iske, Adaptive thinning for bivariate scattered data, preprint, Technische Universität München, 2000.
3. Fasshauer, G. E. and J. W. Jerome, Multistep approximation algorithms: Improved convergence rates through postconditioning with smoothing kernels, Advances in Comp. Math. **10** (1999), 1–27.
4. Floater, M. S. and A. Iske, Multistep scattered data interpolation using compactly supported radial basis functions, J. Comput. Appl. Math. **73** (1996), 65–78.
5. Floater, M. S. and A. Iske, Thinning algorithms for scattered data interpolation, BIT **38** (1998), 705–720.
6. Hochbaum, D. S. (ed.), *Approximation Algorithms for NP-hard Problems*, PWS Publishing Company, Boston, 1997.
7. Hochbaum, D. S. and D. B. Shmoys, A best possible heuristic for the k-center problem, Mathematics of Operations Research **10:2** (1985), 180–184.
8. Iske, A., Reconstruction of smooth signals from irregular samples by using radial basis function approximation, in *Proceedings of the 1999 International Workshop on Sampling Theory and Applications*, Y. Lyubarskii (ed.), The Norwegian University of Science and Technology, Trondheim, 1999, 82–87.

9. Narcowich, F. J., R. Schaback, and J. D. Ward, Multilevel interpolation and approximation, Appl. Comput. Harmonic Anal. **7** (1999), 243–261.
10. Narcowich, F. J. and J. D. Ward, Norms of inverses and condition numbers for matrices associated with scattered data, J. Approx. Theory **64** (1991), 69–94.
11. Shmoys, D. B., Computing near-optimal solutions to combinatorial optimization problems, DIMACS, Ser. Discrete Math. Theor. Comput. Sci. **20** (1995), 355–397.
12. Wendland, H., Piecewise polynomial, positive definite and compactly supported radial functions of minimal degree, Advances in Comp. Math. **4** (1995), 389–396.

Armin Iske
Zentrum Mathematik
Technische Universität München
D-80290 München, Germany
`iske@ma.tum.de`

Bounding the Hausdorff Distance Between Implicitly Defined and/or Parametric Curves

Bert Jüttler

Abstract. This paper is devoted to computational techniques for generating upper bounds on the Hausdorff distance between two planar curves. The results are suitable for pairs of implicitly defined and/or parametric curves. The bounds are computed directly from the control points resp. spline coefficients of the curves. They improve an earlier result of Sederberg [10, Eq. (6.2)]. Potential applications include error bounds for the approximate implicitization of spline curves, for the approximate parameterization of (piecewise) algebraic curves, and for algebraic curve fitting.

§1. Introduction

The notion of distance between two curves (and surfaces) is important for various applications of geometric design; see Bogacki and Weinstein [2] for a detailed discussion of several possible definitions. Parametric distance measures, such as the maximum norm of the difference vector of the parametric representations, are certainly useful in applications. These measures, however, are non–geometrical; they also tend to overestimate the real distance. Moreover, these measures cannot be used if one or both curves are given by an implicit representation, such as for piecewise algebraic spline curves, see [10].

This paper focuses on the well–known Hausdorff distance between two curves. We introduce the auxiliary notion of the footpoint distance, which is closely related to it, and develop a computational technique for generating upper bounds, directly from the control points resp. spline coefficients of the curves. The results are suitable for pairs of implicitly defined and/or parametric curves. They improve an earlier result of Sederberg [10], see the end of Section 3.

The potential applications include error bounds for the approximate implicitization of spline curves and surfaces (cf. [4]), for the approximate parameterization of algebraic curves and surfaces (cf. [1]), and for curve and surface fitting with algebraic spline curves and surfaces.

Mathematical Methods for Curves and Surfaces: Oslo 2000
Tom Lyche and Larry L. Schumaker (eds.), pp. 223–232.

ISBN 0-8265-1378-6

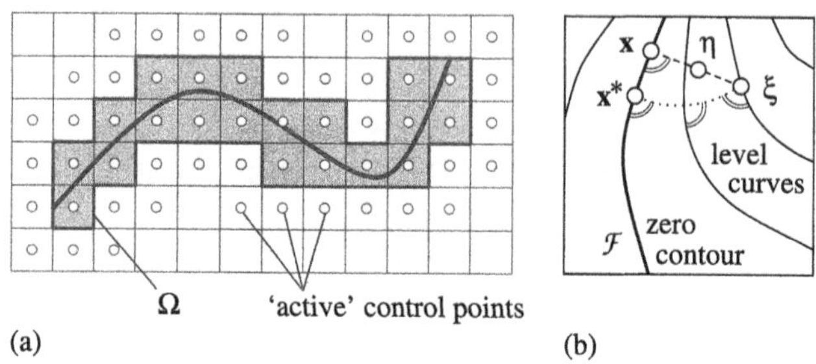

Fig. 1. (a) A planar curve can be defined as the zero contour of a biquadratic tensor–product spline function. (b) Footpoint $\mathbf{x} \in \mathcal{F}$ of a point $\boldsymbol{\xi}$ and the path of steepest descent (dotted line). See Lemma 1 and Corollary 2.

§2. Implicitly Defined Algebraic Spline Curves

Consider a closed bounded set $\Omega \subset \mathbb{R}^2$. The zero contour of a bivariate function $f(\mathbf{x}) = f(x_1, x_2) \not\equiv 0$ with the domain Ω,

$$\mathcal{F} = \{\, \mathbf{x} \mid f(\mathbf{x}) = 0 \,\wedge\, \mathbf{x} \in \Omega \,\}, \tag{1}$$

defines a planar curve $\mathcal{F}$, possibly consisting of more than one segment. If the function f is a bivariate spline function, then the curve $\mathcal{F}$ will be a piecewise algebraic curve, i.e., an algebraic spline curve. For instance, one may choose the function f as a tensor–product spline function of degree (d, d),

$$f(\mathbf{x}) = f(x_1, x_2) = \sum_{(i,j)\in\mathcal{I}} M_i^d(x_1)\, N_j^d(x_2)\, p_{i,j}, \qquad \mathbf{x} \in \Omega, \tag{2}$$

where the B-splines $M_i^d(x_1)$ and $N_j^d(x_2)$ are defined over suitable knot sequences. In this situation, the spline coefficients (control points) $p_{i,j} \in \mathbb{R}$ can be associated with the rectangular grid in the x_1x_2–plane, which is obtained from the Greville abscissas of the knot sequences, see e.g. [9]. The index set $\mathcal{I}$ contains the indices of all 'active' control points,

$$(i,j) \in \mathcal{I} \iff \exists\, \mathbf{x}^* = (x_1^*, x_2^*) \in \Omega : M_i^d(x_1^*)\, N_j^d(x_2^*) \neq 0. \tag{3}$$

An example is shown in Fig. 1a. The curve $\mathcal{F}$ is defined as the zero contour of a biquadratic tensor-product spline function with a regular grid of knot lines. Then, in the biquadratic case, the control points are associated with the centers of the cells. The domain Ω of the spline function consists of all grey cells. The active control points are marked by circles.

Alternatively, algebraic spline curves can be defined by piecing together the zero contours of triangular Bézier surface patches, see e.g. [10]. This representation is particularly useful if the implicit representation of the curve is generated by implicitizing a parametric one.

Throughout this paper we assume that the function f is at least C^1, and that its gradient satisfies $\nabla f(\mathbf{x}) \neq (0,0)$ for all $\mathbf{x} \in \Omega$. Consequently, the zero contour (1) is differentiable and has no singularities, such as multiple points or cusps.

§3. Distance Between Points

Consider a point $\boldsymbol{\xi} \in \Omega$. Any point $\mathbf{x} \in \mathcal{F}$ satisfying $(\boldsymbol{\xi} - \mathbf{x}) \times \nabla f(\mathbf{x}) = 0$ will be called a footpoint of $\boldsymbol{\xi}$, see Fig. 1b. Here, $\times$ is the two–dimensional exterior product, $\mathbf{v} \times \mathbf{w} = v_1 w_2 - v_2 w_1$. The inner product of vectors will be denoted with $\mathbf{v} \cdot \mathbf{w}$, and the Euclidean distance between the corresponding points with $\mathrm{dist}(\mathbf{v}, \mathbf{w}) = \|\mathbf{v} - \mathbf{w}\|$.

Lemma 1. *Consider a point $\boldsymbol{\xi} = (\xi_1, \xi_2) \in \Omega$. Let $\mathbf{x} = (x_1, x_2) \in \mathcal{F}$ be a footpoint of $\boldsymbol{\xi}$. If the line segment $\overline{\mathbf{x}\boldsymbol{\xi}}$ is contained in Ω, then the distance between the points $\boldsymbol{\xi}$ and $\mathbf{x}$ can be expressed as*

$$\mathrm{dist}(\mathbf{x}, \boldsymbol{\xi}) = \frac{\|\nabla f(\mathbf{x})\|}{|\nabla f(\mathbf{x}) \cdot \nabla f(\boldsymbol{\eta})|}\, |f(\boldsymbol{\xi})|, \tag{4}$$

where $\boldsymbol{\eta}$ is a certain point on the line segment connecting the points $\mathbf{x}$ and $\boldsymbol{\xi}$, cf. Fig. 1b.

Proof: Let $g(t)$ be the restriction of the function f to the normal of $\mathcal{F}$ at $\mathbf{x}$,

$$g(t) = f(\mathbf{x} + t\, \frac{\nabla f(\mathbf{x})}{\|\nabla f(\mathbf{x})\|}). \tag{5}$$

As $\mathbf{x} \in \mathcal{F}$ is a footpoint of $\boldsymbol{\xi}$, this function satisfies one of the equations $g(\mathrm{dist}(\mathbf{x}, \boldsymbol{\xi})) = f(\boldsymbol{\xi})$ or $g(-\mathrm{dist}(\mathbf{x}, \boldsymbol{\xi})) = f(\boldsymbol{\xi})$, depending on the orientation of the gradients. Moreover, $g(0) = 0$. Using the mean value theorem, we obtain in the first case

$$\frac{f(\boldsymbol{\xi}) - 0}{\mathrm{dist}(\mathbf{x}, \boldsymbol{\xi})} = g'(\lambda) = \nabla f(\boldsymbol{\eta}) \cdot \frac{\nabla f(\mathbf{x})}{\|\nabla f(\mathbf{x})\|} \tag{6}$$

for some $\lambda \in [0, \mathrm{dist}(\mathbf{x}, \boldsymbol{\xi})]$, and with $\boldsymbol{\eta} = \mathbf{x} + \lambda\, \frac{\nabla f(\mathbf{x})}{\|\nabla f(\mathbf{x})\|}$. The assertion follows by solving this equation for $\mathrm{dist}(\mathbf{x}, \boldsymbol{\xi})$. The second case can be dealt with analogously. □

This lemma can be used for bounding the distance of points $\boldsymbol{\xi} \in \Omega$ from their footpoints. With the help of the control points of the spline functions, we are able to generate an upper bound C on the length of the gradients,

$$\|\nabla f(\mathbf{x})\| \le C \quad \text{for} \quad \mathbf{x} \in \Omega. \tag{7}$$

In addition, we may generate a lower bound D_h on the inner product of the gradients of any two neighbouring points whose distance does not exceed a certain constant h,

$$\|\nabla f(\mathbf{x}) \cdot \nabla f(\mathbf{y})\| \ge D_h \ \text{ holds for all } \ \mathbf{x}, \mathbf{y} \in \Omega \ \text{ with } \ \mathrm{dist}(\mathbf{x}, \mathbf{y}) \le h. \tag{8}$$

The methods used for computing these bounds are described in Section 6. As a consequence of Lemma 1 we obtain the following result, which bounds the distance *without* computing the footpoint $\mathbf{x}$ of $\boldsymbol{\xi}$.

Corollary 2. *Consider again the situation of Lemma 1, and assume (7), (8). Let the parameter h be chosen such that $h \geq C/D_h\, f(\boldsymbol{\xi})$. The distance of the point $\boldsymbol{\xi}$ from its footpoint $\mathbf{x}$ on the curve $\mathcal{F}$ is then bounded by*

$$\operatorname{dist}(\mathbf{x}, \boldsymbol{\xi}) \leq \frac{C}{D_h}\, f(\boldsymbol{\xi}). \tag{9}$$

This result (and similarly the Theorems 3 and 4) can be used only if the parameter h is not too small. On the other hand, the smaller the parameter h, the bigger the lower bound D_h. The choice of a suitable constant h is addressed in Section 7.

If the point $\boldsymbol{\xi}$ approaches its footpoint $\mathbf{x}$, the bound (9) converges to zero, as $\boldsymbol{\xi} \to \mathbf{x}$ implies $f(\boldsymbol{\xi}) \to 0$.

This corollary improves an erroneous result in [10]. According to inequality (6.2) of [10], the distance can be bounded by $C\, f(\boldsymbol{\xi})$. (The original formula is slightly different, but it is in fact equivalent to this one.) However, this formula is valid only if additional assumptions about the domain Ω are satisfied, which have so far not been specified. More precisely, the domain Ω has to contain both the point $\boldsymbol{\xi}$ and the point $\mathbf{x}^*$ on the curve $\mathcal{F}$ which is obtained by following the path of steepest descent, starting at $\boldsymbol{\xi}$, see Fig. 1b. Generally, the latter point is *not* the footpoint $\boldsymbol{\xi}$.

§4. Distance Between Two Implicitly Defined Curves

Let $\mathcal{F}$ be a planar curve defined by (1) and (2), and let the second curve $\mathcal{G}$ be defined by

$$\mathcal{G} = \{\, \mathbf{x} \mid g(\mathbf{x}) = 0 \ \wedge\ \mathbf{g} \in \Omega \,\}, \tag{10}$$

where $g(\mathbf{x})$ is a another bivariate spline function

$$g(\mathbf{x}) = g(x_1, x_2) = \sum_{(i,j) \in \mathcal{I}} M_i^d(x_1)\, N_j^d(x_2)\, q_{i,j}, \qquad \mathbf{x} \in \Omega, \tag{11}$$

with control points $q_{i,j} \in \mathbb{R}$ and domain Ω. The knots of $f(\mathbf{x})$ and $g(\mathbf{x})$ are assumed to be identical. Let $\mathcal{G}_0$ be the segment of $\mathcal{G}$ which consists of all points which have at least one footpoint on $\mathcal{F}$,

$$\mathcal{G}_0 = \{\mathbf{y} \mid \mathbf{y} \in \mathcal{G} \vee \exists \mathbf{x} \in \mathcal{F} : (\mathbf{y} - \mathbf{x}) \cdot \nabla f(\mathbf{x}) = 0\}, \tag{12}$$

see Fig. 2. We consider the maximum distance of the points of G_0 from their footpoints,

$$\operatorname{dist}^{\mathrm{F}}(\mathcal{F}, \mathcal{G}) = \sup_{\mathbf{y} \in \mathcal{G}_0} \ \inf_{\substack{\mathbf{x} \in \mathcal{F} \text{ is a} \\ \text{footpoint of } \mathbf{y}}} \operatorname{dist}(\mathbf{x}, \mathbf{y}). \tag{13}$$

This measure will be called the one–sided footpoint distance. By symmetrizing it we obtain the footpoint distance

$$\operatorname{Dist}^{\mathrm{F}}(\mathcal{F}, \mathcal{G}) = \max\{\, \operatorname{dist}^{\mathrm{F}}(\mathcal{F}, \mathcal{G}),\ \operatorname{dist}^{\mathrm{F}}(\mathcal{G}, \mathcal{F}) \,\}, \tag{14}$$

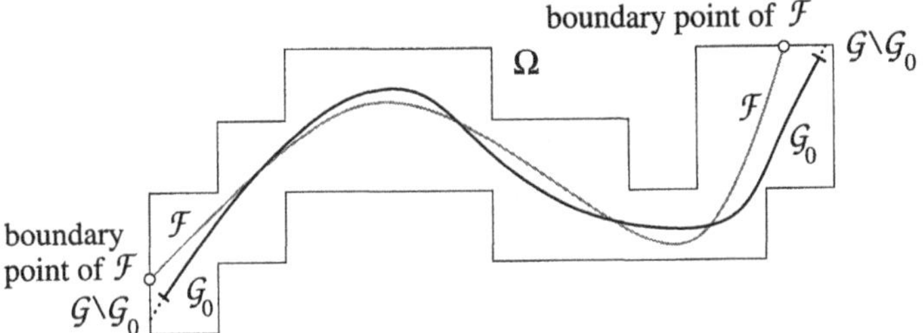

Fig. 2. The curve segments $\mathcal{F}$ (grey), $\mathcal{G}\backslash\mathcal{G}_0$ (dotted), $\mathcal{G}_0$ (solid), and the boundary points $\partial\mathcal{F}$ (marked by circles).

which is closely related to the Hausdorff distance (see [2])

$$\mathrm{Dist}^{\mathrm{H}}(\mathcal{F},\mathcal{G}) = \max\{\ \mathrm{dist}^{\mathrm{H}}(\mathcal{F},\mathcal{G}),\ \mathrm{dist}^{\mathrm{H}}(\mathcal{G},\mathcal{F})\ \}, \tag{15}$$

where $\mathrm{dist}^{\mathrm{H}}(\mathcal{F},\mathcal{G}) = \sup_{\mathbf{y}\in\mathcal{G}} \inf_{\mathbf{x}\in\mathcal{F}} \mathrm{dist}(\mathbf{x},\mathbf{y})$, as follows. The left part of the Hausdorff distance can be bounded by

$$\mathrm{dist}^{\mathrm{H}}(\mathcal{F},\mathcal{G}) \le \max\{\mathrm{dist}^{\mathrm{F}}(\mathcal{F},\mathcal{G}),\ \underbrace{\sup_{\mathbf{x}\in\mathcal{G}\backslash\mathcal{G}_0}\ \inf_{\mathbf{y}\in\partial\mathcal{F}}\ \mathrm{dist}(\mathbf{x},\mathbf{y})\}}_{(*)}, \tag{16}$$

where $\partial\mathcal{F}$ is the set of all (finitely many) boundary points of $\mathcal{F}$, cf. Fig. 2. If the minimum distance of a point $\mathbf{y} \in \mathcal{G}_0$ from $\mathcal{F}$ always occurs at one of its footpoints (and not at the boundary points $\partial\mathcal{F}$), then the inequality (16) becomes an equation. This is a realistic assumption in applications. Consequently, if one ignores the contributions $(*)$ at the boundary, the Hausdorff distance between the curves $\mathcal{F}$ and $\mathcal{G}$ is essentially equal to the footpoint distance (14).

Theorem 3. *Let $K = \max_{(i,j)\in\mathcal{I}} |p_{i,j} - q_{i,j}|$, and assume (7), (8). The domain Ω is assumed to contain the line segments connecting the points in $\mathcal{G}_0$ with their footpoints on $\mathcal{F}$. Let h be chosen such that $h \ge C/D_h\, K$. The one–sided footpoint distance is then bounded by*

$$\mathrm{dist}^{\mathrm{F}}(\mathcal{F},\mathcal{G}) \le \frac{C}{D_h}\,K. \tag{17}$$

Proof: Consider a point $\mathbf{y} \in \mathcal{G}_0$, hence $g(\mathbf{y}) = 0$. Using the convex hull property of B-splines we obtain $|f(\mathbf{y})| = |f(\mathbf{y}) - g(\mathbf{y})| \le K$. Inequality (17) now follows from Corollary 2. □

If the curves $\mathcal{F}$ and $\mathcal{G}$ are sufficiently close to each other, then the difference $f(\mathbf{x}) - g(\mathbf{x})$ can be expected to be small, provided that the sign distributions of f and g are similar. Consequently, Theorem 3 can be expected to give a tight upper bound for the distance between the curves.

Using this theorem, we are now able to derive bounds on $\mathrm{Dist}^{\mathrm{F}}(\mathcal{F},\mathcal{G})$, directly from the control points of the bivariate spline functions f and g. An example is given in Section 7.

§5. Distance Between Implicitly Defined and Parametric Curves

Once again, let $\mathcal{F}$ be a planar curve defined by (1) and (2), and let $\mathcal{H}$ be a parametric B-spline curve of degree k (see [7]),

$$\mathcal{H} = \{ \mathbf{h}(t) \mid t \in [0,1] \}, \tag{18}$$

with the parametric representation $\mathbf{h}(t)$ and parameter domain $[0,1]$. Consider the spline function of degree $2dk$ which is obtained by restricting $f(\mathbf{x})$ to the curve $\mathbf{h}(t)$,

$$f(\mathbf{h}(t)) = \sum_{i=0}^{m} P_i^{2dk}(t)\; h_i, \quad t \in [0,1], \tag{19}$$

where the B-splines $P_i^{2dk}(t)$ of degree $2dk$ are defined over an appropriate knot sequence. In addition to the original knots of $\mathbf{h}(t)$, it contains the parameter values of the intersections of $\mathcal{H}$ with the knot lines of the bivariate spline function $f(\mathbf{x})$, both with sufficient multiplicity. For instance, if $\mathbf{h}(t)$ is a cubic spline curve, then the parameter values of the intersections with the knot lines can be found by solving a couple of cubic equations.

The coefficients $h_i \in \mathbb{R}$ can be computed with the help of algorithms for composing spline functions. A blossoming–based approach has been described in [3]. Alternatively, one may construct the B-spline representation of $f(\mathbf{h}(t))$ with the help of interpolation techniques.

The one–sided footpoint distance $\operatorname{dist}^{\mathrm{F}}(\mathcal{F},\mathcal{H})$ can be defined as in (13). It is a useful distance measure in several applications, such as the approximate implicitization of the parametric B-spline curves $\mathcal{H}$, see [4], or for the approximate parameterization of the implicit curve $\mathcal{F}$. With the help of the control points h_i we obtain the following upper bound.

Theorem 4. *Let $H = \max_{i=0,\dots,m} |h_i|$, and assume (7), (8). The domain Ω is assumed to contain the line segments connecting the points in $\mathcal{H}_0$ (which is defined as $\mathcal{G}_0$) with their footpoints on $\mathcal{F}$. Let h be chosen such that $h \geq C/D_h\, K$. The distance of the points of $\mathcal{H}_0$ from their footpoints is then bounded by*

$$\operatorname{dist}^{\mathrm{F}}(\mathcal{F},\mathcal{H}) \leq \frac{C}{D_h}\, H. \tag{20}$$

Proof: Consider a point $\mathbf{y} \in \mathcal{H}_0$, hence $\mathbf{y} = \mathbf{h}(t_0)$ for some $t_0 \in [0,1]$. Using the convex hull property of B-splines we obtain $|f(\mathbf{y})| = |f(\mathbf{h}(t_0))| \leq H$. Inequality (20) now follows from Corollary 2. □

§6. Generating the Bounds C and D_h

In order to generate the constants C and D_h in the inequalities (7) and (8), we split the spline function (1) into polynomial segments with the subdomains $\Omega^{(k)}$,

$$f(\mathbf{x}) = f^{(k)}(\mathbf{x}) \quad \text{for} \quad \mathbf{x} \in \Omega^{(k)}, \quad k = 1,\dots,K. \tag{21}$$

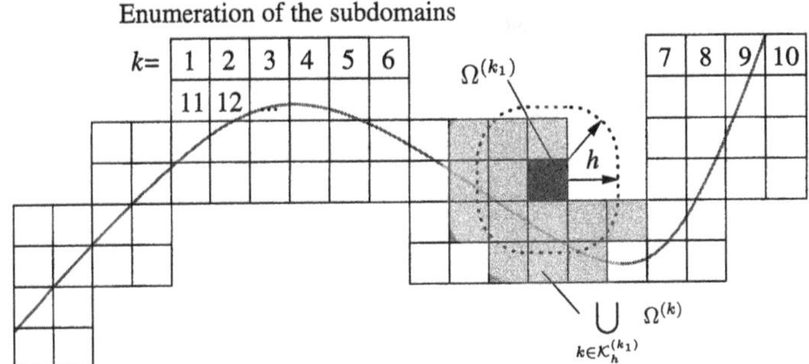

Fig. 3. Splitting the tensor–product spline function into polynomial (tensor–product Bézier) segments, and definition of the index sets $\mathcal{K}_h^{(k)}$.

The K subdomains $\Omega^{(k)}$ of the polynomial pieces are either the cells of the original spline function, or they are obtained by splitting them further into smaller rectangles. Using smaller subdomains $\Omega^{(k)}$, instead of the cells of the spline function, we may obtain tighter bounds C and D_h.

As an example, Fig. 3 shows the enumeration of the subdomains $\Omega^{(k)}$ which are used for splitting the spline function of Fig. 1a into polynomial pieces. Here, they are obtained by subdividing each of the original quadratic cells into four smaller squares.

Using knot insertion we obtain for each subdomain a tensor–product Bézier representation

$$f^{(k)}(\mathbf{x}) = \sum_{i=0}^{d} \sum_{j=0}^{d} B_i^d(x_1)\, B_j^d(x_2)\, b_{i,j}^{(k)}, \quad \mathbf{x} \in \Omega^{(k)}, \tag{22}$$

with certain coefficients $b_{i,j}^{(k)}$. Moreover, we may compute a tensor–product Bézier representation of the associated gradients

$$\nabla f^{(k)}(\mathbf{x}) = \sum_{i=0}^{d} \sum_{j=0}^{d} B_i^d(x_1)\, B_j^d(x_2)\, \mathbf{c}_{i,j}^{(k)} \quad \mathbf{x} \in \Omega^{(k)}, \tag{23}$$

where the control points $\mathbf{c}_{i,j}^{(k)}$ are obtained from the formulas for differentiation and degree elevation of tensor–product polynomials in Bernstein–Bézier form. Note that the first (resp. second) component is a polynomial of degree $(d-1,d)$ (resp. $(d,d-1)$). Thus, degree elevation is needed in order to obtain the representation of the form (23).

Inequality (8) involves gradients at two points with a certain maximum distance h. For each subdomain $\Omega^{(k)}$, we denote with $\mathcal{K}_h^{(k)}$ the indices of all subdomains that are within distance h of it,

$$i \in \mathcal{K}_h^{(k)} \iff \exists \mathbf{x} \in \Omega^{(k)}\ \exists \mathbf{y} \in \Omega^{(i)} : \ \text{dist}(\mathbf{x}, \mathbf{y}) \le h. \tag{24}$$

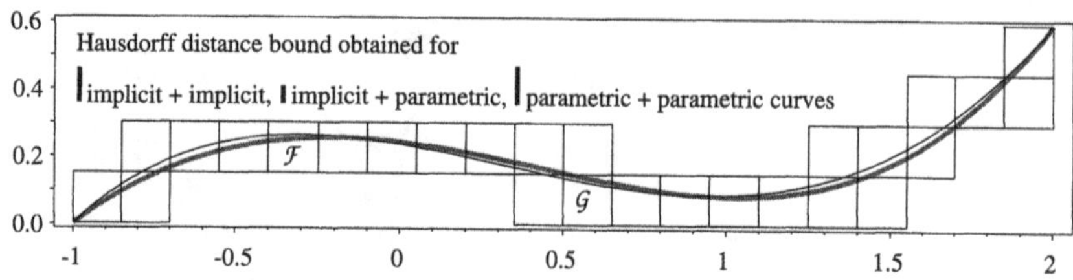

Fig. 4. The Hausdorff distance bounds which are obtained for two cubic Bézier curves by representing them in implicit and/or parametric form.

Geometrically, the set $\mathcal{K}_h^{(k)}$ contains the indices of all subdomains which have points within the offset curve (which consists of line segments and circular arcs) at distance h of the boundary of $\Omega^{(k)}$, see Fig. 3.

Lemma 5. *The inequalities (7) and (8) are valid with the following constants:*

$$C = \max_{\substack{k=1,\ldots,K \\ i,j=0,\ldots,d}} \|\mathbf{c}_{i,j}^{(k)}\| \quad \text{and} \quad D_h = \min_{\substack{k_1=1,\ldots,K;\ k_2\in\mathcal{K}_h^{(k_1)} \\ i_1,i_2,j_1,j_2=0,\ldots,d}} \mathbf{c}_{i_1,j_1}^{(k_1)} \cdot \mathbf{c}_{i_2,j_2}^{(k_2)}. \tag{25}$$

Proof: These constants are obtained by applying the convex hull property of polynomials in Bézier form to the gradient (23) and to the inner product of gradients

$$\nabla f(\mathbf{x}_1) \cdot \nabla f(\mathbf{x}_2), \quad \mathbf{x}_1 \in \Omega,\ \mathbf{x}_2 \in \Omega,\ \mathrm{dist}(\mathbf{x}_1, \mathbf{x}_2) \le h. \quad \square \tag{26}$$

§7. Examples and Conclusions

We apply the theoretical results to the two cubic Bézier curves $\mathcal{F}$ and $\mathcal{G}$ which are shown in Fig. 4. Both curves have identical segment–end points and are fairly close together. Thus we have $\mathcal{F} = \mathcal{F}_0$, $\mathcal{G} = \mathcal{G}_0$, and the minimum distance from a point on one curve to the other curve always occurs at its footpoint. Consequently, the Hausdorff distance (15) is equal to both one–sided footpoint distances $\mathrm{dist}^{\mathrm{F}}(\mathcal{F},\mathcal{G})$ and $\mathrm{dist}^{\mathrm{F}}(\mathcal{G},\mathcal{F})$.

Originally, the curves are given as cubic Bézier curves $\mathbf{f}(t)$ and $\mathbf{g}(t)$, both with the domain [0,1]. After implicitizing them, we obtain (bi-) cubic tensor–product polynomials $f(\mathbf{x})$ and $g(\mathbf{y})$. In order to obtain suitable constants C and D_h, we choose the domain Ω as the union of all squares (size 0.15×0.15) which are shown in the figure. Within each square, the functions f and g are represented in tensor–product Bézier form.

As outlined in Section 6, we generate the constants C and D_h for both curves, where $h = 0.15$. This eventually gives the following bounds on the Hausdorff distance of both curves:

(a) Theorem 3 (pair of implicitly defined curves): $\mathrm{Dist}^{\mathrm{H}}(\mathcal{F},\mathcal{G}) \le 0.108$.

(b) Theorem 4 (implicit and parametric curve): $\mathrm{Dist}^{\mathrm{H}}(\mathcal{F},\mathcal{G}) \le 0.052$. Here, the composition $f(\mathbf{g}(t))$ yields a polynomial of degree 9. The upper bound H is found as the maximum absolute value of the control points which are obtained after splitting it uniformly into four Bézier segments.

Both bounds should be compared with the one which results directly from the parametric representations:

(c) From the convex–hull property (pair of parametric curves):

$$\mathrm{Dist}^{\mathrm{H}}(\mathcal{F}, \mathcal{G}) \leq \max_{t\in[0,1]} \|\mathbf{f}(t) - \mathbf{g}(t)\| \leq 0.119. \tag{27}$$

The upper bound is obtained as the maximum length of the difference vectors of the corresponding control points which are obtained after splitting both curves uniformly into four Bézier segments.

The three bounds are shown in Fig. 4. The second bound is fairly close to the exact Hausdorff distance of both curves.

The above bounds depend on the choice of the constant h, which is an initial estimate of the Hausdorff distance. If a parametric representation of both curves is available, then the corresponding bound (c) can serve as an initial value for h. Alternatively one may use discretization–based or heuristical techniques. This should then be combined with an iterative adaptation technique.

The bound (b), obtained by combining implicit and parametric representations is the tightest one. Theorem 4 will be useful to obtain error bounds for the approximate implicitization and approximate parameterization of algebraic spline curves and surfaces, cf. [1,4].

The bounds (a) and (c), obtained by using either the implicit representations or the parametric ones, are almost identical. However, it should be noted that Theorem 3 provides distance bounds for the more general class of algebraic spline curves, whereas the parameterization–based techniques can deal only with rational ones. Also, the parameterizations of the two curves are relatively similar, as they have identical segment end points. In the general situation the parameterization–based approach is expected to give less accurate results.

The techniques presented in this paper can be used only if, at least within a neighbourhood of the curve, the gradients $\nabla f(\mathbf{x})$, $\nabla g(\mathbf{x})$ of the functions $f(\mathbf{x})$, $g(\mathbf{x})$ satisfy certain regularity conditions. Ideally, these gradients would all be unit vectors. Then, the function $f(\mathbf{x})$ would be simply the signed distance function, or 'normal form', of the curve. See [5,6] for further information on normal forms and their applications. Theorems 3 and 4 can be applied to functions whose gradient field is not too different from the ideal situation. In particular, points with vanishing gradients (extrema and saddle points) of the functions are $f(\mathbf{x})$, $g(\mathbf{x})$ have to be excluded.

In order to obtain tighter bounds, the gradient fields could be improved by multiplying the defining functions with suitable bivariate polynomials.

The curve fitting algorithms described in [8] and [11] produce bivariate spline functions whose gradients approximate unit vectors along its zero contour (i.e., along the corresponding algebraic spline curve). In addition to avoiding singularities, this makes these functions well suited for applying Theorems 3 and 4.

As a matter of future research, we plan to extend the results of this paper to algebraic spline surfaces.

References

1. Bajaj, C. L. and G. Xu, Spline approximations of real algebraic surfaces, J. Symb. Comput. **23** (1997), 315–333.
2. Bogacki, P. and S. E. Weinstein, Generalized Fréchet distance between curves, in *Mathematical Methods for Curves and Surfaces II,* Morten Dæhlen, Tom Lyche, Larry L. Schumaker (eds), Vanderbilt University Press, Nashville & London, 1998, 25–32.
3. De Rose, T., R. N. Goldman, H. Hagen, and S. Mann, Functional composition algorithms via blossoming, ACM Trans. on Graphics **12.2** (1993), 113–135.
4. Dokken, T., Approximate implicitization, in *Mathematical Methods in CAGD: Oslo 2000,* T. Lyche and L. L. Schumaker (eds), Vanderbilt University Press, Nashville, TN, 2001, 81–102.
5. Hartmann, E., The normal form of a planar curve and its application to curve design, in *Mathematical Methods for Curves and Surfaces II,* Morten Dæhlen, Tom Lyche, Larry L. Schumaker (eds), Vanderbilt University Press, Nashville & London, 1998, 237–244.
6. Hartmann, E., On the curvature of curves and surfaces defined by normalforms, Comput. Aided Geom. Design **16** (1999), 355–376.
7. Hoschek, J. and D. Lasser, *Fundamentals of Computer Aided Geometric Design,* AK Peters, Wellesley MA, 1993.
8. Jüttler, B., Least–squares fitting of algebraic spline curves via normal vector estimation, in *The Mathematics of Surfaces IX,* R. Cipolla, R. R. Martin (eds), Springer, London, 2000, 263–280.
9. Schumaker, L. L., *Spline Functions: Basic Theory,* Wiley, New York, 1981.
10. Sederberg, T. W., Planar piecewise algebraic curves, Computer Aided Geom. Des. **1** (1984), 241–255.
11. Taubin, R., Estimation of planar curves, surfaces, and nonplanar space curves defined by implicit equations with applications to edge and range image segmentation, IEEE Trans. Pattern Analysis and Machine Intelligence **13** (1991), 1115–1138.

Bert Jüttler
Johannes Kepler University Linz
Institute of Analysis and Computational Mathematics
Dept. of Applied Geometry
Altenberger Str. 69, 4040 Linz, Austria
`bert.juettler@jk.uni-linz.ac.at`

Biangle Surface Patches

Kęstutis Karčiauskas

Abstract. General construction of a two-sided surface patch of degree $2n$ is introduced and its main properties (degree elevation, subdivision, etc.) are described. As an important special case the biangle patches suitable for inclusion into spline surfaces are presented. They are generalization of Sabin's two-sided patch. Applications are illustrated by freeform spline surfaces representing a sphere exactly.

§1. Introduction

In this paper the general construction of two-sided patch of degree $2n$ is presented. As a special case it gives the biangle patches [4]. Our new construction uses the base point approach for investigation of rational surfaces. From the algebraic point of view, the biangle patches of degree $2n$ as entire surfaces generalize tensor product surfaces of bidegree (n, n). From the geometric point of view, more interesting and important are so called tensor-border patches (a term due to J. Peters). Tensor-border biangles constructed in this paper form a special class of two-sided patches of degree $2n$. They might be considered as the generalization of Sabin's biangle patch [10] – the pioneering work in the two-sided case. Inserting biangle patches into a spline surface makes it more flexible. For example, the freeform spheroids representing a sphere exactly are constructed.

This paper is organized as follows. In Section 2 the definition of a biangle patch of degree $2n$ is given. Its main properties are presented in Section 3. Section 4 is devoted to biangle tensor-border patches. Some applications are briefly described in Section 5.

§2. Basis Functions and Definitions

The surface patches are defined as rational patches in this paper. Only in special cases they do become polynomial. Hence it is convenient to introduce homogeneous control points, as it simplifies many formulations. Namely, for any control point $P \in \mathbb{R}^3$ and its weight w, we set $\underline{P} = (wP, w)$ and call it a

Mathematical Methods for Curves and Surfaces: Oslo 2000
Tom Lyche and Larry L. Schumaker (eds.), pp. 233–242.

ISBN 0-8265-1378-6

homogeneous control point. Infinite control points $\underline{P} = (x_1, x_2, x_3, 0)$ are also used.

Let $D = \Delta V_0 V_1 V_2$ be a domain triangle in a parameter plane (u, v). Any point V is uniquely represented via barycentric coordinates l_0, l_1, l_2 in a form $V = l_0 V_0 + l_1 V_1 + l_2 V_2$. Without loss of generality, we assume $V_0 = (0,0)$, $V_1 = (1,0)$, $V_2 = (0,1)$, and hence $l_0 = 1-u-v$, $l_1 = u$, $l_2 = v$. The functions h_0, h_1, h_2, h_3 are defined for any real number $a > -1$ by the formulas

$$h_0 = l_0^2, \quad h_1 = l_0 l_1, \quad h_2 = l_0 l_2, \quad h_3 = l_1^2 + 2a l_1 l_2 + l_2^2. \tag{1}$$

The functions h_i vanish simultaneously at the intersection points (complex conjugate if $|a| < 1$) of a degenerate conic $h_3 = 0$ and a line $l_0 = 0$. Auxiliary prebasis functions $\tilde{f}_{ij}^n$, $n \geq 1$, $i, j = 0, \ldots, n$, are defined by the formula

$$\tilde{f}_{ij}^n = \begin{cases} h_0^{n-i-j} h_1^i h_2^j, & i+j \leq n, \\ h_3^{i+j-n} h_1^{n-j} h_2^{n-i}, & i+j \geq n. \end{cases} \tag{2}$$

The coefficients k_{ij}^n, $n \geq 1$, i, $j = 0, \ldots, n$, are defined recurrently. We assume $k_{ij}^n = 0$ if $i \notin \{0, \ldots, n\}$ or $j \notin \{0, \ldots, n\}$. For $n = 1$, let $k_{00}^1 = 1$, $k_{10}^1 = 2$, $k_{01}^1 = 2$, $k_{11}^1 = 1$. For $n \geq 1$, the coefficients k_{ij}^{n+1} are defined by the following formula (3) considering separately the cases: $i + j < n + 1$ (first equation); $i + j = n + 1$ (second equation); $i + j > n + 1$ (third equation)

$$\begin{aligned} k_{ij}^{n+1} =& k_{ij}^n + 2k_{i-1,j}^n + 2k_{i,j-1}^n + k_{i,j-2}^n + k_{i-2,j}^n + 2a k_{i-1,j-1}^n, \\ k_{i,n+1-i}^{n+1} =& 2k_{i,n-i}^n + 2k_{i-1,n-i+1}^n + k_{i-1,n-i+2}^n + k_{i+1,n-i}^n + k_{i-2,n-i+1}^n \\ &+ k_{i,n-i-1}^n + 2a k_{i,n-i+1}^n + 2a k_{i-1,n-i}^n, \\ k_{ij}^{n+1} =& k_{i-1,j-1}^n + 2k_{i-1,j}^n + 2k_{i,j-1}^n + k_{i+1,j-1}^n + k_{i-1,j+1}^n + 2a k_{ij}^n. \end{aligned} \tag{3}$$

The coefficients k_{ij}^n satisfy the symmetry relations $k_{ij}^n = k_{ji}^n$, $k_{ij}^n = k_{n-j,n-i}^n$. The basis functions f_{ij}^n, $n \geq 1$, $i, j = 0, \ldots, n$ are defined via the formula

$$f_{ij}^n = k_{ij}^n \tilde{f}_{ij}^n \tag{4}$$

Note that the basis functions f_{ij}^n are linearly independent.

Proposition 1. $\sum_{i,j=0}^n f_{ij}^n = (h_0 + 2h_1 + 2h_2 + h_3)^n$.

Proof: For $n = 1$ this is obvious. By induction, we have

$$(h_0 + 2h_1 + 2h_2 + h_3)^{n+1} = (\sum f_{ij}^n)(h_0 + 2h_1 + 2h_2 + h_3).$$

Expanding the right side of this equation, using formulas (2)–(4) and an identity $h_0 h_3 = h_1^2 + 2a h_1 h_2 + h_2^2$ we get the desired result. This arithmetic procedure is actually a reason why the coefficients k_{ij}^n are defined by the formula (3). □

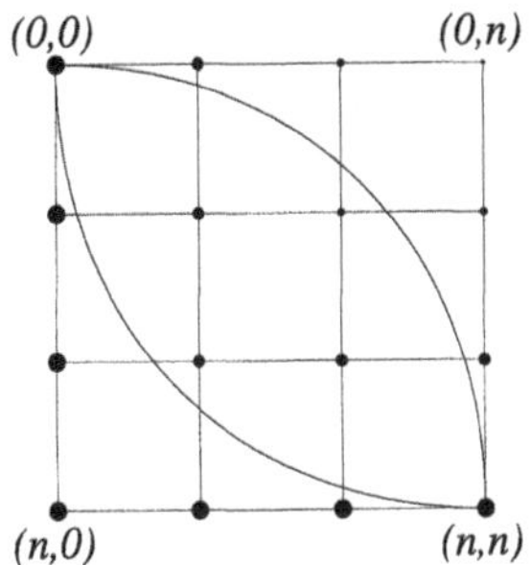

Fig. 1. Control point scheme of a biangle patch.

It is easy to check that $h_0 + 2h_1 + 2h_2 + h_3 = 1$ if and only if $a = 1$. Hence the basis functions sum to 1 only if $a = 1$.

For a fixed n we set $d^0_{ij} = \min(n - i, j)$, $d^1_{ij} = \min(i, n - j)$. An integer d^0_{ij} might be considered as a combinatorial distance from a grid point (i, j) to a half boundary $\{(0,0),\ (1,0), \ldots, (n,0),\ (n,1), \ldots, (n,n)\}$ of a grid rectangle with the corner vertices $(0,0)$, $(n,0)$, (n,n), $(0,n)$. (see Fig. 1). The grid points with $d^0_{ij} = 0$ are displayed in Fig. 1 as the biggest disks, with $d^0_{ij} = 1$ as a little bit smaller, and so on, and similarly for the integer d^1_{ij}. It is easy to check that

$$k^n_{ij} = \binom{2n}{i+j} \quad \text{if} \quad d^0_{ij} = 0 \quad \text{or} \quad d^1_{ij} = 0. \tag{5}$$

Definition 2. *The* rational biangle surface patch *of* degree $2n$ *with the* homogeneous control points $\underline{P}_{ij}$, $i, j = 0, \ldots, n$, *is an image of a mapping* $F : D \to \mathbb{R}^3$, $F = (\tilde{F}_1/\tilde{F}_4, \tilde{F}_2/\tilde{F}_4, \tilde{F}_3/\tilde{F}_4)$, *where* $\tilde{F}$ *is defined by the formula*

$$\tilde{F}(p) = \sum_{i,j=0}^{n} \underline{P}_{ij} f^n_{ij}(p). \tag{6}$$

A mapping F contracts an edge $\overline{V_1V_2}$ of a domain triangle D to a point P_{nn}. Hence $F(D)$ is two-sided surface patch with the corner points P_{00} and P_{nn}. Moreover, it follows from the definition of the basis functions and from the formula (5), that an edge $\overline{V_0V_1}$ maps to a rational Bézier curve of degree $2n$ with the control points $\underline{Q}_s = \underline{P}_{ij}$, $s = i + j$, $s = 0, \ldots, 2n$, $d^0_{ij} = 0$. Similarly, an edge $\overline{V_0V_2}$ maps to Bézier curve with the control points $\underline{P}_{ij}$, $d^1_{ij} = 0$.

Generally, contraction of an edge causes singularity at the image point. Nonsingularity of $F(D)$ at the point P_{nn} follows from a geometric symmetry of a biangle patch.

Proposition 3. *(Symmetry). Let* $\underline{P}'_{ij} = \underline{P}_{n-j,n-i}$. *The homogeneous control points* $\underline{P}'_{ij}$ *and* $\underline{P}_{ij}$ *define the same biangle patch.*

The symmetry of a patch with respect to the change $\underline{P}'_{ij} = \underline{P}_{ji}$ is obvious.

Symmetric two-sided domain. Let $n = 1$, $P_{00} = (1,0,0)$, $P_{10} = (0,1,0)$, $P_{01} = (0,0,1)$, $P_{11} = (-1,0,0)$, $w_{ij} = 1$. Then $F(D)$ is a two-sided domain on the quadric Q, defined by the equation $x^2 + 2(a-1)yz + 2y + 2z - 1 = 0$, and satisfying $y \geq 0$, $z \geq 0$. We denote by H_0, H_1, H_2, H_3 the restrictions to Q of the polynomials $x - y - z + 1$, y, z, $1 - x - y - z$, respectively. It is easy to check that

$$H_0 \circ F/h_0 = H_1 \circ F/h_1 = H_2 \circ F/h_2 = H_3 \circ F/h_3.$$

This relation means that we get the same biangle patch substituting h_i by H_i in (2) and D by $F(D)$ in Definition 2.

If $|a| < 1$, the quadric Q is elliptic; if $a = 1$, parabolic; if $a > 1$, hyperbolic. Biangle patches are called elliptic patches [4], parabolic patches and hyperbolic patches, respectively.

§3. Main Properties of the Biangle Patch

1) *Convex hull property.* A patch lies in a convex hull of the control points P_{ij} if the weights w_{ij} are positive (nonnegative) and $a \geq 0$. It seems to be true also if $-1 < a < 0$, but at the moment we have checked it only for $n \leq 8$.

2) *Cross-derivatives.* The cross-derivative of order m along the boundary curve defined by the the control points $\underline{P}_{ij}$ with $d_{ij}^k = 0$ does not depend of the control points with $d_{ij}^k > m$.

3) *Polynomial patches.* The parabolic patch is polynomial if all weights are equal to 1.

4) *Re-scaling.* A patch remains the same if the weights are changed via the formula $w'_{ij} = \lambda^{i+j} w_{ij}$ for arbitrary $\lambda > 0$.

5) *Degree elevation.* We denote by $\underline{P}_{ij}^n$, $i,j = 0,\ldots,n$, the homogeneous control points of a patch of degree $2n$. We set $\underline{R}_{ij}^n = k_{ij}^n \underline{P}_{ij}^n$, $i,j = 0,\ldots,n$, and define $\underline{R}_{ij}^{n+1}$, $i,j = 0,\ldots,n+1$, via formulas (3), where k is replaced by $\underline{R}$. Finally we define $\underline{P}_{ij}^{n+1} = \underline{R}_{ij}^{n+1}/k_{ij}^{n+1}$.

Proposition 3. *The homogeneous control points $\underline{P}_{ij}^{n+1}$ define a biangle patch of degree $2n+2$ which coincides with the initial patch of degree $2n$.*

6) *Plotting and subdivision.* It follows directly from the definition that a biangle surface patch of degree $2n$ can be represented as a triangular Bézier patch of degree $2n$ with one degenerate side. The mapping $C : (s,t) \to (s(1-t), st)$ maps the unit rectangle to a domain triangle D contracting an edge $s = 0$ to the point V_0. The composition $F \circ C$ represents a biangle patch $F(D)$ as a tensor product patch of bidegree $(2n, 2n)$. Moreover, if $a = 1$ this bidegree is reduced to $(2n, n)$ and a tensor product patch is polynomial if a biangle patch is polynomial. Though a patch $F(D)$ is nonsingular, a mapping $F \circ C$ is degenerate on the edges $s = 0$ and $s = 1$. This might be considered as undesirable feature. Therefore, we also represent $F(D)$ as a collection of nondegenerate Bézier patches of possibly low degree. Various possibilities can

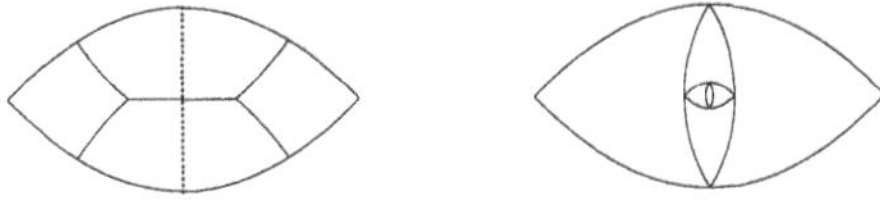

Fig. 2. Subdivision.

be explained via properties of quadratic and biquadratic patches on quadrics (see [1]).

(i) Two "corner" patches of bidegree $(2n, 2n)$ and two "inner" patches of bidegree $(2n, 4n)$ (see Fig. 2, left without dotted line). If $a = 1$ the bidegree of the "inner" subpatches can be reduced to $(2n, 2n)$, but by this reduction a polynomiality is lost.

(ii) Two triangular Bézier patches of degree $4n$ (separated by a dotted line in Fig. 2, left). For $a = 1$ the patches are of degree $2n$ and polynomial if a biangle patch is polynomial. Triangular patches may be subdivided into quadrangular subpatches.

(iii) If a biangle patch is elliptic or hyperbolic, we have in any nondegenerate representation degree $4n$. It is possible to keep the degrees $2n$ by the following recursive procedure (see Fig. 2, right): a biangle patch of degree $2n$ is composed using two Bézier triangles of degree $2n$ and a smaller biangle of degree $2n$ with the same parameter a; if $|a| < 1$ the biangle subpatch fills a hole between triangular subpatches; if $a > 1$ triangular subpatches overlaps a biangle subpatch.

(iv) Subdividing a domain triangle D by a line $l_1 - l_2 = 0$ into two triangles, we get a representation of a biangle of degree $2n$ via two subbiangles of degree $2n$ with parameter $a' = \sqrt{(a+1)/2}$.

7. *Representation of quadrics.*

Proposition 4. *Let $n = 1$ and x_0, x_1, x_2, x_3 be the barycentric coordinates associated with the vertices P_{00}, P_{10}, P_{01}, P_{11}, respectively. Then the biangle patch lies on a quadric defined by the equation*

$$\frac{x_1^2}{w_{10}^2} + \frac{x_2^2}{w_{01}^2} + \frac{2ax_1x_2}{w_{10}w_{01}} - \frac{4x_0x_3}{w_{00}w_{11}} = 0.$$

Conversely, any nonsingular two sided domain on any quadric bounded by two conics can be represented as a biangle patch of degree 2.

A proof of this proposition is a sequence of straightforward computations. The case of higher degree curves on quadrics is more subtle. Using universal parametrization of a sphere from [5], we get the following result.

Proposition 5. *Any two-sided domain on a sphere bounded by Bézier curves of degree $2n$, $n \geq 2$, can be represented as a biangle patch of degree $2[n/2]+4$.*

8) *Implicit equation.* Using standard arguments of algebraic geometry, we get the following result.

Proposition 6. *The implicit degree of a biangle patch of degree $2n$ does not exceed $2n^2$.*

The transformation $T(u,v) = ((\tilde{a}v - u)/A, (\tilde{a}u - v)/A)$, where $\tilde{a} = a + \sqrt{a^2-1}$, $A = (\tilde{a}-1)(\tilde{a}+u+v+1)$, of a parameter plane transforms a hyperbolic biangle patch (as an entire algebraic surface) of degree $2n$ to a tensor product surface of bidegree (n,n). Applying the Dixon resultant (see [11]) we get the equation of a biangle patch. Taking a deeper look at this procedure, we also derive an equation of an elliptic patch (transformation T is complex) and of a parabolic patch (transformation T is undefined!).

In many applications, the boundary of a biangle patch consists of degree raised curves, and hence the implicit degree is $< 2n^2$. For example, the boundary of a biangle of degree 4 are degree raised conics. In this case the implicit degree is ≤ 4. We derive an efficient implicization algorithm, based on the following observation: a biangle patch contains a family of plane sections (maybe complex) composed from two conics. This allows us to reduce constructively the implicization of a surface to a well investigated case of the plane curves.

§4. Tensor-border Nets and Patches

4.1. Terminology and Combinatorial Structure

Sabin [8] introduced multisided patches that behave like tensor product surfaces along the boundary conics. This idea was generalized to a different number of boundary curves and boundary curves of higher degree (see [2,3,6 8,9,10,12]). In retrospect, it appears that the construction of a spline surface of complex topology can be subdivided into two steps. First, data (the control points,weights) representing a smooth join of tensor product patches is created. Second, the multisided patches that behave along the boundary curves like tensor product surfaces are constructed. Corresponding multisided patches were called Sabin–Hosaka– Kimura (SHK) patches in [3]. It is a little bit funny to consider "two-sided" as "multisided", but actually the main principles from [3] were successfully applied also to the biangle surfaces. During a discussion at the conference, Jörg Peters observed that an acronym SHK was being attached to a variety of different constructions leading to misunderstandings; instead he suggested to name the surfaces for their construction principle, e.g. tensor-border patches.

For the fixed integers $m \geq 2$, n, $k \leq [n/2]$ we denote by $\mathcal{L}$ a set of triples (s,i,j), $s = 0,\ldots,m-1$, $i = 0,\ldots,n$, $j = 0,\ldots,k$, where the triples are identified via $(s,i,j) = (s+1,j,n-i)$, $s = 0,\ldots,m-1$, $i \geq n-k$ (index s is treated in a cyclic fashion); if $m = 2$ and $n = 2k$ there are additionally identified $(s,i,k) = (s,n-i,k)$. The m-sided tensor-border net of order n and of depth k is a set of control points $P_q \in \mathbb{R}^3$ and their weights w_q, $q \in \mathcal{L}$. Tensor-border nets have natural combinatoric structure as a collection of m overlapping quadrangular nets (see Fig. 3).

The m-sided tensor-border patch of order n and of depth k is defined via following property: for each boundary curve with the control points $\underline{P}_{s00}$,

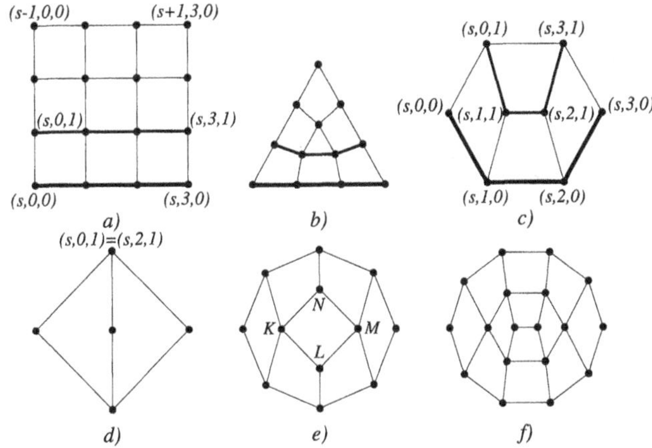

Fig. 3. Schemes of tensor-border patches.

$\underline{P}_{s10}, \dots, \underline{P}_{sn0}$ a patch can be so reparametrized that the crossderivatives up to order k coincide along this curve with the cross-derivatives of tensor product surface patch of bidegree (n, n) with the control points $\underline{Q}_{ij} = \underline{P}_{sij}$, $0 \leq i \leq n$, $0 \leq j \leq k$. To prevent further misunderstandings, we distinguish between tensor-border *net* and tensor-border *patch*.

4.2. Outline of the Construction of Tensor-border Biangles

The principle of construction of two-sided tensor-border patches of order n and of depth 1 is actually the same as in [3], where multisided patches are considered. The patches of depth 2 are handled in a similar fashion. By $B_i^n(t)$ we denote Bernstein polynomials.

(i) The basis functions g_q, $q \in \mathcal{L}$, are defined as linear combinations of prebasis functions $\tilde{f}_{ij}^r$ for some r.

(ii) Reparametrization maps $R_s(u, v) : [0, 1] \times [0, 1] \to D$, $s = 0, 1$, are constructed.

(iii) Let $\overline{g}_q = g_q / \sum_{q' \in \mathcal{L}} g_{q'}$, $q \in \mathcal{L}$. It is proved: $\overline{g}_q \circ R_s|_{v=0} = B_i^n(u)$ if $q = (s, i, 0)$ and 0 else; $(\overline{g}_q \circ R_s)'_v|_{v=0} = (B_i^n(u) B_j^n(v))'_v|_{v=0}$ if $q = (s, i, j)$ and 0 else (in case $n = 2$, $q = (s, 0, 1) = (s, 2, 1)$ it is proved $(\overline{g}_q \circ R_s)'_v|_{v=0} = ((B_0^2(u) + B_2^2(u)) B_1^2(v))'_v|_{v=0}$).

(iv) Rational biangle tensor-border patch of order n and of depth 1 with the homogeneous control points $\underline{P}_q$, $q \in \mathcal{L}$, is defined as an image of a mapping $G : D \to \mathbb{R}^3$, $G = (\tilde{G}_1/\tilde{G}_4, \tilde{G}_2/\tilde{G}_4, \tilde{G}_3/\tilde{G}_4)$, where $\tilde{G}(p) = \sum_{q \in \mathcal{L}} \underline{P}_q g_q(p)$.

(v) Since $G = (\overline{G}_1/\overline{G}_4, \overline{G}_2/\overline{G}_4, \overline{G}_3/\overline{G}_4)$, where $\overline{G}(p) = \sum_{q \in \mathcal{L}} \underline{P}_q \overline{g}_q(p)$, it follows from the properties (iii) of the functions $\overline{g}_q$, that after reparametrisation the biangle patch behaves along each boundary curve like tensor product surface.

4.3. Some Examples of Tensor-border Biangles

We give the explicit formulas for the basis functions of some important two-sided tensor-border patches. Since the patches are symmetric, only necessary formulas are given. The rest can be written by symmetry. Note that basis functions of tensor-border patches are linear combinations of $\tilde{f}^r_{ij}$ for some r. Hence they can be represented as the biangle patches of degree $2r$.

The explicit formulas of the other patches and the proofs can be found in a preprint of the author (in preparation). There also a piecewise polynomial (rational) construction of two-sided tensor-border patches is described.

(i) Tensor-border patch of order 2 and of depth 1 with arbitrary parameter a. The corresponding control point scheme – Sabin net – is shown in Fig. 3d.

$$g_{000} = \tilde{f}^2_{00} + \tilde{f}^2_{20} + \tilde{f}^2_{02},\ g_{010} = 2\tilde{f}^2_{10} + 2\tilde{f}^2_{21},\ g_{011} = 4\tilde{f}^2_{11}. \tag{7}$$

If $a = 1$ (parabolic biangle) it is Sabin's two-sided patch [10].

(ii) Tensor-border patch of order 3 and of depth 1 with arbitrary parameter a and additional shape parameter e (Fig. 3c). Both parameters control a flatness of the biangle patch.

$$\begin{aligned} g_{000} =& \tilde{f}^3_{00} + e(\tilde{f}^3_{10} + \tilde{f}^3_{20} + \tilde{f}^3_{01} + \tilde{f}^3_{02}) + \tilde{f}^3_{30} + \tilde{f}^3_{03} \\ &+ (3 - 2e - 2ea)\tilde{f}^3_{11}, \\ g_{010} =& 3(\tilde{f}^3_{10} + e(\tilde{f}^3_{20} + \tilde{f}^3_{30}) + \tilde{f}^3_{31} + \tilde{f}^3_{11} + (2 - e)\tilde{f}^3_{21}), \\ g_{011} =& 9(\tilde{f}^3_{11} + \tilde{f}^3_{21} + \tilde{f}^3_{12}). \end{aligned} \tag{8}$$

(iii) Elliptic tensor-border patch of order $2r$ and of depth 1 with $a = 0$.

$$\begin{aligned} g_{000} =& \tilde{f}^{r+1}_{00} + \tilde{f}^{r+1}_{20} + \tilde{f}^{r+1}_{02},\ g_{0r0} = \binom{2r}{r}(\tilde{f}^{r+1}_{r,0} + \tilde{f}^{r+1}_{r+1,1}), \\ g_{0i0} =& \binom{2r}{i}(\tilde{f}^{r+1}_{i,0} + \tilde{f}^{r+1}_{i+2,0}),\ 0 < i < r, \\ g_{0i1} =& 2r\binom{2r}{i}\tilde{f}^{r+1}_{i1},\ 1 \le i \le r. \end{aligned} \tag{9}$$

If $r > 2$ we can add in a symmetric fashion to g_q the linear combinations of $\tilde{f}^{r+1}_{ij}$ with $d^0_{ij}, d^1_{ij} > 1$.

(iv) Polynomial tensor-border biangles. The patches of order n defined via formulas (7), (8), (9) are rational and are represented as biangle patches of degree $2[(n+1)/2]+2$. Polynomial tensor-border patches are constructed as the parabolic biangle patches of degree $2[(n+1)/2]+4$.

(v) Tensor-border patch of order 5 and of depth 2 (Fig. 3f) can be represented as an elliptic patch of degree 12 with $a = 0$ or as a parabolic patch of degree 10.

(vi) Convex hull property. If all weights w_q, $q \in \mathcal{L}$, are positive, then for a default range of free parameters the derived tensor-border patches lie in the convex hull of the control points P_q.

(vii) Piecewise polynomial (rational) biangles. The two-sided tensor-border net of order $2k+1$ and depth k is replaced by two triangular tensor-border nets of order $2k+1$ and depth k compatible with each other and with the data of the initial biangle net. As an exception, a similar procedure is valid for Sabin nets (order 2 and depth 1). Corresponding triangular tensor-border patches have a lower degree r as entire biangle patch would have. For example: $n = 2$, $k = 1$ then $r = 4$ (rational), 5 (polynomial); $n = 3$, $k = 1$ then $r = 5$ (rational), 6 (polynomial); $n = 5$, $k = 2$ then $r = 9$ (rational), 11 (polynomial).

§5. Applications

For any (topologically correct) closed triangulation, a smooth spline surface is built using m-sided tensor-border patches of order 2 and of depth 1. It has the following combinatorial structure: corner points of the patches are centroids of the triangles; midpoints of the boundary conics are on the edges; m-sided patch corresponds to vertex of valency m; a biangle patch corresponds to an edge. (For a tetrahedron, Steiner patches can be used instead of triangular tensor-border patches.) This scheme has two global shape parameters controlling input data for tensor-border patches. Local shape parameters depend on the type of tensor-border patches. For a default range of free shape parameters, the surface satisfies convex hull property locally and globally. Taking a piecewise polynomial version of tensor-border patches, a spline surface can be represented via biquartic patches. Using tensor-border patches [3], we get a freeform spline surface representing a sphere exactly for the default position of tetrahedron, octahedron or icosahedron and for default shape values. The construction [7] has a different combinatorial structure, but applied to a triangulation produces a similar surface (it also interpolates the centroids of triangles) which can be represented via bicubic patches.

The input for the representation of a sphere via a tetrahedron, octahedron or icosahedron scheme involves irrational values. Using four tensor-border biangle patches, we get freeform spheroids representing a sphere exactly for the default *rational* values. Here is an example.

Let $K_r = P^r_{011}$, $L_r = P^r_{021}$, $M_r = P^r_{031}$, $N_r = P^r_{121}$, $r = 0, \ldots, 3$, be the inner control points of tensor-border biangle patch of order 4 and of depth 1 (see Fig. 3e). The other control points are defined via the formulas: $P^r_{000} = \sum_{r=0}^{3} K_r/4$; $P^r_{040} = \sum_{r=0}^{3} M_r/4$; $P^r_{010} = (K_{r-1} + K_r)/2$; $P^r_{130} = (K_r + K_{r+1})/2$; $P^r_{030} = (M_{r-1} + M_r)/2$; $P^r_{110} = (M_r + M_{r+1})/2$; $P^r_{020} = (N_{r-1} + L_r)/2$; $P^r_{120} = (N_r + L_{r+1})/2$. The weights are defined by the formulas $w^r_{s00} = 1$, $w^r_{s10} = w^r_{s30} = e_0$, $w^r_{s20} = e_1$, $w^r_{s11} = e_0^2$, $w^r_{s21} = e_0 e_1$, $s = 0, 1$, where e_0, e_1 are any positive numbers. The tensor-border biangle patches are defined via formula (9). We get the smooth spline spheroid with control points K_r, L_r, M_r, N_r and shape parameters e_0, e_1.

Let $K_0 = (1,1,1)$, $L_0 = (2,1,0)$, $M_0 = (1,1,-1)$, $N_0 = (1,2,0)$ and $e_0 = 1/2$, $e_1 = 1/3$. The points K_r, L_r, M_r, N_r, $r = 1,2,3$, are defined by rotating K_0, L_0, M_0, N_0, respectively, around the z-axis by the angle $r\pi/2$. The spline surface is the unit sphere.

References

1. Dietz, R., Hoschek, J., and B. Jüttler, An algebraic approach to curves and surfaces on the sphere and other quadrics, Comput. Aided Geom. Design **10** (1993), 211–229.
2. Hosaka, M. and F. Kimura, Non-four-sided patch expressions with control points, Comput. Aided Geom. Design **1** (1984), 75–86.
3. Karčiauskas K., On five- and six-sided rational surface patches; Rational m-sided Sabin–Hosaka–Kimura like surface patches,preprints, 1999 (http://www.mif.vu.lt/katedros/cs2/publicat/public.htm).
4. Karčiauskas, K. and R. Krasauskas, Rational biangle surface patches, in *Proc. of the Sixth International Conf. in Central Europe on Comp. Graph. and Visualisation*, V. Skala (ed.), 1998, 165–170.
5. Krasauskas R., Universal parametrization of some rational surfaces, in *Curves and Surfaces with Applications in CAGD*, A. Le Méhauté, C. Rabut, and L. L. Schumaker (eds.), Vanderbilt University Press, Nashville, 1997, 231–238.
6. Loop, Ch. and T. DeRose, Generalized B-spline surfaces of arbitrary topology, Computer Graphics **24** (1990), 347–356.
7. Peters J., C^1-surface splines, SIAM J. Numer. Anal., **32**(2), 1995, 645–666.
8. Sabin M., Non rectangular surfaces for inclusion in B-spline surfaces, in *Eurographics'83*, T. Hagen (ed.), 1983, 57–69.
9. Sabin M., A symmetric domain for 6-sided patches, in *The Mathematics of Surfaces IV*, A. Bowyer (ed.), Clarendon Press, 1991, 185–193.
10. Sabin M., Two-sided patches suitable for inclusion in B-spline surface, in *Mathematical Methods for Curves and Surfaces II*, M. Dæhlen, T. Lyche, and L. L. Schumaker (eds.), Vanderbilt University Press, Nashville, 1998, 409–416.
11. Sederberg, T. W., Anderson, D. C., and R. N. Goldman, Implicit representation of parametric curves and surfaces, Computer Vision, Graphics and Image Processing **28** (1984), 72–84.
12. Zheng J. J. and A. A. Ball, Control point schemes over non-four-sided areas, Comput. Aided Geom. Design **14** (1997), 807–821.

K. Karčiauskas
Dept. of Mathematics and Computer Science
Vilnius University
Naugarduko 24, 2600 Vilnius, Lithuania
kestutis.karciauskas@maf.vu.lt

Some Examples of Quasi-Interpolants Constructed from Local Spline Projectors

Byung-Gook Lee, Tom Lyche, and Knut Mørken

Abstract. We give a recipe for deriving local spline approximation methods which reproduce the whole spline space. The methods are obtained by solving a series of local approximation problems. Examples of specific quadratic and cubic approximation methods are given.

§1. Introduction

Many applications of splines make use of some approximation method to produce a spline function from given discrete data. Popular methods include interpolation and least squares approximation. However, both of these methods require solution of a linear system of equations with as many unknowns as the dimension of the spline space, and are therefore not suitable for real-time processing of large streams of data. For this purpose local methods, which determine spline coefficients by using only local information, are more suitable. To ensure good approximation properties it is important that the methods reproduce polynomials and preferably the functions in the given spline space. A method based on derivative information was constructed in [2], while a more general class was studied in [3]. In order to reproduce the spline space, the local information of the methods in [3] was restricted to lie in one knot interval. In this paper we remove this restriction. We then discuss some specific approximation methods for quadratic and cubic splines.

We use B-splines as a basis for splines and denote the ith B-spline of degree d with knots $\boldsymbol{t}$ by $B_{i,d} = B_{i,d,\boldsymbol{t}}$, and the linear space spanned by these B-splines by $\mathbb{S}_{d,\boldsymbol{t}}$.

Mathematical Methods for Curves and Surfaces: Oslo 2000
Tom Lyche and Larry L. Schumaker (eds.), pp. 243–252.

ISBN 0-8265-1378-6

§2. A General Construction of Quasi-interpolants

Given a function f, the basic problem of spline approximation is to determine B-spline coefficients $(c_i)_{i=1}^n$ such that

$$Pf = \sum_{i=1}^{n} c_i B_{i,d}$$

is a reasonable approximation to f. The basic challenge is therefore to devise a procedure for determining the B-spline coefficients. We assume that f is defined on an interval $[a, b]$, and that we have selected a space of splines $\mathbb{S}_{d,\boldsymbol{t}}$ defined on $[a, b]$ (i.e., so that $\boldsymbol{t} = (t_j)_{j=1}^{n+d+1}$ is nondecreasing with $t_{d+1} = a$ and $t_{n+1} = b$). We fix k and propose the following procedure for determining c_k:

(i) Choose a local interval $I = (t_\mu, t_\nu)$ with the property that I intersects the (interior of the) support of $B_{k,d}$:

$$I \cap (t_k, t_{k+d+1}) \neq \emptyset.$$

Denote the restriction of the space $\mathbb{S}_{d,\boldsymbol{t}}$ to the interval I by $\mathbb{S}_{d,\boldsymbol{t},I}$, *i.e.*,

$$\mathbb{S}_{d,\boldsymbol{t},I} = \operatorname{span}\{B_{\mu-d,d}, \ldots, B_{\nu-1,d}\}.$$

(ii) Choose some local approximation method P_I with the property that

$$P_I g = g, \qquad \text{for all } g \text{ in } \mathbb{S}_{d,\boldsymbol{t},I}. \tag{1}$$

(iii) Let f_I denote the restriction of f to the interval I. Then there exist B-spline coefficients $(b_i)_{i=\mu-d}^{\nu-1}$ such that $P_I f_I = \sum_{i=\mu-d}^{\nu-1} b_i B_{i,d}$. Note that $\mu - d \leq k \leq \nu - 1$ since $\operatorname{supp} B_{k,d}$ intersects I.

(iv) Set $c_k = b_k$.

When determining c_k, this procedure gives us the freedom to restrict our attention to a local subinterval $I = [t_\mu, t_\nu]$ of our choice. By doing this we may reduce the complexity of the problem. Secondly, we have the freedom to choose the local approximation method P_I. Typical choices will be interpolation, least squares approximation, or a smoothing spline. As we shall see in Lemma 1, the local condition (1) ensures that if f is a spline in $\mathbb{S}_{d,\boldsymbol{t}}$, it will be reproduced by Pf. In certain situations, other conditions may be more natural, but we will not pursue this any further here.

We first ascertain that the local reproduction condition leads to global reproduction of $\mathbb{S}_{d,\boldsymbol{t}}$.

Lemma 1. *The spline approximation Pf determined by steps (i)–(iv) above has the property that $Pf = f$ for all f in the spline space $\mathbb{S}_{d,\boldsymbol{t}}$.*

Proof: Suppose that $f = \sum_{i=1}^{n} \hat{c}_i B_{i,d}$ for certain coefficients $(\hat{c}_i)_{i=1}^{n}$; we must show that if $Pf = \sum_{i=1}^{n} c_i B_{i,d}$ then $c_i = \hat{c}_i$. We note that $f_I = \sum_{i=\mu-d}^{\nu-1} \hat{c}_i B_{i,d}$, so f_I is clearly in $\mathbb{S}_{d,\boldsymbol{t},I}$. Therefore, by assumption, we have

$$\sum_{i=\mu-d}^{\nu-1} b_i B_{i,d} = P_I f_I = f_I = \sum_{i=\mu-d}^{\nu-1} \hat{c}_i B_{i,d},$$

so $b_i = \hat{c}_i$ for $i = \mu - d, \ldots, \nu - 1$, and in particular $b_k = \hat{c}_k$. But remember that c_k is chosen equal to b_k so we therefore have $c_k = \hat{c}_k$, as required. □

To emphasize the dependence on f, the coefficient c_k is often written $c_k = \lambda_k f$, with λ_k some linear functional. The following lemma gives an explicit formula for the coefficient $\lambda_k f$ in the case where it is a combination of given linear functionals $\lambda_{k,1}, \ldots, \lambda_{k,\nu-\mu+d}$.

Lemma 2. *Suppose that the coefficient c_k of Pf is chosen as*

$$c_k = \frac{\det\bigl(\boldsymbol{\lambda} B_{\mu-d}, \ldots, \boldsymbol{\lambda} B_{k-1}, \boldsymbol{\lambda} f, \boldsymbol{\lambda} B_{k+1}, \ldots, \boldsymbol{\lambda} B_{\nu-1}\bigr)}{\det\bigl(\boldsymbol{\lambda} B_{\mu-d}, \ldots, \boldsymbol{\lambda} b_{\nu-1}\bigr)}, \tag{2}$$

where $\boldsymbol{\lambda} B_j$ denotes the column vector

$$\boldsymbol{\lambda} B_j = (\lambda_{k,1} B_{j,d}, \ldots, \lambda_{k,\nu-\mu+d} B_{j,d})^T$$

and $\lambda_{k,1}, \ldots, \lambda_{k,\nu-\mu+d}$ are linear functionals defined on $\mathbb{S}_{d,\boldsymbol{t}}$ such that the denominator in (2) is nonzero. Then $Pf = f$ for all f in $\mathbb{S}_{d,\boldsymbol{t}}$.

Proof: We choose a special local approximation operator in step (ii) of the construction procedure, namely the one that maps f_I to the spline that solves the generalized interpolation problem

$$\lambda_{k,i}(P_I f_I) = \lambda_{k,i} f_I, \qquad \text{for } i = 1, \ldots, \nu - \mu + d. \tag{3}$$

Expressing $P_I f_I$ in terms of B-splines as $P_I f_I = \sum_{j=\mu-d}^{\nu-1} b_j B_{j,d}$ and inserting this in (3) leads to the linear system of equations

$$\sum_{j=\mu-d}^{\nu-1} (\lambda_{k,i} B_{j,d}) b_j = \lambda_{k,i} f_I, \qquad i = 1, \ldots, \nu - \mu + d. \tag{4}$$

Solving this system for b_k by Cramer's rule and setting $c_k = b_k$ yields the formula (2), and the solution is unique since the denominator in (2) is nonzero. The uniqueness also implies that (1) holds. □

A general class of approximation methods are obtained by letting P_I be given as point functionals of the form

$$\lambda_{k,j} f = f(x_{k,j}), \quad j = 1, \ldots, m_k, \tag{5}$$

where $m_k = \nu - \mu + d$ and $x_{k,1}, \ldots, x_{k,m_k}$ are given points. With this choice, it is well known (see page 200 of [1]) that if

$$B_{\mu-d-1+j,d}(x_{k,j}) > 0, \qquad j = 1, \ldots, m_k, \tag{6}$$

then the denominator in (2) is nonzero and Lemma 2 can be applied. Expanding the numerator in (2), we obtain c_k in the form

$$c_k = \lambda_k f = \sum_{j=1}^{m_k} w_{k,j} f(x_{k,j}), \tag{7}$$

for some vector $\boldsymbol{w}_k = (w_{k,j})$. Equivalently, we can find $\boldsymbol{w}_k$ by solving the linear system

$$\delta_{i,k} = \lambda_k(B_{i,d}) = \sum_{j=1}^{m_k} w_{k,j} B_{i,d}(x_{k,j}) \qquad \text{for } i = \mu - d, \ldots, \nu - 1, \tag{8}$$

where $\delta_{i,k} = 1$ if $i = k$ and zero otherwise, as usual. In practice one would usually determine c_k numerically, either from (2), (4), or (8), except in special cases where the formulas are particularly simple.

Quasi-interpolants of this kind were studied in [3]. However, there the data points $\{x_{k,j}\}_{j=1}^{m_k}$ are restricted to all lie in one subinterval $[t_l, t_{l+1}]$ of $[t_k, t_{k+d+1}]$.

There are standard ways to obtain error estimates for the kind of approximation methods developed here. Let us denote the total approximation by Pf, and suppose we have found a constant C (that may depend on the knots, but not on f) such that

$$\|Pf\| \le C\|f\|. \tag{9}$$

Here $\|f\|$ denotes the uniform norm on the interval $[a, b]$,

$$\|f\| = \max_{x\in[a,b]} |f(x)|.$$

From (9) it follows by a standard argument that

$$\|f - Pf\| \le (1 + C)\,\mathrm{dist}(f, \mathbb{S}_{d,\boldsymbol{t}}), \tag{10}$$

where $\mathrm{dist}(f, \mathbb{S}_{d,\boldsymbol{t}})$ denotes the quantity

$$\mathrm{dist}(f, \mathbb{S}_{d,\boldsymbol{t}}) = \inf_{g\in\mathbb{S}_{d,\boldsymbol{t}}} \|f - g\|.$$

§3. Examples: Projection Into a Given Spline Space

In this section we consider some examples in the case where the knots and the degree of the spline are given.

Example 3. *A quadratic spline.* Suppose $d = 2$ and the knots $\boldsymbol{t} = (t_j)$ are given. To determine c_k we choose a point $(x_{k,2})$ in the middle subinterval of the B-spline $B_{k,2}$. Thus $I = [t_\mu, t_\nu] = [t_{k+1}, t_{k+2}]$ so that we need $m_k = \nu - \mu + d = 3$ local interpolation points in I. The space $\mathbb{S}_{d,\boldsymbol{t},I}$ (quadratic polynomials on I) is spanned by the three B-splines $(B_{i,2})_{i=k-2}^{k}$ and the local operator P_I is the interpolation operator at the three points $(x_{k,j})_{j=1}^{3}$. Since $t_{k+1} = x_{k,1} < x_{k,2} < x_{k,3} = t_{k+2}$, it is clear that (6) holds and the system (4) becomes a simple 3×3 linear system of equations. The coefficient c_k can either be determined as the second of the three resulting coefficients, which is b_k with our labelling, or from (2). The result is

$$c_k = \lambda_k f = \frac{1}{2}\Big(-\theta_k^{-1} f(x_{k,1}) + \theta_k^{-1}(1+\theta_k)^2 f(x_{k,2}) - \theta_k f(x_{k,3})\Big), \qquad (11)$$

where

$$\theta_k = \frac{x_{k,3} - x_{k,2}}{x_{k,2} - x_{k,1}}.$$

The spline approximation $P_2 f = \sum_j \lambda_j f B_{j,2}$ reproduces the quadratic spline space, and from (10) we obtain

$$||f - P_2 f|| \le (3+\rho)\,\mathrm{dist}(f, \mathbb{S}_{2,\boldsymbol{t}}),$$

where

$$\rho = \max_k \{\theta_k, \theta_k^{-1}\}.$$

This holds for any function f and with the special choice $x_{k,2} = (x_{k,1} + x_{k,3})/2$ then $\|f - P_2 f\| \le 4\,\mathrm{dist}(f, \mathbb{S}_{2,\boldsymbol{t}})$. With this special choice of $x_{k,2}$, this operator is classical, and the corresponding approximation method has been used to approximate functions on the sphere ([5]).

Example 4. *A quadratic spline based locally on 5 points.* Another possibility is to choose $[t_\mu, t_\nu] = [t_k, t_{k+3}]$. Then the local spline space $\mathbb{S}_{2,\boldsymbol{t},I}$ has dimension 5 and is spanned by the five B-splines $(B_{i,d})_{i=k-2}^{k+2}$. If we choose three extra points

$$x_{k,1} \in (t_k, t_{k+1}), \qquad x_{k,3} \in (t_{k+1}, t_{k+2}), \qquad x_{k,5} \in (t_{k+2}, t_{k+3})$$

in addition to the two interior knots $x_{k,2} = t_{k+1}$ and $x_{k,4} = t_{k+3}$, we can take P_I to be the operator corresponding to interpolation at the five points $(x_{k,i})_{i=1}^{5}$. Again it is easy to see that (6) holds and we choose c_k as the middle coefficient in (4).

Example 5. *A cubic spline based locally on 5 points.* Similar constructions are possible in the cubic case ($d = 3$). With k fixed, we choose the interval $I = [t_{k+1}, t_{k+3}]$ which means that the local spline space has dimension 5. Again we determine $P_I f$ by interpolation, this time at the three knots $x_{k,1} = t_{k+1}$, $x_{k,3} = t_{k+2}$ and $x_{k,5} = t_{k+3}$ plus the two additional points $x_{k,2} \in (t_{k+1}, t_{k+2})$ and $x_{k,4} \in (t_{k+2}, t_{k+3})$. As before, (6) holds and c_k is chosen as the middle coefficient.

In most cases the formulas for the B-spline coefficients are rather complicated, but in the following case the expressions become quite simple: We choose $x_{k,2}$ as the midpoint between t_{k+1} and t_{k+2}, and $x_{k,4}$ as the midpoint between t_{k+2} and t_{k+4}. If we introduce the knot ratio $\theta = (t_{k+3} - t_{k+2})/(t_{k+2} - t_{k+1})$, the formula for $c_k = \lambda_k f$ becomes

$$\begin{aligned} c_k = & \frac{1}{9}\Big(\frac{1+2\theta}{\theta(1+\theta)} f(x_{k,1}) - 8\frac{1+2\theta}{\theta(1+\theta)} f(x_{k,2}) + (16 + 7\theta^{-1} + 7\theta) f(x_{k,3}) \\ & - 8\frac{\theta(2+\theta)}{1+\theta} f(x_{k,4}) + \frac{\theta(2+\theta)}{1+\theta} f(x_{k,5})\Big). \end{aligned}$$

§4. Approximation of Discrete Data by Quasi-interpolants

In the examples of quasi-interpolants in the previous section it was assumed that a spline space was given, and that we were to construct an approximation method that would project functions onto the spline space. In practice, a more common situation is that a set of discrete data points are given, and the challenge is to construct a spline approximation. In this case, we need to determine a suitable spline space (spline degree and knot vector) before we can compute the approximation, and the knots must be chosen somehow. We consider a class of methods which illustrate how quasi-interpolants of the above type can be used to solve approximation problems of this kind.

We start with an odd number m of data points $(x_j, y_j)_{j=1}^m$, where $m \geq 2d - 1$. We assume that the points are sampled from a function f defined on an interval $[a, b]$, and that the abscissae $\boldsymbol{x} = (x_j)_{j=1}^m$ satisfy $a = x_1 < x_2 < \ldots < x_m = b$. From $\boldsymbol{x}$ we form the knot vector

$$\boldsymbol{t} = (t_j)_{j=1}^{n+d+1} = (\overbrace{x_1, \ldots, x_1}^{d+1}, x_3, x_5, \ldots, x_{m-2}, \overbrace{x_m, \ldots, x_m}^{d+1}),$$

where $n = (m-1)/2 + d$. Note that the knots are related to the abscissae of the data points via the formula $t_j = x_{2(j-d)-1}$ for $j = d+1, \ldots, n+1$. We shall design a quasi-interpolant

$$P_d f = \sum_{k=1}^{n} \lambda_k f B_{k,d}$$

such that each $\lambda_k f$ depends on at most $2d - 1$ data points. To compute $c_k = \lambda_k f$ for k in the range $d \leq k \leq n - d + 1$, we set $I = [t_{k+1}, t_{k+d}]$. The restriction $\mathbb{S}_{d,\boldsymbol{t},I}$ of $\mathbb{S}_{d,\boldsymbol{t}}$ to this interval has dimension $m_k = 2d-1$, but we also have $2d-1$ data points in I, namely the points $x_{k,j} = x_{2(k-d)+j}$ for $j = 1, \ldots, 2d-1$. We can therefore determine a local approximation by forcing a spline in $\mathbb{S}_{d,\boldsymbol{t},I}$ to interpolate these points and then choose the middle coefficient as c_k. The ends must be treated specially. One possibility is to use derivative information, but we do not consider this here. Instead, for $k = 1, \ldots, d-1$ we use the same local interval as for $k = d$, i.e., $I = [t_{d+1}, t_{2d}]$, which means that $\mathbb{S}_{d,\boldsymbol{t},I}$ has dimension $2d-1$ and we have $2d-1$ data points in I. We solve the local interpolation problem and choose $(c_k)_{k=1}^{d-1}$ as the first $d-1$ coefficients of the local interpolant. The right end can be handled similarly.

To be more specific, we work out some of the details in the cubic case (the quadratic case corresponds to the construction in Example 3). We determine the first two coefficient functionals by considering the interval $I = [t_4, t_6] = [x_1, x_5]$. The nonzero B-splines on this interval are $(B_{i,3})_{i=1}^5$ so we can enforce five local interpolation conditions, namely interpolation at the data points $(x_i)_{i=1}^5$. To make sure that we reproduce the local spline space, it is sufficient to require

$$\delta_{i,k} = \lambda_k(B_{i,3}) = \sum_{j=1}^{5} w_{k,j} B_{i,3}(x_j), \qquad \text{for } i = 1, \ldots, 5 \text{ and } k = 1, 2.$$

Since $B_{1,3}(x_1) = 1$ and all other B-splines are 0 at x_1, we see immediately that $\lambda_1 f = f(x_1)$. The other system of equations can be solved numerically to determine the $w_{k,j}$. The right end of the interval $[a, b]$ is treated similarly.

For $k = 3, \ldots, n-2$, we use the five data points $\{x_{k,j}\}_{j=1}^5$ belonging to the subinterval $[t_{k+1}, t_{k+3}]$ to calculate $\lambda_k f$, where $x_{k,j} = x_{2k+j-6}$ for $j = 1, 2, \ldots, 5$. Proceeding as above, we end up with a simple 5×5 system of equations for each set of unknowns $(w_{k,j})_{j=1}^5$.

It is possible to find explicit expressions for the weights by using a computer algebra system, and this can be useful for an analysis of the resulting approximation method, see the next section. When implementing the approximation method however, it is usually more efficient to solve the equations numerically, as the explicit formulas for the weights are rather complicated and expensive to evaluate. However, in certain special cases it is worth precomputing the weights. One such case is when the data points are uniformly spaced when we find

$$\begin{aligned}
\lambda_1(f) &= f(x_1), \\
\lambda_2(f) &= \frac{1}{18}\big(-5f(x_1) + 40f(x_2) - 24f(x_3) + 8f(x_4) - f(x_5)\big), \\
\lambda_3(f) &= \frac{1}{6}\big(f(x_1) - 8f(x_2) + 20f(x_3) - 8f(x_4) + f(x_5)\big).
\end{aligned}$$

The weights for $k = 4, \ldots, n-2$ are the same as those for $k = 3$, while the weights for $k = n-1$ agree with those for $k = 2$, but in reverse order.

Note that if x lies in an interval $[t_\mu, t_{\mu+1})$ for some integer μ in the range $4 \le \mu \le n$, then $P_3 f$ reduces to

$$(P_3 f)(x) = \sum_{j=\mu-3}^{\mu} (\lambda_j f) B_{j,3}(x),$$

and therefore depends on (at most) 11 data values in the vicinity of $[t_\mu, t_{\mu+1})$.

§5. Example of Error Analysis

We give the result of an error analysis for the cubic quasi-interpolant P_3 that we considered in Section 4; the other operators can be analyzed similarly. To state the results, we use the following mesh-ratios:

$$\theta_j = \frac{x_{j+1} - x_j}{x_j - x_{j-1}}, \quad j = 2, \ldots, m-1.$$

Proposition 6. *For each positive integer j with $2j+1 \le m$, we have*

$$||P_3 f||_{\infty,[x_{2j-1},x_{2j+1}]} \le K(\rho) ||f||_{\infty,[x_{2j-5},x_{2j+5}]\cap[x_1,x_m]},$$

where $K(\rho)$ is a polynomial in $\rho = \max_k\{\theta_k, \theta_k^{-1}\}$.

Proof: We deduce the estimate on the first interval $[x_1, x_3]$; the treatment of the other intervals is similar. For $x \in [x_1, x_3]$ we have

$$P_3 f(x) = \sum_{j=1}^{4} \lambda_j f B_{j,3}(x),$$

where $\lambda_1 f = f(x_1)$, $\lambda_2 f = \sum_{j=1}^{5} w_{2,j} f(x_j)$, $\lambda_k f = \sum_{j=1}^{5} w_{k,j} f(x_{2k-6+j})$ for $3 \le k \le m-2$, and where the $w_{k,j}$ can be computed from $\lambda_k f = c_k$ given by (2). Note that only the 7 first x-values are used to define $P_3 f(x)$ for $x \in [x_1, x_3]$. Therefore, since the B-splines form a nonnegative partition of unity, we obtain

$$|(P_3 f)(x)| \le \max_{1\le k\le 4} \sum_{j=1}^{5} |w_{k,j}| \, ||f||_{\infty,[x_1,x_7]}, \quad x \in [x_1, x_3].$$

Since the $w_{k,j}$ for fixed k are computed by forming minors in the numerator in (2), it is clear that they must alternate in sign. This follows since all minors are nonnegative by total positivity properties of B-splines, see *e.g.* page 201 of [1]. Moreover, since P_3 reproduces constants, we have $\sum_{j=1}^{5} w_{k,j} = 1$ for all k. Combining this identity with the fact that the $w_{k,j}$ oscillate in sign, we find that

$$\sum_{j=1}^{5} |w_{k,j}| \le 2C_k + 1,$$

where

$$C_k = |w_{k,2}| + |w_{k,4}|.$$

It then follows that

$$|(P_3 f)(x)| \leq (2 \max_{1\leq k\leq 4} C_k + 1)||f||_{\infty,[x_1,x_7]}, \quad x \in [x_1, x_3].$$

Now $C_1 = 0$ and using *Mathematica* we find

$$C_2 = \frac{(1+\theta_2)^2\left((1+\theta_4)^2 + \theta_3\theta_4(1+\theta_2)\nu_2\right)}{3\theta_2^2\theta_3\theta_4\nu_2},$$

while for $3 \leq k+1 \leq m-2$

$$C_{k+1} = \frac{(1+\theta_{2k-2})^2(1+\theta_{2k-1})(1+\theta_{2k})^2\left(1+\theta_{2k-1}\theta_{2k}(1+\theta_{2k-2}\theta_{2k-1})\right)}{3\theta_{2k-2}\theta_{2k-1}\theta_{2k}(\theta_{2k-1}(1+\theta_{2k}) + \nu_k)},$$

where

$$\nu_k = 2 + \theta_{2k} + \theta_{2k-2}(2 + \theta_{2k} + 2\theta_{2k-1}(1+\theta_{2k})).$$

It follows that these expressions can be bounded by a polynomial in ρ. □

If the data are evenly spaced, or more generally, the even indexed data-points are located midway between their neighbours, the estimate for the constant $K(\rho)$ simplifies.

Corollary 7. *Suppose that $x_{2k} = (x_{2k-1} + x_{2k+1})/2$ for $k \leq (m-1)/2$. For any positive integer j with $2j+1 \leq m$ we have*

$$||P_3 f||_{\infty,[x_{2j-1},x_{2j+1}]} \leq K(\rho)||f||_{\infty,[x_{2j-5},x_{2j+5}]\cap[x_1,x_m]},$$

where

$$K(\rho) = (16\rho + 41)/9.$$

Proof: In this case $\theta_{2k} = 1$ for all k, and the expressions for C_2 and C_{k+1} in the proof of Proposition 6 simplify to

$$C_2 = \frac{8}{9}\left(2 + \frac{1}{\theta_3}\right),$$

$$C_{k+1} = \frac{8}{9}\left(\frac{1}{\theta_{2k-1}} + 1 + \theta_{2k-1}\right), \qquad \text{when } 3 \leq k+1 \leq m-2.$$

Since $\theta_{2k-1}^{-1} + \theta_{2k-1} \leq 1 + \rho$ we find $K(\rho) \leq 2C + 1$, where $C = \max_k C_k \leq \frac{8}{9}(2+\rho)$ for all k. □

§6. Remarks

Remark 6.1. From Proposition 6 and (10) we can deduce that the error $P_3 f - f$ will be of the same order of magnitude as the error in best approximation by cubic splines. In fact if the data is uniform, then $\rho = 1$ and from Corollary 7 we find $K(\rho) = 19/3$. Of course $K(\rho)$ is larger for nonuniform data. In any case, if f has a bounded fourth derivative we see that $P_3 f$ is a fourth order accurate approximation to f.

Remark 6.2. The approximation procedure in this paper can be extended to trigonometric splines [4] and other more general classes of functions.

Remark 6.3. We have given the three formulae (2), (4), and (8) for deriving quasi-interpolants reproducing the whole spline space. In addition formulas based on blossoming could be used. We refer to [4].

Acknowledgments. This work was done during the first authors's visit to Oslo under support of a Korea Science and Engineering Foundation (KOSEF) postdoctoral fellowships program.

References

1. de Boor, C., *A Practical Guide to Splines*, Springer Verlag, New York, 1978.
2. de Boor, C. and G. J. Fix, Spline approximation by quasiinterpolants, J. Approx. Theory **8** (1973), 19–45.
3. Lyche, T. and L. L. Schumaker, Local spline approximation methods, J. Approx. Theory **15** (1975), 294–325.
4. Lyche, T., L. L. Schumaker, and S. Stanley, Quasi-interpolants based on trigonometric splines, J. Approx. Theory **95** (1998), 280–309.
5. Schumaker, L. L. and C. Traas, Fitting scattered data on spherelike surfaces using tensor products of trigonometric and polynomial splines, Numer. Math. **60** (1991), 133–144.

Department of Applied Mathematics
Dongseo University
Pusan, 617-716 Korea
lbg@dongseo.ac.kr

Universitetet i Oslo
Istititt for informatikk
P.O. Box 1080, Blindern
0316 Oslo, Norway
tom@ifi.uio.no
knutm@ifi.uio.no

Single-Valued Tubular Surface Intersection Using Interval Arithmetic

Hélio Lopes and Sinésio Pesco

Abstract. Single-valued tubular surfaces are built from a 3D directrix curve. Given a point on the directrix curve and an angular direction, a point on the surface is obtained by a function that corresponds to the distance between these two points along the given direction. This work presents a new algorithm for intersecting and trimming single-valued tubular surfaces based on interval arithmetic. Applications of this algorithm on boolean operations and on petroleum reservoir modeling are here discussed as well.

§1. Introduction

The main geometrical object of this work is the single-valued tubular surface, which was first introduced by Sánches-Reyes in [8].

A single-valued directrix curve $\mathbf{d} : [t_i, t_f] \to \mathbb{R}^3$ is a function that associates for a given parameter t in the domain, the point $\mathbf{d}(t) = (t, x(t), y(t))$, in which y and z are scalar continuous functions of one variable defined on the same domain $[t_i, t_f]$. The first coordinate function of $\mathbf{d}$, which corresponds to the identity, could be replaced by any inversible scalar continuous function. For a given angle $\theta \in [\theta_i, \theta_f]$, the angular direction $\mathbf{q}_\theta$ is defined to be the vector $\mathbf{q}_\theta = (0, -cos(\theta), sin(\theta))$. A point $\mathbf{s}(t,\theta)$ on the surface is then obtained by choosing a scalar continuous function $r : [t_i, t_f] \times [\theta_i, \theta_f] \to \mathbb{R}$ that corresponds to the distance between the points $\mathbf{d}(t)$ and $\mathbf{s}(t,\theta)$ along the direction $\mathbf{q}_\theta$ (Fig. 1 illustrates an example). In other words, the single-valued tubular surface parametric function $\mathbf{s} : [t_i, t_f] \times [\theta_i, \theta_f] \to \mathbb{R}^3$ assigns for a given ordered pair (t, θ) the point

$$\mathbf{s}(t,\theta) = \mathbf{d}(t) + r(t,\theta)\mathbf{q}_\theta.$$

A tubular solid in $\mathbb{R}^3$ is a volume derived from this kind of surface, whose parametrization $\mathbf{v} : [0,1] \times [t_i, t_f] \times [\theta_i, \theta_f] \to \mathbb{R}^3$ is given by

$$\mathbf{v}(u,t,\theta) = \mathbf{d}(t)(1-u) + \mathbf{s}(t,\theta)u.$$

Mathematical Methods for Curves and Surfaces: Oslo 2000
Tom Lyche and Larry L. Schumaker (eds.), pp. 253–262.

ISBN 0-8265-1378-6

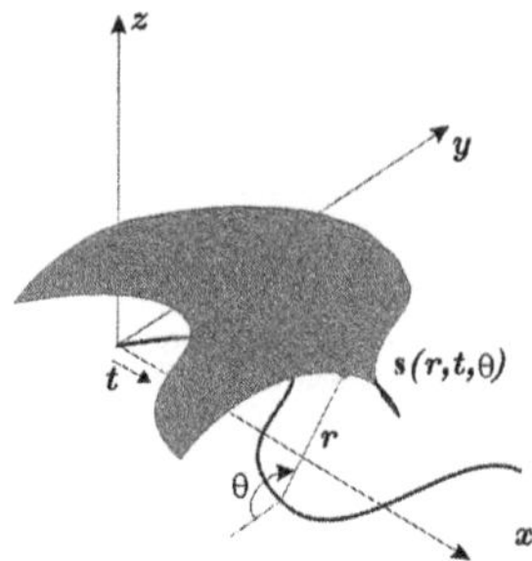

Fig. 1. Single-valued tubular surface example.

The main advantage of this type of solid is that it is very simple to build an algorithm that decides if a point in $\mathbb{R}^3$ is inside, outside or on the boundary of the volume. As a consequence, it can be included as a CSG primitive in a solid modeling system. To do so, it is necessary to develop algorithms that perform boolean operations for this kind of solid. The starting step on these algorithms is the intersection curve calculation.

Algorithms to intersect parametric surfaces are divided into two groups. Algorithms of the first group use continuation methods to obtain the intersection curves [1]. Algorithms of the second group use subdivision methods instead [4].

The initial step of algorithms based on continuation methods is the search of a common point on the surfaces. The intersection curve is then obtained by an iterative method that approximates the next point in the curve by using the curve tangent vector direction. Some problems with this strategy should be mentioned: the surfaces could intersect in several curve components, it is hard to numerically identify closed intersection curve components, and the intersection curve could have singular points.

In algorithms based on subdivision methods, the domain of both surfaces are divided until each surface patch could be replaced by a simple approximation. The intersection problem is then solved for each pair of patches. As a consequence, the number of intersection calculation could be huge. Also, in the algorithms of this group, some connected intersection curve components might not be identified when the surfaces are not sufficiently subdivided. Algorithms based on *range analysis* has been proposed [9] in order to solve this problem.

The main objective of this work is to present a new subdivision algorithm for intersecting and trimming single-valued tubular surfaces based on *Interval Arithmetic*, which has been recognized as a natural tool to perform range analysis computation [7].

This paper is organized as follows. Section 2 defines some interval arithmetic basic concepts. Section 3 presents a point membership classification algorithm for single-valued tubular solids and introduces an interval extension of such algorithm. Section 4 introduces the surface-to-surface intersection algorithm. Finally, Section 5 shows some examples and applications.

§2. Basic Concepts of Interval Arithmetic

The objective of this section is to define the notations and the basic concepts of Interval Arithmetic (IA). The theory of IA can be found in more detail in [7]. The applications of IA to geometric modeling and computer graphics are discussed in [11].

The set of all compact real intervals will be denoted by $\mathbb{IR}$. An element $[x]$ in $\mathbb{IR}$ corresponds to an interval $[\underline{x}, \overline{x}]$, where $\underline{x}$ and $\overline{x}$ are two real numbers such that $\underline{x} \leq \overline{x}$.

For a given interval $[x] = [\underline{x}, \overline{x}] \in \mathbb{IR}$, its two boundary points $\underline{x}$ and $\overline{x}$ will be called, respectively, the infimum and the supremum of $[x]$ and they will be denoted by $\inf([x])$ and $\sup([x])$.

The basic binary arithmetic operation $\circ \in \{+, -, \times, \div\}$ for real numbers can be extended for two intervals $[x]$ and $[y] \in \mathbb{IR}$ in the following way:

$$[x] \circ [y] := \{x \circ y | x \in [x] \text{ and } y \in [y]\}.$$

Of course, the arithmetic operator $\div$ can only be applied if $0 \notin [y]$. The result of any interval arithmetic operation is an interval, and by using its monotonicity properties, they can be expressed in terms of the boundary points of its operands, for example, $[\underline{x}, \overline{x}] + [\underline{y}, \overline{y}] = [\underline{x} + \underline{y}, \overline{x} + \overline{y}]$.

All those binary arithmetic operations $\circ \in \{+, -, \times, \div\}$ are inclusion isotonic, which means: $[z] \subseteq [x]$ and $[w] \subseteq [y] \Rightarrow [z] \circ [w] \subseteq [x] \circ [y]$.

The domain of a real function f will be denoted by $Dom(f)$. The image of a real continous function f over all points x in the compact real interval $[x] \subseteq Dom(f)$ is defined to be the interval $Im(f, [x])$ such that

$$Im(f, [x]) := [\min\{f(x) | x \in [x]\}, \max\{f(x) | x \in [x]\}].$$

The interval extension of an elementary real function $\phi \in \{\exp, \ln, \cos, \sin, \tan, \cos^{-1}, \sin^{-1}, \tan^{-1}, \cosh, \sinh, \tanh, \cosh^{-1}, \sinh^{-1}, \tanh^{-1}\}$ over a given interval $[x] \subseteq Dom(\phi)$ is defined to be the image of ϕ over $[x]$, i.e., $\phi([x]) := Im(\phi, [x])$. The power and rational functions are defined in the same way.

An interval function γ is said to be inclusion isotonic when for all pairs of intervals $[x]$ and $[y]$ contained in its domain, the proposition $[x] \subseteq [y] \Rightarrow \gamma([x]) \subseteq \gamma([y])$ is always true.

By definition, the interval extension of all elementary functions cited above have this important property.

Since each elementary real function is continous and its interval extensions is inclusion isotonic, the interval $\phi([x])$ can also be expressed in terms of the boundary points of $[x]$, for example

$$\exp([x]) := [\exp(\inf([x])), \exp(\sup([x]))],$$

and

$$([x])^n := \begin{cases} [\underline{x}^n, \overline{x}^n] & \text{if } 0 < \underline{x} \text{ or } n \text{ is odd,} \\ [0, \max\{\underline{x}^n, \overline{x}^n\}] & \text{if } 0 \in [x] \text{ and } n \text{ is even,} \\ [\overline{x}^n, \underline{x}^n] & \text{if } \overline{x} < 0 \text{ and } n \text{ is even.} \end{cases}$$

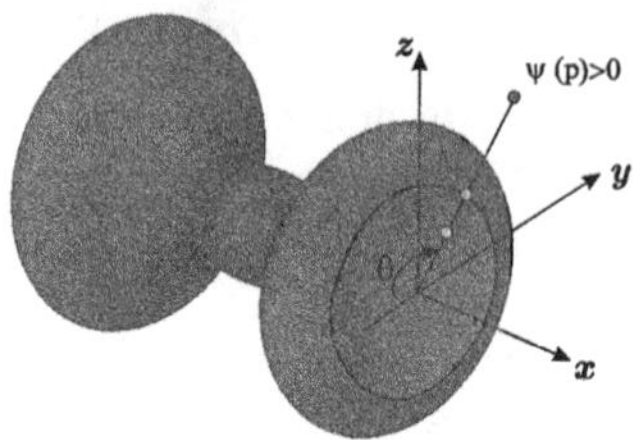

Fig. 2. Point membership classification.

Let f be a scalar function of one real variable x defined by an expression that contains only arithmetic operations and elementary functions. The interval extension of f, denoted by $f_{[]}$, over a compact real interval $[x] \subseteq Dom(f)$ is defined to be the interval function built from the replacement of every ocorrence of the real variable x in the expression of f by the interval variable $[x]$. The interval $f_{[]}([x])$ is then obtained by the use of the corresponding interval arithmetic operators and interval elementary functions on the evaluation of its expression.

When the interval $[x]$ is a single point ($[x] = [a, a]$), the interval extension $f_{[]}$ satisfies $f_{[]}([a, a]) = f(a)$. Since all interval operations and elementary functions are inclusion isotonic, the interval extension of a real function also is inclusion isotonic.

The next theorem is the main tool used in this work. It links IA to range analysis. Its proof will be omitted since it can be found in [7].

Theorem 1. *Let f be a real continous function of one variable and $f_{[]}$ its interval extension. If $[x] \subseteq Dom(f)$, then $Im(f, [x]) \subseteq f_{[]}([x])$.*

When a point $y \notin f_{[]}([x])$, this theorem guarantees that the point y is not in the set $Im(f, [x])$. Unfortunately, the interval $f_{[]}([x])$ is, in general, an overestimation of the range of f over the interval $[x]$. To get a narrower enclosure, several other tools has been proposed. For example, centered forms [3] and affine arithmetic [2] are to be cited.

The standard algebraic operations on the vector space $\mathbb{R}^n$ are also extended to interval vector operations in $\mathbb{IR}^n$.

§3. The Point Membership Classification Algorithm

As introduced above, the volume of a tubular solid is described by the parametric function $\mathbf{v} : [0, 1] \times [t_i, t_f] \times [\theta_i, \theta_f] \to \mathbb{R}^3$, whose expression is given by $\mathbf{v}(u, t, \theta) = \mathbf{d}(t)(1 - u) + \mathbf{s}(t, \theta)u$. The main characteristic of this kind of solid is that it is very simple to build a function that classifies whether a given point $\mathbf{p}_0 = (x_0, y_0, z_0) \in \mathbb{R}^3$ is inside, outside or on the boundary of its volume. The next algorithm is an implementation of such a function.

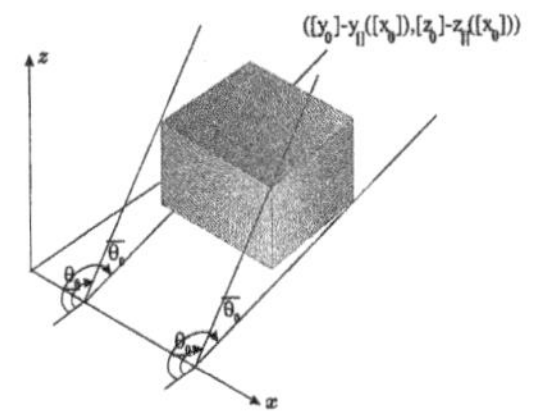

Fig. 3. The $[\theta_0]$ interval calculation.

FUNCTION Classify $(p_0, \mathbf{v}, [t_i, t_f], [\theta_i, \theta_f])$
1) $t_0 = x_0$;
2) IF $(t_0 \notin [t_i, t_f])$ THEN RETURN ($\mathbf{p}_0$ is *outside*);
3) $(d_x, d_y, d_z) = \mathbf{d}(t_0)$; $\theta_0 = \mathrm{ARG}((y_0, z_0) - (d_y, d_z))$;
4) IF $(\theta_0 \notin [\theta_i, \theta_f])$ THEN RETURN ($\mathbf{p}_0$ is *outside*);
5) $r_0 = \sqrt{(d_y - y_0)^2 + (d_z - z_0)^2}$; $\psi_0 = r_0 - r(t_0, \theta_0)$;
6) IF $(\psi_0 < 0)$ THEN RETURN ($\mathbf{p}_0$ is *inside*);
   ELSE IF $(\psi_0 > 0)$ THEN RETURN ($\mathbf{p}_0$ is *outside*);
   ELSE RETURN ($\mathbf{p}_0$ is *on the boundary*);

END Classify.

The function Classify uses the coordinate x_0 of $\mathbf{p}$ to find the directrix curve parameter, since $\mathbf{d}(x_0) = (x_0, y(x_0), z(x_0))$ if $x_0 \in [t_i, t_f]$. After that, the clockwise oriented angle θ_0 between the vectors $(-1, 0)$ and $(y_0 - y(x_0), z_0 - z(x_0))$ is computed by the use of the function ARG. If $\theta_0 \in [\theta_i, \theta_f]$, the distance function r is evaluated in (x_0, θ_0) and the distance r_0 between the points $(y(x_0), z(x_0))$ and (y_0, z_0) is calculated. A new function ψ is now defined by $\psi(p_0) = r_0 - r(x_0, \theta_0)$. In order to perform the membership classification for the given point p_0, the sign of $\psi(p_0)$ is tested (Fig. 2 shows an example).

We now introduce an interval extension for the function ψ. The unique parameter of $\psi_{[]}$ is an interval vector (or simply a box) $B = ([x_0], [y_0], [z_0]) \in \mathbb{IR}^3$, such that $[x_0] \subseteq [t_i, t_f]$. The directrix curve interval extension $\mathbf{d}_{[]}$ is evaluated in $[x_0]$ in order to obtain an estimation for its range. The interval $[\theta_0]$ is calculated by applying an interval extension of the ARG function over the interval vector $([y_0] - y([x_0]), [z_0] - z([x_0])) \in \mathbb{IR}^2$, (Fig. 3 illustrates an example). Then, the value of $\psi_{[]}$ over the interval vector $([x_0], [y_0], [z_0])$ is defined to be the interval

$$\psi_{[]}([x_0], [y_0], [z_0]) = \sqrt{([y_0] - y_{[]}([x_0]))^2 + ([z_0] - z_{[]}([x_0]))^2} - r_{[]}([x_0], [\theta_0]).$$

There are three important facts to observe about the proposed interval extension of the function ψ that follows from its definition.

Proposition 2. *The interval function $\psi_{[]}$ is inclusion isotonic.*

Corollary 3. *If $B = ([a, a], [b, b], [c, c])$, then $\psi_{[]}(B) = \psi(a, b, c)$.*

Corollary 4. *If $0 \notin \psi_{[]}(B)$, the box B does not intersect the single-valued tubular surface* $\mathbf{s}$.

§4. The Surface-to-surface Intersection Algorithm

To include tubular solid as a CSG primitive in a solid modeling system, it is necessary to develop algorithms that perform boolean operations on it.

A tubular solid is bounded by a single-valued tubular surface together with two planar faces. Those faces are parametrized, respectively, by $\mathbf{v}(u, t_i, \theta)$ and $\mathbf{v}(u, t_f, \theta)$ for all $u \in [0,1]$ and $\theta \in [\theta_i, \theta_f]$. Since the intersection curve computation is an important step in the boolean operations, this section will introduce an algorithm based on range analysis to intersect a single-valued tubular surface $\mathcal{S}_1$ with another surface $\mathcal{S}_2$ given by a parametric function $\mathbf{m} : [u_i, u_f] \times [v_i, v_f] \subset \mathbb{R}^2$.

This algorithm will decompose the domain of $\mathcal{S}_2$, and by the use of range analysis will decide whether the $\mathcal{S}_2$ patches intersect $\mathcal{S}_1$. To implement such an algorithm, the following function has been devised:

`FUNCTION` Explore $([u], [v], level)$

1) $([x_0], [y_0], [z_0]) = \mathbf{m}_{[]}([u], [v])$;
2) $[t_0] = [x_0] \cap [t_i, t_f]$;
3) `IF` ($[t_0]$ is empty) `THEN RETURN` (The $\mathcal{S}_2$'s patch doesn't intersect $\mathcal{S}_1$);
4) $[\psi_0] = \psi_{[]}(([t_0], [y_0], [z_0]))$;
5) `IF` ($0 \notin [\psi_0]$) `THEN RETURN` (The $\mathcal{S}_2$'s patch doesn't intersect $\mathcal{S}_1$);
   `ELSE IF` ($level$ = max. level of recursion) `THEN`
      `CALL` Intersect$([u], [v])$;
   `ELSE`
      Explore$([\inf(u), \mathrm{mid}(u)], [\inf(v), \mathrm{mid}(v)], level + 1)$;
      Explore$([\mathrm{mid}(u), \sup(u)], [\inf(v), \mathrm{mid}(v)], level + 1)$;
      Explore$([\inf(u), \mathrm{mid}(u)], [\mathrm{mid}(v), \sup(v)], level + 1)$;
      Explore$([\mathrm{mid}(u), \sup(u)], [\mathrm{mid}(v), \sup(v)], level + 1)$;

`END` Explore.

The function Explore uses a recursion to decompose the domain of $\mathcal{S}_2$ by subdividing each current rectangle into four rectangles with the same size. This algorithm stops when it is possible to assert that the current patch of $\mathcal{S}_2$ couldn't intersect $\mathcal{S}_1$ or when the level of recursion reachs a chosen maximum level.

Suppose that the box $B = ([x_0], [y_0], [z_0])$ corresponds to the estimated range of the $\mathcal{S}_2$ current patch on the recursion. If the interval $[x_0] \cap [t_i, t_f]$ is empty, the $\mathcal{S}_2$ current patch couldn't intersect $\mathcal{S}_1$ since the interval $[t_i, t_f]$ is the exact first coordinate range of $\mathcal{S}_1$. The current patch on the recursion could also be discarded when $0 \notin \psi_{[]}(B)$ (see Corollary 4).

When the maximum level of recusion is reached, the $\mathcal{S}_2$ current patch may intersect $\mathcal{S}_1$. In this case, the function Explore calls the function Intersect in order to verify the existence and to determine the intersection curves. The function Intersect subdivides the rectangle $[u] \times [v]$ into two triangles and verifies for each edge if there is a point p on the edge such that $\psi(\mathbf{m}(p)) = 0$. If such point exists it will correspond to the point where the edge intersects the surface $\mathcal{S}_1$. In the algorithm, the maximum level of recursion is chosen in such a way to have on each transversal edge at most one solution for $\psi(\mathbf{m}(p)) = 0$. Non-transversal edges are specially treated. Taylor Expansions [5] and Automatic Differentiation [3] techniques can be used in order to guarantee that a closed intersection curve couldn't occur inside a rectangle generated on the maximum level of subdivision.

Having the point $\mathbf{m}(p) = (p_x, p_y, p_z)$, such that $\psi(\mathbf{m}(p)) = 0$, the parameters t and θ from the tubular surface $\mathcal{S}_1$ are calculated by the following expressions: $t = p_x$ and $\theta = \text{ARG}((p_y - y(p_x), p_z - z(p_x))$.

When a subdivided triangle in the domain of $\mathcal{S}_2$ has two points p_1 and p_2 such that $\psi(\mathbf{m}(p_1)) = 0$ and $\psi(\mathbf{m}(p_2)) = 0$, the edge p_1p_2 is inserted on the $\mathcal{S}_2$ triangulated surface. The points $\mathbf{m}(p_1)$ and $\mathbf{m}(p_2)$ are then used to calculate the ordered pairs (t_1, θ_1) and (t_2, θ_2) in the parametrization of $\mathcal{S}_1$. Once the two pairs are computed, the edge that connects these two points on the domain is included on the $\mathcal{S}_1$ triangulated surface.

The main advantages of this algorithm are that it is robust and very simple to implement. The use of Interval Arithmetic is essential to guarantee the elimination of $\mathcal{S}_2$ patches that couldn't intersect $\mathcal{S}_1$. However, it could require a large amount of computation because IA may give an overestimation for the range of the functions. When this overestimation is too large, the function Explore will not be able to discard rectangles in the $\mathcal{S}_2$ domain by using only the conditions proposed above. As a consequence, the function Intersect will be called unnecessarilly. Since the use of IA in this algorithm is localized, the interval arithmetic could be replaced by centered forms [3] or affine arithmetic [2] in order to get a narrower approximation for the range of the functions evaluated in the algorithm.

§5. Examples and Applications

This section illustrates the above algorithm with two examples.

Example 1. The directrix curve and distance function of the single-valued tubular surface $\mathcal{S}_1$ are given by: $\mathbf{d}(t) = (t, 0.7 - 0.5\sin(t), 0.4 + 0.2\sin(t))$, $r(t,\theta) = 1$, and the single-valued tubular surface $\mathcal{S}_2$ is defined by the functions $\mathbf{d}(t) = (t, 1, 1)$ and $r(t, \theta) = 1.1 - (4.9^2t^2\cos(4.9t/5))/150$. The domain of both surfaces is $[-3, 3] \times [0, 2\pi]$. Fig. 4 illustrates the 3D surfaces. Fig. 5a shows how the algorithm has decomposed the $\mathcal{S}_2$ domain in order to calulate the intersection curve. Fig. 5b shows the corresponding intersection curve inserted in the $\mathcal{S}_1$ triangulated domain.

Example 2. In this example, the solids are bounded by single-valued tubular surfaces whose directrix curves and r functions are Bézier curves. Fig. 6a illustrates a projection of the two surfaces, in which $\mathcal{S}_1$ is the surface in the left. Fig. 6b shows the chosen position of those surfaces for the intersection. Fig. 7 shows how the algorithm decompose the $\mathcal{S}_2$ domain. Note that, in this example, the IA gives a large overestimation for the functions range, caused by the Bézier curve interval evaluation. As a consequence, the function Intersect is unnecessarily called on the majority of the rectangles generated in

Fig. 4. The two surfaces of the Example 1.

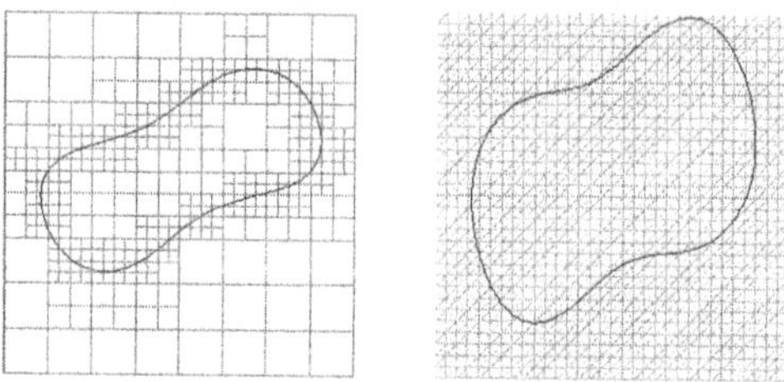

Fig. 5. The surfaces domain of $\mathcal{S}_2$ (in the left) and $\mathcal{S}_1$.

Fig. 6. The surfaces of the second example.

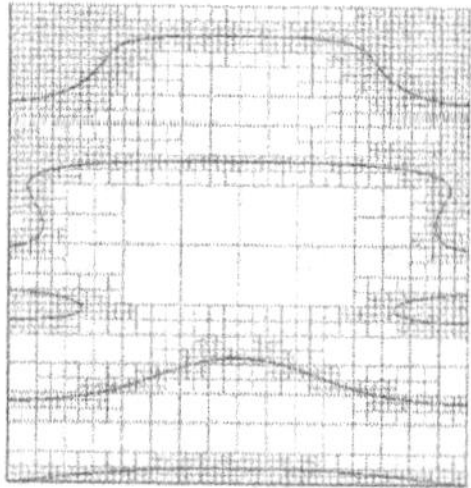

Fig. 7. The $\mathcal{S}_2$ domain decomposition.

the maximum level of recusion. Fig. 8 illustrates the two surfaces cut along the intersection curves. Fig. 9 shows the result of the intersection and union boolean operations on the corresponding solids.

Fig. 8. The surfaces cutted along the intersection curve.

Fig. 9. The result of the intersection and union boolean operations.

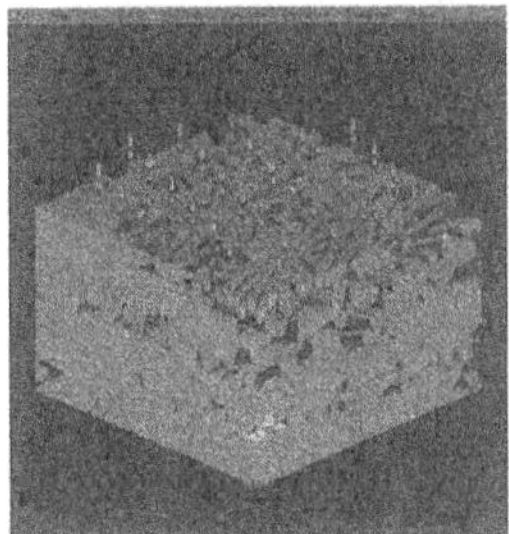

Fig. 10. Fluvial channels in a reservoir model.

The algorithm described here has been implemented in C++. To determine the intersection curve computation it takes about 1 second for the first example and 7 seconds for the second (in a PC Pentium III 500Mhz). In terms of applications, tubular solids turn out to be very suitable for the representation of fluvial sand channels in a petroleum reservoir model [6]. Fig.. 10 shows an example in which the boolean union of several channels corresponds to the total volume of sand in a given domain.

§6. Conclusions

The main contributions of this paper are a robust algorithm for intersecting and trimming single-valued tubular surfaces based on Interval Arithmetic, and the interval extension of the function ψ, since it plays an essential role in the identification of patches that cannot intersect a single-valued tubular surface.

Although this work deals only with directrix curves in Cartesian coordinates, it can easily be extended to include directrix curves in cylindrical coordinates ([9],[10]). The authors also plan to extend this algorithm to adapt the intersection curve by the use of some curvature range analysis criteria.

Acknowledgments. The authors would like to thank Prof. C. M. Hoffmann for introducing them to the area of Interval Analysis, and Prof. G. Tavares for his important ideas in the reservoir modeling application. We would like to thank also the referee for his suggestions. This work was partially supported by FINEP and PETROBRAS.

References

1. Barnhill, R., G. Farin, M. Jordan and B. Piper, Surface/Surface Intersection. Comput. Aided Geom. Design **4** (1987), 3–16.
2. Figueiredo, L. H., Surface intersection using affine arithmetic, Proceedings of the Graphics Interface'96, (1996), 168–175.
3. Hammer, R., M. Hocks, U. Kulisch and D. Ratz, *C++ Toolbox for Verified Computing - Basic Numerical Problems*, Springer Verlag, 1995.
4. Houghton, E. G., R. F. Emnett, J. D. Factor and Ch. L. Sabharwal, Implementation of a divide-and-conquer method for intersection of parametric surfaces, Comput. Aided Geom. Design **2** (1985), 173–183.
5. Huber, E. and W. Barth, Surface-to-surface intersection with complete and guaranteed results, in *Developments in Reliable Computing*, T. Csendes (ed.), Kluwer, 1999, 189–202.
6. Lanzarini, W.L., C.A. Poletto, G. Tavares, S. Pesco and H. Lopes, Stochastic modeling of geometric objects and reservoir heterogeneites, Paper SPE 38953 in the Proc. of the SPE 5th Latin American and Caribbean Petroleum Eng. Conf., Rio de Janeiro, 1997.
7. Moore, R. E., *Interval Analysis*, Prentice-Hall, 1996.
8. Sánches-Reyes, J., Single-valued tubular patches, Comput. Aided Geom. Design **11** (1994), 565–592.
9. Sánches-Reyes, J., Single-valued curves in polar coordinates, CAD **22**(1) (1990), 19–26.
10. Sánches-Reyes, J., Single-valued surfaces in cylindrical coordinates, CAD **23**(8) (1991), 561–568.
11. Snyder, J. M., *Generative Modeling for Computer Graphics and CAD*, Academic Press, 1992.

Hélio Lopes and Sinésio Pesco
Departamento de Matemática
Pontifícia Universidade Católica do Rio de Janeiro
Rua Marquês de São Vicente, 225, Gávea
Rio de Janeiro, RJ, Brazil, CEP:22.453-900
{lopes,pesco}@mat.puc-rio.br

Surface Animation for Flower Growth

Zhaoying Lu, Claire Willis, and Derek Paddon

Abstract. This paper presents a free-form model for surface deformation and growth. A general framework with bicubic patches is developed to generate a growing surface. The key component of this method is to enable the control points to move such that we achieve satisfactory results when the surface growth is constrained by an evolutionary formulation. The evolution theory must take account of natural and artificial perturbations in the growth cycle. This implies that the transient of the control points must be flexible and adaptable. A simple real-time collision detection method is developed to trace the surface growth correctly in an efficient manner. The application we are interested in involves the animation of flower development. Therefore, the modelling must be accurate and suitable for all possible growth patterns. Petals are arranged according to sound biological principles, therefore, any growth must demonstrate collision detection and deformation constraints. We show that the model presented here upholds these principles.

§1. Introduction

Smooth surfaces are of great importance in geometric modelling and computer graphics. Simulation of surface deformation and growth is a key challenge in the virtual environment. A realistic interactive surface model has been studied to make it possible to manipulate and control such surfaces. It is usual to generate surfaces using mathematical representations based upon polynomial functions of two parameters, such as Bézier surfaces, B-splines or rational B-splines. Such surfaces can be defined by an array of control points. However, it is a difficult task to enable the control points to move in such a way that we create a desired shape in an interactive environment.

Recently, many attempts have been made to manipulate surfaces. Some interesting results have been obtained using implicit equations [9], mathematical frameworks [10], convex parametric surface patch fitting [6], partial differential equations [11], and boundary element methods [1]. However, all of the methods require many surface details and are not easily controlled. The interactive design technique we defined here would allow the user to manipulate the surface effectively with a minimal number of control points.

Mathematical Methods for Curves and Surfaces: Oslo 2000
Tom Lyche and Larry L. Schumaker (eds.), pp. 263–272.

ISBN 0-8265-1378-6

The application we are interested in involves the animation of flower development. Biological processes are difficult to simulate because they grow and move according to highly complex natural principles. This paper focuses on the simulation of flower petals through all transient stages from a bud to the fully developed flower. Thus, we need to develop methodologies that support this simulation, to ensure that we achieve realistic results for time transient behaviours. Flowering, or more precisely, reproductive development, is composed of many independent but highly coordinated processes.

Simulation of collisions between mature organs is an important problem in the visualisation of structures with densely packed organs such as flowers. In nature, individual flowers touch each other, which modifies their positions and shapes. Consequently, the mature organs must be carefully modelled and sized to avoid intersections. This is feasible while modelling static structures, but proper simulation of collisions becomes crucial in the realistic animation of plant development. This technique will be presented in this paper. Since it is necessary to obtain a growing surface for representing petals or leaves, it is important to construct a suitable dynamic surface model.

Paper Objectives and Overview

The objectives of this paper are to:

- Present an interactive surface deformation and growth model, the purpose of which is to control the surface to meet the perturbations in the growth period.
- Introduce a method for controlling surface collision and deformation.
- Illustrate step by step the petal development controls with analysis of the shape changes.
- Describe the shape control by the implementation of the plant growth function, which gives a good description of plant development.

In this paper, we propose a method for generating flower growth animation in which petal surface area and shape can be changed simultaneously in real time. We represent a flower petal as a set of control points defining a bicubic patch. In addition, we expand growth function theory to enable the growth rate to vary as the petal develops and control points to move to meet the desired surface.

§2. Description of the Model

2.1. Surface Representation

To achieve a realistic result for the development of a bud into a flower, it is inevitable that the petal model must possess some measure of the complexity of a real petal. This presents problems for the construction of the 3D model for our simulation. The model needs to provide satisfactory surface continuity and smoothness. Building a model of this complexity using traditional polygonalization techniques would involve prohibitive storage and processing overheads, therefore alternative methods must be used.

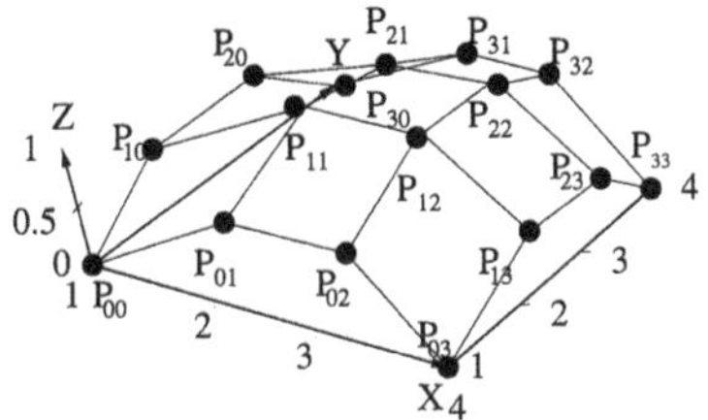

Fig. 1. Control points for bicubic Bézier patch.

2.2. Bézier Patch

Prusinkiewicz and Lindenmayer showed how plant components [8], such as stamens, petals, leaves, seeds, can be built out of bicubic patches. A patch is defined by three polynomials with degree three, with respect to parameters s and t. Just as with two dimensional curves, three dimensions patches may use a variety of control strategies, including Bézier, Hermite, and B-Spline bases. A Bézier patch requires sixteen control points (shown in Figure 1). The Bézier form of the bicubic parametric patch has a very concise matrix formulation specifying a vector point, $P(s,t)$, on the surface in terms of the sixteen control points (more details see [7]).

For a regular and symmetrical petal, we can keep $P_{00}, P_{01}, P_{02}, P_{03}$ as one control point, thus making the control easier. To make it simpler, P_{30}, P_{31}, P_{32}, P_{33} could be put together as one point if the top of the petal is a discontinuity point. As shown in Figure 2, four control points are located at the bottom of the petal while the base portion of the petal is simple. The other two groups of four control points usually control the middle part of the petal. The last four control points play an important role in shaping the petal if they are located at the top of the petal. When we consider the structure for the petal surface, we need to pay attention to where the discontinuity points are, in which the two groups of the four control points can be joined, forming a corner control point. The part of the surface formed by the same group of four control points will generate a smooth and continuous curve. So the surface structure will be changed if the discontinuity points are different.

Bicubic Bézier patches have become a popular tool for surface modelling. The obvious advantages include:

- *Ease of interactivity* – the control point effects are readily observed and understood, and the control points themselves are easily modified, either numerically or interactively;
- *Representational efficiency* – complex surfaces are represented by a very small set of numbers. So this approach is applied here as the surface representation for our flower petal modelling.

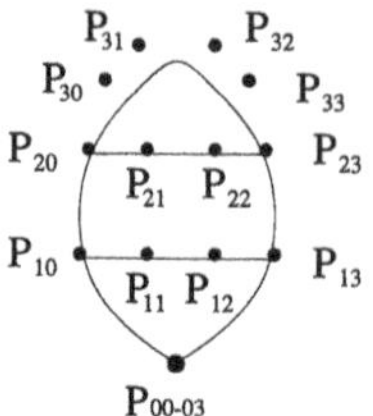

Fig. 2. Petal structure with sixteen control points.

§3. Collisions

Surface collision detection is always a challenge process in interactive surface animation. The simplest approach to detecting collision in a physical simulation is to test each geometric element, such as point, edge, face, against each other geometric element for a possible collision. To achieve practical running times for large simulations, the number of possible collisions must be culled as rapidly as possible using some type of spatial data structure. If the surface connectivity does not change, a two-dimensional surface-based data structure can be applied for the elements. Another method is to distribute the elements into a three-dimensional volume-based data structure. However, it is too expensive in the computation and subdivision of the bounding boxes of all the elements, especially when all the elements are surfaces.

Our method focuses on how to construct a suitable surface-based data structure for the petal surface. As each surface patch has sixteen control points, we can use each group of four control points as a data structure. An efficient way for constructing the hierarchy of boxes is to compute the bounding surfaces using the convex hull property.

A Bézier surface has an attractive convex-hull property which is very useful for our surface intersection test [3]. Each of the boundary curves of a Bézier surface is a Bézier curve. Considering the defining polygon net for the bicubic Bézier surface shown in Figure 1, it is easy to see that the tangent vectors at the patch corners are controlled both in direction and magnitude by the position of adjacent points along the edges of the net. Consequently, the user can control the shape of the surface patch without an intimate knowledge of tangent or twist vectors.

A Bézier curve is determined by a defining polygon. It has some properties, such as the curve generally follows the shape of the defining polygon; the first and last points on the curve are coincident with the first and last points of the defining polygon; and the curve is contained with the convex hull of the defining polygon. We use the convex hull of the control polygon as a convex hull for the curve. For 3D curves, the convex hull is the convex polyhedron formed by the control points.

Analysing the petal growth process from a theoretical viewpoint, we can see the control points moving through space and thereby changing the shape for the petal by the intuitive definition of a surface. We will now formalise this intuitive concept in order to arrive at a mathematical description of a

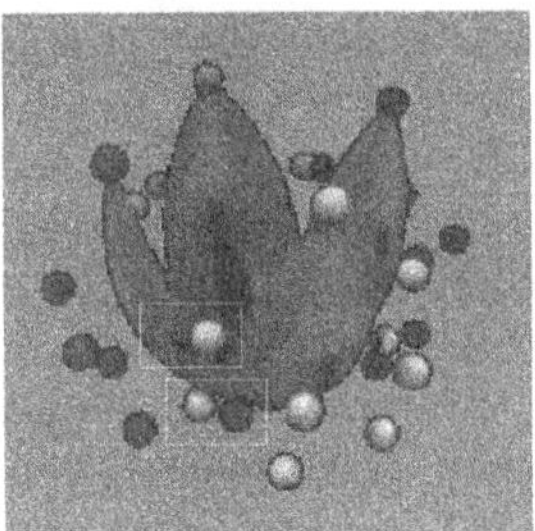

Fig. 3. Control points crossed over.

petal surface. First, we assume that the moving curve is a Bézier curve. At any time, the moving curve is then determined by a set of control points. Each original control point will move through space according to the growth function, and this curve is contained within the convex hull. In Figure 3, small spheres are shown to represent the control points for the petal patch surface. The control points will be moving apart when the petal surfaces separating. The control points of the adjacent petals in the upper rectangle are moving apart, which implies that the upper part of the petals do not collide. However, in the lower rectangle the control points have crossed each other and are out of order, indicating that the lower parts of the petals are intersecting.

As we can see, the four control points in the vertical boundary of the petal surface also form a convex polyhedron. So the collision detection between the two adjacent petal surfaces can be converted to two horizontal convex polyhedron tests (from the two middle sets of control points), and one vertical convex polyhedron intersection test.

If there is no intersection, there is no collision between these two surfaces. If there is an intersection, we can divide the convex polyhedron to two child polyhedrons, then we repeat the test on the child convex polyhedron which is close to the side of the adjacent petal surface and recurse. As we consider both petal surfaces are growing, we do not calculate the exact points of the collision and just need to know which part of the surface caused the collision and avoid it.

§4. Constraints and Deformation

In our flower petal surface application, the petal surfaces intend to keep their own shape even when the unavoidable collisions happen. The surface growth constraints come from the surface growth function described in the next section. Therefore, the surface constrained deformation model needs to keep the petal surface growing features in the length, width and curvature.

4.1. The Deformation Model

Let P be the corner control point of the collided convex polyhedron which is to be deformed. When a function f applied to reposition P, this function will pull this control point away from the collision situation. At the moment, we use a function to pull the control point toward the centre and avoid most of the collision by overlapping the petal surfaces. Some random and different functions can be applied to generate more realistic results, like curving petals.

However, the petal surface will change the original growing length and width rate if only part of the control points have been deformed. It is very important that the surface deformation satisfies all the constraints by a proper choice of the control point matrix. This matrix keeps all the information on the control points, such as related positions, distances and curvatures. Therefore, we can ensure the shape of the curve remains the same by moving all four control points at the same time. An efficient way for moving control points is described below.

4.2. Moving Control Points

It is necessary to determine the optimal method for moving control points such that we avoid collisions between petal surfaces. As we use two sets of four control points to manipulate the body of the petal surface, we must first generate the two Bézier curves, one for each set of control points.

A basic way to draw a parametric cubic is by interactive evaluation of $x(u)$, $y(u)$ and $z(u)$ for incrementally spaced values of u, and plotting lines between successive points [4]. A much more efficient method for evaluating polynomial equations is to recursively generate each succeeding value of the function by incrementing the previously calculated value for the cubic equation through the use of finite differences.

Calculations for successive points are then efficiently carried out as a series of additions. To apply this incremental procedure to Bézier curves, three sets of calculations are needed for the coordinates $x(u)$, $y(u)$, and $z(u)$. For surfaces, incremental calculations are applied for both values of u and v.

In our mathematical petal surface framework, two Bézier curves formed by $P_{10} - P_{13}$ and $P_{20} - P_{23}$ in the petal surface body (as shown in Figure 2) play an important role in the shape control. Therefore, the whole set of control points for the curves are moving together to maintain the same growth function and avoid collision with adjacent petals. Through the method described above, the middle point of the curve can be generated by choosing the value for u and v. It is relatively easy to adjust the position of the curve by moving the middle points only. The other control points could be calculated back from the relative positions to the reference middle point. The surface deformation therefore is controlled by the growth process and the collision detection.

§5. Implementation

5.1. General Growth Function

Continuous processes such as plant growth need to be described by growth functions. A popular example of the growth function is a sigmoidal type,

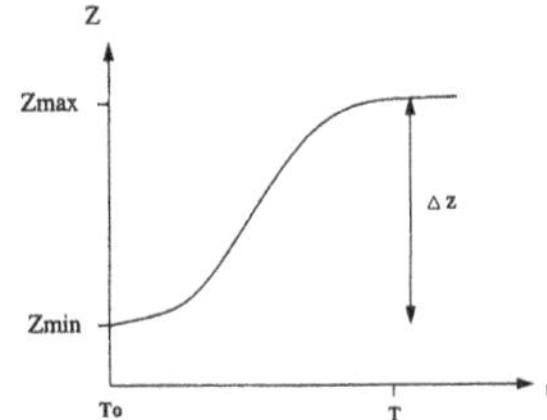

Fig. 4. General growth function.

Fig. 5. Flowers with centre.

monotonically increasing from minimum to maximum with growth rates of zero at both ends of time interval (To, T), as shown in Figure 4. The growth is slow initially, accelerating to reach a maximum, then slowing again and eventually ceasing. We apply the popular Velhurst's logistic function, defined by the equation [1]:

$$\frac{dz}{dt} = l * z(1 - \frac{z}{z_{max}}), \tag{1}$$

(t: time; l: length growth rate; z: the growth in z axis direction).

Usually, bicubic parametric patch modelling suffers from lack of a high-level modelling abstraction for shape control. So the growth function is applied in the x, y, z directions for each control point. It is presented in equation (1) for the z axis direction growth.

The time interval is chosen by the user, according to the smoothness of the desired animation sequence. This time interval is then fixed during the recursive process to determine the surface features. The flexible growth rate curve allows variation for individual control points. It means that we can apply a different growth function for one control point or some of the control points. A unique irregular petal will be generated when one or some control points have different growth tendencies.

The growth of the seeds on the flower centre is applied with the phyllotaxis theory, which detects and eliminates collisions between the organs while optimising their packing (see [5] for details). The divergence angle between consecutively formed organs (measured from the centre of the structure) is close to the Fibonacci angle of $360^o \mathcal{T}^{-2} \approx 137.5^o$, where $\mathcal{T} = (1 + \sqrt{5})/2$. As shown in Figure 5, the flower centre is expanding with the growth of the petals. The phyllotaxis principle is also applied when we place the petals and has direct effect on the petal overlap priority.

5.2. Factors Controlling Growth

With the theory for the growth rate, we need to consider what parameters are necessary for controlling petal shape. However, in order to allow a non-expert to use our system, we have reduced the number of control functions to the minimum while still retaining the benefit of the system. Experience has shown that just three control factors are needed, that is, the length, width and curvature growth factors.

For a regular and symmetrical petal, we can evenly locate the sixteen control points. The growth between two groups of four control points will be considered as having the same growth rate. That is, we apply a linear control function. Of course, more complex growth function can be applied if necessary.

In summary, if a fixed petal structure is provided for the petal surface, the main three growth factors play an important role in surface change for the growth process. These three factors allow the user to control a bud growing into a flower with petals of different length, width and curvature after any time.

5.3. Implementation

We need to set up the growth equation by setting an appropriate growth function. The growth rate value in length, width and depth and the growth factors for all the control points must be given to calculate the new positions after the specified time interval. The selections of all the values in these two steps depend upon the desired petal shape. The growth factors for some control points should be reset if their desired surface growth tendency is changed. When the petals are formed via the growth functions, the petal collision detection will be applied and the relevant control points will be moved to deform the shape of the petal. Then the new control points matrix will be generated, and the patch surface thus formed, will be rendered by ray-tracing.

§6. Results and General Comparison

As we can see in Figure 5, the petals separate and do not collide when the flower grows. This is only because the centre is relatively large compared to the size of the petals. Petal surfaces will be in collision positions when growing if the flower has a small centre, relative to the size of the petals, as shown in Figure 6. In this example, petals have collisions and the adjacent petals occupy the same space which is impossible. Comparing Figure 7, applying the deformation functions to reposition the control points, the petal surfaces are shown to overlap on the other petals and avoid the collision. The flower surfaces are developing according to the growth rates and the perturbations caused by the other petals.

In summary, our method provides a direct method of surface control. Compared to using predefined surfaces or generating surfaces from fixed original and final shapes, this method is more flexible and creative.

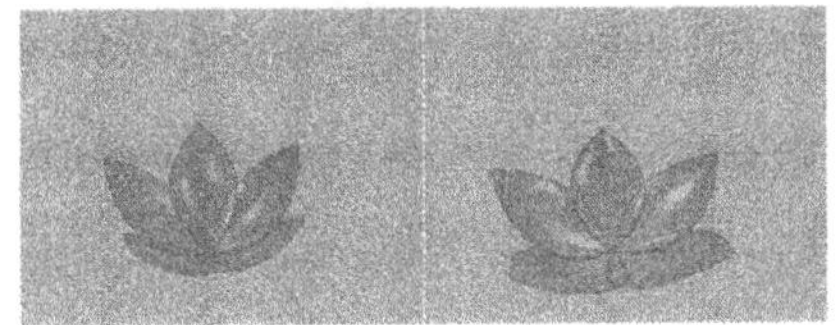

Fig. 6. Intersecting flower petals (petals occupy the same space).

Fig. 7. Overlapped flower petals (the deformation applied to avoid the collision).

§7. Conclusion

Many previous models of plant growth avoid the modelling of surface detail, often resorting to imposing detail such as leaves and petals from a library of simplistic elements. Here, we have demonstrated that surface detail can be effectively represented by dynamic elements that grow and deform naturally as the plant evolves through its life cycle.

We have introduced a smooth surface model using a bicubic patch to represent each petal. This patch is defined by surface and temporal parameters. In this way, the evolution of elements such as petals can be accurately and realistically represented and controlled.

Collision detection of moving objects is essential, otherwise these objects simply pass through each other. Here we have implemented an efficient method to detect these collisions and ensure that elements adapt their shape and position to avoid each other, while maintaining the appropriate growth.

In summary, we believe that the proposed modelling method and its extensions will prove useful in many applications of surface modelling, from research in plant development and ecology to the surface design of plant organs and in the production of animated surface models for use in virtual environments.

References

1. Doug, J. and P. Dinesh, Accurate Real Time Deformable Objects, SIGGRAPH, 1999.
2. Edelstein-Keshet, L., *Mathematical Models in Biology*, 1988.
3. Farin, G., *Curves and Surfaces for Computer Aided Geometric Design*, Academic Press, 1988.

4. Foley, J., A. Dam, S. Feiner, and J. Hughes, *Computer Graphics: Principles and Practice*, second edition, 1990.
5. Fowler, D., P. Prusinkiewicz, and J. Battjes, A Collision-based model of spiral phyllotaxis, Computer Graphics **26** (2) (1992), 361–368.
6. Juettler, B., Convex surface fitting with parametric Bézier surfaces, *Mathematical Methods for Curves and Surfaces II*, M. Dæhlen, T. Lyche, and L. L. Schumaker (eds.), Vanderbilt University Press, Nashville, 1998, 263–270.
7. Lu, Z., C. Willis, and D. Paddon, Perceptually realistic flower generation, WSCG, Plzen, Czech Republic, 2000.
8. Prusinkiewicz, P., and A. Lindenmayer, *The Algorithmic Beauty of Plants*, Springer-Verlag, 1990.
9. Sederberg, T., and F. Chen, Implicitization using moving curves and surfaces, SIGGRAPH, 1995.
10. Skalak, R., D. Farrow, and A. Hoger, Kinematics of surface growth, Journal of Mathematical Biology **35** (1997), 869–907.
11. Ugail, H., M. Bloor, and M. Wilson, Techniques for interactive design using the PDE method, ACM Transactions on Graphics **18** (2) (1999), 195–212.

Zhaoying Lu, Claire Willis and Derek Paddon
Computing Group
Dept. of Maths. Sci.
University of Bath
BATH BA2 7AY
UK
`mapzl@bath.ac.uk`
`cpw@maths.bath.ac.uk`
`derek@maths.bath.ac.uk`

Multiresolution Editing of Pasted Surfaces

Marryat Ma and Stephen Mann

Abstract. Surface pasting allows the insertion of local detail to a tensor product surface without changing the structure of the underlying surface. It works by applying feature surfaces on top of a base surface to create a composite surface. Previous modelling systems for pasted surfaces only allowed users to translate, rotate, and resize pasted features, and did not support direct manipulation. In this paper, we describe a method for the direct manipulation of pasted surfaces that allows the user to edit a surface at any level in the pasting hierarchy.

§1. Introduction

Hierarchical modelling is currently an active area for research. Many surfaces have varying levels of detail, and modelling techniques that explicitly represent these levels of detail are useful both in terms of reduced storage and in interactive modelling paradigms where users want to interact with their models at different levels of detail.

Tensor product B-spline surfaces are commonly used in the computer industry because they can be represented by little information and have attractive continuity properties. However, it is difficult to add detail to these surfaces without globally increasing the complexity of the surfaces, and thus they are poorly suited to multiresolution editing.

Hierarchical B-splines were developed by Forsey and Bartels [7] for adding areas of local detail to a tensor product B-spline surface. A parametrically aligned region of the surface is locally refined to increase its control point density. The control points in the refined region are displaced to create the local detail. Forsey implemented a limited direct manipulation scheme for hierarchical B-splines that allows a user to manipulate a surface at predefined surface points. If the user decides to manipulate the surface at a lower resolution level, then the higher levels of detail are removed from the display. Unfortunately, with the details hidden, the user is unable to see the effect of the manipulation on the entire surface. Hierarchical B-splines allow multiresolution editing and maintain a high level of continuity, but the local details cannot be translated, rotated, or resized.

Mathematical Methods for Curves and Surfaces: Oslo 2000
Tom Lyche and Larry L. Schumaker (eds.), pp. 273–282.

ISBN 0-8265-1378-6

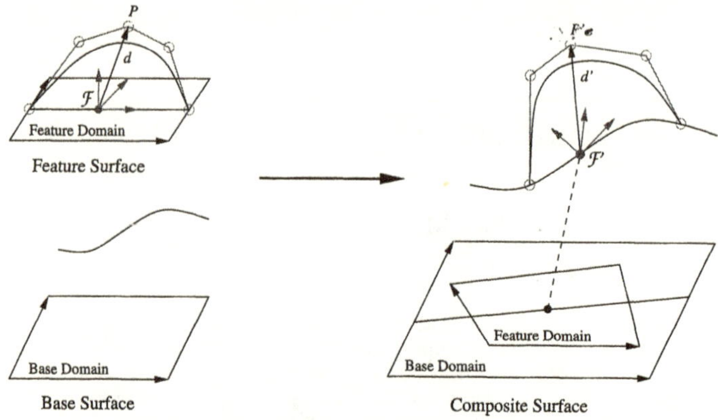

Fig. 1. Pasting a feature on a base surface.

Surface pasting, developed by Bartels and Forsey [3], is a generalization of hierarchical B-splines that allows the insertion of local detail to a tensor product surface without changing the structure of the underlying surface. In surface pasting, the area of local detail is represented as a tensor product surface, called a feature. The feature is placed on an existing surface, called the base, to produce a composite surface. Additional features can be pasted hierarchically on the composite surface to create more complex composite surfaces. Surface pasting has been integrated into Side Effects' *Houdini* software, where it has been successfully used for character animation.

Although surface pasting is a hierarchical modelling method, the user interfaces implemented for previous research have concentrated on positioning the features upon the base surfaces, and adjusting—i.e., translating, rotating, and resizing—the features once they have been pasted. In this paper, we present a technique for the direct manipulation of pasted surfaces that allows the user to edit the surfaces at any resolution in the hierarchy.

§2. Background

A tensor product B-spline surface is a piecewise polynomial surface that is defined by a grid of control points $\{P_{i,j}\}$ and a set of basis functions $\{N_{i,j}\}$:

$$S(u,v) = \sum_{i=1}^{M}\sum_{j=1}^{N} N_{i,j}(u,v)P_{i,j}.$$

Here $N_{i,j}(u,v) = N_i^m(u)N_j^n(v)$, and the N_is and N_js are the degree m and n B-splines for the two parametric domain directions. For a more detailed introduction to B-splines, see any introductory spline text, such as [6].

The surface pasting process (illustrated in Figure 1) is a computationally inexpensive method for adding local detail to tensor product surfaces. First, the feature's domain is embedded into its range space. Next, a local coordinate frame $\mathcal{F}_{i,j} = \{\mathcal{O}_{i,j}, \vec{r}_{i,j}, \vec{s}_{i,j}, \vec{t}_{i,j}\}$ is constructed for each feature control

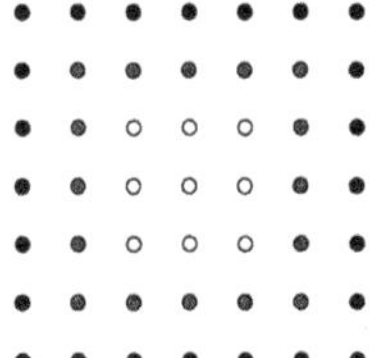

Fig. 2. Outer two layers of feature control points located at Greville points.

point $P_{i,j}$. The origin of the coordinate frame $\mathcal{O}_{i,j}$ is the embedded Greville point corresponding to $P_{i,j}$. The vectors $\vec{r}_{i,j}$ and $\vec{s}_{i,j}$ are the two parametric domain directions for the feature, and $\vec{t}_{i,j} = \vec{r}_{i,j} \times \vec{s}_{i,j}$. Each feature control point is represented as a displacement vector $\vec{d}_{i,j}$ expressed relative to $\mathcal{F}_{i,j}$ by subtracting $\mathcal{O}_{i,j}$ from $P_{i,j}$:

$$\vec{d}_{i,j} = P_{i,j} - \mathcal{O}_{i,j} = \alpha_{i,j}\vec{r}_{i,j} + \beta_{i,j}\vec{s}_{i,j} + \gamma_{i,j}\vec{t}_{i,j}.$$

Next, the feature's domain is mapped into the base's domain. For each feature control point, the corresponding Greville point is mapped into the base's domain. The base surface is evaluated at this domain point to form a new local coordinate frame $\mathcal{F}'_{i,j} = \{\mathcal{O}'_{i,j}, \vec{r}'_{i,j}, \vec{s}'_{i,j}, \vec{t}'_{i,j}\}$. The control point's new origin on the base surface is $\mathcal{O}'_{i,j}$, the vectors $\vec{r}'_{i,j}$ and $\vec{s}'_{i,j}$ are the two partial derivatives at $\mathcal{O}'_{i,j}$, and $\vec{t}'_{i,j} = \vec{r}'_{i,j} \times \vec{s}'_{i,j}$. The feature control point is placed by expressing its displacement vector relative to this new local coordinate frame:

$$P'_{i,j} = \mathcal{O}'_{i,j} + \vec{d}'_{i,j},$$

where

$$\vec{d}'_{i,j} = \alpha_{i,j}\vec{r}'_{i,j} + \beta_{i,j}\vec{s}'_{i,j} + \gamma_{i,j}\vec{t}'_{i,j}. \tag{1}$$

To ensure that the boundary of the feature lies near the base surface, we place the first layer of the feature's control points (the black points of Figure 2) at the Greville points so they have zero displacement vectors. After pasting, these feature control points will lie on the base surface, and the boundary of the feature will lie near the base. By inserting knots into the feature surface, the discontinuity between the feature and the base can be made as small as desired.

By placing the second layer of the feature's control points (the grey points of Figure 2) at the Greville points, we achieve an approximate C^1 join between the feature and the base. Conrad [5] gives a further discussion of continuity issues of pasted surfaces, and shows how to use quasi-interpolation to further reduce both the C^0 and C^1 discontinuity between the feature and the base.

§3. Direct Manipulation of Tensor Product Surfaces

Traditionally, B-spline curves and surfaces were manipulated by moving their control points. This method is unintuitive and requires that the control points be displayed, thus increasing the clutter on the screen.

Bartels and Beatty [2] developed a technique for the direct manipulation of spline curves in which users could pick any point on a curve, move it to a new location, and have the shape of the curve change appropriately. Their method found a set of control points that had maximal influence over the picked point. The amount that each control point moved was proportional to its influence over the picked point.

A number of researchers have investigated techniques for the direct manipulation of tensor product surfaces. For example, Fowler proposed a method for directly manipulating positions, normal vectors, and partial derivatives at any surface point [8]. He also found that the system of equations that must be solved to perform direct manipulation of tensor product surfaces is underdetermined.

We have chosen to calculate new control point locations using a generalization of Bartels and Beatty's curve manipulation technique [2], where the extra degrees of freedom are used to reduce the overall change in the position of the surface's control points. We have extended and altered their method so that it can be applied to the direct manipulation of tensor product B-spline surfaces.

Given a surface $S(u,v) = \sum_i \sum_j N_{i,j}(u,v)P_{i,j}$, suppose we want to move a surface point $S(\overline{u},\overline{v})$ by a vector $\overrightarrow{\Delta P}$, i.e., we want a surface S' such that

$$S'(\overline{u},\overline{v}) = S(\overline{u},\overline{v}) + \overrightarrow{\Delta P}. \tag{2}$$

Then a block of control points that has the most influence over the picked surface point is found. For each control point $P_{i,j}$ in this block, we calculate a weight $w_{i,j}$ that is proportional to the control point's contribution to the surface point $S(\overline{u},\overline{v})$:

$$w_{i,j} = \frac{N_{i,j}(\overline{u},\overline{v})}{\sum\limits_k \sum\limits_l \left(N_{k,l}(\overline{u},\overline{v})\right)^2}, \tag{3}$$

where the double summation is over the block of control points that we are modifying. Each of these control points is updated as

$$P'_{i,j} = P_{i,j} + w_{i,j}\overrightarrow{\Delta P}. \tag{4}$$

The net result is for the vector $\overrightarrow{\Delta P}$ to be distributed over the control points of S so that when S' is evaluated at $(\overline{u},\overline{v})$, the sum of the $w_{i,j}$s weighted by the basis functions is 1.

Bartels and Beatty showed that in the curve case, moving the single most influential control point produced unstable results. This instability occurs because there will be a point on the curve to the left of which control point P_i will have the most influence, and to the right of which control point P_{i+1} will have the most influence. Picking near this division can have markedly

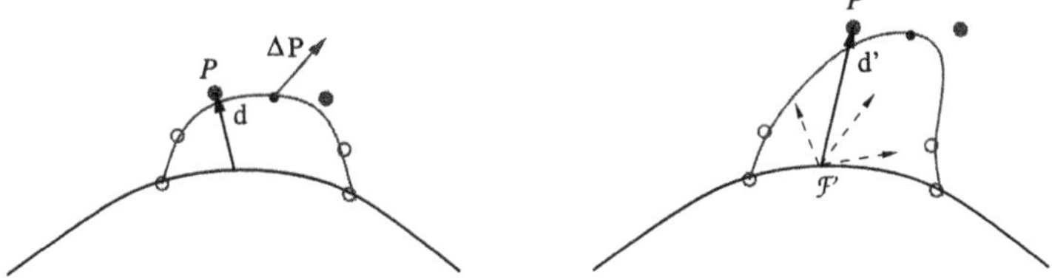

Fig. 3. Updating a control point's displacement vector.

different results depending on which side is picked. A minimum of two control points must be adjusted for their method to be stable.

Just as we need to move two control points in the curve case, we need to move at least two control points in each of the two parametric domain directions to get a stable direct manipulation method for surfaces. We choose to adjust a 2×2 block of control points since this small block size restricts the locality of change. A larger block may be used to modify a larger area of the surface.

§4. Direct Manipulation of Pasted Surfaces

When directly manipulating pasted surfaces, we would like to edit any point on the surface, and to edit the surface at any resolution in the hierarchy regardless of where in the hierarchy the selected point lies. In this section, we begin with a discussion of the extension of the direct manipulation technique to the top level of the pasting hierarchy, i.e., the level in which the selected point lies, and then we describe some problems with this simple approach. In the next section, we will extend this method to solve these problems, which will allow us to edit the surface at any resolution at or below the level of the selected point.

The basic technique for directly manipulating tensor product B-spline surfaces carries over to pasted surfaces with only minor modifications. The first step is to update the control points of the surface we wish to modify using direct manipulation of a tensor product surface as described in §3. Then we update the displacement vector $\vec{d}_{i,j}$ for each control point $P_{i,j}$.

To recalculate each control point's displacement vector, the local coordinate frame $\mathcal{F}'_{i,j} = \{\mathcal{O}'_{i,j}, \vec{r}'_{i,j}, \vec{s}'_{i,j}, \vec{t}'_{i,j}\}$ on the base surface must be reconstructed. The difference $\vec{d}'_{i,j}$ between the new control point location $P'_{i,j}$ and the origin of the coordinate frame $\mathcal{O}'_{i,j}$ is found and expressed in terms of $\mathcal{F}'_{i,j}$:

$$\vec{d}'_{i,j} = \alpha'_{i,j}\vec{r}'_{i,j} + \beta'_{i,j}\vec{s}'_{i,j} + \gamma'_{i,j}\vec{t}'_{i,j}.$$

This gives a 3×3 system of equations to determine the new $\alpha'_{i,j}$, $\beta'_{i,j}$, and $\gamma'_{i,j}$ for each updated displacement vector. After the new displacement vectors are calculated, the feature may be translated or the underlying surface may be changed, and the results of the direct manipulation are preserved.

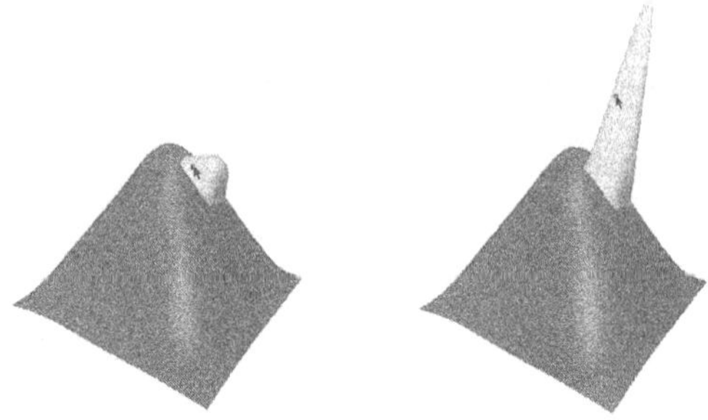

Fig. 4. Example of distortion.

In the remainder of this paper, we will talk about selecting a surface point $S(\overline{u},\overline{v})$ and then manipulating points $S_i(\overline{u},\overline{v})$, i.e., points at other levels in the pasting hierarchy with the same parameter values as the selected point. In effect, we are manipulating S_i at the image of $(\overline{u},\overline{v})$ mapped into the domain of that surface according to the invertible domain transformations defined by the surface pasting operation.

If we select a point on one surface in a pasting hierarchy, and move the selected surface's control points using this direct manipulation technique, we often get the results we want. However, simply using this basic direct manipulation technique on the component surfaces in a pasting hierarchy results in several problems. The first problem is that if we modify the boundary control points of a surface, then we may lose the desired approximate continuity properties, i.e., the composite surface may stop looking smooth or the surface may detach from its underlying base. Thus, we need to fix the boundary and cross boundary derivatives of the manipulated surface by ensuring that the two outermost rings of the surface's control points (indicated in Figure 2) do not move.

If the user attempts to move a surface point whose most influential 2×2 block of control points intersects the two outermost rings of control points, we are faced with two choices. We can disallow the direct manipulation, or we can find the closest block that does not overlap the two outermost rings. In the latter case, the control points we would change have less influence on the selected surface point, and thus they must be displaced farther to move the picked surface point to its new location. This can cause unsightly distortions, similar to those described by Bartels and Beatty [2], in the area of the surface over which these alternative control points have a higher influence. These distortions would likely confuse the user since the maximal change in the surface would not occur at the picked point. An example of such a distortion can be seen in Figure 4. We choose to disallow direct manipulation of a pasted surface when the most influential 2×2 block of control points intersects the two outermost rings of the surface's control points.

A second problem with this simple method for directly manipulating pasted surfaces occurs when we try to edit at a different level in the pasting hierarchy than at the level containing the picked point. To truly edit the pasted surface at any resolution, we need to be able to select a point in a region of high resolution, and have the direct manipulation changes occur at a lower level of resolution. One simple way to implement this type of manipulation is to find the corresponding point $S_i(\overline{u},\overline{v})$ on a lower level surface when the user selects the surface point $S(\overline{u},\overline{v})$, and to use the $\overrightarrow{\Delta P}$ vector to directly manipulate $S_i(\overline{u},\overline{v})$. We would then reapply (1) to each surface above S_i to update the composite surface.

Unfortunately, (1) is not accounted for in the direct manipulation equations, and this method does not result in direct manipulation. To achieve direct manipulation of the composite surface, we will have to make additional adjustments to the control points of the surfaces between S_i and S, as described in the next section.

§5. Direct Manipulation of Hierarchical Pasted Surfaces

To manipulate a pasted surface at a lower resolution than the level at which the selected point lies, we need to modify the direct manipulation method. A first idea is to derive for the pasting hierarchy formulas similar to equations (2), (3), and (4). When we expand (2) for a pasted surface, we get

$$\sum_{i,j} N_{i,j}(\overline{u},\overline{v})\left(\vec{d}''_{i,j} + \mathcal{O}''_{i,j}\right) = \sum_{i,j} N_{i,j}(\overline{u},\overline{v})\left(\vec{d}'_{i,j} + \mathcal{O}'_{i,j}\right) + \overrightarrow{\Delta P}.$$

To derive the direct manipulation equations, we need to expand $\vec{d}''_{i,j}$ (1) and $\mathcal{O}''_{i,j}$ (an evaluation of the base surface). Unfortunately, $\vec{d}''_{i,j}$ depends upon the control points of the base surface in a non-linear manner as we see from (1) and the formula for $\vec{t}'_{i,j}$. Thus, this method of direct manipulation of a hierarchical pasted surface is more expensive than we would like, and as we increase the depth of the hierarchy, the equations become more complicated.

As an alternative method, we chose to modify the control points of multiple surfaces in the hierarchy. While this method is not truly hierarchical, it applies most of the change to a single surface, with other surfaces receiving only minor updates.

The idea behind our method is to push the work down the pasting hierarchy, make a large change at the desired level, and then ascend the hierarchy making small adjustments as needed. Suppose we have a hierarchy of pasted surfaces, $S_0,\ldots,S_H$, with S_0 being the coarsest resolution. Given a point $S(\overline{u},\overline{v}) = S_h(\overline{u},\overline{v})$ on a pasted surface at resolution h, suppose we wish to edit at resolution r, with $h \geq r$. Our method descends the hierarchy of surfaces under the picked point until we reach surface S_r. We adjust S_r so that $S_r(\overline{u},\overline{v})$ is moved by $\overrightarrow{\Delta P}$. This causes S_{r+1} to change (giving a new surface, S'_{r+1}), although $S_{r+1}(\overline{u},\overline{v})$ will not necessarily move by $\overrightarrow{\Delta P}$. We now compute the difference between the desired change and the actual change in S_{r+1}, giving a correction factor $\overrightarrow{\Delta P}' = \overrightarrow{\Delta P} - [S'_{r+1}(\overline{u},\overline{v}) - S_{r+1}(\overline{u},\overline{v})]$. Then

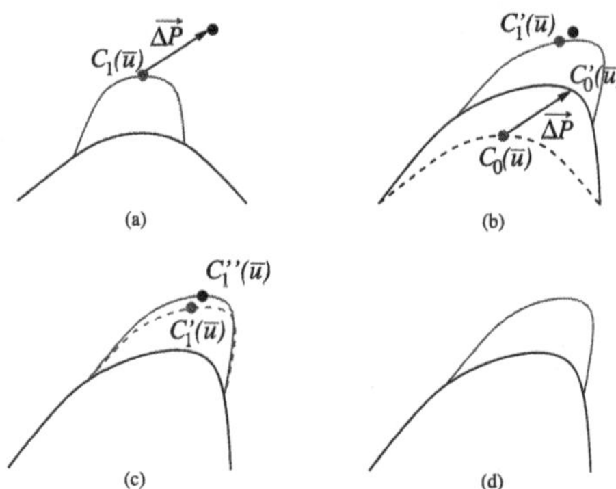

Fig. 5. Example of hierarchical direct manipulation.

we directly manipulate S'_{r+1} by $\overrightarrow{\Delta P}'$, giving a new surface S''_{r+1} such that $S''_{r+1}(\overline{u},\overline{v}) = S_{r+1}(\overline{u},\overline{v}) + \overrightarrow{\Delta P}$. Note that if $S_i(\overline{u},\overline{v})$ lies in the unmodifiable region of S_i (area of the surface where every point has a 2×2 block of control points that intersects the surface's two outermost rings of control points), then we skip it. This procedure is repeated for each surface up the hierarchy until S_h is reached.

We illustrate our method for a curve in Figure 5, in which a point on the top surface is selected and the desired resolution level is the base level; the construction for surfaces is analogous. Initially, the user has chosen a point $C_1(\overline{u})$ to manipulate and the amount $\overrightarrow{\Delta P}$ by which to move the point. In diagram (b), the point $C_0(\overline{u})$ is located and moved by $\overrightarrow{\Delta P}$. The shape of the base changes, resulting in the change of the feature's shape and location of the picked point, as shown in (c). Since $C_1'(\overline{u})$ is not at its desired position, a correction factor is applied, giving $C_1''(\overline{u})$. The last diagram shows the resulting composite curve.

Using our method, the user can select any point on the surface, and directly manipulate the surface, with modifications occurring at or below the level in the hierarchy on which the selected point lies. However, if the user selects a point in the unmodifiable region of the top level surface, then we do not allow the user to edit at the level of the selected surface, and only allow the user to edit at lower levels in the hierarchy. This restriction prevents the type of distortions discussed in §4.

Figure 6 illustrates a sequence of events that occurs when performing hierarchical direct manipulation on the composite surface. The selected point is located on the smallest and lightest coloured surface in the first image. The dark coloured surface(s) in the middle three images are affected in each stage of the manipulation. In each case, the bottom most dark surface undergoes the greatest change, while the other dark surfaces only have correction factors applied to them. The more surfaces that are dark coloured, the broader the change in the composite surface. In the second image, only the top most

Fig. 6. Example of the direct manipulation of hierarchical pasted surfaces.

surface is affected and the selected point was moved to the right and down. In the next image, two surfaces are affected and the selected point was moved up and to the right. Three surfaces are affected and the selected point is moved up and to the right in the fourth image. The last image shows the result of the manipulation.

Our direct manipulation method provides the user with complete feedback no matter what resolution level the user is editing. In contrast, Forsey's hierarchical B-splines editor strips the upper levels of detail out of the display as the user descends the hierarchy, making modelling more difficult.

§6. Conclusions and Future Work

Surface pasting is a method for constructing multiresolution surfaces by hierarchically composing tensor product B-spline surfaces. In this paper, we have shown how to implement direct manipulation of pasted surfaces, allowing the user to edit the composite surface at any level in the hierarchy, either at the selected level, or at a lower level.

Our direct manipulation method is not truly hierarchical since changes are made to multiple resolution levels. However, empirical testing has shown that the correction factors applied to the higher level surfaces tend to be much smaller than the initial $\overrightarrow{\Delta P}$ assigned to the selected resolution level. More research is needed to determine if there are any limits on the size of the correction factors as the pasting hierarchy is ascended.

An improvement to our hierarchical direct manipulation method would be to allow the user to increase the edit resolution for a picked point. For example, suppose there are n levels in the pasting hierarchy and the user wishes to manipulate the picked point at resolution $n+1$. Then a null surface with increased control point density could be pasted under the picked point, and this new surface, rather than the original one, would be manipulated. The ability to edit a surface at a higher resolution would allow the user to add local detail to a region without explicitly pasting a feature surface.

Acknowledgments. We would like to thank Richard Bartels for his contributions to this work. Financial support was provided by NSERC and CITO.

References

1. Barghiel, C., R. Bartels, and D. Forsey, Pasting spline surfaces, in *Mathematical Methods for Curves and Surfaces,* Morten Dæhlen, Tom Lyche, Larry L. Schumaker (eds), Vanderbilt University Press, Nashville & London, 1995, 31-40.
2. Bartels, R. and J. Beatty, A technique for the direct manipulation of spline curves, in *Proceedings of Graphics Interface '89,* S. MacKay, E. M. Kidd (eds), Canadian Human-Computer Communications Society, London ON, 1989, 33–39.
3. Bartels, R. and D. Forsey, Spline overlay surfaces, University of Waterloo Report CS-92-08, 1991.
4. Chan, L., S. Mann, and R. Bartels, World space surface pasting, in *Proceeding of Graphics Interface '97,* W. A. Davis, M. Mantei, R. V. Klassen (eds), Canadian Human-Computer Communications Society, Kelowna, BC, 1997, 146–154.
5. Conrad, B. and S. Mann, Better pasting via quasi-interpolation, in *Curve and Surface Design: Saint-Malo 1999,* P.-J. Laurent, P. Sablonnière, and L. L. Schumaker (eds), Vanderbilt University Press, Nashville TN, 2000, 27–36.
6. Farin, G., *Curves and Surfaces for Computer Aided Geometric Design,* Fourth Edition, Academic Press, NY, 1997.
7. Forsey, D. and R. Bartels, Hierarchical B-spline refinement, Computer Graphics **22(4)** (1988), 205–212.
8. Fowler, B., Geometric manipulation of tensor product surfaces, 1992 Symposium on Interactive 3D Graphics **25(2)** (1992), 101–108.
9. Ma, M., The direct manipulation of pasted surfaces, University of Waterloo Report CS-00-15, 2000.

Marryat Ma and Stephen Mann
Computer Science Department
University of Waterloo
Waterloo, Ontario N2L 3G1, Canada
`mma@cgl.uwaterloo.ca`
`smann@cgl.uwaterloo.ca`

Knot Insertion Algorithms and Local Linear Independence

Esmeralda Mainar and Juan Manuel Peña

Abstract. Knot insertion algorithms for general spaces admitting shape preserving representations were provided in [10]. Given a finite dimensional vector space of univariate functions $\mathcal{U}$, in the knot insertion procedure we obtain a chain of spaces with increasing dimensions and normalized B-bases. Here we study the local linear independence of these bases.

§1. Introduction

A space with a normalized totally positive basis admits shape preserving representations (see [5,12]). The basis with optimal shape preserving properties is called the normalized B-basis. In [9] the authors derive a corner cutting algorithm associated with any normalized B-basis (called B-algorithm) which is an evaluation algorithm and satisfies subdivision properties. On the other hand, the use of knot insertion techniques when working with splines has been very useful in several fields. For instance, in Computer Aided Geometric Design, it is useful that a curve expressed in terms of a B-spline basis of a space $\mathcal{U}$ can be also expressed in terms of other B-spline bases of spaces containing $\mathcal{U}$ in order to increase the flexibility for the interactive design of the curve. In [10], we studied when a B-algorithm also provides a knot insertion algorithm and we also found a long list of spaces such that the corresponding B-algorithms provide a knot insertion algorithm. In fact, many spline spaces possess totally positive bases (see [7] about total positivity), including the spaces of generalized splines of [6,8,14]. The results of [10] can provide knot insertion algorithms for the generalized splines of [6], and therefore for the Tchebycheffian and trigonometric splines (see [13,1]), which are examples of these generalized splines. Let us also recall that the class of Tchebycheffian splines contains many classes of splines such as polynomial, exponential and hyperbolic splines.

Mathematical Methods for Curves and Surfaces: Oslo 2000
Tom Lyche and Larry L. Schumaker (eds.), pp. 283–292.

ISBN 0-8265-1378-6

In the knot insertion procedure we obtain a chain of spaces with increasing dimensions and normalized B-bases. Here we study the local linear independence of these spaces and analyze some related properties. In Section 2 we introduce the basic definitions and auxiliary results. In Section 3 we prove that if the normalized B-basis of a space is locally linearly independent, then all normalized B-bases of the spaces obtained in the mentioned knot insertion procedure are also locally linearly independent and we derive some consequences of this fact.

§2. Basic Definitions and Auxiliary Results

Let $\mathcal{U}$ be a vector space of real functions defined on an interval $I \subseteq \mathbb{R}$, and let $(u_0, \ldots, u_n)$ be a basis of $\mathcal{U}$. Given $D \subseteq I$ and $u \in \mathcal{U}$, we shall denote by $\mathcal{U}\big|_D$ the space $\{u(t)|u \in \mathcal{U}, t \in D\}$. The system of functions $(u_0, \ldots, u_n)$ is normalized if $\sum_{i=0}^n u_i(t) = 1$, $\forall\, t \in I$. The collocation matrix of $(u_0(t), \ldots, u_n(t))$ at $t_0 < \cdots < t_m$ in I is given by

$$M\begin{pmatrix} u_0, \ldots, u_n \\ t_0, \ldots, t_m \end{pmatrix} := (u_j(t_i))_{i=0,\ldots,m; j=0,\ldots,n}. \tag{2.1}$$

Clearly, $(u_0, \ldots, u_n)$ is normalized and formed by nonnegative functions if and only if all its collocation matrices are stochastic (that is, nonnegative and such that the sum of each row is one).

A matrix is totally positive if all its minors are nonnegative, and a system of functions is totally positive when all its collocation matrices (2.1) are totally positive. In CAGD, it is well known (cf. [5,12]) that shape preserving representations of curves by means of control polygons must be associated with normalized totally positive bases. The normalized totally positive basis with optimal shape preserving properties was called in [2] a normalized B-basis. By Proposition 3.12 of [2], a B-basis can be defined by:

Definition 2.1. Let $(u_0, \ldots, u_n)$ be a totally positive basis of a space $\mathcal{U}$ defined on I. Then $(u_0, \ldots, u_n)$ is a B-basis if and only if the following conditions hold

$$\inf\left\{\frac{u_i(t)}{u_j(t)} : t \in I, u_j(t) \neq 0\right\} = 0,$$

for all $i \neq j$.

Examples of B-bases are the Bernstein basis in the case of the space $\mathcal{P}_k([a,b])$ of polynomials of degree less than or equal to k on an interval $[a,b]$, and the B-spline basis in the case of the corresponding polynomial spline space.

Existence of B-bases and normalized B-bases follows from Remark 3.8 and Theorem 4.2 (i) of [2], respectively:

Proposition 2.2. *If a vector space of functions has a totally positive (resp., normalized totally positive) basis, then it has a B-basis (resp., a unique normalized B-basis).*

Let us introduce the concept of local linear independence. Our definition is equivalent with other classical definitions of local linear independence in spaces of spline functions.

Definition 2.3. A finite collection B of functions on some topological space Ω is locally linearly independent if, for any open set $D \subseteq \Omega$ and any $\alpha \in \mathbb{R}^B$, $\sum_{b\in B} \alpha_b b = 0$ on D implies that $\alpha_b b = 0$ on D for all $b \in B$.

The following result corresponds to Corollary 4.3 of [3].

Theorem 2.4. *Let $\mathcal{U}$ be a finite dimensional vector space which has a locally linearly independent TP basis. Then any B-basis is also locally linearly independent.*

By supp(f) we denote the support of a function f. By Proposition 3.2 of [3] we can state the following characterization of locally linearly independent systems:

Proposition 2.5. *Let $u_0, \ldots u_n$ be functions defined on Ω. Then $u_0, \ldots, u_n$ are locally linearly independent if and only if for any $\alpha \in \mathbb{R}^n$,*

$$\mathit{supp}\left(\sum_{i=0}^{n} \alpha_i u_i\right) = \bigcup \{\mathit{supp}\ u_i :\ \alpha_i \neq 0\}.$$

We shall introduce the basic definitions for knot insertion in a given space $\mathcal{U}$ with shape preserving representations. Given $t_0 \in \mathrm{Int}(I)$, let $I' := (-\infty, t_0] \cap I$ and $I'' := (t_0, \infty) \cap I$. Let us recall that, for an $(n+1)$-dimensional space of functions $\mathcal{U}$ and a parameter t_0 such that $\dim(\mathcal{U}\big|_{I'}) = r+1$ and $\dim(\mathcal{U}\big|_{I''}) = s+1$, by formula (2.11) of [10] the normalized B-basis $(u_0, \ldots, u_n)$ of $\mathcal{U}$ satisfies

$$u_i(t) = 0,\ \forall t \in I',\ i = r+1, \ldots, n; \quad u_i(t) = 0,\ \forall t \in I'',\ i = 0, \ldots, n-s-1. \tag{2.2}$$

Therefore, $\dim(\mathcal{U}\big|_{I'}) + \dim(\mathcal{U}\big|_{I''}) \geq \dim(\mathcal{U})$.

Definition 2.6. Let $\mathcal{U}$ be a space of functions defined on I with a normalized totally positive basis and $t_0 \in \mathrm{Int}(I)$ (the interior of I). We say that $k := \dim(\mathcal{U}\big|_{I'}) + \dim(\mathcal{U}\big|_{I''}) - \dim(\mathcal{U}) (\geq 0)$ is the potential knot multiplicity of t_0 in $\mathcal{U}$.

The following formula follows from the previous definition:

$$r + s - n = k - 1. \tag{2.3}$$

Remark 2.7. If $(u_0, \ldots, u_n)$ is a normalized B-basis of functions continuous at t_0 then, by Remark 4.4 of [10], the number of basis functions among $u_0, \ldots, u_n$ which do not vanish at t_0 coincides with the potential knot multiplicity of t_0 in $\mathcal{U}$.

Definition 2.8. Let $\mathcal{U}^{n+1}$ be an $(n+1)$-dimensional space of functions defined on I with a normalized totally positive basis and $t_0 \in \mathrm{Int}(I)$ whose potential knot multiplicity in $\mathcal{U}$ is k. Then we say that we can perform an elementary knot insertion with t_0 if there exits an $(n+2)$-dimensional space $\mathcal{U}^{n+2} \supseteq \mathcal{U}^{n+1}$

with a normalized totally positive basis such that $\mathcal{U}^{n+2}\big|_{I'} = \mathcal{U}^{n+1}\big|_{I'}$ and $\mathcal{U}^{n+2}\big|_{I''} = \mathcal{U}^{n+1}\big|_{I''}$.

Remark 2.9. Let us observe that if we perform an elementary knot insertion with t_0 (whose potential knot multiplicity in $\mathcal{U}^{n+1}$ was k) then $k-1$ is the potential knot multiplicity in $\mathcal{U}^{n+2}$ of t_0. So, we can perform at most k consecutive knot insertions with t_0 and the potential knot multiplicity of a parameter in the space obtained when performing the maximum number of knot insertions is zero.

Let $\mathcal{U}$ be a vector space of functions defined on I with normalized B-basis $(u_0, \ldots, u_n)$. Since $u_0(t), \ldots, u_n(t)$, $t \in I'$ (resp., $t \in I''$), form a totally positive system, by Proposition 2.2 they generate the space $\mathcal{U}\big|_{I'}$ (resp., $\mathcal{U}\big|_{I''}$) which has a normalized B-basis $(\bar{v}_0, \ldots, \bar{v}_r)$ (resp., $(\bar{w}_0, \ldots, \bar{w}_s)$). Therefore there exist matrices L, U such that

$$(u_0(t), \ldots, u_r(t)) = (\bar{v}_0(t), \ldots, \bar{v}_r(t))L, \quad \forall t \in I', \tag{2.4}$$

and

$$(u_{n-s}(t), \ldots, u_n(t)) = (\bar{w}_0(t), \ldots, \bar{w}_s(t))U, \quad \forall t \in I''. \tag{2.5}$$

The matrix L corresponds to the left B-algorithm and the matrix U corresponds to the right B-algorithm.

We say that a space $\mathcal{U}$ of functions is $\mathcal{C}^j$ ($j \geq 0$) if $u^{(j)}$ is continuous for all $u \in \mathcal{U}$, where $u^{(0)} := u$ and $u^{(1)} := u'$. By Remark 2.7, if $(u_0, \ldots, u_n)$ is a normalized B-basis of a space $\mathcal{U}$ of functions continuous at t_0 then the number of basis functions nonvanishing at t_0 is $k = \dim(\mathcal{U}\big|_{I'}) + \dim(\mathcal{U}\big|_{I''}) - \dim(\mathcal{U})$. In order to guarantee the possibility of performing elementary knot insertions, the following concept was introduced in [10]:

Definition 2.10. Let $\mathcal{U}$ be a vector space of functions defined on I with a normalized B-basis $(u_0, \ldots, u_n)$, let $t_0 \in \text{Int}(I)$, $r := \dim(\mathcal{U}\big|_{I'}) - 1$, $s := \dim(\mathcal{U}\big|_{I''}) - 1$ and let $k(>0)$ be the number of basis functions which do not vanish at t_0. We say that t_0 is a k-admissible parameter in $\mathcal{U}$ if $(u_0, \ldots, u_n)$ satisfies the following properties:

(A) There exists $\varepsilon > 0$ such that $\mathcal{U}$ is $\mathcal{C}^{k-1}$ in $(t_0 - \varepsilon, t_0 + \varepsilon) \subseteq I$.

(B) $\det(u_i^{(j)}(t_0))_{i=n-s,\ldots,r;j=0,\ldots,k-1} \neq 0$.

Taking into account Definition 2.10 and Remark 2.7, we deduce that if t_0 is a k-admissible parameter in $\mathcal{U}$, then its potential knot multiplicity in $\mathcal{U}$ is k.

If t_0 is k-admissible ($k \in \mathbf{N}^*$) then by Theorem 6.1 of [10], its corresponding B-algorithm is symmetric, and by Theorem 5.1 of [10], we can perform k elementary knot insertions with t_0 leading to the chain of spaces

$$\mathcal{U}^{n+1} \subset \mathcal{U}^{n+2} \subset \cdots \subset \mathcal{U}^{n+k+1}.$$

In addition, by Corollary 3.4 of [11], there exists a unique knot insertion process at t_0.

§3. Main Result

This section is devoted to proving the following result and deriving some consequences.

Theorem 3.1. *Let $\mathcal{U}^{n+1}$ be an $(n+1)$-dimensional vector space of functions defined on $I \subseteq \mathbb{R}$ with a locally linearly independent normalized B-basis $(u_0^{n+1}, \ldots, u_n^{n+1})$. Let $t_0 \in \text{Int}(I)$ be a k-admissible parameter ($k \in \mathbf{N}^*$). Then the normalized B-bases of the spaces $\mathcal{U}^{n+1} \subset \mathcal{U}^{n+2} \subset \cdots \subset \mathcal{U}^{n+k+1}$ obtained when inserting t_0 (in the corresponding knot insertion process) are locally linearly independent on I.*

Proof: Let us prove that the normalized B-bases of the spaces $\mathcal{U}^{n+p+2}$ ($p = k-1, \ldots, 0$) are locally linearly independent by induction on the dimension of the subspaces. As the first induction step, we prove that the normalized B-basis $(u_0^{n+k+1}, \ldots, u_{n+k}^{n+k+1})$ of $\mathcal{U}^{n+k+1}$ is locally linearly independent (the case $p = k-1$). The functions of $(u_0^{n+1}, \ldots, u_n^{n+1})$ are locally linearly independent on I, and by (2.2), we can deduce that the functions $u_0^{n+1}, \ldots, u_r^{n+1}$ form a locally linearly independent normalized totally positive basis of $\mathcal{U}^{n+1}\big|_{I'}$ and the functions $u_{n-s}^{n+1}, \ldots, u_n^{n+1}$ form a locally linearly independent totally positive basis of $\mathcal{U}^{n+1}\big|_{I''}$. We can then apply Theorem 2.4 and deduce that all B-bases of $\mathcal{U}^{n+1}\big|_{I'}$ (of $\mathcal{U}^{n+1}\big|_{I''}$, respectively) are locally linearly independent on I' (on I'', respectively), and hence its normalized B-basis $(\bar{v}_0, \ldots, \bar{v}_r)$ ($(\bar{w}_0, \ldots, \bar{w}_s)$, respectively) is locally linearly independent on I' (on I'', respectively). Let us define

$$v_i(t) := \begin{cases} \bar{v}_i(t) & \text{if } t \in I', \\ 0 & \text{if } t \in I'' \end{cases}, \quad i = 0, \ldots, r$$

and

$$w_i(t) := \begin{cases} 0 & \text{if } t \in I' \\ \bar{w}_i(t) & \text{if } t \in I'' \end{cases}, \quad i = 0, \ldots, s.$$

These functions generate a vector space of functions

$$\mathcal{V} := \text{span}\{v_0(t), v_1(t), \ldots, v_r(t), w_0(t), w_1(t), \ldots, w_s(t)\}, \quad t \in I.$$

From Corollary 2.12 of [10] we can deduce that

$$t_0 \in \text{supp}\, \bar{v}_{r-i} = \text{supp}\, v_{r-i} \quad \text{and} \quad t_0 \in \text{supp}\, \bar{w}_i = \text{supp}\, w_i \quad \forall i = 0, \ldots, k-1. \tag{3.1}$$

In Theorem 3.3 of [9] it was shown that if $(u_0^{n+1}, \ldots, u_n^{n+1})$ is a normalized B-basis of a space $\mathcal{U}$ then $(v_0, \ldots, v_r, w_0, \ldots, w_s)$ is a normalized B-basis of the vector space $\mathcal{V}$. By Proposition 4.5 of [10], $\mathcal{U}^{n+k+1}$ coincides with the space $\mathcal{V}$, and so

$$(u_0^{n+k+1}, \ldots, u_{n+k}^{n+k+1}) = (v_0, \ldots, v_r, w_0, \ldots, w_s). \tag{3.2}$$

Let us suppose that $D \subseteq \mathbb{R}$ is an open set and $\alpha_0, \ldots, \alpha_r, \beta_0, \ldots, \beta_s$ are real values such that $\sum_{i=0}^{r} \alpha_i v_i + \sum_{i=0}^{s} \beta_i w_i = 0$ on $D \cap I$. By considering

I' we have $\sum_{i=0}^{r} \alpha_i v_i = \sum_{i=0}^{r} \alpha_i \bar{v}_i = 0$ on $D \cap I'$ and by the local linear independence on I' of $(\bar{v}_0, \ldots, \bar{v}_r)$ we deduce that $\alpha_i v_i = \alpha_i \bar{v}_i = 0$ for all $i = 0, \ldots, r$. If we focus now on I'', we have $\sum_{i=0}^{s} \beta_i w_i = \sum_{i=0}^{s} \beta_i \bar{w}_i = 0$ on $D \cap I''$, and by the local linear independence on I'' of $(\bar{w}_0, \ldots, \bar{w}_s)$ we conclude that $\beta_i w_i = \beta_i \bar{w}_i = 0$ for all $i = 0, \ldots, s$. Therefore the normalized B-basis (3.2) of $\mathcal{U}^{n+k+1}$ is locally linearly independent on I.

Let us now assume that the normalized B-basis $(u_0^{n+p+2}, \ldots, u_{n+p+1}^{n+p+2})$ is locally linearly independent and let us prove that the normalized B-basis $(u_0^{n+p+1}, \ldots, u_{n+p}^{n+p+1})$ of $\mathcal{U}^{n+p+1}$ is locally linearly independent on I ($p = k-1, \ldots, 0$). Let $I_i^j := \operatorname{supp}(u_i^j)$. By Proposition 4.1 of [2], $I_i^{n+p+2} = [\alpha_i, \beta_i]$ is an interval ($0 \le i \le n+p+1$) and

$$\alpha_0 \le \cdots \le \alpha_{n+p+1}, \quad \beta_0 \le \cdots \le \beta_{n+p+1}. \tag{3.3}$$

Since $\mathcal{U}^{n+p+1}\big|_{I'} = \mathcal{U}^{n+1}\big|_{I'}$ and $\mathcal{U}^{n+p+1}\big|_{I''} = \mathcal{U}^{n+1}\big|_{I''}$, by formula (2.2),

$$\begin{aligned} &u_i^{n+p+1}(t) = 0, \ \forall t \in I'', \quad i = 0, \ldots, n+p-s-1; \\ &u_i^{n+p+1}(t) = 0, \ \forall t \in I', \quad i = r+1, \ldots, n+p. \end{aligned}$$

By Theorem 6.1 of [10], we have a symmetric B-algorithm, a concept introduced in that paper. Then Theorem 5.1 of [10] holds, and as shown in its proof, the normalized B-bases of $\mathcal{U}^{n+p+1}$ and $\mathcal{U}^{n+p+2}$ are related by

$$(u_0^{n+p+1}, \ldots, u_{n+p}^{n+p+1}) = (u_0^{n+p+2}, \ldots, u_{n+p+1}^{n+p+2}) M_p,$$

where $M_p = (m_{ij}^{(p)})_{0 \le i \le n+p+1; 0 \le j \le n+p}$ is a bidiagonal matrix with

$$m_{ii}^{(p)} \ne 0, \quad m_{i+1,i}^{(p)} \ne 0, \quad i = n+p-s, \ldots, r. \tag{3.4}$$

Then we can write

$$\begin{aligned} u_i^{n+p+1} &= u_i^{n+p+2} \quad \text{for} \quad i = 0, \ldots, n+p-s-1, \\ u_i^{n+p+1} &= m_{ii}^{(p)} u_i^{n+p+2} + m_{i+1,i}^{(p)} u_{i+1}^{n+p+2} \quad \text{for} \quad i = n+p-s, \ldots, r, \\ u_i^{n+p+1} &= u_{i+1}^{n+p+2} \quad \text{for} \quad i = r+1, \ldots, n+p. \end{aligned} \tag{3.5}$$

For any $p = k-1, \ldots, 0$, we derive from (3.5) and (3.2)

$$u_{r+1}^{n+p+2} = u_{r+2}^{n+p+3} = \cdots = u_{r+1+i}^{n+k+1} = w_i, \tag{3.6}$$

with $i = (n+k+1) - (n+p+2) = k-p-1$ and so $i \in \{0, \ldots, k-1\}$. Analogously, by (3.5) and (3.2),

$$u_{n+p-s}^{n+p+2} = u_{n+p-s}^{n+p+3} = \cdots = u_{n+p-s}^{n+k+1} = v_{n+p-s} = v_{r-i} \tag{3.7}$$

for $i = r+s-n-p$ and so by (2.3) $i = k-1-p$ and $i \in \{0, \dots, k-1\}$. We can apply (3.1) and derive from (3.6) $t_0 \in [\alpha_{r+1}, \beta_{r+1}]$, and $t_0 \in [\alpha_{n+p-s}, \beta_{n+p-s}]$ from (3.7). Therefore $\alpha_{r+1} \le \beta_{n+p-s}$. Taking into account (3.3), we have

$$(\alpha_i \le)\alpha_{i+1} \le \beta_i(\le \beta_{i+1}) \quad \forall i = n+p-s, \dots, r. \tag{3.8}$$

If $J \subseteq \{n+p-s, \dots, r+1\}$, by (3.8) we have

$$\cup_{i\in J}\mathrm{supp}(u_i^{n+p+2}) = [\alpha_{\min J}, \beta_{\max J}]. \tag{3.9}$$

Given any linear combination $\sum_{i=n+p-s}^{r+1} \lambda_i u_i^{n+p+2}$, let us define

$$C = \{i|\, \lambda_i \neq 0,\, n+p-s \le i \le r+1\}, \quad h := \min C, \quad j := \max C. \tag{3.10}$$

Taking into account that the functions $u_{n+p-s}^{n+p+2}, \dots, u_{r+1}^{n+p+2}$ are locally linearly independent we can deduce from (3.9) that

$$\mathrm{supp}\left(\sum_{i=n+p-s}^{r+1} \lambda_i u_i^{n+p+2}\right) = \cup_{i\in C}\mathrm{supp}(u_i^{n+p+2}) = [\alpha_h, \beta_j]. \tag{3.11}$$

Since the functions $u_0^{n+p+2}, \dots, u_{n+p+1}^{n+p+2}$ are locally linearly independent, by Proposition 2.5 and (3.4) we can write

$$\begin{aligned}
&\mathrm{supp}(u_i^{n+p+1}) = \mathrm{supp}(u_i^{n+p+2}), \quad 0 \le i \le n+p-s-1;\\
&\mathrm{supp}(u_i^{n+p+1}) = \mathrm{supp}(u_i^{n+p+2}) \cup \mathrm{supp}(u_{i+1}^{n+p+2}), \quad n+p-s \le i \le r;\\
&\mathrm{supp}(u_i^{n+p+1}) = \mathrm{supp}(u_{i+1}^{n+p+2}), \quad r+1 \le i \le n+p.
\end{aligned} \tag{3.12}$$

Let us consider any linear combination $\sum_{i=0}^{n+p} c_i u_i^{n+p+1}$ of the system $(u_0^{n+p+1}, \dots, u_{n+p}^{n+p+1})$. By Proposition 2.5, it is sufficient to see that

$$\mathrm{supp}\left(\sum_{i=0}^{n+p} c_i u_i^{n+p+1}\right) = \cup\{\mathrm{supp}(u_i^{n+p+1})|\, c_i \neq 0,\, 0 \le i \le n+p\} \tag{3.13}$$

in order to prove that $(u_0^{n+p+1}, \dots, u_{n+p}^{n+p+1})$ is locally linearly independent. By (3.5) we can write

$$\sum_{i=0}^{n+p} c_i u_i^{n+p+1} = \sum_{i=0}^{n+p-s-1} c_i u_i^{n+p+2} + \sum_{i=n+p-s}^{r+1} \lambda_i u_i^{n+p+2} + \sum_{i=r+2}^{n+p+1} c_{i-1} u_i^{n+p+2}, \tag{3.14}$$

where $\lambda_{n+p-s} = c_{n+p-s} m_{n+p-s,n+p-s}^{(p)}$, $\lambda_i = c_{i-1} m_{i,i-1}^{(p)} + c_i m_{ii}^{(p)}$ $(i = n+p-s+1, \dots, r)$ and $\lambda_{r+1} = c_r m_{r+1,r}^{(p)}$. Hence, taking into account (3.4) and according to (3.10), we can deduce that

$$\begin{aligned}
\min\{i|\, c_i \neq 0,\, n+p-s \le i \le r\} = h,\\
\max\{i|\, c_i \neq 0,\, n+p-s \le i \le r\} = j-1.
\end{aligned} \tag{3.15}$$

Using again that the functions $u_0^{n+p+2}, \ldots, u_{n+p+1}^{n+p+2}$ are locally linearly independent, we can derive from (3.14) and Proposition 2.5

$$\operatorname{supp}\Big(\sum_{i=0}^{n+p} c_i u_i^{n+p+1}\Big) = S_1 \cup S_2$$

where

$$S_1 = \bigcup\{\operatorname{supp}(u_i^{n+p+2}) |\, c_i \neq 0,\ 0 \leq i \leq n+p-s-1 \text{ or } r+2 \leq i \leq n+p+1\}$$
$$S_2 = \bigcup\{\operatorname{supp}(u_i^{n+p+2}) |\, \lambda_i \neq 0,\ n+p-s \leq i \leq r+1\}.$$

Using (3.11),

$$S_2 = \bigcup[\alpha_h, \beta_j].$$

Applying (3.5), we get

$$S_1 = \bigcup\{\operatorname{supp}(u_i^{n+p+1}) |\, c_i \neq 0,\ 0 \leq i \leq n+p-s-1 \text{ or } r+1 \leq i \leq n+p\}.$$

On the other hand, by (3.5), (3.9) and (3.15) we have

$$\begin{aligned}
&\bigcup\{\operatorname{supp}(u_i^{n+p+1}) |\, c_i \neq 0,\ n+p-s \leq i \leq r\} \\
&\quad = \bigcup\{\operatorname{supp}(u_i^{n+p+2}) |\, c_i \neq 0,\ n+p-s \leq i \leq r\} \cup \\
&\qquad \bigcup\{\operatorname{supp}(u_{i+1}^{n+p+2}) |\, c_i \neq 0,\ n+p-s \leq i \leq r\} = [\alpha_h, \beta_j].
\end{aligned}$$

We conclude (3.13), and the result follows. □

We now collect information on the normalized B-bases of the spaces $\mathcal{U}^{n+p+1}$ $(0 \leq p \leq k)$ which can be derived from Theorem 3.1.

A first application deals with the supports of the basis functions. For numerical purposes it is convenient that the supports of the basis functions of a spline space are small. However, even common spaces such as the space of univariate polynomial splines with multiple knots do not possess bases whose basis functions have minimal support. This motivated the following definition introduced in [3]: a basis $(u_0, \ldots, u_n)$ is **least supported** if for *every* basis $(v_0, \ldots, v_n)$ for this space, there is some permutation σ so that

$$\operatorname{supp}(u_j) \subseteq \operatorname{supp}(v_{\sigma(j)}), \quad j = 0, \ldots, n.$$

In Theorem 3.4 of [3] it was proved that a basis is locally linearly independent if and only if it is a least supported basis.

On the other hand, a nonsingular totally positive matrix is called **almost strictly totally positive** (ASTP) if it satisfies that any minor if and only if its diagonal entries are all positive. By Remark 6.2 of [10] , under the hypotheses of the previous theorem the matrices L and U (see (2.4) and (2.5)) corresponding to the left and right B-algorithm are both ASTP. A system of functions is called ASTP if all its collocation matrices (2.1) are ASTP. By Theorem 3.1 of [4], any collocation matrix (2.1) of a locally linearly independent totally positive system of continuous functions is ASTP.

Therefore, by Theorem 3.1, if $\mathcal{U}^{n+1}$ has a locally linearly independent normalized B-basis of continuous functions, then the collocation matrices of the normalized B-bases of the spaces $\mathcal{U}^{n+1} \subset \cdots \subset \mathcal{U}^{n+k+1}$ obtained when inserting t_0 (in the corresponding knot insertion process) are ASTP and the bases are least supported. Finally, let us recall that in [4] it was shown that an ASTP basis also satisfies a Schoenberg-Whitney interpolation property.

Acknowledgments. This research has been partially supported by the Spanish Research Grant DGES PB96-0730.

References

1. Bister, D. and H. Prautzsch, A new approach to Tchebycheffian B-Splines, in *Curves and Surfaces with applications in CAGD*, A. Le Méhauté, C. Rabut and L. L. Schumaker (eds.), AKPeters, Boston, 1997, 35–41.
2. Carnicer, J. M. and J. M. Peña, Totally positive bases for shape preserving curve design and optimality of B-splines, Comput. Aided Geom. Design **11** (1994), 635–656.
3. Carnicer, J. M. and J. M. Peña, Least supported bases and local linear independence, Numer. Math. **67** (1994), 289–301.
4. Carnicer, J. M. and J. M. Peña, Spaces with almost strictly totally positive bases, Math. Nachrichten **169** (1994), 69–79.
5. Goodman, T. N. T., Shape preserving representations, in *Mathematical Methods in CAGD*, T. Lyche and L. L. Schumaker (eds.), Academic Press, New York, 1989, 333–357.
6. Goodman, T. N. T. and S. L. Lee, Interpolatory and variation diminishing properties of generalized B-splines, Proc. Royal Soc. Edinburgh **96** (1984), 249–259.
7. Karlin, S., *Total Positivity*, Stanford University Press, Stanford, 1968.
8. Lyche, T. and L. L. Schumaker, Total Positivity properties of LB-splines, in *Total positivity and its applications*, M. Gasca and C. A. Micchelli (eds.), Kluwer Academic Press, Dordrecht, 1996, 35–46.
9. Mainar, E. and J. M. Peña, Corner cutting algorithms associated with optimal shape preserving representations, Comput. Aided Geom. Design **16** (1999), 883–906.
10. Mainar, E. and J. M. Peña, Knot insertion and totally positive systems, J. Approx. Theory **104** (2000), 45–76.
11. Mainar, E. and J. M. Peña, On the uniqueness of knot insertion procedures, East J. Approx. **6** (2000), 277–294.

12. Peña, J. M., *Shape preserving representations in Computer-Aided Geometric Design* (ed.), Nova Science Publishers, Commack (New York), 1999.

13. Schumaker, L. L., *Spline Functions: Basic Theory*, John Wiley and Sons, New York, 1981.

14. Sommer, M., and H. Strauss, Weak Descartes systems in generalized spline spaces, Constr. Approx. **4** (1988), 133–145.

Esmeralda Mainar and Juan Manuel Peña
Departamento de Matemática Aplicada
Universidad de Zaragoza,
50009 Zaragoza, Spain
`esme@posta.unizar.es` and `jmpena@posta.unizar.es`

Local Tension Methods for Bivariate Scattered Data Interpolation

Carla Manni

Abstract. The paper surveys some local tension methods which can be used for constrained scattered data interpolation. The graphical performances of various methods are compared with the aid of a classical numerical test.

§1. Introduction

The purpose of this paper is to survey some methods in the area of bivariate scattered data interpolation. The problem of fitting a smooth function to arbitrarily located positions in a planar domain arises often in scientific and engineering applications, see for example [27,56,60].

It is well known that for data points belonging to $\mathbb{R}^d$, $d \geq 2$, it is not possible, in general, to solve the interpolation problem using spaces not depending on the location of the given data [23]. Thus, various methods have been proposed which take into account the position of the data. Among these we will limit our survey to methods which construct the interpolant as a smooth collection of triangular patches over a given triangulation of the data. In addition, it is generally required that the interpolant avoid inflections and oscillations extraneous to the expected behavior of the data. Hence, we will restrict our attention to interpolation schemes with **tension** properties; that is to interpolants, depending on a set of **shape parameters**, which approach the piecewise linear interpolant of the data for limit configurations of the above mentioned parameters. To be more specific, let the data

$$(\mathbf{P}_i,\ f_i = f(\mathbf{P}_i)), \qquad \mathbf{P}_i \in \mathbb{R}^2, \qquad i = 1, \cdots, N,$$

be given. We will discuss some methods for the construction of a smooth function s defined in the convex hull of $\mathbf{P}_1, \cdots, \mathbf{P}_N$ which admits a local construction, interpolates the data, that is

$$s(\mathbf{P}_i) = f_i, \qquad i = 1, \cdots, N,$$

Mathematical Methods for Curves and Surfaces: Oslo 2000
Tom Lyche and Larry L. Schumaker (eds.), pp. 293–314.

ISBN 0-8265-1378-6

and has tension properties.

In the construction of such an interpolant in the bivariate setting, difficulties which do not exist in the one-dimensional case arise. First of all, in the one-dimensional case the x coordinates are naturally ordered, and the minimum set to be considered for the local construction of the interpolant is a single sub-interval of the data. In the bivariate setting, triangles correspond to intervals. However, there is no unique way to arrange the given $\mathbf{P}_i$ as vertices of triangles; that is, to triangulate the data. Different triangulations provide different piecewise linear interpolants of the data and, consequently, different shapes of the data.

Whenever the elementary domain has been determined, derivatives at the boundary of the domain are needed to allow a local construction. Derivatives are required up to an order depending on the global smoothness of the interpolant. For the bivariate case, derivatives must be provided at the vertices and along and across the edges of the given triangulation. This introduces a further dependence on the triangulation. Often derivatives are not explicitly given, so they must be estimated.

Here we will not discuss the construction of the triangulation and the estimation of the derivatives, even if we anticipate that for all the local schemes we are going to describe, the quality of the final interpolant heavily depends on these choices.

Thus, in the following, we will assume that a triangulation of the data is given and the needed derivatives are also given. For the sake of brevity we will basically consider only C^1 schemes.

We emphasize that a crucial point dealing with shape parameters consists in their choice. Preferably, they should be chosen according to an efficient automatic procedure. A simple geometric interpretation of the shape parameters is often essential in order to allow such a procedure.

In the next section we briefly recall some well-known tension methods used for the one-dimensional problem. Then we will describe in Section 3 some extensions of these methods to the bivariate setting. Section 4 is devoted to illustrating a particular strategy for obtaining a class of tension schemes in one or two dimensions. Finally, we conclude in Section 5 with some graphical examples.

§2. One Dimensional Tension Methods

All the one-dimensional methods we will consider can be seen as a generalization of the classical cubic splines. Let the data (x_i, f_i, f_i'), $i = 1, \cdots, N$, be given. Let

$$t = \frac{x - x_i}{h_i}, \qquad h_i = x_{i+1} - x_i.$$

2.1 Exponential Splines

Exponential splines have been introduced in the well-known paper [61] and, since then have continuously received considerable attention from both the

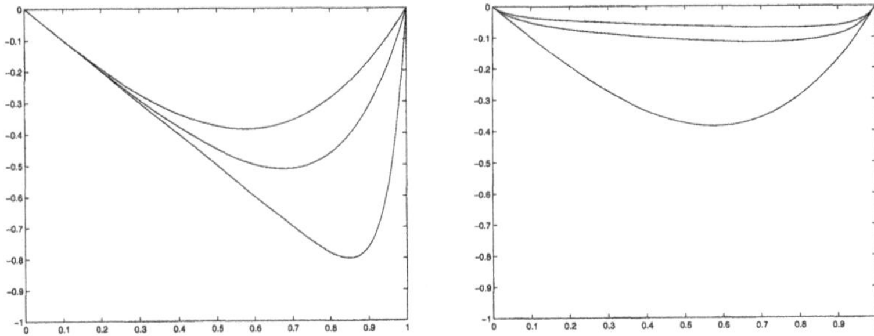

Fig. 1. Left: behavior of $\hat{\varphi}_i$ for $\lambda_i = 0.1,\ 5,\ 20$. Right: behavior of $\tilde{\varphi}_i$ for $w_i = 0,\ 10,\ 20$.

theoretical and the algorithmic point of view, see *e.g.* [13,30,36,38,50,57] and references quoted therein. Exponential splines are probably the first and most well-known method for introducing tension effect in the classical cubic splines. In the sub-interval $[x_i, x_{i+1}]$, the interpolating spline is represented in the form

$$s(x) = f_i(1-t) + f_{i+1}t + \hat{a}_i\hat{\varphi}_i(1-t) + \hat{b}_i\hat{\varphi}_i(t), \tag{1}$$

where

$$\hat{\varphi}_i(t) = \frac{\sinh(\lambda_i t) - t\sinh(\lambda_i)}{\sinh(\lambda_i) - \lambda_i}$$

and $\hat{a}_i$, $\hat{b}_i$ have to be determined, for example, to ensure interpolation of the first derivatives which implies C^1 continuity. Exponential splines tend to a piecewise linear function as the λ_i go to ∞, while they reduce to classical cubic splines as the λ_i approach 0, see Fig. 1 left. This behavior can be easily understood by observing that the exponential splines satisfy the following differential equation which reduces to the equation characterizing the cubic splines if $\lambda_i = 0$:

$$s^{(4)}(x) - \lambda_i^2 s^{(2)}(x) = 0, \quad x \in (x_i, x_{i+1}).$$

Moreover, exponential splines generalize the well-known minimum property of cubics: in fact, it can be seen that they solve the following problem [52]:

$$\min_{g\in\mathcal{E}} \sum_{i=1}^{N-1} \int_{x_i}^{x_{i+1}} [g''(x)]^2 dx + \lambda_i^2 \int_{x_i}^{x_{i+1}} [g'(x)]^2 dx$$
$$\mathcal{E} := \{g : g' \text{ abs. cont. }, g'' \in L^2[x_1, x_N], g(x_i) = f_i\}.$$

2.2 Rational Splines

A second class of functions which allow tension are rational functions. There are several forms used for rational functions, mainly depending on the degrees for the numerator and the denominator. Here we recall the following form for rational cubics, used by Gregory, Delbourgo and others, (see for example [24]):

$$s(x) = f_i(1-t) + f_{i+1}t + \tilde{a}_i\tilde{\varphi}_i(1-t) + \tilde{b}_i\tilde{\varphi}_i(t), \tag{2}$$

where

$$\tilde{\varphi}_i(t) = \frac{t^3 - t}{1 + w_i t(1-t)}.$$

The tension effect can be immediately achieved as the w_i increase, see Fig. 1 right. Rational splines have a more recent history than exponential splines, but nevertheless there is an enormous literature on them. Rational splines have been introduced as a possible solution for shape preserving problems in the functional setting, but recently several authors have successfully used them in the area of constrained interpolation of planar and space curves, see for example [8,28,31,49,59,63] and references quoted therein.

2.3 Variable Degree Splines

Finally, a third possible extension of classical cubic splines which incorporates the tension effect is given by variable degree splines. Each segment of the interpolating spline is obtained as a linear combination of a linear part and two polynomials of degree n_i as follows:

$$s(x) = f_i(1-t) + f_{i+1}t + \bar{a}_i\bar{\varphi}_i(1-t) + \bar{b}_i\bar{\varphi}_i(t), \tag{3}$$

where

$$\bar{\varphi}_i(t) = t^{n_i} - t, \quad n_i \geq 3.$$

The degrees n_i act as tension parameters: as they approach ∞ only the linear part remains in the linear combination. The variable degree polynomial splines have been introduced and used independently by P. Costantini and P. D. Kaklis, see [15,16,17,18,34,35]. The first author favored a geometric approach emphasizing the geometric meaning of the tension parameters. This geometric approach allowed to use variable degree splines as a very efficient tool not only for the interpolation of univariate functions or curves, but also for tensioned interpolation in the bivariate setting [19,21,22]. The above mentioned geometric approach simply consists of representing the function $\bar{\varphi}_i$ with respect to the classical Bernstein basis (see for example [33]) and considering its control points (that is basically its coefficient with respect to the Bernstein basis). From (3) all but one of these control points belong to a line, see Fig. 2 left. Similarly, each segment of the interpolating spline in (3), when expressed with respect to the Bernstein basis, has all the $n_i - 1$ interior control points belonging to a segment which approaches the segment interpolating (x_i, f_i), (x_{i+1}, f_{i+1}) as n_i increases. Thus, the shape (positivity, monotonicity, convexity, etc.) of the interpolant can be immediately controlled by the shape of the polygonal line connecting its control points, that is by its control polygon, see Fig. 2 right. In addition, the proposed variable degree splines are isomorphic to classical cubic splines so that they do not present unpleasant phenomena, such as numerical instability, common for polynomials with high degree.

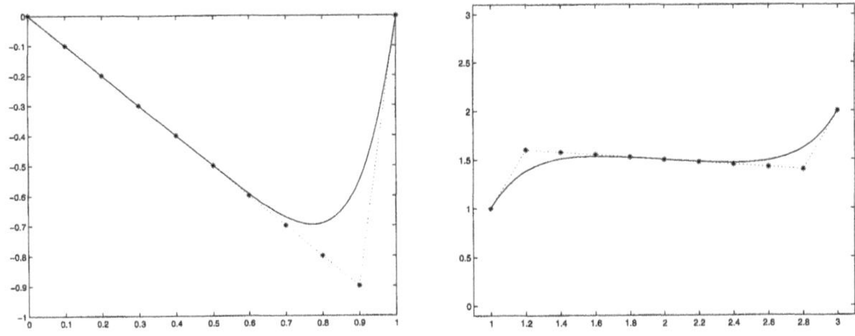

Fig. 2. $\bar{\varphi}_i$ (left), s (right) and their control polygons for $n_i = 10$.

§3. Bivariate Tension Schemes

In this section we will describe how the previously mentioned one-dimensional tension methods have been employed to construct tensioned bivariate interpolants.

First of all, we note that in the bivariate setting we cannot try to introduce tension modifying a cubic polynomial defined in each triangle of the given triangulation. In fact, in general it is not possible to locally construct a C^1 function patching together cubic polynomials interpolating data and derivatives at the three vertices of each triangle of the given triangulation. This is due to the fact that the conditions necessary to guarantee C^1 continuity across the edges impose constraints on the derivatives across the edges and inhibit a local construction (see for example [5,33,54]). Then, the basic element we have to consider cannot be a simple cubic element. We have to consider more complex structures such as those obtained with blending, or the so called macro-elements which require the splitting of each triangle in sub-triangles.

3.1 Side-Vertex Element

Among the various blending techniques proposed to describe a triangular interpolating element, we will briefly recall the side-vertex element proposed in [51] (see Fig. 3). For each triangle, let the values of s and of its derivatives at the vertices be given. The values of s along the edges are constructed according to a one-dimensional Hermite scheme. The normal derivatives across the edge are estimated (in general assuming a linear variation of the derivative along the edge). For each vertex of the triangle, say $\mathbf{P}_i$, a function C_i is constructed defining for each point $\mathbf{P}$ in the interior of the triangle the segment through $\mathbf{P}_i$ and $\mathbf{P}$ and considering a one-dimensional Hermite scheme along this segment (see Fig. 3). Finally, the element is obtained blending the above constructed functions as

$$s(\mathbf{P}) := \frac{u_j^2 u_k^2 C_i(\mathbf{P}) + u_i^2 u_k^2 C_j(\mathbf{P}) + u_i^2 u_j^2 C_k(\mathbf{P})}{u_j^2 u_k^2 + u_i^2 u_k^2 + u_i^2 u_j^2}, \tag{4}$$

where u_l denotes the barycentric coordinate of $\mathbf{P}$ with respect to $\mathbf{P}_l$, $l = i, j, k$.

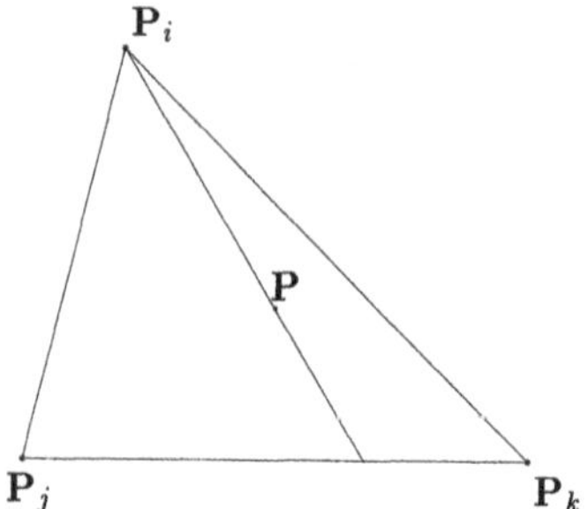

Fig. 3. The construction of the side-vertex element.

Nielson and Franke obtained a side-vertex element with tension properties by using a one-dimensional Hermite scheme based on exponentials as in (1), [52]. Figure 4 (left) shows the tension effect obtained by increasing the values of the parameters λ_i (see (1)) associated to the edges and to the interior of the triangle.

More recently, a side-vertex element based on cubic rational functions has been proposed in [53]. Figure 4 (right) shows how, when increasing the values of the parameters w_i (see (2)), a tensioned element with performance similar to that of the exponential side-vertex element is produced.

The side-vertex element is implemented in a procedural approach and requires the weighted combination of three partial interpolants, see (4). This lack of closed form is the main disadvantage of the method and makes it difficult to control the shape via the tension parameters.

3.2 Variable Degree Splines: Bivariate Extension

Let us consider now a second possibility for the construction of triangular interpolating elements: the splitting of the given triangle. Probably the most well-known splitting is the so called Clough–Tocher (CT) split which consists of dividing the given triangle in three sub-triangles by means of an interior point ([12,14]), see also Fig. 5. The CT splitting allows the construction of a cubic macro-element. The idea of variable degree splines makes it possible to generalize such a construction and to obtain a polynomial macro-element with tension properties [19]. Referring to the classical Bernstein representation with respect to barycentric coordinates, we describe the macro-element in terms of its Bézier control points in each sub-triangle (see for example [33]). Fig. 5 shows the projection of the control points onto the (x, y) plane.

As is well known, (see Fig. 5, top) for the cubic CT macro-element the Bézier points $\odot$ are determined from interpolation conditions (position and derivatives) at the vertices; the $*$ from the normal derivatives across the edges, while the $\otimes$ are obtained by imposing C^1 continuity conditions across the interior edges. It turns out that the macro-element has 12 degrees of freedom. We will deal with the reduced macro-element where we assumed a linear variation of the normal derivatives across the boundary edges. This assumption reduces to 9 the number of degrees of freedom.

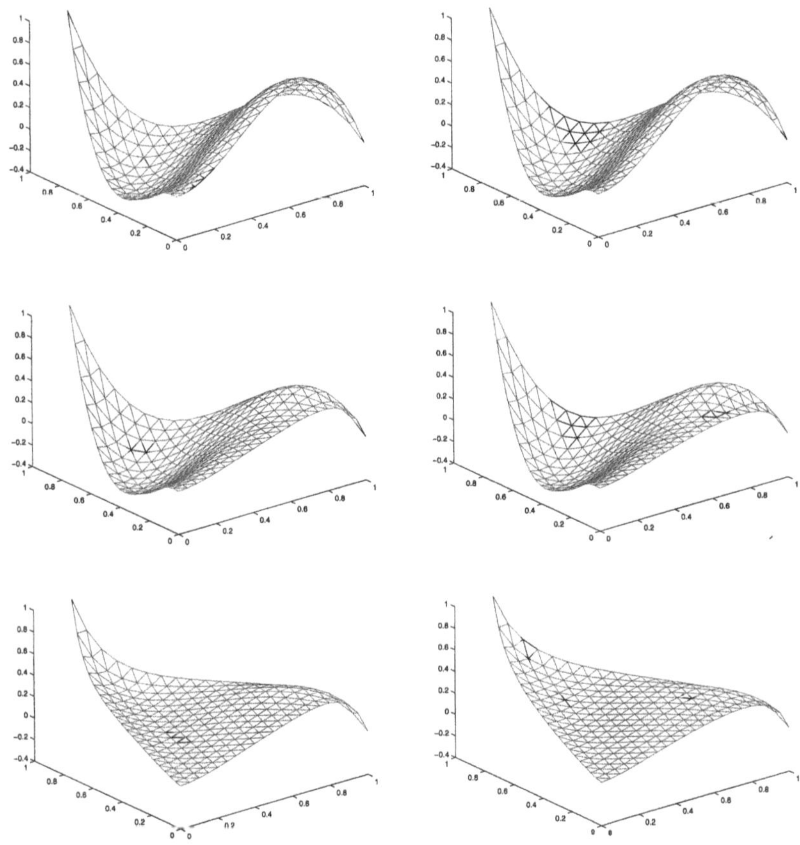

Fig. 4. Side-vertex element based on exponentials (left) and rationals (right).
Top left: $\lambda_i = .1$ for the three edges and for the interior of the triangle.
Center left: $\lambda_i = 10, .1, .1$ for the three edges and $\lambda_i = .1$ for the interior.
Bottom left: $\lambda_i = 10$ for the three edges and for the interior.
Top right: $w_i = 0$ for the three edges and for the interior.
Center right: $w_i = 7, 0, 0$ for the three edges and $w = 0$ for the interior.
Bottom right: $w_i = 7$ for the three edges and for the interior.

In order to introduce a tension effect in the CT macro–element, in [19] a macro-element of degree $n \geq 3$ which still depends only on 9 degrees of freedom has been considered. In each sub-triangle a polynomial of degree n is defined by means of its Bézier control points (see Fig. 5 bottom for $n = 6$). The $\odot$ are determined from interpolation conditions at the vertices; the $\oplus$ are obtained by applying the one-dimensional variable degree scheme as in (3) along the edge; the $*$ are determined from the normal derivatives across the boundary edges (assuming a linear variation of such derivatives), while all the

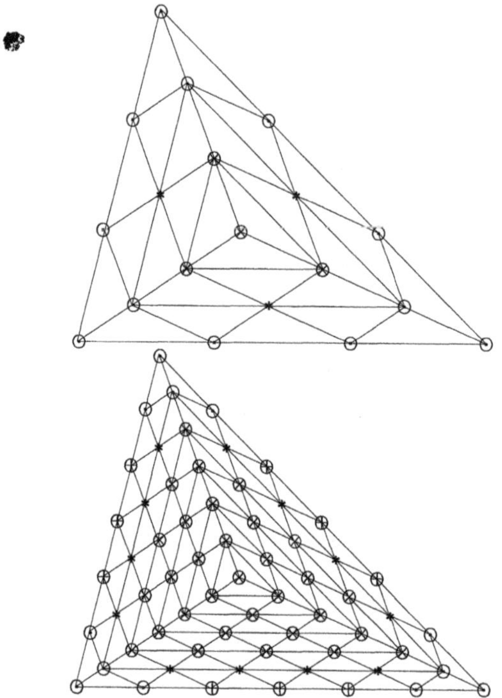

Fig. 5. Variable degree CT polynomial macro-element. Projection of the macro-net onto the x, y plane. Top the CT macro-element. Bottom $n = 6$.

remaining Bézier control points (⊗) are required to belong to the same plane (obtained by imposing C^1 continuity conditions across the interior edges). Due to the construction the resulting macro-element is of class C^1.

Fig. 6 shows the macro-net and the polynomial macro-element for different values of n. The figure on the left demonstrates how the central plane of the ⊗ points invades all of the triangle and approaches the plane interpolating the data as the degree increases. The surface inherits the same behavior due to the convex hull property of the Bézier-Bernstein representation.

To construct the final interpolant, it is necessary to patch together the various macro-elements of the given triangulation. To obtain C^0 and C^1 continuity, the same degree has to be associated with the two triangles sharing a common edge in two neighboring macro-elements (see [19,33]). As a consequence, the same degree for all the triangles must be used. It turns out that the highest degree necessary to control the shape in "difficult" triangles diffuses throughout the whole domain [19]. So, the construction provides a completely global tension scheme, see also Fig. 14.

In [21] a strategy has been proposed to localize the scheme. In each triangle three possibly different degrees $n_i \geq 3$ associated with each edge are considered (see Fig. 7 where the degrees used are 3, 6, and 10). Then the

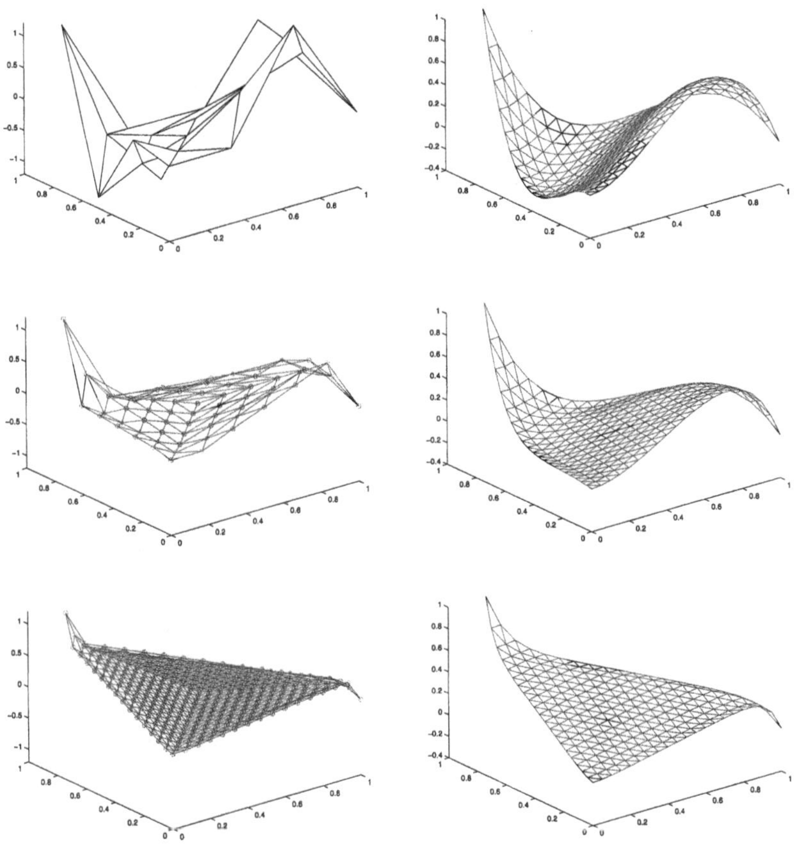

Fig. 6. Variable degree CT polynomial macro-element. Left the macro-net, right the macro-element. Top to bottom $n = 3, 6, 15$.

macro-nets of three variable degree macro-elements are constructed according to the above scheme, with the previously chosen degrees, and the restriction of each macro-net to the corresponding sub-triangle is considered. Of course, the three macro-nets may be discontinuous across the interior edges of the split (see Fig. 7, top). Next, the constructed macro-net is modified in order to obtain a new one defining a C^1 macro-element, which basically retains, in the various sub-triangles, the shape of the corresponding initial macro-element. Let n denote the maximum of the three degrees n_i. As a first step, the degrees in the three sub-triangles are formally equalized by means of degree raising ([33]) from n_i to n. This allows us to express the polynomial of degree n_i as a polynomial of degree n. The degree raising produces a macro-net still discontinuous across the interior edges (see Fig. 7, center). The final step

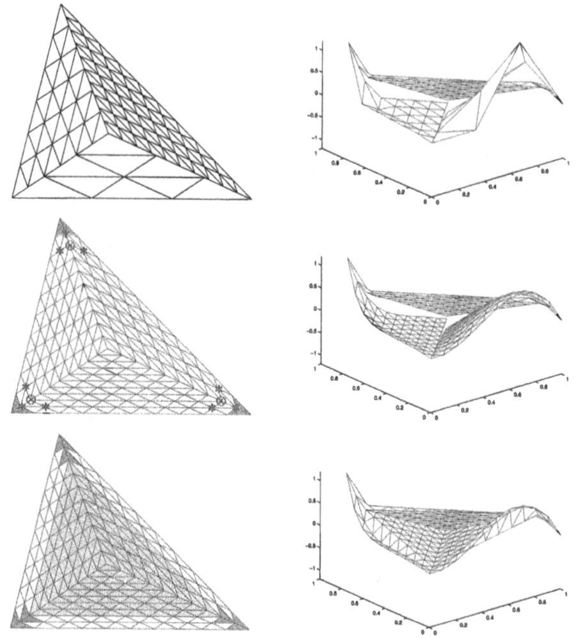

Fig. 7. Variable degree polynomial macro-element: local construction.

consists of modifying the degree raised macro-net in order to restore the C^1 continuity across the interior edges. This is done without changing the points which determine the smoothness behavior of the macro-element across the boundary edges of the triangle. In fact, the triangles of the Bézier net sharing the same vertex of the given triangle are coplanar (see Fig. 7 center, dark shaded triangles) due to the interpolation of the same tangent plane at the vertices and to the properties of the degree raising process. Then it suffices to require that the three $\otimes$ points (see Fig. 7, center) belong to the plane defined by the three surrounding $*$ points, and next that all the points in the light shaded triangle belong to the same plane (see Fig. 7, bottom). This ensures C^1 continuity across the interior edges of the split. We remark that the Bézier control points in the "first" two rows parallel to each boundary edge $\mathbf{e}_i$ remain unchanged. So the macro-element reduces to a polynomial of degree n_i along $\mathbf{e}_i$, and it has the same normal derivative across $\mathbf{e}_i$ as the "initial" macro-element. Therefore, the macro-element (see Fig. 8) has *formally* degree n but provides a C^1 join across $\mathbf{e}_i$ with a macro-element of degree n_i (compare Fig. 8 and Fig. 6 left). So, a completely local scheme is obtained (see also Fig. 15, top).

Such an approach allows an easy control of the shape of the resulting element by means of its control net. Generalizations have been considered in order to obtain variable degree macro-elements, where the tension parameters, that is the degrees, are associated with the vertices rather than the edges [22].

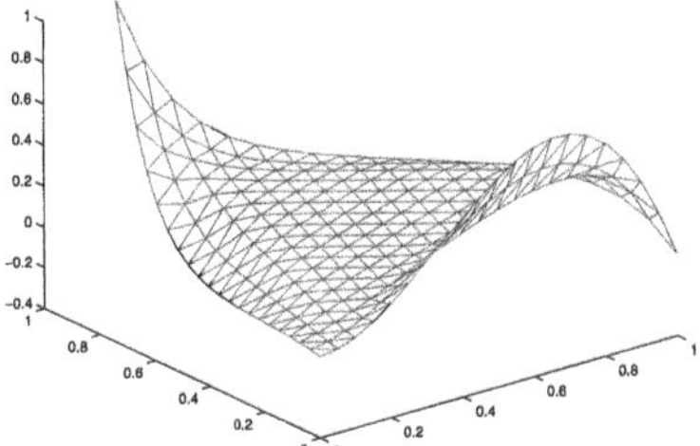

Fig. 8. Local construction of the variable degree polynomial macro-element: the final macro-element. Data and degrees as in Fig. 7.

This makes the tension effect less dependent on the given triangulation.

§4. Parametric Techniques

The last method for the construction of tensioned element we shall describe is the so called parametric approach. We briefly go back to the one-dimensional case in order to describe the basic idea of the approach [45,46,47].

In a given interval $[x_i, x_{i+1}]$ a cubic spline can be expressed in different forms. The Bernstein–Bézier form has been used to introduce the variable degree splines. A very well-known alternative is the Hermite form

$$s(x(t)) = f_i H_0^{(0)}(t) + f_{i+1} H_1^{(0)}(t) + h_i f_i' H_0^{(1)}(t) + h_i f_{i+1}' H_1^{(1)}(t),$$

where $x(t) = x_i + h_i t,\ \ t \in [0,1]$, and $H_i^{(j)},\ i,j = 0,1$, denote the elements of the cardinal basis for cubic Hermite interpolation. Noting that the Hermite interpolation cubic scheme reproduces straight lines, we have that $x(t)$ can be written in a less transparent form as

$$x(t) = x_i H_0^{(0)}(t) + x_{i+1} H_1^{(0)}(t) + h_i H_0^{(1)}(t) + h_i H_1^{(1)}(t).$$

Thus, the graph of $s(x)$, $x \in [x_i, x_{i+1}]$, can be seen as the image of a parametric curve

$$\mathcal{C}(t;1,1) = \begin{cases} x(t) = x_i H_0^{(0)}(t) + x_{i+1} H_1^{(0)}(t) + h_i H_0^{(1)}(t) + h_i H_1^{(1)}(t), \\ y(t) = f_i H_0^{(0)}(t) + f_{i+1} H_1^{(0)}(t) + h_i f_i' H_0^{(1)}(t) + h_i f_{i+1}' H_1^{(1)}(t). \end{cases}$$

This simple remark suggests to consider a family of cubic parametric curves depending on two parameters $\lambda_i^{(0)}$ and $\lambda_i^{(1)}$ each associated with the end-points of the interval as follows (see Fig. 9):

$$\mathcal{C}(t;\lambda_i^{(0)},\lambda_i^{(1)}) = \begin{cases} x(t) = x_i H_0^{(0)}(t) + x_{i+1} H_1^{(0)}(t) + \\ \qquad h_i \lambda_i^{(0)} H_0^{(1)}(t) + h_i \lambda_i^{(1)} H_1^{(1)}(t), \\ y(t) = f_i H_0^{(0)}(t) + f_{i+1} H_1^{(0)}(t) + \\ \qquad h_i \lambda_i^{(0)} f_i' H_0^{(1)}(t) + h_i \lambda_i^{(1)} f_{i+1}' H_1^{(1)}(t). \end{cases} \tag{5}$$

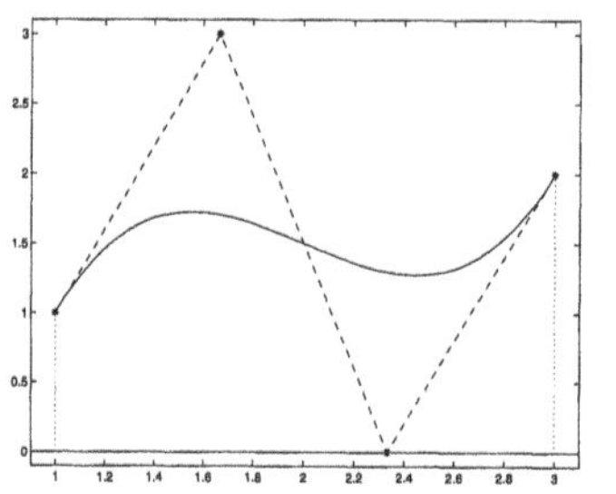
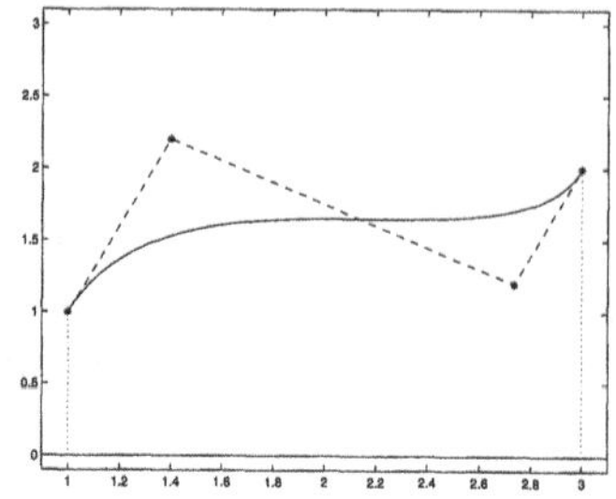

Fig. 9. Behavior of $\mathcal{C}(t;\lambda_i^{(0)},\lambda_i^{(1)})$ and of its control polygon.
Left: $\lambda_i^{(0)}=\lambda_i^{(1)}=1$. Right: $\lambda_i^{(0)}=0.6$, $\lambda_i^{(1)}=0.4$.

From the construction it immediately follows that the curve (5) passes through $\begin{pmatrix} x_{i+j} \\ f_{i+j} \end{pmatrix}$ and has tangent vectors $\lambda_i^{(j)} h_i \begin{pmatrix} 1 \\ f'_{i+j} \end{pmatrix}$, $j=0,1$, at the end-points. Moreover, since the first two elements of the cardinal Hermite basis are non-negative and sum to one, if $\lambda_i^{(0)}=\lambda_i^{(1)}=0$ the curve (5) reduces to a convex combination of the two end-points, that is to a segment. Then $\lambda_i^{(j)}$ determine the amplitude of the tangent vectors at the end-points, and act as **tension parameters** stretching the curve from the classical cubic Hermite interpolant $(\lambda_i^{(j)}=1)$ to the segment interpolating the data $(\lambda_i^{(j)}=0)$.

Since we are interested in functional and not in parametric interpolation, we note that $t \to x(t)$ is *invertible*, $t \in [0,1]$, $\forall\ 0<\lambda_i^{(0)},\lambda_i^{(1)} \leq 1$. Thus, for $0<\lambda_i^{(0)},\lambda_i^{(1)} \leq 1$, the image of $\mathcal{C}(t;\lambda_i^{(0)},\lambda_i^{(1)})$ is the graph of a *function*

$$s(x) := y(t(x)), \quad x \in [x_i, x_{i+1}].$$

Due to this construction, the function $s(x)$ is of class C^1 in the whole interval $[x_1, x_N]$, it interpolates the data and has tension properties. The shape of s is immediately controlled by the control points of the curve $\mathcal{C}$.

Summarizing, looking at the interpolating function as a cubic parametric curve allows us to obtain a simple interpolating functional tension scheme.

Such an approach has various advantages similar to those of the variable degree splines, such as immediate geometric meaning of the tension parameters and easy control of the graph via the control points. Moreover, it provides good approximation properties because the tension parameters can assume a continuous set of values.

On the other hand, the parametric setting itself is the main drawback of the scheme: the analytic expression of $s(x)$ is not explicitly known, (s lies on a branch of an algebraic curve of third degree and genus zero [64]). Thus, the evaluation of $s(\bar{x})$ at a given $\bar{x}$ requires the solution of a third degree equation. Due to the very simple structure of the x component this can be efficiently done by classical methods. Nevertheless, for a tabulation or a plot of s, the parametric form can be directly used with all the computational advantages of the parametric cubics.

This approach can be applied to any Hermite interpolation scheme reproducing $\mathbb{P}_1$ (the space of polynomials of degree less than or equal to 1).

4.1 Parametric Techniques: Bivariate Extension

The parametric approach can be immediately applied in the multivariate setting [48]. Let us consider a bivariate application. Assume λ_i are given parameters associated with the vertices $\mathbf{P}_i$. In order to apply the parametric approach to construct an element with tension properties it is enough:

a) to choose a C^1 interpolating (macro-)element reproducing $\mathbb{P}_1$,

b) to apply it componentwise and to consider the resulting parametric surface,

c) to check when the obtained surface is the graph of a function.

Following [20], let us consider for example the reduced CT cubic macro-element of Sect. 3.2 which reproduces the planes. As previously mentioned, this macro-element is completely determined by the values of the function and of its gradients at the three vertices. Such values determine the $\odot$ Bézier control points around the vertices (see Fig. 5, top). In order to use the parametric approach, the scheme must be applied componentwise, looking also at the x, y components as a function of the parameters which describe the surface. Then, labeling the vertices of the triangle as in Fig. 3 and introducing tension parameters λ_l, the 9 Bézier control points denoted by $\odot$ will be changed according to the following rule [20]:

$$\begin{pmatrix} \mathbf{P}_l \\ f_l \end{pmatrix} \to \begin{pmatrix} \mathbf{P}_l \\ f_l \end{pmatrix},$$

$$\begin{pmatrix} \mathbf{P}_l \\ f_l \end{pmatrix} + \frac{1}{3}\begin{pmatrix} \mathbf{e}_r \\ \langle \nabla f_l, \mathbf{e}_r \rangle \end{pmatrix} \to \begin{pmatrix} \mathbf{P}_l \\ f_l \end{pmatrix} + \frac{\lambda_l}{3}\begin{pmatrix} \mathbf{e}_r \\ \langle \nabla f_l, \mathbf{e}_r \rangle \end{pmatrix},$$

where $l = i, j, k$ and $\mathbf{e}_r$ denotes any edge of the triangle crossing $\mathbf{P}_l$. The remaining control points are determined assuming a linear variation of the normal derivative and imposing C^1 continuity (see also Fig. 10, top).

It is immediate to see that if $\lambda_i = 1$, the classical CT cubic macro-element is obtained. If $\lambda_i = 0$, all the control points belong to the plane interpolating the data at the vertices of the triangle, so the surface reduces to the same plane. Hence, $\lambda_i^{(j)}$ act as tension parameters stretching the surface from the classical CT cubic macro-element $(\lambda_i^{(j)} = 1)$ to a plane $(\lambda_i^{(j)} = 0)$.

The values of the tension parameters λ_i can vary at the vertices of the same triangle allowing a complete localization of the tension around the considered vertex (see Fig. 10).

In the bivariate setting the approach exhibits the same advantages and drawbacks as in the one-dimensional case. However, due to the increased complexity, we can say that the advantages become more interesting. Similarly, the drawbacks become more troublesome.

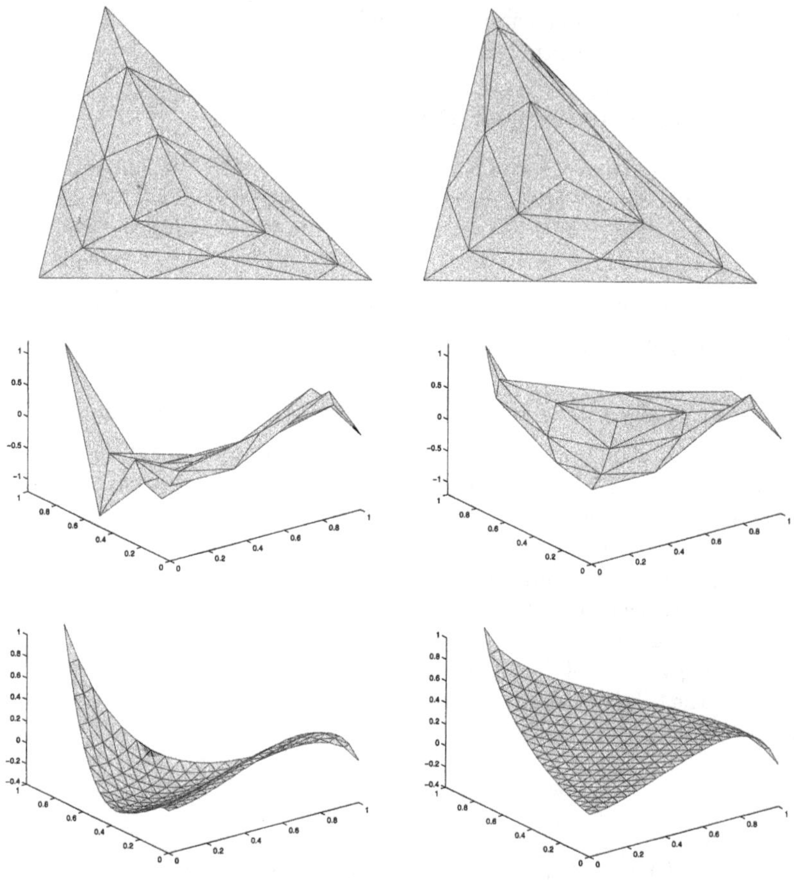

Fig. 10. The parametric CT macro-element. Top to bottom: the projection onto the x, y plane of the Bézier net, the Bézier net and the macro-element for different values of the tension parameters at the various vertices. The vertices are labeled as in Fig. 3. Left: $\lambda_i = \lambda_j = 1,\ \lambda_k = 0.5$. Right: $\lambda_i = 0.3,\ \lambda_j = 1,\ \lambda_k = 0.5$.

So far we have described elements based on the classical CT macro-element in order to stress the analogy to cubics in the one-dimensional case. As far as the parametric approach is concerned, we remark that a construction based on the classical quadratic macro-element over the Powell–Sabin split ([55]) of a triangle offers a very interesting alternative. In particular, it is interesting from a computational point of view, since only quadratics are involved, even though its graphical performances are completely comparable with those of the parametric cubic macro-element [48].

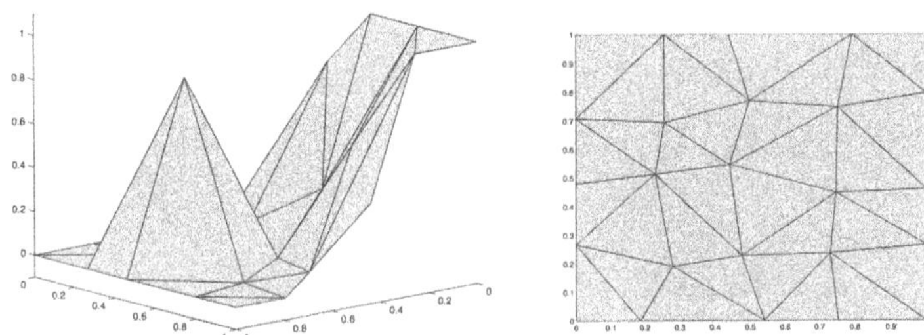

Fig. 11. The data and the considered triangulation.

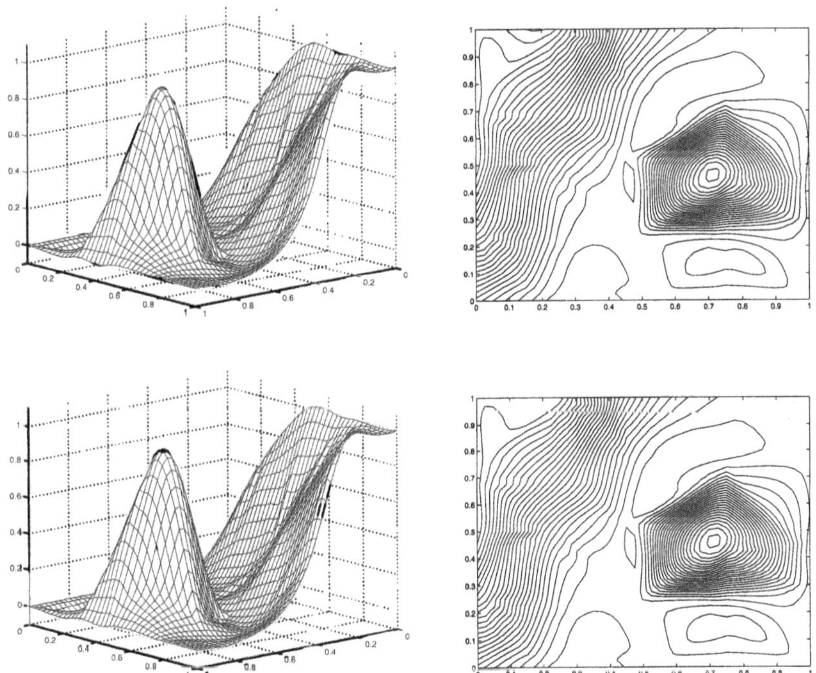

Fig. 12. Interpolant based on the side-vertex element and its contour lines. Top: exponentials with $\lambda_i = .1$ for the edges and the interior. Bottom: rationals with $w_i = 0$ for the edges and the interior.

Local tension schemes with higher smoothness can be obtained via the parametric approach ([48]) from classical macro-elements of class C^r, $r > 1$ as those proposed in [3,39,40].

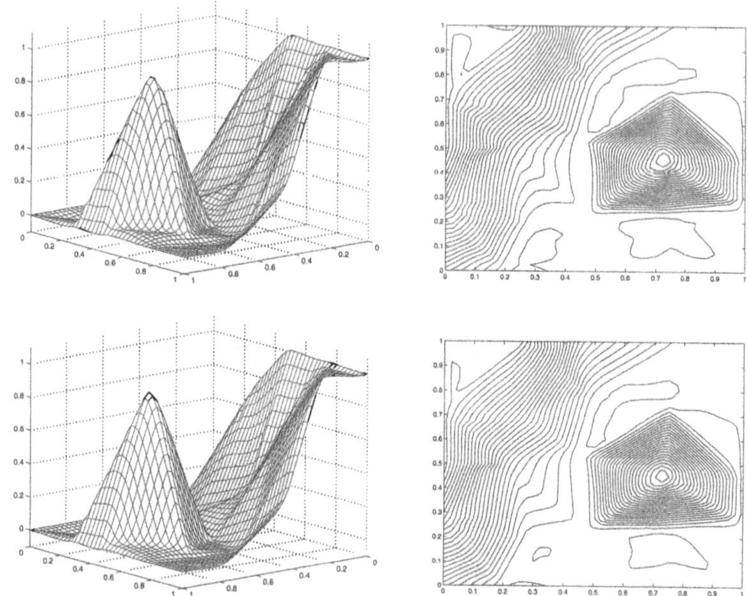

Fig. 13. Interpolant based on the side-vertex element and its contour lines. Top: exponentials with $\lambda_i = 10$ for the edges and $\lambda_i = 2$ in the interior. Bottom: rationals with $w_i = 7$ for the edges and the interior.

§5. Conclusions and Examples

In this section we will present the behavior of the described tension methods over a classical test in bivariate interpolation (see Figures 12–15). The data is obtained by sampling at 25 irregularly distributed points the so called Ritchie function [58] (see Fig. 11). The derivatives at the data points have been estimated according to [19]. The triangulation is depicted in Fig. 11. We emphasize that:

- for all local interpolation schemes the quality of the final interpolant strongly depends on the considered triangulation and the prescribed derivatives at the data points. This has motivated various papers on construction of data dependent triangulations and on estimation of derivatives (see for example [1,2,4,10,11,25,29]). Construction of suitable triangulations and efficient estimation of derivatives is beyond the goal of this paper, which is focused on the tension methods. We only recall that tension methods based on exponential splines have also been used to construct derivatives from the data points [9,37,52].

- As can be seen in Figures 12–15, the various tension schemes have similar graphical performances because all approach the piecewise linear interpolant of the data for limit configuration of the tension parameters.

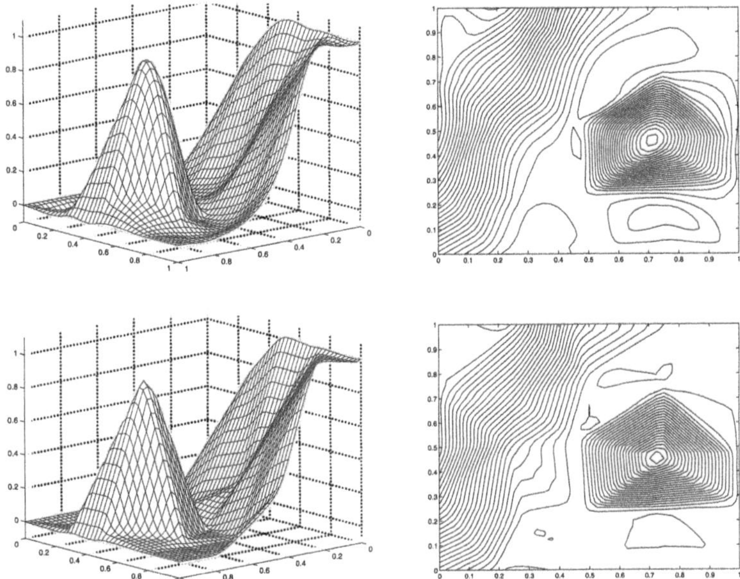

Fig. 14. Interpolant based on the variable degree CT macro-element and its contour lines. The same degree is associated to all the triangles according to [19]. Top: $n_i = 3$. Bottom: $n_i = 9$.

- The various tension schemes have different computational costs.

- Not all tension schemes allow an easy control of the shape via tension parameters: a geometric meaning of the tension parameters is a very important property in this context. Tension schemes based on variable degree polynomials or on the parametric approach possess an immediate geometric interpretation of the tension parameters. This allows us to easily control, via automatic algorithms, the shape (monotonicity, convexity, etc.) of the resulting interpolant along (lines parallel to) the edges of the *given* triangulation [19,20,21,48].

- Global convexity is a more arduous task: it requires the construction of a suitable triangulation and a consistent estimation of derivatives which require a global process (see for example [7,26,41,42,43,62]).

- The methods surveyed in the present paper are devoted to the construction of a *bivariate function* which interpolates (functional) scattered data points. For the enormous literature concerning the problem of constructing a valid two-manifold surface interpolating scattered data in three-dimensional space we refer for example to [10,32,44] and references quoted therein. For multivariate scattered data interpolation see for example [6].

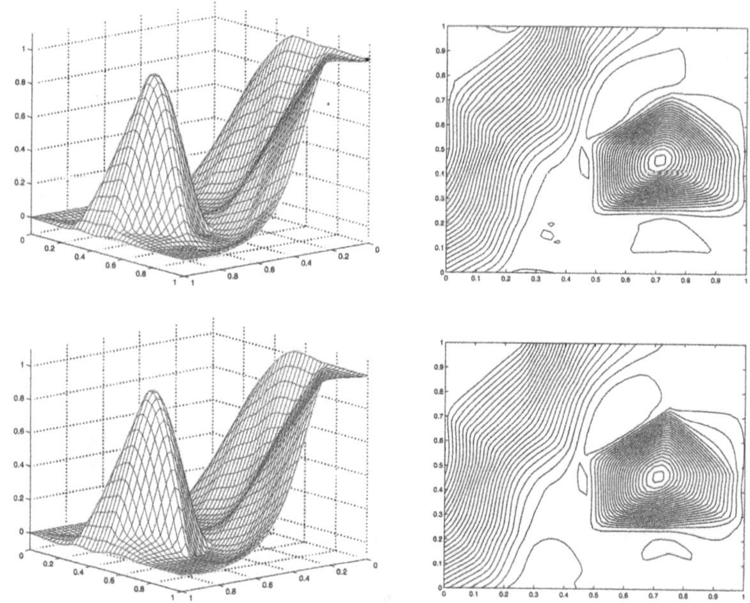

Fig. 15. Top: interpolant based on the variable degree CT macro-element and its contour lines. To each edge of the given triangulation is associated a possibly different degree ($n_i = 9$ or $n_i = 3$) according to [21] (see also Fig. 7, top). The local choice of the degrees is depicted in Fig. 16, left. Bottom: interpolant based on the parametric cubic CT macro-element with $\lambda_i = 1$ except for the circled points (Fig. 16 right) where $\lambda_i = .5$.

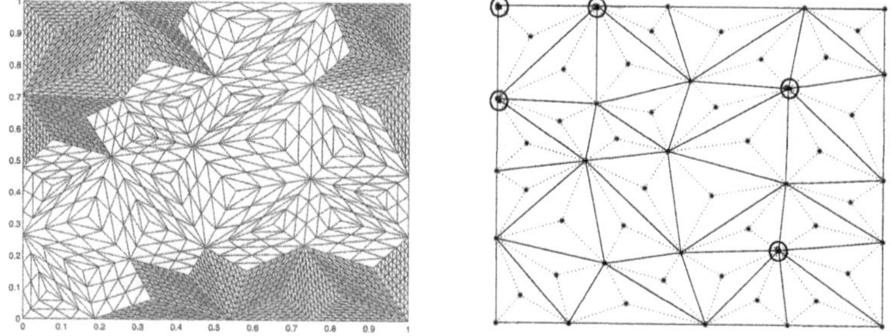

Fig. 16. Local choice of the tension parameters for Fig. 15. Left: projection onto the x, y plane of the "initial" macro net of the variable degree CT macro–element (see also Fig. 7, top). Right: distribution of the tension parameters for the parametric CT macro-element.

Acknowledgments. The author thanks Professor P. Costantini and Professor P. Sablonnière for their helpful comments and remarks.

References

1. Akima, H., On estimating partial derivatives for bivariate interpolation of scattered data, Rocky Mountain J. Math. **14** (1984), 41–52.
2. Alboul L., G. Kloosterman, C. Traas and R. van Damme, Best data-dependent triangulations, J. Comput. Appl. Math. **119** (2000), 1–12.
3. Alfeld, P., A bivariate C^2 Clough-Tocher scheme, Comput. Aided Geom. Design **1** (1984), 257–267.
4. Alfeld, P., Derivative generation from multivariate scattered data by functional minimization, Comput. Aided Geom. Design **2** (1985), 281–296.
5. Alfeld, P. and L. L. Schumaker, The dimension of bivariate spline spaces of smoothness r for degree $d \geq 4r+1$, Constr. Approx. **3** (1987),189–197.
6. Alfeld, P., Scattered data interpolation in three or more variables, in *Mathematical Methods in Computer Aided Geometric Design*, T. Lyche and L. L. Schumaker (eds.), Academic Press, London, 1989, 1–33.
7. Anderson, L. E., T. Elfving, G. Iliev and K. Vlachkova, Interpolation of convex scattered data in $\mathbb{R}^3$ based upon an edge convex minimum norm network, J. Approx. Theory **80** (1995), 299–320.
8. Baumgarten, C. and G. Farin, Smooth rational cubic spline interpolation, in *Mathematical Methods for Curves and Surfaces II*, M. Dæhlen, T. Lyche, and L. L. Schumaker (eds.), Vanderbilt University Press, Nashville, 1998, 17–24.
9. Bacchelli Montefusco, L. and G. Casciola, Algorithm 677: C^1 surface interpolation, ACM Trans. Math. Software **15** (1989), 365–374.
10. Bertram, M., J. C. Barnes, B. Hamann, K. J. Joy, H. Pottmann and D. Wushour, Piecewise optimal triangulation of scattered data in the plane, Comput. Aided Geom. Design **17** (2000), 767–787.
11. Bozzini, M. T. and M. Rossini, On a method of numerical differentiation, in *Curve and Surface Fitting: Saint-Malo 1999*, A. Cohen, C. Rabut, and L. L. Schumaker (eds.) Vanderbilt University Press, Nashville, 2000, 85–94.
12. Ciarlet, P.G., Interpolation error estimates for the reduced Hsieh–Clough–Tocher triangle, Math. Comp. **32** (1978), 335–344.
13. Cline, A. K., Scalar- and planar-valued curve fitting using splines under tension, Commun. ACM **17** (1974), 218–220.
14. Clough, R. W. and J. L. Tocher, Finite element stiffness matrices for analysis of plates in bending, in *Proc. Conf. on Matrix Methods in Structural Mechanics*, Wright–Patterson A.F.B., Ohio, 1965.
15. Costantini, P., On monotone and convex spline interpolation, Math. Comp. **46** (1986), 203–214.
16. Costantini, P., Boundary-valued shape-preserving interpolating splines, ACM Trans. Math. Software **23** (1997), 229–251.

17. Costantini, P., Curve and surface construction using variable degree polynomial splines, Comput. Aided Geom. Design **17** (2000), 419–446.

18. Costantini P., T. N. T. Goodman and C. Manni, Constructing C^3 shape–preserving interpolating space curves, Adv. Comp. Math. (2000) to appear.

19. Costantini, P. and C. Manni, On a class of polynomial triangular macro-elements, J. Comput. Appl. Math. **73** (1996), 45–64.

20. Costantini, P. and C. Manni, A parametric cubic element with tension properties, SIAM J. Numer. Anal. **36** (1999), 607–628.

21. Costantini, P. and C. Manni, A local shape–preserving interpolation scheme for scattered data, Comput. Aided Geom. Design **16** (1999), 385–405.

22. Costantini, P. and C. Manni, Interpolating polynomial macro-elements with tension properties, in *Curve and Surface Fitting: Saint-Malo 1999*, A. Cohen, C. Rabut, and L. L. Schumaker (eds.), Vanderbilt University Press, Nashville, 2000, 143–152.

23. Davis, P. J., *Interpolation and Approximation*, Blaisdell, Waltham, 1963.

24. Delbourgo, R. and J. A. Gregory, Shape preserving piecewise rational interpolation, SIAM J. Sci. Statist. Comput. **6** (1985), 967–976.

25. De Marchi, S., On computing derivatives for C^1 interpolating schemes: an optimization, Computing **60** (1998), 29–53.

26. Floater, M. S., Local and global convexity preservation, in *Mathematical Methods for Curves and Surfaces II*, M. Dæhlen, T. Lyche, and L. L. Schumaker (eds.), Vanderbilt University Press, Nashville, 1998, 183–190.

27. Foley, T. A. and H. Hagen, Advances in scattered data interpolation, Surv. Math. Ind. **4** (1994), 71–84.

28. Goodman, T. N. T., B. H. Ong, and M. L. Sampoli, Automatic interpolation by fair, shape preserving, G^2 space curves, Computer-Aided Design **30** (1998), 813–822.

29. Goodman, T. N. T., H. B. Said and L. H. T. Chang, Local derivative estimation for scattered data interpolation, Appl. Math. Comput. **68** (1995), 41–50.

30. Grandison, C., Behaviour of exponential splines as tensions increase without bound, J. Approx. Theory **89** (1997), 289–307.

31. Gregory, J. A. and M. Sarfraz, A rational cubic spline with tension, Comput. Aided Geom. Design **7** (1990), 1–13.

32. Hahmann, S. and G. P. Bonneau, Triangular G^1 interpolation by 4-splitting domain triangles, Comput. Aided Geom. Design **17** (2000), 731–757.

33. Hoschek, J. and D. Lasser, *Fundamentals of Computer Aided Geometric Design,* A. K. Peters, Wellesley, Massachusetts, 1993.

34. Kaklis, P. D. and D. G. Pandelis, Convexity-preserving polynomial splines of non-uniform degree, IMA J. Numer. Anal. **10** (1990), 223–234.

35. Kaklis, P. D. and M. I. Karavelas, Shape-preserving interpolation in $\mathbf{R}^3$, IMA J. Numer. Anal. **17** (1997), 373–419.

36. Koch, P. E. and T. Lyche, Interpolation with exponential B-splines in tension, in *Geometric modelling*, G. Farin et al. (eds), Springer-Verlag. Comput. Suppl. 8, 1993, 173–190.

37. Kolb, A. and H. P. Seidel, Interpolating scattered data with C^2 surfaces, Computer-Aided Design **27** (1995), 277–282.

38. Kvasov, B. I., Algorithms for shape preserving local approximation with automatic selection of tension parameters, Comput. Aided Geom. Design **17** (2000), 17–37.

39. Laghchim-Lahlou, M. and P. Sablonnière, Triangular finite elements of HCT type and class C^ρ, Advances in Comp. Math. **2** (1994), 101–122.

40. Laghchim-Lahlou, M. and P. Sablonnière, C^r finite elements of Powell-Sabin type on three direction mesh, Advances in Comp. Math. **6** (1996), 191–206.

41. Leung, N. K. and R. J. Renka, C^1 convexity-preserving interpolation of scattered data, SIAM J. Scient. Computing **20** (1999), 1732–1752.

42. Lorente-Pardo, J., P. Sablonnière, and M. C. Serrano-Perez, Subharmonicity and convexity properties of Bernstein polynomials and Bézier nets on triangles, Comput. Aided Geom. Design **16** (1999), 287–300.

43. Lorente-Pardo, J., P. Sablonnière, and M. C. Serrano-Perez, On the convexity of Bézier nets of quadratic Powell-Sabin splines on 12-fold refined triangulations, J. Comput. Appl. Math. **115** (2000), 383–396.

44. Mann, S., C. Loop, M. Lounsbery, D. Meyers, J. Painter, T. DeRose, and K. Sloan, A survey of parametric scattered data fitting using triangular interpolants, in *Curve and Surface Design* H. Hagen, (ed.), SIAM. Geometric Design Publications, Philadelphia, 1992, 145–172.

45. Manni, C., C^1 comonotone Hermite interpolation via parametric cubics, J. Comput. Appl. Math. **69** (1996), 143–157.

46. Manni, C. and M. L. Sampoli, Comonotone parametric Hermite interpolation, in *Mathematical Methods for Curves and Surfaces II*, M. Dæhlen, T. Lyche, and L. L. Schumaker (eds.), Vanderbilt University Press, Nashville, 1998, 343–350.

47. Manni, C., On shape preserving C^2 Hermite interpolation, BIT **14** (2001), to appear.

48. Manni, C., A general parametric framework for functional tension schemes, J. Comput. Appl. Math. **119** (2000), 275–300.

49. Marion, B. W. and J. W. Schmidt, Range restricted interpolation using Gregory's rational cubic splines, J. Comput. Appl. Math. **103** (1999), 221–237.

50. Marusic, M. and M. Rogina, Sharp error bounds for interpolating splines in tension, J. Comput. Appl. Math. **61** (1995), 205–223.

51. Nielson, G. M., The side-vertex method for interpolation in triangles, J. Approx. Theory **25** (1979), 318–336.

52. Nielson, G. M. and R. Franke, A method for construction of surfaces under tension, Rocky Mountain J. Math. **14** (1984), 203–220.

53. Ong B. H. and H. C. Wong, A C^1 positivity preserving scattered data interpolation scheme, in *Advanced Topics in Multivariate Approximation*, F. Fontanella, K. Jetter and P. J. Laurent (eds.), Series in Approximation and Decomposition, 8, World Scientific, Singapore, 1996, 259–274.

54. Pfluger, P. R. and R. H. J. G. Meyling, An algorithm for smooth interpolation to scattered data in $\mathbb{R}^2$, in *Mathematical Methods in Computer Aided Geometric Design*, T. Lyche and L. L. Schumaker (eds.), Academic Press, London, 1989, 469–480.

55. Powell, M. J. D. and M. A. Sabin, Piecewise quadratic approximations on triangles, ACM Trans. Math. Software **3** (1977), 316–325.

56. Powell, M. J. D., A review of methods for multivariable interpolation at scattered data points, in *The State of the Art in Numerical Analysis* I. S. Duff, et al. (eds.), Oxford Clarendon Press, 1997, 283–309.

57. Renka, R. J., Algorithm 716: TSPACK: Tension spline curve-fitting package, ACM Trans. Math. Software **19** (1993), 81–94.

58. Ritchie, S. I. M., Surface representation by finite elements, Ms thesis, University of Calgary, 1978.

59. Schaback, R., Adaptive rational splines, Constr. Approx. **6** (1990), 167–179.

60. Schumaker, L. L., Fitting surfaces to scattered data, in *Approx. Theory II*, G. G. Lorentz, C.K. Chui and L. L. Schumaker (eds.), Academic Press, New York, 1976, 203–268.

61. Schweikert, D. G., An interpolation curve using a spline in tension , J. Math. Phys. **45** (1966), 312–317.

62. Scott, D. S., The complexity of interpolating given data in three space with a convex function of two variables, J. Approx. Theory **42** (1984), 52–63.

63. Seymour, C. and K. Unsworth, Interactive shape preserving interpolation by curvature continuous rational cubic splines, J. Comput. Appl. Math. **102** (1999), 87–117.

64. Walker, R. J., *Algebraic Curves*, Dover, New York, 1962.

Carla Manni
Dipartimento di Matematica
Via Carlo Alberto 10
10123 Torino, Italy
manni@dm.unito.it

Data Reduction in Surface Approximation

Rossana Morandi and Alessandra Sestini

Abstract. Data reduction is a basic tool when a huge amount of data is given, and so it is extremely useful in surface approximation. In this paper we consider such a problem in the case of gridded data, and we propose the use of a two-stage approach in order to have a remarkable data reduction and, at the same time, a good approximation of the given data. In the first stage, a significant set of points is selected through a sampling strategy solving a suitable nonlinear system by means of the iterative Jacobi method. In the second stage, the least-squares approximating surface is computed using a "generalization" of the well known inverse multiquadric basis functions. Numerical results are given to show the performance of the method.

§1. Introduction

Sometimes people use the term "data reduction" for referring to the replacement of a given large set of data with a smaller significant one (preprocessing phase) which can be handled much more economically (e.g. [4]). Thus, any sampling strategy [8] which can be extended to the discrete case can be considered a data reduction tool. However, the term is often used with a more general meaning referring to any approximation scheme determining a good data approximation using few free parameters (e.g. [9,10,12]).

In this paper both these meanings are considered using a two-stage method to determine cheaply a good surface approximation of a large set of gridded data in $\mathbb{R}^3$ (in the general case of nonfunctional data, we assume that also a reasonable set of associated parameter values is assigned).

In the first stage, a generalization to discrete data sets of the sampling strategy proposed in [8] for the continuous case is considered. This strategy allows us to sample highly curved areas more densely than almost flat ones, and it requires a shape function which is used to measure the local curvedness of the surface underlying the data. Even if there are several possibilities [8], we have always used a suitable mean value of the principal curvatures for its definition because it seems suited also to the general parametric case here considered. The nonlinear Jacobi method is used for approximating the

Mathematical Methods for Curves and Surfaces: Oslo 2000
Tom Lyche and Larry L. Schumaker (eds.), pp. 315–324.

ISBN 0-8265-1378-6

solution of a nonlinear system based on this shape function and a result about its convergence is given.

In the second stage, a least-squares approximation scheme is tested for approximating the previously selected set of data using a "generalization" of the inverse multiquadrics [5]. In this approach, each basis function depends on three additional parameters as well as on the position of the associated knot. One of these allows us to associate a different decay to each basis function, and it is analogous to the variable shape parameter which has been profitably used in [6], relating to an interpolation problem. The other two parameters allow us to obtain a different stretching and rotation for each basis function. The numerical experiments have shown that the first parameter is useful almost always, but the other two parameters can be really useful only for sets of data which suggest highly stretched and rotated peaks or depths in the underlying surface. In such situations these "generalized" inverse multiquadric basis functions permit the determination of good approximations without using too many knots. This can facilitate using the suggestions given by the reference surface and by the shape function defined in the first stage for determining the number n of basis functions to be used for a set of data and also for selecting suitable values for the corresponding $5n$ parameters. The approach presented here can be considered as an alternative to that proposed in [3], where the 11 auxiliary free parameters used to obtain a parametric domain distortion and the knot locations ($2n$ parameters) are determined using a nonlinear optimization procedure.

The outline of the paper is as follows. In Section 2 the problem is illustrated, in Section 3 the sampling strategy is introduced and some theoretical results are given and in Section 4 the "generalized" inverse multiquadric basis functions used for computing the least-squares approximation are presented. Finally, in Section 5 some numerical results are displayed to confirm the interest of the method.

§2. The Problem

A large set Ω of gridded data points is given as

$$\Omega := \{\boldsymbol{D}_{ij} \in \mathbb{R}^3, i = 1, \ldots, Nu, \;\; j = 1, \ldots, Nv\},$$

together with an associated parameter set

$$\Omega_P := \{(u_i, v_j), i = 1, \ldots, Nu, \;\; j = 1, \ldots, Nv\}$$

defining a rectangular grid in the reference parameter domain $I_2 := [0,1]^2$. We want to construct an approximating parametric surface $\boldsymbol{S}^* = \boldsymbol{S}^*(\boldsymbol{W})$, $\boldsymbol{W} = (u, v)$ with the shape suggested by the data set Ω using a restricted set of data and few free parameters. To select a significant set of $N \ll Nu \times Nv$ data, we utilize the sampling strategy introduced in the next section.

§3. The Sampling Strategy

As well known, in surface sampling highly curved areas should be sampled more densely than almost flat ones, in order to obtain as much information as possible about the shape of the original surface using a sample of smaller size. Thus, in the first stage we construct a set $\bar{\Omega}$ of $N \ll Nu \times Nv$ points

$$\bar{\Omega} := \{\boldsymbol{P}_l \in \mathbb{R}^3, l = 1, \dots, N\},$$

which are denser where the data suggest highly curved areas of the underlying surface. For this aim, after computing a suitable set $\bar{\Omega}_P \subset I_2$ of N locations,

$$\bar{\Omega}_P := \{\boldsymbol{W}_l = (\bar{u}_l, \bar{v}_l) \in I_2, l = 1, \dots, N\},$$

each point is defined as $\boldsymbol{P}_l := \tilde{\boldsymbol{S}}(\boldsymbol{W}_l)$, where $\tilde{\boldsymbol{S}}(\boldsymbol{W})$ is a reference surface interpolant at the data. We remark that it is not necessary that the reference surface is very smooth but it has to preserve the shape of the data in order to obtain a significant sample. So an adequate reference surface is given for instance by the local bilinear interpolant at the data. Now, we assume that the four corners of I_2 belong to $\bar{\Omega}_P$ and that some locations are assigned on each edge of I_2. Let us denote

Λ_C the set of four indices associated to the corner locations,

Λ_B the set of indices associated to the other locations constrained to some edge of I_2,

Λ_I the set of indices associated to the inner positions,

Λ_l a suitable set of indices of locations neighboring to $\mathbf{W}_l$.

Then, the locations $\boldsymbol{W}_l$, $l = 1, \dots, N$, can be obtained by satisfying the boundary conditions and a particular system of nonlinear equations which can be considered a generalization of a lever system [8], that is

$$\sum_{i \in \Lambda_l} (1 + q\rho(\boldsymbol{W}_i))(\boldsymbol{W}_l - \boldsymbol{W}_i) = 0, \; l \in \Lambda_I, \tag{1}$$

$$\sum_{i \in \Lambda_l} (1 + q\rho(\boldsymbol{W}_i))(\bar{w}_l - \bar{w}_i) = 0, \; l \in \Lambda_B, \tag{2}$$

where q is a non negative parameter, $\bar{w}_l, \bar{w}_i, i \in \Lambda_l$ in (2) can be $\bar{u}_l, \bar{u}_i, i \in \Lambda_l$ or $\bar{v}_l, \bar{v}_i, i \in \Lambda_l$, (it depends on the edge to which $\boldsymbol{W}_l$ is constrained) and the function $\rho(\boldsymbol{W})$ is a shape function. This is defined as

$$\rho(\boldsymbol{W}) := \frac{\tilde{\rho}(\boldsymbol{W}) - \min \tilde{\rho}(\boldsymbol{W})}{\max \tilde{\rho}(\boldsymbol{W}) - \min \tilde{\rho}(\boldsymbol{W})},$$

where $\tilde{\rho}(\boldsymbol{W})$ is an interpolant at the data $(u_i, v_j, \tilde{\rho}_{ij})$, $i = 1, \dots, Nu$, $j = 1, \dots, Nv$, with

$$\tilde{\rho}_{ij} := \sqrt{\frac{(\tilde{\kappa}_1)_{ij}^2 + (\tilde{\kappa}_2)_{ij}^2}{2}}$$

and $(\tilde{\kappa}_1)_{ij}$ and $(\tilde{\kappa}_2)_{ij}$ discrete approximations of the principal curvatures at the data location (u_i, v_j) which can be computed using a standard finite difference approach. We note that $\rho(\boldsymbol{W})$ allows us to control the degree of spacing versus the curvedness in the locations of $\bar{\Omega}_P$ [8]. For other possible definitions of $\tilde{\rho}(\boldsymbol{W})$ see [7].

We can solve the system (1)-(2) by using the nonlinear Jacobi method. To prove its convergence we have used a matrix formulation of the system and a suitable ordering of the unknowns. For this purpose, we note that the number N_1 (N_2), of the u (v) unknowns in the system (1)-(2), is given by

$$N_1 := N - 4 - e^1, \quad (N_2 := N - 4 - e^2),$$

where $e^1 := |\{l \in \Lambda_B : u_l = 0, 1\}|$, ($e^2 := |\{l \in \Lambda_B : v_l = 0, 1\}|$). Thus, in matrix form, we can denote with $\mathbf{X}$ the vector of the unknowns in the system (1)-(2) which has $N_1 + N_2$ components and can be ordered as

$$\boldsymbol{X} := (\boldsymbol{U}^T, \boldsymbol{V}^T)^T, \quad \boldsymbol{U} := (u_{i_1^1}, \cdots, u_{i_{N_1}^1})^T, \quad \boldsymbol{V} := (v_{i_1^2}, \cdots, v_{i_{N_2}^2})^T,$$

where $1 \le i_k^r \le N, k = 1, \ldots, N_r, r = 1, 2$. Thus, after defining the following sets of indices

$$\delta\Lambda_p^1 := \{l \in \Lambda_p \cap (\Lambda_B \cup \Lambda_C) : u_l = 0, 1\},$$
$$\delta\Lambda_p^2 := \{l \in \Lambda_p \cap (\Lambda_B \cup \Lambda_C) : v_l = 0, 1\},$$

the system (1)-(2) can be rewritten as

$$A(\boldsymbol{X})\boldsymbol{X} = \boldsymbol{t}, \tag{3}$$

where $\boldsymbol{t} := ((\boldsymbol{t}^1)^T, (\boldsymbol{t}^2)^T)^T$, with $t_k^r := \sum_{p \in \delta\Lambda^r_{i_k^r}} (1 + q\rho(\boldsymbol{W}_{i_p^r})), k = 1, \ldots, N_r$, $r = 1, 2$. The matrix $A(\boldsymbol{X})$ is a block diagonal matrix, $A(\boldsymbol{X}) := diag(A^1(\boldsymbol{X}), A^2(\boldsymbol{X}))$ and $A^r(\boldsymbol{X})$ are the following square matrices of order $N_r, r = 1, 2$:

$$A_{kp}^r(\boldsymbol{X}) := \begin{cases} m_{i_k^r}^r + q\sum_{l \in \Lambda_{i_k^r} \setminus \delta\Lambda^r_{i_k^r}} \rho(\boldsymbol{W}_{i_l^r}), & \text{if } p = k, \\ -(1 + q\rho(\boldsymbol{W}_{i_p^r})), & \text{if } i_p^r \in \Lambda_{i_k^r} \setminus \delta\Lambda^r_{i_k^r}, \\ 0, & \text{otherwise}, \end{cases} \tag{4}$$

where $m_p^r := |\Lambda_p \setminus \delta\Lambda_p^r|$.

Now, we are ready to prove the convergence of the iterative nonlinear Jacobi method for q sufficiently small. For this we first need to prove the following result

Lemma 1. *For any $q \ge 0$, the matrix $A(\boldsymbol{X}) = A(q, \boldsymbol{X})$ is an $M-$matrix, provided that the sets of indices $\Lambda_l, l = 1, \ldots, N$ are suitably selected and the unknowns are suitably ordered.*

Proof: A suitable choice of the sets of indices Λ_l, $l = 1, \ldots, N$, and of the ordering of the unknowns makes $A(\boldsymbol{X})$ an irreducible diagonally-dominant matrix for any $q \geq 0$. In addition, $A^r(\boldsymbol{X})_{kk} > 0$, and $A^r(\boldsymbol{X})_{kp} \leq 0$, $\forall p \neq k, r = 1, 2$. As any irreducible diagonally-dominant matrix with positive diagonal entries and non-positive extra diagonal entries is an $M-$matrix [11], the lemma is proved. □

Now, the $k-$th iteration of the Jacobi method can be written as

$$\boldsymbol{X}^{(k+1)} = G(\boldsymbol{X}^{(k)}) = B(\boldsymbol{X}^{(k)})\boldsymbol{X}^{(k)} + \boldsymbol{t}, \tag{5}$$

where $B(\boldsymbol{X}) = I - D^{-1}(\boldsymbol{X})A(\boldsymbol{X})$ and I denotes the identity matrix of order $N_1 + N_2$ and $D(\boldsymbol{X}) := diag(A(\boldsymbol{X}))$. So, $B(\boldsymbol{X})$ is a block diagonal matrix, $B(\boldsymbol{X}) := diag(B^1(\boldsymbol{X}), B^2(\boldsymbol{X}))$ where $B^r(\boldsymbol{X}), r = 1, 2$ are matrices of order $N_r, r = 1, 2$ defined as

$$B^r_{kp}(\boldsymbol{X}) := \begin{cases} \frac{(1+q\rho(\boldsymbol{W}_{i^r_p}))}{m^r_{i^r_k} + q\sum_{l \in \Lambda_{i^r_k} \setminus \delta\Lambda^r_{i^r_k}} \rho(\boldsymbol{W}_{i^r_l})}, & \text{if } i^r_p \in \Lambda_{i^r_k} \setminus \delta\Lambda^r_{i^r_k}, \\ 0, & \text{otherwise.} \end{cases} \tag{6}$$

Thus, the following convergence theorem can be stated

Theorem 1. *If q is sufficiently small, the convergence of the vector sequence defined in (5) is guaranteed, provided that the sets of indices $\Lambda_l, l = 1, \ldots, N$, are suitably selected and the unknowns are suitably ordered.*

Proof: Lemma 1 states that, under the hypotheses of the theorem, the matrix A is an $M-$matrix for any $q \geq 0$ and this implies that $B(\boldsymbol{X}) = B(q, \boldsymbol{X})$ is a convergent matrix for any $q \geq 0$, [11]. Now, considering that $B(0, \boldsymbol{X})$ is a constant matrix, let us say $\bar{B}$, if q is sufficiently small we can write

$$B(q, \boldsymbol{X}) \simeq \bar{B} + q\delta B(\boldsymbol{X}),$$

where $\delta B(\boldsymbol{X})$ is a suitable variation matrix non depending on q. So, if q is sufficiently small, (5) implies that

$$G'(\boldsymbol{X}) \simeq \bar{B} + q\epsilon(\boldsymbol{X}),$$

where $\epsilon(\boldsymbol{X})$ is a suitable variation matrix non depending on q. Thus, the theorem is proved. □

§4. "Generalized" Inverse Multiquadric Least-squares Approximation

After determining $\bar{\Omega}_P$, and $\bar{\Omega}$, the approximating surface $\boldsymbol{S}^* = \boldsymbol{S}^*(\boldsymbol{W})$ is determined minimizing the quantity $\sum_{l=1}^{N} \| \boldsymbol{P}_l - \boldsymbol{S}(\boldsymbol{W}_l) \|_2^2$, among all the surfaces $\boldsymbol{S}(\boldsymbol{W}) := \boldsymbol{Q}_0 + \sum_{j=1}^{n} \boldsymbol{Q}_j b_j(\boldsymbol{W})$, with basis functions

$$b_j(\boldsymbol{W}) := \frac{1}{\sqrt{1 + \frac{\|\boldsymbol{W} - \boldsymbol{W}^*{}_j\|^2_{F_j}}{f_j^2}}},$$

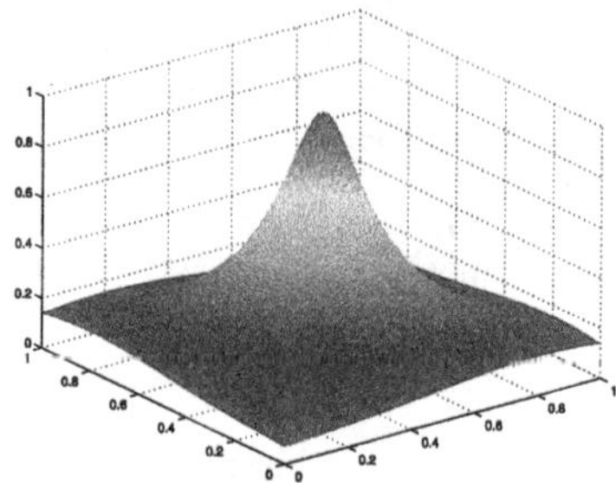

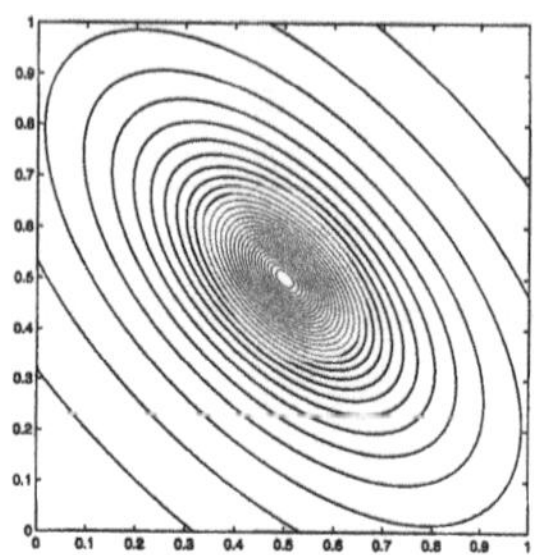

Fig. 1. A basis function and its contour lines. $\bar{\boldsymbol{W}}_j = (0.5, 0.5)^T, f_j = 0.1, \phi_j = \frac{\pi}{4}, \gamma_j = 2.$

where $\boldsymbol{W}^*{}_j \in \mathbb{R}^2$ is a knot associated to $b_j(\boldsymbol{W}), f_j$ is a positive parameter used to control its decay, and F_j is a symmetric and positive definite matrix of order 2 defined as

$$F_j := R_j^T \Gamma_j^2 R_j,$$

with

$$R_j := \begin{pmatrix} \cos\phi_j & \sin\phi_j \\ -\sin\phi_j & \cos\phi_j \end{pmatrix}, \quad \Gamma_j := \begin{pmatrix} 1 & 0 \\ 0 & \gamma_j \end{pmatrix}.$$

Each basis function $b_j(\boldsymbol{W})$ is then defined by 5 parameters, that is the two components of the knot $\boldsymbol{W}^*{}_j$, the coefficient f_j which is used to control its radial decay, and the coefficients γ_j and ϕ_j used for stretching and rotating, respectively. So, we deal with a "generalization" of the usual inverse multiquadric functions [5] which can be obtained as a particular case for $\gamma_j = 1$ and f_j equal to a constant value (related to the usual multiquadric parameter), $\forall\, j$.

It should be noted that, looking at the shape function $\rho(\boldsymbol{W})$, at its contour lines, and at the reference surface $\tilde{\boldsymbol{S}}(\boldsymbol{W})$ defined in the first stage, very useful suggestions for selecting an adequate n and suitable values of the necessary $5n$ coefficients can be given. Some more remarks about this important point are given at the end of Section 5.

We would like to conclude this section by observing that, using our "generalized" inverse multiquadric basis functions, we cannot be certain that the collocation matrix associated with the corresponding least-squares problem is of full rank. However, this is true also when standard inverse multiquadric basis functions are used if the knot locations are not constrained to $\bar{\Omega}_P$ (see e.g. [1,2]). Even if we have not yet considered this important theoretical aspect, we believe that the least-squares problem is much more robust than the corresponding interpolation problem and, in all the numerical tests, using reasonable values of the coefficients, we have had no trouble related to the rank of the collocation matrix.

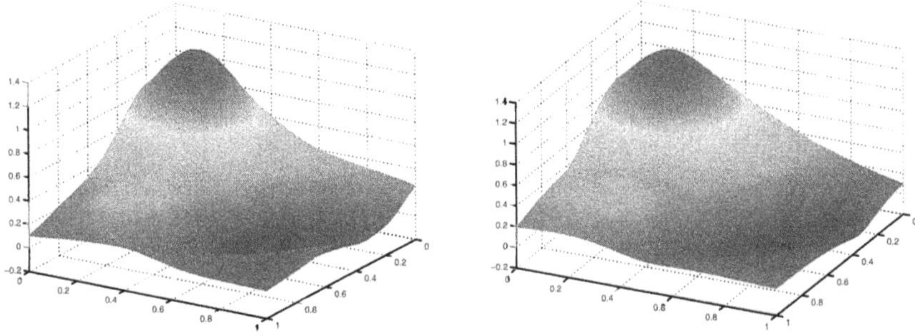

Fig. 2. Test 1. The parent surface (left) and its approximation (right).

§5. Numerical Results

In this section the results obtained with the proposed approach for three sets of data are presented. In all cases Ω is defined by sampling a test surface at 101×101 uniformly gridded locations in the reference domain I_2. As the data are gridded, in the first stage a coarser uniformly spaced grid of $N = mu \times mv$ locations is always used to initialize the vector $\mathbf{X}$. Both the u and v unknowns are ordered from the left to the right and from the bottom upwards and $\Lambda_i := \{i-1, i+1, i-mu, i+mu\}, i \in \Lambda_I$ (and is suitably modified if $i \in \Lambda_B$). Two functional surfaces are used in the first and in the second test, and a parametric one is used for the last. In the first stage always 30 iterations of the nonlinear Jacobi method have been enough to reach convergence, using $q = 10$ for the first and second test and $q = 10/3$ for the third. The reference surface $\tilde{\mathbf{S}}(\mathbf{W})$, necessary to define the set of data $\bar{\Omega}$, is always defined as the local bilinear interpolant at the data.

For each test, the figure on the left shows the parent surface used to define the set Ω, while on the right the final approximating surface is presented. The corresponding sets $\bar{\Omega}_P$ and the knot locations are also reported. In addition, for the second test the shape function and its contour lines are represented in Figure 5.

In the first test the parent surface is almost everywhere analogous to the standard Franke function except for the local depth on the right which is stretched to get a significant case for testing the benefit obtainable using the additional free parameters associated with the basis functions. The 100 locations of the set $\bar{\Omega}_P$ and the 5 positions of the knots are shown in Figure 3.

The parent surface related to the second test is the same used in [1] and [3], and is particularly suited to check the features of the method. In this case the number of the locations of the set $\bar{\Omega}_P$ is 196. The knots used to define the approximating surface are 7, and some of them are out of the domain I_2.

The third test refers to a parametric surface. $\bar{\Omega}_P$ has 100 elements, the knots are 13, 4 out of I_2. The approximating surface has been obtained using the parametric approach, that is the same values of the $5n$ necessary param-

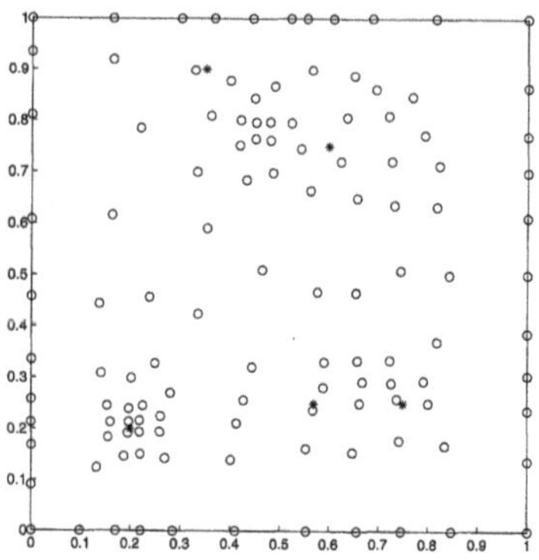

Fig. 3. Test 1. The locations belonging to $\bar{\Omega}_P$ ('O') and the knot locations ('*').

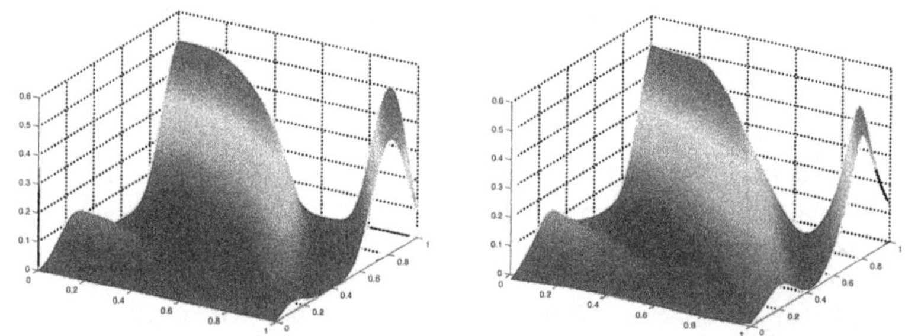

Fig. 4. Test 2. The parent surface (left) and its approximation (right).

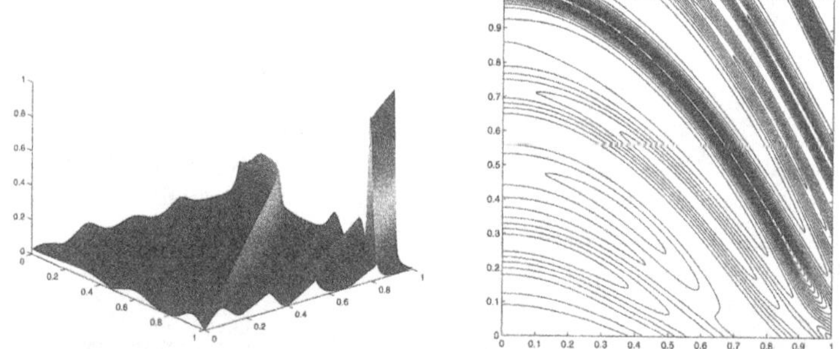

Fig. 5. Test 2. The shape function (left) and its contour lines (right).

eters have been used for all the components. Further improvements can be reached looking at an independent approximation of each component, even if clearly this implies a triple number of free parameters.

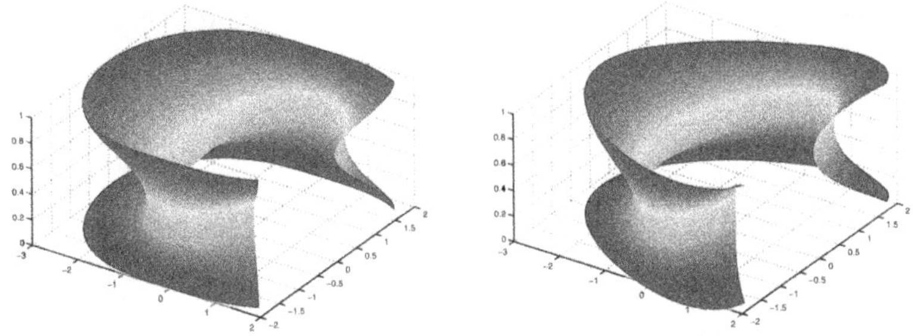

Fig. 6. Test 3. The parent surface (left) and its approximation (right).

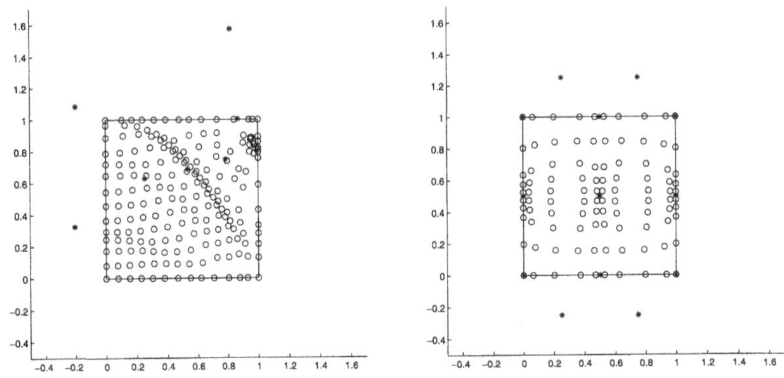

Fig. 7. The locations belonging to $\bar{\Omega}_P$ ('O') and the knot locations ('*'). Test 2 (left) and Test 3 (right).

References

1. Franke, R., Hagen H., and G. Nielson, Least squares surface approximation to scattered data using multiquadric functions, Advances in Comp. Math. **2** (1994), 81–99.
2. Franke, R., Hagen H., and G. Nielson, Repeated knots in least squares multiquadric functions, Computing Suppl. **10** (1995), 177–187.
3. Franke, R. and H. Hagen, Least squares surface approximation using multiquadrics and parametric domain distortion, Comput. Aided Geom. Design **16** (1999), 177–196.
4. Hamann, B. and J. L. Chen, Data point selection for piecewise linear curve approximation, Comput. Aided Geom. Design **11** (1994), 289–301.
5. Hardy R. L., Theory and applications of the multiquadric - biharmonic method, Comput. Math. Appl. **19** (1990), 163–208.
6. Kansa, E. J. and R. E. Carlson, Improved accuracy of multiquadric interpolation using variable shape parameters, Comput. Math. Appl. **24** (1992), 99–120.

7. Koenderink J. J. and A. J. van Doorn, Surface shape and curvature scales, Image and Vision Comput. **10** (1992), 557–564.
8. Li S. Z., Adaptive sampling and mesh generation, Computer-Aided Design **27** (1995), 235–240.
9. Lyche, T. and K. Mørken, A data reduction strategy for splines with applications to the approximation of functions and data, IMA J. Numer. Anal. **8** (1988), 185–208.
10. Morandi R., D. Scaramelli, and A. Sestini, A geometric approach for knots selection in convexity-preserving spline approximation, in *Curve and Surface Fitting: Saint-Malo 1999*, Albert Cohen, Christophe Rabut, and Larry L. Schumaker (eds.), Vanderbilt University Press, Nashville, 2000, 287–296.
11. Ortega J. M., *Matrix Theory*, Plenum Press, New York, 1988.
12. Saux, E. and M. Daniel, Data reduction of polygonal curves using B-splines, Computer-Aided Design **31** (1999), 507–515.

Rossana Morandi
Department of Energetics "Sergio Stecco"
Via Lombroso 6/17, 50134 Firenze
Italy
`morandi@de.unifi.it`

Alessandra Sestini
Department of Energetics "Sergio Stecco"
Via Lombroso 6/17, 50134 Firenze
ITALY
`sestini@de.unifi.it`

The Analytic Blossom

Géraldine Morin and Ron Goldman

Abstract. Blossoming is a powerful tool for studying and computing with Bézier and B-spline curves and surfaces – that is, for the investigation and analysis of polynomials and piecewise polynomials in geometric modeling. In this paper we define a notion of the blossom for Poisson curves. Poisson curves are to analytic functions what Bézier curves are to polynomials – a representation adapted to geometric design. As in the polynomial setting, the blossom provides a simple, powerful, elegant and computationally meaningful way to analyze Poisson curves. Here, we define the analytic blossom and interpret all the known algorithms for Poisson curves – subdivision, trimming, evaluation of the function and its derivatives, and conversion between the Taylor and the Poisson basis – in terms of this analytic blossom.

§1. Introduction

The blossom, or polar form, presented by Ramshaw [17,18,19] and de Casteljau [5] is a simple, elegant and powerful tool, very adapted to working with the Bézier representation of a polynomial. Not only can the dual functionals of the Bernstein basis be expressed very simply in terms of the blossom, but also algorithms like subdivision or change of basis can be explained and justified very easily using the multi-affinity and symmetry of the blossom [9]. The blossom is related as well to other effective tools for analyzing Bézier and B-splines curves and surfaces like the de Boor-Fix formula [9,1,4] and the Marsden identity [11,20,6]. More recently, blossoming, in various forms, has provided a method for extending Bézier curves to non-polynomial settings [2,16,7,10,12,13]. In this paper, we define the equivalent of the polynomial blossom for analytic functions.

Poisson curves are analytic functions expressed in the Poisson basis, a representation adapted to geometric modeling. With a blossom for analytic functions, the algorithms and tools available for Poisson curves have a new, simple and elegant interpretation, and, as in the polynomial setting, many of these algorithms follow simply from the multi-affinity and symmetry of the blossom. The purpose of this paper is to introduce the analytic blossom, and to interpret the algorithms and tools available for Poisson curves in terms of this blossom.

Mathematical Methods for Curves and Surfaces: Oslo 2000
Tom Lyche and Larry L. Schumaker (eds.), pp. 325–346.

ISBN 0-8265-1378-6

§2. The Polynomial Setting

In this section we shall briefly recall the classical definition of the blossom [17]. We shall also see that the blossom can be characterized by replacing the diagonal property with the dual functional property. In the next section, this result will be extended to the analytic setting.

To simplify our notation and to emphasize that a polynomial and its blossom are just different representations of the same underlying object, we will denote a function and its blossoms by the same name. For example, if P is a polynomial of degree n, then $P[x_1,\ldots,x_n]$ will denote the blossom of P evaluated at $(x_i)_{i=1}^n$. More generally, $P[x_1,\ldots,x_k]$ for $k \geq n$ will denote the blossom of P considered as a polynomial of degree k evaluated at $(x_i)_{i=1}^k$. (Note that when using the standard notation p for the blossom of P, the distinction between the polynomial blossoms of different degrees is also determined implicitly by the number of arguments, or domain of the function.) Later, we shall use $P[x_1,\ldots,x_n,0\ldots] = P[x_1,\ldots,x_n,0^\infty]$ to denote the analytic blossom of P.

2.1. The Polynomial Blossom and its Relation to Dual Functionals

Definition 1. *The blossom of a polynomial P of degree n is the unique symmetric, multi-affine, n-ary function that satisfies the diagonal property:* $P[\underbrace{x,\ldots,x}_{n\ times}] = P(x)$ *[17]. That is,*

- $P[x_1,\ldots,x_i,\ldots,x_j,\ldots,x_n] = P[x_1,\ldots,x_j,\ldots,x_i,\ldots,x_n]$,
- $P[..,(1-a)u+av,..] = (1-a)P[..,u,..] + aP[..,v,..]$,
- $P[\underbrace{x,\ldots,x}_{n\ times}] = P[x^n] = P[x]$.

In this paper, we use the notation a^kb^j to represent $\underbrace{a,a,\ldots,a}_{k\ times},\underbrace{b,\ldots,b}_{j\ times}$.

The blossom is a linear operator which is easy to compute. For example, the blossom of $P(t) = \binom{n}{k}t^k$ is $P[x_1,\ldots,x_n] = \sum_{i_a\neq i_b} x_{i_1}\ldots x_{i_k}$.

The dual functional property links Bézier curves to the blossom. Let $B_k^n(t)$ denote the kth Bernstein polynomial of degree n and P_k, $k = 0\ldots n$, the coefficients (or Bézier points) of the polynomial P in the Bernstein basis of degree n. Then the dual functional property asserts that

$$P[0^{n-k}1^k] = P_k.$$

The dual functional property is known to be an easy consequence the three blossoming axioms (see Definition 1). What is not generally appreciated is that this dual functional property can actually replace the diagonal property in the blossoming axioms.

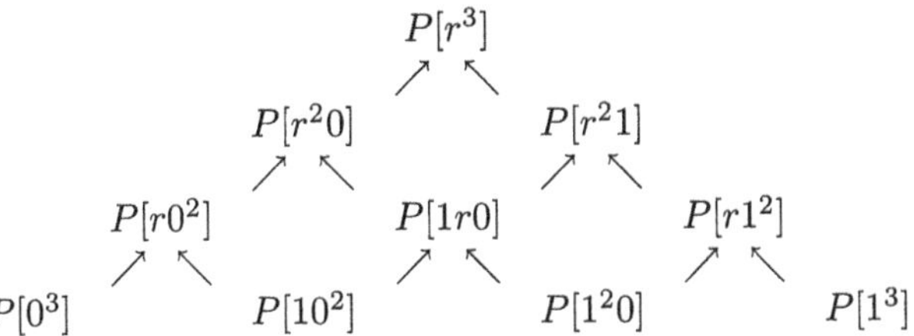

Fig. 1. The de Casteljau algorithm for a cubic Bézier curve P at a parameter r in $(0,1)$: The de Casteljau algorithm is both an evaluation and a subdivision algorithm. The parameter r is inserted at each level using the identity $r = (1-r)0 + r1$ and invoking the symmetry and multi-affinity property of the blossom. The input, the values of the blossom of P over the knot vector $(0^3 1^3)$, is at the base of the diagram. The value $P(r) = P[r^3]$ appears at the apex and the control points after one round of subdivision, appear along the left and right lateral edges.

Proposition 2. *The blossom of a degree n polynomial P is the unique symmetric, multi-affine, n-ary function characterized by the dual functional property:*

$$P_k = P[\underbrace{0,\ldots,0}_{n-k \text{ times}}, \underbrace{1,\ldots,1}_{k \text{ times}}] = P[0^{n-k}1^k], \text{ for all } k = 0,\ldots,n,$$

where the points $P_0,\ldots,P_n$ are the Bézier control points of the polynomial P over the interval $[0,1]$ – that is, the coefficients of P in the Bernstein basis.

Proof: By the definition of the Bézier control points,

$$P(t) = \sum_{k=0}^{n} P_k B_k^n(t). \tag{1}$$

Moreover, by the symmetry and multi-affinity of the blossom (Fig. 1 taking $r = t$)

$$P[t^n] = \sum_{k=0}^{n} P[0^{n-k}1^k] B_k^n(t). \tag{2}$$

If we assume that the blossom satisfies the diagonal property, then from (2)

$$P(t) = \sum_{k=0}^{n} P[0^{n-k}1^k] B_k^n(t).$$

So, from the uniqueness of the coefficients of P in the Bernstein basis,

$$P[0^{n-k}1^k] = P_k, \text{ for all } k = 0,\ldots,n.$$

Hence the diagonal property implies the dual functional property.

Conversely, if the blossom satisfies the dual functional property, then

$$P[0^{n-k}1^k] = P_k, \text{ for all } k = 0, \ldots, n.$$

Using these identities in (1) and (2), we obtain $P[t^n] = P(t)$. Hence the dual functional property implies the diagonal property. □

The knot vector $(0^n 1^n)$ corresponds to the Bernstein basis – that is, the blossom evaluated over this knot vector (at every n consecutive coordinates of the knot vector) provides the dual functionals for the Bernstein basis. The choice of the interval $[0,1]$ is arbitrary. The Bézier curve P can be parameterized over any interval $[a,b]$, in which case the dual functionals are given by $P_k = P[a^{n-k}b^k]$.

2.2. Polynomial Blossom and Algorithms for Bézier Curves

Not only can the control points be interpreted in terms of the blossom, but so too can algorithms for Bézier curves: in particular, subdivision, conversion between monomial and Bernstein form, and evaluation of the function and its derivatives [9]. For all these algorithms, both input and output can be expressed by the blossom of n-tuples. An algorithm, a way to compute the output from the input, then follows by applying the properties of the blossom.

For example, one step of subdivision splits a Bézier curve into two curves. The control points of these new curves are computed from the control points of the original curve. From the dual functional property, the original control points $(P_0, \ldots, P_n)$ over the interval $[0,1]$ are given by the blossom evaluated over the knot vector $(0^n 1^n)$ at every n consecutive coordinates of the knot vector. For subdivision at r, we compute the control points associated with the knot vectors $(0^n r^n)$ and $(r^n 1^n)$ from the control points associated with the knot vector $(0^n 1^n)$. Using the multi-affinity and the symmetry of the blossom, we can generate a dynamic programming algorithm for subdivision: this is the de Casteljau subdivision algorithm (Fig. 1).

2.3. Polynomial Blossom, Marsden Identity and de Boor-Fix Formula

The blossom expresses Bézier control points and algorithms in a simple and elegant way. Moreover, some other powerful tools for studying Bézier and B-splines curves, such as the Marsden identity [11] and the de Boor-Fix formula [4] are closely linked to the blossom [1,20]. We shall briefly review these results and their connections here, and we will extend these identities to the analytic blossom in Section 5.

Given two polynomials $P(u)$ and $Q(u)$ of degree n, define the bilinear form [6]:

$$[P(u), Q(u)]_n = \frac{1}{n!}\sum_{k=0}^{n}(-1)^{n-k}P^{(k)}(\tau)Q^{(n-k)}(\tau).$$

Notice that the right hand side is a constant independent of τ since its derivative is zero.

The next proposition expresses evaluation of the function, its derivatives and its blossom in terms of the bracket operator.

Proposition 3.

$$P(x) = [P(u), (x-u)^n]_n \tag{3}$$

$$P^{(k)}(x) = [P(u), \frac{d^k}{dx^k}(x-u)^n]_n \tag{4}$$

$$P[x_1, \ldots, x_n] = [P(u), (X-u)^n[x_1, \ldots, x_n]]_n = [P(u), (x_1-u)\ldots(x_n-u)]_n \tag{5}$$

Proof: The explicit computation of $[P(u), \frac{d^k}{dx^k}(x-u)^n]_n = \frac{n!}{(n-k)!}[P(u), (x-u)^{n-k}]_n$ leads to the Taylor expansion of $P^{(k)}(x)$ at τ. Equation (5) holds since the right hand side satisfies the three blossoming axioms: symmetry and multi-affinity in the x_i's are immediate from the definition and the diagonal property follows from (3). □

Equation (5) is exactly the de Boor-Fix expression for the blossom derived by Barry [1]. This explicit expression for the blossom provides an alternative proof of the existence of the blossom.

The function $(x-u)^n$ plays a key role with respect to this bracket operator [6]. Bracketing a polynomial P with $(x-u)^n$ not only reproduces P, but also bracketing P with the derivative or the blossom of $(x-u)^n$, reproduces the derivative or the blossom of P. The expression of this same function $(x-u)^n$ in the Bernstein basis is exactly the Marsden identity for the Bernstein basis [20]

$$(x-u)^n = \sum_{k=0}^{n}(-1)^{n-k}u^{n-k}(1-u)^k B_k^n(x).$$

This identity follows directly from the dual functional property of the polynomial blossom.

§3. The Analytic Blossom

Let F be an analytic function at zero. Then F admits not only a Taylor development at zero, but also a Poisson development at zero converging on $D(0,R)$, where R is the radius of convergence of the Taylor series and $D(0,R)$ is the open disk in $\mathbb{C}$ of center 0 and radius R. That is,

$$F(t) = \sum_{k\geq 0} P_k b_k(t), \quad t \in D(0,R),$$

where $b_k(t) = e^{-t}\frac{t^k}{k!}$ [14]. The coefficients (P_k) of F in the Poisson basis $(b_k(t))$ are called the Poisson control points of the curve F. Here, for simplicity, we always consider the Poisson development at zero. This is not a real restriction, since a Poisson curve can be trimmed [15]. That is, the Poisson representation of $F(a+\cdot)$ can be generated from the Poisson representation of F for an arbitrary parameter a in the interior of the domain of convergence.

The Poisson representation is to analytic functions what the Bézier representation is to polynomials. Many properties and algorithms generalize from Bézier to Poisson curves, e.g., Poisson curves follow the shape of the Poisson control polygon, since the convex hull and variation diminishing properties hold in the Poisson basis. There is also a subdivision algorithm for analytic functions, based on the Poisson representation [14].

The main goals of this paper are to define a blossom for analytic functions, establish its existence and uniqueness, investigate its main properties, and apply it to the study of algorithms for Poisson curves. In the remainder of this section, we shall first define the blossom of an analytic function, using three axioms similar to the axioms of the polynomial blossom, and we shall provide as well some simple examples of blossoms of analytic functions. Second, we shall prove the existence of the analytic blossom. We then show that this analytic blossom provides the dual functionals for the Poisson basis, and we prove that, just as in the polynomial setting, this dual functional property can replace the diagonal property in the definition of the blossom. Finally, we will use the dual functional property to establish the uniqueness of the analytic blossom.

3.1. Definition and Examples of the Analytic Blossom

Definition 4. *The blossom of a function F, analytic at zero, is the unique function that takes an infinite number of arguments almost all of which are zero, is symmetric, multi-affine, and satisfies a diagonal property. Thus, we have*

- $F[..,x_i,..,x_j,..] = F[..,x_j,..,x_i,..]$,
- $F[..,(1-a)u+av,..] = (1-a)F[..,u,..] + aF[..,v,..]$,
- $F[\underbrace{\frac{t}{n},\ldots,\frac{t}{n}}_{n \text{ times}},0,\ldots]$ *converges uniformly to $F(t)$ as n goes to ∞ on any disk $\bar{D}(0,b)$, where $b < R$, R is the radius of convergence of the Taylor expansion of F at zero, and $\bar{D}(0,b)$ is the closed disk of center 0 and radius b.*

Moreover, these three properties completely characterize the blossom – existence is proved in the next subsection and uniqueness at the end of the section. Here are some examples of analytic blossoms:

- if $F(t) = t$, then $F[x_1,\ldots,x_n,0,\ldots] = \sum_i x_i$,
- if $F(t) = \frac{t^2}{2}$, then $F[x_1,\ldots,x_n,0,\ldots] = \sum_{i\neq j} x_i x_j$.

More generally

- if $F(t) = \frac{t^k}{k!}$, then $F[x_1,\ldots,x_n,0,\ldots] = \sum_{i_a\neq i_b} x_{i_1}\ldots x_{i_k}$.

Finally, we have

- if $F(t) = e^t$, then
$F[x_1,\ldots,x_n,0,\ldots] = \sum_{k\geq 0}\sum_{i_a\neq i_b} x_{i_1}\ldots x_{i_k} = \prod_{i=1}^n (1+x_i)$.

Notice that, whereas the polynomial blossom of degree n is defined on $\mathbb{C}^n$, the analytic blossom is defined on the direct sum $\bigoplus_{i=1}^{\infty} C_i$, where $C_i = \mathbb{C}$ for all i and not on the direct product $\times_{i=1}^{\infty} C_i$.

3.2. Existence of an Analytic Blossom

The main point of this section is to establish the existence of an analytic blossom satisfying the three axioms of the definition. This proof of existence is based on a relation between the analytic and polynomial blossoms. In particular, a polynomial P has both an analytic and polynomial blossoms.

We start with the following lemma. This result is crucial to proving both the existence and the uniqueness of the analytic blossom.

Lemma 5. *Let $F(t)$ be an analytic function and let R be the radius of convergence of its Poisson series $\sum P_k b_k(t)$ at zero. Then as $n \to \infty$,*

$$\sum_{k=0}^{n} P_k B_k^n(\frac{t}{n}) \text{ converges uniformly to } \sum_{k\geq 0} P_k b_k(t) \text{ on } \bar{D}(0,b),$$

where $b < R$.

Proof: First let us prove that $B_k^n(\frac{t}{n})$ converges uniformly to $b_k(t)$ on $\bar{D}(0,b)$. By definition

$$\begin{aligned} B_k^n(\frac{t}{n}) &= \binom{n}{k} (\frac{t}{n})^k (1-\frac{t}{n})^{n-k} \\ &= \frac{n\dots(n-k+1)}{n^k}\frac{t^k}{k!}(1-\frac{t}{n})^n(1-\frac{t}{n})^{-k}. \end{aligned}$$

But, $\frac{n\dots(n-k+1)}{n^k}$ converges to 1, and the functions $(1-\frac{t}{n})^n$ and $(1-\frac{t}{n})^{-k}$, for any fixed k, converge respectively to e^{-t} and 1 uniformly on $\bar{D}(0,b)$. Therefore $B_k^n(\frac{t}{n})$ converges uniformly to $b_k(t)$ on $\bar{D}(0,b)$. Now, consider

$$\begin{aligned} &\left|\sum_{k=0}^{n} P_k B_k^n(\frac{t}{n}) - \sum_{k\geq 0} P_k b_k(t)\right| \\ &\leq \underbrace{\left|\sum_{k=0}^{m} P_k \left(B_k^n(\frac{t}{n}) - b_k(t)\right)\right|}_{A} + \underbrace{\left|\sum_{k>m} P_k b_k(t)\right|}_{B} + \underbrace{\left|\sum_{k=m+1}^{n} P_k B_k^n(\frac{t}{n})\right|}_{C}. \end{aligned} \tag{6}$$

Because both $F(t)$ and $F(t)e^t$ are defined by their Taylor and Poisson series at 0 on $\bar{D}(0,b)$, for any ε we can choose m so large that

$$\sum_{k\geq m} |P_k \frac{t^k}{k!}| < \varepsilon \text{ and } \sum_{k\geq m} P_k |b_k(t)| < \varepsilon \tag{7}$$

for all $t \in \bar{D}(0,b)$. Thus, in particular, $B < \varepsilon$. Moreover, since $B_k^n(\frac{t}{n}) \longrightarrow b_k(t)$ as $n \to \infty$, for any fixed $0 \leq k \leq m$, there exists n_0 such that for all $n \geq n_0$: $A < \varepsilon$.

Finally, we need to bound C. First observe that for all $n \geq 0$, $0 \leq i \leq n$ and $t \in \bar{D}(0,b)$,

$$\left|(1-\frac{t}{n})^i\right| \leq e^b, \tag{8}$$

because

$$\left|(1-\frac{t}{n})^i\right| = \left|\sum_{j=0}^{i} \binom{i}{j} (\frac{-t}{n})^j\right| = \left|\sum_{j=0}^{i} \frac{i\ldots(i-j+1)}{n^j}\frac{(-t)^j}{j!}\right| \leq \sum_{j\geq 0}\left|\frac{(-t)^j}{j!}\right| \leq e^b.$$

Now

$$C = \left|\sum_{k=m+1}^{n} P_k \frac{t^k}{k!}\frac{n\ldots(n-k+1)}{n^k}(1-\frac{t}{n})^{n-k}\right|.$$

But $0 \leq n-k \leq n-m+1 \leq n$, so from (8) and (7)

$$C \leq e^b \sum_{m+1}^{n} \left|P_k \frac{t^k}{k!}\right| \leq e^b \varepsilon.$$

Since A, B and C can each be bounded uniformly on $\bar{D}(0,b)$, the result follows by (6). □

The next Lemma relates the analytic blossom of Poisson and Taylor basis functions respectively to the polynomial blossom of the Bernstein polynomials and the Taylor monomials.

Lemma 6.

$$b_k[x_1,\ldots,x_n,0^\infty] = B_k^n[x_1,\ldots,x_n] \tag{9}$$

$$\frac{X^k}{k!}[x_1,\ldots,x_n,0^\infty] = \binom{n}{k} X^k[x_1,\ldots,x_n] \tag{10}$$

Proof: First, we need to check that the right hand sides of these two identities are well defined. That is, for $x_n = 0$, we must have

$$B_k^{n-1}[x_1,\ldots,x_{n-1}] = B_k^n[x_1,\ldots,x_{n-1},0], \tag{11}$$

$$\binom{n-1}{k} X^k[x_1,\ldots,x_{n-1}] = \binom{n}{k} X^k[x_1,\ldots,x_{n-1},0]. \tag{12}$$

By the dual functional property, (11) is satisfied for $[x_1,\ldots,x_{n-1}] = [0^{n-1-j}1^j]$. It is then satisfied on any $[x_1,\ldots,x_n]$ by symmetry and multi-affinity. Moreover, since

$$X^k[x_1,\ldots,x_n] = \frac{1}{\binom{n}{k}} \sum_{i_a \neq i_b} x_{i_1}\ldots x_{i_k},$$

equation (12) holds. This proves that the two identities are well defined.

Now we need to check that the right hand sides of equations (9) and (10) are symmetric, multi-affine, and satisfy the diagonal property of the analytic blossom. The symmetry and multi-affinity properties follow from the corresponding properties of the polynomial blossom.

From the diagonal property of the polynomial blossom

$$B_k^n[\left(\frac{t}{n}\right)^n] = B_k^n(\frac{t}{n}).$$

But, from Lemma 5, $\lim_{n\to\infty} B_k^n(\frac{t}{n}) = b_k(t)$ and the convergence is uniform on $\bar{D}(0,b)$ for any $b > 0$. Thus the diagonal property holds for equation (9).

For the right hand side of (10), the diagonal property holds because

$$[\binom{n}{k} t^k][\left(\frac{t}{n}\right)^n] = \frac{n\ldots(n-k+1)}{k!}\left(\frac{t}{n}\right)^k,$$

which converges uniformly to $\frac{t^k}{k!}$ on $\bar{D}(0,b)$. □

The next lemma is the main result of this section; existence of the analytic blossom follows directly from it. Equations (13) and (14) express an analytic blossom of Poisson or Taylor series as the polynomial blossoms of Bernstein or Taylor polynomials.

Lemma 7.

$$(\sum_{k\geq 0} P_k b_k)[x_1,\ldots,x_n,0^\infty] = (\sum_{k=0}^{n} P_k B_k^n)[x_1,\ldots,x_n] \tag{13}$$

$$(\sum_{k\geq 0} A_k\frac{X^k}{k!})[x_1,\ldots,x_n,0^\infty] = \left(\sum_{k=0}^{n} A_k\binom{n}{k}X^k\right)[x_1,\ldots,x_n] \tag{14}$$

Proof: The proof of these two identities mimics the proof of Lemma 6. From Lemma 6 and by linearity, the right hand sides of these identities are well defined. Moreover, from the symmetry, multi-affinity and linearity of the polynomial blossom, the right hand sides of these equations are symmetric and multi-affine.

From the diagonal property of the polynomial blossom

$$(\sum_{k=0}^{n} P_k B_k^n)[\left(\frac{t}{n}\right)^n] = \sum_{k=0}^{n} P_k B_k^n(\frac{t}{n}),$$

and from Lemma 5, the sum on the right hand side converges uniformly to $\sum_{k\geq 0} P_k b_k(t)$ on $\bar{D}(0,b)$, where $b < R$ and R is the radius of convergence of the Poisson series. Thus, the diagonal property for the analytic blossom is satisfied, and therefore (13) holds. Finally, (14) follows from (13). From [14]

$$\sum A_k\frac{t^k}{k!} = \sum P_k b_k(t) \text{ where } P_k = \sum_{i=0}^{k}\binom{k}{i}A_k,$$

and from [8]

$$\binom{n}{i} t^i = \sum_{k=i}^{n} \binom{k}{i} B_k^n(t).$$

Thus

$$\begin{aligned}(\sum A_k \frac{X^k}{k!})[x_1, \ldots, x_n, 0^\infty] &= (\sum_{k=0}^{n} \left(\sum_{i=0}^{k} \binom{k}{i} A_i \right) B_k^n)[x_1, \ldots, x_n] \\ &= (\sum_{i=0}^{n} A_i \sum_{k=i}^{n} \binom{k}{i} B_k^n)[x_1, \ldots, x_n] \\ &= (\sum_{i=0}^{n} A_i \binom{n}{i} X^i)[x_1, \ldots, x_n]. \quad \square\end{aligned}$$

Lemma 7 gives explicit expressions for the analytic blossom of Poisson and Taylor series at zero in terms of the polynomial blossom. Thus, since any analytic function F at zero admits both a Poisson and a Taylor development, an analytic blossom certainly exists.

Note that the blossom of F could as well be defined by another family of polynomials (F_n) if the following two conditions hold:

- the uniform convergence property

$$F(t) = \lim_{n \to \infty} F_n(\frac{t}{n})$$

- the compatibility property verified in the proof of Lemma 6:

$$F_{n-1}[x_1, \ldots, x_{n-1}] = F_n[x_1, \ldots, x_{n-1}, 0].$$

We shall encounter another such family of functions in Section 5, where we provide an alternative proof of existence by exploiting the functions $F(t) = e^{-ut}$ and $F_n(t) = (1 - ut)^n$. The next lemma shows that the analytic blossom in n non-zero arguments is the blossom of the nth Poisson or Taylor partial sum.

Lemma 8.

$$(\sum_{k \geq 0} P_k b_k)[x_1, \ldots, x_n, 0^\infty] = (\sum_{k=0}^{n} P_k b_k)[x_1, \ldots, x_n, 0^\infty] \tag{15}$$

$$(\sum_{k \geq 0} A_k \frac{X^k}{k!})[x_1, \ldots, x_n, 0^\infty] = \sum_{k=0}^{n} A_k (\frac{X^k}{k!})[x_1, \ldots, x_n, 0^\infty] \tag{16}$$

Proof: These results follow immediately from Lemmas 6 and 7 by linearity. $\square$

Lemma 8 says that if we want to compute the analytic blossom of an infinite Poisson or Taylor expansion, we can always truncate to a finite sum. In effect, then, the linearity of the analytic blossom holds even for infinite series, since

$$b_k[x_1, \ldots, x_n, 0^\infty] = 0, \quad k > n,$$

$$X^k[x_1, \ldots, x_n, 0^\infty] = 0, \quad k > n.$$

3.3. The Dual Functional Property

We will now prove that, as in the polynomial case, the diagonal property can be replaced by a dual functional property in the definition of the blossom. This second characterization of the blossom in terms of the dual functional property will be very useful for many reasons; uniqueness will follow very easily from the diagonal property and, in the analytic case, the dual functional property is simpler, more intuitive, and often easier to apply than the diagonal property. The dual functional property also allows us to avoid issues of uniform convergence over subsets of $\mathbf{C}$.

The following lemma is the first step in proving the equivalence of the two definitions of the analytic blossom.

Lemma 9. *Let $F_n(t) = \sum_{k=0}^n P_k B_k^n\left(\frac{t}{n}\right)$ and $F(t) = \sum_{k\geq 0} P'_k b_k(t)$ on the open disk $D(0,R)$. If*

$$(F_n)_{n\geq 0} \text{ converges uniformly to } F(t) \text{ on } \bar{D}(0,b),\ 0 < b < R$$

then

$$P_k = P'_k, \text{ for all } k.$$

Proof: Since we know from Lemma 5 that $\sum_{k=0}^n P'_k B_k^n\left(\frac{t}{n}\right)$ converges uniformly to F on $\bar{D}(0,b)$, it is sufficient to prove that if $G_n(t) = \sum_{k=0}^n (P_k - P'_k) B_k^n\left(\frac{t}{n}\right)$ converges uniformly to $G(t) = 0$ on $\bar{D}(0,b)$, then $Q_k = P_k - P'_k = 0$ for all k.

The functions G_n are polynomials on $\mathbf{C}$ and therefore analytic on $\mathbf{C}$. The uniform convergence of a series of analytic functions (G_n) to an analytic function G (here the zero function) on $\bar{D}(0,b)$ implies the uniform convergence of the derivatives $(G_n^{(k)})$ to the derivative $G^{(k)}$ on the same domain $\bar{D}(0,b)$ [3].

Using this strong convergence property, we shall show by induction that $Q_i = 0$ for all i. First $Q_0 = 0$ because $G_n(0) = Q_0$ for any n and $G(0) = 0$. Thus, from the convergence hypothesis of the lemma, Q_0 converge to 0 – that is, $Q_0 = 0$. Suppose that $Q_i = 0$ for $i \leq m-1$. Then,

$$G_n(t) = \sum_{k=m}^{n} Q_k B_k^n\left(\frac{t}{n}\right) = \sum_{k=m}^{n} Q_k \binom{n}{k} (1 - \frac{t}{n})^{n-k} \left(\frac{t}{n}\right)^k$$

$$= t^m Q_m \binom{n}{m} (1 - \frac{t}{n})^{n-m} \frac{1}{n^m} + (\frac{t}{n})^{m+1} \sum_{k=m+1}^{n} Q_k \binom{n}{k} (1 - \frac{t}{n})^{n-k} \left(\frac{t}{n}\right)^{k-(m+1)}$$

Thus,

$$G_n^{(m)}(0) = Q_m m! \binom{n}{m} \frac{1}{n^m} = Q_m \frac{n \ldots (n-m+1)}{n^m}.$$

Since the sequence $\left(G_n^{(m)}(0)\right)$ converges to $G^{(m)}(0) = 0$ and

$$\lim_{n\to\infty} \frac{n \ldots (n-m+1)}{n^m} = 1,$$

$\lim_{n\to\infty} Q_m = 0$. Thus $Q_m = 0$. □

Proposition 10. *The blossom of an analytic function $F(t) = \sum_{k\geq 0} P_k b_k(t)$ is a symmetric, multi-affine function over an infinite number of arguments almost all of which are zero, characterized by the following dual functional property*

$$P_k = F[1^k 0^\infty] \text{ for all } k \geq 0.$$

Proof: Let G be a symmetric, multi-affine function. Then

$$G\left[\left(\frac{t}{n}\right)^n 0^\infty\right] = \sum_{k=0}^{n} G[1^k 0^\infty] B_k^n\left(\frac{t}{n}\right). \tag{17}$$

Now suppose that G is a blossom of F. Then from the diagonal property

$$G\left[\left(\frac{t}{n}\right)^n 0^\infty\right] \text{ converges to } F(t) = \sum_{k\geq 0} P_k b_k(t) \tag{18}$$

uniformly on any disk $\bar{D}(0,b)$ where $b < R$. From (17) and (18)

$$\lim_{n\to\infty} \sum_{k=0}^{n} G[1^k 0^\infty] B_k^n\left(\frac{t}{n}\right) = \sum_{k\geq 0} P_k b_k(t), \tag{19}$$

and this convergence is uniform on any disk $\bar{D}(0,b)$, where $b < R$ and R is the radius of convergence of $\sum P_k b_k(t)$. Thus, from (19) and Lemma 9,

$$P_k = G[1^k 0^\infty], \qquad \text{for all } k.$$

Conversely, if G satisfies the hypotheses of the proposition, we need to prove that G is a blossom of F – that is, G satisfies the diagonal property. Since $G[1^k 0^\infty] = P_k$ are the Poisson coefficients of the analytic function F, by Lemma 5, $\sum_{k=0}^n G[1^k 0^\infty] B_k^n(\frac{t}{n})$ converges to $\sum_{k\geq 0} G[1^k 0^\infty] b_k(t) = \sum_{k\geq 0} P_k b_k(t) = F(t)$ uniformly on $\bar{D}(0,b)$.

Thus, $G\left[(\frac{t}{n})^n 0^\infty\right] = \sum_{k=0}^n G[1^k 0^\infty] B_k^n(\frac{t}{n})$ (equation (17)) converges uniformly to $F(t)$ on $\bar{D}(0,b)$, so G satisfies the diagonal property and by definition G is a blossom of F. □

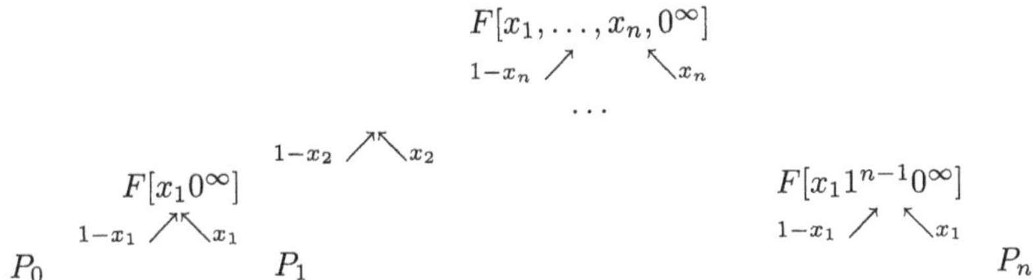

Fig. 2. The analytic blossom $F[x_1, \ldots, x_n, 0^\infty]$ can be computed from the Poisson control points of F by inserting x_i at level i and using the symmetry and multi-affinity properties.

Corollary 11. $b_k[1^j 0^\infty] = \delta_{jk}$.

3.4. Uniqueness of the Analytic Blossom

Now we have all the necessary tools to prove one of the main results of this paper: as in the polynomial case, the analytic blossom exists and is unique.

Theorem 12. *For any analytic function* $F(t) = \sum P_k b_k(t)$ *there exists a unique function that takes an infinite number of arguments almost all of which are zero, is symmetric, multi-affine and satisfies the diagonal property* $\lim_{n\to\infty} F\left[\left(\frac{t}{n}\right)^n 0^\infty\right] = F(t)$, *(or equivalently the dual functional property* $F[1^k 0^\infty] = P_k$*).*

Proof: Existence has been proved already, as a direct consequence of Lemma 7. Uniqueness follows from Proposition 10. The control points (P_k) are uniquely defined; therefore the values of any blossom at $[1^k 0^\infty]$ are fixed. Moreover, defining the blossom at $[1^k 0^\infty]$ determines it everywhere: by the multi-affinity and symmetry properties, the values of the blossom at $[x_1 1^k 0^\infty]$ are easily computed. By induction, so are the values at $[x_1, \ldots, x_n, 1^k 0^\infty]$ and hence too at $[x_1, \ldots, x_n, 0^\infty]$ for arbitrary x_i's (see Fig. 2). Therefore, the blossom of an arbitrary analytic function F exists and is unique. □

The next two results follow from Theorem 12. We shall use these results in the next section.

Corollary 13. *Let* $G(x) = F(rx)$. *Then*

$$G[x_1, \ldots, x_n, 0^\infty] = F[rx_1, \ldots, rx_n, 0^\infty]. \tag{20}$$

Proof: By uniqueness, it is sufficient to verify that the right hand side of (20) satisfies the three properties characterizing the blossom. The symmetry and multi-affinity properties of G follow from those of F. Moreover, for x in $D(0, R)$

$$G[\left(\frac{x}{n}\right)^n 0^\infty] = F[\left(\frac{rx}{n}\right)^n 0^\infty],$$

which by the diagonal property of F converges uniformly to $F(rx) = G(x)$ on any closed disk $\bar{D}(0,b)$, $\frac{b}{r} < R$. □

Corollary 14. *If $(P_k(r))$ denotes the set of control points of the function $G(t) = F(rt)$, then*

$$P_k(r) = F[r^k 0^\infty], \qquad \textit{for all } k.$$

Proof: $P_k(r) = G[1^k 0^\infty] = F[(r1)^k 0^\infty]$ by Corollary 13. □

As in the polynomial case, one can define a knot vector corresponding to the Poisson basis: $(\ldots 0\ 1 \ldots)$, the infinite vector containing first an infinite number of 0's, and then an infinite number of 1's. The Poisson control points are generated by the blossom evaluated over this knot vector; the first point is the blossom evaluated over all the zeros, and the following points are given successively by moving one position to the right which adds each time one more 1 to the set of arguments. Similarly, the points $(P_k(r))$ correspond to the knot vector $(\ldots 0\ r \ldots)$, which corresponds to the basis $b_k(\frac{t}{r})$. Indeed by definition

$$F(rt) = \sum P_k(r) b_k(t),$$

so by Corollary 14, we have

$$F(t) = \sum F(r^k 0^\infty) b_k(\frac{t}{r}).$$

This change of basis will be treated in the subdivision section. Other change of basis algorithms are treated in the subsequent sections.

§4. Algorithms for Poisson Curves

As in the polynomial setting, many of the algorithms acting on a Poisson curve can be expressed in terms of blossoming. When both the input and the output of the algorithm are blossom expressions, then the algorithm itself often follows from the multi-affinity and the symmetry of the blossom. Some convergence results, but not all, follow from the diagonal property.

4.1. Subdivision

The subdivision algorithm for Poisson curves given in [14] has a very elegant blossoming interpretation. At the first step of subdivision, the control points $(P_k(r))_{k\geq 0}$ of $F(rx)$, where $0 < r < 1$, are generated from the initial Poisson control points $(P_k)_{k\geq 0}$ of F; similarly, at the next stage of subdivision, the points $(P_k(\rho r))_{k\geq 0}$ are computed from the points $(P_k(r))_{k\geq 0}$, for any r and ρ in $(0,1)$. Since $P_k(r) = F[r^k 0^\infty]$ (Corollary 14), the first step of subdivision consists of getting from the knot vector $(\ldots 0\ 1 \ldots)$ to the knot vector $(\ldots 0\ r \ldots)$ – that is, from the control points of F in the Poisson basis $(b_k(t))$, to the control points of F in the basis $\left(b_k(\frac{t}{r})\right)_{k\geq 0}$; the next step goes from the knot vector $(\ldots 0\ r \ldots)$ to the knot vector $(\ldots 0\ \rho r \ldots)$. These steps can

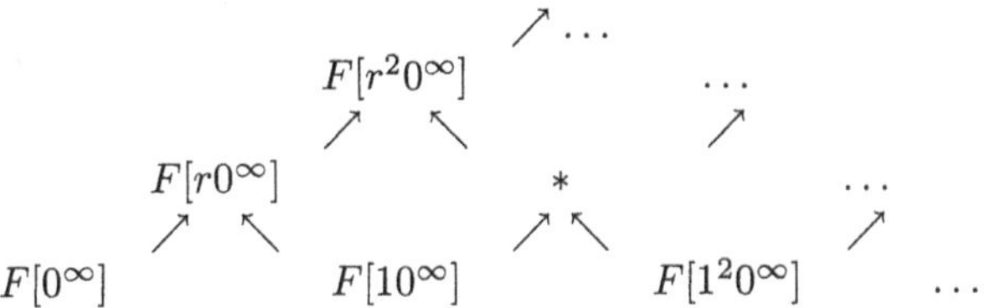

Fig. 3. The subdivision algorithm for Poisson curves. The Poisson control points are placed at the base, and the new control points, after one step of Poisson subdivision, appear along the left lateral edge of the diagram. This process can be iterated to execute further subdivision steps.

be computed easily using the multi-affinity and symmetry properties of the blossom (Fig. 3). This computation is exactly the algorithm proposed in [14] which corresponds, in the polynomial setting, to the left hand side of the de Casteljau algorithm (Fig. 1). It is shown in [14] that on any finite interval $[0, b]$, $b < R$, and R is the radius of convergence of F at 0, the control polygons defined by the control points $(P_k(r))_{k\geq 0}$ converge uniformly to the Poisson curve F as r goes to zero, although, as in the polynomial case, the convergence of the subdivision process does not follow from the blossom interpretation.

4.2. Trimming

The trimming algorithm for Poisson curves proposed in [15] also has a very nice blossom interpretation, from which, using the multi-affine and symmetry properties of the blossom, the trimming algorithm is retrieved. The aim of the trimming algorithm is to compute the Poisson control points $(P_0(a), P_1(a), \ldots)$ of the function $F(a+\cdot)$, where a is a fixed but arbitrary number in $[0, R)$ [15]. Each control point $P_k(a)$ is the limit of a family of control points $(P_k^n(a))_{n\geq 0} = (F[1^k(\frac{a}{n})^n 0^\infty])_{n\geq 0}$. The control points $(P_k^n(a))_{k\geq 0}$ not only converge to the control points of $F(a+\cdot)$, but also characterize a collection of functions F_a^n that converge uniformly to $F(a+\cdot)$ on any compact subinterval of the domain as n goes to infinity. The convergence of the points follows from the blossom interpretation. First, from the diagonal property,

$$\lim_{n\to\infty} P_0^n(a) = \lim F[(\frac{a}{n})^n 0^\infty] = F(a) = F(a+0) = P_0(a).$$

Second, $(P_1, P_2, \ldots)$ and $(P_1(a), P_2(a), \ldots)$ are the control points respectively of the functions $F + F'$ and $F(a+\cdot) + [F(a+\cdot)]'$ [15]. Since, by Proposition 10, $P_k = F[1^k 0^\infty] = (F + F')[1^{k-1}0^\infty]$, the blossom of $F + F'$ is

$$(F + F')[x_1, \ldots, x_n, 0^\infty] = F[1, x_1, \ldots, x_n, 0^\infty].$$

Thus, from the diagonal property

$$\lim_{n\to\infty} P_1^n(a) = \lim F[1(\frac{a}{n})^n 0^\infty] = \lim(F+F')[(\frac{a}{n})^n 0^\infty] = (F+F')(a) = P_1(a).$$

Finally, by induction, $\lim_{n\to\infty} P_k^n(a) = P_k(a)$. The convergence of the functions F_a^n, however, does not follow from the blossom interpretation. Since

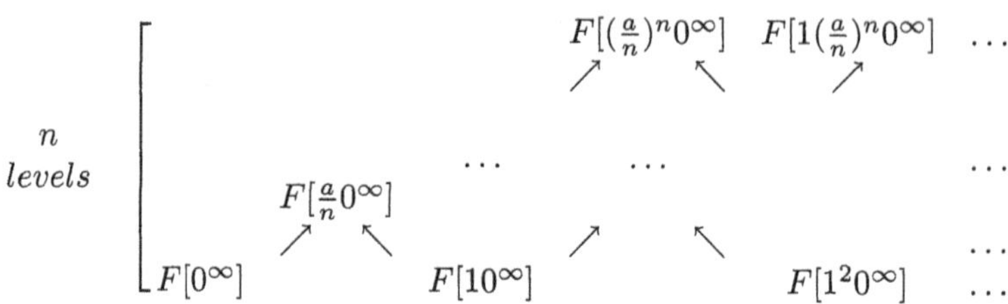

Fig. 4. The trimming algorithm for Poisson curves. Here, just like for subdivision, the computation is justified by the multi-affinity and symmetry of the blossom.

$P_k^n(a) = F[1^k \left(\frac{a}{n}\right)^n 0^\infty]$, the knot vector corresponding to the function F_a^n approximating $F(a+\cdot)$ is $(\ldots 0 \ \left(\frac{a}{n}\right)^n \ 1 \ldots)$. The algorithm for computing the control points of F_a^n is illustrated by Figure 4.

4.3. Evaluation Algorithms for Functions and Derivatives

The blossom can also characterize the control points of the derivatives $F^{(m)}$ of F, as in the polynomial setting [9]. We shall see, though, that the corresponding equalities from the polynomial setting differ by constant factors.

The analytic blossom is a polynomial in the non-zero parameters, so we can homogenize the analytic blossom; indeed the homogenization of the analytic blossom is similar to the homogenization in the polynomial setting. (Detailed descriptions of the homogeneous blossom can be found in [17,9].) Formally, the homogeneous blossom of an analytic function F is the unique symmetric, multi-linear function that when dehomogenized reduces to the multi-affine blossom of F, that is,

$$F[(x_1,1),\ldots,(x_n,1),(0,1)\ldots] = F[x_1,\ldots,x_n,0\ldots].$$

Let (P'_k) denote the Poisson control points of F'. Since $b'_k = b_{k-1} - b_k$,

$$\begin{aligned} P'_k &= P_{k+1} - P_k \\ &= F[1^{k+1}0^\infty] - F[1^k 0^\infty] \end{aligned} \tag{21}$$

From (21) and the linearity of the homogenized analytic blossom

$$P'_k = F[\delta 1^k 0^\infty] \text{ where } (1,1)-(0,1)=(1,0)=\delta.$$

By induction, if $(P_k^{(m)})_{k\geq 0}$ denotes the control points of $F^{(m)}$, then

$$P_k^{(m)} = F[\delta^m 1^k 0^\infty]. \tag{22}$$

Thus, more generally

$$F^{(m)}[x_1,\ldots,x_n,0^\infty] = F[\delta^m, x_1,\ldots,x_n,0^\infty]. \tag{23}$$

$F[\delta 0^\infty]$ $\quad$ $F[\delta 1 0^\infty]$ $\quad \ldots$

P_0' $\quad$ P_1'

$- \nearrow\nwarrow +$ $\quad$ $- \nearrow\nwarrow +$ $\quad \ldots$

P_0 $\quad$ P_1 $\quad$ P_2

$F[0^\infty]$ $\quad$ $F[10^\infty]$ $\quad$ $F[1^2 0^\infty]\ldots$

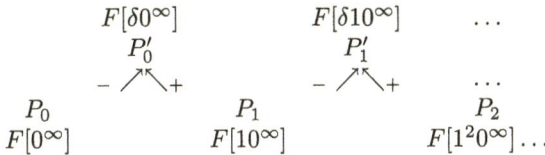

Fig. 5. The control points of the first derivative (upper level) are generated from the original control points (lower level). This process can be iterated to generate the control points of arbitrary derivatives of F.

Using equation (22) and the linearity of the homogeneous blossom, we can compute the control points of a derivative of any order of F from the Poisson control points of F (Fig. 5).

By first finding the control points of $F^{(m)}$ and then using the trimming algorithm, the sequence of points $F[\delta^m \left(\frac{t}{n}\right)^n 0^\infty]$ is generated. From (23) and the diagonal property

$$F^{(m)}(t) = \lim_{n\to\infty} F[\delta^m \left(\frac{t}{n}\right)^n 0^\infty]. \tag{24}$$

Thus we have a de Casteljau like algorithm to approximate F and any of its derivatives.

Thanks to the blossom expression for the derivative, the blossom characterizes the degree of continuity of two analytic curves meeting at 0, as in the polynomial setting [12]. That is,

Proposition 15. *If G and F are two analytic functions at 0, then the following statements are equivalent:*

- $F^{(k)}(0) = G^{(k)}(0)$ *for all* $0 \le k \le j$,
- $F[x_1, \ldots, x_j, 0^\infty] = G[x_1, \ldots, x_j, 0^\infty]$.

Proof: By (23), the first statement is equivalent to: $F[\delta^k 0^\infty] = G[\delta^k 0^\infty]$, for all $0 \le k \le j$. By the linearity of the homogenized blossom, this equality is equivalent to $F[1^k 0^\infty] = G[1^k 0^\infty]$, for all $0 \le k \le j$, since $\delta + 0 = 1$. This last equality is a particular case of and induces the second statement, since j new parameters can be introduced using the symmetry and multi-affinity of the blossom as in the proof of Theorem 12. □

4.4. Conversion between Poisson and Taylor Basis

In the polynomial setting, algorithms to convert from the Bernstein basis to the monomial basis and back are given in [9]. As in the polynomial setting, a scaled monomial basis, the Taylor basis $(\frac{x^k}{k!})$ corresponds to the knot vector $(\ldots 0\, \delta \ldots)$. Indeed, from (24), taking $t = 0$: $F^{(k)}(0) = F[\delta^k 0^\infty]$. Thus, if we denote by T_k the k-th Taylor coefficient of F, $T_k = F[\delta^k 0^\infty]$. Note that the scaling of the monomial basis in the analytic setting differs from the scaling in the polynomial setting, where the knot vector $(0^n\, \delta^n)$ corresponds to the basis $\left(\binom{n}{k} x^k\right)$.

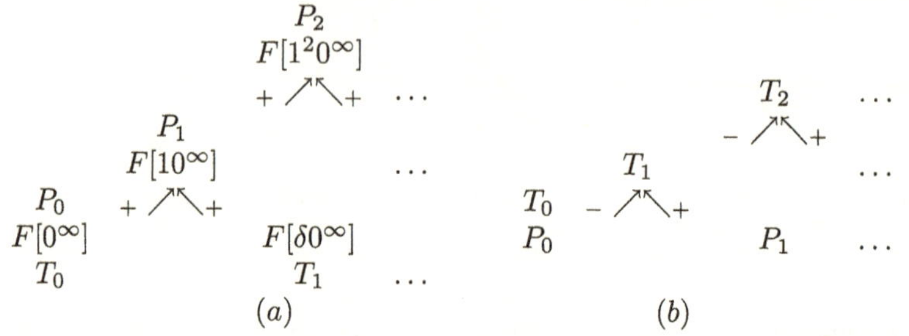

$$
\begin{array}{ccccc}
 & & P_2 & & \\
 & & F[1^2 0^\infty] & & \\
 & & + \nearrow\nwarrow + & \cdots & \\
 & P_1 & & & \\
 & F[10^\infty] & & \cdots & \\
P_0 & + \nearrow\nwarrow + & & & \\
F[0^\infty] & & F[\delta 0^\infty] & & \\
T_0 & & T_1 & \cdots & \\
 & & (a) & &
\end{array}
\qquad
\begin{array}{ccccc}
 & & & T_2 & \cdots \\
 & & & - \nearrow\nwarrow + & \\
 & & T_1 & & \cdots \\
T_0 & - \nearrow\nwarrow + & & & \\
P_0 & & & P_1 & \cdots \\
 & & (b) & &
\end{array}
$$

Fig. 6. On the left (a), the change of basis from Taylor coefficients (T_k) to Poisson coefficients (P_k). On the right (b) the change of basis from Poisson coefficients (P_k) to Taylor coefficients (T_k). The coefficients are computed using the linearity of the blossom because $\delta + 0 = 1$ i.e. $(1,0) + (0,1) = (1,1)$ and $1 - 0 = \delta$.

We can now apply the linearity and the symmetry of the homogenized blossom to compute the Taylor coefficients from the Poisson coefficients, and conversely, the Poisson coefficients from the Taylor coefficients (Fig. 6). Figure 6(a) is exactly the algorithm proposed in [14] to compute Poisson coefficients from Taylor coefficients.

§5. de Boor-Fix Formula, Marsden Identity and Blossoming for Analytic Functions

In [8] the authors present a Marsden identity and a bilinear bracket operator characterizing de Boor-Fix dual functionals for analytic functions. Here we will derive expressions for an analytic function and its derivatives, as well as for the blossom in term of this bracket operator.

Definition 16. *Let F be an analytic function at zero, with radius of convergence R, and let $G(u) = P(u)e^{-xu}$, where P is a polynomial and x is in $[0, R)$. Then*

$$[F(u), G(u)] = \sum_{k \geq 0} \frac{(-1)^k}{k!} F^{(k)}(0) G^{(k)}(0).$$

Proposition 18 will establish that the bracket operator on such functions is well defined – that is, that the right-hand side always converges.

The following proposition introduces a de Boor-Fix like expression for the analytic blossom. As in the polynomial case, this new expression provides an alternative proof of the existence of the analytic blossom.

Proposition 17.

$$F[x_1, \ldots, x_n, 0^\infty] = [F(u), \prod_{i=1}^{n} (1 - x_i u)] \tag{25}$$

Proof: First, note that the bracket operator between F, an analytic function at 0, and a polynomial P is well defined, since $[F(u), P(u)]$ is a finite sum.

To prove (25), we check the three blossoming axioms: symmetry, multi-affinity and the dual functional property. The expression $\prod_{i=1}^{n}(1 - x_i u)$ is certainly symmetric and multi-affine in the x_i's. Thus, so is the expression $[F(u), \prod_{i=1}^{n}(1 - x_i u)]$ by the linearity of the bracket operator. For the dual functional property, we need to prove

$$P_k = [F(u), (1-u)^k]. \tag{26}$$

In [14], an explicit expression for the Poisson control points in terms of the Taylor coefficients is given

$$P_k = \sum_{j=0}^{k} \tbinom{k}{j} F^{(j)}(0). \tag{27}$$

But

$$[F(u), (1-u)^k] = \sum_{j=0}^{k} \frac{(-1)^j}{j!} \frac{(-1)^j k!}{(k-j)!} F^{(j)}(0) = \sum_{j=0}^{k} \tbinom{k}{j} F^{(j)}(0). \tag{28}$$

Equation (26) follows from equations (27) and (28). □

From the blossom examples in Section 3, or by direct verification of the blossom axioms,

$$e^{-Xu}[x_1, \ldots, x_n, 0^\infty] = \prod_{i=1}^{n}(1 - u x_i). \tag{29}$$

Thus, the function e^{-xu} plays the same role for the analytic bracket operator $[.,.]$ as the function $(x-u)^n$ plays for the polynomial bracket operator $[.,.]_n$. The next proposition states this result more precisely.

Proposition 18. *If F is an analytic function with radius of convergence R at 0, then for any x in $[0, R)$,*

- $F(x) = [F(u), e^{-xu}]$,
- $F^{(p)}(x) = [F(u), \frac{d^p}{dx^p}(e^{-xu})]$,
- $F[x_1, \ldots, x_n, 0^\infty] = [F(u), e^{-Xu}[x_1, \ldots, x_n, 0^\infty]]$.

Proof: The right-hand side of the first two equations is exactly the Taylor development of $F(x)$ and $F^{(p)}(x)$ at 0. The last equation follows immediately from (25) in Proposition 17 and (29). □

By linearity and from Proposition 18 the series defining $[F(u), G(u)]$ for $G(u) = P(u)e^{-xu}$ converges, so the bracket operator is well defined.

Similar to the polynomial setting, the expansion of the reproducing function e^{-xu} in the Poisson basis generates a Marsden identity for analytic functions [8]

$$e^{-xu} = \sum_{k\geq 0}(1-u)^k b_k(x).$$

This Marsden identity follows easily from the dual functional property of the analytic blossom, since $e^{-xu} = \sum_{k\geq 0} E_k b_k(x)$, where $E_k = (e^{-Xu})[1^k 0^\infty] = (1-u)^k$ from (29).

§6. Conclusions and Open Questions

In this paper we have defined the analytic blossom and established its existence and uniqueness. Like the polynomial blossom for Bézier curves, the analytic blossom is a simple, powerful and elegant tool for analyzing Poisson curves. All the algorithms presented in Section 4 had already been developed in previous work, but each required a different approach. From the blossoming interpretation given here, these algorithms all follow directly from the multi-affinity and the symmetry of the blossom (or the linearity of the homogenized blossom).

Nevertheless, our definition of the blossom, and in particular the diagonal property, requires us to consider the domain $\mathbb{C}$, since the diagonal property requires uniform convergence on $\bar{D}(0,b)$. This convergence is used to prove the uniqueness of the blossom. However, proving directly that

$$\lim_{n\to\infty}\sum_{k=0}^{n} P_k B_k^n(\frac{t}{n}) = 0 \text{ (uniformly) on } [0,b], b>0 \Leftrightarrow P_k = 0 \text{ for all } k$$

would be sufficient to establish uniqueness and avoid variables in $\mathbb{C}$. We would like to know if this result is true.

We hope that the analytic blossom will help in developing new algorithms and novel tools for analyzing Poisson curves – that is, for investigating analytic functions. By considering the analytic blossom over more general knot vectors, it may even be possible to generate an analogue of B-splines for analytic functions. In order to generalize B-splines to the analytic setting, we may need first to extend the domain of the analytic blossom from the direct sum $\bigoplus_{i=1}^{\infty} C_i$, where $C_i = \mathbb{C}$, to the direct product $\times_{i=1}^{\infty} C_i$. We hope to address this extension of the analytic blossom in a future paper.

Acknowledgments. This work has been supported by NSF grant CCR-9971004 and DFG grant DFG Ka 477-22/1. The authors would like to thank the referee, Marie-Laurence Mazure, for pointing out the importance of uniform convergence in the diagonal property of the analytic blossom.

References

1. Barry, P. J., de Boor-Fix functionals and polar forms, Comput. Aided Geom. Design **7** (1990), 425–430.
2. Barry, P. J., de Boor-Fix dual functionals and algorithms for Tchebycheffian B-spline curves, Constr. Approx. **12(3)** (1996), 385–408.
3. Davis, P. J., *Interpolation & Approximation*, Dover Publications, Inc., New York, Second edition, 1975.
4. de Boor, C., and G. Fix, Spline approximation by quasi-interpolant, J. Approx. Theory **8** (1973), 19–45.
5. de Casteljau, P., Courbes et Surfaces à Pôles, André Citroën Automobiles SA, Paris, 1963.
6. Goldman, R., Dual polynomial bases, J. Approx. Theory **79(3)** (1994), 311–346.
7. Goldman, R., The rational Bernstein bases and the multirational blossoms, Comput. Aided Geom. Design **16** (1999), 701–738.
8. Goldman, R., and G. Morin, Poisson approximation, Proceedings of Geometric Modeling and Processing 2000, 141–149.
9. Goldman, R. N. and T. Lyche, *Knot Insertion and Deletion Algorithms for B-spline Curves and Surfaces*, SIAM, Philadephia, 1993.
10. Gonsor, D., and Neamtu, M., Non-polynomial polar forms, *Curves and Surfaces in Geometric Design*, P.-J. Laurent, A. Le Méhauté, and L. L. Schumaker (eds.), A. K. Peters, Wellesley MA, 1994, 193–200.
11. Marsden, M. J., An identity for spline functions with applications to variation-diminishing spline approximation, J. Approx. Theory **3** (1970), 7–49.
12. Mazure, M. L. and P. J. Laurent, Affine and non affine blossoms, in *Computational Geometry*, A. Conte, V. Demichelis, F. Fontanella, I. Galligani (eds.), World Scientific, 1993, 201–230.
13. Mazure, M. L., Blossoming of Chebychev Splines, *Mathematical Methods for Curves and Surfaces*, M. Dæhlen, T. Lyche, and L. L. Schumaker (eds.), Vanderbilt University Press, Nashville, 1995, 353–364.
14. Morin, G. and R. Goldman, A subdivision scheme for Poisson curves and surfaces, Comput. Aided Geom. Design **17(9)** (2000), 813–833.
15. Morin, G. and R. Goldman, Trimming analytic functions using right sided Poisson subdivision, Computer-Aided Design, to appear.
16. Pottman, H., The geometry of Tchebycheffian splines, Comput. Aided Geom. Design **10** (1993), 181–210.
17. Ramshaw, L., Blossoming: a connect-the-dots approach to splines, Techn. Rept. No.19, Digital Systems Research Center, Palo Alto, CA, 1987.

18. Ramshaw, L., Bézier and B-splines as multiaffine maps, Theoretical Foundations of Computer Graphics and CAD, NATO Advanced Study Institute Series, R. A. Earnshaw (ed), Springer-Verlag (Heidelberg), 1988, 757–776.

19. Ramshaw, L., Blossoms are polar forms, Comput. Aided Geometric Design **6(4)** (1989), 323–358.

20. Stephanus, Y. and R. Goldman, Blossoming Marsden's identity, Comput. Aided Geom. Design **9** (1992), 73–84.

Géraldine Morin and Ron Goldman
Computer Science Department, Rice University
Houston, TX 77251-1892, USA
`{gege,rng}@cs.rice.edu`

Bézier Curves: Topological Convergence of the Control Polygon

Manuela Neagu, Emmanuelle Calcoen, and Bernard Lacolle

Abstract. In terms of distance (e.g., Hausdorff distance), the control polygon of a Bézier curve converges to the curve via de Casteljau subdivision. This convergence is widely studied in the literature. In this paper, we adopt a different point of view for the convergence problem. We study whether the control polygon preserves the topology of the associated curve. For this purpose, we deal with two main topological features of the Bézier curve, the existence of multiple points and the convexity.

§1. Introduction

The geometric properties of the Bézier curves need no longer be introduced in CAGD. For many applications, the convergence of the control polygon to the associated curve via de Casteljau subdivision is the most important of these properties. Namely, in many cases it is not necessary to deal with the curve itself, a "good enough" polygonal approximation suffices.

We are thus led to study how good an approximation the control polygon can become by subdivision. If this question is asked in terms of distance (e.g., Hausdorff distance), the problem has been well studied in the literature and the answer is satisfying, especially for convex Bézier curves. But the question above can be also addressed from a topological point of view, that is, we can study whether the control polygon has the same topological characteristics as its associated curve. This problem is an important one among the emerging challenges in computational topology, and as far as we know, it has never been treated.

The two topological features that we shall deal with in this paper are the existence of multiple points and the convexity (we recall that a convex curve is a curve which lies on the boundary of its convex hull). Thus, there are three questions to be answered:

- can one associate a self-intersecting control polygon to any self-intersecting Bézier curve?

Mathematical Methods for Curves and Surfaces: Oslo 2000
Tom Lyche and Larry L. Schumaker (eds.), pp. 347–354.

ISBN 0-8265-1378-6

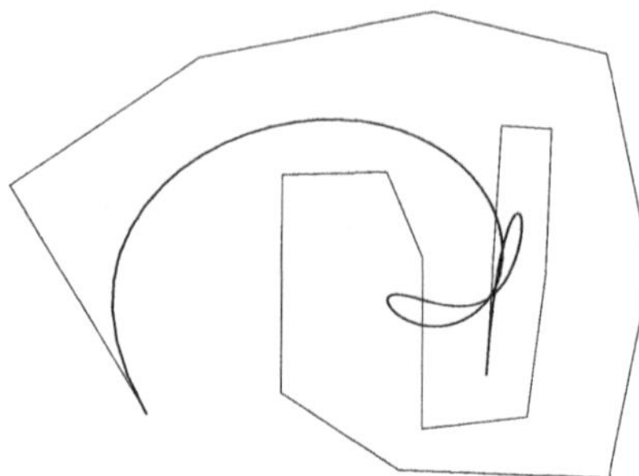

Fig. 1. Self-intersecting Bézier curve with non-self-intersecting control polygon.

- can one associate a non-self-intersecting control polygon to any non-self-intersecting Bézier curve?
- can one associate a convex control polygon with any convex Bézier curve?

We note that a non-convex Bézier curve cannot have a convex control polygon.

The topology-preserving control polygon that we are looking for should be obtained by successive de Casteljau subdivisions of the initial control polygon. Moreover, any control polygon obtained from it by subdivision must also be topology-preserving. We note that in this paper, we restrict our study to the case where the subdivision parameter is fixed and equal to $1/2$.

The layout of the paper is the following: In Section 2, we answer the first one of the three questions above. In Section 3, we briefly recall the definition and main properties of the hodograph of a Bézier curve. In Sections 4 and 5, we respectively address the two remaining questions. We remark that the answer to the second question depends on the regularity of the Bézier parameterisation of the curve. In Section 6, we give a few conclusions, as well as some remarks on the use of an adaptive subdivision parameter.

§2. Self-Intersecting Bézier Curve

From this point on, we shall say that a topological property of a control polygon $\mathcal{P}$ is subdivision invariant, or subdivision invariantly satisfied, if any control polygon obtained by subdivision from $\mathcal{P}$ satisfies this property.

In Figure 1, we present an example of a Bézier curve which has multiple points, and such that its control polygon is not self-intersecting. As the control polygon converges to the curve in terms of distance, the proof of the following lemma is straightforward:

Lemma 1. *Let $\mathcal{B}$ be a self-intersecting Bézier curve. Then we can subdivide $\mathcal{B}$ until its control polygon becomes subdivision invariantly self-intersecting.*

For the other two questions of the introduction, our discussion is based on the connection between the Bézier curve and its hodograph. Therefore, let us briefly recall the definition and main properties of the latter.

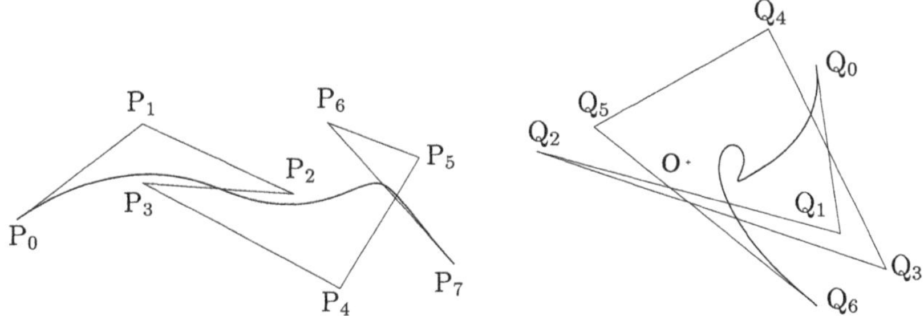

Fig. 2. A Bézier curve and its hodograph. O denotes the origin.

§3. Hodograph of a Bézier Curve

Let $\boldsymbol{c}$ be a $\mathcal{C}^1$ curve. Then its **hodograph** is the curve described by the endpoints of the tangent vectors of $\boldsymbol{c}$ translated to a common origin. Of course, the hodograph depends on the parameterisation of the curve. It is easy to see that the hodograph of a Bézier curve $\mathcal{B}$ of degree n is a Bézier curve $\mathcal{H}(\mathcal{B})$ of degree $n-1$. The control points of $\mathcal{H}(\mathcal{B})$ are the points

$$n\,(\mathrm{P}_{i+1} - \mathrm{P}_i) \qquad \text{for } i \in \{0, \ldots, n-1\},$$

where P_i are the control points of $\mathcal{B}$, $i \in \{0, \ldots, n\}$.

The properties of the hodograph that we are going to use do not depend on the scale at which it is represented. Thus, we can consider that the control polygon of $\mathcal{H}(\mathcal{B})$ is $\mathcal{Q} = \mathrm{Q}_0 \ldots \mathrm{Q}_{n-1}$, where $\mathrm{Q}_i = \mathrm{P}_{i+1} - \mathrm{P}_i$ for all $i \in \{0, \ldots, n-1\}$. In Figure 2, we present an example of Bézier curve and its associated hodograph.

Let us now suppose that we subdivide the curve $\mathcal{B}$ using a parameter t_0. We denote by $\mathcal{B}'$ and $\mathcal{B}''$ the two resulting Bézier curves. Then we can see that the hodographs of $\mathcal{B}'$ and $\mathcal{B}''$ are respectively the two Bézier curves obtained from $\mathcal{H}(\mathcal{B})$ by subdivision of parameter t_0.

Now let us answer the two remaining questions on the topological properties of the control polygon.

§4. Non-Self-Intersecting Bézier Curve

As in the previous case, it is well known that there exist Bézier curves which have no multiple points and such that their control polygon is self-intersecting. Such a curve is presented in Figure 3.

As mentioned in the introduction, the existence of a control polygon with no multiple points depends on the regularity of the Bézier parameterisation: If the derivative does not vanish, then there exists a topology-preserving control polygon; otherwise, there are cases where any control polygon obtained by subdivision of parameter $1/2$ is self-intersecting.

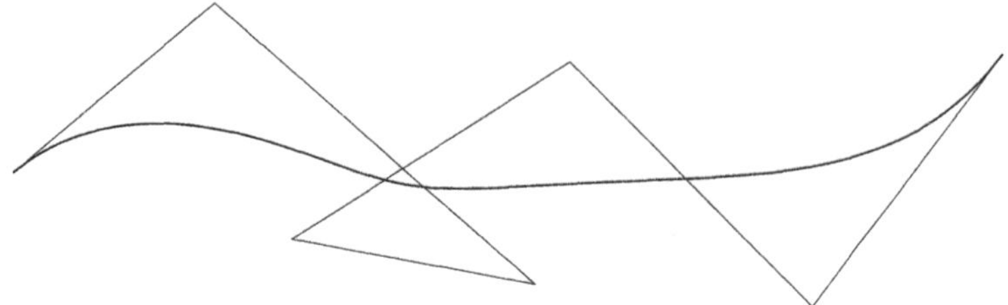

Fig. 3. Non-self-intersecting Bézier curve with self-intersecting control polygon.

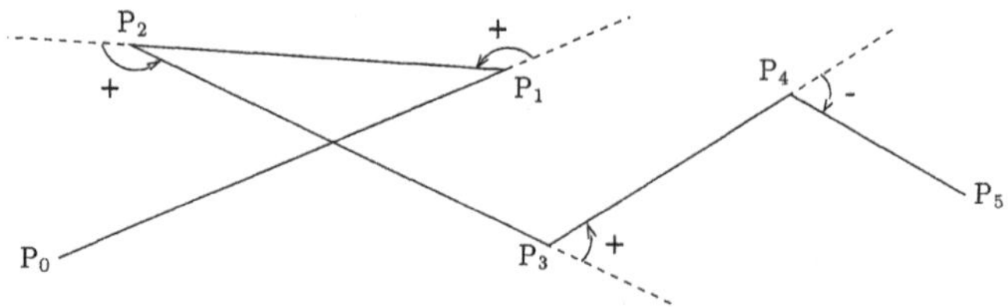

Fig. 4. Angles along a polygonal chain.

4.1. Regular Bézier Parameterisation

In order to prove the existence of a non-self-intersecting control polygon, let us first take a look at the angular properties of a self-intersecting polygonal chain.

Let $\mathcal{P} = \mathrm{P}_0\mathrm{P}_1 \dots \mathrm{P}_n$ be a polygonal chain. For $i \in \{1, \dots, n-1\}$, the angle $\widehat{\mathrm{P}}_i$ is the oriented angle between the oriented lines $\overrightarrow{\mathrm{P}_{i-1}\mathrm{P}_i}$ and $\overrightarrow{\mathrm{P}_i\mathrm{P}_{i+1}}$. If the segments $[\mathrm{P}_{i-1}\mathrm{P}_i]$ and $[\mathrm{P}_i\mathrm{P}_{i+1}]$ overlap, we set $\widehat{\mathrm{P}}_i = \pi$; thus, $\widehat{\mathrm{P}}_i \in (-\pi, \pi]$. In Figure 3, we show the signs of the angles associated with the vertices of a polygonal chain. Using this notation, it is easy to prove the following lemma.

Lemma 2. *If $\mathcal{P}$ is self-intersecting, then at least one of the two following properties is fulfilled:*

1) *there exists $k \in \{1, \dots, n-1\}$ such that $\widehat{\mathrm{P}}_k = \pi$;*
2) *there exist $k_1, k_2 \in \{1, \dots, n-1\}$, $k_1 < k_2$, such that $\sum_{i=k_1}^{k_2} \widehat{\mathrm{P}}_i \notin [-\pi, \pi]$.*

Now we can state the main result of this subsection.

Proposition 3. *Let $\mathcal{B}$ be a Bézier curve with no multiple points and regular Bézier parameterisation. Then we can subdivide it until its control polygon becomes subdivision invariantly non-self-intersecting.*

Proof: In this proof, all the subdivisions that we perform are concurrently applied to the curve $\mathcal{B}$ and to its hodograph. It is obvious that we can subdivide $\mathcal{B}$ until any two distinct Bézier curves that we obtain have subdivision invarintly disjoint control polygons. Once this property is achieved, we subdivide until the control polygon of each sub-Bézier curve of the hodograph can be separated from the origin by a line. As the first derivative of $\mathcal{B}$ does not vanish, the hodograph does not pass through the origin, so this is possible. It is also clear that this property is subdivision invariant.

It is easy to see that if the control polygon of the hodograph is separated from the origin by a line, then the control polygon of the curve cannot be self-intersecting. Indeed, the segments that join the origin to the control points of the hodograph are parallel to the edges of the control polygon of the curve. Thus, two such edges cannot overlap (the segment joining the corresponding control points of the hodograph should pass through the origin, which is impossible) and the sum of any successive angles belongs to the interval $(-\pi, \pi)$. Using the previous lemma, we deduce that the control polygon of the curve has no multiple points.

So, each control polygon of the sub-curves is non-self-intersecting, and any two such polygons are disjoint. We conclude that the control polygon of $\mathcal{B}$ has no multiple points, and this is a subdivision invariant property. □

We remark that in order to be sure of the subdivision invariability of the non-self-intersecting property, we really have to subdivide the curve until its hodograph satisfies the criterion of the proof above. Indeed, there are cases where an initially non-self-intersecting control polygon can become self-intersecting after subdivision. In Figure 5, we present such an example. The first picture represents a Bézier cubic. The second one shows the upper half of the previous after subdivision and scaling. The third is obtained in the same way from the second and finally the fourth shows the third after subdivision.

In Figure 6, we can see the hodograph of the (first) curve of Figure 5, with a zoom on the origin, which is very close to the upper endpoint of the hodograph, and inside the region delimited by the hodograph and its control polygon up to the fourth subdivision.

4.2. Non-regular Bézier Parameterisation

If the first derivative of the Bézier curve is allowed to vanish, there exist examples of non-self-intersecting curves with subdivision invariantly self-intersecting control polygon. Such an example is shown in Figure 7.

The derivative of the curve vanishes for $t = 1/3$. The point of the hodograph corresponding to this value of the parameter is the origin, where the hodograph presents a cusp. It is obvious that this point will never become the common endpoint of two successive curves if we use the subdivision parameter $t_0 = 1/2$. Then the sub-Bézier curve containing this point will have a self-intersecting control polygon, no matter how many subdivisions are applied to the curve.

To prove this statement, let us study the evolution of the hodograph. All the angles of its control polygon are negative and this property is obviously subdivision invariant. Therefore, as a Bézier curve is a subset of the convex hull of its control polygon, the control polygon of the sub-curve containing the origin must "turn around" this one, as shown in Figure 8. But if the control polygon of the hodograph has this configuration, then the sum of the angles of the control polygon of the curve (which are also all negative) is smaller

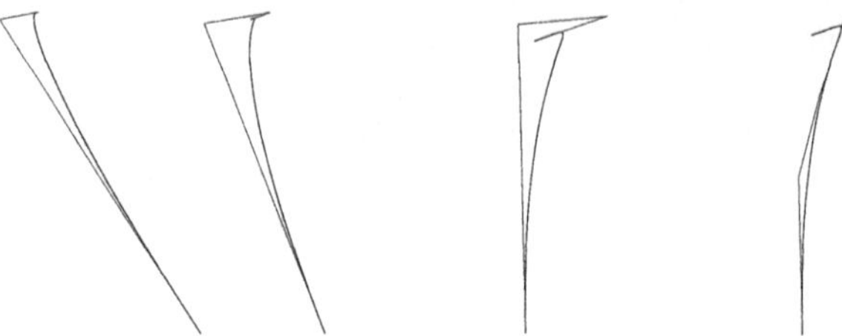

Fig. 5. A non-self-intersecting control polygon can become self-intersecting.

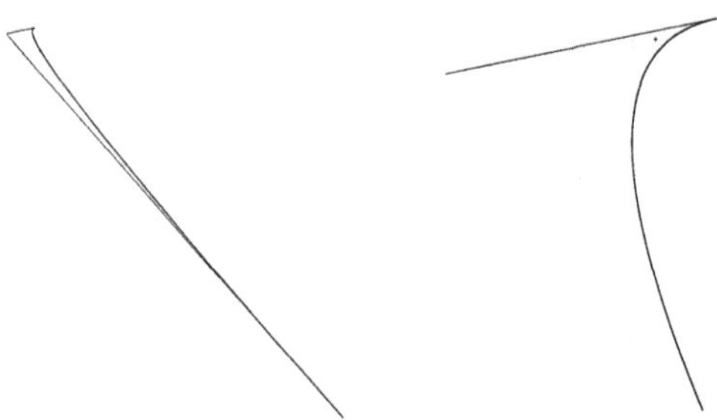

Fig. 6. The hodograph of the curve of Figure 5, with a zoom on the origin.

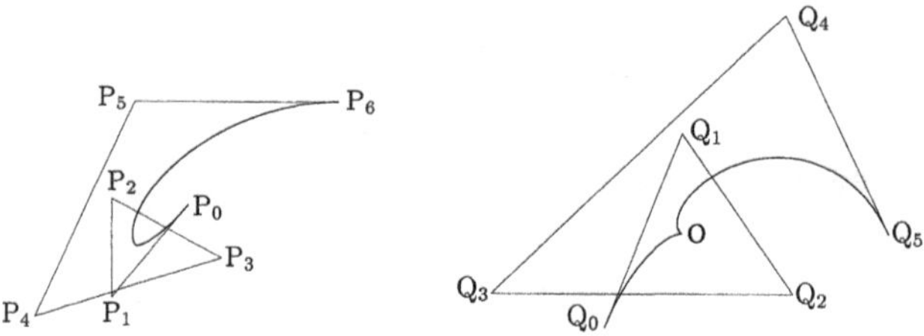

Fig. 7. Bézier curve with self-intersecting control polygon and its hodograph.

Fig. 8. Scaled representation of the sub-curve of the hodograph containing the origin after three subdivisions.

than -2π. It results that the control polygon of the curve is either a spiral or self-intersecting. As the (sub-)curve is convex after the first subdivision, its control polygon cannot be a spiral, and thus it is self-intersecting.

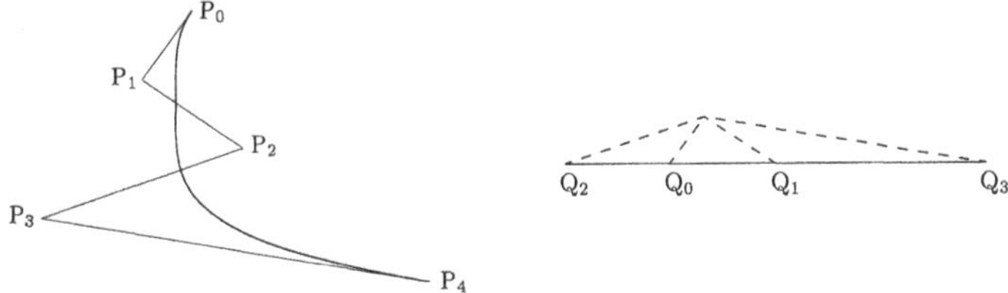

Fig. 9. Convex Bézier curve with non-convex control polygon and its hodograph.

§5. Convex Bézier Curve

It is well known that there exist convex Bézier curves which have non-convex control polygon. It is less known (and less intuitive) that there exist convex Bézier curves which have *subdvision invariantly* non-convex control polygon. Nevertheless, such curves exist, and an example is given in Figure 9.

The point corresponding to the parameter $t = 1/3$ is a flat point of the curve, the second and third derivatives vanish there. As we have also seen in the previous section, this point will always belong to the interior of a sub-Bézier curve if we use the subdivision of parameter $t_0 = 1/2$. Then the curve containing this point will have a non-convex control polygon, no matter how many subdivisions we apply.

Indeed, let us take a look at the corresponding point of the hodograph. It is not a multiple point (the parameterisation does not go back and forth on the segment that represents the curve), but the first derivative of the hodograph vanishes for $t = 1/3$. That implies that the points Q_0, Q_1, Q_2, and Q_3 cannot be well-ordered (if they were, the y-coordinate of the hodograph should be strictly increasing, and thus the y-derivative greater than 0). This property will be fulfilled by all the sub-curves of the hodograph which contain the point initially corresponding to $t = 1/3$. But if the control points of the hodograph are not well-ordered, the slope of the edges of the control polygon of the curve cannot vary monotonically. Consequently, the latter is not convex.

§6. Conclusion

The results that we have presented in this paper are somehow surprising. As the control polygon has very good convergence properties in terms of distance, we might be prone to expect equally good topological convergence. Perhaps the most unexpected is the possible behaviour of the control polygon of a convex Bézier curve. Of course, all the "bad" examples that we have given correspond to some kind of degeneracy of the Bézier parameterisation, which usually are not encountered in practice. Nevertheless, it is interesting to note that these cases exist.

These problems of preservation of topology cannot be solved by the use of an adaptive parameter subdivision. The reason is that in order to find the adapted parameter, we must solve algebraic equations corresponding to the special points of the Bézier curve (where the derivatives vanish). As the only way of solving such equations is the use of numerical methods, we generally do not find *exactly* the good parameter. We can still prove that for convex Bézier curves the use of the suitable subdivision parameter would generate convex control polygons.

As a last remark, let us note that all the proofs that he have presented also hold if we replace the repeated de Casteljau subdivision by repeated degree elevation.

References

1. Andersson, L.-E., T. J. Peters, and N. F. Stewart, Self-intersection of composite curves and surfaces, Comput. Aided Geom. Design **15** (1998), 507–527.
2. Bern, M., D. Eppstein, et al., Emerging challenges in computational topology, http://arxiv.org/abs/cs.CG/9909001, 1999.
3. Calcoen, E., Approximations polygonales d'objets convexes du plan pour la géométrie algorithmique, Ph.D. Thesis, University Joseph Fourier, Grenoble, 1996.
4. Farin, G., *Curves and Surfaces for Computer Aided Geometric Design: a Practical Guide*, London Academic Press, 1997.
5. Fiorot, J. C., Courbes de Bézier, in *Colloque Courbes et Surfaces Bézier/ B-splines*, Rennes, 1987, 1–31.
6. Hoschek, J., and D. Lasser, *Fundamentals of Computing Aided Geometric Design*, AK Peters, Wellesley, 1993.

Manuela Neagu
LICN-IUT
Allée André Maurois
87065 Limoges Cedex, France
manuela.neagu@unilim.fr

Emmanuelle Calcoen
ESA
55, rue Rabelais
BP 748
49007 Angers Cedex, France
e.calcoen@esa-angers.educagri.fr

Bernard Lacolle
LMC-IMAG
BP 53
38041 Grenoble Cedex 9, France
Bernard.Lacolle@imag.fr

What is the Natural Generalization of Univariate Splines to Higher Dimensions?

Marian Neamtu

Abstract. In the first part of the paper, the problem of defining multivariate splines in a natural way is formulated and discussed. Then, several existing constructions of multivariate splines are surveyed, namely those based on simplex splines. Various difficulties and practical limitations associated with such constructions are pointed out. The second part of the paper is concerned with the description of a new generalization of univariate splines. This generalization utilizes the novel concept of the so-called Delaunay configurations, used to select collections of knot-sets for simplex splines. The linear span of the simplex splines forms a spline space with several interesting properties. The space depends uniquely and in a local way on the prescribed knots and does not require the use of auxiliary or perturbed knots, as is the case with some earlier constructions. Moreover, the spline space has a useful structure that makes it possible to represent polynomials explicitly in terms of simplex splines. This representation closely resembles a familiar univariate result in which polar forms are used to express polynomials as linear combinations of the classical B-splines.

§1. Introduction

This paper describes the material I presented at the *Fifth International Conference on Mathematical Methods for Curves and Surfaces*, held in Oslo in June 2000. In my talk, I addressed the topic of a meaningful generalization of the classical univariate splines to higher dimensions. The quest for finding such generalizations is certainly not new. Many researchers in the theory of multivariate splines have addressed this or similar questions. Indeed, there is a great variety of multivariate generalizations of splines available today. However, there does not seem to exist a generalization that is commonly agreed to be the "right" one. This fact perhaps prompted Carl de Boor to conclude his survey "What is a multivariate spline?" [8] with a touch of irony: "*If this leaves you a bit wondering what multivariate splines might be, I am pleased. For I don't know myself.*"

Given the large number of possible approaches to multivariate splines, it is clear that the term "natural generalization" in the title is necessarily vague

Mathematical Methods for Curves and Surfaces: Oslo 2000
Tom Lyche and Larry L. Schumaker (eds.), pp. 355–392.

ISBN 0-8265-1378-6

unless we agree on its precise meaning. Therefore, let us first discuss which properties a multivariate spline should ideally possess. My own understanding of "spline" in this paper is that it should be a *piecewise polynomial* of a given fixed degree. Second, the individual polynomial pieces of the spline should be associated with polygonal (polyhedral) regions. This means that the domain of definition of the spline should be partitioned by segments of straight lines (hyperplanes), *i.e.*, it should be a *rectilinear partition.* In particular, we will assume that the partition is a union of polytopes, each of which is the convex hull of a finite set of points, called *knots.* The third important property of splines is that they are "local". Thus, not only should the spline space have a finite local dimension, but also the space should contain splines of *compact support.* Ideally, a good spline space should be such that *every* spline in the space can be written as a combination of compactly supported splines.

Mathematicians usually endeavor to obtain results in their greatest possible generality. Therefore, it is desirable to have a generalization of splines that applies to all degrees, (almost) arbitrary knot locations, and all spatial dimensions. Moreover, in the one-dimensional case the spline construction should reduce to the familiar univariate splines. We must also postulate that the obtained spline space is large enough so that it can be used to effectively approximate other functions. Namely, the spline space should contain constants and, more generally, all polynomials up to a given degree, ideally equal to the spline degree. The requirement of polynomial reproduction is now standard, since it is well known that the degree of approximation is closely related to the degree of local polynomial reproduction, see *e.g.* [6,131].

Before we formulate the above requirements more rigorously, we need some notation and a few technical assumptions. We will be concerned with s-variate splines *i.e.*, functions whose domain is $\mathbb{R}^s$, $s = 1, 2, \ldots$ The **degree** of a polynomial or a spline will be denoted by n, $n = 0, 1, \ldots$ The symbol K will stand for the set of **knots**, *i.e.*, points in $\mathbb{R}^s$. To avoid the consideration of "boundary conditions", we assume throughout that the convex hull $[K]$ of K equals $\mathbb{R}^s$. Moreover, we only allow knot-sets without accumulation points, or, equivalently, K must be such that $K \cap \Omega$ is finite for every compact $\Omega \subset \mathbb{R}^s$. Lastly, K will be "generic" in the sense (to be made more precise later) that certain properties of the corresponding spline space will be required to hold only for "almost all" knot-sets, not necessarily all knot-sets. This will help us avoid various special cases (such as knot-multiplicities), which, while deserving attention in practical considerations of splines, do not have direct bearing on the central issue of existence of a natural generalization of splines. The set K, together with a set of rules of how the knots in K are connected, will give rise to a partition of $\mathbb{R}^s$ and hence also to a spline space, denoted by $\mathcal{S}_n$, of all real-valued piecewise polynomials of a given degree n and smoothness C^{n-1}. This smoothness is also sometimes called **optimal** since a higher than optimal smoothness would necessarily imply $\mathcal{S}_n \subset \Pi_n$, where, as usual, Π_n stands for the space of polynomials of total degree at most n. This, however, is not a spline space in our terminology since clearly Π_n does not contain functions of compact support, and hence $\mathcal{S}_n$ cannot be spanned by such functions. The above assumption that all splines in $\mathcal{S}_n$ are optimally smooth is motivated by the fact that univariate splines with generic knots (*i.e.*, knots without multiplicities) also have optimal smoothness.

We are now ready to define more precisely the problem of finding a natural generalization of splines, which we will refer to as the *fundamental problem.*

Fundamental Problem. *For any integers $s \geq 1$ and $n \geq 0$, and any generic set of knots K, construct a spline space $\mathcal{S}_n$ on $\mathbb{R}^s$, such that*

(A) *Each spline in $\mathcal{S}_n$ is a piecewise polynomial of degree n, associated with a rectilinear partition determined by K;*
(B) *Each spline in $\mathcal{S}_n$ has optimal smoothness i.e., $\mathcal{S}_n \subset C^{n-1}(\mathbb{R}^s)$;*
(C) *Splines in $\mathcal{S}_n$ reproduce polynomials i.e., $\Pi_n \subset \mathcal{S}_n$;*
(D) *$\mathcal{S}_n$ is locally finite dimensional, i.e., $\dim \mathcal{S}_n|_\Omega < \infty$, for all compact $\Omega \subset \mathbb{R}^s$;*
(E) *There exists a countable collection $\mathcal{B}_n$ of compactly-supported functions which forms a basis for $\mathcal{S}_n$, in the sense that for every $x \in \mathbb{R}^s$, all but a finite number of functions in $\mathcal{B}_n$ vanish at x, and every spline in $\mathcal{S}_n$ can be uniquely represented as a linear combination of the form $\sum_{B \in \mathcal{B}_n} c_B B$, where $c_B \in \mathbb{R}, B \in \mathcal{B}_n$;*
(F) *For $s = 1$, the space $\mathcal{S}_n$ reduces to the classical space of univariate splines and the functions in $\mathcal{B}_n$ are the ordinary univariate B-splines.*

Several remarks on the above formulation are in order. Condition (E) implies, among other things, that the functions in $\mathcal{B}_n$ are linearly independent. Moreover, a consequence of (D) and (E) is that for every compact $\Omega \subset \mathbb{R}^s$, there is a finite number of functions in $\mathcal{B}_n$ with support in Ω. The elements of $\mathcal{B}_n$ will be called (multivariate) B-splines. Requirement (F) expresses the desire, discussed earlier, that the construction of $\mathcal{S}_n$ must reduce to the familiar splines in one dimension. Strictly speaking, this condition is somewhat vague since in principle one could have different constructions for $s \geq 2$ and $s = 1$. However, our implicit assumption is that the sought-for construction of splines operates essentially in the same way for *all* dimensions s. Thus, in this sense the well-known multivariate box splines do not satisfy (F) since for $s = 1$, such splines are different from the usual univariate splines (except for special knot-sets). In Section 3, we will give examples of other multivariate spline spaces that satisfy (A)–(E), but are not "natural" since they fail (F).

The reader will notice that there are other important requirements that one might want to impose on the space $\mathcal{S}_n$. For example, next to the linear independence of the functions in $\mathcal{B}_n$, it is often imperative to have a stronger property of stability of $\mathcal{B}_n$. Also, it is advantageous to have efficient algorithms for numerical manipulation with the splines, such as recurrence relations for their evaluation. No doubt, the wish list of additional properties of splines could be much longer. However, there is no point in contemplating such a list when in fact it is not clear whether the fundamental problem has a solution. Indeed all types of multivariate splines that come to my mind are established by relaxing some of the requirements (A)–(F). For example, it is known that the classical piecewise polynomials on triangulations do not in general satisfy

(B) and (E) unless the smoothness degree is strictly less than the optimal degree. The already mentioned box splines also do not solve the fundamental problem, since the space $\mathcal{S}_n$ is not even defined for non-uniform knots K. These and several other examples of splines (see *e.g.* Section 3) show that it is conceivable that requirements (A)–(F) may be too restrictive. In fact, one reason why I became interested in the subject of this paper was a realization that *none* of the available spaces of multivariate splines seemed to provide a solution to the fundamental problem.

In this paper I will address the fundamental problem by first explaining that the most obvious candidates for spaces $\mathcal{S}_n$ satisfying (A)–(F) are those spanned by *simplex splines*. In the next section, I briefly discuss the history of these functions and recall their definition, along with some of their properties. In Section 3, I review several constructions of simplex spline spaces that are currently available and explain why they do not satisfy all the above six properties. Then, in Sections 4 and 5, a new type of simplex spline spaces is described, consistent with the requirements of the fundamental problem. The main idea of the construction is based on a new concept of a *Delaunay configuration*, introduced in this paper, which is a generalization of the classical Delaunay triangulation.

§2. Simplex Splines

There are two basic approaches to multivariate splines. We can first define the space $\mathcal{S}_n$ and then check whether it has the desired properties, *e.g.* whether it possesses a compactly-supported basis. Examples of spaces defined by this principle are the classical piecewise polynomials on planar triangulations. The alternative approach is to start with appropriate compactly-supported piecewise polynomials, and then form $\mathcal{S}_n$ as the linear span of these functions.

The first approach is not appropriate in our situation. This is because by requirement (B), the underlying rectilinear partition for the spline space will likely be very complicated. Note that we cannot expect the sought-for space $\mathcal{S}_n$ to coincide with the usual space of piecewise polynomials on triangulations or other relatively simple partitions of $\mathbb{R}^s$, since it is known that there may be no nontrivial compactly-supported splines of optimal smoothness on such partitions, let alone a basis consisting of such splines. Thus, it may be difficult or impossible to "guess" what a suitable partition should be. Note that this is trivial in the univariate case since partitions of $\mathbb{R}$ are unions of intervals.

The second approach seems much more tractable since it suggests that we first design appropriate individual "B-splines" and only then proceed with defining the corresponding spline space. The reason why this may be an easier path to follow is that there is already a well-understood class of multivariate compactly-supported splines, the so-called polyhedral splines. In particular, let $n \geq 1$, σ be a bounded convex polyhedron in $\mathbb{R}^{n+s}$ with positive volume, and let $X \subset \mathbb{R}^s$ be the canonical projection of the vertices of σ onto $\mathbb{R}^s$. Recall that the canonical projection of a point $v = (v_1, \ldots, v_s, \ldots, v_{s+n}) \in \mathbb{R}^{n+s}$ onto $\mathbb{R}^s$ is defined as $v|_{\mathbb{R}^s} := (v_1, \ldots, v_s)$. The polyhedral spline M_σ is defined

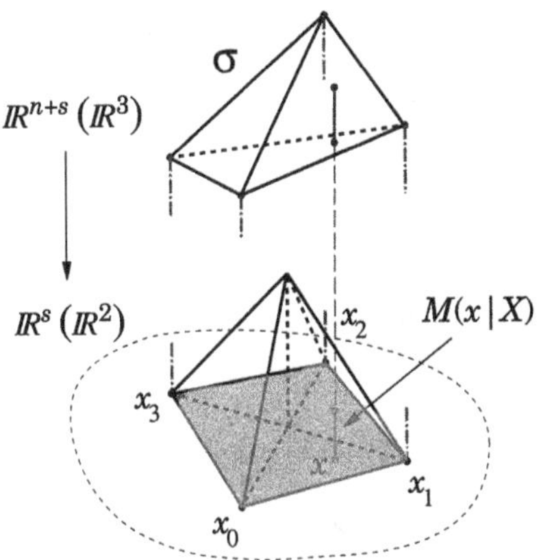

Fig. 1. Geometric interpretation of simplex splines ($s = 2, n = 1$).

as the volumetric projection of σ onto $\mathbb{R}^s$, that is

$$M_\sigma(x) := \frac{\text{vol}_n\{v \in \sigma : \; v|_{\mathbb{R}^s} = x\}}{\text{vol}_{n+s}\sigma}, \tag{1}$$

where $\text{vol}_k(A)$ stands for the k-dimensional volume of a set A. In the special case $n = 0$, σ is a polyhedron in $\mathbb{R}^s$, and one defines

$$M_\sigma := \frac{\chi_\sigma}{\text{vol}_s\sigma}, \tag{2}$$

where χ_σ is the characteristic function of σ. Polyhedral splines were introduced in [11]. They are known to be piecewise polynomials of degree n and of optimal smoothness if X is in generic position. By definition, polyhedral splines are obviously compactly supported with support $[X]$, non-negative, and the normalization in the above two identities is such that the integral of M_σ is one. Special cases of polyhedral splines are the already mentioned box splines and simplex splines, obtained when σ is a parallelepiped and a simplex, respectively.

Since every polyhedron can be decomposed as a union of simplices, every polyhedral spline is a linear combination of simplex splines. Therefore, from now on we will restrict ourselves, without loss of generality, to the special case where σ is a simplex. It turns out that in this situation $M_\sigma(x)$ only depends on the projection X, which justifies a new and more appropriate notation $M(x|X) := M_\sigma(x)$, used henceforth. The geometric idea of the definition of $M(\cdot|X)$ via volumetric projection is depicted in Figure 1. In this figure, a bivariate simplex spline of degree one is obtained by volumetrically projecting a three-dimensional simplex σ onto the plane. This gives a pyramid-like function whose support is the convex hull of the four knots $X = \{x_0, x_1, x_2, x_3\}$, being the projections of the vertices of σ onto $\mathbb{R}^2$. A typical bivariate quadratic simplex spline is displayed in Figure 2, along with its support, the convex hull

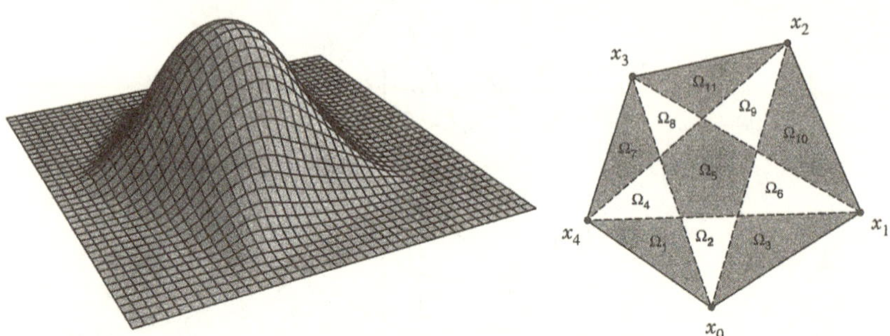

Fig. 2. A bivariate quadratic simplex spline and its support ($s = 2, n = 2$).

of five knots (since in the quadratic case, σ is a simplex in $\mathbb{R}^4$). In general, a collection X of $n+s+1$ knots in $\mathbb{R}^s$ gives rise to an s-variate simplex spline of degree n.

For a number of reasons, simplex splines are an excellent point of departure for attempting to solve the fundamental problem. First, they are defined for all s and n, and also they are optimally smooth (C^{n-1}) for generic knot-sets X. In fact, "generic" for simplex splines means that the knots in X must be in general position, *i.e.*, no $s+1$ of them are allowed to lie in a common hyperplane in $\mathbb{R}^s$. For example, the quadratic spline in Figure 2 has its knots in general position (since no three of them are collinear), hence the spline is tangent-plane continuous (C^1). Thus, simplex splines satisfy (A) and (B) since any linear combination of simplex splines is a piecewise polynomial of degree n and optimal smoothness. Moreover, the definition of $M(\cdot|X)$ is consistent with (F) in the sense that $M(\cdot|X)$ is just the univariate B-spline for $s = 1$. Simplex splines and univariate B-splines share many useful properties, see *e.g.* [30,87]. For example, simplex splines can be evaluated by the Micchelli recurrence [85], expressing simplex splines in terms of their lower-degree versions. In particular,

$$M(x|X) = \frac{n+s}{n} \sum_{y \in X} \lambda_y M(x|X \backslash \{y\}), \quad x \in \mathbb{R}^s, \tag{3}$$

where the numbers $\lambda_y \in \mathbb{R}, y \in X$, are chosen so that $\sum_{y \in X} \lambda_y y = x$ and $\sum_{y \in X} \lambda_y = 1$. Note that this is a generalization of the familiar B-spline recurrence. Beside the original proof of Micchelli, alternative proofs of the recurrence were discovered in [11,15,18,65,68,84,91,105] (and possibly elsewhere).

There are also other reasons, described in more detail in the next section, why simplex splines are an ideal starting point for our investigation. In particular, one can show that simplex splines have the striking property that if their linear combinations reproduce constants, then this will automatically imply that they can represent *all* polynomials up to the degree of the simplex splines. Another fact worth mentioning is that *every* piecewise polynomial on a rectilinear partition of $\mathbb{R}^s$ can be expressed as a linear combination of appropriately chosen simplex splines. Thus, assuming that piecewise polynomials in the sought-for space $\mathcal{S}_n$ are combinations of simplex splines is not a real restriction.

We finish this section with a few historical remarks and comments on references. Simplex splines were formally introduced in 1976 by de Boor [5], who followed a suggestion of Schoenberg [112]. However, simplex splines were already known to statisticians several years earlier (see *e.g.* [126]). We refer the reader to the surveys [6,7,10,30,47,70,113,115,123], for many results on simplex splines. The paper [31] dwells on the origins of simplex splines in statistics. There are several recent books containing chapters on simplex splines, including [1,4,87]. Since the introduction of simplex splines, a steady stream of papers has been written each year on this subject, perhaps with the exception of the last few years, during which there seemed to be less visible activity in the area. We include here a list of references that contains an up-to-date set of 130 research articles, surveys, dissertations, and unpublished papers. To my best knowledge, the list is complete. However, in case it is not, the reader is encouraged to send me additions.

§3. Available Constructions of Simplex Spline Spaces

While individual simplex splines are very appealing functions mathematically, it is their linear combinations that are of main interest in applications. To solve the fundamental problem, we need to choose appropriate collections $\mathcal{C}_n$ of knot-sets $X \subset K$ of size $n+s+1$, so that the associated simplex spline spaces

$$\mathcal{S}_n := \operatorname{span}\{M(\cdot|X) : \ X \in \mathcal{C}_n\} := \left\{ \sum_{X \in \mathcal{C}_n} c_X M(\cdot|X) : \ c_X \in \mathbb{R} \right\} \tag{4}$$

will satisfy conditions (A)–(F). Before we address this issue in more detail, it is tempting to review the existing constructions of simplex spline spaces, to see whether they do not already provide a solution to the fundamental problem. We shall see that this is not the case.

3.1. Complete Configurations of Simplex Splines

For the purpose of this subsection only, suppose that K is finite. A possible way to construct a space $\mathcal{S}_n$ is to consider the so called complete configurations of knot-sets $X \subset K$ of size $n+s+1$, that is

$$\mathcal{C}_n := \{X \subset K : \ \#X = n+s+1\},$$

where $\#X$ denotes the size of the set X. Complete configurations were investigated in [28] and also in [62,66]. From the practical point of view, the space $\mathcal{S}_n$ corresponding to such collections $\mathcal{C}_n$ is not very useful. For example, the simplex splines generated by the knot-sets in $\mathcal{C}_n$ are linearly dependent and also can have "large" supports *i.e.*, can be non-local. Nevertheless, the space $\mathcal{S}_n$ is

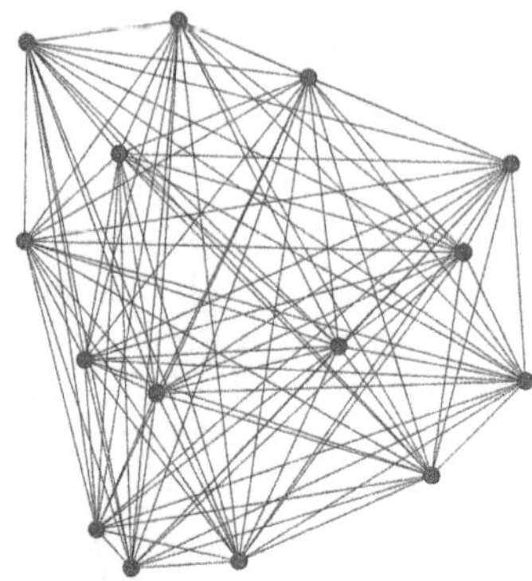

Fig. 3. The partition corresponding to a complete configuration of a knot-set.

helpful in gaining theoretical insight into simplex splines. In particular, it is known [66] that $\mathcal{S}_n$ is precisely the space of optimally smooth piecewise polynomials of degree n on the partition of $[K]$, generated by all $(s-1)$-dimensional faces (or simplices) of the form $[Y]$, where $Y \subset K, \#Y = s$. Figure 3 shows a typical example of such a partition in $\mathbb{R}^2$. The above implies that *every* piecewise polynomial of optimal smoothness and compact support in $[K]$, which reduces to a polynomial of degree at most n in each region bounded but not intersected by the faces $[Y]$, is a linear combination of simplex splines.

Another well-known fact about $\mathcal{S}_n$, proved in [28] and independently in [66], is that

$$\dim \mathcal{S}_n = \binom{\#K - n - 1}{s}. \tag{5}$$

A simple consequence of this identity, obtained by setting $\#K = n+s+1$, is that a simplex spline with $n+s+1$ knots in general position is the *unique* (up to normalization) piecewise polynomial of degree n and (global) smoothness C^{n-1}, supported on the partition of $[K]$, generated by all $(s-1)$-dimensional faces $[Y]$, where $Y \subset K, \#Y = s$. This property, which is sometimes referred to as the **minimal support property**, is reminiscent of a familiar property of univariate B-splines, and can be used as an alternative definition of simplex splines. The term "minimal" refers to the fact that taking $\#K < n+s+1$ implies that $\mathcal{S}_n$ contains the zero function only. Thus, for example, the quadratic simplex spline in Figure 2 is the unique C^1 piecewise quadratic function supported on the convex hull of the knots $\{x_0, \ldots, x_4\}$, which is a single polynomial in each of the displayed regions Ω_j. There is no such nontrivial spline if we only take four knots, instead of five.

The above dimension formula also shows why choosing $\mathcal{C}_n$ as the complete configuration associated with K is not a good idea. Namely, clearly, $\#\mathcal{C}_n = \binom{\#K}{s}$, which is larger than the dimension of $\mathcal{S}_n$. Consequently, the simplex splines corresponding to $\mathcal{C}_n$ are linearly dependent. This explains why researchers looked for better ways to define suitable sets $\mathcal{C}_n$.

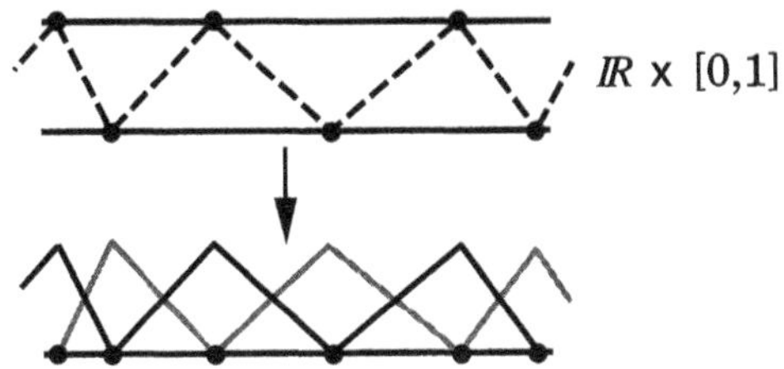

Fig. 4. A geometric construction of univariate splines ($s = 1, n = 1$).

3.2. A Geometric Construction

The geometric definition (1) of simplex splines generated considerable interest among researchers since it immediately gives a recipe for selecting suitable collections $\mathcal{C}_n$. Specifically, de Boor [5] proposes to consider the cylinder or "slab" $\mathbb{R}^s \times \Omega$, where $\Omega \subset \mathbb{R}^n$ is a suitable convex polytope of unit volume, and subdivide this slab into nontrivial simplices. The corresponding simplex splines will then automatically form a partition of unity. This is because the volumetric projection of the above slab, which is the union of the mentioned simplices, is obviously the constant function with value $\mathrm{vol}_n(\Omega) = 1$. To obtain a *linearly independent* collection of simplex splines, we must choose Ω to be a simplex [21,27]. For example, to define linear univariate splines, we can set $\Omega = [0, 1]$ and then triangulate the slab $\mathbb{R} \times [0, 1]$, as shown in Figure 4.

Similarly, to obtain bivariate quadratic splines, one can take the four-dimensional slab $\mathbb{R}^2 \times \Omega \subset \mathbb{R}^4$, where Ω is a 2-simplex, and triangulate it, *i.e.*, decompose it into four-dimensional simplices, each of which gives rise to a quadratic simplex spline (such as the one in Figure 2).

One can show that by the above geometric construction, the associated spline space $\mathcal{S}_n$ will contain not only constants, but also *all* polynomials of degree n. This is yet another surprising property of simplex splines, which is in general not satisfied by other polyhedral spline spaces, *e.g.* box splines. The first proof of the fact that polynomial reproduction is obtained "for free" from the partition of unity was given in [16] and is also explained in many other papers, including [6,27,30,47,50,69,70]. The proof is based on the already mentioned fact that if $\sigma \subset \mathbb{R}^{n+s}$ is a simplex, then M_σ is determined by the projection X of the vertices of σ onto $\mathbb{R}^s$ *i.e.*, the actual shape of σ does not matter as long as X stays the same. Note that this property is unique to simplex splines and does not hold for other polyhedra σ.

The above geometric idea was elaborated on in detail in [50] (see also [47]). There, explicit formulae were given for representing bivariate polynomials as linear combinations of simplex splines, closely resembling similar identities for univariate splines. However, despite such results and the apparent elegance of the geometric construction, this idea did not gain popularity, perhaps as a consequence of the need to triangulate four or higher-dimensional polyhedra. Therefore, other means of obtaining appropriate collections $\mathcal{C}_n$ were proposed.

3.3. Triangulations of Simploids

We have seen that a serious disadvantage of the geometric construction is that the user has to construct triangulations in $\mathbb{R}^{n+s}$, a task that is practically not feasible for $n + s > 3$. A computationally more attractive method of decomposing the slab $\mathbb{R}^s \times \Omega$ into simplices was proposed in [27] and independently in [69].

This method proceeds in two stages. In the first stage, we start with a triangulation Δ of the knots K *i.e.*, a decomposition of $\mathbb{R}^s$ into simplices $T \in \Delta$, whose vertices are the given knots. This gives rise to a decomposition of the slab into polyhedra of the form $T \times \Omega$, called **simploids** in [27]. Hence, to find a triangulation of $\mathbb{R}^s \times \Omega$, it is sufficient to triangulate the individual simploids $T \times \Omega$, provided that this triangulation is done in such a way that it results in a global triangulation of $\mathbb{R}^s \times \Omega$, see [27,69]. This turns out to be a comparatively simpler task since simploids can be triangulated in a canonical way, see [6,27,29,47,69,123]. In fact, the knot-sets corresponding to the simplices of this canonical triangulation can be obtained using simple combinatorial rules, and hence computing an actual triangulation of the simploids is not necessary. One can show that in this way one obtains $\binom{n+s}{s}$ simplex splines per simplex T, spanning Π_n on each T. This is because the simplex splines obtained in this way have multiple knots and in fact they are equal to the **Bernstein polynomials** associated with the simplices $T \in \Delta$. Thus, this construction simply gives rise to a space $\mathcal{S}_n$ that is precisely the space of all piecewise polynomials of degree n, associated with the triangulation Δ. In particular, the space $\mathcal{S}_n$ satisfies (C). However, $\mathcal{S}_n$ also contains functions that are discontinuous along the boundaries of the simplices $T \in \Delta$, hence $\mathcal{S}_n$ fails to satisfy condition (B). This is a consequence of the fact that the individual simplex splines that span $\mathcal{S}_n$ are themselves discontinuous, due to their multiple knots.

The second stage of the method consists in modifying or "smoothing" the space $\mathcal{S}_n$ so that all of its elements are optimally smooth. Roughly, this can be achieved by removing the multiplicities of the knots of the above simplex splines, by perturbing them or "pulling them apart". This is why sometimes this construction is referred to as the **pulling-apart method**. Thus, the smoothed-out space $\mathcal{S}_n$ is spanned by modified simplex splines obtained as perturbed versions of Bernstein polynomials. We refer the reader to the original papers [27,69], for a detailed description of this method.

Examples of the pulling-apart process and the resulting knot-sets are displayed in Figures 5-8. In particular, Figure 5 shows the situation in the univariate case, where the simplices T are just intervals. In this case the simploids are rectangles partitioning the slab $\mathbb{R} \times [0, 1]$, each giving rise to two discontinuous linear Bernstein polynomials, *i.e.*, simplex splines with double knots. By perturbing the triangles of the depicted triangulation of $\mathbb{R} \times [0, 1]$, and then projecting volumetrically those triangles onto $\mathbb{R}$, we obtain *continuous* simplex splines, shown in the figure, which are simply hat functions. Thus, the pulling-apart step removed the multiplicity of the double knots (shown as filled circles in the figure) and replaced them with pairs of *distinct*

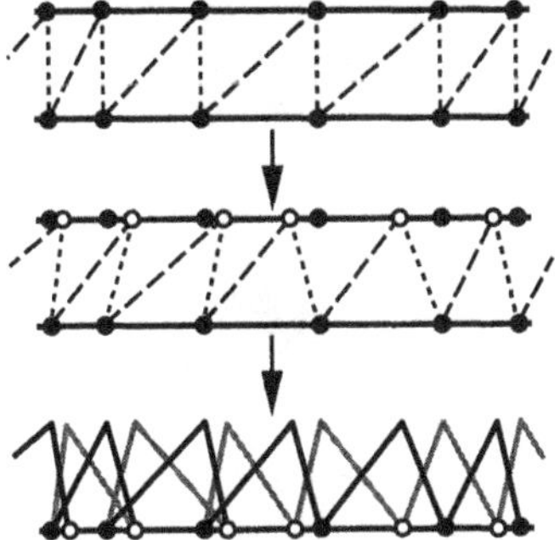

Fig. 5. C^0 linear splines obtained by pulling apart the knots ($s = 1, n = 1$).

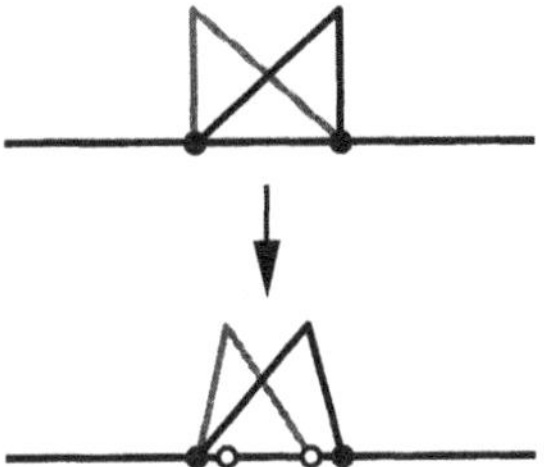

Fig. 6. Two discontinuous simplex splines and their perturbations ($s = 1, n = 1$).

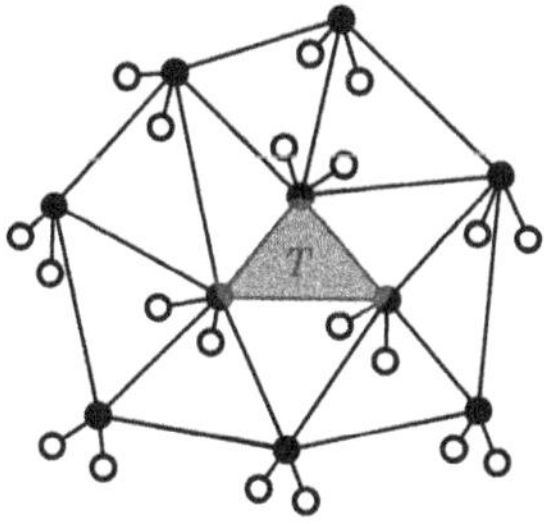

Fig. 7. The pulling-apart method for bivariate quadratic splines ($s = 2, n = 2$).

knots, consisting of the original knots (filled circle) and their perturbations (empty circles). The smoothing effect of the pulling-apart process on the linear Bernstein polynomials corresponding to a single interval can be better seen in Figure 6.

Figure 7 shows an analogous situation for bivariate quadratic simplex splines. Here, a planar region is first triangulated and then each knot, which is thought of as a triple knot, is separated into three distinct knots, the original one (filled circle), and two "auxiliary" knots (empty circles). Figure 8 displays

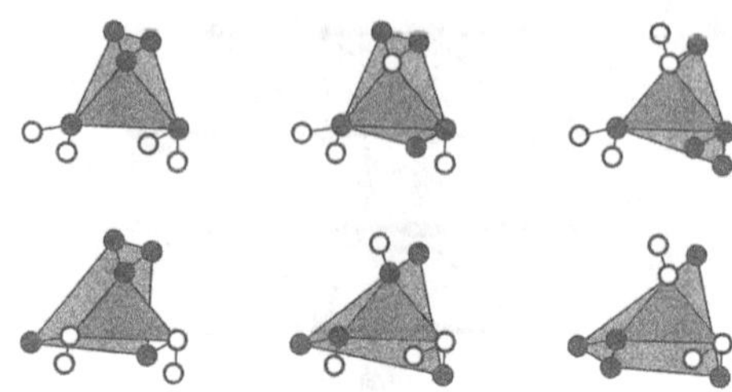

Fig. 8. Six knot-sets of size five corresponding to the shaded triangle in Figure 7.

the collection of six knot-sets of size five, associated with the shaded triangle T from Figure 7. In Figure 8, the filled circles represent the "active" knots forming the given knot-set of size five, whereas the empty circles are the remaining knots associated with the shaded triangle, not belonging to the knot-set.

It should be noted that by the described construction, the resulting space $\mathcal{S}_n$ will still contain Π_n *i.e.*, the perturbation of the knots will not compromise property (C). It is also worth stressing that, as we have seen, to obtain $\mathcal{S}_n$ it is *not* necessary to triangulate the mentioned $(n+s)$-dimensional slab, we only have to deal with triangulations in $\mathbb{R}^s$. The collections of simplex splines constructed by the described method can be shown to have several additional properties. In particular, the simplex splines in such collections are linearly independent and also, under some mild conditions, stable. Nevertheless, as will be discussed in Section 3.5, the spline space $\mathcal{S}_n$ does not satisfy all properties (A)–(F). Moreover, there are several other shortcomings of the presented construction that necessitated a search for a better solution.

3.4. DMS-Splines

Except for the method proposed later in this paper, the most recent construction of simplex spline spaces, often called DMS-splines or triangular B-splines, was given in [32].

Unlike the previous two methods, this construction is not based on the geometric principle. However, one can still think of the simplex splines that one obtains as being generated by a perturbation of Bernstein polynomials or, more precisely, perturbation of the de Casteljau algorithm for these polynomials. In particular, this construction also employs auxiliary or pulled-apart knots. The difference with the method of Section 3.3 is that the knots are used to construct collections of simplex splines in a "more symmetric" way. For details on this construction, the interested reader is referred to [32] and the survey [113].

Figures 9–11 give examples of collections of knot-sets obtained by this method. The univariate case is explained in Figure 9. Here, the filled circles along the real axis represent the original knots and the empty circles are the auxiliary knots (there is only one per original knot in the linear case $n = 1$). Two simplex splines corresponding to a single interval are depicted in

Fig. 9. Construction of DMS-splines ($s = 1, n = 1$).

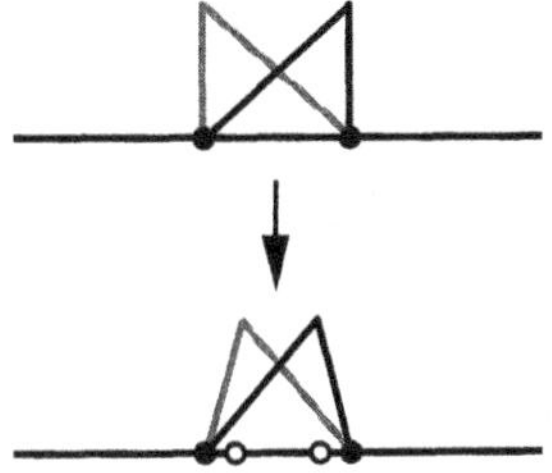

Fig. 10. Perturbations of two discontinuous simplex splines by the DMS-method.

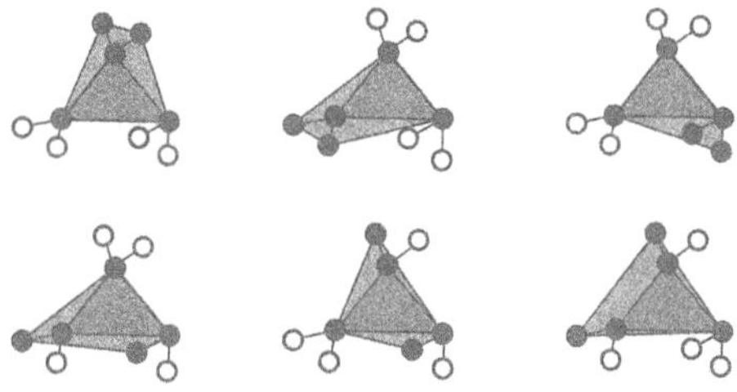

Fig. 11. Construction of knot-sets for DMS-splines ($s = 2, n = 2$).

Figure 10, for the original (not pulled-apart) knots and also for their perturbations. Clearly, compared to the analogous figure for the method of Section 3.3 (*cf.* Figure 6), this construction is more symmetric. This can also be seen in Figure 11, which is a counterpart of Figure 8. The figure displays the collection of six knot-sets of size five, associated with the shaded triangle T from Figure 7, giving rise to six quadratic simplex splines. In contrast with Figure 8, the collections in Figure 11 are more symmetric. For example, renumbering the three vertices (the original knots) of the shaded triangle will result in the same configuration of knot-sets, as opposed to the construction of the previous subsection.

Superficially, the difference between DMS-splines and the splines of Section 3.3 may seem marginal and unimportant. However, a closer inspection reveals a significant difference. The present construction, due to its inherent symmetry, has more interesting and useful structure. Among other things, it is possible to derive elegant explicit formulae for the representation of polynomials as combinations of simplex splines. It will be instructive to dwell on this aspect of DMS-splines in more detail. The space $\mathcal{S}_n$ of DMS-splines can be written as

$$\mathcal{S}_n = \text{span}\{M_{j,T} : \ j = 1, \ldots, J(n, s), T \in \Delta\},$$

where $J(n, s) := \binom{n+s}{s}$, Δ is a triangulation of $\mathbb{R}^s$, and $\{M_{j,T}\}_{j=1}^{J(n,s)}$ are the simplex splines corresponding to a given $T \in \Delta$. Let $\Pi_n(\Delta)$ denote the space of piecewise polynomials of degree n and optimal smoothness C^{n-1}, associated with the triangulation Δ. It was shown in [114], that every $f \in \Pi_n(\Delta)$ can be expressed as a linear combination of the simplex splines, namely

$$f = \sum_{T \in \Delta} \sum_{j=1}^{J(n,s)} F_T(K_{j,T}) N_{j,T}, \tag{6}$$

where $N_{j,T}$ are appropriate normalizations of the simplex splines $M_{j,T}$. We also used the following notation. Since f is a piecewise polynomial on the triangulation Δ, its restriction f_T to a simplex $T \in \Delta$ is a single polynomial. In the above formula, F_T stands for the *polar form* of f_T (see Section 4, for a precise definition of polar forms) and $K_{j,T}$ are appropriately chosen collections of knots of size n. Since the details of formula (6) will not be needed here, we refer the reader to the original article [114] and the surveys [113,115], for a discussion. However, it is important to note that (6) closely resembles a well-known univariate formula, discussed in Section 4.6. Moreover, (6) and $\Pi_n \subset \Pi_n(\Delta)$ imply $\Pi_n \subset \mathcal{S}_n$, which was already proved in [32]. This means that the construction of DMS-splines is consistent with requirement (C). Finally, identity (6) also proves the aforementioned fact, that every optimally smooth piecewise polynomial on a rectilinear partition can be written as a linear combination of simplex splines. This is because every rectilinear partition can be triangulated in such a way that any piecewise polynomial f of optimal smoothness on this partition is also a piecewise polynomial on the triangulation. Hence by (6), f is a combination of simplex splines. We note in passing that this assertion is true, even without the assumption of optimal smoothness, if we allow simplex splines with multiple knots.

3.5. Existing Constructions do not Solve the Fundamental Problem

Unfortunately, all of the above constructions suffer from various shortcomings that are not encountered in the univariate situation. We have already explained that the construction via complete configurations is not acceptable. As for the other three methods, they generate, strictly speaking, a family of spline spaces, not a *single* space for a given collection of knots K. This family either depends on how we triangulate the slab in Section 3.2, or on how we choose the set of auxiliary (or pulled-apart) knots. The least elegant ingredient of these constructions is probably that it is not clear how one can choose the auxiliary knots in a natural way. This complication may explain why so far simplex splines have not gained popularity in applications.

Of course, in principle one could make a generic (*e.g.* random) choice of the auxiliary knots. However, not surprisingly, this may result in spline spaces with "less structure". For example, for the method of Section 3.3, a generic choice of the auxiliary knots will in the univariate case not lead to the classical splines. On the other hand, a clever selection of these knots does give rise to the ordinary univariate splines, see [27,29,69]. Unfortunately, such selection does not seem to have an analog in higher dimensions. This led de Boor [7] to the grim conclusion that it is "*not likely that simplex splines will be used as a basis for a good subspace of a given smooth [piecewise polynomial] space of functions*". This should be contrasted with his more optimistic view on DMS-splines. In his survey [10], de Boor writes: "*Unfortunately, the first scheme [of Section 3.3] did not lead to a spline space with easily constructed quasi-interpolant schemes. However, very recently, a scheme has become available, in [32], that, in hindsight, appears to be the 'right' one.*

Despite the above praise and a considerable appeal of DMS-splines, *e.g.* exhibited by formula (6), these splines are not a solution to the fundamental problem either. In contrast to the geometric construction of Section 3.3, where there is not enough symmetry, in the case of DMS-splines there is "too much" symmetry in some sense. This symmetry makes it *impossible* to choose the auxiliary knots so as to obtain the classical univariate splines in the case $s = 1$. Thus, DMS-splines are not consistent with requirement (F), except in the special case when the original knots K are not pulled apart, *i.e.*, when the splines remain discontinuous and hence not optimally smooth.

To conclude, we see that in the context of our criteria (A)–(F), none of the four described constructions of multivariate splines can be viewed as "natural". We need to go back to the drawing board and explore alternative ways of defining simplex spline spaces.

§4. A New Construction

Below, we describe a new construction of spline spaces that does not require the introduction of auxiliary knots, and which nevertheless still exhibits many desirable properties. We begin with a glimpse at univariate splines.

4.1. Univariate Splines without Total Ordering of the Real Line

In the univariate case, as we know, the issue of selecting knots for B-splines is resolved trivially by choosing knot-sets formed by *consecutive* knots. Thus, in the one-dimensional case, suitable collections of knots arise naturally as a byproduct of the *total ordering* of the real line. Since there is no such useful ordering in higher-dimensional Euclidean spaces, another principle of selecting appropriate collections of knot-sets $\mathcal{C}_n$ for simplex splines must be found. However, then, in view of property (F), a similar principle should also apply in the univariate case. Hence, there should be a way to dispense with the total ordering and still be able to group the knots into appropriate collections. Moreover, this new selection principle should give rise to the usual consecutive knots for $s = 1$.

Figure 12 displays a set of knots in the plane, along with two collections of knot-sets of size five. There is little doubt which of the two should be picked

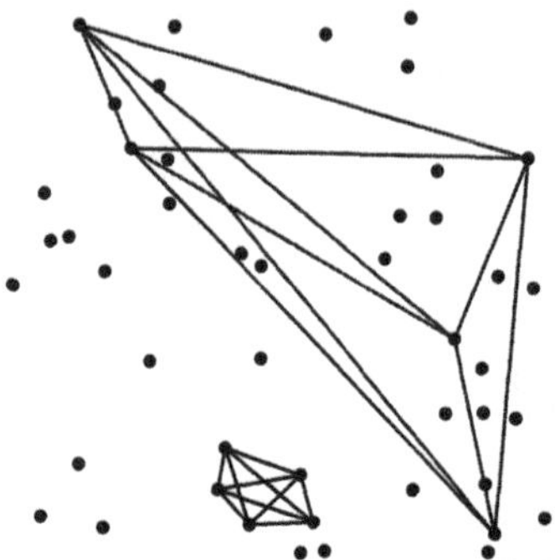

Fig. 12. Two knot-sets of size five.

by any reasonable selection principle. It is clear that the knot-set with smaller convex hull is the obvious choice. As for the other knot-set, its knots seem "too far" from each other, which is unacceptable if we want to avoid simplex splines with "large" supports. This discussion suggests the following heuristic selection principle for constructing the collections $\mathcal{C}_n$.

Heuristic Principle. *Given a knot-set $X \subset K$ of size $n + s + 1$, X should be an element of $\mathcal{C}_n$ if and only if X contains "nearby" knots.*

While obviously very vague, intuitively this principle expresses a natural requirement that $\mathcal{C}_n$ should be defined by choosing knot-sets corresponding to simplex splines with "small" supports. The following two examples show that this heuristic principle can be formulated rigorously.

Example 1. *Define $\mathcal{C}_n$ as the set of all knot-sets $X \subset K$ of size $n + s + 1$ such that $[X] \cap K = X$ i.e., the convex hull of X does not contain any knots from K not belonging to X.*

Example 2. *Define $\mathcal{C}_n$ as the set of all knot-sets $X \subset K$ of size $n+s+1$ such that* $\mathrm{diam}(X \cup \{x\}) > \mathrm{diam}(X)$, *for every $x \in K \backslash X$, where* $\mathrm{diam}(A)$ *denotes the diameter of a set A.*

It is easy to see that in both examples, the construction of $\mathcal{C}_n$ is well defined in any dimension s, since to determine $\mathcal{C}_n$ we need only to compute the convex hull of a finite number of points and its diameter, which clearly can be done for all $s \geq 1$. Also note that for $s = 1$, both principles give rise to knot-sets of consecutive knots, which confirms that the use of total ordering of the real line can be avoided.

It is another question whether the collections $\mathcal{C}_n$ from Examples 1 and 2 are actually suitable for constructing meaningful simplex spline spaces $\mathcal{S}_n$ for $s > 1$. The answer is negative in both cases and for different reasons, even though these collections are identical and the "correct" ones for $s = 1$. The collection $\mathcal{C}_n$ from the first example is not appropriate in higher dimensions since there are too many knot-sets in this collection. Namely, observe that

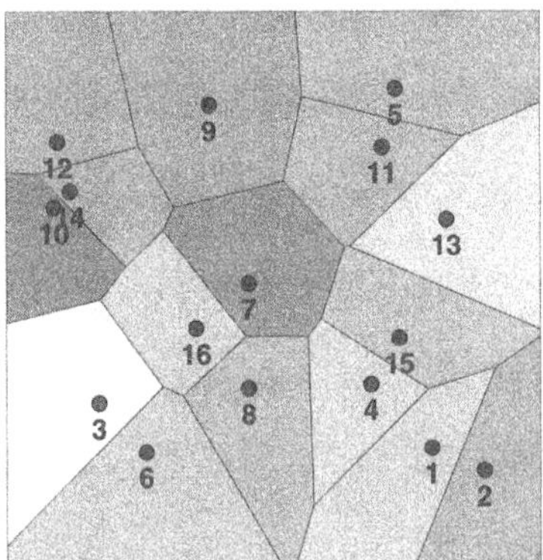

Fig. 13. Voronoi diagram of a set of points.

the condition $[X] \cap K = X$ does not prevent $\mathcal{C}_n$ from containing knot-sets whose convex hulls are "long" and "thin". However, a more detailed analysis shows that the space $\mathcal{S}_n$, corresponding to $\mathcal{C}_n$ via relation (4), is rich enough so that $\Pi_n \subset \mathcal{S}_n$. The collection $\mathcal{C}_n$ obtained in Example 2 has the opposite drawback. This time, $\mathcal{C}_n$ is too small since $\mathcal{S}_n$ does not contain all polynomials of degree n.

There are many other similar selection criteria that lead to knot-sets of consecutive knots in the univariate case, but fail to give rise to good collections in the multivariate case (see Example 4 below). Nevertheless, the above examples illustrate that the idea of selecting knots according to whether the knots are nearby or not, makes sense, at least in principle. We only need to define accurately what "nearby" means. A natural place to look for a good definition is in the theory of Voronoi diagrams, which are well-known concepts from computational geometry. Voronoi diagrams record information about proximity relations between point-sets and can help us determine whether points are close or far from each other.

4.2. Voronoi Diagrams and Higher Order Voronoi Diagrams

The Voronoi diagram of a set of knots K is defined as the partition of $\mathbb{R}^s$ into subsets that are closest to one of the points in K. Thus, for each $x \in K$ there is a closed set $c(x) \subset \mathbb{R}^s$, called a Voronoi cell, such that the distance of every $y \in c(x)$ from x is less than or equal the distance of y from any other point in K. Voronoi cells are known to be polyhedral, *i.e.*, intersections of finitely many closed half-spaces. Figure 13 shows the Voronoi diagram of a set of points in the plane. Voronoi diagrams belong to standard tools in many areas of mathematics and computational geometry. For more on Voronoi diagrams, we refer the reader to [133,134].

Voronoi diagrams can be generalized to the so-called higher order Voronoi diagrams, introduced in 1975 by Shamos and Hoey [137] (see also [133,134]).

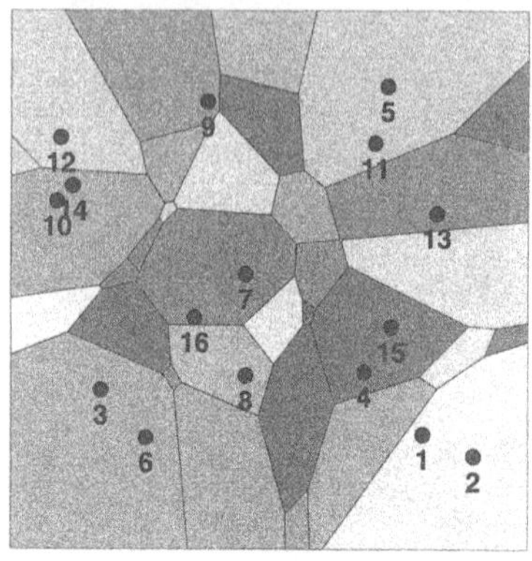

Fig. 14. Voronoi diagram of order two.

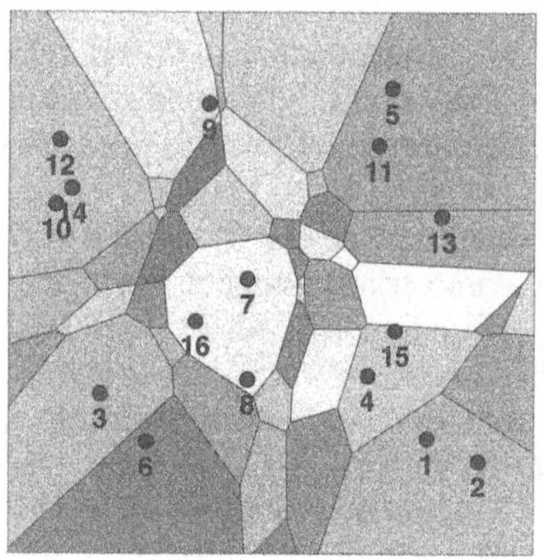

Fig. 15. Voronoi diagram of order three.

The usefulness of these generalizations of Voronoi diagrams stems from the fact that they allow us to find, for any given point $x \in \mathbb{R}^s$ and any number k, a set of k points from K closest to x. The precise definition is as follows.

Definition 3. *The Voronoi diagram of order k of the set K is the subdivision of $\mathbb{R}^s$ into regions, called* Voronoi cells (of degree k), *such that each cell contains the same k nearest points from K. In particular, let $X \subset K$ be such that $\#X = k$. The Voronoi cell $c(X)$ is defined as*

$$c(X) := \{x \in \mathbb{R}^s : \ d(x, X) \leq d(x, Y), Y \subset K, \#Y = k\},$$

where $d(x, Y) := \max_{y \in Y} \|x - y\|$, and $\| \cdot \|$ is the Euclidean norm.

Note that some Voronoi cells could be empty sets. Voronoi diagrams of order one are the usual Voronoi diagrams. Just as in the case of ordinary

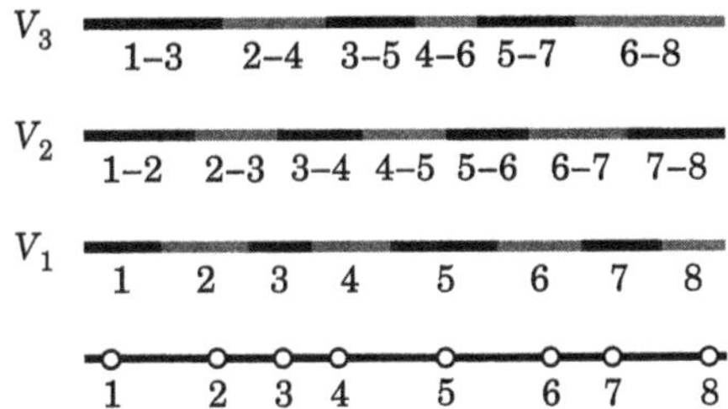

Fig. 16. Sabin's suggestion works well for univariate knot-sets.

diagrams, the cells $c(X)$ are convex polyhedral sets. Figures 14 and 15 show the Voronoi diagrams of orders two and three, respectively, for the point-set from Figure 13.

Let us now return to the problem of constructing appropriate collections of knot-sets $\mathcal{C}_n$, using the heuristic principle mentioned earlier. Definition 3 motivates the following alternative to the constructions in Examples 1 and 2.

Example 4. *Define $\mathcal{C}_n$ as the set of all knot-sets $X \subset K$ of size $n + s + 1$, which correspond to nonempty Voronoi cells of order $n + s + 1$.*

This suggestion makes sense since if $c(X)$ is nonempty, then all knots in X are "nearby" in that there is a region in $\mathbb{R}^s$, namely $c(X)$, such that the points in this region are closer to the knots X than to any other knots from K. Clearly, this means that points in X cannot be too far apart.

The above interesting choice for $\mathcal{C}_n$ was proposed in 1988 by Malcolm Sabin at the *First Oslo Conference on Curves and Surfaces.* In his paper for the proceedings of that conference [107], Sabin discussed the problem of using bivariate quadratic simplex splines to define useful spline spaces. In that paper, he first considers splines with knots on a uniform triangular grid, after which he suggests using higher order Voronoi diagrams to choose knot-sets for quadratic simplex splines. He writes: "*The fifth order Voronoi diagram ... identifies quintuples of vertices on the basis of locality*". However, he continues on a pessimistic note: "*I thought this was a lovely idea until I discovered that it would not produce the regular grid example. Indeed it is difficult to imagine any plausible locality-based selection of fives ... which would omit the vertex in the middle of the regular grid ...*"

It is not difficult to see that in the univariate case, Sabin's idea works well in that it leads to collections of consecutive knots. This is illustrated in Figure 16, displaying Voronoi diagrams of orders one, two, and three for a set of eight knots on the real line. For example, the interval marked as 3–5 in this figure is the Voronoi cell of order three, corresponding to the knots numbered 3,4,5.

Unfortunately, Sabin's suggestion does not give rise to appropriate collections of knot-sets in higher dimensions. This is explained in Figure 17, representing the Voronoi diagram of order three for a set of four knots (numbered 1 through 4 in the figure). It can be seen that in this specific example,

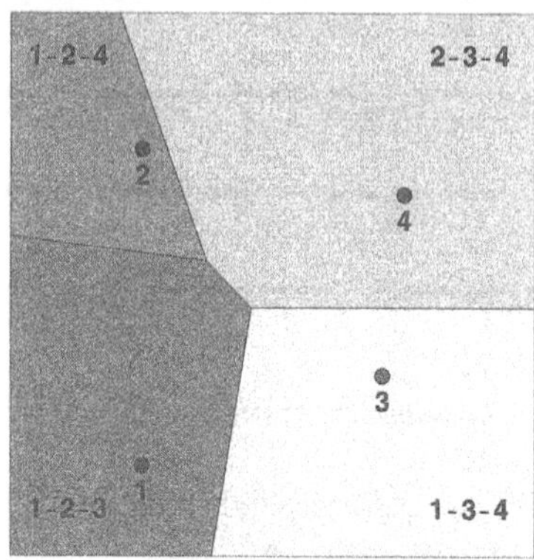

Fig. 17. Voronoi diagram of order three of four knots — Sabin's idea fails.

the Voronoi diagram consists of four nonempty Voronoi cells of order three. For example, the Voronoi cell marked 1-2-4 is the region of points in $\mathbb{R}^2$ that are closest to knots 1, 2, and 4 (or, equivalently, farthest to knot 3). Thus, according to Sabin, the collection $\mathcal{C}_0$ should consist of the four triples $\mathcal{C}_0 = \{\{1,2,3\},\{1,3,4\},\{2,3,4\},\{1,2,4\}\}$. The space $\mathcal{S}_0$ in this case is therefore spanned by the characteristic functions of the triangles whose vertices are these four triples of knots. It is easy to see that these four functions are linearly dependent. Hence, Sabin's recipe leads to a collection of "too many" simplex splines.

In the next section we will see that Sabin's suggestion can be modified to give rise to more appropriate collections $\mathcal{C}_n$.

4.3. Delaunay Triangulations

Returning to the example in Figure 17, it is clear that instead of considering the collection $\mathcal{C}_0$ of all four triples of knots, it would be more natural to consider just two of those knot-sets, for example $\mathcal{C}_0 = \{\{1,2,3\},\{2,3,4\}\}$. These knot-sets give rise to two simplex splines that are clearly linearly independent (since their supports are two non-overlapping triangles) and whose linear combinations reproduce constants on the convex hull of the four knots. More generally, if one wants to construct collections $\mathcal{C}_0$ corresponding to an arbitrary set of knots K, then instead of using the approach of Example 4, it is preferable to start with a *triangulation* Δ of the knots K. This leads to a space $\mathcal{S}_0$, spanned by the lowest-degree simplex splines, namely the characteristic functions of the triangles (simplices) of the triangulation Δ.

This brings us to the concept of the dual graph of the Voronoi diagram, obtained by connecting two points from K by an edge if their associated Voronoi cells are neighbors, *i.e.*, share a common $(s-1)$-dimensional face. From now on, we will assume that the knots K are such that no $s+2$ of them are co-spherical, *i.e.*, lying on a common $(s-1)$-dimensional sphere. For

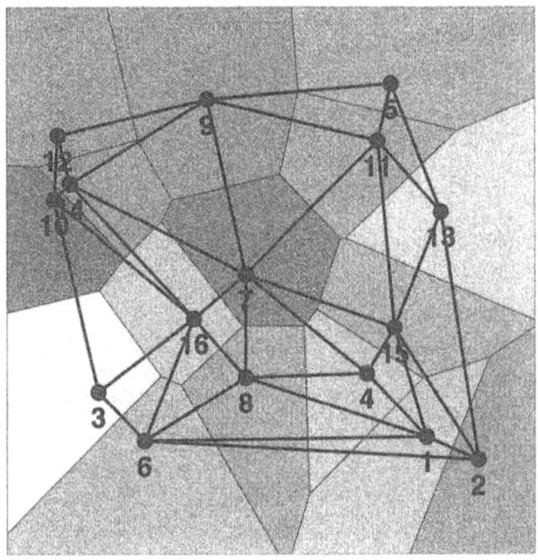

Fig. 18. Delaunay triangulation.

$s = 2$, this assumption means that no four knots from K are co-circular. A consequence of this slight restriction is that any given point in $\mathbb{R}^s$ can belong to at most $s+1$ Voronoi cells. In his 1934 paper [132], B. N. Delaunay proved that the dual graph of the Voronoi diagram is a triangulation of the convex hull $[K]$, now called the Delaunay triangulation. Note that in our case $[K] = \mathbb{R}^s$, by an earlier assumption on K. The Delaunay triangulation corresponding to the point-set from Figure 13 is shown in Figure 18.

For our purposes, it will be convenient to interpret Delaunay's result in function-theoretic language. Namely, denoting the Delaunay triangulation by Δ, this result can be restated as

$$\sum_{T \in \Delta} \mathrm{vol}_s(T) M(x|V(T)) = 1, \quad x \in \mathbb{R}^s, \tag{7}$$

where $V(T)$ are the vertices of the simplex $T \in \Delta$. This follows from formula $M(x|V(T)) = \chi_T / \mathrm{vol}_s(T)$, which is a consequence of (2). Hence, identity (7) asserts that appropriate linear combinations of the lowest-degree simplex splines, corresponding to the collection of knot-sets

$$\mathcal{C}_0 = \{V(T), T \in \Delta\}, \tag{8}$$

reproduce constants. Strictly speaking, equality (7) is true only for all x in the interior of the simplices of Δ, but one can redefine the lowest-degree simplex splines on the boundary of the simplices so as to achieve pointwise equality everywhere on $\mathbb{R}^s$.

To summarize, in the lowest-degree case $n = 0$, it is a better idea to utilize Delaunay triangulations to construct collections $\mathcal{C}_0$, instead of Voronoi diagrams of order $s+1$, as suggested by Sabin. This is because the collection (8) clearly gives rise to independent simplex splines and yet their number is large enough so that $\Pi_0 \subset \mathcal{S}_0$.

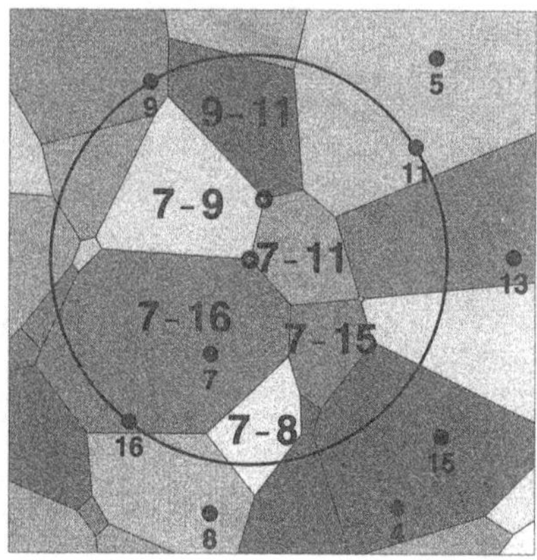

Fig. 19. Voronoi diagram of order two and Delaunay configuration of degree one.

4.4. Delaunay Configurations

Motivated by (7) and (8), the main idea of our new construction can be explained as follows. Since the lowest-degree splines are obtained from Delaunay triangulations, it is conceivable that splines of degrees $n \geq 1$ can be generated using appropriate higher-degree analogs of such triangulations, called here Delaunay configurations. Just as Delaunay triangulations can be derived from Voronoi diagrams, Delaunay configurations will be intimately related to the already mentioned higher order Voronoi diagrams.

To motivate the definition of such configurations, let us consider the particular planar Voronoi diagram of order two, displayed in Figure 19. The large-font pairs of numbers in some of the Voronoi cells represent the pairs of knots associated with a given cell. Thus, for example, the cell marked 7−16 consists of all points in the plane that are closer to knots 7 and 16 than to any other pair from the given set of knots. To explain the idea behind Delaunay configurations, let us recall that a Voronoi vertex is a point in the plane (or in $\mathbb{R}^s$) which is common to exactly three (or $s+1$) Voronoi cells (recall that by our assumption, there are no points in $\mathbb{R}^2$ ($\mathbb{R}^s$) belonging to four ($s+2$) or more Voronoi cells). It is not difficult to see that every Voronoi vertex is the incenter of a circle circumscribed to a triple of knots. In fact, there are precisely two types of such Voronoi vertices, depicted in the figure. The first type corresponds to the Voronoi vertex common to cells 7−9,7−11,9−11, hence it is the center of the circle passing through knots 7,9,11. The second depicted Voronoi vertex is common to cells 7−9,7−11,7−16 and is the center of the circle passing through knots 9,11,16. The essential difference between these two types of vertices is that the first mentioned circle contains no knots in its interior, whereas the second circle contains one knot, namely the knot 7. Therefore, by a well-known fact about Delaunay triangulations, in the first case the triangle with vertices 7,9,11 is a Delaunay triangle, while in

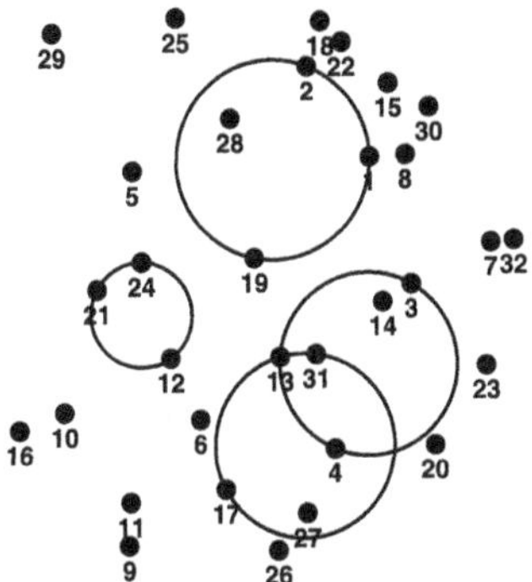

Fig. 20. Delaunay configurations and circumscribed circles.

the second case, the triangle with vertices 9,11,16 is not a Delaunay triangle. The Delaunay triangle gives rise, as we mentioned earlier, to a simplex spline of degree zero (a constant simplex spline). On the other hand, the quadruple $(7, 9, 11, 16)$ corresponds to a (linear) simplex spline of degree one. This quadruple will be called a Delaunay configuration of degree one ("one" refers to the number of knots inside the circle circumscribed to the triangle with vertices 9,11,16) or simply a Delaunay quadruple. We shall see that simplex splines corresponding to such collections will give rise to a spline space with several striking properties.

The above discussion motivates the following definition.

Definition 5. *The family of pairs*

$$\Delta_n := \{X = (X_B, X_I)\},$$

such that

$$X_B, X_I \subset K, \quad \#X_B = s + 1, \quad \#X_I = n,$$

and such that the closed ball $\Omega \subset \mathbb{R}^s$*, circumscribed to* X_B*, contains* X_I *in its interior and no other knots from* K*, i.e.,* $X_B \cup X_I = \Omega \cap K$ *and* $X_I = \operatorname{int}(\Omega) \cap K$*, is called the* (oriented) Delaunay configuration of degree n associated with the knots K.

Figure 20 shows examples of Delaunay configurations of varying degrees. In particular, the triple of points $\{12, 21, 24\}$ is a Delaunay triangle, $(\{1, 2, 19\}, \{28\})$ is a Delaunay quadruple, and both $(\{3, 4, 13\}, \{14, 31\})$ and $(\{13, 17, 31\}, \{4, 27\})$ are Delaunay quintuples (*i.e.*, Delaunay configurations of degree two). Figure 21 shows a Delaunay configuration of degree three.

Before we embark on discussing the relevance of the above notion for constructing multivariate splines, a few remarks on Definition 5 are in order. The concept of a Delaunay configuration seems to be new. Although it is implicit in several papers on higher order Voronoi diagrams, see *e.g.* [138], it

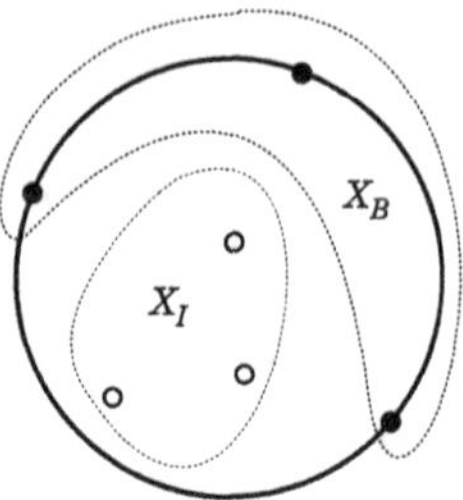

Fig. 21. Boundary and interior knots of a Delaunay configuration of degree three.

does not seem to have been studied as a separate entity. It is immediately clear from the well-known properties of Delaunay triangulations that these are Delaunay configurations of degree $n = 0$, *i.e.*, $\Delta_0 = \Delta$. It should be noted that Delaunay configurations for $n \geq 1$ do not form a partition of $\mathbb{R}^s$ *i.e.*, a Delaunay configuration is not a tessellation. Clearly, for every $X \in \Delta_n$, the sets X_B and X_I are disjoint. In our notation, the subscript "B" stands for "boundary" and "I" for "interior" knots.

4.5. Partition of Unity and Reproduction of Polynomials

We now return to the central issue of this section and present a construction of appropriate collections $\mathcal{C}_n$ of knot-sets, along with the accompanying simplex spline spaces $\mathcal{S}_n$.

First recall that a Delaunay configuration of degree n corresponds to a collection of knot-sets of size $n + s + 1$. This suggests that we define $\mathcal{S}_n$ as the space spanned by the simplex splines whose knot-sets are precisely the elements of the Delaunay configuration Δ_n. As we have pointed out earlier, this certainly makes sense in the lowest-degree case $n = 0$, since then $\mathcal{S}_0$ contains constant functions, as a consequence of (7). In fact, this property carries over to all degrees n.

To formulate the above more precisely, let us first recall the definition of polar forms of polynomials. The concept of a polar form or "blossom", while known for quite some time in an algebraic context, has been introduced into the spline theory by de Casteljau and independently by Ramshaw (see [135], for an introduction). Polar forms have proven to be a convenient mathematical tool for describing (piecewise) polynomial functions and for analyzing various spline algorithms. Given an s-variate polynomial $p \in \Pi_n$, the polar form P of p is defined as the unique function of n vector variables $x_1, \ldots, x_n \in \mathbb{R}^s$, which is symmetric, affine in each of these variables, and such that p is equal to P on the diagonal, *i.e.*, $P(x, \ldots, x) = p(x), x \in \mathbb{R}^s$. For $n = 0$, we define $P := p$.

We are now ready to state the announced result, that every polynomial $p \in \Pi_n$ can be expressed as a linear combination of the simplex splines whose knot-sets are Delaunay configurations of degree n.

Theorem 6. *Let $p \in \Pi_n$ and let P be the polar form of p. Also, let Δ_n be the Delaunay configuration of degree n of the set of knots $K \subset \mathbb{R}^s$. Then*

$$p = \sum_{X \in \Delta_n} P(X_I) N(\cdot|X), \tag{9}$$

where $N(\cdot|X)$ are normalized simplex splines, defined by

$$N(\cdot|X) := \binom{n+s}{s}^{-1} \mathrm{vol}_s[X_B] M(\cdot|X), \tag{10}$$

and $M(\cdot|X)$ is the simplex spline with knots $X_B \cup X_I$. In particular,

$$\Pi_n \subset \mathcal{S}_n,$$

where $\mathcal{S}_n$ is the spline space corresponding to Δ_n, i.e.,

$$\mathcal{S}_n := \mathrm{span}\{M(\cdot|X), X \in \Delta_n\}.$$

Proof: The theorem is proved in [95]. However, let us remark that for $n = 0$, assertion (9) follows from the fact that $N(\cdot|X)$ is the characteristic function of the Delaunay simplex $[X_B]$ and that such simplices form a partition of $\mathbb{R}^s$. Hence, (9) is just a restatement of (7).

For $n \geq 1$, the proof employs the identity

$$\sum_{X \in \Delta_n} P(X_I) N(\cdot|X) = \sum_{X' \in \Delta_{n-1}} P(X'_I, x) N(\cdot|X'). \tag{11}$$

This identity can be proved using recurrence (3) along with some combinatorial properties of Delaunay configurations. Namely, there is a natural way to use (3) in our setting to evaluate $N(\cdot|X)$, which is a scalar multiple of $M(\cdot|X)$. In particular, we can choose the coefficients $\lambda_y, y \in X$, in (3) to be such that all but $s+1$ of them are zero, where the nonzero values are associated with the $s+1$ knots $y \in X_B$. Since these knots are in general position, it is well known that the coefficients λ_y are uniquely determined. The recurrence relation allows us to rewrite the left-hand side of (11) as a linear combination of simplex splines of degree $n-1$. The proof of (11) is completed by a judicious manipulation of this linear combination. Next, iterating identity (11) n times, one obtains

$$\begin{aligned}
\sum_{X \in \Delta_n} P(X_I) N(x|X) &= \sum_{X'' \in \Delta_{n-2}} P(X''_I, x, x) N(\cdot|X'') = \dots \\
&= \sum_{X''' \in \Delta_0} P(x, \dots, x) N(x|X''') \\
&= p(x) \sum_{X''' \in \Delta_0} \chi_{[X''']}(x) = p(x),
\end{aligned}$$

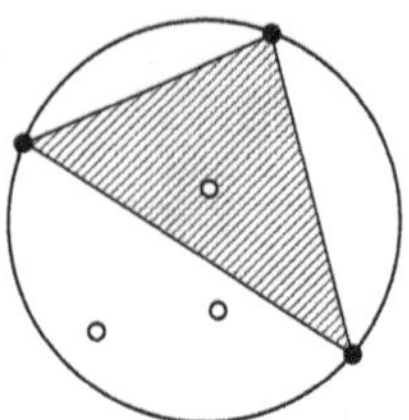

Fig. 22. The area of the triangle enters the normalizing factor in (10).

for all $x \in \mathbb{R}^s$, where we again used that Δ_0 is a tessellation of $\mathbb{R}^s$. □

An immediate consequence of (10) is that the normalized simplex splines add up to one. This follows from (9) by setting $p \equiv 1$, or $P \equiv 1$. The geometric nature of the normalization is illustrated for a Delaunay configuration of degree three in Figure 22.

Corollary 7. *The normalized simplex splines form a partition of unity, i.e.,*

$$\sum_{X \in \Delta_n} N(\cdot|X) \equiv 1.$$

4.6. The Univariate Case

It will be interesting to compare identities (9) and (10) with their univariate counterparts. Consider an increasing sequence of knots $K = \{x_i\}_{i \in \mathbb{Z}} \subset \mathbb{R}$. The univariate normalized B-spline (see [136]) with knots $x_i, \ldots, x_{i+n+1}$ is given by

$$N(\cdot|x_i, \ldots, x_{i+n+1}) = \frac{x_{i+n+1} - x_i}{n+1} M(\cdot|x_i, \ldots, x_{i+n+1}), \tag{12}$$

where M is the univariate B-spline of degree n, normalized to have unit integral. Observe that (12) is the *exact* analog of (10). This is because the knot-set $\{x_i, \ldots, x_{i+n+1}\}$ can be thought of as an element of the Delaunay configuration of K of degree n, with $X = (X_B, X_I)$, where the "boundary" knots are $X_B = \{x_i, x_{i+n+1}\}$ and the "interior" knots are given as $X_I = \{x_{i+1}, \ldots, x_{i+n}\}$. Note that the Delaunay configuration of degree n of the knot-set K is precisely the set of all $(n+2)$-tuples of consecutive knots from K. Thus, setting $s = 1$ and noticing that $\mathrm{vol}_1[X_B] = x_{i+n+1} - x_i$, shows that (10) reduces to (12).

The univariate counterpart of (9) is

$$p = \sum_{i \in \mathbb{Z}} P(x_{i+1}, \ldots, x_{i+n}) N(\cdot|x_i, \ldots, x_{i+n+1}), \tag{13}$$

where p is now a univariate polynomial of degree at most n, and P is the polar form of p. This is a well-known polynomial reproduction formula for univariate splines, see [135].

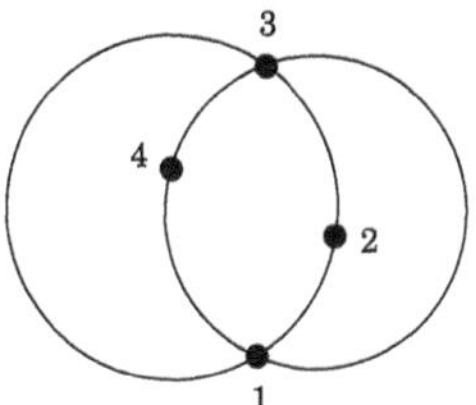

Fig. 23. Two different Delaunay configurations of degree one with identical knots.

§5. Are Simplex Splines B-Splines?

Does the spline space $\mathcal{S}_n$ introduced in the previous section provide a solution to the fundamental problem? We have seen in Section 4.6 that $\mathcal{S}_n$ reduces to the univariate splines for $s = 1$ and hence requirement (F) is met. Conditions (A) and (B) are consequences of the property of simplex splines that, for generic knots, they are piecewise polynomials of optimal smoothness. Requirement (C) is the assertion of Theorem 6. As for (D), this follows from the fact that for any given compact set $\Omega \subset \mathbb{R}^s$ there is a finite number of simplex splines overlapping Ω, which in turn is a consequence of the local nature of Delaunay configurations [95].

Unfortunately, the collection of simplex splines $\mathcal{B}_n := \{M(\cdot|X), X \in \Delta_n\}$, that spans $\mathcal{S}_n$, may not satisfy property (E). This is because the simplex splines in $\mathcal{B}_n$ can be dependent, *i.e.*, these splines are not necessarily B-splines. The reason for this is that some Delaunay configurations might be different as pairs and yet the same as sets. In particular, there could be two different configurations (X_B^1, X_I^1) and (X_B^2, X_I^2), corresponding to the same set of points, *i.e.*, $X_B^1 \cup X_I^1 = X_B^2 \cup X_I^2$, but $X_B^1 \neq X_B^2$, $X_I^1 \neq X_I^2$. An example of this situation is displayed in Figure 23, depicting four knots from a set K (the remaining knots from K are not shown in the figure, but are assumed to lie outside both circles). In this figure, two Delaunay configurations are formed from the four knots, namely, $X^1 = (X_B^1, X_I^1) = (\{1,2,3\},\{4\})$ and $X^2 = (X_B^2, X_I^2) = (\{1,3,4\},\{2\})$. Hence, $M(\cdot|X^1)$ and $M(\cdot|X^2)$ are trivially dependent since they are identical.

One way to deal with the above difficulty is to identify all such dependent splines in $\mathcal{B}_n$. Equivalently, instead of using the Delaunay configurations introduced in Definition 5, one could consider "un-oriented" Delaunay configurations, consisting of knot-sets of the form $X_B \cup X_I$, instead of the pairs (X_B, X_I). Indeed, it turns out that by removing multiple entries in the collection $\mathcal{B}_n$, the resulting simplex splines are B-splines in the sense of property (E) and, in particular, these splines are independent. Thus, with this straightforward modification of $\mathcal{B}_n$, the space $\mathcal{S}_n$ is a solution to the fundamental problem. Note, however, that for certain purposes it may still be important to treat the configurations of knots that are the same as sets, as being different, for example when employing identity (9).

Before we address another possibility of how to get rid of the described dependencies in the collection $\{M(\cdot|X), X \in \Delta_n\}$, let us recall an interesting identity for univariate splines, which is a generalization of the reproduction formula (13), see [135]. Let f be a univariate spline of degree n with knots $K = \{x_i\}_{i\in\mathbb{Z}}$. Then

$$f = \sum_{i\in\mathbb{Z}} F_i(x_{i+1}, \ldots, x_{i+n}) N(\cdot|x_i, \ldots, x_{i+n+1}). \tag{14}$$

Here, the symbol F_i stands for the polar form of the polynomial $f|_{(x_j,x_{j+1})}$, the restriction of f to an interval (x_j, x_{j+1}), where j is any integer such that $i \le j \le i+n$. Equivalently, F_i can be taken as the polar form of any polynomial piece of f that "lives" inside the interval $[x_i, x_{i+n+1}]$, the support of the B-spline $M(\cdot|x_i, \ldots, x_{i+n+1})$. In spite of the freedom that we have to choose F_i, formula (14) is well defined since one can show that the value of F_i at the knots $x_{i+1}, \ldots, x_{i+n}$ is the same, no matter which interval (x_j, x_{j+1}) is picked for the definition of F_i.

Identity (14) allows us to express any univariate piecewise polynomial explicitly as a linear combination of B-splines. We have already encountered a multivariate analog of this identity in this paper. Namely, we have seen in Section 3.4 that formula (6) makes it possible to express every $f \in \Pi_n(\Delta)$, *i.e.*, every optimally smooth piecewise polynomial on a triangulation Δ, as a linear combination of simplex splines. However, formula (6) is not a full analog of (14) since it holds only for single polynomials or piecewise polynomials on triangulations, but not for *all* DMS-splines.

Thus, a natural question is whether identity (14) has a generalization for our new type of splines. It turns out (see Theorem 9 below) that, just as with the space of DMS-splines, the space $\mathcal{S}_n$ introduced in Section 4 also contains $\Pi_n(\Delta)$, where $\Delta = \Delta_0$. As a consequence, (9) can be extended to all piecewise polynomials from $\Pi_n(\Delta)$. However, again, there does not seem to be an extension of (9) that would express *every* spline in $\mathcal{S}_n$ as a combination of simplex splines from $\mathcal{B}_n$. An additional aesthetic flaw of (9) is that the arguments X_I in (9) are equal for all simplex splines with the same interior knots, and hence those simplex splines have identical coefficients in this formula. Therefore, the simplex splines in $\mathcal{B}_n$ are "redundant", in some sense.

This brings us to a second possibility of how one can handle the linear dependence of the simplex splines $\{M(\cdot|X), X \in \Delta_n\}$. We could define a new set of compactly-supported splines by adding all simplex splines in $\{N(\cdot|X), X \in \Delta_n\}$ sharing the same set of interior knots. The linear span of these new functions, denoted by $\mathcal{S}'_n$, will be a smaller space of splines, *i.e.*, $\mathcal{S}'_n$ will be a proper subspace of $\mathcal{S}_n$. Nevertheless, $\mathcal{S}'_n$ will still be large enough since it will contain all polynomials of degree n.

Definition 8. *Let* $n \ge 1$, $\mathcal{I}_n := \{X_I : X = (X_B, X_I) \in \Delta_n\}$, *and* $I \in \mathcal{I}_n$. *A* multivariate B-spline B_I *is the function defined as*

$$B_I := \sum_{X\in\Delta_n, X_I = I} N(\cdot|X).$$

It is an immediate consequence of (9) that the B-splines $\mathcal{B}'_n := \{B_I, I \in \mathcal{I}_n\}$ form a partition of unity and the reproduction formula (9) can be written as

$$p = \sum_{I \in \mathcal{I}_n} P(I) B_I, \quad p \in \Pi_n,$$

which confirms that the new space $\mathcal{S}'_n := \operatorname{span}\{B_I, I \in \mathcal{I}_n\} \subset \mathcal{S}_n$ still contains Π_n. In fact, this space, together with the collection $\mathcal{B}'_n$, also solves the fundamental problem. Note that in the univariate case, there is only one simplex spline per set I of interior knots. Hence, the new multivariate B-splines reduce to the univariate B-splines for $s = 1$, which establishes property (F).

We finish this section with a striking generalization of formula (9), which, unlike (6), is a *complete* analog of (14). Let $f \in \mathcal{S}'_n$ and $I \in \mathcal{I}_n$. Just as in the univariate case, it can be shown that if f_I is a single polynomial piece of f on a region in the support of B_I, bounded but not intersected by any hyperplane containing s knots, then the polar form F_I of f_I, evaluated at X_I, does not depend on which polynomial piece f_I we pick. That is, the number $F_I(X_I)$ is well defined as long as the polynomial f_I corresponds to a region inside the support of B_I. With this notation, we obtain

Theorem 9 [95]. *Every spline in the space $\mathcal{S}'_n$ can be written uniquely as a linear combination of B-splines. Namely,*

$$f = \sum_{I \in \mathcal{I}_n} F_I(I) B_I,$$

for every $f \in \mathcal{S}'_n$. Moreover, the space $\mathcal{S}'_n$ contains all optimally smooth piecewise polynomials of degree n, associated with the Delaunay triangulation Δ, i.e., $\Pi_n(\Delta) \subset \mathcal{S}'_n$.

§6. Conclusions

The results of the previous two sections show that simplex splines can give rise to interesting spaces of multivariate splines. In particular, our construction yields spline spaces for which polynomials can be represented explicitly in an elegant and simple way. This representation is virtually identical to the univariate case, which is in contrast to earlier constructions, such as the construction of DMS-splines in [32].

On the other hand, we leave many obvious questions unanswered, such as: How can one compute Delaunay configurations?, Can the splines be efficiently and stably evaluated by means of a "de Boor-like" algorithm?, Do the splines have good approximation properties?

While the new splines seem very natural in many respects, it is also not clear at this time how useful they might become for applications. In this paper, I viewed the described fundamental problem as a purely mathematical problem, whose consideration may or may not have practical significance. In my opinion a natural construction of splines should be mathematically elegant and conceptually simple. However, this does not necessarily mean simplicity in the computational sense. Therefore, at this stage of the investigation it is too early to tell whether the splines constructed here can compete successfully with other types of multivariate splines currently in use. Nevertheless, I hope that the described results will spark sufficient interest to continue the research along the new directions outlined in this paper.

Acknowledgments. I would like to thank Tom Lyche, Knut Mørken, Ewald Quak, and Larry Schumaker, for organizing this successful conference and for generously inviting me to present a lecture. This research was supported by the National Science Foundation under grant DMS-9803501.

References on Simplex Splines

1. Atteia, M., *Hilbertian Kernels and Spline Functions*, Studies in Computational Mathematics, 4, North-Holland Publishing Co., Amsterdam, 1992.
2. Auerbach, S., R. H. J. Gmelig Meyling, M. Neamtu, and H. Schaeben, Approximation and geometric modeling with simplex B-splines associated with irregular triangles, Comput. Aided Geom. Design **8** (1991), 67–87.
3. Boehm, W., Multivariate spline algorithms, in *The Mathematics of Surfaces*, J. A. Gregory (ed.), Clarendon Press, Oxford, 1986, 197–215.
4. Bojanov, B. D., H. A. Hakopian, and A. A. Sahakian, *Spline functions and multivariate interpolations*, Mathematics and its Applications, 248, Kluwer, Dordrecht, 1993.
5. Boor, C. de, Splines as linear combinations of B-splines. A survey, in *Approximation Theory, II*, G. G. Lorentz, C. K. Chui, and L. L. Schumaker (eds.), Academic Press, New York, 1976, 1–47.
6. Boor, C. de, Topics in multivariate approximation theory, in *Topics in Numerical Analysis*, P. R. Turner (ed.), Lecture Notes in Math., 965, Springer, Berlin-New York, 1982, 39–78.
7. Boor, C. de, Multivariate approximation, in *The State of the Art in Numerical Analysis*, A. Iserles and M. J. D. Powell (eds.), Oxford University Press, New York, 1987, 87–109.
8. Boor, C. de, What is a multivariate spline?, in *Proc. First Intern. Conf. Industr. Applied Math., Paris 1987*, J. McKenna and R. Temam (eds.), SIAM, Philadelphia, 1988, 90–101.
9. Boor, C. de, *Splinefunktionen*, Lectures in Mathematics ETH Zürich, Birkhäuser, Basel, 1990.
10. Boor, C. de, Multivariate piecewise polynomials, Acta Numerica **2** (1993), 65–109.
11. Boor, C. de and K. Höllig, Recurrence relations for multivariate B-splines, Proc. Amer. Math. Soc. **85** (1982), 397–400.

12. Carlson, B. C., B-splines, hypergeometric functions, and Dirichlet averages, J. Approx. Theory **67** (1991), 311–325.

13. Cohen, E. T., T. Lyche, and R. Riesenfeld, Cones and recurrence relations for simplex splines, Constr. Approx. **3** (1987), 131–141.

14. Dæhlen, M., On the evaluation of box splines, in *Mathematical Methods in Computer Aided Geometric Design*, T. Lyche and L. L. Schumaker (eds.), Academic Press, New York, 1989, 167–179.

15. Dahmen, W., Multivariate B-splines — recurrence relations and linear combinations of truncated powers, in *Multivariate Approximation Theory* (Proc. Conf., Math. Res. Inst., Oberwolfach, 1979), Internat. Ser. Numer. Math., 51, Birkhäuser, Basel-Boston, 1979, 64–82.

16. Dahmen, W., Polynomials as linear combinations of multivariate B-splines, Math. Z. **169** (1979), 93–98.

17. Dahmen, W., Approximation by smooth multivariate splines on nonuniform grids, in *Quantitative Approximation*, R. A. DeVore and K. Scherer (eds.), Academic Press, New York, 1980, 99–114.

18. Dahmen, W., On multivariate B-splines, SIAM J. Numer. Anal. **17** (1980), 179–191.

19. Dahmen, W., Konstruktion mehrdimensionaler B-splines und ihre Anwendungen auf Approximationsprobleme, in *Numerical Methods of Approximation Theory*, L. Collatz, G. Meinardus, and H. Werner (eds.), Birkhäuser, Basel, 1980, 84–110.

20. Dahmen, W., Multivariate B-splines — ein neuer Ansatz im Rahmen der konstruktiven mehrdimensionalen Approximationstheorie, Habilitationsschrift, Bonn, 1980.

21. Dahmen, W., Approximation by linear combinations of multivariate B-splines. J. Approx. Theory **31** (1981), 299–324.

22. Dahmen, W., Adaptive approximation by multivariate smooth splines, J. Approx. Theory **36** (1982), 119–140.

23. Dahmen, W. and C. A. Micchelli, Numerical algorithms for least squares approximation by multivariate B-splines, in *Numerical Methods of Approximation Theory*, vol. 6, L. Collatz, G. Meinardus, and H. Werner (eds.), Birkhäuser, Basel, 1981, 85–114.

24. Dahmen, W. and C. A. Micchelli, On limits of multivariate B-splines, J. Analyse Math. **39** (1981), 256–278.

25. Dahmen, W. and C. A. Micchelli, Computation of inner products of multivariate B-splines, Numer. Func. Anal. Optim. **3** (1981), 367–375.

26. Dahmen, W. and C. A. Micchelli, Some remarks on multivariate B-splines, in *Multivariate Approximation Theory II (Oberwolfach, 1982)*, Internat. Ser. Numer. Math., 61, Birkhäuser, Basel-Boston, 1982, 81–87.

27. Dahmen, W. and C. A. Micchelli, On the linear independence of multivariate B-splines. I. Triangulations of simploids, SIAM J. Numer. Anal. **19** (1982), 993–1012.

28. Dahmen, W. and C. A. Micchelli, On the linear independence of multivariate B-splines. II. Complete configurations, Math. Comp. **41** (1983), 143–163.

29. Dahmen, W. and C. A. Micchelli, Multivariate splines — A new constructive approach, in *Surfaces in Computer Aided Geometric Design*, R. E. Barnhill and W. Boehm (eds.), North-Holland, Amsterdam-New York, 1983, 191–215.

30. Dahmen, W. and C. A. Micchelli, Recent progress in multivariate splines, *Approximation Theory IV*, C. Chui, L. Schumaker, and J. Ward (eds.), Academic Press, New York, 1983, 27–121.

31. Dahmen, W. and C. A. Micchelli, Statistical encounters with B-splines, in *Function Estimates (Arcata, Calif., 1985)*, Contemp. Math., 59, Amer. Math. Soc., Providence, 1986, 17–48.

32. Dahmen, W., C. A. Micchelli, and H.-P. Seidel, Blossoming begets B-splines built better by B-patches, Math. Comp. **59** (1992), 97–115.

33. Dickey, J., P. H. Garthwaite, and G. Bian, An elementary continuous-type nonparametric distribution estimate, Int. J. Math. Stat. Sci. **4** (1995), 193–247.

34. Farwig, R., Multivariate truncated powers and B-splines with coalescent knots, SIAM J. Numer. Anal. **22** (1985), 592–603.

35. Fleet, van P. J., Some recurrence formulas for box splines and cone splines, 1994, preprint.

36. Fong, P., Shape control for multivariate B-spline surfaces over arbitrary triangulations, M.Sc. thesis, University of Waterloo, 1992.

37. Fong, P. and H.-P. Seidel, Control points for multivariate B-spline surfaces over arbitrary triangulations, Computer Graphics Forum **10** (1991), 309-317.

38. Fong, P. and H.-P. Seidel, An implementation of multivariate B-spline surfaces over arbitrary triangulations, in *Proceedings of Graphics Interface '92*, Canadian Human-Computer Communications Society, Morgan Kaufmann, 1992, 1–10.

39. Fong, P. and H.-P. Seidel, An implementation of triangular B-spline surfaces over arbitrary triangulations, Comput. Aided Geom. Design **10** (1993), 267–275.

40. Franssen, M. G. J., Evaluation of DMS-splines, M.Sc. thesis, Eindhoven University, 1995.

41. Gasparini, M., Exact multivariate Bayesian bootstrap distributions of moments, Ann. Statist. **23** (1995), 762–768.

42. Gmelig Meyling, R. H. J., Polynomial spline approximation in two variables, Dissertation, University of Amsterdam, 1986.

43. Gmelig Meyling, R. H. J., An algorithm for constructing configurations of knots for bivariate B-splines, SIAM J. Numer. Anal. **24** (1987), 706–724.

44. Gmelig Meyling, R. H. J., On algorithms and applications for bivariate B-splines, in *Algorithms for Approximation*, J. C. Mason and M. G. Cox (eds.), Oxford University Press, 1987, 83–93.

45. Goodman, T. N. T., Interpolation in minimum seminorm and multivariate B-splines, J. Approx. Theory **37** (1983), 212–223.

46. Goodman, T. N. T., Some properties of bivariate Bernstein-Schoenberg operators, Constr. Approx. **3** (1987), 123–130.

47. Goodman, T. N. T., Polyhedral splines, in *Computation of Curves and Surfaces*, W. Dahmen, M. Gasca, and C. A. Micchelli (eds.), Kluwer, Dordrecht, 1990, 347–382.

48. Goodman, T. N. T., Asymptotic formulas for multivariate Bernstein-Schoenberg operators, Constr. Approx. **11** (1995), 439–453.

49. Goodman, T. N. T., Bernstein-Schoenberg operators, in *Mathematical Methods for Curves and Surfaces*, M. Dæhlen, T. Lyche, and L. L. Schumaker (eds.), Vanderbilt University Press, Nashville, 1995, 161–175.

50. Goodman, T. N. T. and S. L. Lee, Spline approximation operators of Bernstein-Schoenberg type in one and two variables, J. Approx. Theory **33** (1981), 248–263.

51. Gormaz, R., Floraisons polynomiales: Applications et à l'étude des B-splines à plusieurs variables, Dissertation, L'Universite Joseph Fourier, 1993.

52. Gormaz, R., B-spline knot-line elimination and Bézier continuity conditions, *Curves and Surfaces in Geometric Design*, P.-J. Laurent, A. Le Méhauté, and L. L. Schumaker (eds.), A. K. Peters, Wellesley MA, 1994, 209–216.

53. Gormaz, R., Polar cone splines, in *Multivariate Approximation: Recent Trends and Results*, W. Haußman, K. Jetter, and M. Reimer (eds.), Math. Res. **101** (1997), Akademie Verlag, Berlin, 83–94.

54. Gormaz, R. and P.-J. Laurent, Some results on blossoming and multivariate B-splines, in *Multivariate Approximation: From CAGD to Wavelets*, K. Jetter and F. Utreras (eds.), Ser. Approx. Decompos., 3, World Sci. Publishing, River Edge, 1993, 147–165.

55. Gormaz, R. and C. A. Micchelli, On the convergence of B-patch subdivision, Annals of Numerical Mathematics **3** (1996), 105–115.

56. Grandine, T. A., Computing with multivariate simplex splines, *Approximation Theory V*, C. Chui, L. Schumaker, and J. Ward (eds.), Academic Press, New York, 1986, 359–362.

57. Grandine, T. A., The evaluation of inner products of multivariate simplex splines, SIAM J. Numer. Anal. **24** (1987), 882–886.

58. Grandine, T. A., The computational cost of simplex spline functions, SIAM J. Numer. Anal. **24** (1987), 887–890.

59. Grandine, T. A., The stable evaluation of multivariate simplex splines, Math. Comp. **181** (1988), 197–205.

60. Greiner, G. and H.-P. Seidel, Modeling with triangular B-splines, IEEE Comp. Graphics Appl. **14** (1994), 56–60.

61. Ha, K. V., On multivariate simplex B-splines, Dissertation, University of Oslo, 1988.

62. Hakopian, H. A., Multivariate spline functions, divided differences and polynomial interpolations, Dissertation, University of Gdansk, 1981.

63. Hakopian, H. A., Les différences divisées de plusieurs variables et les interpolations multidimensionnelles de types lagrangien et hermitien, C. R. Acad. Sci. Paris Sér. I Math. **292** (1981), 453–456.

64. Hakopian, H. A., Multivariate divided differences and multivariate interpolation of Langrange and Hermite type, J. Approx. Theory **34** (1982), 286–305.

65. Hakopian, H., Multivariate spline functions, B-spline basis and polynomial interpolations, SIAM J. Numer. Anal. **19** (1982), 510–517.

66. Hakopian, H., Multivariate spline functions, B-spline bases and polynomial interpolations. II, Studia Math. **79** (1984), 91–102.

67. Hakopian, H. A. and A. A. Sahakian, Multidimensional splines and polynomial interpolation, Uspekhi Mat. Nauk **48** (1993), 3–76.

68. Höllig, K., A remark on multivariate B-splines, J. Approx. Theory **33** (1981), 119–125.

69. Höllig, K., Multivariate splines, SIAM J. Numer. Anal. **19** (1982), 1013–1031.

70. Höllig, K., Multivariate splines, in *Approximation Theory* (New Orleans, La., 1986), Proc. Sympos. Appl. Math., 36, AMS, Providence, 1986, 103–127.

71. Höllig, K. and C. A. Micchelli, Divided differences, hyperbolic equations, and lifting distributions, Constr. Approx. **3** (1987), 143–156.

72. Hu, C. L. and S. L. Lee, Multivariate B-splines in the joint distributions of the circular serial correlation coefficients, preprint, 1983.

73. Ignatov, Z. G. and V. K. Kaishev, Multivariate B-splines, analysis of contingency tables and serial correlation, Mathematical statistics and probability theory, Vol. B (Bad Tatzmannsdorf, 1986), Reidel, Dordrecht-Boston, MA-London, 1987, 125–137.

74. Ignatov, Z. G. and V. K. Kaishev, Some properties of generalized B-splines, Constructive theory of functions (Varna, 1987), Bulgar. Acad. Sci., Sofia, 1988, 233–241.

75. Ignatov, Z. G. and V. K. Kaishev, A probabilistic interpretation of multivariate B-splines and some applications, Serdica **15** (1989), 91–99.

76. Karlin, S., C. A. Micchelli, and Y. Rinott, Some probabilistic aspects in multivariate splines, in *Multivariate analysis VI (Pittsburgh, Pa., 1983)*, North-Holland, Amsterdam-New York, 1985, 355–360.

77. Karlin, S., C. A. Micchelli, and Y. Rinott, Multivariate splines: A probabilistic perspective, J. Multivariate Anal. **20** (1986), 69–90.

78. Koch, P. E., Multivariate trigonometric B-splines, J. Approx. Theory **54** (1988), 162–168.

79. Kochevar, P., An application of multivariate B-splines to computer aided geometric design, Rocky Mountain J. Math. **14** (1984), 159–175.

80. Lautsch, M., A spline inversion formula for the Radon transform, SIAM J. Numer. Anal. **26** (1989), 456–467.

81. Lee, S. L., The use of homogeneous coordinates in spline functions and polynomial interpolation, in *Multivariate Approximation and Interpolation (Duisburg, 1989)*, Internat. Ser. Numer. Math., 94, Birkhäuser, Basel, 1990, 167–178.

82. Lokar, M., How to compute a multivariate B-spline, in *VI Conference on Applied Mathematics* (Tara, 1988), Univ. Belgrade, 1989, 106–112.

83. Micchelli, C. A., Smooth multivariate piecewise polynomials: A method for computing multivariate B-splines, Instituto per le Applicazioni del Calcolo, Rome, 1979, preprint.

84. Micchelli, C. A., On a numerically efficient method for computing multivariate B-splines, in *Multivariate Approximation Theory*, W. Schempp and K. Zeller (eds.), Birkhäuser, Basel, 1979, 211–248.

85. Micchelli, C. A., A constructive approach to Kergin interpolation in $\mathbb{R}^k$: multivariate B-splines and Lagrange interpolation, Rocky Mountain J. Math. **10** (1980), 485–497.

86. Micchelli, C. A., Recent progress in multivariate splines, in *Proceedings of the International Congress of Mathematicians, Vol. 1, 2 (Warsaw, 1983)*, PWN, Warsaw, 1984, 1523–1524.

87. Micchelli, C. A., *Mathematical Aspects of Geometric Modeling*, CBMS-NSF Regional Conference Series in Applied Mathematics, 65, SIAM, Philadelphia, 1995.

88. Neamtu, M., Multivariate B-splines and their evaluation, Memorandum #598, University of Twente, 1986.

89. Neamtu, M., Multivariate divided differences and B-splines, in *Approximation Theory VI*, C. Chui, L. Schumaker, and J. Ward (eds.), Academic Press, New York, 1989, 445–448.

90. Neamtu, M., Subdividing multivariate polynomials in Bernstein-Bézier form without de Casteljau algorithm, in *Curves and Surfaces*, P.-J. Laurent, A. Le Méhauté, and L. L. Schumaker (eds.), Academic Press, New York, 1991, 359–362.

91. Neamtu, M., Multivariate Splines, Dissertation, University of Twente, 1991.

92. Neamtu, M., Multivariate divided differences. I. Basic properties, SIAM J. Numer. Anal. **29** (1992), 1435–1445.

93. Neamtu, M., On discrete simplex splines and subdivision, J. Approx. Theory **70** (1992), 358–374.

94. Neamtu, M., Homogeneous simplex splines, J. Comput. Appl. Math. **73** (1996), 173–189.

95. Neamtu, M., Delaunay configurations and multivariate splines, in preparation.

96. Neamtu, M. and C. R. Traas, On computational aspects of simplicial splines, Constr. Approx. **7** (1991), 209–220.

97. Neuman, E., Inequalities involving multivariate convex functions. II, Proc. Amer. Math. Soc. **109** (1990), 965–974.

98. Neuman, E. and J. Pečarić, Inequalities involving multivariate convex functions, J. Math. Anal. Appl. **137** (1989), 541–549.

99. Neuman, E. and P. J. Van Fleet, Moments of Dirichlet splines and their applications to hypergeometric functions, J. Comput. Appl. Math. **53** (1994), 225–241.

100. Pfeifle, R. N., Approximation and interpolation using quadratic triangular B-splines, Dissertation, University of Erlangen, 1995.

101. Pfeifle, R. N. and H.-P. Seidel, Faster evaluation of quadratic bivariate DMS spline surfaces, in *Proceedings of Graphics Interface '94*, Canadian Human-Computer Communications Soc., Morgan Kaufmann, 1995, 182–189.

102. Pfeifle, R. N. and H.-P. Seidel, Spherical triangular B-splines with application to data fitting, in *Proc. EUROGRAPHICS '95*, Blackwell, 1995, 89–96.

103. Pfeifle, R. N. and H.-P. Seidel, Fitting triangular B-splines to functional scattered data, in *Proceedings of Graphics Interface '95*, Canadian Human-Computer Communications Soc., Morgan Kaufmann, 1995, 80–88.

104. Pfeifle, R. N. and H.-P. Seidel, Triangular B-splines for blending and filling of polygonal holes, in *Proceedings of Graphics Interface '96*, Canadian Human-Computer Communications Soc., Morgan Kaufmann, 1995, 186–193.

105. Prautzsch, H., Unterteilungsalgorithmen für multivariate splines, Dissertation, Technische Universität Braunschweig, 1984.

106. Prautzsch, H., Degree elevation of B-spline curves, Comput. Aided Geom. Design **1** (1984), 193–198.

107. Sabin, M., Open questions in the application of multivariate B-splines, in *Mathematical Methods in Computer Aided Geometric Design*, T. Lyche and L. L. Schumaker (eds.), Academic Press, New York, 1989, 529–537.

108. Sauer, T., Multivariate B-splines with (almost) arbitrary knots, in *Approximation Theory VIII, Vol. 1: Approximation and Interpolation*, Charles K. Chui and Larry L. Schumaker (eds.), World Scientific Publishing Co., Inc., Singapore, 1995, 477–484.

109. Sauer, T., Polynomial interpolation of minimal degree, Numer. Math. **78** (1997), 59–85.

110. Sauer, T. and Y. Xu, On multivariate Lagrange interpolation, Math. Comp. **64** (1995), 1147–1170.

111. Sauer, T. and Y. Xu, On multivariate Hermite interpolation, Advances in Comp. Math. **4** (1995), 207–259.

112. Schoenberg, I. J., A letter to P. J. Davis, 1965 (published in [87]).

113. Seidel, H.-P., Polar forms and triangular B-spline surfaces, in *Blossoming: The New Polar-Form Approach to Spline Curves and Surfaces SIGGRAPH '91*, Course Notes #26, ACM SIGGRAPH, 1991.

114. Seidel, H.-P., Representing piecewise polynomials as linear combinations of multivariate B-splines, in *Mathematical Methods in Computer Aided Geometric Design II*, T. Lyche and L. L. Schumaker (eds.), Academic Press, New York, 1992, 559–566.

115. Seidel, H.-P., Polar forms and triangular B-spline surfaces, in *Computing in Euclidean Geometry*, D. Z. Du and F. Hwang (eds.), Lecture Notes Ser. Comput., 1, World Sci. Publishing, River Edge, 1992, 235–286.

116. Seidel, H.-P., Simplex splines, polar simplex splines and triangular B-splines, in *The Mathematics of Surfaces, VI* (Uxbridge, 1994), Inst. Math. Appl. Conf. Ser. New Ser., 58, Oxford Univ. Press, New York, 1996, 535–547.

117. Seidel, H.-P., Functional data fitting and fairing with triangular B-splines, in *Surface Fitting and Multiresolution Methods*, A. Le Méhauté, C. Rabut, and L. L. Schumaker (eds.), Vanderbilt University Press, Nashville, 1997, 319–328.

118. Seidel, H.-P. and A. H. Vermeulen, Simplex splines support surprisingly strong symmetric structures and subdivision, in *Curves and Surfaces in Geometric Design*, P.-J. Laurent, A. Le Méhauté, and L. L. Schumaker (eds.), A. K. Peters, Wellesley MA, 1994, 443–455.

119. Strøm, K., On convolutions of B-splines, J. Comput. Appl. Math. **55** (1994), 1–29.

120. Sun, J. C., The Fourier transform approach to multivariate B-splines, Math. Numer. Sinica **8** (1986), 191–199.

121. Traas, C. R., Boundary conditions with bivariate quadratic B-splines, in *Approximation Theory V*, C. Chui, L. Schumaker, and J. Ward (eds.), Academic Press, New York, 1986, 595–598.

122. Traas, C. R., Approximation of surfaces constrained by a differential equation using simplex splines, in *Mathematical Methods in Computer Aided Geometric Design*, T. Lyche and L. L. Schumaker (eds.), Academic Press, New York, 1989, 593–599.

123. Traas, C. R., Practice of bivariate quadratic simplicial splines, in *Computation of Curves and Surfaces*, W. Dahmen, M. Gasca, and C. A. Micchelli (eds.), Kluwer, Dordrecht, 1990, 383–422.

124. Traas, C. R., Constructing bivariate simplicial spline spaces of class C^1, Memorandum #870, University of Twente, 1990.

125. Waldron, S., A multivariate form of Hardy's inequality and L_p-error bounds for multivariate Lagrange interpolation schemes, SIAM J. Math. Anal. **28** (1997), 233–258.

126. Watson, G. S., On the joint distribution of circular serial correlation coefficients, Biometrika **43** (1956), 161–168.

127. Wenz, H.-J., On the control net of certain multivariate spline functions, Acta Math. Inform. Univ. Ostraviensis **2** (1994), 113–125.

128. Wenz, H.-J., On local approximation methods for multivariate polynomial spline surfaces, Results Math. **31** (1997), 170–179.

129. Wesselink, J.-W., Variational modeling of curves and surfaces, Dissertation, Technische Universiteit Eindhoven, 1996.

130. Wong, Z.-Y., Construction of multivariate B-splines with multiple nodes, J. Math. Res. Exposition **7** (1987), 635–639.

Other References

131. Boor, C. de, Quasiinterpolants and approximation power of multivariate splines, in *Computation of Curves and Surfaces*, W. Dahmen, M. Gasca, and C. A. Micchelli (eds.), Kluwer, Dordrecht, 1990, 313–345.

132. Delaunay, B., Sur la sphère vide, Bull. Acad. Sci. USSR (VII), Classe Sci. Mat. Nat., 1934, 793–800.

133. Okabe, A., B. Boots, and K. Sugihara, *Spatial Tessellations: Concepts and Applications of Voronoi Diagrams*, Wiley, Chichester, England, 1992.

134. Preparata, F. P. and M. I. Shamos, *Computational Geometry*, Springer, 1985.

135. Ramshaw, L., Blossoms are polar forms, Comput. Aided Geom. Design **6** (1989), 323–358.

136. Schumaker, L. L., *Spline Functions: Basic Theory*, Interscience (New York), 1981. Reprinted by Krieger, Malabar, Florida, 1993.

137. Shamos, M. I. and D. Hoey, Closest-point problems, in *IEEE Symposium on Foundations of Computer Science*, 1975, 151–162.

138. Teillaud, M., *Towards Dynamic Randomized Algorithms in Computational Geometry*, Lecture Notes in Computer Science, 758, Springer, 1993.

Department of Mathematics
Vanderbilt University
Nashville, TN 37240, USA
`neamtu@math.vanderbilt.edu`
`http://www.math.vanderbilt.edu/~neamtu`

Local Lagrange Interpolation by Bivariate C^1 Cubic Splines

Günther Nürnberger, Larry L. Schumaker,
and Frank Zeilfelder

Abstract. Lagrange interpolation schemes are constructed based on C^1 cubic splines on certain triangulations obtained from checkerboard quadrangulations.

§1. Introduction

Given a triangulation $\triangle$ of a simply connected polygonal domain Ω, the space of C^1 cubic splines is defined by

$$\mathcal{S}_3^1(\triangle) := \{s \in C^1(\Omega) : \; s|_T \in \mathcal{P}_3, \text{ all } T \in \triangle\},$$

where $\mathcal{P}_3$ is the space of cubic bivariate polynomials.

In this paper we are interested in constructing spline interpolation methods that are based on a given set of Lagrange data and which deliver full approximation power. It is well known that to work with $\mathcal{S}_3^1(\triangle)$ successfully, we have to restrict our attention to special classes of triangulations. Indeed, for general triangulations, at this point it is not known whether interpolation at all of the vertices of $\triangle$ is even possible, and the dimension of $\mathcal{S}_3^1(\triangle)$ is also unknown. Moreover, it is known [3] that the space is defective in the sense that it does not give full approximation power on some triangulations (including the very regular type-I triangulations). This implies that in general it does not have a stable local basis.

There are several classes of triangulations where the situation is simplified. First, one can work on the refined triangulation $\triangle_{CT}$ which is obtained from $\triangle$ by splitting each triangle into three subtriangles. The classical Clough-Tocher C^1 cubic element can then be constructed locally from values and gradients at each of the vertices of $\triangle$. If certain cross-derivative information is also available, the method gives full approximation power, see

Mathematical Methods for Curves and Surfaces: Oslo 2000
Tom Lyche and Larry L. Schumaker (eds.), pp. 393–403.

ISBN 0-8265-1378-6

e.g. [2,12]. Another class of triangulations where Hermite interpolation can be performed easily with $\mathcal{S}_3^1(\triangle)$ are the triangulations $\triangle$ which are obtained from a quadrangulation by drawing in both diagonals, see [9,10,17]. To use these interpolation methods where only Lagrange data is available, we need enough data to estimate the required derivatives accurately.

In this paper we give a direct construction of a C^1 cubic spline interpolant which uses only Lagrange data. For other work on Lagrange interpolation methods based on spline spaces, see [4–5, 13–16].

The paper is organized as follows. In Sect. 2 we present a basic definition and some notation. The new concept of Lagrange minimal determining sets is introduced in Sect. 3 and several useful lemmas are established. Sect. 4 presents the main results for checkerboard triangulations, and in Sect. 5 we establish analogous results for certain reduced checkerboard triangulations. We conclude the paper with a numerical example and remarks.

§2. Notation and Preliminaries

One of the keys to our discussion is the idea of a minimal determining set for a spline space. The concept was introduced in [1], and has since been heavily used in the multivariate spline literature. Here we need a more general form.

Definition 1. *Suppose Λ is a set of linear functionals defined on $\mathcal{S} \subseteq \mathcal{S}_d^0(\triangle)$. Then $\mathcal{M} \subseteq \Lambda$ is called* a determining set for $\mathcal{S}$ *provided that for any $s \in \mathcal{S}$, $\lambda s = 0$ all $\lambda \in \mathcal{M}$ implies $s \equiv 0$. The set $\mathcal{M}$ is called* a minimal determining set (MDS) for $\mathcal{S}$ *provided Λ does not contain any smaller determining set.*

Another way to describe a minimal determining set is to note that it is a set such that setting λs for all $\lambda \in \mathcal{M}$ uniquely determines s. It is easy to see, cf. [1], that if $\mathcal{M}$ is a MDS for $\mathcal{S}$, then $\dim \mathcal{S} = \#\mathcal{M}$. An MDS can also be used to construct a basis for $\mathcal{S}$. Indeed, suppose $\mathcal{M}$ is an MDS. Then for each $\lambda \in \mathcal{M}$, there is a unique spline $B_\lambda \in \mathcal{S}$ such that

$$\gamma B_\lambda = \delta_{\lambda,\gamma}, \qquad \text{all } \gamma \in \mathcal{M}. \tag{1}$$

The splines $\{B_\lambda\}_{\lambda\in\mathcal{M}}$ are called the dual basis splines corresponding to $\mathcal{M}$. We are especially interested in choosing $\mathcal{M}$ so that the dual basis splines have local support.

In [1] and the rest of the subsequent spline literature, Λ was always taken to consist of linear functionals which pick off Bernstein-Bézier coefficients. The essential difference in this paper is that we will use point evaluation functionals instead.

While we intend to work with Lagrange data, it is still useful to write polynomials of degree d in their Bernstein-Bézier form

$$p = \sum_{i+j+k=d} c_{ijk}^T B_{ijk}^d, \tag{2}$$

where B_{ijk}^T are the *Bernstein polynomials* of degree d associated with T. This is called the B-representation of p, and the c_{ijk}^T are called its B-coefficients.

Assuming $T := \langle u_1, u_2, u_3 \rangle$, it is common to associate these coefficients with the domain points $\xi^T_{ijk} := (iu_1 + ju_2 + ku_3)/d$. The point ξ^T_{d00} is at the vertex u_1 while the points $\xi^T_{d-1,1,0}$ and $\xi^T_{d-1,0,1}$ are said to be on the ring $R^T_1(u_1)$. We set $D^T_1(u_1) = \{\xi^T_{d00}, \xi^T_{d-1,1,0}, \xi^T_{d-1,0,1}\}$, with similar definitions for u_2, u_3.

§3. Lagrange Minimal Determining Sets

If $\mathcal{M}$ is a set points such that the corresponding point evaluation functionals form a MDS for $\mathcal{S}$, we call $\mathcal{M}$ a Lagrange MDS for $\mathcal{S}$. We prove our first result for general d, although we intend to apply it for $d = 3$.

Lemma 2. *The set of all domain points in T is a Lagrange minimal determining set for the space $\mathcal{P}_d$.*

Proof: Let $n := (d+2)(d+1)/2$, and let $B_1, \ldots, B_n$ be the Bernstein polynomials B^d_{ijk} written in lexicographical order. Let c be the vector of coefficients in the same order, and let $b := (p(t_1), \ldots, p(t_n))$, where $t_1, \ldots, t_n$ are the values ξ^T_{ijk} in the same order. Then the coefficients must solve the system $Ac = b$, where $A_{ij} := B_i(t_j)$ for $i, j = 1, \ldots, n$. It is easy to show that the determinant of the matrix A is nonzero and depends only on d. This means that c is stably and uniquely determined by b. $\square$

It follows from the proof of Lemma 2 and the fact that $\sum_{i+j+k=d} B^d_{ijk} \equiv 1$ that $\{B^d_{ijk}\}$ is a stable basis for $\mathcal{P}_d$ in the sense that there exists a nonzero constant K_1 such that $\|p\|_\infty \le \|c\|_\infty$ and $\|c\|_\infty \le K_1 \|p\|_\infty$ for all $p \in \mathcal{P}_d$. Indeed, we can take $K_1 := \|A^{-1}\|$.

We give two examples of Lagrange minimal determining sets for C^1 cubic splines.

Lemma 3. *Suppose $\triangle$ consists of two triangles $T_1 := \langle u_1, u_2, u_3 \rangle$ and $T_2 := \langle u_1, u_3, u_4 \rangle$ sharing the edge $\langle u_1, u_3 \rangle$. Then the set*

$$\mathcal{M} := \Big(\bigcup_{i=1}^{3} D^{T_1}_1(u_i) \Big) \cup D^{T_2}_1(u_4) \cup \{\xi^{T_1}_{111}\}$$

is a Lagrange MDS for $\mathcal{S}^1_3(\triangle)$.

Proof: By Lemma 2, the points of $\mathcal{M}$ uniquely determine all ten B-coefficients of $s|_{T_1}$. Writing $s|_{T_2}$ in B-form, we see that by the C^1 smoothness conditions, all of its B-coefficients are determined except for $c := (c_{300}, c_{210}, c_{201})$. But then $Gc = b$, where $b := (p(\xi_{300}), p(\xi_{210}), p(\xi_{201}))$ and

$$G := \begin{pmatrix} B^3_{300}(\xi_{300}) & B^3_{210}(\xi_{300}) & B^3_{201}(\xi_{300}) \\ B^3_{300}(\xi_{210}) & B^3_{210}(\xi_{210}) & B^3_{201}(\xi_{210}) \\ B^3_{300}(\xi_{201}) & B^3_{210}(\xi_{201}) & B^3_{201}(\xi_{201}) \end{pmatrix} = \begin{pmatrix} 1 & 0 & 0 \\ \frac{8}{27} & \frac{4}{9} & 0 \\ \frac{8}{27} & 0 & \frac{4}{9} \end{pmatrix}.$$

Thus, all B-coefficients of s are uniquely determined. $\square$

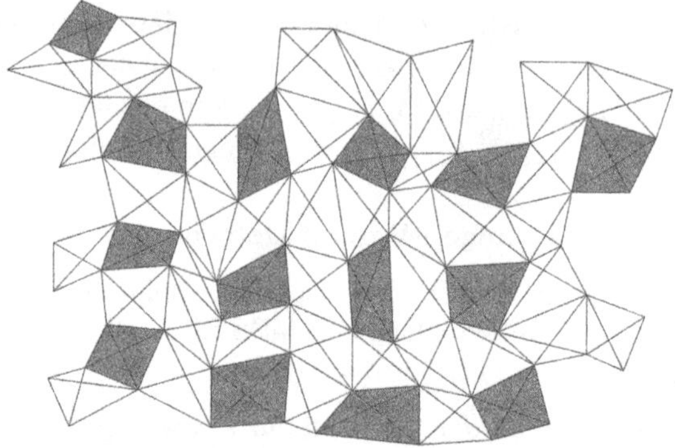

Fig. 1. The set $\mathcal{G}$ for a checkerboard triangulation.

The construction in Lemma 3 is stable in the sense that the maximum coefficient of s is bounded by $K \max_{\xi \in \mathcal{M}} |p(\xi)|$, where K is a constant depending only on the smallest angle in $\triangle$. In this case we say that $\mathcal{M}$ is a stable Lagrange MDS for $\mathcal{S}_3^1(\triangle)$.

Lemma 4. *Suppose $\triangle$ consists of four triangles $T_i := \langle u, u_i, u_{i+1} \rangle$ obtained by inserting both diagonals into a quadrilateral $Q := \langle u_1, u_2, u_3, u_4 \rangle$. Then the set*

$$\mathcal{M} := \Big(\bigcup_{i=1}^{2} D_1^{T_1}(u_i) \Big) \cup D_1^{T_1}(u) \cup D_1^{T_2}(u_3) \cup D_1^{T_4}(u_4) \cup \{\xi_{111}^{T_1}\}$$

is a stable Lagrange MDS for $\mathcal{S}_3^1(\triangle)$.

Proof: Applying Lemma 3 to $T_1 \cup T_2$, it follows that the B-coefficients of s associated with domain points in $T_1 \cup T_2$ are uniquely and stable determined by the data. A similar argument shows that same holds for T_4. Then the coefficients in T_3 can be stably computed from the C^1 smoothness conditions. □

§4. Checkerboard Triangulations

Definition 5. *Suppose $\diamondsuit$ is a quadrangulation consisting of quadrilaterals with largest interior angle less than π. Suppose that the quadrilaterals can be colored black and white in such a way that any two quadrilaterals sharing an edge have the opposite color. Then we call $\diamondsuit$ a* checkerboard quadrangulation. *The triangulation $\triangle$ which is obtained by drawing in both diagonals of all quadrilaterals will be called a* checkerboard *triangulation.*

Suppose $\mathcal{B}$ and $\mathcal{W}$ denote the sets of black and white quadrilaterals of $\diamondsuit$, respectively. Throughout this paper, we assume that all interior vertices of $\diamondsuit$ are of degree four. This assumption implies that there exists $\mathcal{G} \subset \mathcal{B}$ such that for every interior vertex v of $\diamondsuit$, there is a unique quadrilateral $Q \in \mathcal{G}$ sharing the vertex v. For $i = 1, 2, 3$, let $\mathcal{W}_i$ be the set of white quadrilaterals which share i edges with black quadrilaterals. Let $n_B = \#\mathcal{B}$ and $n_i := \#\mathcal{W}_i$ for $i = 1, 2, 3$, and let n_V be the total number of vertices of $\diamondsuit$. Fig. 1 shows a typical checkerboard triangulation in which the quadrilaterals in the set $\mathcal{G}$ are shaded grey. Note that we have not colored the other black quadrilaterals.

Theorem 6. *Suppose $\triangle$ is a checkerboard triangulation. Then*

$$\dim \mathcal{S}_3^1(\triangle) = 3n_V + 4n_B + 3n_1 + 2n_2 + n_3. \tag{3}$$

Moreover, the following set $\mathcal{M}$ of domain points is a stable Lagrange MDS:

1) *if $Q \in \mathcal{G}$, choose points as in Lemma 4,*
2) *if $Q \in \mathcal{B} \setminus \mathcal{G}$, choose points as in Lemma 4, leaving out the points in the sets $D_1^T(v)$ whenever v is a vertex of Q which is interior to $\diamondsuit$,*
3) *Suppose $Q := \langle u_1, u_2, u_3, u_4 \rangle \in \mathcal{W}$ and let $e_i := \langle v_i, v_{i+1} \rangle$ for $i = 1, 2, 3, 4$.*
   a) *if $Q \in \mathcal{W}_3$, choose the point $\xi_{300}^{T_1}$,*
   b) *if Q shares two edges with black quadrilaterals, say e_1, e_2, choose the points $\xi_{300}^{T_1}, \xi_{210}^{T_1}$ and the points in $D_1^{T_3}(v_4)$,*
   c) *if Q shares one edge with black quadrilaterals, say e_1, choose the points $\xi_{300}^{T_1}, \xi_{210}^{T_1}, \xi_{201}^{T_1}$ and $D_1^{T_2}(v_3)$, $D_1^{T_4}(v_4)$.*

Proof: To establish that $\mathcal{M}$ is a Lagrange MDS, suppose $s \in \mathcal{S}_3^1(\triangle)$ and that we are given values for $s(\xi)$ for all $\xi \in \mathcal{M}$. We need to show that all of the B-coefficients of s are uniquely determined. By Lemma 4, all B-coefficients of s associated with domain points lying in quadrilaterals $Q \in \mathcal{G}$ are uniquely determined. Now consider $Q \in \mathcal{B} \setminus \mathcal{G}$. For each vertex $v \in Q$ which is an interior vertex of $\diamondsuit$, the B-coefficients corresponding to domain points in the disk $D_1^T(v)$ are already uniquely determined by C^1 continuity from the neighboring pieces. Leaving the corresponding basis functions out, we can then argue exactly as in Lemma 4 to see that all B-coefficients of s corresponding to the remaining domain points in Q are uniquely determined.

Now suppose $Q \in \mathcal{W}$. If Q shares four edges with black quadrilaterals, then using the C^1 continuity, it is easy to see that all B-coefficients of s corresponding to domain points in Q are uniquely determined. If Q shares the three edges e_1, e_2, e_4 with black quadrilaterals, then all B-coefficients of $s|_{T_1}$ are uniquely determined by C^1 continuity except for $c_{300}^{T_1}$ which is uniquely determined by 3a). If Q shares the two edges e_1 and e_2 with black quadrilaterals, then the C^1 conditions imply that all of the B-coefficients of $s|_{T_1}$ are uniquely determined except for $c_{300}^{T_1}$ and $c_{210}^{T_1}$. These can be determined from the data of 3b) by solving a 2×2 system. In case 3c), all of the B-coefficients

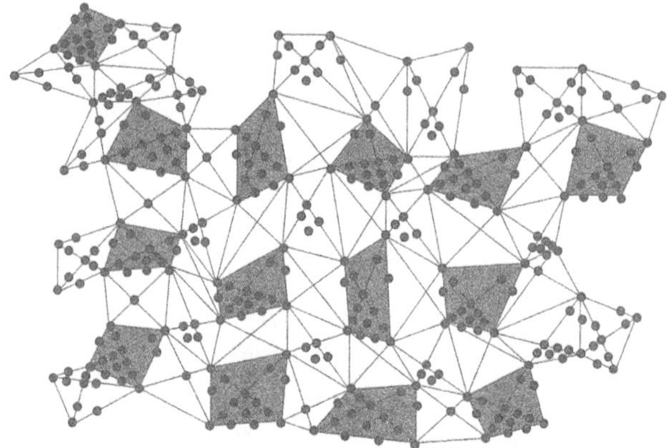

Fig. 2. The Lagrange MDS for the triangulation of Fig. 1.

of $s|_{T_1}$ are uniquely determined by the C^1 conditions except for those associated with domain points in $D^{T_1}(v_Q)$, where v_Q is the crossing point of the two diagonals. Using the data of 3c) and solving the same 3×3 system as in the proof of Lemma 3 shows that these coefficients are also uniquely determined. The C^1 continuity and the additional data of 3c) can be used to uniquely determine the B-coefficients corresponding to the remaining domain points in Q.

Since we have shown that $\mathcal{M}$ is a MDS, it follows that $\dim \mathcal{S}_3^1(\triangle) = \#\mathcal{M}$. We have chosen three points for each vertex. This contributes $3n_V$ to the count. All black quadrilaterals Q contain $\xi_{111}^{T_1}$ and the three points in $D_1^{T_1}(v_Q)$, where v_Q is the crossing point of the two diagonals of Q. For each $Q \in \mathcal{W}_i$ with $1 \le i \le 3$, we have included $4 - i$ additional points.

Finally, we note that all of the above computations are stable in the sense that the size of the computed B-coefficients is bounded by a constant depending only on the smallest angle in the triangulation $\triangle$. This follows from the fact that the computations of Lemmas 2–4 are stable, and the fact that computing coefficients from C^1 smoothness conditions is automatically stable, cf. eg. [7]. □

Fig. 2 shows the Lagrange MDS of Theorem 6 for the checkerboard triangulation of Fig. 1. We now examine the dual basis splines corresponding to $\mathcal{M}$. Given $\xi \in \mathcal{M}$, B_ξ is defined to be the spline in $\mathcal{S}_3^1(\triangle)$ such that $B_\xi(\xi) = 1$ and $B_\xi(\eta) = 0$ for all other points $\eta \in \mathcal{M}$.

Corollary 7. *Let $\triangle$ be a checkerboard triangulation, and let $\mathcal{M}$ be the set defined in Theorem 6. Then the dual basis splines corresponding to $\mathcal{M}$ form a stable local basis for $\mathcal{S}_3^1(\triangle)$.*

Proof: The proof of Theorem 6 shows that all B-coefficients of B_ξ are uniquely and stable determined. It remains to discuss the support of B_ξ. Suppose ξ lies in a quadrilateral Q_ξ. Then we claim

1) $\operatorname{supp}(B_\xi) = Q$ if $\xi \in \mathcal{W}$,
2) $\operatorname{supp}(B_\xi) \subset \operatorname{star}(Q_\xi)$ if $\xi \in \mathcal{B} \setminus \mathcal{G}$,
3) $\operatorname{supp}(B_\xi) \subset \operatorname{star}^2(Q_\xi)$ otherwise.

Here $\operatorname{star}(Q)$ is the union of all quadrilaterals which intersect with Q in at least one point, and $\operatorname{star}^2(Q) := \operatorname{star}(\operatorname{star}(Q))$. These assertions follow immediately from the checkerboard nature of the quadrangulation and the observation that B_ξ vanishes identically on Q whenever

1) $Q \neq Q_\xi$ and $Q \in \mathcal{G}$,
2) $Q \neq Q_\xi$ and $Q \in \mathcal{B}$ does not intersect Q_ξ. □

We are now ready to discuss interpolation. Suppose $\triangle$ is a checkerboard triangulation, and that B_ξ are the dual basis functions of Corollary 7 corresponding to the Lagrange MDS $\mathcal{M}$ for $\mathcal{S}_3^1(\triangle)$ defined in Theorem 6. Given any function defined on Ω, let

$$\mathcal{I}f := \sum_{\xi \in \mathcal{M}} f(\xi)\, B_\xi. \tag{4}$$

By the duality (1) of the basis functions B_ξ, it is clear that the cubic spline $\mathcal{I}f$ interpolates f at all the points of $\mathcal{M}$, *i.e.*,

$$\mathcal{I}f(\xi) = f(\xi), \qquad \xi \in \mathcal{M}. \tag{5}$$

This includes in particular all vertices of $\diamondsuit$. We now give an error bound for this interpolation method.

Theorem 8. *There exists a constant C depending only on the smallest angle in $\triangle$ such that if f is in the Sobolev space $W_\infty^{m+1}(\Omega)$ with $0 \le m \le 3$,*

$$\|D_x^\alpha D_y^\beta (f - \mathcal{I}f)\|_{\infty,\Omega} \le C\, |\triangle|^{m+1-\alpha-\beta}\, |f|_{m+1,\infty,\Omega}, \tag{6}$$

for all $0 \le \alpha + \beta \le m$. Here $|\triangle|$ is the maximum of the diameters of the triangles in $\triangle$.

Proof: We apply Theorem 5.1 of [11]. Clearly, $\mathcal{I}p = p$ for all cubic polynomials. The hypothesis (5.3) of that theorem is trivial since $|f(\xi)| \le \|f\|_{T_\xi}$, where T_ξ is the triangle which contains ξ. □

The result of Theorem 8 can also be established with the weak-interpolation methods described in [6].

§5. Reduced Checkerboard Triangulations

In this section we triangulate a checkerboard quadrangulation in a different way which involves fewer triangles but still leads to Lagrange interpolating C^1 cubic splines with full approximation power.

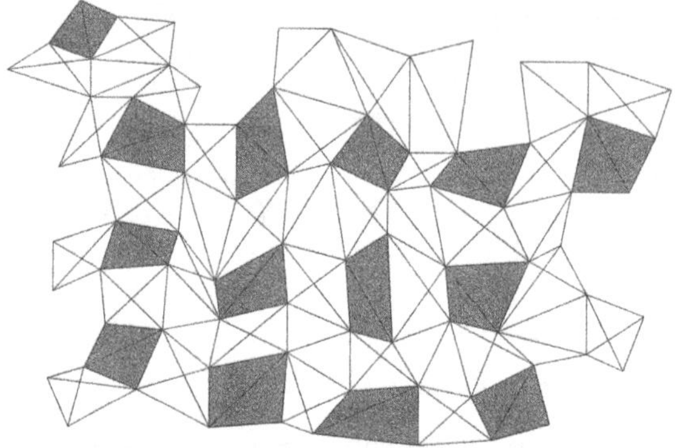

Fig. 3. A reduced checkerboard triangulation.

Definition 9. *Suppose $\diamondsuit$ is a checkerboard quadrangulation, and let $\mathcal{B}$ and $\mathcal{W}$ be the sets of black and white quadrilaterals, respectively. Let $\triangle$ be the triangulation obtained by drawing in one diagonal of each black quadrilateral, and both diagonals of each white quadrilateral. We call $\triangle$ a* reduced checkerboard *triangulation.*

The proof of the following theorem is almost the same as that of Theorem 6 and Corollary 7.

Theorem 10. *Suppose $\triangle$ is a reduced checkerboard triangulation. Then*

$$\dim \mathcal{S}_3^1(\triangle) = 3n_V + n_B + 3n_1 + 2n_2 + n_3, \tag{7}$$

where n_V, n_B, n_1, n_2, n_3 are as in Theorem 6. Moreover, the following set $\mathcal{M}$ of domain points is a stable Lagrange MDS:

1) *if $Q \in \mathcal{G}$, choose points as in Lemma 3,*
2) *if $Q \in \mathcal{B} \setminus \mathcal{G}$, choose points as in Lemma 3, leaving out the points in the sets $D_1^T(v)$ whenever v is a vertex of Q which is interior to $\diamondsuit$,*
3) *if Q is a white quadrilateral, choose points as in 3) of Theorem 6.*

The corresponding dual basis $\{B_\xi\}_{\xi\in\mathcal{M}}$ is a stable local basis for $\mathcal{S}_3^1(\triangle)$.

Using the basis functions B_ξ of this theorem, the C^1 cubic spline $\mathcal{I}f$ defined in (4) clearly satisfies the interpolation conditions (5). As before, $\mathcal{I}p = p$ for all cubic polynomials, and the error bound given in Theorem 8 also holds.

§6. Numerical Example

We illustrate our method by interpolating Franke's test function

$$f(x,y) = \tfrac{3}{4}e^{-\frac{1}{4}((9x-2)^2+(9y-2)^2)} + \tfrac{3}{4}e^{-\frac{1}{49}(9x+1)^2-\frac{1}{10}(9y+1)} + \tfrac{1}{2}e^{-\frac{1}{4}((9x-7)^2+(9y-3)^2)} - \tfrac{1}{5}e^{-(9x-4)^2-(9y-7)^2}$$

n	# Data	$\|\triangle_n\|$	Error	Rate
1	69	3.33E-01	2.12E-01	
2	177	2.00E-01	5.58E-02	2.61
3	549	1.11E-01	1.09E-02	2.78
4	1917	5.88E-02	2.78E-03	2.15
5	7149	3.03E-02	2.38E-04	3.70
6	27597	1.54E-02	2.00E-05	3.65
7	108429	7.75E-03	1.37E-06	3.92
8	429837	3.89E-03	8.83E-08	3.98
9	1711629	1.95E-03	5.57E-09	4.00

Tab. 1. Results for a sequence of uniform quadrangulations.

n	# Data	$\|\triangle_n\|$	Error	Rate
1	69	4.43E-01	3.53E-01	
2	177	2.96E-01	1.16E-01	2.76
3	549	1.70E-01	2.39E-02	2.85
4	1917	8.99E-02	2.31E-03	3.67
5	7149	4.58E-02	2.72E-04	3.17
6	27597	2.36E-02	2.51E-05	3.59
7	108429	1.19E-02	3.56E-06	2.85
8	429837	6.11E-03	2.69E-07	3.87
9	1711629	3.97E-03	1.92E-08	3.84

Tab. 2. Results for a sequence of randomized quadrangulations.

on the unit square $[0,1] \times [0,1]$. The first test was done on a sequence of uniform quadrangulations associated with the vertices $(i/N, j/N)$ for $i,j = 0,N$, where $N = 2^n + 1$ for $n = 1,\ldots,9$. For each n, we computed the maximal error (using 25 points per quadrilateral). In the following table we list the number of data points (which is also the dimension of the corresponding spline space), the size of $|\triangle_n|$, the maximal error E_n, and the rate of convergence $\ln(E_{n-1}/E_n)/\ln(|\triangle_{n-1}|/|\triangle_n|)$. (Note that for this sequence of checkerboard triangulations, $|\triangle_n|$ is not exactly one-half of $|\triangle_{n-1}|$). The table shows that the method achieves the convergence rate of four.

As a second test, we deformed the quadrangulations using a random number generator. Each vertex was deformed by a sufficiently small amount to maintain the topology of the quadrangulation and to insure convext quadrilaterals. This changed the values of $|\triangle_n|$, of course, and also affected the smallest angle in the triangulation which no doubt has some effect on the constant in the error bound. Table 2 shows the corresponding results which also show a convergence rate of four.

§7. Remarks

Remark 11. It is well known from graph coloring theory [8], Theorem 14, that a quadrangulation admits a black/white coloring if and only if all interior vertices are even. However, the set $\mathcal{G}$ does not exist for every such quadrangulation. For a simple example, consider a quadrangulation with one interior vertex of degree six surrounded by interior vertices of degree four. This is the reason why we require that all interior vertices of a checkerboard quadrangulation be of degree four.

Remark 12. In this paper we have focused on Lagrange interpolation. Clearly, our methods can also be used to create C^1 cubic splines which satisfy Hermite interpolation conditions where a function value and gradient values are specified at the vertices of $\triangle$.

Remark 13. The problem of extending the current results to more general classes of quadrangulations is currently under study. It seems to involve some difficult coloring problems which have not been addressed in the graph-theory literature.

Acknowledgments. The second author was supported by the National Science Foundation under grant DMS-9803340 and by the Army Research Office under grant DAAD-19-99-1-0160

References

1. Alfeld, P. and L. L. Schumaker, The dimension of bivariate spline spaces of smoothness r for degree $d \geq 4r+1$, Constr. Approx. **3** (1987), 189–197.
2. Alfeld, P. and L. L. Schumaker, Smooth finite elements based on Clough-Tocher triangle splits, Numer. Math., to appear.
3. Boor, C. de and K. Höllig, Approximation order from bivariate C^1-cubics: a counterexample, Proc. Amer. Math. Soc. **87** (1983), 649–655.
4. Davydov, O. and G. Nürnberger, Interpolation by C^1 splines of degree $q \geq 4$ on triangulations, J. Comput. Appl. Math., to appear.
5. Davydov, O., G. Nürnberger, and F. Zeilfelder, Cubic spline interpolation on nested polygon triangulations, in *Curve and Surface Fitting: Saint-Malo 1999,* A. Cohen, C. Rabut, and L. L. Schumaker (eds), Vanderbilt University Press, Nashville TN, 2000, 161–170.
6. Davydov, O., G. Nürnberger, and F. Zeilfelder, Bivariate spline interpolation with optimal approximation order, Constr. Approx., to appear.
7. Davydov, O. and L. L. Schumaker, On stable local bases for bivariate polynomial spline spaces, Constr. Approx., to appear.
8. Jensen, T. R. and B. Toft, *Graph Coloring Problems,* Wiley, New York, 1995.

9. Fraeijs de Veubeke, B., A conforming finite element for plate bending, J. Solids Structures **4** (1968), 95–108.
10. Lai, M.-J., Scattered data interpolation and approximation using bivariate C^1 piecewise cubic polynomials, Comput. Aided Geom. Design **13** (1996), 81–88.
11. Lai, M.-J. and L. L. Schumaker, On the approximation power of bivariate splines, Advances in Comp. Math. **9** (1998), 251–279.
12. Lai, M.-J. and L. L. Schumaker, Macro-elements and stable local bases for splines on Clough-Tocher triangulations, Math. Comp., to appear.
13. Nürnberger, G. and F. Zeilfelder, Spline interpolation on convex quadrangulations, in *Approximation Theory IX, Vol. 2: Computational Aspects*, Charles K. Chui and Larry L. Schumaker (eds), Vanderbilt University Press, Nashville TN, 1998, 259–266.
14. Nürnberger, G. and F. Zeilfelder, Interpolation by spline spaces on classes of triangulations, J. Comput. Appl. Math. **119** (2000), 347–376.
15. Nürnberger, G. and F. Zeilfelder, Developments in bivariate spline interpolation, J. Comput. Appl. Math. **121** (2000), 125–152.
16. Nürnberger, G. and F. Zeilfelder, Local Lagrange interpolation by cubic splines on a class of triangulations, Trends in Approximation Theory, K. Kopotun, T. Lyche, and M. Neamtu, (eds.), Vanderbilt University Press, Nashville, 2001, to appear.
17. Sander, G., Bornes supérieures et inférieures dans l'analyse matricielle des plaques en flexion-torsion, Bull. Soc. Royale Sciences Liège **33** (1964), 456–494.

Günther Nürnberger and Frank Zeilfelder
Institut für Mathematik
Universität Mannheim, D-618131 Mannheim, Germany
nuernberger@euklid.math.uni-mannheim.de
zeilfeld@mpi-sb.mpg.de

Larry L. Schumaker
Department of Mathematics, Vanderbilt Univerisity
Nashville, TN 37205, USA
s@mars.cas.vanderbilt.edu

Stability of Collocation by Smooth Splines for Volterra Integral Equations

Peeter Oja

Abstract. General results about the numerical stability of the spline collocation method for Volterra integral equations are developed in the case of smooth polynomial splines. Besides stability conditions for linear and quadratic splines, quantitative results are presented concerning the instability of cubic and higher order splines. Numerical tests supporting the theoretical results are also presented.

§1. Introduction

In this paper we mean by stability the boundedness of approximate solutions in the uniform norm when the number of knots is increased. In general such stability is necessary for convergence, and it is also sufficient in the case of a certain test equation. Convergence theory for collocation is well developed for polynomial splines without any continuity conditions in the knots or which are only continuous [2,3,4,6,7]. Our investigations [10] show that in these cases the collocation method is stable for any order of spline and any choice of collocation parameters.

In the case of equidistant collocation points, stability is investigated in [8], and the general distribution of collocation parameters is treated in [5]. In this paper, we will focus our main attention on splines of maximal smoothness. Already in [9] it was shown for cubic splines that collocation at the knots is not convergent for the test equation, and therefore not stable. Here we derive stability conditions for linear and quadratic splines. For higher order smooth splines we prove instability, and explain how the instable behavior depends on the collocation parameter.

Mathematical Methods for Curves and Surfaces: Oslo 2000
Tom Lyche and Larry L. Schumaker (eds.), pp. 405–412.

ISBN 0-8265-1378-6

§2. The Spline Collocation Method

Consider the equation

$$y(t) = \lambda \int_0^t y(s)ds + f(t)\,, \quad t \in [0,T]\,, \tag{1}$$

with a given function f and, in general, any complex number λ. Typically such an equation is considered as the relevant test equation for investigations of numerical methods for general Volterra integral equations of the form

$$y(t) = \int_0^t \mathcal{K}(t,s,y(s))ds + f(t), \quad t \in [0,T]. \tag{2}$$

We are now going to describe the spline collocation method.

A mesh $0 = t_0 < t_1 < \cdots < t_N = T$ will be used, with the choice of knots t_n being of course dependent on N, since we want investigate the process $N \to \infty$. Denote $h_n = t_n - t_{n-1}$ and $\sigma_n = (t_{n-1}, t_n]$, $n = 1, \ldots, N$, $\Delta_N = \{t_1, \ldots, t_{N-1}\}$.

For given integers $m \geq 1$ and $d \geq -1$, define the space of splines

$$S^d_{m+d}(\Delta_N) = \{u \in C^d[0,T] \colon u\big|_{\sigma_n} \in \mathcal{P}_{m+d}, \quad n = 1, \ldots, N\},$$

where $\mathcal{P}_k$ denotes the set of all polynomials with degree not exceeding k.

Suppose also that we have a fixed selection of collocation parameters $0 < c_1 < \cdots < c_m \leq 1$, defining collocation points $t_{nj} = t_{n-1} + c_j h_n$, $j = 1, \ldots, m$, $n = 1, \ldots, N$. In order to determine an approximate solution $u \in S^d_{m+d}(\Delta_N)$ of equation (2), we impose the following collocation conditions

$$u(t_{nj}) = \int_0^{t_{nj}} \mathcal{K}(t_{nj}, s, u(s))ds + f(t_{nj}),$$

for all $n = 1, \ldots, N$ and $j = 1, \ldots, m$.

To be able to start the calculations of this method, we assume that we can use the initial values $u^{(j)}(0) = y^{(j)}(0)$, $j = 0, \ldots, d$ (or at least some approximations of them), which is justified by the requirement $u \in C^d[0,T]$. Thus, on every interval σ_n we have $d+1$ conditions of smoothness and m collocation conditions to determine $m+d+1$ parameters of u as a polynomial of degree $m+d$ on σ_n. This allows us to implement the method step-by-step, progressing from interval σ_n to the next one.

§3. Stability of the Spline Collocation Method

Suppose in the following that the mesh is uniform, i.e. $h_n = h = T/N$ for all n. Denote also $d_1 = \max\{d, 1\}$.

Definition 1. *We say that the spline collocation method is* stable *if for any $\lambda \in \mathbf{C}$ and any $f \in C^{d_1}[0,T]$ the approximate solution u of the test equation (1) remains bounded in $L_\infty(0,T)$ as $h \to 0$.*

The principle of uniform boundedness yields that the spline collocation method is stable if and only if

$$||u||_{L_\infty(0,T)} \leq \text{const}||f||_{C^{d_1}[0,T]}, \qquad \forall f \in C^{d_1}[0,T],$$

where the constant may depend only on T, λ and on the parameters c_j. To describe the stability conditions, we need $(m+d+1) \times (m+d+1)$ matrices V_0 and V as follows:

$$V_0 = \left(\frac{I \,|\, 0}{C}\right), \qquad V = \left(\frac{A}{C}\right),$$

I being the $(d+1) \times (d+1)$ identity matrix, A being the $(d+1) \times (m+d+1)$ matrix

$$A = \begin{pmatrix} 1 & 1 & 1 & 1 & \dots & 1 \\ 0 & 1 & 2 & 3 & \dots & m+d \\ 0 & 0 & 1 & 3 & \dots & \binom{m+d}{2} \\ \dots & \dots & \dots & \dots & \dots & \dots \\ 0 & \dots & \dots & 1 & \dots & \binom{m+d}{d} \end{pmatrix},$$

and

$$C = \begin{pmatrix} 1 & c_1 & \dots & c_1^{m+d} \\ \dots & \dots & \dots & \dots \\ 1 & c_m & \dots & c_m^{m+d} \end{pmatrix}.$$

As the matrix V_0 is invertible, we may introduce $M = V_0^{-1}V$. It turns out that the stability depends only on the eigenvalues of M.

It is clear that $\text{Ker}(M - \mu I) = \text{Ker}(V - \mu V_0)$. Thus for our stability considerations we need to solve the equation

$$\det(V - \mu V_0) = 0\,. \tag{3}$$

We see that the geometric multiplicity of $\mu = 1$ is $\dim \text{Ker}(V - V_0)$, but $\dim \text{Ker}(V - V_0) = m+d+1-\text{rank}(V-V_0) = m$ because $\text{rank}(V-V_0) = d+1$. Thus, the matrix M has eigenvalue $\mu = 1$ with geometric multiplicity m.

In [10] we have established the following results:

1) If all eigenvalues of M are in the closed unit disk and if those which lie on the unit circle have equal algebraic and geometric multiplicities, then the spline collocation method is stable.
2) If M has an eigenvalue outside of the closed unit disk, then the spline collocation method is unstable with exponential growth of the norm of u. In fact, the greater the modulus of an eigenvalue, the stronger the resulting instability.
3) If all eigenvalues of M are in the closed unit disk and there exists one which lies on the unit circle and has different algebraic and geometric multiplicity, then the method is unstable with polynomial growth of the norm of u.
4) The method is stable if and only if it is convergent in $L_\infty(0,T)$ for equation (1).

§4. Stability of the Collocation Method for Smooth Splines

We now concentrate our attention on the case $m = 1$, i.e. on the splines of maximal smoothness. For briefness we write the single parameter c_1 as c. First let us analyse two particular cases. They are considered also in [10], but for completeness we present them briefly here because we need them in the following general analysis as a basis of induction.

In the case $d = 0$ (linear splines), we have

$$V_0 = \begin{pmatrix} 1 & 0 \\ 1 & c \end{pmatrix}, \quad V = \begin{pmatrix} 1 & 1 \\ 1 & c \end{pmatrix},$$

and besides $\mu = 1$, the equation (3) has the solution $\mu = 1 - 1/c$. Thus, the method is stable if and only if $1/2 \le c \le 1$.

Consider now the case $d = 1$ (quadratic splines). Here also the geometric multiplicity of $\mu = 1$ as a solution of (3) is 1. Denote $\nu = 1 - \mu$. Then besides $\nu = 0$ (corresponding to $\mu = 1$), equation (3) has the two solutions

$$\nu = (1 + 2c \pm (1 + 4c(1-c))^{\frac{1}{2}}/2c^2.$$

From this we get $\nu > 0$, and thus $\mu < 1$. For $c = 1$ there are eigenvalues $\mu = 0$ and $\mu = -1$ corresponding to $\nu = 1$ and $\nu = 2$. The function

$$\varphi(c) = (1 + 2c + (1 + 4c(1-c))^{\frac{1}{2}}/2c^2$$

is decreasing ($\varphi'(c) < 0$), and hence for $c < 1$ we get $\nu > 2$ and $\mu < -1$. This means that the method is stable if and only if $c = 1$.

In the general case, we also use the variable $\nu = 1 - \mu$ and denote $\varphi_d(\nu) = \det(V - \mu V_0)$, which has the form

$$\varphi_d(\nu) = \left| \begin{array}{cccccccc} \nu & 1 & 1 & 1 & \dots & 1 & 1 & 1 \\ 0 & \nu & 2 & 3 & \dots & d-1 & d & d+1 \\ 0 & 0 & \nu & 3 & \dots & \binom{d-1}{2} & \binom{2}{2} & \binom{d+1}{2} \\ \dots & \dots & \dots & \dots & \dots & \dots & \dots & \dots \\ 0 & \dots & \dots & \dots & \dots & \nu & \binom{d}{d-1} & \binom{d+1}{d-1} \\ 0 & \dots & \dots & \dots & \dots & 0 & \nu & \binom{d+1}{d} \\ \nu & \nu c & \nu c^2 & \dots & \dots & \dots & \dots & \nu c^{d+1} \end{array} \right| .$$

Repeated expansion by the penultimate row leads to the result

$$\varphi_d(\nu) = c^{d+1}\nu^{d+2} - \sum_{i=0}^{d} \binom{d+1}{i} \nu^{d-i}\varphi_{i-1}(\nu),$$

with initial polynomial $\varphi_{-1}(\nu) = \nu$, or

$$\varphi_d(\nu) = \nu^{d+1}(c^{d+1}\nu - 1) - \sum_{i=1}^{d} \binom{d+1}{i} \nu^{d-i}\varphi_{i-1}(\nu). \tag{4}$$

Note that we have already used $\varphi_0(\nu) = \nu(c\nu - 1)$ and $\varphi_1(\nu) = \nu(c^2\nu^2 - (1 + 2c)\nu + 2)$. In fact, φ_d depends also on c, and that is why we write sometimes $\varphi_{d,c}$.

Let $\nu_{d,c}$ be the maximal (real) root of $\varphi_{d,c}$.

Proposition 2. *The maximal (real) root $\nu_{d,c}$ of $\varphi_{d,c}$ satisfies $\nu_{d,c} > \nu_{d-1,c}$.*

Proof: Clearly $\varphi_{d,c}(\nu) \to \infty$ in the limit process $\nu \to \infty$. Then it is sufficient to show that $\varphi_{d,c}(\nu_{d-1,c}) < 0$. As $c^{d+1} \le c^d$ and $\binom{d+1}{i} > \binom{d}{i}$, we get from (4), taking into account $\varphi_{d-1,c}(\nu_{d-1,c}) = 0$, that

$$\varphi_{d,c}(\nu_{d-1,c}) < \nu_{d-1,c}\varphi_{d-1,c}(\nu_{d-1,c}) = 0,$$

which completes the proof. □

We have seen already that $\nu_{1,c} \geq 2$. This yields the following

Corollary 3. *For $d \geq 2$ and any $c \in (0,1]$ the collocation method is unstable.*

A straightforward calculation gives

$$\frac{\partial}{\partial c}\varphi_{d,c}(\nu) = (d+1)\nu\varphi_{d-1,c}(\nu)\,. \tag{5}$$

Proposition 4. *For two different values of c such that $c_1 < c_2$, we have $\nu_{d,c_1} > \nu_{d,c_2}$.*

Proof: From Proposition 3 and (5) we obtain

$$\frac{\partial}{\partial c}\varphi_{d,c}(\nu_{d,c}) = (d+1)\nu_{d,c}\varphi_{d-1,c}(\nu_{d,c}) > 0$$

for any $c \in (0,1]$. As $\varphi_{d,c}(\nu_{d,c}) = 0$, we get $\varphi_{d,c-\epsilon}(\nu_{d,c}) < 0$ for sufficiently small $\epsilon > 0$ (depending of course on c), which in turn yields $\nu_{d,c-\epsilon} > \nu_{d,c}$. Now using the standard compactness argument covering $[c_1, c_2]$ by intervals $(c-\epsilon, c)$ for $c < c_2$ and $(c_2 - \epsilon, c_2]$, which are open in $(-\infty, c_2]$, we get $\nu_{d,c_1} > \nu_{d,c_2}$, completing the proof. □

Proposition 5. *The polynomial φ_d has coefficients alternating in sign.*

Proof: We prove that

$$\varphi_d(\nu) = \nu\big(c^{d+1}\nu^{d+1} + \sum_{i=0}^{d} a_{d,i}\nu^i\big)\,, \tag{6}$$

with $(-1)^{d+i}a_{d,i} < 0$. Arguing by induction, (4) implies

$$\begin{aligned}\varphi_d(\nu) &= \nu\left(c^{d+1}\nu^{d+1} - \nu^d - \sum_{i=1}^{d}\binom{d+1}{i}\nu^{d-i}\left(c^i\nu^i + \sum_{j=0}^{i-1}a_{i-1,j}\nu^j\right)\right)\\ &= \nu\left(c^{d+1}\nu^{d+1} - \nu^d\left(1 + \sum_{i=1}^{d}\binom{d+1}{i}c^i\right)\right.\\ &\quad\left. + \sum_{k=1}^{d}\left(-\sum_{i-j=k}a_{i-1,j}\binom{d+1}{i}\right)\nu^{d-k}\right).\end{aligned}$$

Denoting $b_k = -\sum_{i-j=k}a_{i-1,j}\binom{d+1}{i}$, we have to prove that $(-1)^k b_k < 0$. But

$$\begin{aligned}(-1)^k b_k &= (-1)^{k+1}\sum_{i-j=k}a_{i-1,j}\binom{d+1}{i}\\ &= \sum_{i-j=k}(-1)^{i-1+j}a_{i-1,j}\binom{d+1}{i} < 0,\end{aligned}$$

which completes the proof. □

Proposition 5 implies

Corollary 6. *The polynomial φ_d has no negative roots.*

Thus if $\varphi_{d,c}$ has only real roots (which seems to be the case), then $\nu_{d,c}$ is in fact the root of $\varphi_{d,c}$ with the greatest modulus, and characterizes the unstable behavior of the method.

§5. Numerical Tests

We choose the function $f(t) = \cos t$ and $\lambda = 1$ in the equation (1) on the interval $[0,1]$. This equation has the exact solution $y(t) = \frac{1}{2}(\sin t + \cos t + e^t)$, and was used already in [1] and by us in [10] for several values of d and m as a test equation. As an approximate value of $||u||_\infty$ we actually calculated $\max\limits_{1\le n\le N}\max\limits_{0\le k\le 10}\left|u_n(t_{n-1}+\frac{k}{10}h)\right|$. The results for smooth splines are presented in the following tables.

Case $d = 0$ (linear splines)

N	4	8	16	32	64
$c = 1.0$	2.05146	2.05038	2.05012	2.05005	2.05003
$c = 0.5$	2.04432	2.04856	2.04966	2.04994	2.05000
$c = 0.2$	2.04331	2.01338	$1.29 \cdot 10^2$	$3.32 \cdot 10^{10}$	$3.72 \cdot 10^{28}$

Case $d = 1$ (quadratic splines)

N	4	8	16	32	64
$c = 1.0$	2.05008	2.05003	2.05003	2.05003	2.05003
$c = 0.5$	2.06247	2.96007	$7.58 \cdot 10^4$	$8.40 \cdot 10^{15}$	$1.65 \cdot 10^{39}$
$c = 0.2$	6.85402	$3.14 \cdot 10^5$	$2.37 \cdot 10^{16}$	$2.27 \cdot 10^{39}$	$3.40 \cdot 10^{86}$

Case $d = 2$ (cubic splines)

N	4	8	16	32	64
$c = 1.0$	2.0498	2.0491	33.56	$3.10 \cdot 10^9$	$3.85 \cdot 10^{26}$
$c = 0.5$	3.089	$1.42 \cdot 10^4$	$4.65 \cdot 10^{13}$	$8.33 \cdot 10^{33}$	$4.39 \cdot 10^{75}$
$c = 0.2$	$1.29 \cdot 10^3$	$1.46 \cdot 10^{11}$	$3.30 \cdot 10^{28}$	$2.84 \cdot 10^{64}$	$3.47 \cdot 10^{137}$

Acknowledgments. The author wishes to acknowledge the hospitality provided by Tom Lyche and Ewald Quak as well as by their colleagues at the University of Oslo and SINTEF, where a part of this work was done in June 2000. His visit was supported by the Norwegian Academy of Sciences and Letters, and by the Estonian Science Foundation grant 3926.

References

1. Baker, C. T. H., *The Numerical Treatment of Integral Equations*, Clarendon Press, Oxford, 1977.

2. Brunner, H., The numerical solution of weakly singular Volterra integral equations by collocation on graded meshes, Math. Comp. **45** (1985), 417–437.

3. Brunner, H., and van der Houwen, P. J., *The Numerical Solution of Volterra Equations*, North-Holland, Amsterdam, 1986.

4. Brunner, H., A. Pedas, and G. Vainikko, The piecewise polynomial collocation method for nonlinear weakly singular Volterra equations, Math. Comp. **68** (1999), 1079–1095.

5. Danciu, I., Numerical stability of the spline collocation methods for Volterra integral equations, in *Proceedings of the International Conference on Approximation and Optimization Cluj-Napoca, 1996*, Vol. II, 1997, 69–78.

6. De Hoog, F. R. and R. Weiss, High order methods for a class of Volterra integral equations with weakly singular kernels, SIAM J. Numer. Anal. **11** (1974), 1166–1180.

7. De Hoog, F. R. and R. Weiss, Implicit Runge-Kutta methods for second kind Volterra integral equations, Numer. Math. **23** (1975), 199–213.

8. El Tom, M. A. E., On the numerical stability of spline function approximations to solutions of Volterra integral equations of the second kind, BIT **14** (1974), 136–143.

9. Hung, H.-S., The numerical solution of differential and integral equations by spline functions, MRC Tech. report No. 1053, University of Wisconsin, Madison, 1970.

10. Oja, P., Stability of the spline collocation for Volterra integral equations, J. Integral Equations Appl., to appear.

Peeter Oja
Institute of Applied Mathematics
University of Tartu
Liivi 2-206
50409 Tartu
Estonia
`peeter_o@ut.ee`

G^1-Hermite Interpolation of Ruled Surfaces

Martin Peternell

Abstract. This article discusses two methods for G^1-Hermite interpolation of ruled surfaces with low degree rational ruled surfaces. We will interpret ruled surfaces as one-parameter families of straight lines. Given two generating lines G_1, G_2 and tangent planes at points of these lines, we want to construct polynomial or rational ruled surfaces of low degree interpolating these boundary data.

§1. Introduction and Fundamentals

Ruled surfaces are among the simplest surfaces used for modeling and design, since one family of parameter curves are straight lines. Applications of ruled surfaces include surface modeling, motion design and wire-cut EDM. In the last few years several articles dealing with design of ruled surfaces have appeared, see for instance [2,5,7]. Different viewpoints and techniques can be chosen to study ruled surfaces. Here, they shall be treated as (differentiable) one-parameter families of lines.

A ruled surface Φ in Euclidean space $\mathbb{R}^3$ possesses a parametric representation

$$\mathbf{x}(u,v) = \mathbf{a}(u) + v\mathbf{r}(u), \qquad u \in I, v \in \mathbb{R}, \tag{1}$$

where $\mathbf{a}(u)$ denotes a directrix curve and $\mathbf{r}(u) \neq (0,0,0)$ denotes a vector field. In the following, we assume sufficient differentiability and regularity of the functions involved. The generating lines $G(u)$ of Φ are obtained by inserting a constant u_0 into (1). If $\mathbf{r}(u) = \mathbf{c}$ is a constant vector, (1) parametrizes a general cylinder.

The tangent plane at a regular surface point is spanned by the partial derivative vectors $\mathbf{x}_u = \dot{\mathbf{a}} + v\dot{\mathbf{r}}$ and $\mathbf{x}_v = \mathbf{r}$. Thus, the surface normal at $\mathbf{x}$ is

$$\mathbf{n}(u,v) = \mathbf{x}_u \times \mathbf{x}_v = \dot{\mathbf{a}}(u) \times \mathbf{r}(u) + v(\dot{\mathbf{r}}(u) \times \mathbf{r}(u)) = \mathbf{n}_1(u) + v\mathbf{n}_2(u).$$

A generating line $G(u_0)$ is called non–torsal iff $\det(\dot{\mathbf{a}}(u_0), \mathbf{r}(u_0), \dot{\mathbf{r}}(u_0)) \neq 0$, which expresses linear independence of $\mathbf{n}_1 = \dot{\mathbf{a}} \times \mathbf{r}$ and $\mathbf{n}_2 = \dot{\mathbf{r}} \times \mathbf{r}$.

Mathematical Methods for Curves and Surfaces: Oslo 2000
Tom Lyche and Larry L. Schumaker (eds.), pp. 413–422.

ISBN 0-8265-1378-6

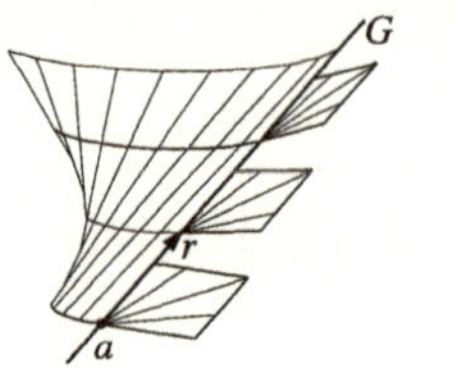

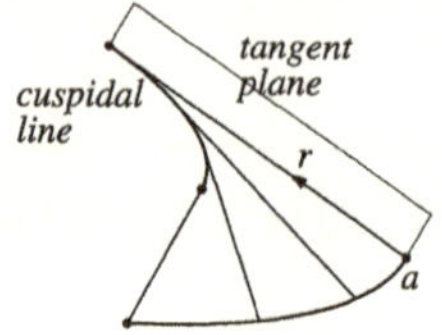

Fig. 1. Non-torsal ruled surface and torsal ruled surface with cuspidal line.

The surface normals along the generator $G(u_0)$ are parametrized by

$$\mathbf{y}(v,w) = \mathbf{a}(u_0) + v\mathbf{r}(u_0) + w(\mathbf{n}_1(u_0) + v\mathbf{n}_2(u_0)).$$

Because of bilinearity in the parameters v and w, this parametrization represents a hyperbolic paraboloid. The points $\mathbf{a} + v\mathbf{r}$ on $G(u_0)$ are in bijective correspondence to the normals $\mathbf{n}_1 + v\mathbf{n}_2$ (or tangent planes) by

$$\mathbf{a}(u_0) + v\mathbf{r}(u_0) \mapsto \mathbf{n}_1(u_0) + v\mathbf{n}_2(u_0). \tag{2}$$

We can extend formula (2) for the parameter value $v = \infty$. This implies that the point at infinity of $G(u_0)$ possesses a tangent plane with normal vector $\mathbf{n}_2(u_0)$. The mapping (2) is called contact projectivity along a non–torsal generator. This leads to the following lemma:

Lemma 1. *Two ruled surfaces $\mathbf{x}_1, \mathbf{x}_2$ are tangent at all points of a common generator G, iff $\mathbf{x}_1, \mathbf{x}_2$ possess same tangent planes at three points of G.*

A generator $G(u_0)$ is called torsal if all regular points of $G(u_0)$ possess the same tangent plane. Analytically, we have

$$\det(\dot{\mathbf{a}}(u_0), \mathbf{r}(u_0), \dot{\mathbf{r}}(u_0)) = 0. \tag{3}$$

There are two cases to be distinguished. First, if $rank(\mathbf{r}(u_0), \dot{\mathbf{r}}(u_0)) = 1$, all points on $G(u_0)$ are regular and the common surface normal along $G(u_0)$ is $\dot{\mathbf{a}} \times \mathbf{r}$. If Φ is a cylinder surface, all rulings are of this type and parallel to each other. Thus, $G(u_0)$ are called cylindrical. Secondly, if $G(u_0)$ is not cylindrical, there exists exactly one singular point on $G(u_0)$, whose parameter value is

$$v_c = -\frac{(\dot{\mathbf{a}} \times \mathbf{r}) \cdot (\dot{\mathbf{r}} \times \mathbf{r})}{(\dot{\mathbf{r}} \times \mathbf{r}) \cdot (\dot{\mathbf{r}} \times \mathbf{r})}. \tag{4}$$

It is called cuspidal point. The surface normal in all other points of $G(u_0)$ is $\dot{\mathbf{a}} \times \mathbf{r}$. A ruled surface Φ is called torsal, if it is developable, which expresses that Φ can be represented as envelope of its one-parameter family of tangent planes

$$(\mathbf{x} - \mathbf{a}) \cdot (\dot{\mathbf{a}} \times \mathbf{r}) = 0.$$

The singular curve $\mathbf{x}(u, v_c(u)) = \mathbf{c}(u)$ on Φ is called line of regression or cuspidal line. If the curve $\mathbf{c}(u)$ consists of one point only, Φ is called a cone.

If $G(u_0)$ is a non-torsal generator, the parameter value $v_s = v_c$ in (4) parametrizes the striction curve $\mathbf{a}(u) + v_s\mathbf{r}(u)$. The distribution parameter

$$\delta(u) = -\frac{\det(\dot{\mathbf{a}}, \mathbf{r}, \dot{\mathbf{r}})}{(\mathbf{r} \times \dot{\mathbf{r}})^2}(u)$$

is a Euclidean differential invariant of first order. It measures how fast the tangent plane turns around $G(u_0)$ if the point $\mathbf{x}(u_0, v)$ travels along $G(u_0)$. It is a signed invariant, and is zero for torsal generators (except for some cylindrical generators of higher order). For more details on Euclidean line geometry, see for instance [1,3].

§3. An Elementary Method for G^1-Hermite Interpolation

Let $R : \mathbf{x}(u, v) = \mathbf{y}(u) + v\mathbf{r}(u)$ be a given ruled surface in $\mathbb{R}^3$, $u \in I$ and $v \in \mathbb{R}$. We assume that there is a cartesian coordinate system such that all generating lines $G(u) = \mathbf{y}(u) + v\mathbf{r}(u)$ intersect two parallel planes $E_1 : z = 0$ and $E_2 : z = 1$. Let $\mathbf{c}_1(u)$, $\mathbf{c}_2(u)$ be the intersection curves of R with E_1, E_2. We pick a sequence of generating lines G_i, $i = 1, \ldots, N$, and compute tangent planes along them. Our task is to determine a ruled surface S which interpolates two adjacent generators $G, H \subset \{G_i\}$ such that S and R are tangent at all points of G and H. Let

$$\mathbf{a} = G \cap E_1, \quad \mathbf{b} = H \cap E_1, \quad \mathbf{p} = G \cap E_2, \quad \mathbf{q} = H \cap E_2.$$

The intersection points of the tangent planes at $\mathbf{a}, \mathbf{b}$ and $\mathbf{p}, \mathbf{q}$ along G, H with E_1, E_2 are the inner points $\mathbf{c}, \mathbf{r}$, see Fig. 2.

We construct a low degree rational (or polynomial) ruled surface S as tensor product surface of degrees $(d, 1)$,

$$S : \mathbf{x}(u, v) = (1 - v)\mathbf{k}_1(u) + v\mathbf{k}_2(u),$$

such that the planar intersection curves $\mathbf{k}_1$ and $\mathbf{k}_2$ interpolate points $\mathbf{a}, \mathbf{b}$ and $\mathbf{p}, \mathbf{q}$ plus the given tangents determined by $\mathbf{c}$ and $\mathbf{r}$. This yields that S and R possess common generating lines G, H and same tangent planes at the points $\mathbf{a}, \mathbf{p}$ and $\mathbf{b}, \mathbf{q}$.

Applying Lemma 1, R and S are tangent at all points of G (or H), if they have the same tangent plane at a *third* point, different from $\mathbf{a}, \mathbf{p}$ (or $\mathbf{b}$, $\mathbf{q}$). For simplicity we choose this third point as midpoint of $\mathbf{a}, \mathbf{p}$ (or $\mathbf{b}, \mathbf{q}$), see Fig. 2.

There is a one parameter family of conics

$$\mathbf{k}_1(u) = \frac{(1-u)^2\mathbf{a} + 2tu(1-u)\mathbf{c} + u^2\mathbf{b}}{(1-u)^2 + 2tu(1-u) + u^2}$$

satisfying the G^1-requirements in the plane E_1. Since we set the weights at $\mathbf{a}, \mathbf{b}$ to 1, we use a special parametrization here. Additionally there is a one parameter family of conics

$$\mathbf{k}_2(u) = \frac{(1-u)^2\mathbf{p}w_1 + 2u(1-u)\mathbf{r} + u^2\mathbf{q}w_2}{w_1(1-u)^2 + 2u(1-u) + w_2u^2}$$

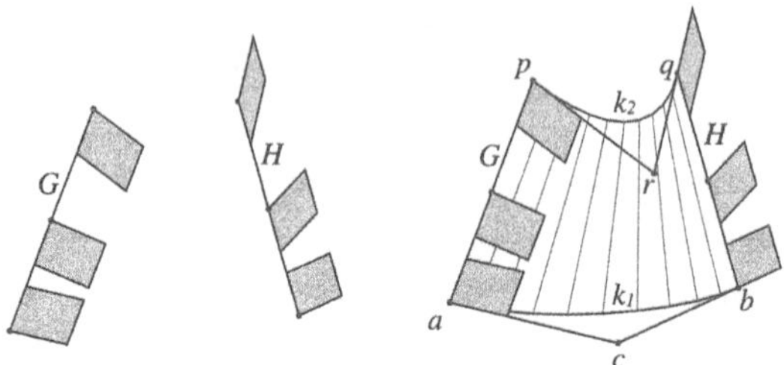

Fig. 2. G^1 Hermite boundary data and solution of the interpolation problem.

satisfying the G^1-requirements in the plane E_2. We put weights w_1, w_2 to points $\mathbf{p}, \mathbf{q}$ to be flexible with the parametrization. Further, let $\mathbf{n}$ and $\mathbf{m}$ be the given surface normals at the midpoints $1/2(\mathbf{a}+\mathbf{p})$ and $1/2(\mathbf{b}+\mathbf{q})$.

Inserting the G^1 condition for the midpoints, we find that the surface S is tangent to R at all points of G and H iff the following conditions on the parameters t, w_1, w_2 hold:

$$tw_1 = -\frac{(\bar{\mathbf{r}}-\bar{\mathbf{p}})\cdot\bar{\mathbf{n}}}{(\bar{\mathbf{c}}-\bar{\mathbf{a}})\cdot\bar{\mathbf{n}}} = \tau_1, \quad tw_2 = -\frac{(\bar{\mathbf{q}}-\bar{\mathbf{r}})\cdot\bar{\mathbf{m}}}{(\bar{\mathbf{b}}-\bar{\mathbf{c}})\cdot\bar{\mathbf{m}}} = \tau_2. \tag{5}$$

Here, $\bar{\mathbf{x}} = (x_1, x_2)$ denotes the vector built by the first two coordinates of the vector $\mathbf{x} \in \mathbb{R}^3$.

Expressing w_1, w_2 in terms of t yields a one parameter family of ruled surfaces $S(t)$ solving the G^1-Hermite interpolation problem.

Useful solutions

The above discussed algorithm results in useful solutions if τ_1, τ_2 possess the same sign. If this is not the case, one of the conics possesses points at infinity. This is caused by a too large difference of the distribution parameters at G and H. To avoid this, we can choose generators G, H 'closer'; or, alternatively, we let E_2 be the plane $z = 0.5$, for instance, which reduces the turning angle of the tangent planes. If the segments on G, H are sufficiently small and G and H are close enough, τ_1, τ_2 will have same sign. For a useful distance measure between lines, see [6].

If G is a torsal generator, $(\bar{\mathbf{c}}-\bar{\mathbf{a}})\cdot\mathbf{n} = 0$ and $(\bar{\mathbf{r}}-\bar{\mathbf{p}})\cdot\bar{\mathbf{n}}$ vanishes too. Thus, the G^1-condition along a torsal generator is already satisfied, and the weight w_1 can be chosen arbitrarily.

If both generators are torsal, all weights can be chosen arbitrarily, for instance, equal to one. The solution S is in general a non–torsal polynomial ruled surface of degree 4 with two torsal generators.

Theorem 2. *Given G^1-Hermite boundary data of a ruled surface, there is a one parameter family $S(t)$ of rational ruled surfaces of degrees (2,1) and order 4 which interpolate the given data. The surfaces $S(t)$ carry a one parameter family of conics, and intersect two given planes E_1, E_2 in conics.*

Our method is general, and works for arbitrary input data, and the planes E_1, E_2 need not be parallel.

In Section 4 we will compute a lowest degree solution of the G^1-Hermite interpolation problem, which consists of a smoothly joined pair of ruled quadrics. But, since quadrics never possess torsal generators, this method is restricted to ruled surfaces without torsal generators. To derive this method, we have to introduce some theory of lines in space.

A modeling scheme using cubic ruled surfaces is difficult because it can be proved that cubic surfaces do not fit all possible data. A combined method consisting of quadric pairs and cubic surface segments is discussed in [5].

§3. Local Coordinates of Lines

A local parametrization of the set of lines in $\mathbb{R}^3$, or at least in a domain of interest, shall be constructed. With some restrictions, a line G can be mapped onto a vector $\mathbf{G} = (g_1, \ldots, g_4) \in \mathbb{R}^4$. This implies that a ruled surface Φ is mapped onto a curve and a two parametric family of lines is mapped onto a surface in $\mathbb{R}^4$. Thus, the G^1-Hermite interpolation of ruled surfaces in $\mathbb{R}^3$ will be translated to G^1-Hermite interpolation with curves in $\mathbb{R}^4$.

For practical purposes, it is often sufficient to consider patches of ruled surfaces, bounded by two planes which enclose the domain of interest in $\mathbb{R}^3$. In the following we will assume that these two planes are parallel and are chosen to be $E_1 : z = 0$ and $E_2 : z = 1$, perpendicular to the z-axis of the coordinate system. The intersection points $\mathbf{g}_1 = (g_1, g_2, 0)$ and $\mathbf{g}_2 = (g_3, g_4, 1)$ of a line G and the planes E_1, E_2 define a parametrization of all *non-horizontal lines* $\mathcal{L}$ by

$$\begin{aligned} \mu : \mathbb{R}^4 = \mathbb{R}^2 \times \mathbb{R}^2 &\to \mathcal{L} \\ \mathbf{G} = (g_1, g_2, g_3, g_4) &\mapsto \mu(\mathbf{G}) = G. \end{aligned} \tag{6}$$

Parametrization μ is a local mapping, and depends essentially on the coordinate system. In applications, the z-axis of the coordinate system can be computed as solution of a regression problem, see [6].

Some Linear Subsets of $\mathbb{R}^4$ and their μ-Images

Given two non–intersecting lines G, H in $\mathbb{R}^3$, there is a unique bilinear tensor product surface (hyperbolic paraboloid)

$$\mathbf{x}(u,v) = (1-v)\left((1-u)\mathbf{g}_1 + u\mathbf{g}_2\right) + v\left((1-u)\mathbf{h}_1 + u\mathbf{h}_2\right).$$

The μ^{-1}-image curve of $\mathbf{x}$ is the straight line segment in $\mathbb{R}^4$ connecting $\mathbf{G} = \mu^{-1}(G)$ and $\mathbf{H} = \mu^{-1}(H)$, see Fig. 3.

If G and H are intersecting, $\mathbf{h}_1 - \mathbf{g}_1$ and $\mathbf{h}_2 - \mathbf{g}_2$ are linearly dependent. Analytically, this says that

$$(h_1 - g_1)(h_4 - g_4) - (h_2 - g_2)(h_3 - g_3) = 0.$$

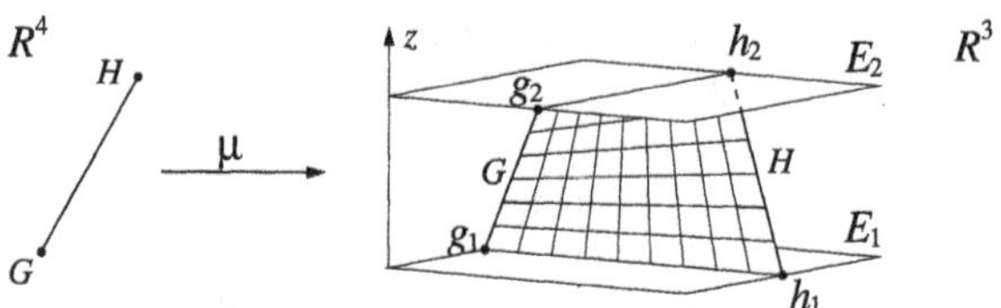

Fig. 3. Hyperbolic paraboloid in $\mathbb{R}^3$ as μ-image of a straight line $\mathbb{R}^4$.

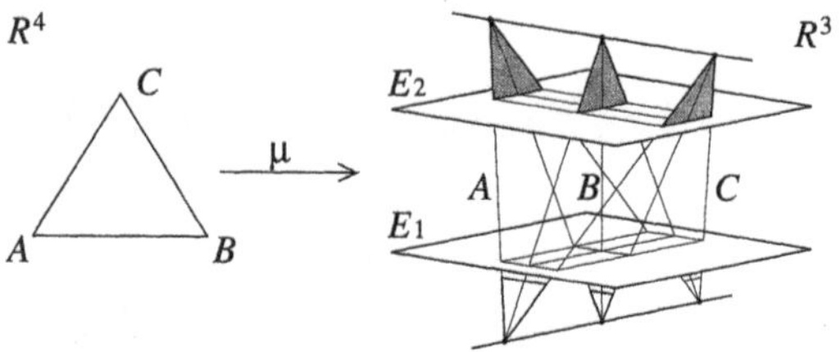

Fig. 4. Hyperbolic net in $\mathbb{R}^3$ as μ-image of a plane in $\mathbb{R}^4$.

The direction vector $\mathbf{H} - \mathbf{G} = (h_1 - g_1, \ldots, h_4 - g_4)$ satisfies the quadratic equation of the indefinite quadratic form

$$I : \langle \mathbf{X}, \mathbf{X} \rangle_i = x_1 x_4 - x_2 x_3 = 0, \mathbf{X} \in \mathbb{R}^4. \tag{7}$$

Vectors satisfying (7) shall be called isotropic. The μ-image of an isotropic line $(1-u)\mathbf{G} + u\mathbf{H}$ (line with isotropic direction vector) is a pencil of lines, spanned by intersecting lines G, H. We summarize.

Corollary 3. *The μ-image of a line G in $\mathbb{R}^4$ is a pencil of lines or a hyperbolic paraboloid, depending on whether G is an isotropic line or not.*

Let E be a plane in $\mathbb{R}^4$ which is parametrized by $E : \mathbf{A} + \sigma(\mathbf{B} - \mathbf{A}) + \tau(\mathbf{C} - \mathbf{A})$. What does the corresponding set of lines in $\mathbb{R}^3$ look like? We compute the isotropic directions in E and obtain the following quadratic equation in the homogeneous parameter $(\sigma : \tau)$:

$$\langle \sigma(\mathbf{B} - \mathbf{A}) + \tau(\mathbf{C} - \mathbf{A}), \sigma(\mathbf{B} - \mathbf{A}) + \tau(\mathbf{C} - \mathbf{A}) \rangle_i = 0. \tag{8}$$

If (8) vanishes identically, the plane E is called isotropic since it contains only isotropic directions. The family of lines $\mu(E)$ in $\mathbb{R}^3$ is a bundle (all lines through a fixed point) or a field (lines in a fixed plane).

Otherwise, the equation (8) has two, one or zero solutions and the planes are of hyperbolic, parabolic or elliptic type. The family of lines $\mu(E)$ in $\mathbb{R}^3$ is called a hyperbolic, parabolic or elliptic net.

A plane E of hyperbolic type carries two 1-parameter families of isotropic lines, parallel to the isotropic directions. The corresponding hyperbolic net $\mu(E)$ consists of two 1-parameter families of pencils of lines. The vertices of these pencils form the two horizontal focal lines or axes of the net, see Figure 4.

A plane E of elliptic type in $\mathbb{R}^4$ contains no real isotropic directions such that the focal lines of the elliptic net $\mu(E)$ are conjugate imaginary. There are no pencils contained in that net $\mu(E)$ and pairwise distinct lines G, H of $\mu(E)$ are skew.

A plane E of parabolic type carries a 1-parameter family of isotropic lines such that the parabolic net $\mu(E)$ consists of a 1-parameter family of pencils of lines. The vertices of these pencils lie on one horizontal focal line and the planes containing the pencils pass through this focal line. The vertices and the planes are in a projective correspondence.

Corollary 4. *The μ-image of a plane E in $\mathbb{R}^4$ is a bundle or field of lines in the case when E contains only isotropic directions and otherwise it is an elliptic, parabolic or hyperbolic net of lines.*

Ruled Surfaces as μ-images of curves $\subset \mathbb{R}^4$

Consider a smooth curve $\mathbf{C}(t)$ in $\mathbb{R}^4$, different from a straight line. The ruled surface $\mu(\mathbf{C}(t))$ is parametrizable in the form

$$\mathbf{x}(t,v) = (1-v)\mathbf{c}_1(t) + v\mathbf{c}_2(t), \tag{9}$$

where $\mathbf{c}_1, \mathbf{c}_2$ are the intersection curves of $\mu(\mathbf{C})$ with the planes E_1, E_2.

The tangent line $\mathbf{C}(t_0) + \lambda\dot{\mathbf{C}}(t_0)$ determines the first order properties of the ruled surface $\mu(\mathbf{C})$ at the generating line $\mu(\mathbf{C}(t_0))$. If $\dot{\mathbf{C}}(t_0)$ is isotropic, $\mu(\mathbf{C}(t_0))$ is a torsal generator and the cuspidal point $\mathbf{v}(t_0)$ is the vertex of the pencil of lines $\mu(\mathbf{C} + \lambda\dot{\mathbf{C}})(t_0)$.

If $\dot{\mathbf{C}}(t_0)$ is not isotropic, the hyperbolic paraboloid determined by the tangent line $\mathbf{C}(t_0) + \lambda\dot{\mathbf{C}}(t_0)$ touches the ruled surface $\mu(\mathbf{C})$ in all points of the generator $\mu(\mathbf{C}(t_0))$.

Let $\mathbf{C}(t)$ be a planar curve in a non–isotropic plane E. The ruled surface $\mu(\mathbf{C}(t))$ is contained in the net of lines $\mu(E)$. If E is isotropic, $\mu(\mathbf{C}(t))$ is a cone or the set of tangent lines of a planar curve, depending on whether $\mu(E)$ is a bundle or field of lines.

In particular, if $\mathbf{C}$ is a conic in a non–isotropic plane, $\mu(\mathbf{C})$ is a rational ruled surface of order ≤ 4. The parametrization (9) is a (2,1) tensor product representation and $\mathbf{c}_1, \mathbf{c}_2$ are conics in the planes E_1, E_2. The degree is less than 4 if $\mathbf{c}_1$ and $\mathbf{c}_2$ possess common points for common parameter values. These parameter values as well as the common points need not to be real. Additionally, common points could lie at infinity.

In general, let $\mathbf{C}(t)$ be a rational curve of degree d in $\mathbb{R}^4$. A $(d,1)$ rational tensor product point representation of $\mu(\mathbf{C})$ is given by (9). The order of the ruled surface is at most $2d$, since the planar curves $\mathbf{c}_1, \mathbf{c}_2$ are in general of degree d.

§4. Interpolation by Pairs of Quadrics

Given two generating lines A, B of a ruled surface R and tangent planes (or surface normals) at points of A and B, we want to construct a ruled surface,

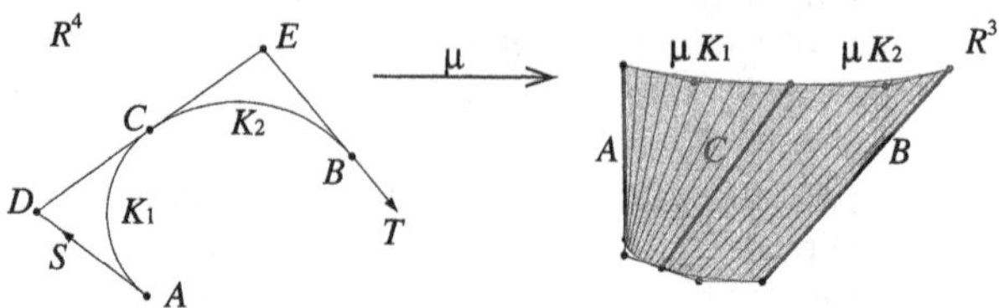

Fig. 5. Isotropic biarc in $\mathbb{R}^4$ and pair of quadrics in $\mathbb{R}^3$.

consisting of a pair of ruled quadrics, which interpolates the given boundary data. Since ruled quadrics do not possess torsal generators, we have to restrict ourselves to ruled surfaces R without torsal generators. This implies that the distribution parameter δ has no sign changes in the interval determined by A and B.

The local parameterization μ maps generating lines A, B with associated tangent planes onto points $\mathbf{A}$, $\mathbf{B}$ with associated tangent lines with direction vectors $\mathbf{S}$ and $\mathbf{T}$, respectively. In general, the data $\mathbf{A}, \mathbf{B}, \mathbf{T}, \mathbf{S}$ span a 3–space in $\mathbb{R}^4$.

In Section 3 we introduced an indefinite quadratic form I in formula (7). Similar to Euclidean space, one can define isotropic circles with respect to I. A curve $\mathbf{K}$ is an isotropic circle if it is a planar intersection of an isotropic sphere $\Sigma : \langle \mathbf{X}, \mathbf{X} \rangle_i = r$. Additionally, we require that the plane carrying $\mathbf{K}$ is not tangent to Σ. This excludes degeneracies where $\mu(\mathbf{K})$ is contained in a bundle or a field or is the union of two pencils.

An isotropic circle (i-circle) in a plane of hyperbolic or elliptic type is a conic $\mathbf{K}$, whose points $\mathbf{X}$ possess constant isotropic distance r from a given center $\mathbf{M}$,

$$\|\mathbf{M} - \mathbf{X}\|_i = \sqrt{\langle \mathbf{M} - \mathbf{X}, \mathbf{M} - \mathbf{X} \rangle_i} = r = const.$$

An isotropic circle in a plane of parabolic type is a parabola $\mathbf{K}$ with isotropic axis. The following theorem holds.

Theorem 5. *The μ-images of isotropic circles $\mathbf{K}$ are ruled quadrics.*

This theorem allows us to apply an isotropic biarc construction in $\mathbb{R}^4$. We will construct two isotropic circles $\mathbf{K}_1(t), \mathbf{K}_2(t)$, which interpolate the given data $\mathbf{A}, \mathbf{T}$ and $\mathbf{B}, \mathbf{S}$ and join smoothly at a point $\mathbf{C}$. Taking into account that no torsal generators occur in the segment under consideration, we can assume that

$$sign(\langle \mathbf{S}, \mathbf{S} \rangle_i) = sign(\langle \mathbf{T}, \mathbf{T} \rangle_i).$$

This allows the normalization $\|\mathbf{S}\|_i = \|\mathbf{T}\|_i = 1$. Further, by letting $\mathbf{Y} = \mathbf{B} - \mathbf{A}$ we have to guarantee that $sign(\langle \mathbf{Y}, \mathbf{Y} \rangle_i) = sign(\langle \mathbf{T}, \mathbf{T} \rangle_i)$, which can always be achieved by an appropriate choice of the data lines A, B.

The biarc $\mathbf{K}_1, \mathbf{K}_2$ possesses inner Bézier points $\mathbf{D}, \mathbf{E}$ and junction point

Fig. 6. Sequence of pairs of ruled quadrics interpolating G^1-Hermite data.

$\mathbf{C}$, see Fig. 5, with

$$\mathbf{D} = \mathbf{A} + \lambda\mathbf{S},\ \mathbf{E} = \mathbf{B} - \mu\mathbf{T},\ \mathbf{C} = \frac{\mu}{\lambda+\mu}\mathbf{D} + \frac{\lambda}{\lambda+\mu}\mathbf{E}.$$

The weights w_1, w_2 of the inner control points $\mathbf{D}, \mathbf{E}$ are computed by

$$w_1 = \frac{\langle \mathbf{C}-\mathbf{A}, \mathbf{D}-\mathbf{A}\rangle_i}{\|\mathbf{C}-\mathbf{A}\|_i \|\mathbf{D}-\mathbf{A}\|_i}, \quad w_2 = \frac{\langle \mathbf{B}-\mathbf{C}, \mathbf{E}-\mathbf{C}\rangle_i}{\|\mathbf{B}-\mathbf{C}\|_i \|\mathbf{E}-\mathbf{C}\|_i}.$$

The points $\mathbf{A}, \mathbf{B}, \mathbf{C}, \mathbf{D}, \mathbf{E}$ and weights w_1, w_2 determine a biarc if and only if

$$\langle \mathbf{E}-\mathbf{D}, \mathbf{E}-\mathbf{D}\rangle_i = (\lambda+\mu)^2.$$

Elaborating this gives a bilinear equation in λ and μ,

$$\langle \mathbf{Y}, \mathbf{Y}\rangle_i - 2\lambda\langle \mathbf{Y}, \mathbf{S}\rangle_i - 2\mu\langle \mathbf{Y}, \mathbf{T}\rangle_i + 2\lambda\mu\left[\langle \mathbf{S}, \mathbf{T}\rangle_i - 1\right] = 0.$$

We summarize the result:

Theorem 6. *Given G^1-Hermite boundary data satisfying the above restrictions, there exists a one-parameter family of isotropic biarcs in $\mathbb{R}^4$ interpolating the given data. Finally, there is a one-parameter family of quadric pairs solving the G^1-Hermite interpolation problem in $\mathbb{R}^3$.*

To obtain a unique solution, we might let $\lambda = \mu$ or $\lambda + \mu \to \min$, see [4]. More generally, optimization in curve design is often done by minimizing a functional involving second derivatives. Thus, we can require that

$$F = \int_{t_0}^{t_1} \langle \ddot{\mathbf{C}}, \ddot{\mathbf{C}}\rangle_e dt \to \min, \tag{10}$$

where $\mu(\mathbf{C}(t))$ is a ruled surface. Here, $\langle \mathbf{X}, \mathbf{X}\rangle_e = x_1^2+x_2^2+x_3^2+x_4^2+x_1x_3+x_2x_4$ denotes a Euclidean scalar product in $\mathbb{R}^4$ which induces a distance measure between lines, see [6]. The minimization of (10) is equivalent to minimizing the functional

$$3\int_{t_0}^{t_1}\int_0^1 [(1-u)\ddot{\mathbf{c}}_1 + u\ddot{\mathbf{c}}_2]^2 du dt, \tag{11}$$

where $(1-u)\mathbf{c}_1 + u\mathbf{c}_2$ are level curves of the ruled surface $\mu(\mathbf{C})$. Thus, we minimize the linearized curvatures of all these level curves.

Theorem 7. *Minimizing the functional (11) over all level curves* $(1-u)\mathbf{c}_1 + u\mathbf{c}_2$ *of the ruled surface* $S = \mu(\mathbf{C}(t))$ *is equivalent to minimizing the linearized bending energy of the image curve* $\mathbf{C}(t)$ *in* $\mathbb{R}^4$ *with respect to the Euclidean scalar product* $\langle , \rangle_e$.

Acknowledgments. This work has been supported by grant No. P13648-MAT of the Austrian Science Fund.

References

1. Hoschek, J., *Liniengeometrie*, Zürich, Bibliographisches Institut, 1971.
2. Hoschek, J. and U. Schwanecke, Interpolation and approximation with ruled surfaces, in *The Mathematics of Surfaces VIII*, Robert Cripps (ed.), Information Geometers, 1998, 213–231.
3. Hlavaty, V., *Differential Line Geometry*, Groningen, P. Nordhoff Ltd., 1953.
4. Nutbourne, A.W. and R.R. Martin, *Differential Geometry Applied to Curve and Surface Design*, Ellis Horwood Ldt., 1988.
5. Peternell, M., H. Pottmann, and B. Ravani, On the computational geometry of ruled surfaces, Computer Aided Design **31** (1999), 17–32.
6. Peternell, M., and H. Pottmann, Interpolating Functions on Lines in 3-Space, in *Curve and Surface Design: Saint-Malo 1999*, Pierre-Jean Laurent, Paul Sablonnière, and Larry L. Schumaker (eds.), Vanderbilt University Press, Nashville, 2000, 351–358.
7. Pottmann, H., M. Peternell, and B. Ravani, Approximation in line space: applications in robot kinematics and surface reconstruction, in *Advances in Robot Kinematics: Analysis and Control*, J. Lenarcic and M. Husty (eds.), Kluwer, 1998, 403-412.
8. Wang, W. and B. Joe, Interpolation on quadric surfaces with rational quadratic spline curves, Computer Aided Geometric Design **14** (1997), 207–230.

Martin Peternell
Institute of Geometry
Vienna University of Technology
Wiedner Hauptstrasse 8–10
A-1040 Wien, Austria
`martin@geometrie.tuwien.ac.at`
http://www.geometrie.tuwien.ac.at/peternell/

Recovering Structural Information from Triangulated Surfaces

Ch. Rössl, L. Kobbelt, and H.-P. Seidel

Abstract. We present a technique for recovering structural information from triangular meshes that can then be used for segmentation, e.g. in reverse engineering applications. In a preprocessing step, we detect feature regions on the surface by classifying the vertices according to some discrete curvature measure. Then we apply a skeletonization algorithm for extracting feature lines from these regions. To achieve this, we generalize the concept of morphological operators to unorganized triangle meshes, providing techniques for noise reduction on the binary feature classification and for skeletonization. The necessary operations are easy to implement, robust, and can be executed efficiently on a mesh data structure.

§1. Introduction

Discrete surface representations are becoming more and more important in geometric design applications, since such surface data typically arises when real-world objects are sampled by some 3D scanning device. The sensor will provide a discrete set of data from which a polygonal mesh is finally generated. There are various technologies for data acquisition and algorithms for the polygonal reconstruction step [2].

In order to make existing CAD systems applicable to input models from 3D scanning, further processing is required: The piecewise linear representation must be converted into a piecewise smooth, continuous surface model as this is the standard representation for current CAD applications.

For this purpose, additional structural information has to be derived from the polygonal representation. This information is used for separating surface regions that can be replaced by one smooth surface patch, such as splines or parts of geometric primitives like cylinders, spheres, etc.

This paper focuses on recovering such structural information from triangulated surfaces as they are typically obtained from a 3D scanner. Such meshes are usually very dense point clouds with stochastical noise that have

Mathematical Methods for Curves and Surfaces: Oslo 2000
Tom Lyche and Larry L. Schumaker (eds.), pp. 423–432.

ISBN 0-8265-1378-6

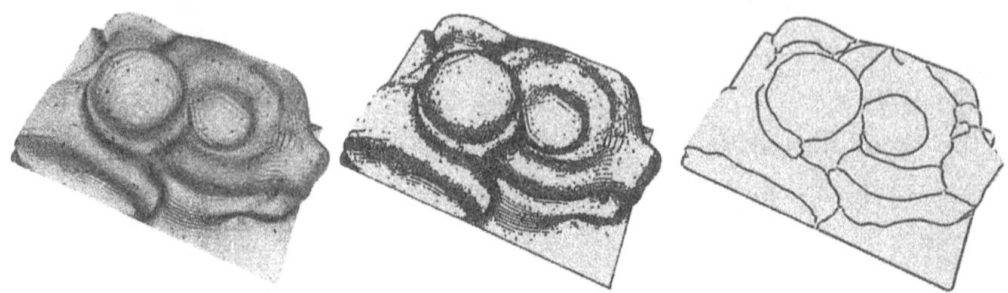

Fig. 1. The figures show a resampled version of the well-known benchmark object from [5]. The maximum curvature on the mesh (left) is thresholded to obtain the initial feature regions (middle). The (smoothed) feature lines are then extracted from these regions by skeletonization (right).

been triangulated by the mentioned algorithms. Deriving structural information is also known as segmentation in the reverse engineering process where suitable computer models are created from surfaces of physical objects. Segmentation is crucial for reverse engineering when an object cannot be approximated by a single surface. A recent overview of this topic is given in [12].

There are two generic approaches to surface segmentation. The first one is region growing or also called face-based approach [11]. A small seed region is grown by adding neighboring regions that are similar with respect to some flatness criterion e.g. [7,10]. Besides the bottom-up region growing, there is also a top-down approach. In [12] a sophisticated cascade of surface fitting steps is used that finally decomposes a surface region in the last step when no surface primitive could be determined for this region. An initial segmentation is obtained by a preprocessing step (planar filtering) similar to the second approach below.

This second approach is called edge-detection or edge-based. Instead of finding explicitly the different surface regions, the boundaries between these regions are estimated e.g. by assuming rapid changes of angular variation at edges [6]. Edge detection schemes may suffer from noise and poor sampling in sharp surface regions. In this paper we try to overcome some of these difficulties.

We use an arbitrary edge detection scheme that is based on surface curvature information and a conservative criterion for extracting "feature regions". The extracted regions will not represent the desired boundaries or surface features exactly, but most of these feature lines can be expected to lie inside the detected regions.

The wide feature regions are narrowed down to feature lines that approximate the structural information and which can be used as natural boundaries for different surface regions (Fig. 1).

In order to implement this skeletonization process, morphological operators are generalized from digital image processing to (bounded manifold) triangle meshes. These operators then work on boolean valued functions on arbitrary meshes rather than on regular domains. Dilation and erosion are used for reducing noise on the feature regions, while a skeletonization operator is introduced for shrinking these regions down to feature lines. For constructing consistent boundary curves, a more sophisticated graph structure is extracted from the feature lines representing the network of boundary curves.

The operators used are simple and can be implemented very efficiently because they only rely on basic operations on the triangle mesh data structures. Of course there is a tradeoff between simplicity and the capabilities of our algorithm as will be explained below.

§2. Preprocessing and Notation

Our algorithm processes triangle meshes as they are typically generated from the combination of a 3D scanning device and surface reconstruction algorithms. The vertices of the mesh are assumed to be more or less uniformly scattered over the surface such that the edge lengths do not vary too much. Otherwise the input mesh should be resampled or remeshed in a way that the condition is fulfilled.

This is necessary for our morphological operators to work reasonably, as they are topological operators. As a consequence, a geometric interpretation is only valid if there is some correlation, i.e. edge lengths are approximately constant. Working just on the mesh topology results in a very efficient algorithm, much in the spirit of morphological operators in digital image processing.

Feature regions are detected by applying some kind of discrete curvature analysis. Curvature information is used to classify feature and non-feature vertices. This may be done with a simple but robust thresholding operation on e.g. the maximum curvature at a vertex (cf. Fig. 1) or a more sophisticated scheme for ridge detection such as [8]. Either way, the criterion for feature vertices is relaxed such that wide feature regions are extracted.

Here, the technique suffers from a principal disadvantage of edge-based methods as smooth surface parts cannot be segmented reliably. So the output of our algorithm may need some refinement, e.g. using traditional methods.

Consider a triangle mesh with vertices $\{V_1, \ldots, V_n\}$ and edges $\mathbb{E} := \{(V_i, V_j)\}$. Then we describe the feature region as the vector $\boldsymbol{F} \in \{0,1\}^n$ with

$$\boldsymbol{F}_i = \begin{cases} 1, & V_i \text{ is a feature vertex,} \\ 0, & \text{otherwise.} \end{cases}$$

Vertices V_i with $\boldsymbol{F}_i = 1$ will be called marked. For convenience we introduce an alternative notation. Let $\boldsymbol{F}$ be assigned the set

$$\mathbb{F} := \{i \in \{1, \ldots, n\} \,|\, \boldsymbol{F}_i = 1\},$$

with $\overline{\mathbb{F}} := \{1, \ldots, n\} \setminus \mathbb{F}$.

There is only one operation on the triangle mesh that is needed: the enumeration of the 1-neighborhood of a vertex. We define this neighborhood

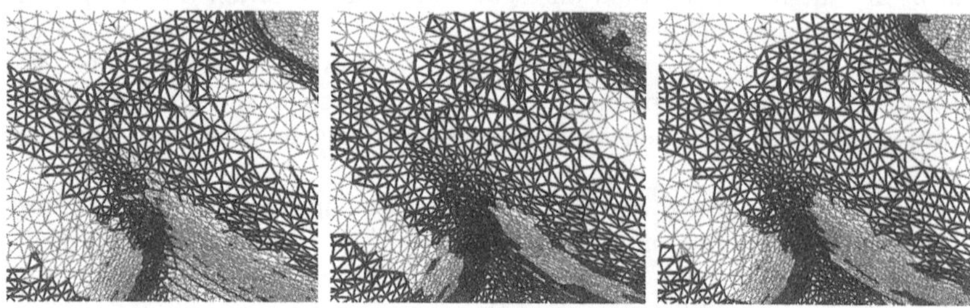

Fig. 2. Left: Initial feature region (close up view from Fig. 1); middle: after dilation; right: and subsequent erosion (=closing).

relation as follows

$$\mathbf{nhd}\{i\} := \{i\} \cup \{j|(V_i, V_j) \in \mathbb{E}\}.$$

For the sake of a simple notation, **nhd** maps vertex indices i rather than vertices V_i. Using this definition, the radius of a neighborhood can be recursively enlarged by defining a d-neighborhood $\mathbf{nhd}^d$ as

$$\mathbf{nhd}\{i_1, \ldots, i_k\} := \bigcup_{1\le\mu\le k} \mathbf{nhd}\{i_\mu\}$$

$$\mathbf{nhd}^1\{i\} := \mathbf{nhd}\{i\}, \quad \mathbf{nhd}^{d+1}\{i\} := \mathbf{nhd}\left(\mathbf{nhd}^d\{i\}\right) \quad (d > 0)$$

§3. Morphological Operators

Mathematical morphology has been used in digital image analysis for quite a long time. Morphological operators are particularly interesting and often preferred to convolution operators because of their simplicity and the fact that they can be efficiently implemented in hardware [4].

We adapt morphological operators to operate on the binary feature vector $\boldsymbol{F}$. As we are using discrete curvature for setting up the initial feature region, high frequency noise may be a problem. Apart from some prefiltering of the input data, we use the dilation and erosion operators to suppress "classification noise" in $\boldsymbol{F}$.

The classical definitions of dilation and erosion are based on addition in a (m-dimensional) Euclidean vector space E^m. E.g. the dilation operator generates from two given sets $A, B \subset E^m$ the union $A \oplus B = \{c \in E^m \,|\, c = a + b, a \in A, b \in B\}$. Here A is the image or pattern to be dilated, and B denotes the so called structure element. Note that we are dealing with binary values in contrast to gray scale operators as used in [6] for edge detection.

Such definitions for a vector space E^n cannot be used directly on general triangle meshes. Since the connectivity of the mesh is irregular, there is no reasonable definition for an addition. Therefore we generalize morphological operators for triangle meshes, even though in a limited way, i.e. only for special types of structure elements.

Definition 1. *Let* $\mathbb{F} \subseteq \{1, \ldots, n\}$. *The* dilation *of* $\mathbb{F}$ *by* $\mathbf{nhd}^d$ *is defined as*

$$\mathbf{dilate}^d(\mathbb{F}) := \left\{ j \mid \exists i \in \mathbb{F} : j \in \mathbf{nhd}^d \{i\} \right\}.$$

The d-neighborhood $\mathbf{nhd}^d$ is used as structure element for every vertex. Thus it adapts in a way to the local mesh connectivity. One can utilize neighborhoods with different radii d as structure element, but anisotropic dilation is not possible. For that reason our operators resemble the classical ones defined in Euclidean vector space restricted to a disk-like structure element $\{(x, y) \mid -d \leq x, y \leq d\}$. As one single structure element $\mathbf{nhd}^d$ will be employed we use a unary notation.

The dilation operator adds vertices to the feature, $\#\mathbf{dilate}^d(\mathbb{F}) \geq \#\mathbb{F}$. $\mathbb{F}$ is grown in a way roughly preserving its "shape" on the mesh. The dilation operator can therefore be effectively used to fill "holes" of unmarked vertices inside and at the boundary of the feature.

Another operation is needed to reverse the effect of dilation and to, ideally, recover the original shape. Therefore we have to shrink $\mathbb{F}$. Apart from that, this shrinking or erosion operator cuts off undesired branches.

Definition 2. *Let* $\mathbb{F} \subseteq \{1, \ldots, n\}$. *The* erosion *of* $\mathbb{F}$ *by* $\mathbf{nhd}^d$ *is defined as*

$$\mathbf{erode}^d(\mathbb{F}) := \left\{ j \mid \mathbf{nhd}^d \{j\} \subseteq \mathbb{F} \right\}.$$

Dilation and erosion can be applied with a d-neighborhood as structure element. For implementation, it might be helpful to note that the effect of using a larger structure element can also be achieved by applying the respective operator with smaller structure element several times. Hence $\mathbf{dilate}^d = \mathbf{dilate} \circ \cdots \circ \mathbf{dilate}$ (d times). The same is true for erosion.

Dilation and erosion remove classification artifacts that may be left even after prefiltering the geometry, but they do not preserve the overall size of the feature. This can be avoided by combining both operators.

First eroding and then dilating $\mathbb{F}$ will cut branches while preserving the original shape. The combined operator is called opening. Furthermore, small isolated regions of marked vertices are just removed as small branches. By changing the order of the operations, we obtain the closing operator that fills holes in the interior of the feature and cuts along the boundary.

Opening and closing can effectively be used to reduce noise on the feature region $\mathbb{F}$. Fig. 2 shows the effects of closing. Both operators are very easy to implement, and can be applied efficiently if appropriate data structures are used [1]. The resulting feature region approximately preserves the shape of the initial feature, and thus it is still a region rather than lines, i.e. it is too wide to be useful. So further operators are defined for narrowing and extracting feature lines.

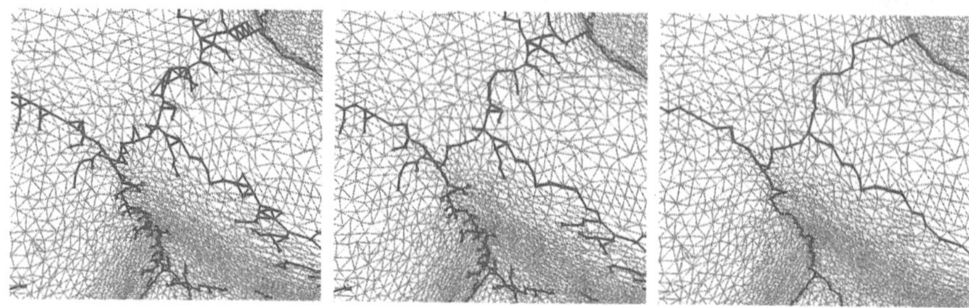

Fig. 3. Results after **skeletonize** (left), **preprune** (middle) and **prune** (right).

§4. Skeletonization

Here, feature lines are polygons defined by the triangle edges between feature vertices. A suitable removal criterion (see below) ensures that with the exception of regions where multiple feature lines meet, there are no triangles in the mesh that contain more than two such feature vertices.

For the extraction of these feature lines, we use a new skeletonization or thinning algorithm similar to techniques used in digital image processing [3]. Mathematically, the skeleton of the feature region can be defined via the medial axis transformation. The skeleton or medial axis of a region R with boundary B contains all points in R that have more than one single closest neighbor in B. This can be imagined as the set of centers of the largest disks that lie completely in R and are not included in any larger disk in R.

The medial axis transformation is relatively expensive to compute even for a planar 2D image. This is why iterative algorithms have been developed that efficiently produce a reasonable approximation of the skeleton. These thinning or skeletonization algorithms typically thin the region by peeling off one layer after another until the skeleton finally remains.

We propose a similar technique for the problem of refining the feature regions on a triangle mesh. This is some kind of controlled erosion. So we will obtain the topological medial axis rather than the geometric one. Defining an additional criterion, whether a marked vertex may be removed from the feature or not, is the essential difference between the skeletonization and the erosion operator. This guarantees topology preservation. Hence, we define a special class of vertices that must not be removed from the feature vector $\boldsymbol{F}$:

Definition 3. *Let* $(u^i_\mu)_{0\le\mu<v_i}$ *be the ordered sequence of vertex indices of the* v_i *neighbors of vertex* V_i. *Let* $c_i := \sum_{0\le\mu<v_i} |\boldsymbol{F}_{u^i_\mu} - \boldsymbol{F}_{u^i_{\mu+1 \bmod v_i}}|$. *A vertex* V_i *is* complex *iff* $\boldsymbol{F}_i = 1$ *and* $c_i \ge 4$. *The number* c_i *is said to be the* complexity *of* V_i.

In order to determine if a vertex V_i is complex, one enumerates its neighbors $V_{u^i_\mu}$ $(0 \le \mu < v_i)$ while counting the number c_i of transitions from marked to unmarked vertices and vice versa. This is the same basic operation on the triangle mesh data structure as used for enumerating **nhd**. The resulting complexity c_i is always even. Complex vertices are either part of the feature line (arc, $c_i = 4$) or they belong to a node, where several of such lines meet ($c_i > 4$). For practical reasons, it might be useful to define all outer boundary vertices of the mesh as complex, i.e. the boundary of the surface is always part of the feature lines.

Definition 4. *Let $i \in \mathbb{F}$. The vertex i is said to be a* center *if* $\mathbf{nhd}\{i\} \in \mathbb{F}$. *The set $O_i := \{j \,|\, i \,\text{is center} \wedge j \in \mathbf{nhd}\,\{i\} \setminus \{i\}\}$ is called* ring *around the center i.*

As $i \in \mathbf{nhd}\{i\}$, only vertices V_i with $i \in \mathbb{F}$ can be centers. If all neighbors of such a marked vertex are also marked, then this vertex is called a center and its neighbors form the surrounding ring. Note that vertices on rings may also be centers themselves. We now use the last two definitions to construct the skeletonization operator.

Definition 5. *Let $C = \{i \in \mathbb{F} \,|\, c_i \geq 4\} \subseteq \mathbb{F}$ be the set of all complex vertices in $\mathbb{F}$. Let $\bigcirc := \bigcup_{1 \leq i \leq n} O_i \subset \mathbb{F}$ be the union of rings and $\odot := \{i \in \mathbb{F} \,|\, i \,\text{is center}\}$ the set of centers. The* skeletonization *operator is defined as*

$$\mathbf{skeletonize}(\mathbb{F}) := \mathbb{F} \setminus (\bigcirc \cap \overline{C \cup \odot}).$$

The **skeletonize** operator is iteratively applied to the feature $\mathbb{F}$. With every iteration the outmost layer of feature vertices is peeled off. These vertices can be characterized as being part of rings while not being centers themselves. This is equivalent to erosion. By additionally respecting complex vertices, the resulting thin parts of the feature do not vanish, but will form the final feature skeleton. The topology of the feature region is preserved, and connected parts will remain connected.

It is obvious, that $\mathbb{F}' := \mathbf{skeletonize}(\mathbb{F}) \subseteq \mathbb{F}$. If $\odot = \emptyset$ or if $\odot$ contains only centers with all ring-vertices being complex, then $\mathbb{F}' = \mathbb{F}$. This is when the skeletonization terminates. As the number of centers and rings can only decrease with every iteration, the algorithm always terminates after a finite number of iterations. The number of complex vertices usually increases with every iteration.

The resulting feature contains complex vertices and vertices that cannot be removed, because they do not have a center as neighbor. These are vertices at the end of feature-line-branches, and vertices that are close to the feature lines and hence violate the previously demanded property of a maximum of two feature vertices per triangle in "non-node" regions (cf. Figs. 4/5, left). Both classes of vertices are disturbing and will be eliminated in the next step.

The first so called pre-pruning step will remove the second class of vertices so that the skeleton will consist of feature lines only. Therefore, all non-complex vertices are removed unless they are not situated at the end of branches, i.e. that have more than one marked neighbor.

Definition 6. *Let $\mathbb{F}_S$ denote the feature after skeletonization, and let $C_S \subseteq \mathbb{F}_S$ be the set of its complex vertices. Then the* pre-pruning operator *is defined as*

$$\mathbf{preprune}(\mathbb{F}_S) = \mathbb{F}_S \setminus \left\{\min \left\{i \in \overline{C_S} \,|\, \#(\mathbf{nhd}\,\{i\} \cap \mathbb{F}_S) \geq 1\right\}\right\}.$$

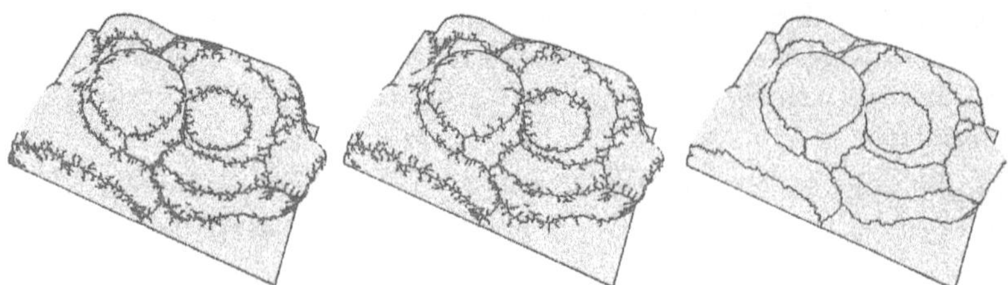

Fig. 4. Results after **skeletonize** (left), **preprune** (middle) and **prune** (right).

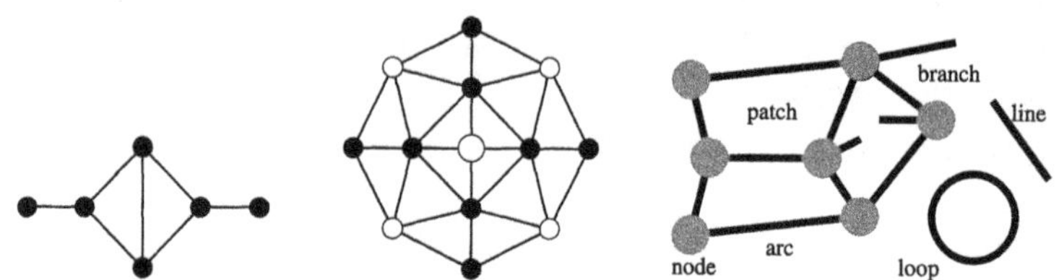

Fig. 5. Left: The **preprune** operator must not change the feature topology, so only one non-complex vertex is removed at a time. Middle: The non-complex center vertex is considered to be part of the node. Right: elements of the high-level graph structure.

The **preprune** operator is designed to remove one single feature vertex at a time (i.e. the one with the smallest index). This is necessary to handle configurations of complex vertices as shown in Fig. 5 (left) correctly. Note that C_S may change with every application of **preprune**. **preprune** is iterated k times until $\textbf{preprune}^k(\mathbb{F}_S) = \mathbb{F}_S$. The resulting feature $\mathbb{F}_{SP} := \textbf{preprune}^k(\mathbb{F}_S)$ represents the feature lines just as demanded (cf. Figs. 4/5, middle), but there are still small branches that are regarded as unwanted artifacts. The ends of these branches are the only non-complex vertices in $\mathbb{F}_{SP}$. This makes it simple to define an iterative pruning scheme $\textbf{prune}(\mathbb{F}_{SP}) := \mathbb{F}_{SP} \setminus C_{SP}$.

Of course the pruning operator cannot clearly distinguish wanted feature lines from unwanted artifacts. A simple heuristic is that unwanted branches are short, say of length $\leq m$ vertices. So **prune** is iterated m times. In order to prevent open ended feature lines from being shortened, we store the pruning history i.e. the vertices that have been removed. So after pruning, we can back off and restore the removed vertices for branches with more than m vertices. The backtracking is started from all open ends (non-complex vertices) that survived the application of $\textbf{prune}^m$.

The resulting feature lines represent the topological medial axis of the initial wide feature regions (cf. Figs. 4/5, right). They roughly follow the center of these regions as the mesh is assumed to have very uniform edge lengths allowing a reasonable geometric interpretation. In order to use the feature lines in a comfortable way and to be able to implement more high-level operations, some postprocessing will be done for embedding the lines into the mesh structure.

§5. Postprocessing

After skeletonization and pruning, $\mathbb{F}$ represents the wanted feature lines as a set of vertices. It may be of great advantage to finally convert this set into a more high-level data structure i.e. into a graph where the vertices are associated with elements of that graph. We distinguish between the following elements (cf. Fig. 5, right):

- Nodes are clusters of complex vertices V_i with complexity $c_i > 4$. They are the first elements to be identified by finding a seed vertex for the next node and then iteratively adding adjacent node vertices. This is repeated until no more seed vertices can be found. Prepruning may remove non-complex feature vertices that will cause "holes" in a single node like the center vertex Fig. 5 (middle). Such vertices that have only node-vertices as neighbors are considered to be part of the node.
- Arcs are ordered sequences of complex vertices with $c_i = 4$ that start and end from a node. They are traced from node to node.
- Branches are just like arcs, but open ended feature lines. They start at a node and end with a non-complex vertex.
- Loops are closed feature lines that do not touch any node.
- The remaining feature vertices – if any – are isolated lines, i.e. feature lines with two non-complex vertices at the ends.
- All other vertices of the triangular mesh in $\overline{\mathbb{F}}$ can now be grouped into patches that are bounded by or include some of the previous elements.

Extracting the different elements in this order simplifies the implementation. The polygons associated with arcs, etc. are non-smooth and are subject to aliasing effect. So they are smoothed or approximated by smooth curves (cf. Fig. 1, right).

The new data structure allows some more advanced operations on the feature, e.g. the feature lines can be edited manually, or small patches can be removed by first marking/filling them and then reskeletonizing the feature region. Connecting branches automatically is straightforward for two branches in the same patch.

§6. Conclusions

We presented a technique for extracting feature lines on a triangular mesh. On preprocessing, an edge-detection scheme is used to extract a wide feature region. This can be done by a single thresholding operation. Then we use easy-to-implement and efficient morphological operators and a skeletonization algorithm to suppress "classification noise" and to extract their topological medial axis. So we preserve the topology of the original feature region. The result can be used directly for segmentation or as initial guess e.g. for dynamic techniques like [9] for optimization.

For now, our operators are purely based on the mesh connectivity. The requirements on the input mesh could probably be relaxed by also taking into account geometric information, e.g. **nhd** could be defined in terms of Euclidean or geodesic distances instead of topological distance. While the operations would get more complex and less efficient, we expect that this could even enable us to process point clouds directly rather than first reconstructing a triangular mesh.

References

1. Campagna, S., L. Kobbelt, and H.-P. Seidel, Directed edges – a scalable representation for triangle meshes, ACM Journal of Graphics Tools **3** (4) (1998), 1–12.
2. Curless, B. and S. Seitz, 3D Photography, SIGGRAPH 2000, Course Notes.
3. Gonzales, R. C. and R. E. Woods, *Digital Image Processing*, Addison-Wesley, 1993.
4. Haralick, R. M., S. R. Sternberg, and X. Zhuang, Image analysis using mathematical morphology, IEEE Transactions on Pattern Analysis and Machine Intelligence, **PAMI-9** (4) (1987), 532–560.
5. Hoschek, J. and W. Dankwort (eds.), *Reverse Engineering*, Teubner, 1996.
6. Hoschek, J., U. Dietz, and W. Wilke, A geometric concept of reverse engineering of shape: Approximation and feature lines, in *Mathematical Methods for Curves and Surfaces II*, M. Dæhlen, T. Lyche, and L. L. Schumaker (eds.), Vanderbilt University Press, Nashville, 1998, 253–262.
7. Isselhard, F., G. Brunnet, and Th. Schreiber, Extraction of first-order feature lines from a discretized surface, in *Mathematics of Surfaces*, R. Cripps (ed.), Birmingham, 1998, 125–137.
8. Lukács, G. and L. Andor, Computing natural division lines on free-form surfaces based on measured data, in *Mathematical Methods for Curves and Surfaces II*, M. Dæhlen, T. Lyche, and L. L. Schumaker (eds.), Vanderbilt University Press, Nashville, 1998, 319–326.
9. Milroy, M. J., C. Bradley, and G. W. Vickers, Segmentation of a wrap-around model using an active contour, Computer-Aided Design **29** (1997), 299–320.
10. Sapidis, N. S. and P. J. Besl, Direct construction of polynomial surfaces from dense range images through region growing, ACM Transactions of Graphics **14** (2) (1995), 171–200.
11. Várady, T., R. R. Martin, and J. Cox, Reverse engineering of geometric models — an introduction, Computer-Aided Design **29** (1997), 255–268.
12. Várady, T. and P. Benkő, Reverse engineering B-rep models from multiple point clouds, in *Geometric Modeling and Processing 2000*, 3–12.

Christian Rössl, Leif Kobbelt, and Hans-Peter Seidel
Max-Planck-Institut für Informatik
Stuhlsatzenhausenweg 85, 66123 Saarbrücken, Germany
{roessl,kobbelt,hpseidel}@mpi-sb.mpg.de

Lower Bounds for Bernstein-Bézier Condition Number

Karl Scherer

Abstract. Recently T. Lyche and the author established the bound

$$\kappa_{n,p} \le \left(\frac{2\pi}{n}\right)^{1/2p} 2^{n-1/2}\bigl(1+\mathcal{O}(1/\sqrt{n})\bigr), \qquad 1 \le p \le \infty$$

for the basis of Bernstein-Bezier polynomials of degree n defined in the L_p-sense on $[-1,1]$. It is exact in the cases $p = 1, 2$ and ∞. Here we complement with lower bounds.

§1. Introduction

The purpose of this note is to establish upper and lower bounds for the condition number

$$\kappa^*_{n,p} := \sup_{a_i} \frac{|\{a_i\}|_p}{\|\sum a_i B_i^n\|_p} \cdot \sup_{b_i} \frac{\|\sum b_i B_i^n\|_p}{|\{b_i\}|_p} \tag{1}$$

of the Bernstein-polynomials $B_i^n(x)$ of degree n defined on $[-1,1]$ by

$$B_i^n(x) := 2^{-n}\binom{n}{i}(1+x)^i(1-x)^{n-i}, \qquad 0 \le i \le n.$$

In some cases these numbers are known exactly; namely with $n = 2m+k$ for $k \in \{0,1\}$ we have (see [3,4])

$$\kappa^*_{n,\infty} = \frac{(2n-1)(2n-3)\cdots(2m+3)(2m+1)^{1-k}}{1\cdot 3\cdots(2m-1)}. \tag{2}$$

From this the asymptotic relation

$$\kappa^*_{n,\infty} = 2^{n-1/2}[1+\mathcal{O}(1/n)], \qquad n \to \infty, \tag{3}$$

Mathematical Methods for Curves and Surfaces: Oslo 2000
Tom Lyche and Larry L. Schumaker (eds.), pp. 433–443.

ISBN 0-8265-1378-6

follows. In case, that $p = 2$, it has been shown in [2] that

$$\kappa_{n,2}^* = \sqrt{\binom{2n+1}{n}}. \tag{4}$$

Then Wallis' inequality yields

$$\kappa_{n,2}^* = \frac{2^{n+1/2}}{\big(\pi n)\big)^{1/4}}[1 + \mathcal{O}(1/n)], \qquad n \to \infty. \tag{5}$$

Recently T. Lyche and the author have obtained [5] quite precise upper and lower bounds for the case $p = 1$. In particular we determined the asymptotic behaviour

$$\kappa_{n,1}^* = 2^n \sqrt{\frac{\pi}{n}}\ [1 + \mathcal{O}(1/n)], \qquad n \to \infty. \tag{6}$$

We also note that the second factor in (1) is known explicitly (see [5]):

$$\sup_{b_i} \frac{\| \sum b_i B_i^n \|_p}{|\{b_i\}|_p} = \Big(\frac{2}{n+1}\Big)^{1/p}. \tag{7}$$

Upper bounds for the intermediate values $1 < p < \infty$ can be obtained from the following version of the (see [1, p.187])

(Interpolation Theorem of Riesz-Thorin). *Let (R, μ) and (S, ν) be two $\sigma-$finite measure spaces, and let T be a linear transformation defined on the space $L^1(\mu) + L^\infty(\mu)$ with values in the space $L^1(\nu) + L^\infty(\nu)$. If for each $f \in L^{p_i}(\mu)$,*

$$||Tf||_{L^{p_i}(\nu)} \leq M_i \ ||f||_{L^{p_i}(\mu)}, \qquad (i = 1, 2),$$

where $1 \leq p_1 < p_2 \leq \infty$, then T is also a bounded linear transformation on $L^p(\mu)$ into $L^p(\nu)$, where

$$1/p = \theta/p_1 + (1 - \theta)/p_2, \qquad 0 < \theta < 1.$$

In addition, the operator norms satisfy the convexity estimate

$$||T||_{L^p(\mu), L^p(\nu)} \leq M_1^\theta \ M_2^{1-\theta}. \tag{8}$$

We apply this theorem to the mapping $T : (\pi_n, L^p[-1, 1]) \to (\mathbb{R}^{n+1}, l^p)$ that assigns to a polynomial f its BB coefficients $T_n(f)$ as well as to its inverse mapping. Choosing $p_1 = 1, p_2 = 2$ with $1/p = (1 + \theta)/2$ on the one hand, and $p_1 = 2, p_2 = \infty$ with $1/p = \theta/2$ on the other hand, leads to

Theorem 1. *Let $n \geq 1$. Then for $1 \leq p \leq 2$,*

$$\kappa_{n,p}^* \leq (\kappa_{n,1}^*)^{-1+2/p}(\kappa_{n,2}^*)^{2-2/p} \leq 2^n \left(\frac{\pi}{2}\right)^{-1+1/p} \left(\frac{\pi}{n}\right)^{1/2p} \left[1+\mathcal{O}(1/\sqrt{n})\right], \quad (9)$$

and for $2 \leq p \leq \infty$,

$$\kappa_{n,p}^* \leq (\kappa_{n,2}^*)^{1/2p}(\kappa_{n,\infty}^*)^{1-2/p} \leq 2^{n-1/2}\left(\frac{4}{\sqrt{\pi n}}\right)^{1/p} \left[1+\mathcal{O}(1/\sqrt{n})\right]. \quad (10)$$

Remark: These bounds are better than those mentioned in the abstract. The reason is that in [5] we applied the Riesz-Thorin theorem for $p = 1, p = \infty$ with $1/p = \theta$, which for $p = 2$ yields the bound $\overline{\lim}_{n\mapsto\infty} 2^{-n} n^{1/4} \kappa_{n,2}^* \leq (\pi/2)^{1/4}$, which is weaker than (5).

§2. A Lower Bound by Extrapolation

We want to complement the last result by sharp lower bounds. To this end we first use an extrapolation version of the Riesz-Thorin -theorem. This is a simple consequence of it, but seems not to be stated explicitly in the literature.

Lemma 2. *Let T be defined as in the preceding theorem with $p_1 < p_2$, and denote by $||T||_{p_i}$ its norm as an operator on $L^{q_i}(\mu)$ into $L^{q_i}(\nu)$.*

a) *If $||T||_{p_1} \geq A_1 > 0$ and $||T||_{p_2} \leq A_2$, then for $\sigma > 1$*

$$||T||_p \geq A_1^{\sigma}\ A_2^{1-\sigma}, \qquad 1/p = \sigma/p_1 + (1-\sigma)/p_2. \quad (11)$$

b) *If $||T||_{p_1} \leq B_1$ and $||T||_{p_2} \geq B_2 > 0$, then for $\sigma > 1$*

$$||T||_p \geq B_1^{1-\sigma}\ B_2^{\sigma}, \qquad 1/p = \sigma/p_2 + (1-\sigma)/p_1. \quad (12)$$

Proof: The idea is simply to consider the inequality (8) of the Riesz-Thorin-theorem in the opposite direction to obtain a lower bound. Solving for $||T||_{p_1}$, we obtain

$$||T||_{p_1} \geq ||T||_p^{1/\theta}\ ||T||_{p_2}^{1-1/\theta}.$$

Now we change the notation, replacing $||T||_{p_1}$ by $||T||_p$ and p_1 by p so that $1/p_1 = \theta/p + (1-\theta)/p_2$. But this is equivalent to the relation in (11) for σ after setting $\sigma := 1/\theta$, and assertion a) follows.

In case b) we proceed analogously. We solve for $||T||_{p_2}$ and obtain (12) after replacing p_2 by p, $||T||_{p_2}$ by $||T||_p$ and setting $\sigma := 1/(1-\theta)$. □

Remark: The fact that $\sigma > 1$ shows that p lies in case a) at the left of the interpolation points p_1, p_2, and in case b) at the right of p_1, p_2, i.e. we have extrapolation.

Now we apply part b) of Lemma 2 to the mapping T_n in Theorem 1 for $p_1 = 1$ and $p_2 = 2$ so that $1/p = 1 - \sigma/2$, $\sigma = 2 - 2/p > 1$ for $p \geq 2$ and $\sigma = (2/p) - 1$. This means that we have to take $B_1 = 2^{n-1}\sqrt{\pi n}\,(1+\mathcal{O}(1/n))$ and $B_2 = 2^n\left(\frac{n}{\pi}\right)^{1/4}(1+\mathcal{O}(1/n))$ in view of the asymptotic relations in (5),(6) and relation (7). Then (12) yields

$$\sup_{a_i} \frac{|\{a_i\}|_p}{\|\sum a_i B_i^n\|_p} \geq \left(\sqrt{\pi n}2^{n-1}\right)^{(2/p)-1}\left(2^{2n}\sqrt{\frac{n}{\pi}}\right)^{1-1/q}(1+\mathcal{O}(1/n))$$

$$\geq 2^n\, n^{1/2p}\, \pi^{3/2p-1}\, 2^{1-2/p}(1+\mathcal{O}(1/n)). \tag{13}$$

For the remaining values of q, we apply part a) of Lemma 2 for $p_1 = 2$ and $p_2 = \infty$. Then $\sigma = 2/p$ and $p < 2$ for $\sigma > 1$. We have to take $A_1 = 2^n\left(\frac{n}{\pi}\right)^{1/4}[1+\mathcal{O}(1/n)]$, $A_2 = 2^{n-1/2}(1+\mathcal{O}(1/n))$, and get from (11)

$$\sup_{a_i} \frac{|\{a_i\}|_p}{\|\sum a_i B_i^n\|_p} \geq \left(2^{2n}\sqrt{\frac{n}{\pi}}\right)^{1/p}\left(2^{n-1/2}\right)^{1-2/p}[1+\mathcal{O}(1/n)]$$

$$\geq \frac{2^{n-1/2+1/p}\, n^{1/2p}}{\pi^{1/2p}}\left[1+\mathcal{O}(1/n)\right]. \tag{14}$$

Now we multiply the inequalities (13),(14) according to (7) by $(2/(n+1))^{1/p}$. This establishes

Theorem 3. *For* $1 \leq p \leq 2$,

a)

$$\kappa_{n,p}^* \geq 2^{n-1/2}\left(\frac{4}{\sqrt{\pi n}}\right)^{1/p}[1+\mathcal{O}(1/n)],$$

and for $2 \leq p \leq \infty$,

b)

$$\kappa_{n,p}^* \leq (\kappa_{n,1}^*)^{-1+2/p}(\kappa_{n,2}^*)^{2-2/p} \leq 2^n\left(\frac{\pi}{2}\right)^{-1+1/p}\left(\frac{\pi}{n}\right)^{1/2p}\left[1+\mathcal{O}(1/\sqrt{n})\right].$$

If we compare the inequalities of Theorem 3 with those of Theorem 1, we see that they are the same except that they hold on the complementary intervals of each other. This is due to the extrapolation principle of Lemma 2. As a consequence the lower bounds cannot match up with the upper bounds.

A little computation shows that the upper bound in case a) is (asymptotically) only by a factor $(\pi/\sqrt{8})^{2/p-1}$ larger than the lower bound. In case b), where $2 \leq p \leq \infty$, the upper bound is larger by the reciprocal factor $(\pi/\sqrt{8})^{1-2/p}$ than the lower one. The largest factor occurs for the endpoints $p = 1$ and $p = \infty$, respectively, where it equals to $\pi/\sqrt{8}$. At the middle point, the factor is 1 since upper and lower bound match up.

Remark: With some additional care, with the same method one could have computed bounds which hold not only in the asymptotic sense, but for each $n \leq 1$.

§3. Lower Bounds by Means of Orthogonal Polynomials

It has been shown in [2,3] that for $p = 2, \infty$, the supremum for the first quotient in $\kappa_{n,p}$ is realized by the Legendre- and Chebychev-polynomials, respectively. Moreover in case $p = 1$, it is asymptotically achieved by the Chebychev-polynomials of the second kind. These polynomials are special cases of the Jacobi-polynomials $P_n^{(\alpha,\beta)}$ which are defined e.g. by the Rodrigues-formula $(\alpha, \beta > -1)$

$$(1-x)^\alpha(1+x)^\beta P_n^{(\alpha,\beta)}(x) = \frac{(-1)^n}{2^n\ n!}\left(\frac{d}{dt}\right)^n [(1-t)^{n+\alpha}(1-t)^{n+\beta}]. \qquad (15)$$

They are orthogonal with respect to the weight $w(x) := (1-x)^\alpha(1+x)^\beta$, and it is well known (see [6]) that they can be expanded in terms of Bernstein-polynomials:

$$P_n^{(\alpha,\beta)}(x) = \sum_{i=0}^{n}[\binom{n+\alpha}{i}\binom{n+\alpha}{n-i}/\binom{n}{i}](-1)^{n-i}B_i^n(x) := \sum_{i=0}^{n} c_i B_i^n(x). \qquad (16)$$

In accordance with the special cases above, for the L_p norm now we consider the polynomials

$$P_n^{(\alpha,\alpha)}(x), \qquad \text{with} \quad \alpha := \frac{1}{p} - \frac{1}{2}, \quad 1 \le p \le \infty. \qquad (17)$$

In order to compute the relevant supremum in (1), we need exact bounds for the quantitities $||P_n^{(\alpha,\alpha)}||_{p,(-1,1)}$ and $||\{c_i\}||_p$ of the coefficients c_i in the expansion (16) for α in (17).

In the first case, our basic tool is the formula of Darboux (1878, cf. [6, p.196]) which yields (for $\alpha = \beta$)

$$P_m^{(\alpha,\alpha)}(\cos\theta)(\sin\theta)^{\alpha+1/2} = \frac{2^{\alpha+1/2}}{\sqrt{\pi n}}\left[\cos(M\theta+\eta) + |\sin\theta|^{-1}\mathcal{O}(n^{-1}),\right. \qquad (18)$$
$$cn^{-1} \le \theta \le \pi - cn^{-1},$$

where $\mathcal{O}$ depends only on c and

$$M := n + \alpha + 1/2, \qquad \eta := -(\alpha + 1/2)\pi/2.$$

There is a variant (cf. [6, p.194]) which gives a better description of the behaviour in the interior of $[-1,1]$,

$$P_m^{(\alpha,\alpha)}(cos\theta) = \frac{1}{\sqrt{\pi n}}\left(\frac{2}{\sin\theta}\right)^{\alpha+1/2} cos(M\theta+\eta) + O(n^{-3/2}), \qquad (19)$$

uniformly on each subinterval $[\epsilon, \pi-\epsilon]$. Here $\mathcal{O}$ depends only on ϵ, but may depend on n if it is taken as in (18). We first establish

Lemma 4. *There holds*

$$|P_n^{(\alpha,\alpha)}(x)|(1-x^2)^{1/2p} \le \frac{2^{1/p}}{\sqrt{\pi n}}[1+\mathcal{O}(n^{-1})], \tag{20}$$

where $\mathcal{O}$ does not depend on n and x.

Proof: In [6, p.166] it is shown by a theorem of Sonin that for the related ultraspherical polynomials,

$$P_n^{(\lambda)}(x) := \frac{\Gamma(\alpha+1)}{\Gamma(2\alpha+1)}\frac{\Gamma(n+2\alpha+1)}{\Gamma(n+\alpha+1)}P_n^{(\alpha,\alpha)}(x), \qquad \lambda := \alpha+1/2 = 1/p,$$

the sequence of the relative maxima of $u(\theta) := (\sin\theta)^\lambda P_n^{(\lambda)}(\cos\theta)$ increases for $0 \le \theta \le \pi/2$ and decreases for $\pi/2 \le \theta \le \pi$. Thus, for even n the absolute maximum is attained for $\theta = \pi/2$. For odd n, the distance between the two maxima nearest to $\pi/2$ is less than $\pi/(2n+1)$. This follows from the well-known distribution of the zeros of Jacobi-polynomials (see [6, p.121]). Therefore, we have for any n

$$|P_n^{(\alpha,\alpha)}(\cos\theta)(\sin\theta)^{\alpha+1/2}| \le \max_{|\theta-\pi/2|\le\pi/(2n+1)} |P_n^{(\alpha,\alpha)}(\cos\theta)|. \tag{21}$$

Now we apply formula (19) for $\epsilon = \pi/4$ in order to get a definite bound in (21), and obtain

$$|P_n^{(\alpha,\alpha)}(\cos\theta)(\sin\theta)^{\alpha+1/2}| \le |P_n^{(\alpha,\alpha)}(0)| + \mathcal{O}(n^{-3/2}) = \frac{2^{1/p}}{\sqrt{\pi n}} + \mathcal{O}(n^{-3/2}),$$

whence (20) follows. □

An immediate consequence of Lemma 4 is the inequality

$$||P_n^{(\alpha,\alpha)}||_{p,(-1,1)} \le \frac{(2\pi)^{1/p}}{\sqrt{\pi n}}[1+\mathcal{O}(n^{-1})] \tag{22}$$

for the L_p norm of $P_n^{(\alpha,\alpha)}$. But this bound can be sharpened using the first formula (18) of Darboux.

Lemma 5. *For $1 \le p < \infty$, the polynomials in (17) satisfy*

$$||P_n^{(\alpha,\alpha)}||_{p,(-1,1)} = \frac{2^{1/p}}{\sqrt{\pi n}}\,\gamma_p\,[1+\mathcal{O}(n^{-1/2p}], \tag{23}$$

where

$$\gamma_p := ||\cos\theta||_{p,(0,\pi)} = \left(\frac{\Gamma(1/2+p/2)\Gamma(1/2)}{\Gamma=(1+p/2)}\right)^{1/p}. \tag{24}$$

Proof: First observe that

$$||P_n^{(\alpha,\alpha)}||_{p,(-1,1)} = ||P_n^{(\alpha,\alpha)}(\cos\,\theta)(\sin\theta)^{1/p}||_{p,(0,\pi)}.$$

Then we split this integral into the interval $U := [\eta_n, \pi - \eta_n]$ and its complement in $[-1, 1]$, respectively. The latter can be estimated by

$$||P_n^{(\alpha,\alpha)}(\cos\,\theta)(\sin\theta)^{1/p}||_{p,(0,\pi)/U} \leq \frac{2^{1/p}}{\sqrt{\pi n}}[1 + \mathcal{O}(n^{-1})]\Big(\int_0^{\eta_n} + \int_{\pi-\eta_n}^{\pi} d\theta\Big)^{1/p}$$

in view of the bound (20). If we choose $\eta_n = 1/\sqrt{n}$, this gives a contribution of order $\mathcal{O}(n^{-1/2-1/2p})$ which is of the order of the remainder in (23).

Next we integrate formula (18) over the interval $U := [\eta_n, \pi - \eta_n]$. Together with the triangle inequality, this gives

$$\left|||P_n^{(\alpha,\alpha)}(\cos\,\theta)(\sin\theta)^{1/p}||_{p,U} - \frac{2^{1/p}}{\sqrt{\pi n}}||\cos(M\theta + \eta)||_{p,U}\right| \leq ||R(\theta)||_{p,U},$$

where we write $R(\theta) := C(\theta)/n\sin\theta)$ for the remainder in formula (18). We can assume $C(\theta) \leq C$ for $\theta \in U$, where C does not depend on θ and n, if we choose $\eta_n = \sqrt{n}$. In order to estimate the L_p -norm of $R(\theta)$, we set $\cos\theta = x$ so that

$$||R(\theta)||_{p,U} \leq \frac{2C}{n}\left(\int_0^{1-c_n}(1-x^2)^{-(p+1)/2}dx\right)^{1/p}, \tag{25}$$

where $c_n = \mathcal{O}(n^{-1})$. Here the integral restricted to the interval [0,1/2] is bounded. For the remaining interval, after transformation $x = 1 - t$ we have

$$\left(\int_{1/2}^{1-c_n}(1-x^2)^{-(p+1)/2}dx\right)^{1/p} \approx \left(\int_{c_n}^{1/2} t^{-(p+1)/2}dt\right)^{1/p} \approx n^{1/2-1/2p},$$

where $\approx$ means equality up to an absolute constant. Hence $||R(\theta)||_{p,U} \leq \mathcal{O}(n^{-1/2-1/2p})$ which yields together with the previous estimate

$$||P_n^{(\alpha,\alpha)}(\cos\,\theta)(\sin\theta)^{1/p}||_{p,U} \leq \frac{2^{1/p}}{\sqrt{\pi n}}||\cos(M\theta + \eta)||_{p,U} + \mathcal{O}(n^{-1/2-1/2p}).$$

Now it is easy to see that for $U = [\sqrt{n}, 1 - \sqrt{n}]$,

$$\big|\ ||\cos(M\theta + \eta)||_{p,U} - ||\cos(M\theta + \eta)||_{p,[0,\pi]}\ \big| = \mathcal{O}(n^{-1/2p}).$$

Furthermore,

$$||\cos(M\theta + \eta)||_{p,[0,\pi]} = ||\cos\,\theta||_{p,[0,\pi]},$$

and the lemma is proved. □

Next we need a lower bound for the coefficients c_i in the expansion (16) for $\alpha = \beta$. First observe that

$$|c_i| = \frac{\binom{n+\alpha}{i}\binom{n+\alpha}{n-i}}{\binom{n}{i}} = \frac{\Gamma(n+\alpha+1)}{\Gamma(n+\alpha+1-i)}\frac{\Gamma(n+\alpha+1)}{\Gamma(i+\alpha+1)\Gamma(n+1)}.$$

Then use Stirling's formula

$$\Gamma(z) = z^{z-1/2}e^{-z}(\sqrt{2\pi}+\phi(z)), \qquad |\phi(z)| < 1/|12z|,$$

and consider only values of i satisfying

$$n-i+\alpha+1 \geq \sqrt{n}, \qquad i+\alpha+1 \geq \sqrt{n}.$$

This gives

$$|c_i| = \frac{e(n+\alpha+1)^{2n+2\alpha+1}[1+\mathcal{O}(\sqrt{1/n})]}{(n-i+\alpha+1)^{n-i+\alpha+1/2}(i+\alpha+1)^{i+\alpha+1/2}(n+1)^{n+1/2}\sqrt{2\pi}},$$

where $\mathcal{O}$ denotes a constant independent of n, i and $\alpha \in [-1/2, 1/2]$. Now write

$$\frac{(n+\alpha+1)^{2n+2\alpha+1}}{(n+1)^n} = \left(\frac{n}{n+1}\right)^n\left(\frac{n+\alpha+1}{n}\right)^n\left(\frac{n+\alpha+1}{2}\right)^{n+2\alpha+1}2^{n+2\alpha+1}$$

so that

$$|c_i| = \frac{e2^{n+2\alpha+1}}{\sqrt{2\pi}}\left(\frac{n}{n+1}\right)^n\left(\frac{n+1+\alpha}{n}\right)^n(n+1)^{-1/2}[1+\mathcal{O}(\sqrt{1/n})] \times F,$$

with

$$F := \frac{((n+\alpha+1)/2)^{n+2\alpha+1}}{(n-i+\alpha+1)^{n-i+\alpha+1/2}(i+\alpha+1)^{i+\alpha+1/2}}.$$

This expression can be simplified to

$$|c_i| = e^{\alpha+1}2^{n+2\alpha+1}(2\pi n))^{-1/2}[1+\mathcal{O}(\sqrt{1/n})] \times F. \tag{26}$$

Then introduce m and y by

$$m := \alpha+1+n/2, \qquad i = y\sqrt{n}\ + n/2,$$

so that $n-i+\alpha+1 = m - y\sqrt{n}$, and $i+\alpha+1 = m+\sqrt{n}$. After division by m^{2m-1} in the numerator and denominator, F transforms to

$$\begin{aligned} F &= \frac{\left(1-\frac{1+\alpha}{2m}\right)^{2m-1}}{[(1-y\sqrt{n}/m)(1+y\sqrt{n}/m)]^{m-1/2}}\left(\frac{1-y\sqrt{n}/m}{1+y\sqrt{n}/m}\right)^{y\sqrt{n}} \\ &= \frac{e^{-\alpha-1}[1+\mathcal{O}(1/n)]}{(1-y^2n/m^2)^{m-1/2}}\left(1-\frac{2y}{y+m/\sqrt{n}}\right)^{y\sqrt{n}}. \end{aligned}$$

Then observe

$$(m-1/2)\log(1-\frac{y^2n}{m^2}) \leq -(m-1/2)\frac{y^2n}{m^2} = -2y^2[1+\mathcal{O}(1/n)],$$

so that one obtains for the denominator

$$(1-y^2n/m^2)^{m-1/2} \leq e^{-2y^2+\mathcal{O}(\sqrt{1/n})}, \qquad y^2 \leq \sqrt{n}, \tag{27}$$

where $\mathcal{O}$ is independent of y and n. For the second factor in y observe that for $a>0, N>0$ with $a/N<1$,

$$N\log(1-\frac{a}{N}) = -a\left(1+\frac{1}{2}\frac{a}{N}+\frac{1}{3}(\frac{a}{N})^2+\cdots\right) \leq \frac{-a}{1-a/N} \leq -a(1+2\frac{a}{N}).$$

Applying this with $N := y\sqrt{n}$ and $a := 2y^2n/(y\sqrt{n}+m)$, we obtain

$$(1-\frac{a}{N})^N \geq e^{-a(1+4y\sqrt{n}/(y\sqrt{n}+m))}.$$

Now $a = -4y^2/[1+2y/\sqrt{n}+(2\alpha+2)/n] = -4y^2+\mathcal{O}(n^{-1/4})$ under the asssumption $y \leq n^{1/4}$ and $y\sqrt{n}/(y\sqrt{n}+m) \leq y\sqrt{n}/m \leq n^{-1/4}$, so that

$$\left(1-\frac{2y}{y+m/\sqrt{n}}\right)^{y\sqrt{n}} \geq e^{-4y^2+\mathcal{O}(n^{-1/4})}, \qquad y \leq n^{1/4}, \tag{28}$$

where $\mathcal{O}$ is independent of y and n. Combining (27) and (28) then gives the following lower estimate

$$F \geq e^{-\alpha-1}e^{-2y^2+\mathcal{O}(n^{-1/4})}, \qquad y \leq n^{1/4}.$$

We remark that one can establish by the same argument a lower bound complementary to (27) and an upper bound complementary to (28), the latter however under the stronger restriction $y \leq n^{1/8}$. The result is the upper estimate

$$F \leq e^{-\alpha-1}e^{-2y^2+\mathcal{O}(n^{-1/8})}, \qquad y \leq n^{1/8}.$$

Then, inserting these bounds into (26) we have proved

Lemma 6. *For α in (17) the coefficients in the expansion (16) satisfy*

$$|c_i| \geq \frac{2^{n+2/p}[1+\mathcal{O}(n^{-1/4})] \times d_i}{\sqrt{2\pi(n+1)}}, \qquad d_i := e^{-2y_i^2}, \quad i = y_i\sqrt{n}+n/2, \tag{29}$$

where $\mathcal{O}$ is independent of n and y provided $y \leq n^{1/4}$. An upper bound of the same form holds for the c_i with $y \leq n^{-1/4}$ replaced by $y \leq n^{-1/8}$.

Next we estimate

$$\sum_{i=0}^{n}|d_i|^p \geq \sum_{|j|<n^{3/4}}|d_{j+n/2}|^p = \sum_{|j/\sqrt{n}|<n^{1/4}} e^{-2p(j/\sqrt{n})^2}.$$

Since $f(t) := e^{-2pt^2}$ is monotone, the value $e^{-2p(j/\sqrt{n})^2}$ can be bounded by the integrals $\sqrt{n}\int_{I_{j+1}} f(t)dt$ and $\sqrt{n}\int_{I_j} f(t)dt$ from below and above, respectively, where $I_j := [(|j|-1)/\sqrt{n}, |j|/\sqrt{n}]$. Summing this, we see that

$$\left(\sum_{i=0}^{n} |d_i|^p\right)^{1/p} \geq \left(\sqrt{n}\int_{\infty}^{\infty} e^{-2pt^2}dt\right)^{1/p} - E_n,$$

where

$$E_n = \left(\int_{|t|\geq n^{1/4}} e^{-2pt^2}dt\right)^{1/p} \leq \frac{n^{1/2p}}{e^{2\sqrt{n}}}\left(\int_{|t|\geq n^{1/4}} t^{-2}dt\right)^{1/p} = 2n^{1/4p}e^{-2\sqrt{n}}.$$

Since

$$\int_{\mathbf{R}} e^{-2pt^2}dt = \sqrt{\frac{\pi}{2p}},$$

in combination with Lemma 6 this yields

Lemma 7. *There holds*

$$\left(\sum_{i=0}^{n} |c_i|^p\right)^{1/p} \geq \frac{2^n}{\sqrt{2\pi n}} \cdot \left(\frac{8\pi n}{p}\right)^{1/2p} [1 + \mathcal{O}(n^{-1/4})]. \tag{30}$$

Now we combine this with the upper bound (23) for $||P_n^{(\alpha,\alpha)}||_{p,(-1,1)}$ and multiply the result with the factor in (7), giving us

Theorem 8. *For* $1 \leq p < \infty$,

$$\kappa_{n,p} \geq \frac{2^{n-1/2}}{\gamma_p\ n^{1/2p}}\left(\frac{8\pi}{p}\right)^{1/2p} [1 + \mathcal{O}(n^{-\min(1/4,1/2p)})].$$

It is interesting to compare this lower bounds asymptotically with the upper ones of Theorem 1. A little computation yields

$$\frac{\text{upper bound}}{\text{lower bound}} = \begin{cases} (\pi^2/8)^{1/2p-1/2}(p/4)^{1/2p}\gamma_p, & 1 \leq p \leq 2, \\ (2p/\pi^2)^{1/2p}\gamma_p, & 2 \leq p \leq \infty. \end{cases}$$

The numerical calculation of the functions in p on the right hand side shows that they are convex on the intervals $[1, 2]$ and $[2, \infty)$, respectively, with maximal values approximately equal to 1.01158 at $p \approx$1.343 and to 1.01483 at $p \approx$ 4.05.

Thus the question arises whether the lower bounds given by ultraspherical polynomials $P_n^{(\alpha,\alpha)}$ are (asymtotically) equal to the L_p condition number for *all* p, $1 \leq p \leq \infty$. In this respect, we remark that the lower bound for $\left(\sum_{i=0}^{n} |c_i|^p\right)^{1/p}$ in Lemma 7 is sharp as is seen by the second statement in Lemma 6. The same is true for the upper bound of Lemma 5, so that the lower bound of Theorem 8 is sharp. In spite of this, the above question is not settled, since the upper bounds of Theorem 1 do not match up with it. To this end, one has to show that the ultraspherical polynomials are indeed asymtotically extremal, which seems to be a quite difficult task.

Acknowledgments. The author would like to thank the referee for his suggestions which lead to an improvement of the results in the first version.

References

1. Butzer, P. L. and H. Berens, *Semigroups of Operators and Approximation*, Springer, New York, 1967.
2. Ciesielskii, Z. and J. Domsta, The degenerate B-Spline basis in the space of algebraic polynomials, Annales Polonici Mathematici **XXVI**(1985), 71-79.
3. Lyche, T., A note on the condition numbers of the B-spline bases, J. Approx. Theory **22** (1978), 202–205.
4. Lyche, T. and K. Scherer, On the sup-norm condition number of the multivariate triangular Bernstein basis, in *Multivariate Approximation and Splines*, G. Nürnberger, J. W. Schmidt, and G. Walz (eds.), ISNM Vol. 125. Birkhäuser Verlag, Basel, 1997, 141– 151.
5. Lyche, T. and K. Scherer, On the 1-norm condition number of the univariate Bernstein basis. preprint.
6. Szegö, G., *Orthogonal Polynomials*, American Mathematical Society, New York 1959.

Karl Scherer
Institut f. Angewandte Mathematik
University of Bonn
Wegelerstr. 6, 53115 Bonn, Germany
scherer@iam.uni-bonn.de

Improved Bi-Laplacian Mesh Fairing

Robert Schneider, Leif Kobbelt and Hans-Peter Seidel

Abstract. Algorithms to create fair meshes can be divided into two categories, depending on whether they are linear or nonlinear. Linear methods have the advantage of being fast, robust and easy to implement, but the results depend highly on the chosen parameterization strategy. Nonlinear methods usually are based on intrinsic surface properties that only depend on the surface geometry and hence lead to surfaces that show high quality fairness and are free from parameterization artifacts. But such methods are considerably slower, more involved to implement, and their convergence depends on the quality of the initial surface that is used in the iterative construction process. In this paper we present a nonlinear mesh fairing algorithm enabling G^1 boundary conditions that lies between the linear and completely intrinsic methods, leading to a construction process that has many advantages of the linear approach while producing a mesh quality that is superior to the results of strictly linear methods.

§1. Introduction

When fairing triangular meshes, one has to take into account that there are two types of fairness. First, the mesh has to satisfy an outer fairness, i.e. the geometry of the mesh should look aesthetic. Second, the mesh has to satisfy an inner fairness, which means that the distribution of the mesh vertices *within* the surface and the shape of the individual faces should satisfy some prescribed requirements. Linear fairing algorithms [9,13,6] cannot separate inner and outer fairness of a mesh. The chosen local discretization strategy will determine both fairness types. And since the results depend highly on the chosen parameterization, there is no guarantee that there will be no parameterization artifacts in the solution. The solution to these problems is to use intrinsic mesh fairing algorithms as presented in [14,11], which also lead to superior outer fairness quality. These strategies are however much more complex than linear approaches. Moreover, to achieve convergence, one cannot use an arbitrary initial mesh to start the construction process. In fact, one needs a mesh that is smooth and not too far away from the final solution.

The fairing technique that we present in this paper lies somewhat between the linear and the intrinsic algorithms, while still sharing many properties

Mathematical Methods for Curves and Surfaces: Oslo 2000
Tom Lyche and Larry L. Schumaker (eds.), pp. 445–454.

ISBN 0-8265-1378-6

of the linear approach. The basic idea is to solve a Bi-Laplacian equation $\Delta\Delta f = 0$ that is not completely linear, but has some degrees of freedom left that are used during the iteration process to construct a mesh that is closer to the result of an intrinsic mesh fairing scheme.

§2. Notation

We partition the vertices of a mesh M into two classes, denoting the set of all border vertices with $V_B(M)$, and the set of all vertices in the interior of M with $V_I(M)$. Let us further assume we can extend our mesh at the boundary vertices to a mesh $\tilde{M}$ such that every boundary vertex has a complete neighborhood in $\tilde{M}$, and let $V_O(M)$ denote that vertices of $\tilde{M}$ that are not in M (Fig. 1). Let the number of interior vertices be n and the number of vertices in $V_B(M)$ resp. $V_O(M)$ be m resp. o. We assume that our vertices are numbered from 1 to $n+m+o$, where the vertices $q_1, \cdots, q_n$ are in $V_I(M)$, the vertices $q_{n+1}, \cdots, q_{n+m}$ in $V_B(M)$ and $q_{n+m+1}, \cdots, q_{n+m+o}$ in $V_O(M)$. For each vertex q_i in $V_B(M) \cup V_I(M)$, let $N(q_i)$ be the set of vertices q_j in $\tilde{M}$ that are adjacent to q_i and let $D(q_i) = N(q_i) \cup \{q_i\}$ be the according 1-disk. The number of vertices in a neighborhood $N(q_i)$ defines the function $valence(q_j) = |N(q_i)|$.

§3. Discretizing the Bi-Laplacian Equation

As mentioned previously, our final construction algorithm is based on solving a nonlinear Bi-Laplacian equation. In this section we present some necessary formulas, show how to discretize the linear Bi-Laplacian, and then present an intrinsic discretization based on the Laplace-Beltrami operator. The latter serves as starting point for the derivation of our mesh fairing algorithm that is presented in the next section.

Let us first define how to discretize the local Laplacian operator Δq_i at a vertex $q_i \in V_B(M) \cup V_I(M)$. In its generalized form, the discrete Laplacian is

$$\Delta(q_i) = \sum_{q_j \in N(q_i)} \lambda_{ij}(q_j - q_i), \tag{1}$$

where $\lambda_{ij} \in \mathbb{R}$ are scalar weights that are assigned to every vertex $q_j \in N(q_i)$. Discretizations of such a form can be found e.g. in [1,3,9,13]. Although the Laplacian can be discretized with a larger support [6], we restrict the support to $D(q_i)$, since this is sufficient for our mesh fairing scheme.

3.1. Discretized Bi-Laplacian Equation for a General Laplacian

In the following we assume that we have prescribed boundary conditions determined by the vertices $V_B(M) \cup V_O(M)$ and also a prescribed connectivity for all interior vertices $V_I(M)$ and are searching the unknown vertex positions for all $q_i \in V_I(M)$. Later in Section 5 and Section 6 we show how to handle other conditions. The unknown vertex positions are determined by discretizing the Bi-Laplacian equation: that means the vertices q_i have to satisfy

$$\Delta\Delta(q_i) = 0, \qquad \forall q_i \in V_I(M). \tag{2}$$

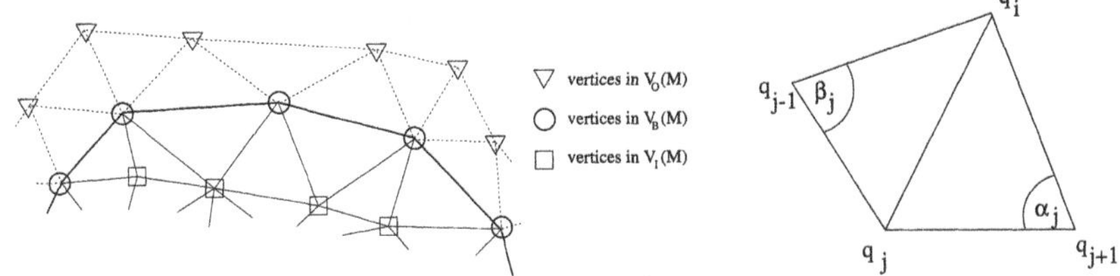

Fig. 1. Left: The mesh M is extended by vertices $V_O(M)$ such that every boundary vertex has a complete 1-neighborhood. Right: The angles α_j and β_j that occur in the discretization of the Laplace-Beltrami operator.

We note that the right Laplacian operator has to be discretized for all vertices $q_i \in V_I(M) \cup V_B(M)$, while the left Laplacian only has to be discretized for the interior vertices. Assuming that we use the same discretization for the left and right Laplacian at interior vertices in (2), we arrive at the following matrix equation

$$\begin{pmatrix} A & B \end{pmatrix} \begin{pmatrix} A & B & 0 \\ C & D & E \end{pmatrix} \vec{q} = 0,$$

where $A \in \mathbb{R}^{n \times n}$, $B \in \mathbb{R}^{n \times m}$, $C \in \mathbb{R}^{m \times n}$, $D \in \mathbb{R}^{m \times m}$ and $E \in \mathbb{R}^{m \times o}$ are sparse submatrices with coefficient entries λ_{ij} given by (1) and $\vec{q}$ is a vector consisting of the vertices of our mesh $\tilde{M}$

$$\vec{q} = (\, q_1, \cdots, q_n, q_{n+1}, \cdots q_{n+m}, q_{n+m+1}, \cdots, q_{n+m+o}\,)^t \,.$$

This leads us to the $n \times n$ linear system

$$\begin{pmatrix} A & B \end{pmatrix} \begin{pmatrix} A \\ C \end{pmatrix} \vec{q}_I = -\begin{pmatrix} A & B \end{pmatrix} \begin{pmatrix} B & 0 \\ D & E \end{pmatrix} \vec{q}_R, \tag{3}$$

for the unknowns $q_i \in V_I(M)$, with

$$\vec{q}_I = (\, q_1, \cdots, q_n\,)^t, \quad \vec{q}_R = (\, q_{n+1}, \cdots q_{n+m}, q_{n+m+1}, \cdots, q_{n+m+o}\,)^t \,.$$

3.2. Discretized Bi-Laplacian Equation for an Intrinsic Laplacian

The outer fairness of the mesh resulting from solving (3) depends highly on the chosen discretization strategy for the local Laplacians given by (1). To resolve this dependency, we have to use a Laplacian that doesn't depend on the underlying local parameterization, but only on the intrinsic surface geometry of the mesh. A generalization of the Laplacian that satisfies this condition is given by the well known Laplace-Beltrami operator Δ_B. Recently Desbrun et al. [1,2] presented a discretization for Δ_B at a vertex q_i that is of the form (1):

$$\Delta_B q_i \;=\; \frac{3}{2A_i} \sum_{q_j \in N(q_i)} (cot\, \alpha_j + cot\, \beta_j)(q_j - q_i). \tag{4}$$

Here A_i is the area sum of the triangles adjacent to q_i and α_j and β_j are the triangle angles as shown in Figure 1.

Using these weights to discretize the Bi-Laplacian (2), (3) becomes

$$D_n S D_{n+m} S^t \vec{q}_I = \vec{b}. \tag{5}$$

The matrix $S \in \mathbb{R}^{n \times n+m}$ is given by

$$S_{ii} = -\sum_{q_j \in N(q_i)} (\cot\alpha_j + \cot\beta_j), \quad S_{ij} = \begin{cases} \cot\alpha_j + \cot\beta_j, & q_j \in N(q_i) \\ 0, & \text{otherwise,} \end{cases}$$

where $q_i \in V_I(M)$. D_n and D_{n+m} are diagonal matrices

$$D_n = diag(\frac{3}{2A_1}, \cdots \frac{3}{2A_n}), \quad D_{n+m} = diag(\frac{3}{2A_1}, \cdots \frac{3}{2A_n}, \frac{3}{2A_{n+1}}, \cdots \frac{3}{2A_{n+m}}).$$

The vector $\vec{b}$ is given by the right side of equation (3) and is independent of the vertices in $V_I(M)$. Multiplying both sides with D_n^{-1}, we can make our linear system symmetric and positive definite. The latter can be seen if we rewrite the matrix $SD_{n+m}S^t$ as $(S\sqrt{(}D_{n+m}))(S\sqrt{(}D_{n+m}))^t$, since matrices of type AA^t are positive definite if A has full row rank. A proof that the matrix S has full rank can be found in a paper by Pinkall and Polthier [10], where one can also find more information about the structure of S.

Discretizing the Bi-Laplacian using the Δ_B operator leads to the nonlinear problem of finding a mesh M^∞ such that (5) is satisfied by its inner vertices, with α_j, β_j and A_i measured on M^∞. A common technique to solve such a problem is to create a sequence of meshes $M^k \to M^{k+1}$ with an initial mesh M^0, such that M^k converges to M^∞. Here M^{k+1} should depend linearly on M^k. In our case we could try to create such a mesh sequence by setting M^{k+1} to the solution of (5) with α_j, β_j and A_i measured on M^k. This would require solving a symmetric positive linear system in every iteration step.

Unfortunately, this algorithm usually doesn't converge. To find a possible explanation for this, let us take a look at an intrinsic mesh fairing scheme that doesn't show such a behavior. In [11] a fairing algorithm was presented that is based on the idea of discretizing the scalar valued partial differential equation $\Delta_B H = 0$, where H is the mean curvature. The outer fairness was controlled by moving the vertices along the surface normal, while the inner fairness was defined independently. However, the technique described above would discretize a vector valued intrinsic equation, leaving no degrees of freedom left for an inner fairness concept. Here also the question arises whether it is possible to find a mesh that satisfies the vector valued intrinsic equation $\Delta_B \Delta_B(q_i) = 0$ for arbitrary $V_B(M)$ and $V_O(M)$ exactly.

§4. Our Discrete Fairing Approach

Because of the failure of the intrinsic fairing algorithm mentioned above, we conclude that we have to integrate an inner fairness force into the construction algorithm. Therefore, the basic idea of our fairing approach is to enforce an inner fairness that controls the vertex distribution inside the surface, while the outer fairness is determined by using the remaining degrees of freedom to approximate the Laplace-Beltrami operator.

4.1. Regular Approximation of the Laplace-Beltrami Operator

Since it seems reasonable to try to distribute the vertices equally across the surface, we decided to choose a regular inner fairness criterion. Following the definition given in [11], a mesh satisfies a **regular inner fairness condition** if all $\Delta(q_i)$ with

$$\Delta(q_i) = s_i \sum_{q_j \in N(q_i)} (q_j - q_i), \quad \text{and} \quad s_i > 0 \tag{6}$$

have vanishing tangential components, *i.e.*, there are scalar values t_i such that $\Delta(q_i) = t_i \vec{n}_i$, where $\vec{n}_i$ is the discrete normal at q_i.

We note that the scaling factors s_i have no influence on whether the inner fairness is satisfied in this definition or not. With this in mind, we decided to discretize (2) based on discrete Laplacians of type (6). For $s_i = \frac{1}{valence(q_i)}$ we arrive at the discrete fairing technique presented in [9]. However, for other scalar factors s_i we produce different results, so these factors are our remaining degrees of freedom to control the mesh fairness. Since the Laplacian is approximately parallel to the normal $\vec{n}_i$, the scaling factors s_i allow us to influence the shape of the resulting mesh mostly along the normal directions. With this discretization, the resulting solution of the Bi-Laplacian doesn't have to satisfy the regular inner fairness criterion given above exactly, but will still produce a mesh with an inner fairness close to it.

Using (6) the linear system (3) becomes

$$\bar{D}_n U \bar{D}_{n+m} U^t \vec{q}_I = \vec{b}, \tag{7}$$

where $\vec{b}$ doesn't depend on inner vertices and $U \in \mathbb{R}^{n \times n+m}$ is given by

$$U_{ii} = -valence(q_i), \quad U_{ij} = \begin{cases} 1, & q_j \in N(q_i) \\ 0, & \text{otherwise}, \end{cases}$$

with $q_i \in V_I(M)$. The matrices $\bar{D}_n$ and $\bar{D}_{n+m}$ are diagonal matrices

$$\bar{D}_n = diag(s_1, \cdots s_n), \quad \bar{D}_{n+m} = diag(s_1, \cdots s_n, s_{n+1}, \cdots s_{n+m}).$$

Again we can make this linear system symmetric by multiplying with $(\bar{D}_n)^{-1}$. Since the matrix U has full rank (see e.g. [3]), with an analogous argument as in Section 3.2 we see that we arrive at a linear system with a symmetric matrix $U\bar{D}_{n+m}U^t$ that is positive definite.

The remaining question that is left is how to choose our degrees of freedom s_i. Assuming that the chosen regular inner fairness criterion leads to local neighborhoods $D(q_i)$ that will be close to a regular polygon with q_i as its centre, the discrete Δ_B operator at q_i given by (4) can be approximated as

$$\Delta_B q_i \approx \frac{3}{2A_i} \sum_{q_j \in N(q_i)} (2\cot\alpha_i)(q_j - q_i) = \frac{3}{A_i} \cot\alpha_i \sum_{q_j \in N(q_i)} (q_j - q_i),$$

with

$$\alpha_i = \frac{\pi}{2} - \frac{\pi}{valence(q_i)}.$$

Therefore, using the remaining degrees of freedom to approximate the Laplace-Beltrami operator leads us to the scalar weights

$$s_i \;=\; \frac{3}{A_i} cot\,(\frac{\pi}{2} - \frac{\pi}{valence(q_i)}), \tag{8}$$

where A_i is the area sum of the triangles adjacent to q_i.

4.2. Construction Algorithm

Discretizing the Bi-Laplacian as described above leads to the problem of finding a mesh M^∞ such that (7) is satisfied with s_i given by (8). This is again a nonlinear problem since the A_i depend on the vertices q_i, but now creating a mesh sequence $M^k \rightarrow M^{k+1}$ as mentioned above converges to a solution. Since only the vertex positions change during the iteration while the connectivity and topology of the mesh stays the same, the mesh sequence is affected only by the areas A_i on M^k that are used to create the next mesh M^{k+1}. Contrary to the approach without an additional inner fairness condition, we didn't notice any convergence problems with this algorithm. In each iteration step, M^{k+1} can be computed by solving a symmetric and positive definite linear system given by (7). In our implementation this was done by using preconditioned conjugate gradients iterations (see [4]). Because the matrix U is weakly diagonally dominant, we used a diagonal matrix C^2 as preconditioner with

$$C^2 = diag(valence(q_1)^2 s_1, \cdots, valence(q_i)^2 s_i, \cdots, valence(q_n)^2 s_n).$$

We noticed that this preconditioner speeds up the convergence of the conjugate gradient iteration considerably with nearly no extra costs. It is not necessary to exactly solve the linear system in every iteration step. In our implementation, M^{k+1} was created by iterating the preconditioned conjugate gradients algorithm for a small number of steps, which resulted in a fast and stable mesh fairing algorithm.

4.3. Connection to Minimal Energy Surfaces

There is a close connection between creating a mesh sequence $M^k \rightarrow M^{k+1}$ by iteratively solving $\Delta_B \Delta_B f = 0$ with Δ_B discretized on M^k, and an algorithm for smooth surface fairing that was presented by Greiner. In [5] he proposed to create a sequence of smooth spline surfaces $F^k \rightarrow F^{k+1}$ based on minimizing

$$\int_A \langle \Delta_B F, \Delta_B F \rangle dA, \tag{9}$$

assuming that one parameterizes on F^k. If this sequence converges to some surface F^∞, he showed that if the spline space satisfies some additional requirements, the surface F^∞ is close to a minimal energy surface. Minimal energy surfaces minimize $\int_A \kappa_1^2 + \kappa_2^2 dA$, where κ_1 and κ_2 are the principal curvatures.

If we discretize (9) on the mesh $\tilde{M}$ using a piecewise linear approximation, we arrive at

$$\int_{\tilde{M}} \sum_{q_i \in V_I(M) \cup V_B(M)} \langle \Delta_B q_i, \Delta_B q_i \rangle \phi_i dA = \sum_{q_i \in V_I(M) \cup V_B(M)} \langle \Delta_B q_i, \Delta_B q_i \rangle \frac{1}{3} A_i,$$

where ϕ_i is the linear hat function at q_i satisfying $\phi_i(q_i) = 1$ and $\phi_i(q_j) = 0$ for all vertices $q_j \neq q_i$. Requiring that the minimum satisfies $\frac{\partial}{\partial q_i} = 0$ for $q_i \in V_I(M)$ and using (4) to discretize the Δ_B operator, we get

$$-\Delta_B q_i \sum_{q_j \in N(q_i)} (cot\, \alpha_j + cot\, \beta_j) + \sum_{q_j \in N(q_i)} \Delta_B q_j (cot\, \alpha_j + cot\, \beta_j) = 0.$$

Scaling both sides with $\frac{3}{2A_i}$, this results in $\Delta_B \Delta_B q_i = 0$. This is again a justification for our strategy to choose the weights s_i such that we approximate the Δ_B operator. Although in our case not all requirements mentioned in [5] are satisfied since our surfaces are not smooth, one can interpret it as a strategy to be closer to a minimal energy surface.

§5. Multigrid Techniques

The iterative preconditioned conjugate gradients algorithm used in the previous section assumes that the mesh connectivity stays constant and that only the vertex positions change. This works very well for meshes with a few thousand vertices, and is even fast enough to be used in interactive fairing for such meshes. However, for very large triangular meshes, it is well known that the convergence of mesh fairing algorithms can be accelerated if multigrid techniques are integrated into the construction process [8,9,6,11]. To handle large data sets, we also implemented a mesh fairing version that is based on multigrid techniques. Here the necessary hierarchy levels are constructed using the progressive mesh approach [7] with half-edge collapses.

We start with the construction of a discrete solution on the coarsest level of the progressive mesh representation, and then each solution on a coarse level serves as starting point for the iteration algorithm on the next finer hierarchy level. Between two hierarchy levels we need a prolongation operator that predicts the position of the new vertices using the vertex split information of the progressive mesh. This local prolongation operator is used when adding vertices between hierarchy levels. If the next hierarchy level is reached, we again use preconditioned conjugate gradients iterations on that level.

§6. G^1 Boundary Handling

The assumption that we have boundary conditions given by the vertices $V_B(M) \cup V_O(M)$ is convenient, but usually nonlinear fourth order problems

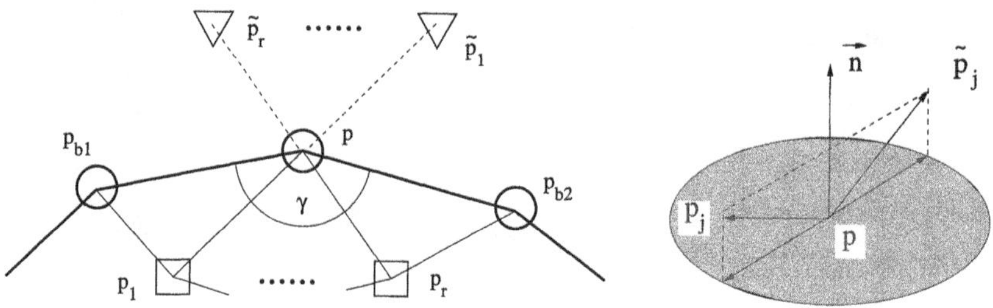

Fig. 2. The 1-neighborhood at the boundary vertex p is completed by reflecting the interior vertices along the line defined by p and $\vec{n}$.

are solved assuming G^1 boundary conditions, that means with prescribed vertices and unit normals along the boundary. Another reason not to use such a boundary condition is that it is not well-suited in combination with multigrid techniques. If we coarsen the mesh M, the local neighborhood of a boundary vertex will no longer be approximately uniformly parameterized, since the 1-ring neighborhood of a boundary vertex becomes unsymmetric. We could restrict the mesh reduction algorithm so that vertices near the boundary are not reduced, but this would considerably reduce the speed gain resulting from the multigrid approach.

The idea behind our G^1 boundary handling strategy is to locally and temporarily complete the neighborhood at the boundary vertices in every step $M^k \to M^{k+1}$. Let p be a boundary vertex, p_{b1} and p_{b2} its adjacent boundary vertices, and let $p_1, \cdots, p_r$ be the r adjacent inner vertices of p (Fig. 2). We complete this neighborhood by assigning a vertex $\tilde{p}_j$ to every interior vertex p_j, $1 \leq j \leq r$, leading to a vertex with the valence $2 + 2r$.

Remembering that the normal curvature distribution $\kappa(\phi)$ at a point on a smooth surface has the property $\kappa(\phi) = \kappa(-\phi)$ - here ϕ is the angle of the normal curvature direction in an arbitrary orthonormal basis of the tangent plane - we determine the new vertices $\tilde{p}_j = 2p - p_j - 2\langle p - p_j, \vec{n}\rangle\vec{n}$ by reflecting the p_j along the line defined by the vertex p and the normal $\vec{n}$. Because of the connection $\Delta_B f = 2H\vec{n}$ between the Δ_B operator and the mean curvature normal $H\vec{n}$, it seems reasonable to assume that the discretization of the Laplacian at the border of the mesh M^k should have no tangential components. Since for every p_j the vector $\tilde{p}_i - p + p_i - p$ is parallel to the normal vector $\vec{n}$ the tangential part of Δp is determined by p_{b1} and p_{b2}. To get a discrete Laplacian that has no tangential component at all, we therefore only use that part of the vector $p_{b1} - p + p_{b2} - p$, that is parallel to the normal vector.

If we project the vertices p_{b1} and p_{b2} into the tangential plane, let γ be the angle that is spanned by the projections $p_{b1} - p$ and $p_{b2} - p$ as shown in Figure 2. If γ is greater than π, it is possible that some $\tilde{p}_i$ lie inside the interior of the mesh. To avoid any special case handling at such concave corners, we determine the area A_i not by using the extended 1-disk of p, but by setting $A_i = A_i(p)\frac{2\pi}{\gamma}$, where $A_i(p)$ is the area sum of the triangles in the interior of

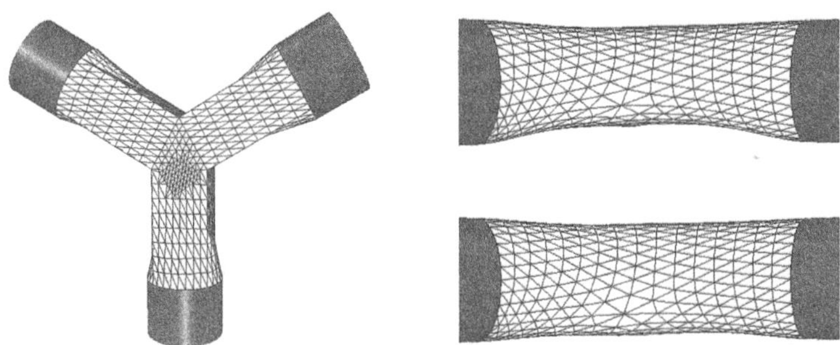

Fig. 3. In this example we compare our mesh fairing scheme with Bi-Laplacian fairing based on a regular linear parameterization [9]. Left: The initial mesh has subdivision connectivity except that at one side we added an additional patch of vertices. The boundary condition is given by 3 cylinders arranged symmetrically. Right: The local uniform parameterization produces a mesh that is clearly not symmetric (top). The additional vertices lead to local as well as global surfaces distortions. However, our nonlinear fairing scheme produces a mesh that is approximately symmetric.

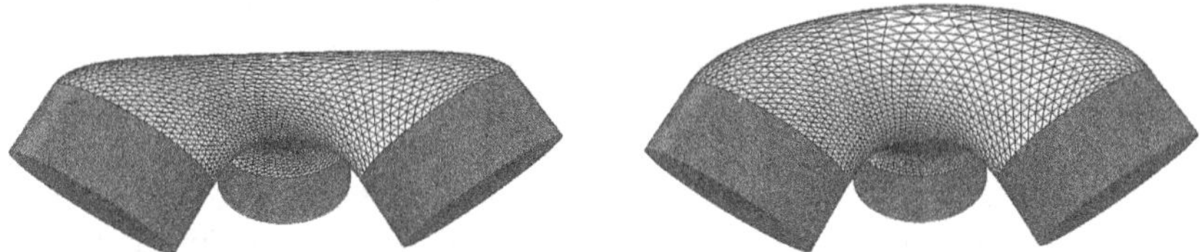

Fig. 4. Another comparison with the regular linear scheme. While the linear method has problems to find a fair surface for this demanding boundary condition (left), our nonlinear scheme produces a surface that looks much more like what one expects.

M^k adjacent to p.

§7. Conclusion

In this paper we presented a nonlinear mesh fairing algorithm that is fast, easy to implement and is able to handle arbitrary initial meshes M^0 even if they are not smooth or are far from the final solution. Because of its speed, this algorithm should be interesting in fields of application where fast fairing methods are needed, e.g. interactive mesh fairing or multiresolution mesh modeling [9]. The resulting mesh quality should be sufficient for many applications. However, if truly intrinsic mesh quality is required, the presented method can serve as preprocessing step that is used to create an initial mesh of high quality.

In future work we will have to study what other inner fairness criteria can be used instead of a regular vertex distribution. We will also have to investigate if a factorization of the Bi-Laplacian as in [12] or more involved preconditioning strategies are worth the extra cost.

References

1. Desbrun, M., M. Meyer, P. Schröder, and A. H. Barr, Implicit fairing of irregular meshes using diffusion and curvature flow, SIGGRAPH 99 Conference Proceedings, 317-324.
2. Desbrun, M., M. Meyer, P. Schröder, and A. H. Barr, Discrete Differential-Geometry Operators in nD, preprint.
3. Floater, M. S., Parametrization and smooth approximation of surface triangulations, Comput. Aided Geom. Design **14** (1997), 231–250.
4. Golub, G. H. and C. F. Van Loan, *Matrix Computations*, Johns Hopkins University Press, Baltimore, 1989.
5. Greiner, G., Blending surfaces with Minimal Curvature, Proc. Dagstuhl Workshop Graphics and Robotics 1994, 163–174.
6. Guskov, I., W. Sweldens, and P. Schröder, Multiresolution Signal Processing for Meshes, SIGGRAPH 99 Conference Proceedings, 325–334.
7. Hoppe, H., Progressive meshes, SIGGRAPH 96 Conference Proceedings, 99–108.
8. Kobbelt, L., Discrete fairing, Proceedings of the Seventh IMA Conference on the Mathematics of Surfaces, 101–131, 1996.
9. Kobbelt, L., S. Campagna, J. Vorsatz, and H-P. Seidel, Interactive Multi-Resolution Modeling on Arbitrary Meshes, SIGGRAPH 98 Conference Proceedings, 105–114.
10. Pinkall, U. and K. Polthier, Computing discrete minimal surfaces and their conjugates, Experimental Mathematics **2** (1993), 15–36.
11. Schneider, R. and, L. Kobbelt, Geometric Fairing of Irregular Meshes for Free-Form Surface Design, Comput. Aided Geom. Design, to appear.
12. Smith, J., The coupled equation approach to the numerical solution of the biharmonic equation by finite differences I., SIAM J. Numer. Anal. **5** (1968), 323–339.
13. Taubin G., A signal processing approach to fair surface design, SIGGRAPH 95 Conference Proceedings, 351–358.
14. Welch, W. and A. Witkin, Free-Form shape design using triangulated surfaces, SIGGRAPH 94 Conference Proceedings, 247–256.

Robert Schneider, Leif Kobbelt and H. P. Seidel
Max-Planck Institute for Computer Science
66123 Saarbrücken, Germany
`{schneider,kobbelt,hpseidel}@mpi-sb.mpg.de`

Approximate Envelope Reconstruction for Moving Solids

U. Schwanecke and L. Kobbelt

Abstract. We present a new approach to approximatively construct the envelope of a moving solid. It is based on dynamically updating an octree that approximates the envelope surface. Our approach guarantees a prescribed error-bound, and is scalable in the sense that it allows a very fast calculation of coarse approximations, while better approximations can be obtained by investing more computation time and memory. Furthermore, the algorithm is robust due to the fact that no badly conditioned surface-surface intersections have to be computed.

§1. Introduction

A solid object undergoing a motion creates a volume, in general. The resulting volume is called a **swept volume**. Swept volumes play an important role in NC (numerical controlled) machining, robotics, and motion planning, e.g. in order to avoid collisions of manipulators. Different approaches to construct the swept volume of a moving solid were developed during the last decades.

One method for representing and analyzing swept volumes is the envelope method (cf. [13]). The main drawback of this method is that there are no really efficient algorithms, due to the essential limitation of efficiently solving the nonlinear envelope equations, and due to the fact that some envelope surfaces tend to resist accurate calculation by both analytical and numerical means.

To overcome the deficiencies of envelope theory, the Sweep Differential Equation (SDE) and Sweep-Envelope Differential Equation (SEDE) method, respectively, were developed (cf. [11]). It subsumes the method of envelopes and is inherently global. The success of the SEDE/SDE method for the numerical computation of swept volumes is largely due to the fact that only a finite set of points (**grazing points**) need to be calculated at each time step.

All of these analytic methods become extremely complex when the topology of the swept volume gets more complicated. Moreover, none of the mentioned methods is able to satisfactory handle self intersections of the swept

Mathematical Methods for Curves and Surfaces: Oslo 2000
Tom Lyche and Larry L. Schumaker (eds.), pp. 455–466.

ISBN 0-8265-1378-6

volume boundary that occur for even fairly simple sweeps. This **trimming problem** for swept volumes is discussed in detail in [2]. The authors give an overview of trimming methods, and develop new trimming strategies for local and global trimming of swept volumes. All of these trimming algorithms consist of computing the candidate set of the boundary of the volume, and in a second step of a test to determine the trim set.

If computational time does not matter, high quality renderings of quite general swept volumes can be obtained by using the Raycasting Engine [17]. For generalized cylinders, i.e. objects defined by sweeping a two-dimensional contour along a three-dimensional trajectory, the computation of the intersection points with a ray can be reduced to the problem of intersecting two two-dimensional curves [6]. A more efficient way of displaying generalized cylinders is the surface scanning algorithm described in [5], which draws contours generated by plane intersections close enough to cover every pixel to which the surface is projected, while avoiding drawing to many superfluous contours. All of these methods only visualize the volume from one specific viewpoint, and cannot address self intersections in an efficient manner.

For the verification of NC machining, several boundary representation and CSG (constructive solid geometry) -based methods have been developed to reconstruct generated objects by moving one object in space. An accurate boundary representation to represent and manipulate solids based on the extension of the octree data structure has been suggested in [7]. This approach can maintain the resulting model, but the processing time per cut, given as a Boolean operation, increases rapidly with part complexity due to the fact that all of the object components have to be cut to each other. Because a part of average complexity requires several thousands and more cuts, the resulting algorithm is too slow for practical use. To overcome this problem, a faster algorithm generating cross sections of the swept profile using extended quadtrees has been given in [15]. However, this approach does not reconstruct the whole object, but only visualizes cross sections, and therefore cannot give a complete impression of the object. In order to construct a three-dimensional model, one has to connect the contours in a post-processing step (cf. [19]).

In this article we present an algorithm for reconstructing a polygonal approximation of the envelope of a swept solid that automatically generates the topologically correct solution within a prescribed error tolerance. The resulting envelope surface is a guaranteed manifold. In our approach we do not have to pay special attention to the trimming of the swept envelope. As a consequence, the presented algorithm is highly eligible for NC milling simulation for example, where typically a lot of self intersections occur due to the fact that the milling tool usually passes the same surface region many times.

Instead of calculating an algebraic expression of the swept solid, we determine a subdivision of the space the swept volume lives in and use linear sweeps between discrete time steps to approximate the target surface. With the help of an octree based data structure, a polygonization of this surface can be determined in a very efficient manner. Dynamically updating the octree in each time step, we incrementally construct the whole piecewise linear approximation of the swept volume.

§2. Functional Representation of Solids

A solid object S can be represented as a closed subset of $\mathbb{R}^3$ with a defining function f. Thereby f is a real continuous implicit function of point coordinates, particularly a generalization of a distance field, with

$$f(\mathbf{p}) \begin{cases} > 0 & \text{for points } \mathbf{p} \in \mathbb{R}^3 \text{ inside the solid,} \\ = 0 & \text{for points } \mathbf{p} \in \mathbb{R}^3 \text{ on the solids boundary,} \\ < 0 & \text{for points } \mathbf{p} \in \mathbb{R}^3 \text{ outside the solid.} \end{cases} \tag{1}$$

Defining functions of complex solids can be created from a finite set of solid primitives or given defining functions with set-theoretic operations by a CSG-like scheme. With the help of the theory of R–functions, the exact analytical definitions of set theoretic operations can be expressed (cf. [20,23]). For example, if solid S_1 is defined as $f_1 \geq 0$ and solid S_2 as $f_2 \geq 0$, then

$$\begin{array}{llll} \text{Intersection} & : S_3 = S_1 \cap S_2 & : & f_3 = f_1 \wedge f_2 = f_1 + f_2 - \sqrt{f_1^2 + f_2^2} \\ \text{Union} & : S_3 = S_1 \cup S_2 & : & f_3 = f_1 \vee f_2 = f_1 + f_2 + \sqrt{f_1^2 + f_2^2} \\ \text{Complement} & : S_3 = S_1 & : & f_3 = -f_1 \\ \text{Subtraction} & : S_3 = S_1 \backslash S_2 & : & f_3 = f_1 \backslash f_2 = f_1 - f_2 - \sqrt{f_1^2 + f_2^2} \end{array} \tag{2}$$

These R-functions have C^1 discontinuities only in points where both arguments equal zero, i.e. at intersections of the belonging solid's surfaces.

Using the concept of defining functions of solids, one can make these solids move by introducing a fourth variable t representing the time. Thus, the general defining function of a moving solid is

$$f(x, y, z, t) \geq 0, \tag{3}$$

where $f(x, y, z, t_1)$ and $f(x, y, z, t_2)$ only differ by a rigid transformation.

Since the elimination of the parameter t in (3) is quite difficult in general, one uses a piecewise constant approximation of the moving solid given by

$$S := S(t_1) \cup S(t_1 + dt) \cup S(t_1 + 2dt) \cup \ldots \cup S(t_2), \tag{4}$$

where $\cup$ is the set-theoretic union defined in (2). For $dt \to 0$ we get a more and more accurate approximation of the swept solid.

In order to achieve a piecewise linear approximation instead of the piecewise constant given by (4), one can define the sweep as the union of two solid samples in initial and final positions, and an envelope swept by the moving solid. Thus, the swept object can be described by

$$f(x, y, z) = f(x, y, z, t_1) \vee f(x, y, z, t_2) \vee E(x, y, z), \tag{5}$$

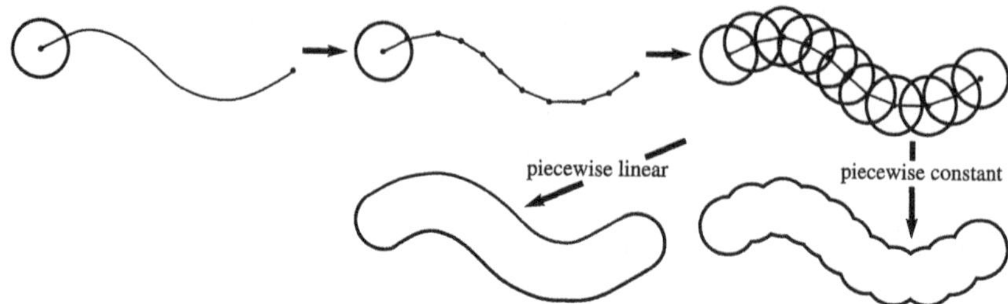

Fig. 1. Piecewise constant or linear approximation respectively.

where E is the envelope, i.e., the object swept during the motion of the generator (cf. Fig. 1).

The envelope surface is the surface which is tangential to all members of the family of surfaces. Consider the two members of the family at times t and $t + dt$. The first one has equation $f(x, y, z, t) = 0$, while the second one has equation $f(x, y, z, t + dt) = 0$. Due to the fact that the envelope must meet both of these surfaces, and thus satisfy both equations, it can be seen by expanding the latter in terms of dt, and letting dt tend to 0, that the envelope surface must simultaneously satisfy

$$f(x, y, z, t) = 0 \quad \text{and} \quad \frac{\partial f(x, y, z, t)}{\partial t} = 0.$$

The implicit form of the envelope surface can be obtained by eliminating t from these two equations. If the surfaces are described by polynomials, the elimination can be performed automatically using methods of *Computer Algebra* such as resultants or Gröbner bases (cf. [9]). Although these methods are restricted to rational functions, many other forms can be converted to them. As an example, we can represent $\sin(s)$ and $\cos(s)$ as the rational functions $2t/(1+t^2)$ and $(1-t^2)(1+t^2)$, with help of the substitution $\tan(s/2) = t$.

All of the proposed methods are computationally expensive, needing exponential or even doubly exponential time in the number of time-steps t_i. Moreover, the resulting implicit functions in general have very high degree which implies hundreds of coefficients.

Moving a rigid solid along the trajectory $\mathbf{g}(\mathbf{t}) : [t_1, t_2] \longrightarrow \mathbb{R}^3$ with the orientation $\mathbf{p}(t) : [t_1, t_2] \longrightarrow \mathbb{R}^{3\times 3}$, the structure of the defining function f is

$$f(x, y, z, t) = s(\mathbf{p}(t)(x, y, z)^T + \mathbf{g}(t))$$

In our actual implementation we only consider spheres, thus we can assume $\mathbf{p}(t)$ to be the identity. The error we make by approximating the trajectory $\mathbf{g}(t)$ by the polygon $\mathbf{l}(t)$ being the piecewise linear interpolation of the points $\mathbf{g}(t_1), \mathbf{g}(t_1 + dt), \ldots, \mathbf{g}(t_2)$ can be estimated by

$$\max_{u\in[t_1,t_2]} \|\mathbf{g}(t) - \mathbf{l}(t)\|_2 \leq \frac{1}{8}\, l_{\max}^2 \cdot \max_{u\in[t_1,t_2]} \|\mathbf{g}''(t)\|_2, \tag{6}$$

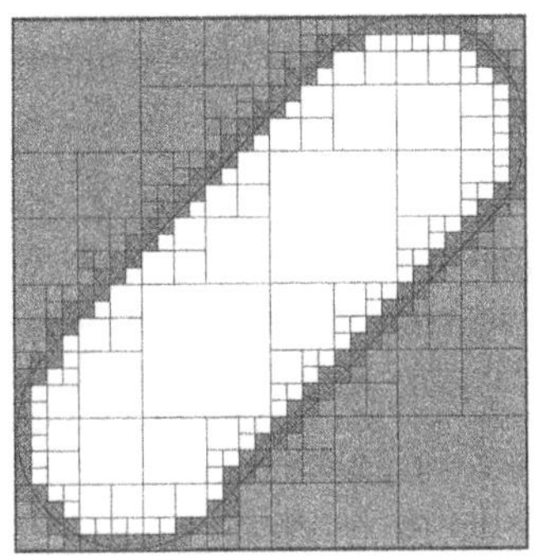

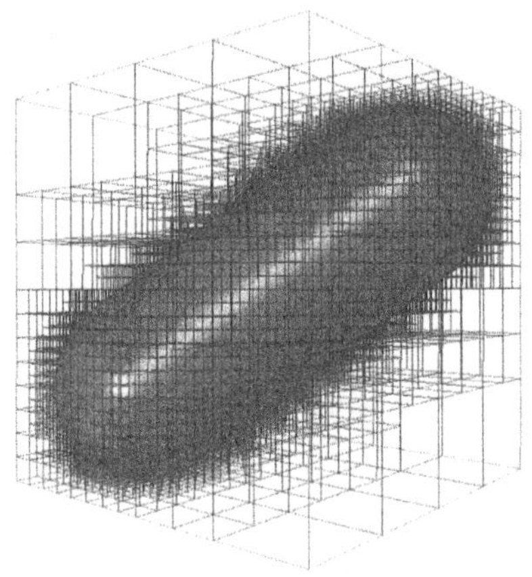

Fig. 2. Quadtree and octree-representation of an implicit function respectively.

where $l_{\max} = \max \|\mathbf{g}(t_1 + idt) - \mathbf{g}(t_1 + (i+1)dt)\|_2$ is the length of the longest polygon edge. If the moving solid is a sphere, than the right hand side of (6) is also an upper bound for the error made by approximating the moving sphere's envelope by a collection of cylinders. This is obvious if we take into account the fact that the tangent to the trajectory $\mathbf{g}(t)$ is parallel to $\mathbf{l}(t)$ at the parameter value $t_{\max}$ where the maximum error occurs. Thus, the envelope constructed by our approach converges quadratically to the desired envelope. Calculating only a piecewise constant approximation instead of a linear one, we increase the error (6) by

$$E(r) = r - \frac{\sqrt{4r^2 - l_{\max}^2}}{2}, \tag{7}$$

where r is the radius of the sphere, resulting in linear instead of quadratic convergence.

§3. Polygonization

One simple way to visualize a solid is to raytrace it directly from its defining implicit function. This only visualizes the object from one specific rendering viewpoint, and needs very high computational resources as we have still mentioned.

For applications beyond mere visualization, it is useful to approximate the surface with an *explicit representation* (indirect visualization), such as a set of triangles or polygons. In fact, any continuous manifold may be approximated by a triangulation (cf. [25]). This process, called polygonization, generally involves sampling the surface. One standard approach for implicitly defined surfaces is to compute the intersection of a three-dimensional grid with the surface in order to determine the location and connectivity of surface points.

Using the divide-and conquer strategy of binary *subdivision* is one of the most efficient ways to generate this three-dimensional grid. It leads to quadtree and octree data-structures, respectively: A quadtree (octree) is derived by successively subdividing a plane (space) in both (three) dimensions to form quadrants (octants). An comprehensive survey of quadtrees, octrees and other forms of subdivision such as KD-trees and bintrees can be found in [21,22]. When an octree is used to represent an implicit function, each cell may be inside, outside, or partially inside and outside (also called white, gray and black respectively) the solid described by the implicit function (cf. Fig. 2).

When building up the octree, we only have to subdivide gray cells (adaptive refinement). To test if a given cell in the octree has to be subdivided, we check the sign of the implicit function at the corners. Different signs at both ends of an edge indicate that the surface must have an intersection with that edge.

Unfortunately, there are configurations where parts of the surface lie in the interior of the cell, yet the implicit function has the same sign at all corners. Such situations are quite hard to detect in general, since many special cases have to be checked [7]. However, in the case of adaptive refinement for isosurface extraction, it is sufficient to find a *conservative* splitting criterion. For correct reconstruction, we have to guarantee that a cell is subdivided if some part of the surface lies in the interior, but we can tolerate wrong decisions where the cell if subdivided, even if the surface does not intersect the cell.

In the case of a moving solid, we can derive such a conservative criterion by computing a bounding box or a bounding sphere that encloses the solid. For such basic primitives, the intersection test with a given cell can be implemented very efficiently (cf. [1]). In oder to reduce the number of erroneous decisions, we can additionally compute an empty box or empty sphere in the interior of the moving solid. For the practically relevant geometries, i.e., the standard shapes of milling tools, the approximation of the moving solid by an outer bounding box and an inner empty box is sufficiently tight.

Another criterion to subdivide given octants in order to achieve an approximation of an object is the variation of the target's distance field over the parent cell (cf. [10]).

Once the octree representation of the solid is given, one can construct a polygonal approximation of the solid's surface by examining the different configurations of the corners of the partially full voxels (border voxels). Therefore, we use the distance values $f(\mathbf{p})$ evaluated at the voxel corners and stored in our hash table. Each of the 8 voxel corners can have a negative or nonnegative value, resulting in 256 different corner configurations. These cases can be handled by a table method (*Marching Cubes*) proposed in [14,18], or an algorithmic method described in [3]. Because our implicit defining functions are continuous, they always intersect an edge of a voxel connecting differently signed vertices. A coarse approximation of the surface samples are the midpoints of these edges. In order to get a better approximation of the desired object, we either approximate the exact intersection using bisection or Newton iteration, or linear interpolate with respect to the values assigned to the corners of the voxels. Figure 3 illustrates the algorithmic approach, and shows two resulting triangular meshes, with and without linear interpolation.

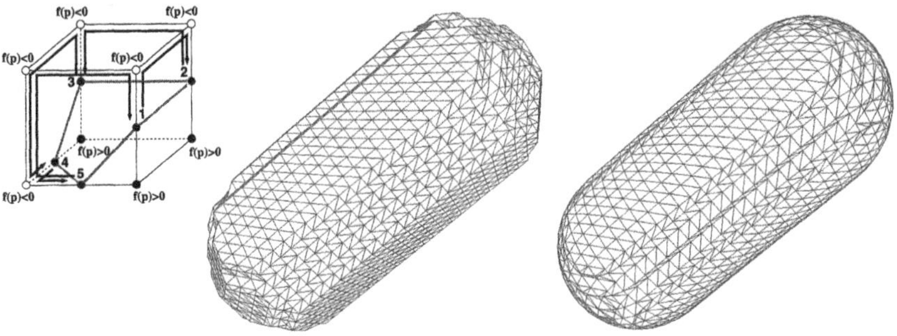

Fig. 3. Left: The algorithmic polygonization method: The surface vertices are ordered by walking from one surface vertex to the next, around a face of a given voxel in clockwise order. When arriving at a new vertex, the face across the vertex's edge from the current face becomes the new current face. Middle and right: Surface obtained by connecting midpoints or linear interpolation respectively.

§4. Dynamic Octree Manipulation

In this section we describe our new approach to approximately reconstructing the envelope of a moving solid. In particular our goal is to construct a polygonal approximation of the target surface given by (4) and (5).

We denote the volume given by the linear interpolation of $S(t_1 + idt)$ and $S(t_1 + (i+1)dt)$ by L_i. Because it is computationally too expensive to determine the implicit function given by (4) and (5), we just consider the volume L_{i+1} with respect to that given by $L_0 \cup \ldots \cup L_i$. In this incremental reconstruction process we choose a bounding box (= root cell of the octree) that is large enough to contain the whole swept volume given by (4), and start computing the octree approximating L_0.

In order to construct the octree approximating that part of the envelope that is generated by $L_0 \cup L_1$, we just traverse the initial octree of L_0 and check in what manner it is changed by L_1. Thereby, we do not have to worry about branches that are marked inside because they can never change status. We only have to update leaves that are marked as border or outside leaves. This makes our approach quite fast.

If a branch of the octree is marked as outside, we have to construct a new suboctree corresponding to L_1. If a branch is marked as border, we have to check whether this branch is inside L_1 or not. If it is not, we go on with the traversal of the initial octree with respect to L_1. Thereby all suboctrees of the initial octree now being inside L_1 can be removed to reduce computation time and memory (cf. Fig. 4).

In order to get the correct results, we also have to update the values of the defining functions evaluated at the voxel corners. Due to the fact that a region once marked as inside alway stays inside, we have to compute the new values and store the higher one of an old and a new value in our hash table.

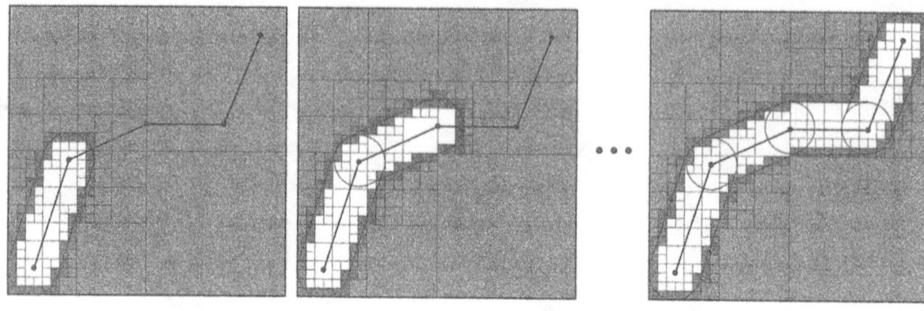

Fig. 4. Dynamically updating a quadtree in order to approximate the envelope of a circle moving along a polygon.

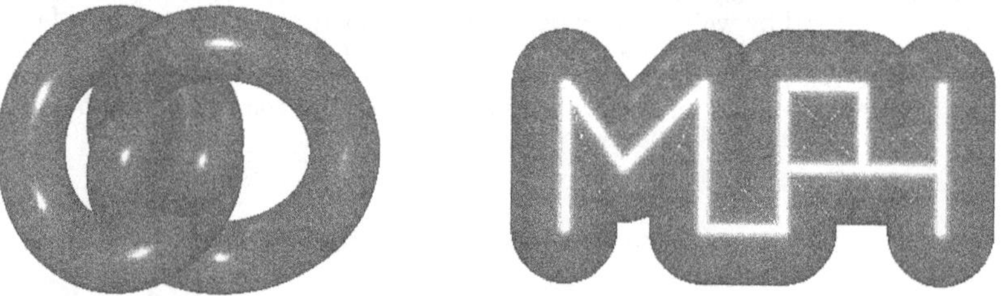

Fig. 5. Some example sweeps. Notice the alias effect along intersections.

If we repeat this procedure until we reach L_{t_2-dt}, we have built an octree representing the whole moving solid given by (4) and (5) up to an error tolerance determined by the size of the voxels or the maximum depth of the octree respectively. Thereby the time step dt has to be chosen sufficiently small, so that the error made by our discretization (4) is smaller than the size of the voxels. Figure 4 illustrates the dynamically growing octree and quadtree, respectively. Notice that voxels of regions that have been subdivided to the maximum refinement level can be combined and deleted during this process.

Figure 5 shows some resulting sweeps. On the left hand side a sphere of radius $r = 2$ is moving along the knot curve defined by

$$k(t) = \begin{pmatrix} 10\cos(t) + \cos(3t) + \cos(2t) + \cos(4t) \\ 6\sin(t) + 10\sin(3t) \\ 4\sin(3t)\sin(\frac{5}{2}t) + 4\sin(4t) - 2\sin(6t) \end{pmatrix}, \quad t \in [0, 2\pi].$$

With $dt = \frac{2\pi}{1000}$ and edge-length of the smallest voxels equals 0.02 (=error tolerance). The resulting polygonal approximation consists of approximately 376000 triangles, and it took about 50 seconds to compute it. On the right hand side a sphere of radius $r = 5$ is moving along 13 points building a polygon describing the letters MPI. The edge-length of the smallest voxels has also been set to 0.02 resulting in a triangular mesh with about 500000 triangles. It took about 20 seconds to calculate this mesh, which is faster than the first example because fewer time steps had to be computed.

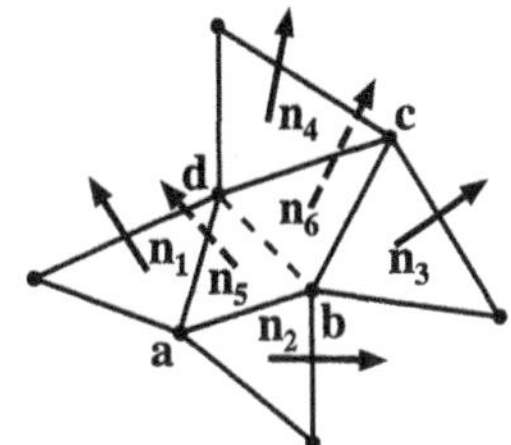

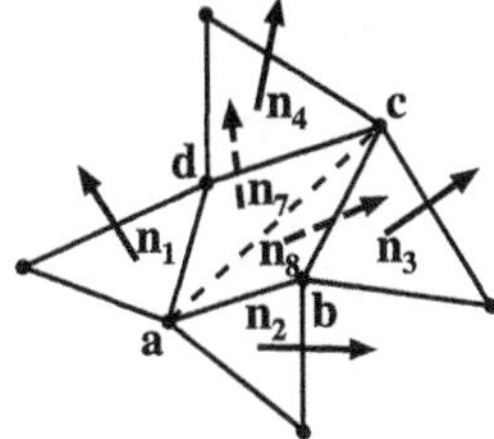

Fig. 6. Flipping an edge according to the deviation of normals.

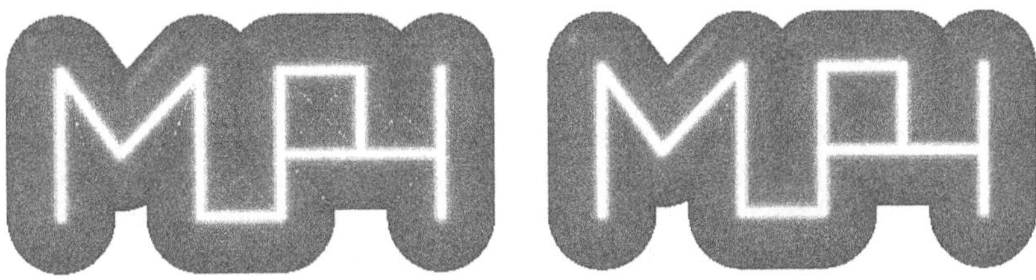

Fig. 7. A Swept surface before and after flipping the edges.

§5. Post Processing

The approach described in the previous section guarantees a prescribed error-bound. The error made by our reconstruction process is at most the length of the edges of the voxels. But due to the discretization we made, and due to the fact that we only choose one possible triangulation of the calculated points, our approach, like every discrete approach, produces alias effects appearing at sharp features (cf. Fig. 5).

In order to reduce these effects, one can improve the triangulation by flipping edges of the mesh representing the solids envelope. Therefore, one has to define a global curvature functional as a criterion for the mesh quality. Such a functional may be the deviation of the normals of adjacent triangles or the discrete Gaussian curvature. Several methods to calculate the discrete Gaussian curvature of a polygonal net exist. A fast and numerically stable method to estimate not only the Gaussian curvature, but also the tensor of curvature of a polygonal net can be found in [24]. A comparison of the errors of several approximations to the surface normal and the Gaussian curvature can be found in [16].

In our implementation we chose a functional based on the deviation of the normals of adjacent triangles which is illustrated in Fig. 6. We examine all triangles of our net and flip an edge **bd** shared by two adjacent triangles into the edge **ac** if $\min(A\setminus \min(A)) > \min(B\setminus \min(B))$ with $A = \{\mathbf{n}_1^T\mathbf{n}_5, \mathbf{n}_4^T\mathbf{n}_5, \mathbf{n}_3^T\mathbf{n}_6, \mathbf{n}_2^T\mathbf{n}_6\}$ and $B = \{\mathbf{n}_1^T\mathbf{n}_7, \mathbf{n}_4^T\mathbf{n}_7, \mathbf{n}_3^T\mathbf{n}_8, \mathbf{n}_2^T\mathbf{n}_8\}$.

Figure 7 shows a surface before and after processing the edge-flipping step. It can be seen that the quality of the mesh has been improved drastically. Unfortunately, as one can also see, the described edge–flipping process does

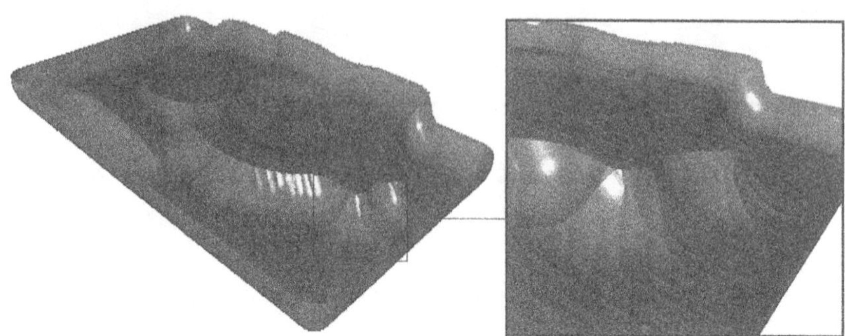

Fig. 8. A NC machining simulation example.

not remove all of the occurring alias effects but at least most of them. This is due to the fact that our greedy approach can become stuck at local minima of the functional measuring the mesh quality. To overcome this problem, one has to choose more sophisticated techniques, e.g., considering the mesh quality not only after performing one edge-flip, but two ore even more, which will be future work.

In Figure 8 an example from an industrial application is shown. The pictured surface resulted from moving a ball cutter of diameter $r = 8$ mm along a path given by 3770 points. The edge-length of the voxels approximating the surface is 0.1 mm. The resulting surface consists of about 1.2 million triangles, and it took about 10 minutes to calculate.

§6. Conclusion and Future Work

We presented a technique to construct a polygonal approximation of the envelope of a moving solid which is based on dynamically updating an octree approximating the envelope surface. Since we use linear sweeps between discrete time steps, our approach is very fast. It can reconstruct the envelope of a moving solid up to a prescribed error bound, while it is scalable in the sense that tolerating a larger approximation error speed up the calculation so that rough approximations even could be obtained in real time. Because no surface-surface intersections have to be computed explicitly, the algorithm is robust and highly eligible for applications like NC milling simulation, where typically a lot of these intersections turn up.

Finally, we showed that flipping the edges of a mesh can improve the mesh-quality drastically. Up to now we only have implemented a simple greedy edge-flipping procedure. Using a more sophisticated optimization technique could improve the mesh-quality even more.

Our actual implementation only allows moving spheres, because there exist very fast and simple algorithms to check whether a sphere or ellipsoid intersects a given box or not. Our algorithm could be generalized to arbitrary solids, as long as we can evaluate the distance function in an efficient manner (e.g. for polygonal meshes cf. [12]). However, already the envelope swept by an arbitrary solid moving along a straight line can become quite complex which has been shown in Section 2. At this point, the piecewise constant approximation might lead to the more efficient reconstruction algorithm.

Acknowledgments. This research was partly supported by Siemens AG München.

References

1. Arvo, J., A simple method for box-sphere intersection testing, in *Graphics Gems*, A. S. Glassner (ed.), Academic Press, New York, 1990, 335–339.
2. Blackmore, D., R. Samulyak and M. C. Leu, Trimming swept volumes, Computer-Aided Design **31** (1999), 215-223.
3. Bloomenthal, J., Polygonization of implicit surfaces, Computer Aided Geom. Design **5** (1988), 341–355.
4. Bloomenthal, J., C. Bajaj, J. Blinn, M. Cani-Gascuel, B. Wyvill and G. Wyvill, *Introduction to Implicit Surfaces*, Morgan–Kaufmann, San Francisco, 1997.
5. Bronsvoort, W., A surface scanning algorithm for displaying generalized cylinders, The Visual Computer **8** (3) (1992), 162–170.
6. Bronsvoort, W. and F. Klok, Ray Tracing generalized cylinders, ACM Trans. on Graphics **4** (4) (1985), 291–303.
7. Brunet, P. and I. Navazo, Solid representation and operation using extended octrees, ACM Trans. on Graphics **9** (2) (1990), 170–197.
8. Carmo do, M., *Differential Geometry of Curves and Surfaces*, Prentice Hall, 1976.
9. Davenport, J. H., Y. Siret and F. Tournier, *Computer Algebra: Systems and Algorithms for Algebraic Computation*, Academic Press, 1993.
10. Frisken, S. F., R. N. Perry, A. P. Rockwood and T. R. Jones, Adaptively sampled distance fields: a general representation of shape for computer graphics, Proc. Siggraph '00, ACM, 2000.
11. Gouping, W., S. Jiaguang and H. Xuanji, The sweep-envelope differential equation algorithm for general deformed swept volumes, Comput. Aided Geom. Design **17** (2000), 399–418.
12. Guéziec, A., "Meshsweeper": Fast closest point on a polygonal mesh and applications, IEEE Trans. on Visualization and Computer Graphics, to appear.
13. HU, Z. and Z. Ling, Swept volumes generated by the natural quadric surfaces, Computer and Graphics **20** (2) (1996), 263–274.
14. Lorensen, W. E. and H. E. Cline, Marching cubes: a high resolution 3D surface construction algorithm, Computer Graphics **21** (3) (1987), 163–169.

15. Liu, C., D. M. Esterling, J. Fontdecaba and E. Mosel, Dimensional verification of NC Machining profiles using extended quadtrees, Computer-Aided Design **28** (11) (1996), 845–852.

16. Meek, D. S. and D. J. Walton, On surface normal and Gaussian curvature approximations given data sampled from a smooth surface, Comput. Aided Geom. Design **17** (2000), 521–543.

17. Menon, J. P. and D. M. Robinson, Advanced NC verification via massively parallel raycasting, ASME DE Manufacturing Review **6** (1993), 141–154.

18. Montani, C., R. Scateni and R. Scopigno, A modified look-up table for implicit disambiguation of marching cubes, The Visual Computer (**10**) (1994), 353–355.

19. Oliva, J-M., M. Perrin and S. Coquillart, 3D Reconstruction of complex polyhedral shapes from contours using a simplified generalized Voronoi diagram, EUROGRAPHICS 96 (1996), 397–408.

20. Rvachev, V. L., *Methods of Logic Algebra in Mathematical Physics*, Naukova Dumka Publishers, Kiev, 1974.

21. Samet, H., *Applications of Spatial Data Structures*, Addison-Wes., 1990.

22. Samet, H., *Design and Analysis of Spatial Data Structures*, Addison-Wesley, 1990.

23. Shapiro, V., Real functions for representation of rigid solid, Comput. Aided Geom. Design **11** (1994), 153–175.

24. Taubin, G., Estimating the tensor of curvature of a surface from a polyhedral approximation, Proc. ICCV '95, 1995, 902–907.

25. Whitney, H., Elementary structure of real algebraic varieties, Annals of Mathematics **66** (1957), 545–556.

Ulrich Schwanecke, Leif Kobbelt
Max-Planck-Institute for Computer Sciences
Stuhlsatzenhausenweg 85
66123 Saarbrücken, Germany
{schwanecke,kobbelt}@mpi-sb.mpg.de

Towards the μ-Basis of a Rational Surface

Thomas W. Sederberg and Jianmin Zheng

Abstract. The method of moving lines and moving curves is a technique for computing the implicit equation of a parametric curve. The efficiency of this method has been significantly enhanced by the introduction of the μ-basis for planar rational curves. This paper makes some initial contributions in defining the μ-basis of a rational surface, and it proves some important properties of the proposed basis. The concepts are based on syzygy modules.

§1. Introduction

In computer aided geometric design, curves and surfaces are usually represented in one of two standard ways: the rational parametric form or the algebraic implicit form. Parametric equations are convenient for generating points along a curve or surface; implicit equations can easily determine whether or not a point lies on a curve or surface. When both representations are available, many algorithms for geometric modeling can be simplified.

It is known from classical algebraic geometry that any polynomial or rational parametric representation admits an implicit form while the converse is not true. The process of finding an implicit equation given a parametric representation is called implicitization. Three major approaches can be used to implicitize a parametric curve or surface [5]: polynomial resultants, Gröbner bases, and Wu-Ritt methods. These approaches typically involve significant computation which limits the practical use of implicitization.

A recent development in implicitization is a method known as the moving curve and surface technique [7,8,9]. A moving curve is defined as

$$C(x,y,w;t) := \sum_{i=0}^{m} f_i(x,y,w)t^i,$$

where $f_i(x,y,w)$ is a homogeneous polynomial of degree d. Thus, $C(x,y,w;t) = 0$ is a family of algebraic curves, with one curve corresponding to each

Mathematical Methods for Curves and Surfaces: Oslo 2000
Tom Lyche and Larry L. Schumaker (eds.), pp. 467–476.

ISBN 0-8265-1378-6

t. In particular, when $d = 1$, $C(x, y, w; t) = 0$ is a family of implicitly defined lines. Therefore we call it a **moving line** of degree m. A moving curve $C(x, y, w; t) = 0$ is said to *follow* a planar rational curve $\mathbf{r}(t) = (x(t), y(t), w(t))$ if $C(x(t), y(t), w(t); t)$ is identically zero, i.e., for all values of t, the point $\mathbf{r}(t)$ lies on the moving curve $C(x, y, w; t) = 0$.

A **moving surface** is defined as

$$S(x, y, z, w; s, t) := \sum_{i=1}^{\sigma} h_i(x, y, z, w)\gamma_i(s, t) = 0,$$

where the equations $h_i(x, y, z, w) = 0$, $i = 1, \cdots, \sigma$, define a collection of implicit surfaces, and where the $\gamma_i(s, t)$, $i = 1, \cdots, \sigma$, are a collection of polynomials in s and t, called the blending functions for the moving surface. We require the blending functions to be linearly independent and to have no nonconstant common factor. A moving surface is said to follow a rational surface $\mathbf{r}(s, t) = (X(s, t), Y(s, t), Z(s, t), W(s, t))$ if $S(X(s, t), Y(s, t), Z(s, t), W(s, t); s, t) \equiv 0$.

The moving curve and surface technique is to find $m + 1$ independent moving curves or σ independent moving surfaces that follow a given rational curve or surface. A square matrix can then be formed from the coefficients of these moving curves or surfaces, and the determinant of the matrix gives the desired implicit equation [7]. Empirical studies show that moving curve and surface methods work well for implicitization of rational curves and surfaces. Usually they give a more compact representation for the implicit equation than do standard resultant based approaches. More importantly, when there are base points in the parametric representation, the moving curve and surface techniques actually simplify while the resultant methods fail [7].

Much research on the moving curve and surface method has focused on efficient algorithms for computing a set of $m + 1$ moving curves or σ moving surfaces. In [7], such a collection of moving curves or surfaces was found by solving a large set of linear equations. In [2], the problem of finding an appropriate set of moving lines was illuminated by the description of the μ-basis. A μ-basis for a degree-n planar rational curve consists of two moving lines, of degree μ and $n - \mu$ respectively, which form an ideal basis for all moving lines that follow the curve. Once the μ-basis of a curve is known, the problem of finding $m + 1$ linearly independent moving curves is greatly simplified. Furthermore, the μ-basis for curves provides a framework for proving that the method of moving lines always works. Further valuable insight into the method of moving lines has resulted from Cox's observation that this method is akin to the theory of modules and syzygies [2].

Motivated by the illumination that the μ-basis for curves has cast on the method of moving lines, the authors along with Ron Goldman, David Cox, and Falai Chen have struggled to devise a similar μ-basis for surfaces. This paper presents some initial results that propose a definition for the μ-basis of a rational surface and prove some important properties of the proposed μ-basis.

The results presented herein are largely the offspring of David Cox's observation that the μ-basis of a planar rational curve is the Gröbner basis of the syzygy module of the curve. These concepts are reviewed in Section 2.

§2. Modules and Syzygies

Let $R[t]^m$ be the set of m-dimensional column vectors with entries in the polynomial ring $R[t]$ where R is the field of real numbers or the field of rational numbers. Let $F_1(t)$, $\cdots$, $F_n(t) \in R[t]^m$, and let M be the collection of all $F \in R[t]^m$ which can be written in the form

$$F = a_1 F_1 + \cdots + a_n F_n$$

with $a_i \in R[t]$ for all i. Then M is a submodule of $R[t]^m$, or a module over $R[t]$. We use the notation $M = \langle F_1, \cdots, F_n \rangle$, and say M is *finitely generated* by $F_1, \cdots, F_n$, or the generating set $\{F_1, \cdots, F_n\}$ *spans* M. Given a finitely generated module M over $R[t]$, define the minimal generating set of the module M to be the generating set with the smallest number, denoted $\mu(M)$, of elements of M. It can be proved that the minimal generating set is also the generating set whose elements have the lowest degrees. If a generating set of M is linearly independent over $R[t]$, we call it a basis for M. If a module M has a basis, one can show that the basis has $\mu(M)$ elements. Thus we have

Lemma 1. *Let M be a submodule of $R[t]^m$. Then its minimal number $\mu(M)$ is at most m.*

Let $N = (F_1, \cdots, F_k)$ be an ordered k-tuple of elements of some $R[t]$-module M. A relation on N is an $R[t]$-linear combination of the F_i which is equal to zero:

$$a_1 F_1 + \cdots + a_k F_k = 0, \quad a_i \in R[t].$$

The k-tuple $(a_1, \cdots, a_k)^T$ is called a syzygy of N. The set consisting of all syzygies $(a_1, \cdots, a_k)^T$ of N itself is a submodule of $R[t]^k$, or a module over $R[t]$, called the syzygy module of N, and denoted $Syz(F_1, \cdots, F_k)$.

Lemma 2. *Let $M = \langle F_1(t), \cdots, F_k(t) \rangle$ be a finitely generated module over the polynomial ring $R[t]$, where $F_i \in R[t]^m$, and let A be the syzygy module of $(F_1, \cdots, F_k)$. Then the minimal numbers of M and A satisfy the relation $\mu(A) = k - \mu(M)$.*

§3. Curve Case: the μ-Basis is the Minimal Generators

For a degree n planar rational curve $\mathbf{r}(t)$ in homogeneous form

$$(x, y, w) = (a(t), b(t), c(t)),$$

where $n = max\{deg(a), deg(b), deg(c)\}$, and $a(t) = \sum_{i=0}^{n} a_i t^i$, $b(t) = \sum_{i=0}^{n} b_i t^i$, and $c(t) = \sum_{i=0}^{n} c_i t^i$ $(\neq 0)$ are relatively prime, Cox, Sederberg and Chen proved [2] that there exist two moving lines p and q of degree $\mu (\leq n/2)$ and $n - \mu$ in t respectively, that satisfy

1) p is the moving line with the lowest degree in t that follows the curve.
2) Any moving line $Ax+By+Cw=0$ that follows the curve can be written uniquely in the form

$$Ax + By + Cw = h_1(t)p + h_2(t)q, \quad h_1, h_2 \in R[t].$$

These two moving lines are referred to as the μ-basis.

The μ-basis is implied by the Hilbert Syzygy Theorem [1]. Each moving line $Ax + By + Cw = 0$ is in one-to-one correspondence to a triple (A, B, C), and the moving line following a curve means $Aa + Bb + Cc = 0$. This means that the μ-basis is, in fact, the minimal generating set of the syzygy module of $(a(t), b(t), c(t))$. It is easy to check that $\mu(\langle a(t), b(t), c(t)\rangle) = 1$. From Lemma 2, we know that the minimal generators consist of two elements. Their degrees can be found using the Hilbert polynomial. This result can be extended to space curves [2]:

Lemma 3. *For a degree n rational space curve $(X(t), Y(t), Z(t), W(t)) = (a(t), b(t), c(t), d(t))$, where $a(t)$, $b(t)$, $c(t)$, and $d(t) \neq 0$ are relatively prime polynomials in t, and n is the maximum of their degrees, let*

$$I_{1,m} = \{A(t)X + B(t)Y + C(t)Z + D(t)W : Aa + Bb + Cc + Dd \equiv 0, \\ m = \max(deg(A), deg(B), deg(C), deg(D))\}.$$

Then there exist three polynomials $p_1 \in I_{1,\mu_1}$, $p_2 \in I_{1,\mu_2}$ and $p_3 \in I_{1,\mu_3}$ such that

1) $\mu_1 + \mu_2 + \mu_3 = n.$
2) *Every $q \in I_{1,m}$ can be written uniquely as $q = h_1(t)p + h_2(t)p_2 + h_3(t)p_3$, where $h_1, h_2, h_3 \in R[t]$ and $deg(h_i) \leq m - \mu_i$.*

The μ-basis is useful for implicitization. A typical resultant based method for implicitizing a curve uses Bezout's resultant. Bezout's resultant for a rational curve of degree n is actually formed from n linearly independent moving lines that follow the curve. However, multiplying the μ-basis by suitable t^k can lead to exactly n linearly independent moving lines of degree $n-1$ following the curve. The matrix formed by the coefficients of these n moving lines is the same as the Bezout matrix up to a constant nonsingular matrix. Therefore, the determinant of the new matrix also gives the implicit equation. Furthermore, if we apply a variant of Bezout's resultant directly to the μ-basis, we obtain an $(n-\mu) \times (n-\mu)$ determinant with μ rows quadratic in x and y, and the remaining $n - 2\mu$ rows linear in x and y. This determinant gives the implicit equation of the rational curve with a compact representation as a determinant of dimension $(n-\mu) \times (n-\mu)$, while the standard Bezout method gives an $n \times n$ determinant [4,6]. Therefore, once the μ-basis is found, the implicitization problem is simplified.

§4. Surface Case: Minimal Generating Set and μ-Basis

In this section we expect to extend the treatment of the μ-basis for curves to the surface case. Consider a degree $m \times n$ rational surface

$$\mathbf{r}(s,t) = (X(s,t), Y(s,t), Z(s,t), W(s,t)).$$

We study the moving plane of degree $n-1$ in t and any degree in s:

$$L(s,t) = \sum_{i=0}^{n-1} L_i(s)t^i = L_0(s) + L_1(s)t + \cdots + L_{n-1}(s)t^{n-1},$$

where $L_i(s) = A_i(s)X + B_i(s)Y + C_i(s)Z + D_i(s)W$. Then each $L(s,t)$ is in one-to-one correspondence with a vector

$$V = (A_0, B_0, C_0, D_0, \cdots, A_{n-1}, B_{n-1}, C_{n-1}, D_{n-1}).$$

The vector V has dimension $4n$ and each element in V is a polynomial in s. In shorthand notation, we also write $V = (L_0, L_1, \cdots, L_{n-1})$.

Rewrite $\mathbf{r}(s,t)$ in the form

$$\mathbf{r}(s,t) = (X(s,t), Y(s,t), Z(s,t), W(s,t)) = \sum_{i=0}^{n}(X_i(s), Y_i(s), Z_i(s), W_i(s))t^i$$

$$= \sum_{i=0}^{n} P_i(s)t^i = P_0(s) + P_1(s)t + \cdots + P_n(s)t^n.$$

If the moving plane $L(s,t)$ follows the surface $\mathbf{r}(s,t)$, then

$$L(s,t)\cdot\mathbf{r}(s,t) = \sum_{i=0}^{n-1} L_i(s)t^i \cdot \sum_{j=0}^{n} P_j(s)t^j$$

$$= \sum_{k=0}^{2n-1}\left(\sum_{i=\max(0,k-n)}^{\min(n-1,k)} L_i(s)\cdot P_{k-i}(s)\right)t^k \equiv 0.$$

Thus,

$$\sum_{i=\max(0,k-n)}^{\min(n-1,k)} L_i(s)\cdot P_{k-i}(s) \equiv 0, \qquad k = 0, 1, \cdots, 2n-1.$$

This creates $2n$ equations and $4n$ unknowns. Rewrite the above equations in matrix form

$$\begin{pmatrix} P_0 & 0 & 0 & \cdots & 0 & 0 & 0 \\ P_1 & P_0 & 0 & \cdots & 0 & 0 & 0 \\ \vdots & \vdots & \vdots & \cdots & \vdots & \vdots & \vdots \\ P_n & P_{n-1} & P_{n-2} & \cdots & 0 & 0 & 0 \\ 0 & P_n & P_{n-1} & \cdots & 0 & 0 & 0 \\ \vdots & \vdots & \vdots & \cdots & \vdots & \vdots & \vdots \\ 0 & 0 & 0 & \cdots & 0 & P_n & P_{n-1} \\ 0 & 0 & 0 & \cdots & 0 & 0 & P_n \end{pmatrix} \begin{pmatrix} L_0 \\ L_1 \\ \vdots \\ L_{n-1} \end{pmatrix} = 0 \qquad (1)$$

or $CM(s)L(s) = 0$.

Define $M = \{(L_0, L_1, \cdots, L_{n-1}) : CM(s)L(s) = 0\}$. Then M is a module over $R[s]$, or a submodule of $R[s]^{4n}$. The geometric meaning of M is that M consists of all moving planes of degree at most $n-1$ in t and any degree in s that follow the surface $\mathbf{r}(s,t)$. In addition, the columns of $CM(s)$ also form a module, and M can be viewed as the syzygies of the columns of $CM(s)$. We now call the minimal generating set of M the μ-basis with regard to s of the surface $\mathbf{r}(s,t)$. In the following we expect to prove

1) the number of minimal generators is $2n$;
2) the sum of the degrees of the minimal generators in s is the implicit degree of the surface.

Lemma 4. *Let $f_1(s,t)$, $f_2(s,t)$, $\cdots$, $f_k(s,t)$ be a set of polynomials in s and t. If for an infinite number of values s_j of s, $f_i(s_j,t)$ have common root(s), then they have a common factor $d(s,t)$ involving t, i.e., $f_i(s,t) = d(s,t)h_i(s,t)$.*

Proof: Consider two planar curves $f_1(s,t) = 0$ and $f_2(s,t) = 0$. Since for an infinite number of values s_j of s, the polynomials $f_1(s_j,t)$, $f_2(s_j,t)$, $\cdots$, $f_k(s_j,t)$ have common roots, $f_1(s,t) = 0$ and $f_2(s,t) = 0$ have an infinite number of intersection points. This implies that the curves $f_1(s,t) = 0$ and $f_2(s,t) = 0$ have a common part, *i.e.*, $f_1(s,t)$ and $f_2(s,t)$ have a non-constant common factor. Let $d_1(s,t)$ be their greatest common factor. Then $f_1(s,t) = d_1(s,t)g_1(s,t)$ and $f_2(s,t) = d_1(s,t)g_2(s,t)$, and the curves $g_1(s,t) = 0$ and $g_2(s,t) = 0$ have only a finite number of intersection points. Therefore among all the intersection points of $f_1(s,t) = 0$ and $f_2(s,t) = 0$, there are still an infinite number of intersection points lying on the curve $d_1(s,t) = 0$. Notice that since these infinite intersection points have different values of s, $d_1(s,t)$ must involve t.

Next consider the planar curves $d_1(s,t) = 0$ and $f_3(s,t) = 0$. They have an infinite number of intersection points. Therefore $d_1(s,t)$ and $f_3(s,t)$ have a common factor. The greatest common factor of $d_1(s,t)$ and $f_3(s,t)$ is also a factor of $f_1(s,t)$ and $f_2(s,t)$. So the above process can be continued. The proof is completed by induction. □

Theorem 1. *The number of moving planes in the μ-basis with regard to s is $2n$ if and only if $X(s,t)$, $Y(s,t)$, $Z(s,t)$ and $W(s,t)$ have no common factor involving t.*

Proof: Let N be the module over $R[s]$ generated by the columns of the coefficient matrix $CM(s)$ in (1). Since the matrix $CM(s)$ has dimension $2n \times 4n$, by Lemma 2, the number of moving planes in the μ-basis with regard to s is $2n$ if and only if $\mu(N) = 4n - 2n = 2n$. And by Lemma 1, $\mu(N) \leq 2n$. So in the following we need only prove that $\mu(N) < 2n$ if and only if $X(s,t)$, $Y(s,t)$, $Z(s,t)$ and $W(s,t)$ have a common factor involving t.

Assume that $\mu(N) < 2n$. Then the columns of $CM(s)$ can be generated by fewer than $2n$ elements. Thus for any fixed value of s, any $2n \times 2n$ minor of $CM(s)$ vanishes. Now fix a value of s, and consider a space curve Γ_s:

$$(X(t), Y(t), Z(t), W(t)) = (X(s,t), Y(s,t), Z(s,t), W(s,t)).$$

Since the surface is degree $m \times n$, except for a finite number of values of s, the curve Γ_s has degree n. Denote $L(t) \in I_{1,n-1}$ by

$$L(t) = L_0 + L_1 t + \cdots + L_{n-1} t^{n-1}.$$

Then $L(t)$ satisfies

$$CM(s)(L_0, \cdots, L_{n-1})^T = 0.$$

Thus the number of linearly independent solutions of $L(t)$ is given by $4n - rank(CM(s)) > 4n - 2n = 2n$. On the other hand, if $X(t)$, $Y(t)$, $Z(t)$ and $W(t)$ have no common factor, i.e. $X(s,t)$, $Y(s,t)$, $Z(s,t)$ and $W(s,t)$ have no common factor involving t, then the curve Γ_s has degree n. By Lemma 3, there exist $p_1(t) \in I_{1,\mu_1}$, $p_2(t) \in I_{1,\mu_2}$ and $p_3(t) \in I_{1,\mu_3}$ such that $\mu_1 + \mu_2 + \mu_3 = n$ and any element $q \in I_{1,m}$ can be uniquely written in the form $q = a_1 p_1 + a_2 p_2 + a_3 p_3$. Thus, the linearly independent solutions of $L(t) \in I_{1,n-1}$ can also be represented by p_1, p_2 and p_3. However, the number of linearly independent elements of degrees at most $n-1$ which are formed from p_1, p_2 and p_3 is $(n-\mu_1) + (n-\mu_2) + (n-\mu_3) = 2n$. This leads to a contradiction. Therefore, $X(t)$, $Y(t)$, $Z(t)$ and $W(t)$ have a common factor. Since s can be arbitrarily chosen except for a finite number of values, by Lemma 4, $X(s,t)$, $Y(s,t)$, $Z(s,t)$ and $W(s,t)$ must have a common factor involving t.

Conversely, if $X(s,t)$, $Y(s,t)$, $Z(s,t)$ and $W(s,t)$ have a common factor involving t, from the above analysis, we know that for any value of s, $rank(CM(s)) < 2n$, i.e. any $2n \times 2n$ minor of $CM(s)$ vanishes. Thus $\mu(N) < 2n$. The proof is completed. □

Conjecture. *If $X(s,t)$, $Y(s,t)$, $Z(s,t)$ and $W(s,t)$ have no common factors and the map sending (s,t) to $\mathbf{r}(s,t) = (X(s,t), Y(s,t), Z(s,t), W(s,t))$ is generically one-to-one, then the sum of degrees of the μ-basis with regard to s is equal to the implicit degree of the rational surface.*

While we believe that we may use the Hilbert polynomial to prove this conjecture; the strict proof needs to be worked out. David Cox proved the special case of the conjecture in which the degree $m \times n$ rational surface $\mathbf{r}(s,t)$ has no basepoints [3]. Thus the sum of degrees of the μ-basis with regard to s is equal to $2mn$ in such a case.

One application of the proposed μ-basis is that we can multiply the μ-basis by suitable s^k to obtain linearly independent moving planes. These moving planes may form a square matrix and the implicit equation may be obtained by computing the determinant of this matrix. For example, consider a degree $m \times n$ rational surface $\mathbf{r}(s,t)$ without base points. The μ-basis has $2n$ elements, whose degrees in s are μ_1, $\cdots$, μ_{2n} with $\sum_{k=0}^{2n} \mu_k = 2mn$. Thus multiplying the k-th element p_k by 1, s, $\cdots$, $s^{2m-1-\mu_k}$ leads to $2mn$ moving planes following the rational surface. The blending functions for these moving planes are $s^i t^j$, $i = 0, \cdots, 2m-1$, $j = 0, \cdots, n-1$. Hence these $2mn$ moving planes can generate a $2mn \times 2mn$ square matrix. Using a different idea, Zhang, Goldman and Chionh proved that this approch to implicitization works in the generic case [10].

§5. Examples

We present several concrete examples of rational surfaces and their μ-bases.

Example 1. (*Bicubic surface without base points*) Consider a bicubic surface

$$\begin{aligned} X(s,t) &= 3600t^3s^3 - 5400t^2s^3 + 3700s^3 - 5400t^3s^2 + 8100t^2s^2 \\ &\quad - 6300s^2 + 3000t^3s - 4500t^2s + 4200s + 1700 \\ Y(s,t) &= -2460t^2s^3 + 2460ts^3 + 3690t^2s^2 - 3690ts^2 - 1980t^2 + 1980t \\ Z(s,t) &= -1800s^3 - 1500t^3s^3 + 2250t^2s^3 + 1800t^3s^2 - 2700t^2s^2 \\ &\quad + 2925s^2 + 1350t^3s - 2025t^2s + 675s - 1650t^3 + 2475t^2 + 600 \\ W(s,t) &= 1000. \end{aligned}$$

The μ-basis consists of 6 moving planes whose degrees in s are all 3. Moreover, if we multiply these moving planes by 1, s and s^2, we obtain 18 linearly independent moving planes of degree 5 in s and degree 2 in t, which form a matrix of dimension 18. Its determinant gives the implicit equation of the surface. (Since the the expressions are complicated, we leave them out here.)

Example 2. (*Bicubic surface with base points*) Consider a bicubic surface

$$\begin{aligned} X(s,t) &= -82t^3s^3 - 96t^2s^3 + 438ts^3 + 150t^3s^2 + 180t^2s^2 - 810ts^2 \\ &\quad - 75t^3s - 90t^2s + 405ts \\ Y(s,t) &= 82t^3s^3 - 342t^2s^3 + 260s^3 - 150t^3s^2 + 630t^2s^2 - 480s^2 \\ &\quad + 75t^3s - 315t^2s + 240s \\ Z(s,t) &= 45s^3 - 90s^2 + 315 \\ W(s,t) &= 100. \end{aligned}$$

This surface has 5 base points, and the implicit degree is 13. The μ-basis consists of 6 moving planes whose degrees in s are respectively 2, 2, 2, 2, 2 and 3. The sum of the degrees is still equal to the implicit degree of the surface. However, it is now not obvious how to produce 13 moving planes from the μ-basis so that they form a matrix whose determinant gives the implicit equation of the surface.

Example 3. (*Surface with common factors in the representation*) A degree 3×2 rational surface is defined in homogeneous form

$$\begin{aligned} X(s,t) &= t + ts^2 + t^2 + 2t^2s + t^2s^2 \\ Y(s,t) &= 1 + s^2 + t + 2ts + ts^2 \\ Z(s,t) &= 2t + 2ts^2 + t^2 + 2t^2s + t^2s^2 + 1 + s^2 + 2ts \\ W(s,t) &= 1 + s^2 + t + 3ts + ts^2 + ts^3 + t^2s + 2t^2s^2 + t^2s^3. \end{aligned}$$

The μ-base with regard to s are the column vectors in the matrix

$$\begin{pmatrix} -1 & -1 & -1 & s & 1 \\ -1 & 0 & 0 & 1 & 0 \\ 1 & 0 & 0 & 0 & 0 \\ 0 & 0 & 0 & -1 & 0 \\ 0 & 0 & -1 & 0 & s \\ 0 & 1 & 0 & 0 & 0 \\ 0 & 0 & 1 & 0 & 0 \\ 0 & 0 & 0 & 0 & -1 \end{pmatrix}.$$

Note that the number of moving planes in the μ-basis has 5 (*not 4*) elements. This is because the surface representation has a common factor $1 + s^2 + t + 2ts + ts^2$.

§6. Future work

The preceding sections report some initial results in defining the μ-basis of a surface. There are many problems that need to be explored further:

1) how to efficiently compute the μ-basis of a rational surface,
2) how to generate approriate linearly independent moving planes from the μ-basis such that they can form a square matrix,
3) when do the linearly independent moving planes form a matrix whose determinant does not vanish?
4) how to prove the degree conjecture (i.e. the sum of the degrees of the μ-basis is the same as the implicit degree of the surface, even if there exist base points).

Acknowledgments. This work was supported in part by NSF grant CCR-9712407. Jianmin Zheng is partially supported by National Natural Science Foundation of China (69973042) and 973 Project on Mathematical Mechanics (G1998030600).

References

1. Cox, D., J. Little, and D. O'Shea, *Using Algebraic Geometry*, Graduate Texts in Mathematics, Vol. 185, Springer-Verlag, New York, 1998.
2. Cox, D., T. Sederberg, and F. Chen, The moving line ideal basis of planar rational curves, Comput. Aided Geom. Design **15** (1998), 803–827.
3. Cox, D., Private communication, 1999.
4. Goldman, R., T. Sederberg, and D. Anderson, Vector elimination: A technique for the implicitization, inversion and intersection of planar parametric rational polynomial curves, Comput. Aided Geom. Design **1** (1984), 327–356.

5. Hoffmann, C., Implicit curves and surfaces in CAGD, IEEE Computer Graphics & Applications **13** (1993), 79–88.
6. De Montaudouin, Y., and W. Tiller, The Cayley method in computer aided geometric design, Comput. Aided Geom. Design **1** (1984), 309–326.
7. Sederberg, T., and F. Chen, (1995), Implicitization using moving curves and surfaces, SIGGRAPH 95 Conference Proceedings, Annual Conference Series, Addison Wesley, 1995, 301–308.
8. Sederberg, T., R. Goldman, and H. Du, Implicitizing rational curves by the method of moving algebraic curves, J. Symbolic Computation **23** (1997), 153–175.
9. Sederberg, T., T. Saito, D. Qi, and K. Klimaszewski, Curve implicitization using moving lines, Comput. Aided Geom. Design **11** (1994), 687–706.
10. Zhang, M., R. Goldman, and E. Chionh, Efficient implicitization of rational surfaces by moving planes, in the Proceedings of ASCM'2000, Chiang Mai, Thailand, December 17-21, 2000.

Thomas W. Sederberg
Department of Computer Science
Brigham Young University
Provo, UT 84602, USA
`tom@cs.byu.edu`

Jianmin Zheng
Institute of Computer Images and Graphics
Department of Mathematics
Zhejiang University
Hangzhou 310027, People's Republic of China
`jm_zheng@sina.com`

A Strategy for the Construction of Piecewise Linear Prewavelets over Type-1 Triangulations in any Space Dimension

Erich Suter

Abstract. We investigate candidates for basis functions for piecewise linear wavelet spaces over Type-1 triangulations of uniformly distributed data in any space dimension.

§1. Introduction

In order to find candidates for basis functions for piecewise linear wavelet spaces over Type-1 triangulations of uniformly distributed data in any space dimension, we propose to use so-called semi-prewavelets for the construction of piecewise linear prewavelets of continuity C^0 with small support. This method is applicable to both unbounded and bounded triangulations of type-1. As a foundation for this, a scheme for stable subdivision of the simplices in a uniform n-dimensional triangulation is described. This is not trivial, as each n-simplex can be subdivided a number of ways, especially in higher dimensions. The proposed approach results in basis functions for the wavelet spaces in the 1- and 2-dimensional cases, but it is not yet known if this is the case also in higher dimensions. This is because three known methods for proving this basis property in similar situations give inconclusive results in higher space dimensions.

The article is based on the papers [9,11] concerning subdivision of an n-dimensional cube into simplices, [1,12] for refinements of tetrahedralizations, [7] for subdivision of n-dimensional simplices into subsimplices, [8] for the construction of a piecewise linear prewavelet basis over an unbounded Type-1 triangulation in $\mathbb{R}^2$, and on [3–6] which describe the construction of prewavelets that constitute a basis for a piecewise linear wavelet space over an arbitrary triangulation in $\mathbb{R}^2$, using semi-prewavelets. In $\mathbb{R}^2$ the basis functions found in this paper are special cases of the prewavelets found in [3–5], and also correspond to one of the prewavelets in [8].

Mathematical Methods for Curves and Surfaces: Oslo 2000
Tom Lyche and Larry L. Schumaker (eds.), pp. 477–486.

ISBN 0-8265-1378-6

§2. Type-1 Triangulations

We describe a method for triangulating uniformly distributed data in any space dimension. It is not a trivial task to split an n-dimensional cube into n-dimensional simplices (n-simplices or simplices for short), and, as in the refinement step explained below, to split each simplex into subsimplices so that the refined triangulation has the same structure as the coarse one.

Let $\mathbf{U}_n^* = \{e_1, e_2, \ldots, e_n\}$ denote the standard orthonormal basis in $\mathbb{R}^n$, and let $\mathcal{P}(\mathbf{U}_n^*)$ be the power set of these vectors, namely the set of all subsets of $\mathbf{U}_n^*$, including the empty set and $\mathbf{U}_n^*$ itself. Furthermore, let the elements in the set $\mathbf{U}_n$ be obtained by taking the vector sum of the elements in each subset of $\mathcal{P}(\mathbf{U}_n^*)$, omitting the empty set. For example, in $\mathbb{R}^3$ we have that $\mathbf{U}_3^* = \{e_1, e_2, e_3\}$, which gives $\mathcal{P}(\mathbf{U}_3^*) = \{\emptyset, \{e_1\}, \{e_2\}, \{e_3\}, \{e_1, e_2\}, \{e_1, e_3\}, \{e_2, e_3\}, \{e_1, e_2, e_3\}\}$, and $\mathbf{U}_3 = \{e_1, e_2, e_3, e_1+e_2, e_1+e_3, e_2+e_3, e_1+e_2+e_3\}$.

Let a Type-1 triangulation $\mathcal{T}^j$ in $\mathbb{R}^n$ be a triangulation induced by the vectors in $2^{-j}\mathbf{U}_n$, namely such that each edge $[v, v^*]$ in the triangulation is parallel to a vector from $2^{-j}\mathbf{U}_n$, where $j \in \mathbb{Z}$ indicates the level of refinement. The edges in $\mathcal{T}^j$ are denoted E^j. The vertices V^j of each triangulation $\mathcal{T}^j$ are thus given by a subset of $2^{-j}\mathbb{Z}^n$. This triangulation can be bounded or unbounded, and in $\mathbb{R}^2$ it is a 3-directional mesh.

Furthermore, we require that the Type-1 triangulation sketched above fulfill the following properties:

- The intersection $T_1 \cap T_2$ between any pair of n-simplices T_1 and T_2 from $\mathcal{T}^j$ is either empty or is itself a simplex of dimension k, for $k \in \{0, \ldots, n-1\}$.
- Each facet ($(n-1)$-dimensional simplex) of an n-simplex is either on the boundary of $\mathcal{T}^j$, or else is a common facet of exactly two n-simplices.
- The intersection between any n-simplex in $\mathcal{T}^j$ and at least one other n-simplex from $\mathcal{T}^j$ is a facet.

§3. Refinement of Type-1 Triangulations

The dyadic refinement of a triangulation in $\mathbb{R}^2$ is described by connecting edge midpoints with new edges. However, in space dimensions larger than two, this method leaves us with many possibilities for choosing edges for each simplex in the triangulation, and several of these choices might result in valid triangulations. The main point of letting the edges in E^j be induced by $\mathbf{U}_n$ is that we thereby make a certain selection of the edges allowed in $\mathcal{T}^j$, and based on this selection the triangulation will be unique for a given set of vertices V^j.

We proceed by describing how a stable subdivision of the simplices in the triangulation can be performed. By selecting some lowest level triangulation $\mathcal{T}^L$ in $\mathbb{R}^n$, where $L \in \mathbb{Z}$, that has vertices V^L from $2^{-L}\mathbb{Z}^n$ and edges induced by the set $2^{-L}\mathbf{U}_n$, we can refine it by inserting vertices dyadically, i.e. on the edge midpoints, and let the refined triangulation $\mathcal{T}^{L+1}$ only have edges as given by the vectors in $2^{-(L+1)}\mathbf{U}_n$. Then $\mathcal{T}^{L+1}$ will have vertices V^{L+1} from $2^{-(L+1)}\mathbb{Z}^n$. This process can be repeated, and all the finer triangulations will have the same overall structure as the coarsest one. Also let $\Omega \in \mathbb{R}^n$ be the convex hull of all n-simplices in $\mathcal{T}^L$. Then we have that the convex hull of the simplices of each triangulation $\mathcal{T}^j$ obtained by refining $\mathcal{T}^L$ equals Ω, for $j = L+1, L+2, \ldots$. Furthermore, each n-simplex $T' \in \mathcal{T}^{j+1}$ is contained in exactly one n-simplex $T \in \mathcal{T}^j$, and any vertex of T' is either a vertex or an edge midpoint of T.

Note that in $\mathbb{R}^1$ and $\mathbb{R}^2$ the simplices in a triangulation will all be of equal shape, but in higher dimensions this is not the case. For example, in $\mathbb{R}^3$ we obtain two n-simplices of different shape, which can be seen as mirrored versions of each other, and for $n > 3$ we obtain even more different shapes. However, all simplices in a triangulation on a given level will have equal volumes.

§4. Nested Piecewise Linear Function Spaces

Given data values $f_v \in \mathbb{R}$ for $v \in V^j$, there is a unique function $f : \Omega \to \mathbb{R}$ which is linear on each simplex in $\mathcal{T}^j$ and interpolates the data: $f(v) = f_v, v \in V^j$. The set of all such piecewise linear functions f constitute a linear space S^j with dimension $|V^j|$.

For each $v \in V^j$, let $\phi_v^j : \Omega \to \mathbb{R}$ be the unique 'hat' function in S^j such that for $w \in V^j$,

$$\phi_v^j(w) = \begin{cases} 1, & \text{if } w = v, \\ 0, & \text{otherwise.} \end{cases}$$

It is well known that the set of functions $\{\phi_v^j\}_{v \in V^j}$ is a basis for the space S^j. The support of ϕ_v^j is the union of all simplices which contain v.

Let V_v^j denote the set of neighbours in V^j of a vertex v in V^j. The hat function centred in v on a level $j-1$ may be expressed in terms of hat functions for the vertices in V_v^j on the finer level j,

$$\phi_v^{j-1} = \sum_{w \in V^j} \phi_v^{j-1}(w) \phi_w^j = \phi_v^j + \frac{1}{2} \sum_{w \in V_v^j} \phi_w^j, \quad v \in V^{j-1}.$$

It follows that S^{j-1} is a subspace of S^j, and therefore we obtain a nested sequence of spaces

$$S^L \subset S^{L+1} \subset S^{L+2} \subset \cdots .$$

Let $\langle \cdot, \cdot \rangle$ be the following inner product, defined for continuous real-valued functions over Ω:

$$\langle f, g \rangle = \int_\Omega f(x) g(x) \, dx, \quad f, g \in C(\Omega), \tag{1}$$

where $C(\Omega)$ denotes the set of all functions with continuity C^0 over Ω. We can make use of an unweighted inner product because all n-simplices in a Type-1 triangulation on a certain level j are of equal size, which is in contrast to the case of arbitrary triangulations, where typically the volumes of the simplices must be taken into consideration.

Let W^{j-1} denote the orthogonal complement of the coarse space S^{j-1} relative to the fine space S^j, so that $S^j = S^{j-1} \oplus W^{j-1}$. The set of 'new' vertices in V^j relative to V^{j-1} is denoted $V_*^j = V^j \setminus V^{j-1}$. The dimension of W^{j-1} is $|V^j| - |V^{j-1}| = |V_*^j| = |E^{j-1}|$.

§5. Prewavelets

The subspace W^{j-1} of S^j is a wavelet space, and any nonzero element in this space is called a prewavelet. With each new vertex u in V_*^j, we associate a prewavelet ψ_u^{j-1} in W_u^{j-1}, a subspace of W^{j-1} consisting of prewavelets with certain small support around the vertex u, and in the following we will explore the set $\{\psi_u^{j-1}\}_{u\in V_*^j}$ of prewavelets. Having in mind how a function $f \in S^j$ can be written as a linear combination of the functions ϕ_w^j, namely as $f(x) = \sum_{v\in V^j} f(v)\phi_v^j(x), x \in \Omega$, we have that $\psi_u^{j-1} \in S^j$, for $u \in V_*^j$, can be written as a linear combination of the same basis functions ϕ_w^j,

$$\psi_u^{j-1}(x) = \sum_{w\in V^j} q_{w,u}\phi_w^j(x), \quad x \in \Omega, \tag{2}$$

where the coefficients of ψ_u^{j-1} are given by $q_{w,u} = \psi_u^{j-1}(w)$, for $w \in V^j$.

5.1. Inner Products of Functions from S^{j-1} and W^{j-1}

Since any prewavelet in S^j has a unique representation as in (2), it is clear that the element ψ_u^{j-1} of S^j belongs to W^{j-1} if and only if

$$\langle \phi_v^{j-1}, \psi_u^{j-1}\rangle = \sum_{w\in V^j} \langle \phi_v^{j-1}, \phi_w^j\rangle\, q_{w,u} = 0, \quad \text{for all } v \in V^{j-1}.$$

Using the inner product from (1), we see that we need to determine the inner product of any pair of basis functions ϕ_v^{j-1} in S^{j-1} and ϕ_w^j in S^j in order to determine the prewavelet coefficients $q_{w,u}$. Due to the local support of these functions, a lot of these inner products are zero, and in fact $\langle \phi_v^{j-1}, \phi_w^j\rangle \neq 0$ if and only if the vertices v and w belong to some common simplex T in $\mathcal{T}^{j-1}$.

In order to calculate the inner product over one simplex, we recall that the integral of every quadratic Bernstein polynomial over a simplex is the same, only depending on the dimension of the simplex. A straightforward calculation shows that the formula for the integral of two linear polynomials $f, g : \mathbb{R}^n \to \mathbb{R}$ over an n-simplex T is thus given by

$$\begin{aligned}&\int_T f(x)g(x)\,dx \\ &\quad = \frac{V(T)}{2\cdot\binom{2+n}{2}}(f_0g_0 + \cdots + f_ng_n + (f_0 + \cdots + f_n)(g_0 + \cdots + g_n)),\end{aligned} \tag{3}$$

where $f_i = f(x_i)$ and $g_i = g(x_i)$ for $i = 0, 1, \ldots, n$, and where x_i are the vertices of the simplex. The volume of the simplex is denoted by $V(T)$.

In order to apply (3) when calculating inner products, we define two sets of simplices in $\mathcal{T}^{j-1}$ and $\mathcal{T}^j$, respectively. First for $v \in V^{j-1}$ and $w \in V^j$, let $T(v,w) = \{T \in \mathcal{T}^{j-1} \mid v, w \in T\}$ denote the set of simplices in $\mathcal{T}^{j-1}$ that contain both v and w, and therefore over which $\langle \phi_v^{j-1}, \phi_w^j \rangle \neq 0$, and secondly, for each simplex $T \in T(v,w)$ let $T^*(v,w,T) = \{T' \in \mathcal{T}^j \mid T' \subset T \in T(v,w),\ w \in T'\}$ be the subsimplices T' of T that contain w.

Now we see how these two sets come into play, as $T(v,w)$ represents the simplices in $\mathcal{T}^{j-1}$ where ϕ_v^{j-1} is nonzero, and $T^*(v,w,T)$ denotes the subsimplices in each simplex T from $T(v,w)$ where ϕ_w^j is nonzero. We use these sets in the intermediate inner product formula

$$\langle \phi_v^{j-1}, \phi_w^j \rangle = \sum_{T \in T(v,w)} \sum_{T' \in T^*(v,w,T)} \int_{T'} \phi_v^{j-1}(x) \phi_w^j(x)\, dx. \tag{4}$$

By substituting (3) in (4) and letting $f = \phi^{j-1}$ and $g = \phi^j$, we arrive at

$$\begin{aligned} \langle \phi_v^{j-1}, \phi_w^j \rangle &= \sum_{T \in T(v,w)} \frac{V(T')}{2 \cdot \binom{2+n}{2}} \sum_{T' \in T^*(v,w,T)} K(T') \\ &= \frac{V(T)}{2^{n+1}\binom{2+n}{2}} \sum_{T \in T(v,w)} \sum_{T' \in T^*(v,w,T)} K(T'), \end{aligned} \tag{5}$$

where $K(T') = (\phi_0^{j-1}\phi_0^j + \cdots + \phi_n^{j-1}\phi_n^j + (\phi_0^{j-1} + \cdots + \phi_n^{j-1})(\phi_0^j + \cdots + \phi_n^j))$ is evaluated in the $n+1$ vertices of each $T' \in T^*(v,w,T) \subset \mathcal{T}^j$, so that we associate function values $\phi_i^{j-1} = \phi_v^{j-1}(x_i)$ and $\phi_i^j = \phi_w^j(x_i)$, for $i = 0, \ldots, n$, where $x_i \in V^j$ are the vertices of T'. Furthermore, we have used the fact that the relationship between the volumes of the simplices on the two different refinement levels is given by $V(T') = 2^{-n}V(T)$.

5.2. Semi-Prewavelets

There exist many possible choices of prewavelets with small support around u, especially in higher dimensions, but we constrain ourselves to discussing a method for selecting only one prewavelet. This method is based on the concept of semi-prewavelets, discussed by Floater and Quak in [4–6]. The strategy is to consider two simpler functions with smaller support than the prewavelet and add these together. The result is a prewavelet that is a candidate for a basis function in W^{j-1}.

Recall that V_v^j denotes the set of neighbours in V^j of a vertex v in V^j. Let $v \in V^{j-1}$ be a vertex from the coarse triangulation $\mathcal{T}^{j-1}$, and let $u \in V^j$ be one of its neighbours in the fine triangulation $\mathcal{T}^j$, so that $u \in V_v^j$. This means that u is the midpoint of some edge $[v, v^*] \in E^{j-1}$, where $v^* \in V_v^{j-1}$.

Definition 1. *A function $\sigma_{v,u}^{j-1}$ in S^j of the following form, with all coefficients being real numbers,*

$$\sigma_{v,u}^{j-1}(x) = B_v \phi_v^j(x) + \sum_{w \in V_v^j} B_w \phi_w^j(x)$$

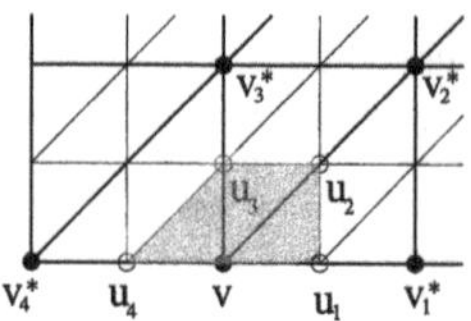

Fig. 1. Four semi-prewavelets centred in v.

is called a semi-prewavelet *if, for any coarse vertex* $a \in V^{j-1}$ *and for some* $\gamma \neq 0$,

$$\langle \phi_a^{j-1}, \sigma_{v,u}^{j-1} \rangle = \begin{cases} -2^{-nj}\gamma & \text{if } a = v, \\ 2^{-nj}\gamma & \text{if } a = v^*, \\ 0 & \text{otherwise.} \end{cases}$$

The semi-prewavelet $\sigma_{v,u}^{j-1}$ is orthogonal to all but two coarse hat functions in S^{j-1}, namely ϕ_v^{j-1} and $\phi_{v^*}^{j-1}$. Observe that since u is also a neighbour of v^* in V^j, there is a corresponding semi-prewavelet $\sigma_{v^*,u}^{j-1}$. A prewavelet can now be constructed by adding the two semi-prewavelets, so that for a fixed γ, the function $\psi_u^{j-1} = \sigma_{v,u}^{j-1} + \sigma_{v^*,u}^{j-1}$ is a prewavelet. This is clear by Definition 1, as we see that $\langle \phi_a^{j-1}, \psi_u^{j-1} \rangle = \langle \phi_a^{j-1}, \sigma_{v,u}^{j-1} \rangle + \langle \phi_a^{j-1}, \sigma_{v^*,u}^{j-1} \rangle = 0,\ a \in V^{j-1}$, and therefore ψ_u^{j-1} belongs to the wavelet space W^{j-1}.

We think of a semi-prewavelet as 'centred' at v or v^*, respectively, though it also depends on u. It is clear that the prewavelets $\{\psi_u^{j-1}\}_{u \in V_*^j}$ have indeed a small local support.

5.3. Semi-Prewavelet Coefficients

Semi-prewavelets are known to exist in $\mathbb{R}^1$ and $\mathbb{R}^2$, but their existence remains to be proven in higher dimensions. However, numerical tests show that they exist in $\mathbb{R}^3$ and $\mathbb{R}^4$, both when

- $v \in V^{j-1}$ is an interior vertex,
- $v \in V^{j-1}$ is any boundary vertex and Ω is of 'rectangular' shape.

Note that we have not checked the existence of semi-prewavelets for other boundary configurations.

Let $V_\sigma(v,u) = \{v\} \cup V_v^j$ be the set of neighbours of any $v \in V^{j-1}$ in V^j, but including v itself. Furthermore, let $C_\sigma(v,u) = \{v\} \cup V_v^{j-1}$ be the set of neighbours in V^{j-1} of v, also including v. This means that $|V_\sigma(v,u)| = |C_\sigma(v,u)|$. We also have corresponding sets for v^*.

We can now express the semi-prewavelet $\sigma_{v,u}^{j-1}$ in terms of its coefficients $q_{w,u}^\sigma$ as

$$\sigma_{v,u}^{j-1}(x) = \sum_{w \in V_\sigma(v,u)} q_{w,u}^\sigma \phi_w^j(x).$$

Our task is to find these coefficients. We have the relation

$$\sum_{w \in V_\sigma(v,u)} \langle \phi_a^{j-1}, \phi_w^j \rangle \, q_{w,u}^\sigma = y_a, \quad \text{for all } a \in C_\sigma(v,u),$$

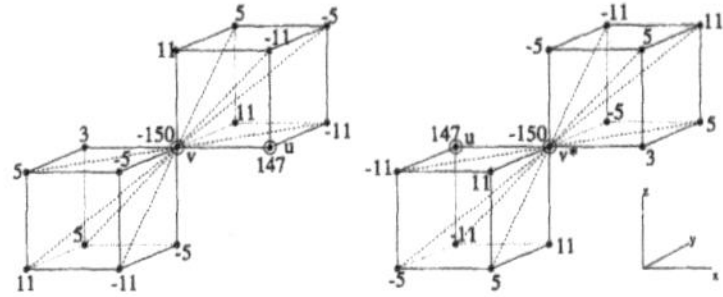

Fig. 2. The masks for the interior semi-prewavelets $\sigma^{-1}_{v_{222},v_{322}}$ and $\sigma^{-1}_{v_{422},v_{322}}$.

where the elements $y_a \in \mathbb{R}$ are set up in correspondence with Definition 1. Each set $V_\sigma(v,u)$ and $C_\sigma(v,u)$ is uniquely related to each specific semi-prewavelet $\sigma^{j-1}_{v,u}$.

Furthermore we can construct the quadratic inner product matrix $A_\sigma = (\langle \phi^{j-1}_a, \phi^j_w \rangle)$ with entries for $a \in C_\sigma(v,u)$ and $w \in V_\sigma(v,u)$, where $C_\sigma(v,u)$ and $V_\sigma(v,u)$ can be arbitrarily ordered. This presents us with the linear system

$$A_\sigma q_\sigma = y, \tag{6}$$

where q_σ is the vector of the semi-prewavelet coefficients $q^\sigma_{w,u}$, and y is the vector of the values y_a. If A_σ is nonsingular, then the system is uniquely solvable for q_σ. By finding the semi-prewavelet coefficients also for the corresponding $\sigma^{j-1}_{v^*,u}$ centred in v^*, we obtain the prewavelet ψ^{j-1}_u by adding the two semi-prewavelets.

Note that the inner product matrix A_σ can be used for finding the coefficients of all semi-prewavelets centred in v. This is done by solving (6) for all $|C_\sigma(v,u)| - 1$ different vectors y that arise when we choose v^* from the different vertices in $C_\sigma(v,u) \setminus v$. An example for a Type-1 triangulation in $\mathbb{R}^2$ is shown in Figure 1 where v is a boundary vertex and v^*_i, for $i = 1, \ldots, 4$, are the vertices in V^{j-1}_v. The grey area in the figure is the support of the fine hat function ϕ^j centred in v.

Due to the regular structure of the Type-1 triangulation it is clear that we can only obtain a limited number of masks for the semi-prewavelets, depending on the space dimension and the structure of the boundary of $\mathcal{T}^{j-1}$.

§6. Prewavelets in $\mathbb{R}^3$

As an example, we will describe a situation in $\mathbb{R}^3$. Note that the vertices are labelled so that $v_{265} = (2,6,5)$, etc. We start by assuming that the vertex $v = v_{222} \in V^{-1}$ is an interior vertcx, and that it therefore has a full set of neighbours in $\mathcal{T}^{-1}$ given by $V^{-1}_v = \{v_{422}, v_{242}, v_{224}, v_{442}, v_{424}, v_{244}, v_{444}, v_{022}, v_{202}, v_{220}, v_{002}, v_{020}, v_{200}, v_{000}\}$, and correspondingly in the refined tetrahedralisation $\mathcal{T}^0$, namely $V^0_v = \{v_{322}, v_{232}, v_{223}, v_{332}, v_{323}, v_{233}, v_{333}, v_{122}, v_{212}, v_{221}, v_{112}, v_{121}, v_{211}, v_{111}\}$. Furthermore, we have the sets related to all semi-prewavelets centred in v, namely $C_\sigma(v,u) = V^{-1}_v \cup v_{222}$ and $V_\sigma(v,u) = V^0_v \cup v_{222}$, where for now the orderings are as indicated, with v as the last member in both cases.

We can now find the inner product matrix A_σ by evaluating $\langle \phi^{j-1}_a, \phi^j_w \rangle$ in all coarse vertices $a \in C_\sigma(v,u)$ and all fine vertices $w \in V_\sigma(v,u)$ and furthermore using the fact that $V(T) = 4/3$ for any $T \in \mathcal{T}^{-1}$. This matrix is nonsingular, and is shown below. Note again that we have evaluated A_σ for many cases, including all boundary cases in a Type-1 triangulation of a rectangular domain, and interior and all boundary cases for a Type-1 triangulation of a rectangular domain in $\mathbb{R}^4$, and in each case A_σ was nonsingular.

$$A_\sigma = \begin{pmatrix}
3 & & & \frac{2}{3} & \frac{2}{3} & & \frac{1}{3} & & \frac{1}{3} & \frac{1}{3} & & & \frac{2}{3} & & \frac{1}{5} \\
& 3 & & \frac{2}{3} & & \frac{2}{3} & \frac{1}{3} & \frac{1}{3} & & \frac{1}{3} & & \frac{2}{3} & & & \frac{1}{5} \\
& & 3 & & \frac{2}{3} & \frac{2}{3} & \frac{1}{3} & \frac{1}{3} & \frac{1}{3} & & \frac{2}{3} & & & & \frac{1}{5} \\
\frac{1}{3} & \frac{1}{3} & & \frac{8}{3} & & & \frac{1}{3} & & & \frac{1}{3} & & & & & \frac{2}{15} \\
\frac{1}{3} & & \frac{1}{3} & & \frac{8}{3} & & \frac{1}{3} & & \frac{1}{3} & & & & & & \frac{2}{15} \\
& \frac{1}{3} & \frac{1}{3} & & & \frac{8}{3} & \frac{1}{3} & \frac{1}{3} & & & & & & & \frac{2}{15} \\
\frac{1}{3} & \frac{1}{3} & \frac{1}{3} & \frac{2}{3} & \frac{2}{3} & \frac{2}{3} & 3 & & & & & & & & \frac{1}{5} \\
& \frac{1}{3} & \frac{1}{3} & & & \frac{2}{3} & & 3 & & & \frac{2}{3} & \frac{2}{3} & & \frac{1}{3} & \frac{1}{5} \\
\frac{1}{3} & & \frac{1}{3} & & \frac{2}{3} & & & & 3 & & \frac{2}{3} & & \frac{2}{3} & \frac{1}{3} & \frac{1}{5} \\
\frac{1}{3} & \frac{1}{3} & & \frac{2}{3} & & & & & & 3 & & \frac{2}{3} & \frac{2}{3} & \frac{1}{3} & \frac{1}{5} \\
& & \frac{1}{3} & & & & & \frac{1}{3} & \frac{1}{3} & & \frac{8}{3} & & & \frac{1}{3} & \frac{2}{15} \\
& \frac{1}{3} & & & & & & \frac{1}{3} & & \frac{1}{3} & & \frac{8}{3} & & \frac{1}{3} & \frac{2}{15} \\
\frac{1}{3} & & & & & & & & \frac{1}{3} & \frac{1}{3} & & & \frac{8}{3} & \frac{1}{3} & \frac{2}{15} \\
& & & & & & & \frac{1}{3} & \frac{1}{3} & \frac{1}{3} & \frac{2}{3} & \frac{2}{3} & \frac{2}{3} & 3 & \frac{1}{5} \\
3 & 3 & 3 & \frac{8}{3} & \frac{8}{3} & \frac{8}{3} & 3 & 3 & 3 & 3 & \frac{8}{3} & \frac{8}{3} & \frac{8}{3} & 3 & \frac{28}{5}
\end{pmatrix}$$

We now fix $\gamma = 384$. By selecting v^* to be for instance $v^* = v_{422} \in C_\sigma(v,u)$, so that $u = v_{322} \in V_\sigma(v,u)$, we obtain $y = (384, 0, 0, 0, 0, 0, 0, 0, 0, 0, 0, 0, 0, 0, -384)^T$, so that we can solve (6) for this case. We find the semi-prewavelet coefficient vector q_σ, giving the coefficients for $\sigma_{v,u}^{-1} = \sigma_{v_{222},v_{322}}^{-1}$, to be $q_\sigma = (147, 11, 11, -11, -11, 5, -5, 3, -5, -5, 5, 5, -11, 11, -150)^T$, which is the mask displayed to the left in Figure 2. Here only the fine vertices in $V_\sigma(v,u)$ are shown.

$$\begin{pmatrix}
147 & 11 & 11 & -11 & -11 & 5 & -5 & 3 & -5 & -5 & 5 & 5 & -11 & 11 & -150 \\
11 & 147 & 11 & -11 & 5 & -11 & -5 & -5 & 3 & -5 & 5 & -11 & 5 & 11 & -150 \\
11 & 11 & 147 & 5 & -11 & -11 & -5 & -5 & -5 & 3 & -11 & 5 & 5 & 11 & -150 \\
-28 & -28 & 4 & 164 & 12 & 12 & -28 & 4 & 4 & -28 & 4 & 12 & 12 & 4 & -120 \\
-28 & 4 & -28 & 12 & 164 & 12 & -28 & 4 & -28 & 4 & 12 & 4 & 12 & 4 & -120 \\
4 & -28 & -28 & 12 & 12 & 164 & -28 & -28 & 4 & 4 & 12 & 12 & 4 & 4 & -120 \\
-5 & -5 & -5 & -11 & -11 & -11 & 147 & 11 & 11 & 11 & 5 & 5 & 5 & 3 & -150 \\
3 & -5 & -5 & 5 & 5 & -11 & 11 & 147 & 11 & 11 & -11 & -11 & 5 & -5 & -150 \\
-5 & 3 & -5 & 5 & -11 & 5 & 11 & 11 & 147 & 11 & -11 & 5 & -11 & -5 & -150 \\
-5 & -5 & 3 & -11 & 5 & 5 & 11 & 11 & 11 & 147 & 5 & -11 & -11 & -5 & -150 \\
4 & 4 & -28 & 4 & 12 & 12 & 4 & -28 & -28 & 4 & 164 & 12 & 12 & -28 & -120 \\
4 & -28 & 4 & 12 & 4 & 12 & 4 & -28 & 4 & -28 & 12 & 164 & 12 & -28 & -120 \\
-28 & 4 & 4 & 12 & 12 & 4 & 4 & 4 & -28 & -28 & 12 & 12 & 164 & -28 & -120 \\
11 & 11 & 11 & 5 & 5 & 5 & 3 & -5 & -5 & -5 & -11 & -11 & -11 & 147 & -150
\end{pmatrix}$$

As A_σ is already found, we can proceed to find the other semi-prewavelets centred in v. As explained earlier, this can be achieved purely by selecting another v^*, i.e. to shift the positive entry of y to the location corresponding to the new $v^* \in C_\sigma(v,u)$, and solve (6) for each new y obtained this way, that is, for all vertices in V_v^{-1}.

Following the ordering of the vertices in V_v^{-1} already indicated, we find the coefficients for the 14 semi-prewavelets centred in v to be as given in the matrix shown above. The row number corresponds to the location of v^* in $C_\sigma(v,u)$, or equivalently, u in $V_\sigma(v,u)$, and the column number corresponds to the location of w in $V_\sigma(v,u)$, for the coefficient $\sigma_{v,u}^{-1}(w)$. Note that the first row contains the coefficients for the semi-prewavelet already found. In Figure 2 we have shown the masks of two semi-prewavelets next to each other, but note that the vertex u is the same point in space for both semi-prewavelets. They thus add up to the prewavelet $\psi_u^{-1} = \psi_{v_{322}}^{-1} = \sigma_{v_{222},v_{322}}^{-1} + \sigma_{v_{422},v_{322}}^{-1}$.

§7. Wavelet Spaces in $\mathbb{R}^3$ and $\mathbb{R}^4$

The square matrix $R^j = \left(q_{w,u}\right) = \left(\psi_u^{j-1}(w)\right)$, evaluated in $u, w \in V_*^j$, is a collocation matrix. The vertices from V_*^j are ordered arbitrarily, but in the same way for the rows and columns of R^j. A set of prewavelets $\{\psi_u^{j-1}\}_{u \in V_*^j}$ in W^{j-1} constitutes a basis of W^{j-1} if R^j is nonsingular.

We will now try to determine whether a collocation matrix based on the elements found in the previous chapter is nonsingular. Recall that this is the case if it is positive definite and symmetric. It can be shown that the collocation matrix for prewavelets defined over a Type-1 triangulation in $\mathbb{R}^3$ is not symmetric, due to the fact that the set of coefficients for any two semi-prewavelets are not equal. Hence this method gives no indication of nonsingularity of R^j.

Another way of proving nonsingularity of R^j is to check whether it is row-wise diagonally dominant. This is equivalent to examining whether $\sigma_{v,u}^{j-1}(u) > \sum_{w \in V_v^j \setminus u} |\sigma_{v,u}^{j-1}(w)|, u \in V_v^j$, for all $v \in V^{j-1}$. However, this condition does not hold.

We have also checked the collocation matrix for columnwise diagonal dominance, which means checking whether $\sigma_{v,u}^{j-1}(u) > \sum_{w \in V_v^j \setminus u} |\sigma_{v,w}^{j-1}(u)|, \quad u \in V_v^j$, for all $v \in V^{j-1}$. For an unbounded Type-1 triangulation, the (infinite) collocation matrix is in fact weakly columnwise diagonally dominant, but this is not the case for a bounded triangulation. For a Type-1 triangulation in $\mathbb{R}^4$, we also cannot decide if the collocation matrix is nonsingular via positive definiteness and symmetry or diagonal dominance, for the same reasons as stated above. Note that here even the infinite collocation matrix is not columnwise diagonally dominant.

In $\mathbb{R}^1$ and $\mathbb{R}^2$ the collocation matrix for any bounded or unbounded Type-1 triangulation is symmetric and diagonally dominant in the setting we have described, and thus results in a Riesz basis for the wavelet space we are interested in.

We still conjecture, however, that the process described here will result in a basis for piecewise linear wavelet spaces over Type-1 triangulations of uniformly distributed data in any space dimension.

Acknowledgments. I wish to thank E. G. Quak and M. S. Floater for their valuable suggestions.

References

1. Bey, J., Tetrahedral grid refinement, Computing **55** (1995), 355–378.
2. de Boor, C., B-form basics, in *Geometric Modeling: Algorithms and New Trends*, G. E. Farin (ed.), SIAM Publications, Philadelphia, 1987, 131–148.
3. Floater, M. S. and E. G. Quak, Piecewise linear prewavelets on arbitrary triangulations, Numer. Math. **82** (1999), 221–252.
4. Floater, M. S. and E. G. Quak, A semi-prewavelet approach to piecewise linear prewavelets on triangulations, *Approximation Theory IX, Vol. 2: Computational Aspects*, Charles K. Chui and Larry L. Schumaker (eds.), Vanderbilt University Press, Nashville, 1998, 63–70.
5. Floater, M. S. and E. G. Quak, Linear independence and L_2 stability of piecewise linear prewavelets on arbitrary triangulations, SIAM J. Numer. Anal., to appear.
6. Floater, M. S., E. G. Quak, and M. Reimers, Filter bank algorithms for piecewise linear prewavelets on arbitrary triangulations, J. Comput. Appl. Math. **119** (2000), 185–207.
7. Freudenthal, H., Simplizialzerlegungen von beschränkter flachheit, Annals of Mathematics **43** (1942), 580–582.
8. Kotyczka, U. and P. Oswald, Piecewise linear prewavelets of small support, in *Approximation Theory VIII, Vol. 2: Wavelets*, Charles K. Chui and Larry L. Schumaker (eds.), World Scientific Publishing Co., Inc., Singapore, 1995, 235–242.
9. Kuhn, H. W., Some combinatorial lemmas in topology, IBM J. Res. Dev. **45** (1960), 518–524.
10. Lawson, C. L., Properties of n-dimensional triangulations, Comput. Aided Geom. Design **3** (1986), 231–246.
11. Mara, P. S., Triangulations for the cube, Journal of Combinatorial Theory (A) **20** (1976), 170–177.
12. Ong, M. E. G., Uniform refinement of a tetrahedron, in SIAM J. Scient. Computing **15** (1994), 1134–1144.
13. Suter, E., Multivariate semi-prewavelets over triangulations of uniform grids, M. Sc. Thesis, Department of Informatics, University of Oslo, Norway, 1999.

Erich Suter
SINTEF Applied Mathematics,
P.O. Box 124, Blindern
N-0314 Oslo
Norway
`Erich.Suter@math.sintef.no`

A Mixed Representation Approach to Offsets of Rational Curves

Salim Taleb

Abstract. The mixed representation of a rational curve describes this geometrical shape as the envelope of a set of tangents given in a parametric form. This representation is naturally related to dual description of curves. The mixed description yields an efficient geometric tool for the design of rational curves with rational offset curves. It leads to remarkably simple results.

§1. Introduction

Offset curves are defined as locus of the points at a constant distance d from a generator curve. Offsets are important in many geometric modelling applications. They are widely used in various applications such as NC machining, geometric tolerancing and the construction of blends and fillets. If the generator curve is rational, then its offset is in general not a rational curve.

Farouki and Sakkalis [2] introduced a class of special planar polynomial curve called Pythagorean hodograph (PH) curve $\boldsymbol{C}(t) = (X(t), Y(t))$, whose hodograph (derivative) components $\dot{X}(t)$, $\dot{Y}(t)$ and a polynomial $\sigma(t)$ form a pythagorean triple $\dot{X}^2(t)+\dot{Y}^2(t) = \sigma^2(t)$. Thus, the PH-curve has polynomial parametric speed $\sigma(t)$; therefore its offset is a rational curve and its arc length parameter is a polynomial function $s(t)$ of the parameter t.

Pottmann [7,8] and independently Fiorot and Gensane [4] extended the notion of PH-curves to the full class of rational curves with rational offsets. The approach in Pottmann [7,8] made an elegant use of the projective dual representation wich has been introduced in CAGD by Hoschek [5]. In dual representation, a curve is considered as an envelope to a family of tangent lines defined by an implicit equation.

As an alternative approach, a curve is considered in mixed representation as an envelope to a family of tangent lines defined by parametric equations.

The structure of the paper is as follows. Mixed (Bézier) representation is presented in Sections 2 and 3. Algorithms for conversion to dual form and to

Mathematical Methods for Curves and Surfaces: Oslo 2000
Tom Lyche and Larry L. Schumaker (eds.), pp. 487–496.

ISBN 0-8265-1378-6

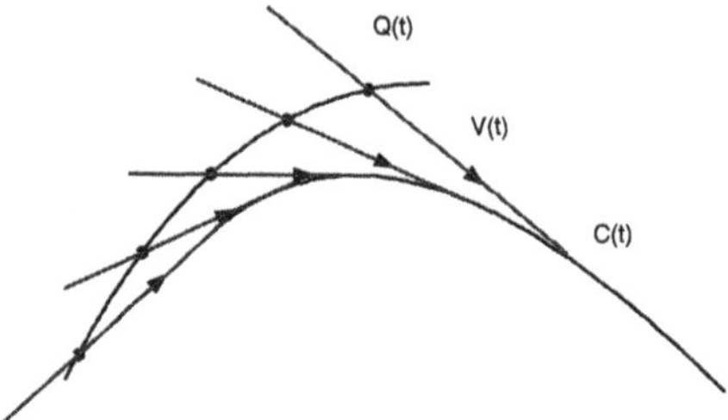

Fig. 1. C is the envelope of a family of parametrized lines $Q(t) + \lambda V(t)$.

Bézier standard form are given in Sections 3 and 4. Section 5 is devoted to the mixed description of rational PH-curves. Applications to rational offsets and to curves with rational arc length parameter are derived subsequently.

In the following, we will work in the projective extension $\mathcal{P}^2$ of the real Euclidean plane $\mathcal{E}^2$. We will denote by (X^P, Y^P) the inhomogeneous Cartesian coordinates of a point $\boldsymbol{P}$. The homogeneous Cartesian coordinates (x^P, y^P, ω^P) of the point $\boldsymbol{P}$ are then collected in the vector $\boldsymbol{p}$. The one-dimensional subspace of $\mathbb{R}^3$ spanned by $\boldsymbol{p}$ is the point $\boldsymbol{P}$ in $\mathcal{P}^2$.

For points not at infinity, i.e. $\omega^P \neq 0$, the corresponding inhomogeneous coordinates are $X^P = x^P/\omega^P$ and $Y^P = y^P/\omega^P$.

Accordingly to the immersion of $\mathbb{R}^2$ in $\mathbb{R}^3$, the vector $\boldsymbol{V} = (\alpha, \beta)^T$ is associated with $\boldsymbol{v} = (\alpha, \beta, 0)^T$.

§2. Mixed Representation of Curves

A mixed representation of a curve $\boldsymbol{C}$ describes this object as the family of its tangent lines defined by the parametric equations

$$\boldsymbol{L}_\lambda(t) = \boldsymbol{Q}(t) + \lambda \boldsymbol{V}(t), \tag{1}$$

where $\boldsymbol{Q}$ is a planar curve and $\boldsymbol{V}$ is a vector field of $\mathbb{R}^2$. The parameter of the family is t, and λ is the parameter along the lines.

It may be seen in Fig. 1 that the lines appear to gather along another curve $\boldsymbol{C}$. The boundary curve $\boldsymbol{C}$ is simultaneously tangent to all the lines in the family. The new curve is called the envelope of the lines $\boldsymbol{L}_\lambda(t)$.

The curve $\boldsymbol{Q}$ will be called the supporting curve (S-curve) of $\boldsymbol{C}$ and the vector field $\boldsymbol{V}$ the tangential direction (T-direction) of $\boldsymbol{C}$.

Let us suppose the existence of the envelope $\boldsymbol{C}$ to the family (1). Each point of $\boldsymbol{C}$ belongs to a tangent line $L_\lambda(t)$. Hence the following relation holds:

$$\boldsymbol{C}(t) = \boldsymbol{Q}(t) + \lambda(t)\boldsymbol{V}(t). \tag{2}$$

Since the curve $\boldsymbol{C}$ is tangent to each line of the family (1), a necessary condition is that the first derivative vector or hodograph $\dot{\boldsymbol{C}}(t)$ has the same direction as the vector $\boldsymbol{V}(t)$. It follows that their determinant vanishes:

$$\det(\dot{\boldsymbol{C}}, \boldsymbol{V}) = 0. \tag{3}$$

From (3), we can compute λ as the function

$$\lambda = -\det(\dot{\boldsymbol{Q}}, \boldsymbol{V}) \; / \; \det(\dot{\boldsymbol{V}} \wedge \boldsymbol{V}). \tag{4}$$

Given $\boldsymbol{Q}$ and $\boldsymbol{V}$, the envelope $\boldsymbol{C}$ is computed as the solution to the linear system consisting of equations (2) and (3).

The equations (2) and (3) ensure the existence of the envelope $\boldsymbol{C}$ if we suppose the additional conditions (see e.g. Ramis et. al. [10]):

(a) $\boldsymbol{Q}$ (resp. $\boldsymbol{V}$) is a C^2-mapping of a real interval $I \subset \mathbb{R}$ into $\mathcal{P}^2$ (resp. $\mathbb{R}^2 - \{0\}$),

(b) $\det(\dot{\boldsymbol{V}}, \boldsymbol{V}) \neq 0$ over I.

In the following, the conditions (a) and (b) are assumed true over $I = [0,1]$. It may be noted that the mixed description (1) of a curve $\boldsymbol{C}$ is not unique. The same curve $\boldsymbol{C}$ can be defined by different equivalent mixed forms. More precisely, we have the following properties: the envelope $\boldsymbol{C}$ is invariant under the following transformations of the mixed representation:

(i) $(\boldsymbol{Q}, \boldsymbol{V}) \to (\boldsymbol{Q}, \alpha \boldsymbol{V})$,

(ii) $(\boldsymbol{Q}, \boldsymbol{V}) \to (\boldsymbol{Q} + \alpha \boldsymbol{V}, \boldsymbol{V})$,

where α is a C^2-continuous function such that $\alpha(t) \neq 0$ for $t \in I$.

§3. The Mixed Bézier Representation

Let us consider a polynomial S-curve $\boldsymbol{Q}$ and a polynomial T-direction $\boldsymbol{V}$. From the linear system (2) and (4), one can derive the following standard description of the curve $\boldsymbol{C}$ in homogeneous coordinates:

$$\boldsymbol{c} = \det(\dot{\boldsymbol{V}}, \boldsymbol{V}) \; \boldsymbol{q} - \det(\dot{\boldsymbol{Q}}, \boldsymbol{V}) \; \boldsymbol{v}. \tag{5}$$

Using the Bernstein basis $B_i^n(t) = \binom{n}{i}(1-t)^{n-i}t^i$ with $i = 0, 1, \ldots, n$, we obtain the mixed Bézier form of the curve $\boldsymbol{C}$

$$\boldsymbol{L}_\lambda(t) = \sum_{i=0}^{n} (\boldsymbol{Q}_i + \lambda \boldsymbol{V}_i) B_i^n(t). \tag{6}$$

For each value of t, we obtain a tangent line to $\boldsymbol{C}$. This line goes through the point of the Bézier S-curve $\boldsymbol{Q}(t) = \sum_{i=0}^{n} \boldsymbol{Q}_i B_i^n(t)$ and is oriented by the tangent T-direction $\boldsymbol{V}(t) = \sum_{i=0}^{n} \boldsymbol{V}_i B_i^n(t)$. Utilizing the usual terminology, we will denote by L_i the i-th control line $\boldsymbol{Q}_i + \lambda \boldsymbol{V}_i$ for $i = 0, 1, \ldots, n$. $\boldsymbol{Q}_i$ is called the i-th control supporting point (S-point) and $\boldsymbol{V}_i$ the i-th control tangent vector (T-vector) for $i = 0, 1, \ldots, n$.

The standard Bézier homogeneous representation of the curve $\boldsymbol{C}$ defined by (6) is given by $\boldsymbol{c}(t) = \sum_{m=0}^{3n-1} \boldsymbol{p}_m B_m^{3n-1}$, where the control points are expressed by the formula

$$\boldsymbol{p}_m = \frac{1}{\binom{3n-1}{m}} \sum_{i+j+k=m} \binom{n-1}{i} \binom{n}{j} \binom{n}{k} \cdot \boldsymbol{r}_{i,j,k},$$

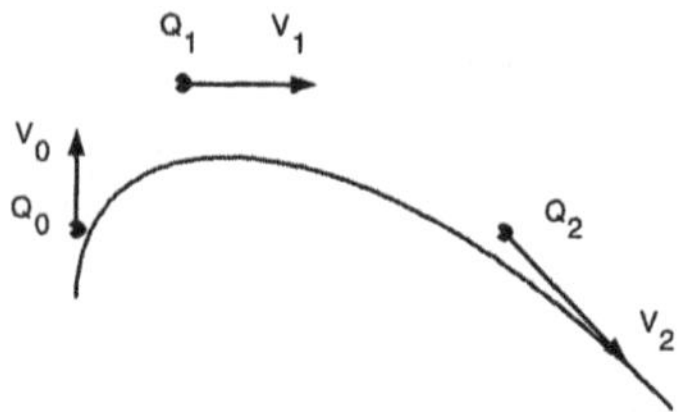

Fig. 2. Quintic defined by a mixed form of degree $n = 2$.

with $\boldsymbol{r}_{i,j,k} = \det(\Delta \boldsymbol{V}_i, \boldsymbol{V}_j)\ \boldsymbol{q}_k - \det(\Delta \boldsymbol{Q}_i, \boldsymbol{V}_j)\ \boldsymbol{v}_k$ for $m = 0, 1, \ldots, 3n-1$; $i = 0, 1, \ldots, n-1$; and $j, k = 0, 1, \ldots, n$. It may be noted that the rational curve $\boldsymbol{C}$ possesses the degree $3n-1$ when the degree of (6) is n. The curve is therefore a rational algebraic curve of maximal order $3n-1$. For instance, we can generate parabolas with $n = 1$ and a subset of the quintics with $n = 2$.

An example of quintic is shown in Fig. 2 for $n = 2$. It may be seen that the points $\boldsymbol{Q}_0$ and $\boldsymbol{Q}_n$ do not belong necessarily to the curve $\boldsymbol{C}$. If we want to fix an end point of the curve segment $\boldsymbol{C}$ to be $\boldsymbol{C}(0) = \boldsymbol{Q}_0$ (resp. $\boldsymbol{C}(1) = \boldsymbol{Q}_n$), a simple geometric condition is that $\boldsymbol{Q}_1$ (resp. $\boldsymbol{Q}_{n-1}$) is the intersection point of the lines L_0 and L_1 (resp. L_{n-1} and L_n). For more details on the geometric properties and further algorithms, the reader is referred to [11].

The use of mixed Bézier representation is suitable for an interactive handling of rational curves. The T-direction can be considered as the guiding system of the curve $\boldsymbol{C}$. Positioning the curve $\boldsymbol{C}$ can be done through the control structure of the supporting curve $\boldsymbol{Q}$.

Given now a rational S-curve $\boldsymbol{Q}$ and a rational T-direction $\boldsymbol{V}$, the envelope $\boldsymbol{C}$ is then computed in homogeneous coordinates as

$$\boldsymbol{c} = \det(\boldsymbol{q}, \dot{\boldsymbol{v}}, \boldsymbol{v})\ \boldsymbol{q} + \det(\boldsymbol{v}, \dot{\boldsymbol{q}}, \boldsymbol{q})\ \boldsymbol{v}.$$

The mixed rational Bézier form of a rational curve $\boldsymbol{C}$

$$L_\lambda(t) = \frac{\sum_{i=0}^{n} \omega_i (\boldsymbol{Q}_i + \lambda \boldsymbol{V}_i) B_i^n(t)}{\sum_{i=0}^{n} \omega_i B_i^n(t)} \tag{7}$$

is similarly converted to the standard description of $\boldsymbol{C}$ by the formula $\boldsymbol{c}(t) = \sum_{m=0}^{4n-2} \boldsymbol{p}_m B_m^{4n-2}$ where the control points are expressed in homogeneous coordinates as

$$\boldsymbol{p}_m = \frac{1}{\binom{4n-2}{m}} \sum_{i+j+k+l=m} \binom{n}{i}\binom{n-1}{j}\binom{n-1}{k}\binom{n}{l} \cdot \boldsymbol{r}_{i,j,k,l},$$

$$\text{with} \quad \boldsymbol{r}_{i,j,k,l} = \omega_{j+1}\omega_k \det(\boldsymbol{q}_i, \boldsymbol{v}_{j+1}, \boldsymbol{v}_k)\ \boldsymbol{q}_l + \omega_i \omega_l \det(\boldsymbol{v}_i, \boldsymbol{q}_{j+1}, \boldsymbol{q}_k)\ \boldsymbol{v}_l,$$

for $m = 0, 1, \ldots, 4n-2$; $i, l = 0, 1, \ldots, n$; and $j, k = 0, 1, \ldots, n-1$. In the rational case, the envelope possesses the degree $4n-2$. The curve is therefore a rational algebraic curve of maximal order $4n-2$.

The order of the envelope gets large very quickly as n is incremented. This leverage effect allows the handling of certain high order curves through a limited amount of control points and vectors.

§4. Connection with Dual Representation

Since both mixed and dual forms make use of families of tangent lines to represent planar curves, one can wonder to what extent the mixed structure properties differ from those of the dual.

The dual representation of curves is a description based on the family of tangent lines defined by the implicit equation

$$A(t)\ X + B(t)\ Y + C(t) = 0.$$

Let $\boldsymbol{u}$ denote the coefficient vector $(A, B, C)^T$ that contains the homogeneous line coordinates. The dual Bézier representation of a planar rational curve segment $\boldsymbol{C}$ constructs a tangent line $\boldsymbol{u}(t)$ to each parameter $t \in [0,1]$:

$$\boldsymbol{u}(t) = \sum_{i=0}^{n} \omega_i \boldsymbol{B}_i^* B_i^n(t). \tag{8}$$

The vectors $\boldsymbol{B}_i^*$ are the line coordinates vectors of the Bézier lines. Since $\boldsymbol{B}_i^*$ and any multiple $\omega_i \boldsymbol{B}_i^*$, $\omega_i \neq 0$, represent the same line, the curve is not determined by the lines $\boldsymbol{B}_i^*$ alone, but also by the weights ω_i. Analogously to the familiar point representation, a more practical description can be based on the use of the Farin lines $\boldsymbol{F}_i^*$. These are the concurrent lines with $\boldsymbol{B}_i^*$ and $\boldsymbol{B}_{i+1}^*$ that are represented by the vectors $\boldsymbol{F}_i^* = \boldsymbol{B}_i^* + \boldsymbol{B}_{i+1}^*$ instead of the weights ω_i (cf. [7,8]). The weights ω_i are then adjusted by the position of the lines $\boldsymbol{F}_i^*$.

The mixed representation is naturally related to the dual one. In projective geometry, each point of the projective plane is described as a one-dimensional linear subspace. We associate to the line $L_\lambda(t)$ defined by (6) or (7) a linear subspace ρ of dimension 2 defined by the equation $A\ X + B\ Y + C\ Z = 0$. The vectors $\boldsymbol{v}$ and $\boldsymbol{q}$ belong to the subspace ρ, and $\boldsymbol{u}$ is an orthogonal vector to ρ. This yields the conversion formula from the mixed form to the dual representation as

$$\boldsymbol{u} = \boldsymbol{q} \wedge \boldsymbol{v},$$

where $\wedge$ is the exterior product.

The above equation implies the formula for conversion from mixed Bézier form to its dual Bézier description $\sum_{k=0}^{2n} \boldsymbol{B}_k^* B_k^n(t)$, where control lines coordinates are given by

$$B_k^* = \frac{1}{\binom{2n}{k}} \sum_{i+j=k} \binom{n}{i}\binom{n}{j} \boldsymbol{q}_i \wedge \boldsymbol{v}_j.$$

For a mixed form of degree n, the algebraic class of $\boldsymbol{C}$ is in general $2n$. From the view point of computational efficiency, it can be suitable to use the mixed description instead of the dual one. Moreover, designing a curve by its mixed Bézier structure, *i.e.*, via the control structure of $\boldsymbol{Q}$ and $\boldsymbol{V}$ seems to be more intuitive than using the Bézier and Farin lines in dual representation.

§5. Mixed Representation of Rational PH-curves

In the sequel, we denote by $\boldsymbol{N}^V$ the orthogonal vector to $\boldsymbol{V}$ expressed by $\boldsymbol{N}^V = (\beta, -\alpha)$ for $\boldsymbol{V} = (\alpha, \beta)$.

Let us consider a rational curve $\boldsymbol{C}$ with the rational parametric representation $\boldsymbol{C}(t)$. With $\boldsymbol{N}(t)$ as normal vector of $\boldsymbol{C}$, the offset at distance d is

$$\boldsymbol{C}^d(t) = \boldsymbol{C}(t) + d\,\frac{\boldsymbol{N}(t)}{\|\boldsymbol{N}(t)\|}.$$

Since we would like $\boldsymbol{C}^d$ to have a rational parametrization, $\boldsymbol{N}(t) = (N_1(t), N_2(t))$ must be of the form (cf. Farouki and Sakkalis [2], Pottmann [7,8])

$$N_1(t) = k(t)\,(2a(t)b(t)), \quad N_2(t) = k(t)\,\left(b^2(t) - a^2(t)\right)$$

with a, b and k polynomials in t.

We deduce that $V = (a^2 - b^2, 2ab)$ is a valid choice as a tangent vector to the curve at $\boldsymbol{C}(t)$. The T-direction provides the major information about the variation of the tangent orientation. Actually, the relation between the hodograph $\dot{\boldsymbol{C}}$ and the T-direction $\boldsymbol{V}$

$$\dot{\boldsymbol{C}}(t) = \mu(t)\boldsymbol{V}(t) \tag{9}$$

is easily established from the equation (3), with proportionality function

$$\mu = \frac{\det(\dot{\boldsymbol{V}}, \dot{\boldsymbol{C}})}{\det(\dot{\boldsymbol{V}}, \boldsymbol{V})} = \dot{\lambda} + \frac{\det(\dot{\boldsymbol{V}}, \dot{\boldsymbol{Q}})}{\det(\dot{\boldsymbol{V}}, \boldsymbol{V})}, \quad \text{with} \quad \lambda = -\frac{\det(\dot{\boldsymbol{Q}}, \dot{\boldsymbol{V}})}{\det(\dot{\boldsymbol{V}}, \boldsymbol{V})}.$$

The choice of Pythagorean T-direction $V(t) = (a^2 - b^2, 2ab)$ induces that the parametric speed is $\|\dot{\boldsymbol{C}}\| = |\mu|\cdot(a^2+b^2)$. Hence, replacing $\boldsymbol{Q}$ by an arbitrary rational curve leads to a rational function $\|\dot{\boldsymbol{C}}\|$; accordingly the envelope $\boldsymbol{C}$ is then a rational PH-curve. Let us summarize our results:

Theorem 1. *Any rational curve $\boldsymbol{C}$ with rational offsets can be expressed in the mixed representation $\boldsymbol{Q}+\lambda\boldsymbol{V}$, where $\boldsymbol{Q}$ is an arbitrary rational S-curve and $\boldsymbol{V} = (a^2 - b^2, 2ab)$ is a T-direction with a, b relatively prime polynomials in t. The mixed form of the offset $\boldsymbol{C}^d$ at distance d is $\boldsymbol{P} + \lambda\boldsymbol{V}$, where*

$$\boldsymbol{P} = \boldsymbol{Q} + d\,\frac{\boldsymbol{N}^V}{\left\|\boldsymbol{N}^V\right\|}.$$

As in the dual approach, the mixed description reveals also a remarkably simple design scheme for rational PH-curves.

The mixed form of Theorem 1 leads to the following explicit parametric representation of the PH-curve $\boldsymbol{C}$:

$$\begin{aligned} X &= X^Q + \lambda(a^2 - b^2), \\ Y &= Y^Q + \lambda(2ab), \\ \lambda &= \frac{(a^2 - b^2)\dot{Y}^Q - (2ab)\dot{X}^Q}{2(a^2 + b^2)(\dot{a}b - a\dot{b})}. \end{aligned} \tag{10}$$

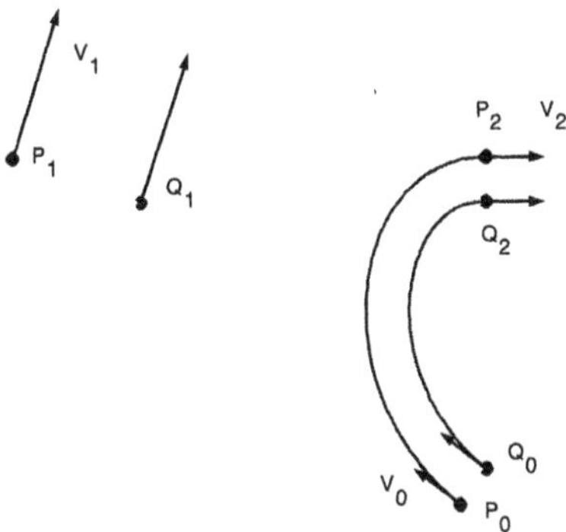

Fig. 3. Rational PH-curve and an offset.

Corollary 2. *All rational curves $\boldsymbol{C}$ with rational offsets $\boldsymbol{C}^d$ can be expressed explicitly in the form (10). Here $\boldsymbol{Q} = (X^Q, Y^Q)$ is an arbitrary rational curve of parameter t, a and b are relatively prime polynomials in t. Replacing $\boldsymbol{Q}$ by*

$$\boldsymbol{Q} + \frac{d}{a^2+b^2}\begin{pmatrix} 2ab \\ b^2-a^2 \end{pmatrix}$$

leads to the explicit form of the offset $\boldsymbol{C}^d$ at distance d.

The explicit form (10) is equivalent to previous results in Pottmann [8], Fiorot and Gensane [4]. Actually, the explicit description derived in [8]

$$X = \frac{2ab}{a^2+b^2}\cdot h + \frac{a^2-b^2}{2(a\dot{b}-\dot{a}b)}\cdot \dot{h},$$
$$Y = \frac{b^2-a^2}{a^2+b^2}\cdot h + \frac{2ab}{2(a\dot{b}-\dot{a}b)}\cdot \dot{h},$$

is equivalent to the special mixed form defined by S-curve

$$\boldsymbol{Q} = \frac{h}{a^2+b^2}\begin{pmatrix} 2ab \\ b^2-a^2 \end{pmatrix}, \quad \text{and T}-\text{direction} \quad \boldsymbol{V} = \begin{pmatrix} a^2-b^2 \\ 2ab \end{pmatrix}.$$

Since the function h is the signed distance of the tangent line to the curve C from the origin, the curve $\boldsymbol{Q}$ is the pedal curve of $\boldsymbol{C}$ with respect to the origin.

Let us study now the connection between the mixed Bézier control structures of the curves $\boldsymbol{C}$ and $\boldsymbol{C}^d$. One can always rewrite a mixed rational Bézier form with a Pythagorean T-direction in such a way that

$$\boldsymbol{V}(t) = \frac{\sum_{i=0}^{n} \omega_i \boldsymbol{V}_i B_i^n(t)}{\sum_{i=0}^{n} \omega_i B_i^n(t)} \tag{11}$$

corresponds to the unit tangent vector field. This transformation can be done by a normalization of the T-direction through property (i) of Section 2, reduction to the same denominator, and degree elevation if necessary. From Theorem 1, one can easily proceed to the following result.

Theorem 3. *Let us consider the mixed rational Bézier form*

$$\boldsymbol{Q} + \lambda \boldsymbol{V} = \frac{\sum_{i=0}^{n} \omega_i (\boldsymbol{Q}_i + \lambda \boldsymbol{V}_i) B_i^n(t)}{\sum_{i=0}^{n} \omega_i B_i^n(t)}$$

of a rational curve $\boldsymbol{C}$ *with* $\boldsymbol{V}$ *a unit Pythagorean T-direction, then the mixed rational Bézier representation of the offset* $\boldsymbol{C}^d$ *at distance* d *is given by*

$$\boldsymbol{P} + \lambda \boldsymbol{V} = \frac{\sum_{i=0}^{n} \omega_i (\boldsymbol{P}_i + \lambda \boldsymbol{V}_i) B_i^n(t)}{\sum_{i=0}^{n} \omega_i B_i^n(t)}$$

with $\boldsymbol{P}_i = \boldsymbol{Q}_i + d\, \boldsymbol{N}_i^V$, $i = 0, 1, \ldots, n$.

This result yields a user friendly scheme. Figure 3 shows a rational PH-curve and its offset of mixed degree $n = 2$. Both are of algebraic order 6. The i-th control lines of $\boldsymbol{C}$ and $\boldsymbol{C}^d$ are parallel, and their distance is d times the norm of $\boldsymbol{V}_i$. The i-th control S-points $\boldsymbol{Q}_i$ and $\boldsymbol{P}_i$ are situated on an orthogonal line to the i-th control line of $\boldsymbol{C}$ or $\boldsymbol{C}^d$.

For an interactive construction of rational PH-curves, the key idea is an appropriate choice of the Pythagorean T-direction $\boldsymbol{V}$. This operation determine the values of the weights. We can afterwards perform arbitrary moves on the control S-points $\boldsymbol{Q}_i$.

Method 4. *To construct a rational PH- curve and its offsets in the mixed rational Bézier frame, perform the following steps :*

(i) Construct a rational Bézier curve on the unit circle. This determine both the weights ω_i *and the control T-vectors* $\boldsymbol{V}_i$ *for* $i = 0, 1, \ldots, n$.

(ii) Choose arbitrary points $\boldsymbol{Q}_i$ *for* $i = 0, 1, \ldots, n$ *such that the corresponding weighted control points* $(\boldsymbol{Q}_i, \omega_i)$ *generate the rational S-curve* $\boldsymbol{Q}$.

(iii) Give a value to the distance d *and apply Theorem 3.*

With this method, the tangential direction is pythagorean. Rational PH-curves and their offsets can then be constructed interactively. Thus, mixed rational Bézier representation provides a simple approach to the design of rational curves with rational offsets.

Figure 4 shows the effect of the move of one control S-point at a time on the shape of a rational PH-curve $\boldsymbol{C}$. The Pythagorean T-direction $\boldsymbol{V}$ and the weights ω_i, $i = 0, 1, 2$ are maintained unchanged during the handling.

For a polynomial PH-curve, the parametric speed $ds/dt = \sqrt{\dot{X}^2 + \dot{Y}^2}$ is polynomial. The arc length parameter $s(t)$ is therefore also polynomial. For rational curves $\boldsymbol{C}(t)$ the situation is more complicated. If ds/dt is rational, the arc length s in not necessarily rational.

Pottmann in [8] proved that the rational curves whose arc length parameter $s(t)$ is a rational function of t are exactly the evolutes of the rational curves with only rational offsets. Since an evolute is the envelope of the family of normal lines, the previous characterization leads us directly to the following result.

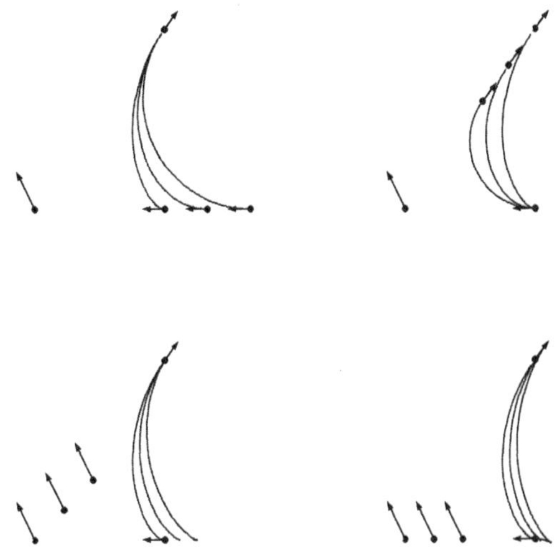

Fig. 4. Interactive handling of a rational PH-curve.

Theorem 5. *Let $\boldsymbol{Q}_1$ denote an arbitrary rational curve. The rational curves $C(t)$ whose arc length parameter $s(t)$ is a rational function of t can be expressed in the mixed rational form $\boldsymbol{Q} + \lambda \boldsymbol{V}$ where $\boldsymbol{V}$ is a Pythagorean T-direction, and $\boldsymbol{Q}$ is itself defined by the mixed rational form $\boldsymbol{Q}_1 + \lambda_1 \boldsymbol{N}^V$.*

References

1. Farin, G., *Curves and Surfaces for Computer Aided Geometric Design*, Academic Press, Boston, MA, 3rd ed., 1992.
2. Farouki, R. T. and T. Sakkalis, Pythagorean hodographs, IBM J. of Research and Development **34** (1990), 736–752.
3. Farouki, R. T., The conformal map $z \to z^2$ of the hodograph plane, Comput. Aided Geom. Design **11** (1994), 363–390.
4. Fiorot, J. C. and T. Gensane, Characterizations of the set of rational curves with rational offsets, in *Curves and Surfaces in Geometric Design*, P.-J. Laurent, A. Le Méhauté, and L. L. Schumaker (eds.), A. K. Peters, Wellesley MA, 1994, 151–158.
5. Hoschek, J., Dual Bézier curves and surfaces, in *Surfaces in Computer Aided Geometric Design*, R. E. Barnhill & W. Boehm (eds.), North Holland, 1983, 147–156.
6. Hoschek, J. and D. Lasser, *Fundamentals of Computer Aided Geometric Design*, AK. Peters, Wellesley, Ma., 1993.

7. Pottmann, H., Applications of the dual Bézier representation of rational curves and surfaces, in *Curves and Surfaces in Geometric Design*, P.-J. Laurent, A. Le Méhauté, and L. L. Schumaker (eds.), A. K. Peters, Wellesley MA, 1994, 377–384.

8. Pottmann, H., Rational curves and surfaces with rational offsets, Comput. Aided Geom. Design **12** (1995), 175–192.

9. Pottmann, H., Curve design with rational Pythagorean-hodograph curves, Advances in Comp. Math. **3** (1995), 147–170.

10. Ramis E., C. Deschamps, and J. Odoux, *Cours de Mathématiques : Applications de l'Analyse à la Géométrie*, Tome 5, Dunod, Paris, 1998.

11. Taleb, S., Mixed Bézier representation of curves, Technical Report, Univ. of Valenciennes, 2000.

Salim Taleb
Laboratoire Macs
Université de Valenciennes
59313 Valenciennes Cedex 9, France
`salim.taleb@univ-valenciennes.fr`

Pedal Curves and Surfaces

Kenji Ueda

Abstract. The positive pedal curve of a given curve is the locus of the feet of perpendiculars from a fixed point to all the tangent lines to the curve. The given curve is called the negative pedal curve of the pedal curve. Higher pedals are defined by induction. In this paper, it is shown that complex numbers are suitable for representing pedal curves. Pedal surfaces are defined in a similar manner to pedal curves, and quaternions are suitable for expressing pedal surfaces.

§1. Introduction

The construction of pedal curves [1,2,4,5,8,11] is one of the classical methods for deriving curves from a given plane curve. Higher pedal curves are defined by induction. A sequence of pedal curves represented as Bézier curves are investigated in [13]. The inverse curve is also derived from a plane curve via inversion. There is a close relationship between pedal curves and inverse curves.

Pedal surfaces [1] or inverse surfaces are constructed from a given surface in a similar manner. There are analogous properties for the derived surfaces to the derived curves. A relationship between pedal curves (surfaces) and dual curves (surfaces) [6] can be found in [13].

Complex numbers [9] and quaternions [7] are tools in computer aided geometric design [3] or in computer graphics [10]. They are also suitable for expressing pedal curves or surfaces. In this paper, (higher) pedal curves and surfaces are investigated using these tools. First, positive and negative pedal curves are introduced and a relationship between pedal curves and inverse curves is presented. Then, higher pedal curves are investigated on the complex plane. Pedal surfaces and their properties are also shown via quaternion calculus.

Mathematical Methods for Curves and Surfaces: Oslo 2000
Tom Lyche and Larry L. Schumaker (eds.), pp. 497–506.

ISBN 0-8265-1378-6

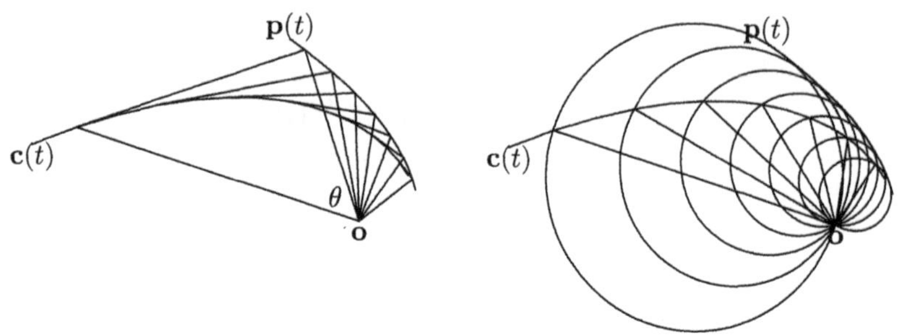

Fig. 1. Pedal curve.

§2. Positive and Negative Pedal Curves

The positive pedal curve of a given curve is the locus of the feet of perpendiculars from a fixed point to all the tangent lines to the curve.

Definition 1. *The* pedal curve *of a regular curve* $\mathbf{c}(t) = (\,x(t) \;\; y(t)\,)^T$ *with respect to the origin is defined by*

$$\pi\,\mathbf{c}(t) = \frac{\mathbf{c}(t)\cdot\mathbf{n}(t)}{\mathbf{n}(t)\cdot\mathbf{n}(t)}\mathbf{n}(t) = \frac{x(t)y'(t) - x'(t)y(t)}{x'(t)^2 + y'(t)^2}\begin{pmatrix} y'(t) \\ -x'(t) \end{pmatrix}, \tag{1}$$

where $\mathbf{n}(t)$ *is the normal vector* $\mathbf{n}(t) = (y'(t), -x'(t))^T$ *to the curve* $\mathbf{c}(t)$.

The operator π derives the pedal curve from a given curve. The pedal curve $\pi\,\mathbf{c}(t)$ is also the envelope of circles of which the diameters are the line segments between $\mathbf{c}(t)$ and the origin $\mathbf{o}$, as illustrated in Figure 1.

Definition 2. *The* inverse curve *of a regular curve* $\mathbf{c}(t) = (\,x(t) \;\; y(t)\,)^T$ *in the unit circle centered at the origin is defined by*

$$\iota\,\mathbf{c}(t) = \iota\begin{pmatrix} x(t) \\ y(t) \end{pmatrix} \equiv \frac{\mathbf{c}(t)}{||\mathbf{c}(t)||^2} = \frac{\mathbf{c}(t)}{\mathbf{c}(t)\cdot\mathbf{c}(t)} = \frac{1}{x(t)^2 + y(t)^2}\begin{pmatrix} x(t) \\ y(t) \end{pmatrix}. \tag{2}$$

The inverse curve is obtained from a curve via inversion in the unit circle centered at the origin. The operator ι derives the inverse curve from a given curve, and it is a self-inverse operator, i.e., $\iota^{-1} = \iota$. The composition of operators, α and β, is denoted by $\beta \circ \alpha$. The composition $\iota \circ \iota$ yields the identity operator $I = \iota \circ \iota = \iota^2$.

The curve for which a given curve is a positive pedal curve is said to be a negative pedal curve. The negative pedal curve of a curve with respect to the origin is the envelope of the line drawn through a point on the curve perpendicular to the position vector of the point.

Theorem 1. *The negative pedal curve of a regular curve* $\mathbf{c}(t) = (\,x(t) \;\; y(t)\,)^T$ *with respect to the origin is obtained by applying the following composition to the curve:*

$$\pi^{-1} = \iota \circ \pi \circ \iota. \tag{3}$$

Proof: Suppose the curve $\mathbf{p}(t)$ is the pedal of a given curve $\mathbf{c}(t)$, the curve $\mathbf{q}(t)$ is the inverse of the curve $\mathbf{p}(t)$, and the curve $\mathbf{r}(t)$ is the pedal of the curve $\mathbf{q}(t)$. These curves are expressed by

$$\mathbf{p}(t) = \pi\,\mathbf{c}(t) = \frac{x(t)y'(t) - x'(t)y(t)}{x'(t)^2 + y'(t)^2}\begin{pmatrix} y'(t) \\ -x'(t) \end{pmatrix}, \tag{4}$$

$$\mathbf{q}(t) = \iota\,\mathbf{p}(t) = (\iota \circ \pi)\,\mathbf{c}(t) = \frac{1}{x(t)y'(t) - x'(t)y(t)}\begin{pmatrix} y'(t) \\ -x'(t) \end{pmatrix}, \tag{5}$$

$$\mathbf{r}(t) = \pi\,\mathbf{q}(t) = (\pi \circ \iota \circ \pi)\,\mathbf{c}(t) = \frac{1}{x(t)^2 + y(t)^2}\begin{pmatrix} x(t) \\ y(t) \end{pmatrix}. \tag{6}$$

As the curve $\mathbf{r}(t)$ is equivalent to the inverse curve (2) of the curve $\mathbf{c}(t)$, the curve $\mathbf{c}(t)$ is represented as $\mathbf{c}(t) = \iota\,\mathbf{r}(t) = (\iota \circ \pi \circ \iota \circ \pi)\,\mathbf{c}(t)$. The relationship among the curves is expressed by the following diagram.

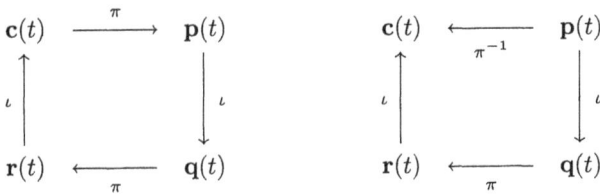

$$\begin{array}{ccc} \mathbf{c}(t) & \xrightarrow{\;\pi\;} & \mathbf{p}(t) \\ \uparrow \iota & & \downarrow \iota \\ \mathbf{r}(t) & \xleftarrow[\;\pi\;]{} & \mathbf{q}(t) \end{array} \qquad\qquad \begin{array}{ccc} \mathbf{c}(t) & \xleftarrow[\;\pi^{-1}\;]{} & \mathbf{p}(t) \\ \uparrow \iota & & \downarrow \iota \\ \mathbf{r}(t) & \xleftarrow[\;\pi\;]{} & \mathbf{q}(t) \end{array}$$

Here, we conclude that $\mathbf{c}(t) = \pi^{-1}\mathbf{p}(t) = (\iota \circ \pi \circ \iota)\,\mathbf{p}(t)$. □

The negative pedal curve $\pi^{-1}\,\mathbf{c}(t)$ is expressed by

$$\frac{1}{x(t)y'(t) - y(t)x'(t)}\begin{pmatrix} (x(t)^2 - y(t)^2)y'(t) - 2x(t)y(t)x'(t) \\ 2x(t)y(t)y'(t) + (x(t)^2 - y(t)^2)x'(t) \end{pmatrix}. \tag{7}$$

While the curve $\mathbf{p}(t)$ is the pedal curve of the curve $\mathbf{c}(t)$, the inverse curve $\iota\,\mathbf{p}(t)$ is the negative pedal curve of the inverse curve $\iota\,\mathbf{c}(t)$, namely, $\iota\,\mathbf{p}(t) = \pi^{-1}(\iota\,\mathbf{c}(t))$. The circles in Figure 1 are transformed into the tangent lines to the curve $\iota\,\mathbf{p}(t)$ via inversion, and the intersecting points between the circles and the curve $\mathbf{c}(t)$ are transformed into the points on the curve $\iota\,\mathbf{c}(t)$ which are the feet of the perpendiculars from the origin to the tangent lines to the inverse curve $\iota\,\mathbf{p}(t)$.

The curve $\pi\,\mathbf{c}(t)$ is the first pedal of the curve $\mathbf{c}(t)$, and the second pedal $\pi^2\,\mathbf{c}(t)$ can be obtained as

$$\frac{(y'(t)x(t) - x'(t)y(t))^2}{(x'(t)^2 + y'(t)^2)^2(x(t)^2 + y(t)^2)}\begin{pmatrix} (y'(t)^2 - x'(t)^2)x(t) - 2x'(t)y'(t)y(t) \\ -2x'(t)y'(t)x(t) + (x'(t)^2 - y'(t)^2)y(t) \end{pmatrix}. \tag{8}$$

Higher negative or positive pedals are obtained by induction.

§3. Pedal Curves in the Complex Plane

By using the complex numbers $\mathbf{z} = x + \mathrm{i}\,y$ and $\mathbf{w} = u + \mathrm{i}\,v$ and their conjugates $\overline{\mathbf{z}} = x - \mathrm{i}\,y$ and $\overline{\mathbf{w}} = u - \mathrm{i}\,v$, we can define a pair of products, the dot product $(\cdot)$ and the cross product $(\times)$ as

$$\mathbf{z} \cdot \mathbf{w} \equiv \frac{\overline{\mathbf{z}}\mathbf{w} + \mathbf{z}\overline{\mathbf{w}}}{2} = xu + yv, \qquad \mathbf{z} \times \mathbf{w} \equiv \frac{\overline{\mathbf{z}}\mathbf{w} - \mathbf{z}\overline{\mathbf{w}}}{2} = \mathrm{i}\,(xv - yu). \tag{9}$$

The properties $\mathbf{z} \cdot \mathbf{z} = x^2 + y^2 = ||\mathbf{z}||^2$ and $\mathbf{z} \times \mathbf{z} = \mathbf{0}$ are easily obtained from the definitions.

In the complex plane, a plane curve $(\,x(t) \;\; y(t)\,)^T$ is expressed as $\mathbf{z}(t) = x(t) + \mathrm{i}\,y(t)$. The derivative $\mathbf{z}'(t)$ and the normal vector $\mathbf{n}(t)$ become $\mathbf{z}'(t) = x'(t) + \mathrm{i}\,y'(t)$ and $\mathbf{n}(t) = -\mathrm{i}\,\mathbf{z}'(t)$.

The positive pedal curve $\pi\,\mathbf{z}(t)$, the inverse curve $\iota\,\mathbf{z}(t)$, and the negative pedal curve $\pi^{-1}\,\mathbf{z}(t)$ are expressed by

$$\pi\,\mathbf{z}(t) = \frac{\mathbf{z}(t)\overline{\mathbf{z}}'(t) - \overline{\mathbf{z}}(t)\mathbf{z}'(t)}{2\overline{\mathbf{z}}'(t)}, \qquad \iota\,\mathbf{z}(t) = \overline{\mathbf{z}}(t)^{-1} = \frac{\mathbf{z}(t)}{||\mathbf{z}(t)||^2}, \tag{10}$$

$$\pi^{-1}\,\mathbf{z}(t) = (\iota \circ \pi \circ \iota)\,\mathbf{z}(t) = \frac{2\mathbf{z}(t)^2}{\mathbf{z}(t)\overline{\mathbf{z}}'(t) - \overline{\mathbf{z}}(t)\mathbf{z}'(t)}\overline{\mathbf{z}}'(t). \tag{11}$$

Higher pedal curves in the complex plane are obtained as follows.

Theorem 2. *For an integer n, the nth pedal curve of a curve $\mathbf{z}(t)$ is expressed by*

$$\pi^n\,\mathbf{z}(t) = \mathbf{\Pi}^n\,\mathbf{z}(t), \tag{12}$$

where $\mathbf{\Pi}$ is a complex number given by

$$\mathbf{\Pi}^{-1} = 1 + \frac{\mathbf{n}(t) \times \mathbf{z}(t)}{\mathbf{n}(t) \cdot \mathbf{z}(t)} = 1 + \frac{\overline{\mathbf{n}}(t)\mathbf{z}(t) - \mathbf{n}(t)\overline{\mathbf{z}}(t)}{\overline{\mathbf{n}}(t)\mathbf{z}(t) + \mathbf{n}(t)\overline{\mathbf{z}}(t)}. \tag{13}$$

Proof: For $n = 1$, (12) becomes $\pi\,\mathbf{z}(t) = \mathbf{\Pi}\,\mathbf{z}(t)$. Hence, the complex number $\mathbf{\Pi}$ is given by

$$\mathbf{\Pi}^{-1} = \frac{\mathbf{z}(t)}{\pi\,\mathbf{z}(t)} = 1 + \frac{\mathbf{z}(t)\overline{\mathbf{z}}'(t) + \overline{\mathbf{z}}(t)\mathbf{z}'(t)}{\mathbf{z}(t)\overline{\mathbf{z}}'(t) - \overline{\mathbf{z}}(t)\mathbf{z}'(t)}. \tag{14}$$

As the derivative and its conjugates are represented by $\mathbf{z}'(t) = \mathrm{i}\,\mathbf{n}(t)$ and $\overline{\mathbf{z}}'(t) = -\mathrm{i}\,\overline{\mathbf{n}}(t)$, we obtain (13).

On the other hand, the multiplication $\mathbf{\Pi}^{-1}\,\mathbf{z}(t)$ is transformed into

$$\mathbf{\Pi}^{-1}\,\mathbf{z}(t) = \frac{2\mathbf{z}(t)^2}{\mathbf{z}(t)\overline{\mathbf{z}}'(t) - \overline{\mathbf{z}}(t)\mathbf{z}'(t)}\overline{\mathbf{z}}'(t). \tag{15}$$

From (11) and (15), the negative pedal curve $\pi^{-1}\,\mathbf{z}(t)$ is expressed by the multiplication $\mathbf{\Pi}^{-1}\,\mathbf{z}(t)$, namely,

$$\mathbf{\Pi}^{-1}\,\mathbf{z}(t) \;\underset{\pi^{-1}}{\overset{\pi}{\rightleftarrows}}\; \mathbf{z}(t) \;\underset{\pi^{-1}}{\overset{\pi}{\rightleftarrows}}\; \mathbf{\Pi}\,\mathbf{z}(t)\,. \tag{16}$$

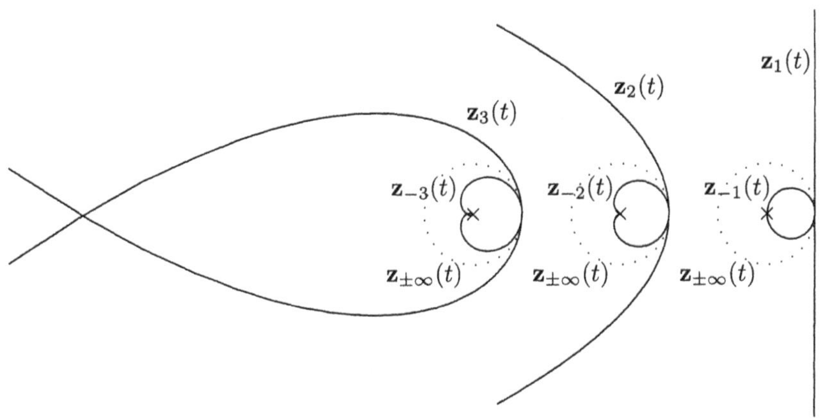

Fig. 2. Rational sinusoidal spirals ($m = \pm 3, \pm 2, \pm 1$) with the limit circle ($m = \pm\infty$).

As this diagram means $\pi^{i+1}\,\mathbf{z}(t)/\pi^i\,\mathbf{z}(t) = \pi^i\,\mathbf{z}(t)/\pi^{i-1}\,\mathbf{z}(t)$ for any curve $\mathbf{z}(t)$, the nth pedal curve of a curve $\mathbf{z}(t)$ is expressed by $\pi^n\,\mathbf{z}(t) = \mathbf{\Pi}^n\,\mathbf{z}(t)$. □

Diagram (16) also shows the property $[\pi^{-1}\,\mathbf{z}(t)]\,[\pi\,\mathbf{z}(t)] = \mathbf{z}(t)^2$. By using the angle θ in Figure 1, the complex number is rewritten as

$$\mathbf{\Pi}^{-1} = 1 + \frac{\mathbf{n}(t) \times \mathbf{z}(t)}{\mathbf{n}(t) \cdot \mathbf{z}(t)} = 1 + \mathrm{i}\,\frac{||\mathbf{n}(t)||\,||\mathbf{z}(t)||\,\sin\theta}{||\mathbf{n}(t)||\,||\mathbf{z}(t)||\,\cos\theta} = 1 + \mathrm{i}\,\tan\theta. \qquad (17)$$

The conjugate $\overline{\mathbf{\Pi}}$ of the complex number $\mathbf{\Pi}$ is transformed as

$$\overline{\mathbf{\Pi}}^{-1} = \overline{\mathbf{\Pi}^{-1}} = \overline{\left(\frac{\mathbf{z}(t)}{\pi\,\mathbf{z}(t)}\right)} = \frac{\overline{\mathbf{z}}(t)}{\overline{\pi\,\mathbf{z}(t)}} = \frac{\iota\,(\pi\,\mathbf{z}(t))}{\iota\,\mathbf{z}(t)} = \frac{\pi^{-1}\,(\iota\,\mathbf{z}(t))}{\iota\,\mathbf{z}(t)}. \qquad (18)$$

Equation (18) shows that the conjugate $\overline{\mathbf{\Pi}}$ is the coefficient for the pedal curves of the inverse curve $\iota\,\mathbf{z}(t)$:

$$\pi^n\,(\iota\,\mathbf{z}(t)) = \overline{\mathbf{\Pi}}^n\,(\iota\,\mathbf{z}(t)) = \left(1 - \frac{\mathbf{n}(t) \times \mathbf{z}(t)}{\mathbf{n}(t) \cdot \mathbf{z}(t)}\right)^{-n} (\iota\,\mathbf{z}(t)). \qquad (19)$$

For example, a family of curves $\mathbf{z}_m(t)$ [13] for an integer m is defined as $\mathbf{z}_m(t) = (1 + \mathrm{i}\,t)^m$. Substituting $t = \tan(\theta/m)$, we can obtain the polar equation of the curve $\rho = ||\mathbf{z}_m(t)|| = \sqrt{1+t^2}^{\,m} = \cos^{-m}(\theta/m)$. As the polar equation of sinusoidal spirals is $\rho^p = a^p \cos p\theta$ [8], the curve $\mathbf{z}_m(t)$ is the case of $a = 1$ and $p = -1/m$, i.e., $\rho = \cos^{-m}(\theta/m)$. The family of curves $\mathbf{z}_m(t)$ can be called rational sinusoidal spirals.

For each m, the curve $\mathbf{z}_m(t)$ is called as in [8]

$\mathbf{z}_1(t)$:	straight line,	$\mathbf{z}_{-1}(t)$:	circle (radius : $\frac{1}{2}$),
$\mathbf{z}_2(t)$:	parabola,	$\mathbf{z}_{-2}(t)$:	cardioid,
$\mathbf{z}_3(t)$:	Tschirnhausen cubic,	$\mathbf{z}_{-3}(t)$:	Cayley's sextic,
$\vdots$		$\vdots$	
$\mathbf{z}_\infty(t)$:	unit circle,	$\mathbf{z}_{-\infty}(t)$:	unit circle.

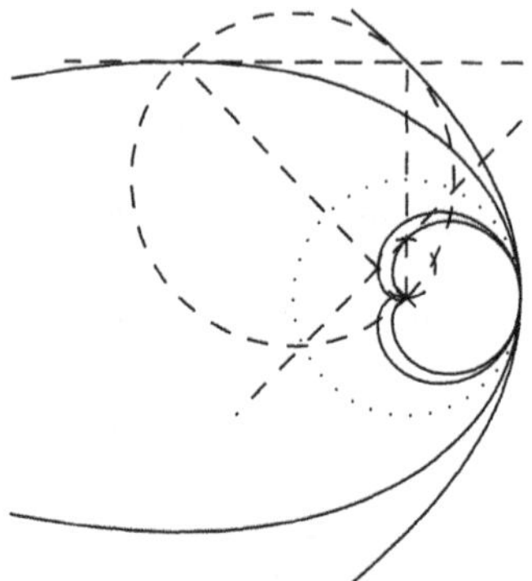

Fig. 3. Relationship between pedals and inverses.

Figure 2 illustrates several curves in this family of curves.

As the complex number $\mathbf{\Pi}$ for the curve $\mathbf{z}_m(t)$ is $(1+\mathrm{i}\,t)^{-1}$, the nth pedal curve is expressed by

$$\pi^n\,\mathbf{z}_m(t) = (1+\mathrm{i}\,t)^{m-n} = \mathbf{z}_{m-n}(t). \tag{20}$$

The following diagram shows the relationship in this family of curves:

$$\begin{array}{ccccccccccc}
\xrightarrow{\pi} & \mathbf{z}_3(t) & \xrightarrow{\pi} & \mathbf{z}_2(t) & \xrightarrow{\pi} & \mathbf{z}_1(t) & \xrightarrow{\pi} & \mathbf{z}_0(t) & \xrightarrow{\pi} & \mathbf{z}_{-1}(t) & \xrightarrow{\pi} \\
 & \updownarrow\iota & & \updownarrow\iota & & \updownarrow\iota & & \updownarrow\iota & & \updownarrow\iota & \\
\xleftarrow[\pi]{} & \overline{\mathbf{z}}_{-3}(t) & \xleftarrow[\pi]{} & \overline{\mathbf{z}}_{-2}(t) & \xleftarrow[\pi]{} & \overline{\mathbf{z}}_{-1}(t) & \xleftarrow[\pi]{} & \overline{\mathbf{z}}_0(t) & \xleftarrow[\pi]{} & \overline{\mathbf{z}}_1(t) & \xleftarrow[\pi]{}
\end{array}$$

Figure 3 illustrates the relationship among the curves $\mathbf{z}_{\pm 2}$ and $\mathbf{z}_{\pm 3}$.

As the shape of the curve $\mathbf{z}_m(t)$ is symmetric about the real axis, the conjugate $\overline{\mathbf{z}}_m(t)$ is equivalent to the curve $\mathbf{z}_m(t)$ geometrically. Hence, the inverse curve $\iota\,\mathbf{z}_m(t) = \overline{\mathbf{z}}_{-m}(t)$ and the pedal curve $\pi^n\,\mathbf{z}_m(t)$ are included in this family of curves.

Equation (20) holds true for a rational number m. The pedal curve and the inverse curve of the rectangular hyperbola $\mathbf{z}_{\frac{1}{2}}(t)$ are the lemniscates of Bernoulli, $\mathbf{z}_{-\frac{1}{2}}(t)$ and $\overline{\mathbf{z}}_{-\frac{1}{2}}(\iota)$, respectively.

§4. Pedal Surfaces and Quaternions

Pedal surfaces are defined in a similar manner to pedal curves. A surface $\mathbf{s}(u,v)$ and its partial derivatives $\mathbf{s}_u(u,v)$ and $\mathbf{s}_v(u,v)$ are given by

$$\mathbf{s}(u,v) = \begin{pmatrix} x(u,v) \\ y(u,v) \\ z(u,v) \end{pmatrix}, \quad \mathbf{s}_u(u,v) = \begin{pmatrix} x_u(u,v) \\ y_u(u,v) \\ z_u(u,v) \end{pmatrix}, \quad \mathbf{s}_v(u,v) = \begin{pmatrix} x_v(u,v) \\ y_v(u,v) \\ z_v(u,v) \end{pmatrix}. \tag{21}$$

The normal vector $\mathbf{n}(u,v)$ to the surface is defined by $\mathbf{n}(u,v) = \mathbf{s}_u(u,v) \times \mathbf{s}_v(u,v)$.

Definition 3. *The* pedal surface $\pi\,\mathbf{s}(u,v)$ *of a surface* $\mathbf{s}(u,v)$ *with respect to the origin is defined by*

$$\pi\,\mathbf{s}(u,v) = \frac{\mathbf{s}(u,v)\cdot\mathbf{n}(u,v)}{\mathbf{n}(u,v)\cdot\mathbf{n}(u,v)}\mathbf{n}(u,v) \tag{22}$$

$$= \frac{x(y_u z_v - z_u y_v) + y(z_u x_v - x_u z_v) + z(x_u y_v - y_u x_v)}{(y_u z_v - z_u y_v)^2 + (z_u x_v - x_u z_v)^2 + (x_u y_v - y_u x_v)^2}\begin{pmatrix} y_u z_v - z_u y_v \\ z_u x_v - x_u z_v \\ x_u y_v - y_u x_v \end{pmatrix}.$$

Definition 4. *The* inverse surface $\iota\,\mathbf{s}(u,v)$ *of a surface* $\mathbf{s}(u,v)$ *in the unit sphere centered at the origin is defined by*

$$\iota\,\mathbf{s}(u,v) = \frac{1}{\mathbf{s}(u,v)\cdot\mathbf{s}(u,v)}\mathbf{s}(u,v) = \frac{1}{x(u,v)^2 + y(u,v)^2 + z(u,v)^2}\begin{pmatrix} x(u,v) \\ y(u,v) \\ z(u,v) \end{pmatrix}. \tag{23}$$

Theorem 3. *The negative pedal surface of a regular surface* $\mathbf{s}(u,v)$ *with respect to the origin is obtained by applying the following composition to the surface:*

$$\pi^{-1} = \iota \circ \pi \circ \iota. \tag{24}$$

Proof: Suppose the surface $\mathbf{p}(u,v)$ is the pedal curve $\pi\,\mathbf{s}(u,v)$ of the surface $\mathbf{s}(u,v)$, and the surface $\mathbf{q}(u,v)$ is the inverse surface $\mathbf{q}(u,v) = \iota\,\mathbf{p}(u,v)$ of the surface $\mathbf{p}(u,v)$. The pedal surface $\mathbf{r}(u,v)$ of the surface $\mathbf{q}(u,v)$ is expressed by

$$\mathbf{r}(u,v) = \left(\frac{\mathbf{p}(u,v)\cdot(\mathbf{p}_u(u,v)\times\mathbf{p}_v(u,v))}{(\mathbf{p}_u(u,v)\times\mathbf{p}_v(u,v))\cdot(\mathbf{p}_u(u,v)\times\mathbf{p}_v(u,v))}\right)(\mathbf{p}_u(u,v)\times\mathbf{p}_v(u,v))$$

$$= \frac{1}{x^2(u,v) + y^2(u,v) + z^2(u,v)}\begin{pmatrix} x(u,v) \\ y(u,v) \\ z(u,v) \end{pmatrix} = \iota\,\mathbf{s}(u,v). \tag{25}$$

As the surface $\mathbf{r}(u,v)$ is also expressed as $\mathbf{r}(u,v) = \pi\,\mathbf{q}(u,v) = \pi\,(\iota\,\mathbf{p}(u,v)) = \pi\,(\iota\,(\pi\,\mathbf{s}(u,v)))$, we obtain the equation $\mathbf{s}(u,v) = (\iota\circ\pi\circ\iota\circ\pi)\,\mathbf{s}(u,v)$, and conclude (24). □

The following diagram similar to that of pedal curves is obtained for pedal surfaces.

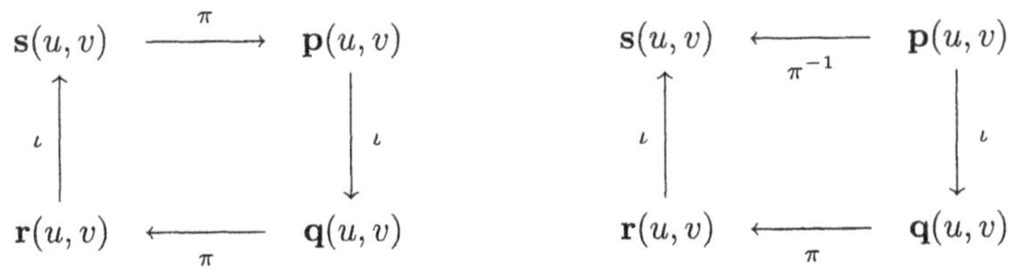

$$\begin{array}{ccc} \mathbf{s}(u,v) & \xrightarrow{\pi} & \mathbf{p}(u,v) \\ \uparrow \iota & & \downarrow \iota \\ \mathbf{r}(u,v) & \xleftarrow[\pi]{} & \mathbf{q}(u,v) \end{array} \qquad \begin{array}{ccc} \mathbf{s}(u,v) & \xleftarrow[\pi^{-1}]{} & \mathbf{p}(u,v) \\ \uparrow \iota & & \downarrow \iota \\ \mathbf{r}(u,v) & \xleftarrow[\pi]{} & \mathbf{q}(u,v) \end{array}$$

Quaternion calculus is suitable for expressing higher pedal surfaces. A quaternion is an extension of a complex number [7] and can be expressed as the elements of the 4-space $\mathbb{R}^4$ of the span $\{1, \mathrm{i}, \mathrm{j}, \mathrm{k}\}$, which follows the rules of production $\mathrm{ij} = -\mathrm{ji} = \mathrm{k}$, $\mathrm{jk} = -\mathrm{kj} = \mathrm{i}$, $\mathrm{ki} = -\mathrm{ik} = \mathrm{j}$ and $\mathrm{i}^2 = \mathrm{j}^2 = \mathrm{k}^2 = -1$. The conjugate $\overline{\mathbf{Q}}$, norm $||\mathbf{Q}||$ and inverse $\mathbf{Q}^{-1}$ of a quaternion $\mathbf{Q} = Q + \mathrm{i}\,Q_x + \mathrm{j}\,Q_y + \mathrm{k}\,Q_z$ are defined as

$$\overline{\mathbf{Q}} = Q - Q_x\mathrm{i} - Q_y\mathrm{j} - Q_z\mathrm{k}, \quad ||\mathbf{Q}||^2 = \mathbf{Q}\overline{\mathbf{Q}}, \quad \mathbf{Q}^{-1} = \frac{\overline{\mathbf{Q}}}{||\mathbf{Q}||^2}. \tag{26}$$

The product of two quaternion, $\mathbf{P} = P + \mathrm{i}\,P_x + \mathrm{j}\,P_y + \mathrm{k}\,P_z$ and $\mathbf{Q}$, is represented as

$$\begin{aligned}\mathbf{P}\,\mathbf{Q} =&(PQ - P_xQ_x - P_yQ_y - P_zQ_z) + \mathrm{i}\,(PQ_x + P_xQ + P_yQ_z - P_zQ_y) \\ &+\mathrm{j}\,(PQ_y - P_xQ_z + P_yQ + P_zQ_x) + \mathrm{k}\,(PQ_z + P_xQ_y - P_yQ_x + P_zQ).\end{aligned} \tag{27}$$

The following properties are easily obtained:

$$(\mathbf{P}\,\mathbf{Q})\,\mathbf{R} = \mathbf{P}\,(\mathbf{Q}\,\mathbf{R}), \qquad (\mathbf{P}\,\mathbf{Q})^{-1} = \mathbf{Q}^{-1}\,\mathbf{P}^{-1}, \qquad \overline{\mathbf{P}\,\mathbf{Q}} = \overline{\mathbf{Q}}\,\overline{\mathbf{P}}. \tag{28}$$

Usually, pure vector quaternions are used to express 3D space $\mathbb{R}^3$. A quaternion $\mathbf{p}$ such that $\overline{\mathbf{p}} = -\mathbf{p}$ is called a pure vector quaternion. The conjugate $\overline{\mathbf{p}}$, norm $||\mathbf{p}||$ and inverse $\mathbf{p}^{-1}$ of a pure vector quaternion $\mathbf{p} = p + \mathrm{i}\,p_x + \mathrm{j}\,p_y + \mathrm{k}\,p_z$ are defined as

$$\overline{\mathbf{p}} = -\mathbf{p}, \qquad ||\mathbf{p}||^2 = \mathbf{p}\,\overline{\mathbf{p}} = -\mathbf{p}^2, \qquad \mathbf{p}^{-1} = -\frac{\mathbf{p}}{||\mathbf{p}||^2} = \frac{\mathbf{p}}{\mathbf{p}^2}. \tag{29}$$

The dot product and cross product in vector calculus can be expressed using pure vector quaternions. The dot product $\mathbf{p}\cdot\mathbf{q}$ and cross product $\mathbf{p}\times\mathbf{q}$ of two pure vector quaternions $\mathbf{p}$ and $\mathbf{q} = \mathrm{i}\,q_x + \mathrm{j}\,q_y + \mathrm{k}\,q_z$ can be defined as

$$\mathbf{p}\cdot\mathbf{q} \equiv -\frac{\mathbf{p}\mathbf{q} + \mathbf{q}\mathbf{p}}{2} = p_xq_x + p_yq_y + p_zq_z, \tag{30}$$

$$\mathbf{p}\times\mathbf{q} \equiv \frac{\mathbf{p}\mathbf{q} - \mathbf{q}\mathbf{p}}{2} = \mathrm{i}\,(p_yq_z - p_zq_y) + \mathrm{j}\,(p_zq_x - p_xq_z) + \mathrm{k}\,(p_xq_y - p_yq_x). \tag{31}$$

Hence, the product $\mathbf{p}\,\mathbf{q}$ of two pure vector quaternions is equal to $-\mathbf{p}\cdot\mathbf{q} + \mathbf{p}\times\mathbf{q}$, and has the following properties.

$$\mathbf{p}\times\mathbf{p} = \mathbf{0}, \quad \mathbf{p}\cdot\mathbf{p} = -\mathbf{p}^2 = ||\mathbf{p}||^2, \quad \mathbf{p}\times\mathbf{q} = -(\mathbf{q}\times\mathbf{p}), \quad \mathbf{p}\cdot\mathbf{q} = \mathbf{q}\cdot\mathbf{p}. \tag{32}$$

A surface $\mathbf{s}(u,v)$, its partial derivatives $\mathbf{s}_u(u,v)$ and $\mathbf{s}_v(u,v)$ and the normal vector $\mathbf{n}(u,v)$ to the surface are given by

$$\mathbf{s}(u,v) = \mathrm{i}\,x(u,v) + \mathrm{j}\,y(u,v) + \mathrm{k}\,z(u,v), \tag{33}$$

$$\mathbf{s}_u(u,v) = \mathrm{i}\,x_u(u,v) + \mathrm{j}\,y_u(u,v) + \mathrm{k}\,z_u(u,v), \tag{34}$$

$$\mathbf{s}_v(u,v) = \mathrm{i}\,x_v(u,v) + \mathrm{j}\,y_v(u,v) + \mathrm{k}\,z_v(u,v), \tag{35}$$

$$\mathbf{n}(u,v) = \mathbf{s}_u(u,v) \times \mathbf{s}_v(u,v). \tag{36}$$

The pedal surface $\pi\,\mathbf{s}(u,v)$ and the inverse surface $\iota\,\mathbf{s}(u,v)$ of the surface $\mathbf{s}(u,v)$ are also given by

$$\pi\,\mathbf{s}(u,v) = \frac{\mathbf{n}(u,v)\cdot\mathbf{s}(u,v)}{\mathbf{n}(u,v)\cdot\mathbf{n}(u,v)}\mathbf{n}(u,v), \quad \iota\,\mathbf{s}(u,v) = \overline{\mathbf{s}}(u,v)^{-1} = \frac{\mathbf{s}(u,v)}{||\mathbf{s}(u,v)||^2}. \tag{37}$$

Theorem 4. *For an integer n, the nth pedal surface of a surface represented by a pure vector quaternion $\mathbf{s}(u,v)$ is expressed by*

$$\pi^n\,\mathbf{s}(u,v) = \mathbf{\Pi}^n\,\mathbf{s}(u,v), \tag{38}$$

where $\mathbf{\Pi}$ is a quaternion given by

$$\mathbf{\Pi}^{-1} = 1 + \frac{\mathbf{n}(u,v)\times\mathbf{s}(u,v)}{\mathbf{n}(u,v)\cdot\mathbf{s}(u,v)} = 1 - \frac{\mathbf{n}(u,v)\,\mathbf{s}(u,v) - \mathbf{s}(u,v)\,\mathbf{n}(u,v)}{\mathbf{n}(u,v)\,\mathbf{s}(u,v) + \mathbf{s}(u,v)\,\mathbf{n}(u,v)}. \tag{39}$$

Proof: From the definitions, we obtain

$$\mathbf{s}(u,v)\,(\pi\,\mathbf{s}(u,v))^{-1} = -\frac{\mathbf{s}(u,v)\,\mathbf{n}(u,v)}{\mathbf{n}(u,v)\cdot\mathbf{s}(u,v)} = 1 + \frac{\mathbf{n}(u,v)\times\mathbf{s}(u,v)}{\mathbf{n}(u,v)\cdot\mathbf{s}(u,v)}. \tag{40}$$

For $n = 1$, the quaternion $\mathbf{\Pi}$ is defined as $\mathbf{\Pi}^{-1} = \mathbf{s}(u,v)\,(\pi\,\mathbf{s}(u,v))^{-1}$. On the other hand, the product of the inverse of the quaternion $\mathbf{\Pi}^{-1}$ and the surface $\mathbf{s}(u,v)$ yields the surface $\mathbf{t}(u,v) = \mathbf{\Pi}^{-1}\,\mathbf{s}(u,v)$. Using a rational function $f(u,v)$, the normal vector $\mathbf{m}(u,v)$ to the surface $\mathbf{t}(u,v)$ can be expressed as

$$\mathbf{m}(u,v) = \mathbf{t}_u(u,v)\times\mathbf{t}_v(u,v) = f(u,v)\mathbf{s}(u,v), \tag{41}$$

and the pedal surface $\pi\,\mathbf{t}(u,v)$ of the surface $\mathbf{t}(u,v)$ with respect to the origin becomes

$$\pi\,\mathbf{t}(u,v) = \frac{\mathbf{t}\cdot\mathbf{m}}{\mathbf{m}\cdot\mathbf{m}}\mathbf{m} = \frac{f(u,v)(x^2+y^2+z^2)}{f^2(u,v)(x^2+y^2+z^2)}f(u,v)\mathbf{s}(u,v) = \mathbf{s}(u,v). \tag{42}$$

As (42) means that the surface $\mathbf{t}(u,v)$ is the negative pedal surface of the surface $\mathbf{s}(u,v)$, there is the following relationship:

$$\mathbf{\Pi}^{-1}\,\mathbf{s}(u,v) \;\underset{\pi^{-1}}{\overset{\pi}{\rightleftarrows}}\; \mathbf{s}(u,v) \;\underset{\pi^{-1}}{\overset{\pi}{\rightleftarrows}}\; \mathbf{\Pi}\,\mathbf{s}(u,v)\,. \tag{43}$$

Hence, the nth pedal surface $\pi^n\,\mathbf{s}(u,v)$ of a surface $\mathbf{s}(u,v)$ is expressed by $\mathbf{\Pi}^n\,\mathbf{s}(u,v)$. □

From $\mathbf{\Pi}\,\mathbf{s}(u,v) = \mathbf{s}(u,v)\,\overline{\mathbf{\Pi}}$, the nth pedal surface is also given by

$$\pi^n\,\mathbf{s}(u,v) = \mathbf{s}(u,v)\,\overline{\mathbf{\Pi}}^n. \tag{44}$$

For the inverse surface $\iota\,\mathbf{s}(u,v)$, the nth pedal surface is expressed by

$$\pi^n\,(\iota\,\mathbf{s}(u,v)) = (\iota\,\mathbf{s}(u,v))\,\mathbf{\Pi}^n = \overline{\mathbf{\Pi}}^n\,(\iota\,\mathbf{s}(u,v)). \tag{45}$$

For example, the pedal surface of the surface $\mathbf{q}_n(u,v) = (\mathrm{i} + \mathrm{j}\,u + \mathrm{k}\,v)^n$ is a surface of revolution which is obtained by rotating the curve $\pi\,\mathbf{z}_n(t)$ about the real axis, because the surface $\mathbf{q}_1(u,v)$ is the plane with a distance of 1 from the origin.

§5. Conclusion

In this paper, a relationship between pedal curves and inverse curves and a property in a family of positive or negative pedal curves derived from a curve have been shown. There are analogous properties for pedal surfaces and inverse surfaces.

It has been also shown that complex numbers and quaternions are suitable for representing pedal curves and surfaces, respectively.

References

1. Berger M., *Geometry*, Springer-Verlag, 1987.
2. Bruce, J. W. and P. J. Giblin, *Curves and Singularities*, 2nd ed., Cambridge University Press, 1992.
3. Farouki, R. T., The conformal map $z \to z^2$ of the hodograph plane, Computer Aided Geometric Design **11** (1994), 363–390.
4. Gray, A., *Modern Differential Geometry of Curves and Surfaces*, CRC Press, 1993.
5. Hilbert, D. and S. Cohn-Vossen, *Geometry and the Imagination*, Chelsea Publishing Co., 1952.
6. Hoschek, J., Dual Bézier curves and surfaces, in *Surfaces in Computer Aided Geometric Design*, R. E. Barnhill and W. Boehm (eds.), North-Holland, 1983, 147–156.
7. Kantor, I. L. and A. S. Solodovnikov, *Hypercomplex Numbers: An Elementary Introduction to Algebras*, Springer-Verlag, 1989.
8. Lawrence, J. D., *A Catalog of Special Plane Curves*, Dover, 1972.
9. Needham, T., *Visual Complex Analysis*, Oxford University Press, 1997.
10. Pletinckx, D., Quarternion calculus as a basic tool in computer graphics, The Visual Computer **5** (1989), 2–13.
11. Shikin, E. V., *Handbook and Atlas of Curves*, CRC Press, 1995.
12. Ueda, K., A sequence of Bézier curves generated by successive pedal-point constructions, in *Curves and Surfaces with Applications in CAGD*, A. Le Méhauté, C. Rabut, and L. L. Schumaker (eds.), Vanderbilt University Press, Nashville, 1997, 427–434.
13. Ueda, K., Polar curves and surfaces, in *The Mathematics of Surfaces IX*, R. Cipolla and R. Martin (eds.), Springer-Verlag, 2000, 372–388.

Kenji Ueda
Ricoh Company, Ltd.
1-1-17, Koishikawa, Bunkyo-ku, Tokyo 112-0002, Japan
ueda@src.ricoh.co.jp

Generalizing the C^4 Four-directional Box Spline to Surfaces of Arbitrary Topology

Luiz Velho

Abstract. In this paper we introduce a new scheme that generalizes the four-directional box spline of class C^4 to surfaces of arbitrary topological type. The scheme is composed of semi-regular binary refinement together with separable two-pass smoothing by repeated convolution.

§1. The $[4.8^2]$ Laves Tiling and Bisection Refinement

Mesh refinement methods are usually based on regular tilings, that is, tessellations of the plane composed by regular n-gons. The basic idea is to start with an initial uniform tessellation, and then apply repeatedly some refinement rule, such that at every step a finer tessellation, similar under scaling to the original, is produced.

There are only three types of plane tilings formed by tiles that are congruent to a single regular polygon. They correspond to uniform tessellations generated by squares, equilateral triangles, and regular hexagons. The most common ones are the triangle and quadrilateral tessellations. Note that the above tilings have the desired property: it is possible to subdivide the tiles obtaining a new tiling made by similar elements of smaller size.

A larger class of refinable tilings are the monohedral tilings with regular vertices, also known as Laves tilings, named after the crystallographer Fritz Laves, who studied them [2].

In a monohedral tiling, every tile is congruent to one fixed tile, called the prototile. This means that all faces in the tessellation have the same shape and size. A vertex v of a tiling is called regular if the angle between each consecutive pair of edges that are incident in v is equal to $2\pi/d$, where d is the valence of v.

There are eleven tilings that satisfy these two conditions. We classify these tilings by listing the degree of the vertices of their prototile in cyclic order. Thus, they are named using the notation $[d_1, d_2 \ldots, d_k]$, where d_i is

Mathematical Methods for Curves and Surfaces: Oslo 2000
Tom Lyche and Larry L. Schumaker (eds.), pp. 507–516.

ISBN 0-8265-1378-6

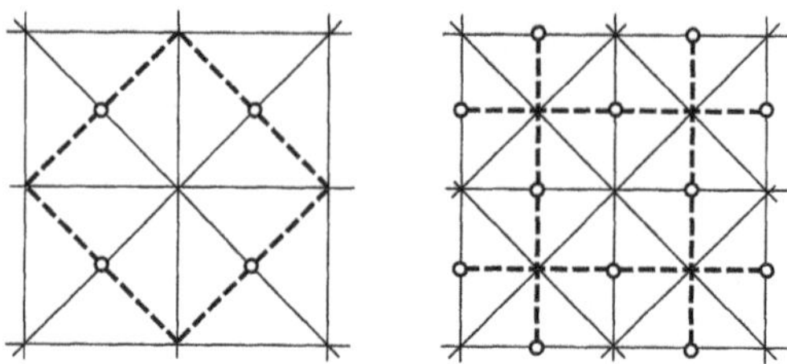

Fig. 1. Two bisection refinement steps of the 4–8 the refinement procedure.

the valence of vertex v_i (we also use superscripts to indicate repetition of symbols). As expected, regular triangle - quadrilateral - and hexagonal tilings also belong to this class. They are, respectively, the Laves tilings of type $[6^3]$, $[4^4]$ and $[3^6]$. Naturally, it is possible to extend all refinement concepts to the Laves tilings.

The $[4.8^2]$ Laves tiling has a rich structure that can be exploited in the context of subdivision with many advantages over the traditional $[6^3]$ and $[4^4]$ tilings. Among other things, refinement is based on bisections, and uniform as well as non-uniform refinement are both supported.

The $[4.8^2]$ tiling is composed of isosceles right triangles. The basic structure is a pair of triangles forming a square block divided along one of its diagonals. We call this structure a basic block.

A regular 4–8 mesh is a cell complex that has the same connectivity as a $[4.8^2]$ Laves tiling. By definition, every face has one vertex of valence 4 and two vertices of valence 8. The 4–8 mesh has edges of two types: 8-8 edges, linking two vertices of valence 8; and 4–8 edges, linking one vertex of valence 4 to one vertex of valence 8.

We would like to devise a refinement procedure based just on topological information. Observe that edges of type 8–8 occur only as the diagonal edges of basic blocks. This follows directly from the regularity condition. Using this observation, we specify a 4–8 bisection refinement procedure.

The procedure is as follows: First, we bisect all edges of type 8–8, by inserting a split vertex. Then, we subdivide all faces into two sub-faces, by linking the vertex of degree 4 to the split vertex of the opposite edge.

Note that, in order to produce a self-similar mesh, 4–8 bisection refinement has to be applied twice (applying just one subdivision step results in a mesh that when rotated by 45 degrees is self-similar to the original). For this reason, the regular 4–8 refinement is defined as a double step of 4–8 bisection refinement This is illustrated in Figure 1.

The regular 4–8 refinement procedure relies on the special topological structure of the mesh. In order to make it widely applicable, particularly for the representation of 2D manifolds, it is necessary to extend the refinement procedure to accept arbitrary initial control meshes.

The generalization of regular 4–8 refinement to non-regular meshes exploits the fact that subdivision operates on quadrilateral blocks. Thus, our strategy is to take a triangulation as input, and, in a pre-processing phase,

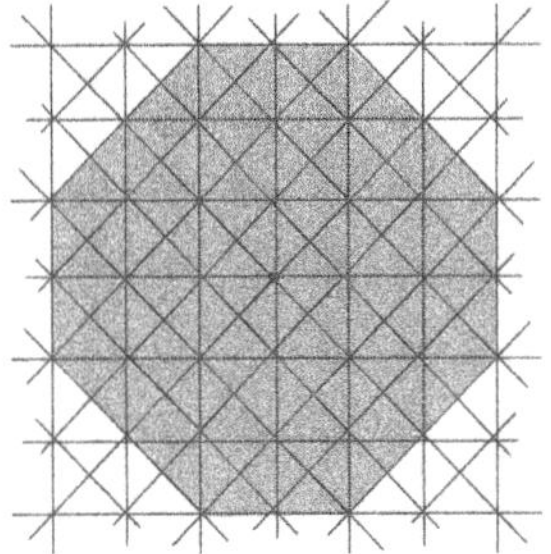
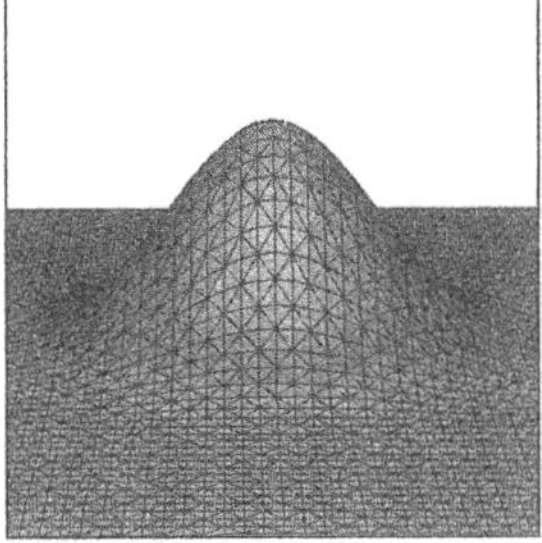

Fig. 2. C^4 Box spline function.

construct a block structure that can be handled subsequently by regular 4–8 refinement. This pre-processing transforms an arbitrary initial mesh into a triangulated quadrangulation [3]. We call the complete procedure a semi-regular 4–8 refinement, and the mesh produced under its action semi-regular 4–8 mesh. Thus, a semi-regular 4–8 mesh is a triangle mesh in which original vertices may have arbitrary valence and all other vertices have either valence 8 or valence 4. First, when a vertex is generated at refinement step i, its valence is 4. Then, at the subsequent refinement step $i+1$, its valence becomes 8.

§2. $\mathbf{C^4}$ Box Spline Subdivision

4–8 meshes are closely related to the four-directional grid, well known in the theory of splines [1]. A box spline is generated by convolutions of the characteristic function of the unit partition, along a prescribed set of directions. They are smooth piecewise polynomial functions with compact support. They are refinable, and their translates form a basis. Box spline basis are usually specified by a set of direction vectors.

Box spline functions can be used to create surfaces that are defined parametrically by a function $g: U \subset \mathbf{R}^2 \to \mathbf{R}^3$. In this setting, a box spline surface is specified by control points $c_{uv} \in \mathbf{R}^3$ that are associated with grid points $(u, v) \in \mathbf{Z}^2$ of the domain U.

The simplest smooth box spline over a four-directional grid is the Zwart-Powell function [7], also known as the ZP element. It is associated with the set of vectors $D = \{e_1, e_2, e_1 + e_2, e_1 - e_2\}$, where $e_1 = (1, 0)$ and $e_2 = (0, 1)$.

A four-directional box spline that exhibits a higher order of smoothness than the ZP element is the function generated by the set of direction vectors

$$D = \begin{pmatrix} 1 & 1 & 0 & 0 & 1 & 1 & 1 & 1 \\ 0 & 0 & 1 & 1 & 1 & 1 & -1 & -1 \end{pmatrix}.$$

This function is a piecewise polynomial of degree 6 and it is C^4 continuous. Figure 2 shows a plot of the function, as well as its support on the four-directional grid.

Using refinement, we express the function on a coarse grid as a linear combination of scaled and translated functions on a finer grid. This two-scale

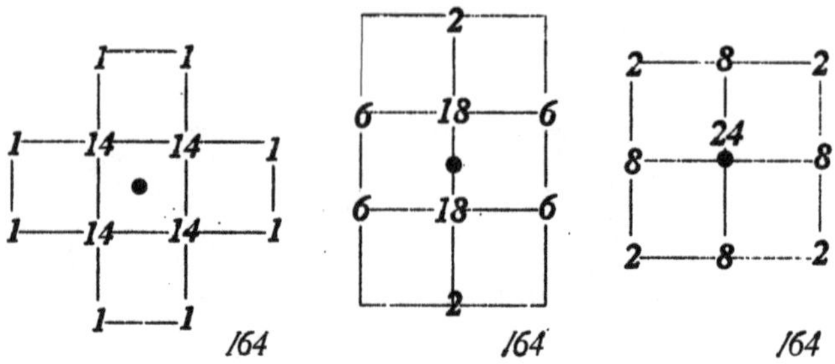

Fig. 3. Face, edge and vertex masks for C^4 box spline.

relation can be computed from the generating function $S(z_1, z_2)$ associated with the induced subdivision scheme (See [6]). The generating function of the C^4 four-directional box spline is given by

$$S(z_1, z_2) = \frac{1}{64}(1+z_1)^2(1+z_2)^2(1+z_1z_2)^2(1+z_1/z_2)^2.$$

Expanding it, we compute the coefficients of the two-scale relation. They are the coefficients of the monomials of $S(z_1, z_2)$, where the weight at grid point (u, v) is the coefficient of $z_1^u z_2^v$. These coefficients are shown below (without normalization by the factor $\frac{1}{64}$).

		1	2	1		
	2	6	8	6	2	
1	6	14	18	14	6	1
2	8	18	24	18	8	2
1	6	14	18	14	6	1
	2	6	8	6	2	
		1	2	1		

As the grid is refined, values at grid points of the finer grid are calculated as linear combinations of values at grid points of the coarse grid. From the subdivision formula we obtain the smoothing masks for face, edge and vertex points, shown in Figure 3. The masks extend over a 2-neighborhood, as expected.

The large support of the masks makes the implementation of a subdivision scheme based on the C^4 box spline difficult. The separability property of semi-regular 4–8 refinement, allows us to factorize such a high order scheme into manageable pieces.

We decompose the C^4 smoothing operator into two masks, shown in Figure 4. It is easy to see that the smoothing filter corresponds to averaging in the horizontal, vertical, and diagonal directions. When applied successively in the two binary steps comprising the 4–8 refinement, the result is a discrete convolution twice in each of the four directions.

First, we use the face mask shown in Figure 3 (left) to compute the values at new vertices $v' \in V'$ (recall that, by construction, their 1-neighborhood

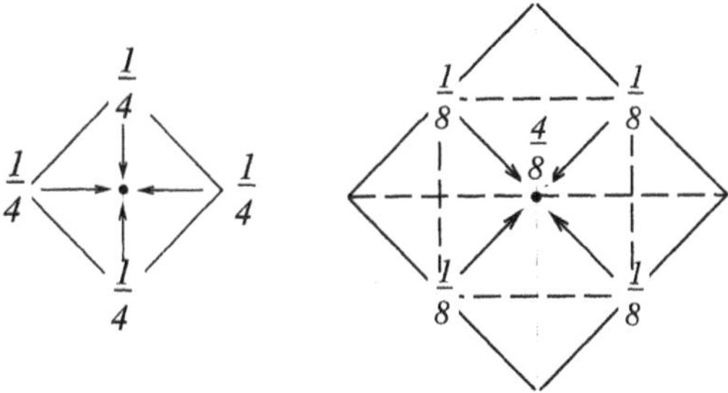

Fig. 4. Factorized face and vertex masks for C^4 Box spline.

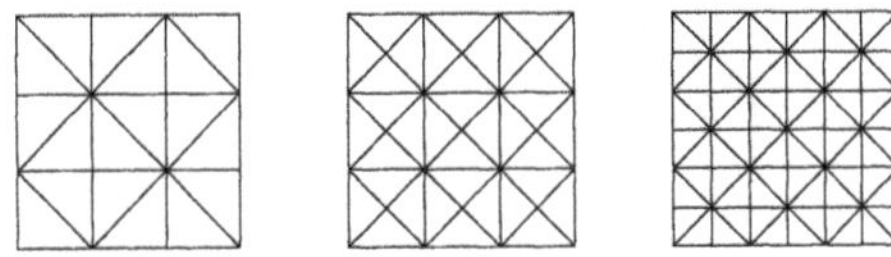

Fig. 5. Three consecutive subdivision levels, $j-1$, j and $j+1$.

$N_1(v')$ consists of exactly 4 vertices). The filter function is an average of the values at the neighbors. Next, we update the values of old ordinary vertices as follows: for any vertex v with valence 8 *after subdivision*, (i.e. old ordinary vertices), update it by weighted average of its neighboring valence-8 vertices, $\{v' \in N_1(v); deg(v') = 8\}$, (always 4 vertices) and v itself. This vertex mask is shown in Figure 4 (right).

The two factorized face and vertex masks in Figure 4 combine to produce the original face, edge, and vertex masks of the C^4 box spline in Figure 3.

We now describe in more detail the factorization of the C^4 subdivision scheme. We will analyze three consecutive levels of subdivision, $j+1$, j and $j-1$. We want to compute the values of the vertices of the mesh, $v_i^{j+1} \in V$ at level $j+1$. The face, vertex, and edge masks in Figure 3 compute new values based on previous values of the vertices of the mesh at level $j-1$. The factorization of these masks will employ a combination of the face and vertex masks in Figure 3 at levels $j+1$ and j, using values of the vertices at levels j and $j-1$, respectively. We remark that the factorized masks appear rotated by 45 degrees at two consecutive levels. In the following figures, the face and vertex masks will be indicated, respectively, by dashed arrows on a gray background and by solid arrows. Figure 5 shows a piece of the mesh at these three consecutive levels of subdivision, which we will use in the analysis. For clarity, in the figures we will only show the grid lines of level j.

Figure 6 shows the labeling scheme that we will use to identify the vertices of the mesh.

The C^4 face mask is decomposed into an application of face and vertex masks at level j, followed by an application of vertex mask at level $j+1$.

$a \quad b \quad c \quad d$

$q \quad r \quad s$

$e \quad f \quad g \quad h$

$t \quad z \quad u \quad v$

$i \quad j \quad k \quad l$

$w \quad x \quad y$

$m \quad n \quad o \quad p$

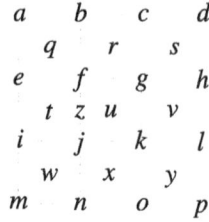

Fig. 6. Labeling scheme for the vertices of the mesh.

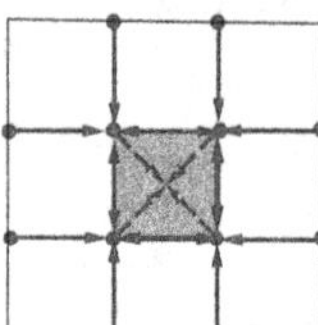

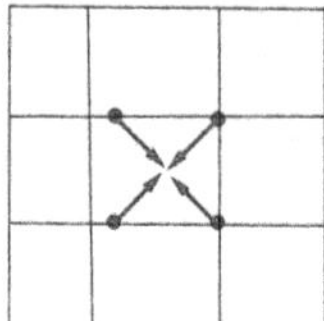

Fig. 7. Decomposition of the center mask.

Figure 7 shows the flow of computations used for updating vertex u.

At level $j+1$ the new value of vertex u is computed using the vertex mask

$$u^{j+1} = \frac{4}{8}u^j + \frac{1}{8}\left(f^j + g^j + k^j + j^j\right).$$

The values of u^j, f^j, g^j, k^j, and j^j are computed at level j using the face mask for computing u^j

$$u^j = \frac{1}{4}\left(f^{j-1} + g^{j-1} + k^{j-1} + j^{j-1}\right),$$

and the vertex mask for computing the values of f^j, g^j, k^j, and j^j. So, we have

$$f^j = \frac{4}{8}f^{j-1} + \frac{1}{8}\left(b^{j-1} + g^{j-1} + j^{j-1} + e^{j-1}\right),$$

and similarly for vertices g^j, k^j, and j^j. Substituting the values of vertices u^j, f^j, g^j, k^j, and j^j into the equation for u^{j+1}, we obtain the C^4 face mask, which gives the value of u^{j+1} in terms of the values of the vertices of the mesh ate level $j-1$.

The C^4 edge mask is decomposed into an application of face and vertex masks at level j, followed by an application of face mask at level $j+1$. Figure 8 shows the flow of computations used for updating vertex z.

At level $j+1$ the new value of vertex z is computed using the face mask

$$z^{j+1} = \frac{1}{4}\left(f^j + u^j + j^j + t^j\right).$$

Fig. 8. Decomposition of the edge mask.

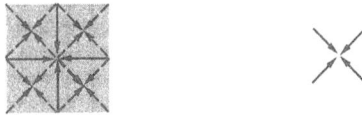

Fig. 9. Decomposition of the vertex mask.

The values of f^j, u^j, j^j, and t^j are computed at level j using the vertex mask for computing f^j and j^j, and the face mask for computing u^j and t^j. The computation for j^j is similar to the computation of f^j, while the computation for t^j is similar to the computation of u^j. Substituting the values of f^j, u^j, j^j, and t^j into equation for z^{j+1}, we obtain the C^4 edge mask.

The C^4 vertex mask is decomposed into an application of face and vertex masks at level j, followed by an application of vertex mask at level $j+1$. Figure 9 shows the flow of computations used for updating vertex f.

At level $j+1$ the new value of vertex f is computed using the vertex mask

$$f^{j+1} = \frac{4}{8}f^j + \frac{1}{8}\left(r^j + u^j + t^j + q^j\right).$$

The values of f^j, r^j ,u^j, t^j, and q^j are computed at level j using the vertex mask for computing f^j and the face mask for computing r^j ,u^j, t^j, and q^j. The computation for r^j, t^j, and q^j is similar to the computation of u^j. Again, substituting the values of f^j, r^j ,u^j, t^j, and q^j, into the equation for f^{j+1}, we obtain the C^4 vertex mask.

The factorization described above applies only for regular vertices of valence 4 and 8. In order to extend the subdivision scheme to arbitrary meshes, we have to devise masks for extraordinary vertices of valence $2n$ (since vertices of a semi-regular 4-8 mesh have even valence). It turns out that is only necessary to generalize the factorized vertex mask. This mask will combine with the face mask to produce new C^4 face, edge and vertex masks for arbitrary meshes.

Figure 10 shows the generalization of the factorized vertex mask. Note that extraordinary vertices have n new neighbors with valence 4 and n old

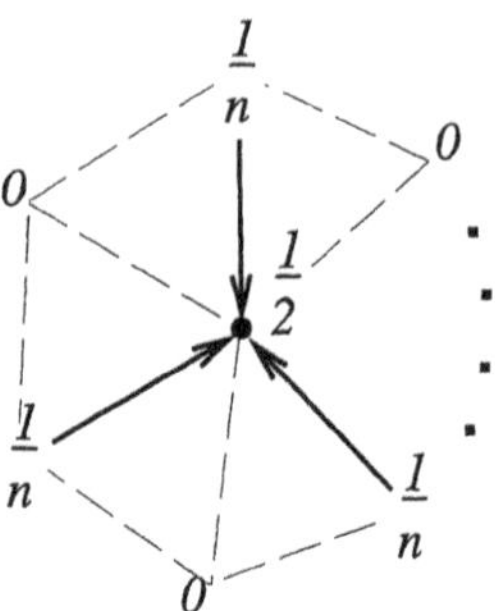

Fig. 10. Generalization of factorized vertex masks for valence $2n$.

neighbors with valence greater that 4. The mask is a weighted average of the extraordinary vertex and its old neighbors.

The smoothing operator is computed by a simple procedure that first applies the face mask to new vertices of valence 4, $v_i' \in V^{l+1} \setminus V^l$,

$$p^{l+1}(v_i') = \frac{1}{4} \sum_{v_j \in N_1(v_i')} p^l(v_j),$$

and then applies the vertex mask to old vertices of arbitrary valence, $v_i \in V^l$,

$$p^{l+1}(v_i) = \frac{1}{2} p^l(v_i) + \frac{1}{2n} \sum_{v_j \in N_1(v_i) \cap V^l} p^l(v_j).$$

Smoothness of the subdivision scheme follows from the regular C^4 four-directional box spline. The limit surface is C^4 continuous almost everywhere, except at isolated extraordinary vertices. In [3] we show that the surface is C^1 continuous at extraordinary vertices.

Figure 11 shows an example of a C^4 continuous box spline surface generated from a mesh of control points with arbitrary connectivity.

§3. Conclusions

We have presented a new subdivision scheme based on the [4.8^2] Laves tiling that extends the four-directional box splines of class C^4 to surfaces of arbitrary topological type. Our scheme is composed of a semi-regular 4–8 refinement operator a separable two-pass smoothing operator. The characteristics of these two operators make 4–8 subdivision a powerful tool for CAGD.

Semi-regular 4–8 refinement employs only bisections, and generates a hierarchical mesh structure that supports adaptive multiresolution. Separable two-pass smoothing allows a simple and efficient implementation of large masks through a decomposition of the subdivision rules.

In addition to the C^4 box spline, the 4–8 subdivision framework makes possible the construction of other schemes based on two-directional and four-directional grids.

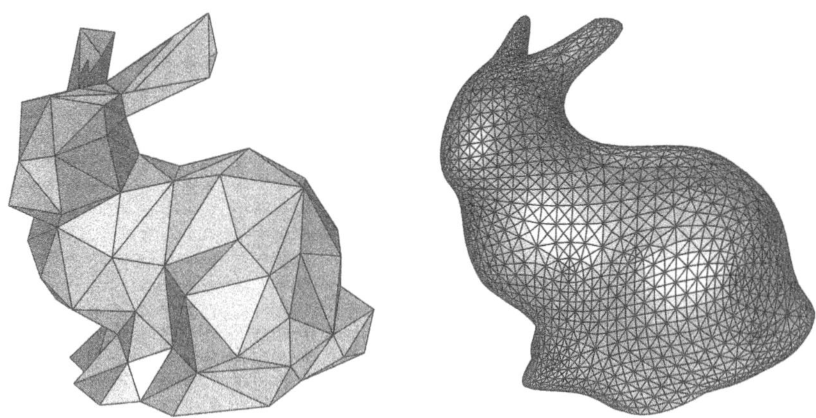

Fig. 11. Stanford Bunny.

We have described in [4] the implementation of Doo-Sabin and Catmull-Clark surfaces using the 4–8 subdivision framework. We have also implemented the Midedge scheme, that is based on the C^1 four-directional box spline.

An important aspect of 4–8 subdivision is its support for adaptive multiresolution. In [5], we discuss the hierarchical structure for 4–k meshes and give an overview of its applications.

References

1. de Boor, C., D. Hollig, and S. Riemenschneider, *Box Splines*, Springer-Verlag, New York, NY, 1994.
2. Grünbaum and G. Shephard, *Tilings and Patterns*, W. H. Freeman, 1987.
3. Velho, L. and D. Zorin, 4–8 subdivision, Comput. Aided Geom. Design, to appear.
4. Velho, L., Using semi-regular 4–8 meshes for subdivision surfaces Journal of Graphics Tools, to appear.
5. Velho, L. and J. Gomes, Variable resolution 4-K meshes: concepts and applications, Computer Graphics Forum, 2000.
6. Warren, J., *Subdivision Methods for Geometric Design*, unpublished monograph, 1995.
7. Zwart, P. B., Multivariate splines with nondegenerate partitions, SIAM J. Numer. Anal. **10** (4) (1973), 665–673.

Luiz Velho
IMPA – Instituto de Matemática Pura e Aplicada
Estrada Dona Castorina 110, Rio de Janeiro, RJ, Brazil, 22460-320
lvelho@visgraf.impa.br

Moving Least Squares Approximation on the Sphere

Holger Wendland

Abstract. We introduce moving least squares approximation as an approximation scheme on the sphere. We prove error estimates and approximation orders. Finally, we show certain numerical results.

§1. Introduction

Recently, approximation on the sphere has become important because of its obvious applications to Meteorology, Oceanography and Geoscience and Geoengineering in general. Over the last years several approaches were made to reconstruct a continuous function on the sphere from a finite number of discrete data (see [4] for a recent overview). One possibility is based on the theory of radial basis functions. Much work in this direction was done by [5,6,9,15] and this list is far from being complete. Even if this approach allows to use scattered data for the reconstruction process, the computation of the approximating function needs the solution of a full $N \times N$ linear system where N denotes the number of centers, and is intolerably expensive when N is large. Another technique [14] is hyperinterpolation approximation. It uses quadrature rules to discretize the L_2-orthogonal projection operator onto the space of spherical harmonics. Here, the advantage is that no linear system has to be solved. On the other hand, this method depends on the chosen quadrature rule, and since it is based on projection onto the space of spherical harmonics, it inherits the problems of projections, i.e., the Lebesgue constants cannot be uniformly bounded. Other local methods are based on spherical splines. These macro-element methods depend on a spherical triangulation (cf. [1]).

In this paper we want to introduce and investigate the method of moving least squares approximation on the sphere. It will turn out that the effort of computing the approximant is bounded by a constant which depends only on the space dimension and the desired approximation order. Furthermore, it can be shown that the involved Lebesgue constants are uniformly bounded.

Mathematical Methods for Curves and Surfaces: Oslo 2000
Tom Lyche and Larry L. Schumaker (eds.), pp. 517–526.

ISBN 0-8265-1378-6

This is achieved by two means: oversampling and only local reconstruction of spherical harmonics. Finally, it is a meshless method.

The paper is organized as follows. In the rest of this section we collect some basic notations. In the next section we describe the method of moving least squares. The third section contains our main result on approximation orders. The final section deals with examples.

We denote the sphere in $\mathbb{R}^n$ by $S^{n-1} = \{x \in \mathbb{R}^n : \|x\|_2 = 1\}$. We have to review some basics about spherical harmonics (see [11]). Let $\pi_\ell(\mathbb{R}^n)$ denote the polynomials of total degree ℓ on $\mathbb{R}^n$. Then $\pi_\ell(S^{n-1})$ denotes the restriction of these polynomials to the sphere S^{n-1}. In the following we need a basis for this space. Let $\pi_{\ell-1}^{\perp}(S^{n-1})$ denote the orthogonal complement of $\pi_{\ell-1}(S^{n-1})$ in $\pi_\ell(S^{n-1})$ with respect to the usual inner product of $L_2(S^{n-1})$. The spherical harmonics $\{Y_{\ell,k} : 1 \le k \le N(n,\ell)\}$ are an orthonormal set which spans $\pi_{\ell-1}^{\perp}(S^{n-1})$. Here,

$$N(n,\ell) = \dim\, \pi_{\ell-1}^{\perp}(S^{n-1}) = \begin{cases} 1, & \text{if } \ell = 0, \\ \frac{2\ell+n-2}{\ell}\binom{\ell+n-3}{\ell-1}, & \text{if } \ell \ge 1. \end{cases}$$

The dimension of the space $\pi_\ell(S^{n-1})$ is therefore given by $N(n+1,\ell)$.

Finally, we denote the distance between two points on the sphere by $d(x,y) = \arccos(x^T y)$.

§2. Moving Least Squares

Moving least squares approximation has been introduced in scattered data approximation on $\mathbb{R}^n$ several years ago (cf. [3,7]). It simplest form coincides with Shepard's interpolation method [13]. Recently, moving least squares have attracted more attention as a so-called meshless method in application to partial differential equations (cf. [2]).

We will formulate the application for the sphere. Suppose a continuous function $f \in C(S^{n-1})$ shall be reconstructed at $x \in S^{n-1}$ from its values $f(x_1), \dots, f(x_N)$ on scattered, pairwise distinct centers $X = \{x_1, \dots, x_N\} \subseteq S^{n-1}$. Then the approximate value $p^*(x)$ is given by the solution p^* of

$$\min \left\{ \sum_{i=1}^{N} (f(x_i) - p(x_i))^2 w(x, x_i) \;: p \in \mathcal{P} \right\},$$

where $\mathcal{P} \subseteq C(S^{n-1})$ is a finite dimensional subspace, usually spanned by spherical harmonics, and $w : S^{n-1} \times S^{n-1} \to [0,\infty]$ is a continuous function with possible exception at the diagonal. This exception is made to include also interpolation, but this will not bother us any further. Note that a local reproduction of any function can be forced simply by adding it to the space $\mathcal{P}$. This might be interesting in case of known singularities.

Under mild conditions (see Theorem 1 and also [8,16]) it can be shown that $p^*(x)$ can be written as $p^*(x) = \sum_{i=1}^{N} a_i^*(x) f(x_i)$ where the $a_i^*(x)$ minimize the quadratic form

$$\frac{1}{2} \sum_{i=1}^{N} a_i^2 \theta(x_i, x)$$

subject to the linear constraints

$$\sum_{i=1}^{N} a_i p(x_i) = p(x), \qquad p \in \mathcal{P},$$

with $\theta(x, y) = 1/w(x, y)$. Thus we will concentrate on this optimization problem.

Since we want to have a local process because of numerical efficiency, we choose $w(x, y)$ as a radial and compactly supported function, even if radiality is not really necessary. To be more precise, we choose a continuous function $\phi : [0, \infty) \to [0, \infty)$ with

- $\phi(r) > 0,\ 0 \le r \le 1/2,$
- $\phi(r) = 0,\ r \ge 1,$

and define

$$\theta_\delta(x, y) := \frac{1}{\phi(\frac{d(x,y)}{\delta})}$$

for $\delta > 0$. If we further define for $X = \{x_1, \ldots, x_N\}$ the set of indices

$$I(x) \equiv I(x, \delta, X) = \{j \in \{1, \ldots, N\} : d(x, x_j) < \delta\}$$

of centers contained in the cap of "radius" δ around x, and use $\mathcal{P} = \pi_\ell(S^{n-1})$ the space of spherical harmonics of order up to ℓ, the moving least squares approximation now takes the form

$$s_{f,X}(x) = \sum_{j \in I(x)} a_j^*(x) f(x_j), \tag{1}$$

where the coefficients $a_j^*(x)$ are determined by minimizing

$$\frac{1}{2} \sum_{j \in I(x)} a_j(x)^2 \theta_\delta(x, x_j) \tag{2}$$

under the constraints

$$\sum_{j \in I(x)} a_j(x) Y(x_j) = Y(x), \qquad Y \in \pi_\ell(S^{n-1}). \tag{3}$$

The proof of the following theorem is the same as the proof of the corresponding result from [16] in $\mathbb{R}^d$.

Theorem 1. *Suppose $Z = \{x_j \in X : j \in I(x, \delta, X)\}$ is $\pi_\ell(S^{n-1})$-unisolvent (i.e., zero is the only function from $\pi_\ell(S^{n-1})$ that vanishes on X). Then the minimization problem (2) with constraints (3) has a unique solution $a_j^*(x)$, $j \in I(x)$. The solution has the representation*

$$a_j^*(x) = \phi(\frac{d(x, x_j)}{\delta}) \sum_{i=0}^{\ell} \sum_{k=1}^{N(n,\ell)} \lambda_{i,k} Y_{i,k}(x_j), \tag{4}$$

where the $\lambda_{i,k}$ are the unique solutions of

$$\sum_{i=0}^{\ell}\sum_{k=1}^{N(n,\ell)}\lambda_{i,k}\sum_{j\in I(x)}\phi(\frac{d(x,x_j)}{\delta})Y_{i,k}(x_j)Y_{\nu,\mu}(x_j)=Y_{\nu,\mu}(x), \tag{5}$$

for $0\le\nu\le\ell$ *and* $1\le\mu\le N(n,\ell)$.

From this we can read off that the computational complexity of the approximant at a single point x is bounded by $\mathcal{O}(N(n+1,\ell)^3+N(n+1,\ell)^2|I(x)|+|I(x)|)$, where $|I(x)|$ denotes the number of elements in $I(x)$. Of course this is only true if we know the set $I(x)$ in advance. But this can be done in a preprocessing step of $\mathcal{O}(N)$ effort based on a boxing strategy. Thus, after the preprocessing step the computational complexity is constant if the number of points in $I(x)$ can be uniformly bounded. The latter can be achieved by choosing δ proportional to the fill distance

$$h_X\equiv h_{X,S^{n-1}}:=\sup_{x\in S^{n-1}}\min_{x_j\in X}d(x,x_j).$$

§3. Error Estimates

We want to measure the error $f(x)-s_f(x)$ in terms of the fill-distance $h_{X,S^{n-1}}$. To achieve our results we have to assume that all set of centers $X=\{x_1,\dots,x_N\}$ are quasi-uniform, i.e., there exists a global constant $c_1>0$, such that $q_X\le h_X\le c_1q_X$, where q_X is the separation distance

$$q_X:=\frac{1}{2}\min_{j\neq k}d(x_j,x_k).$$

Since the method is local, this is actually not a severe restriction. With the computational aspects in mind, we set

$$\delta\equiv\delta_X:=C_3h_X. \tag{6}$$

Before we can give our main result, we need:

Lemma 2. *There exist constants $h_0,C_1,C_2>0$ such that for every set $X=\{x_1,\dots,x_N\}\subseteq S^{n-1}$ with $h_{X,S^{n-1}}\le h_0$ and every $x\in S^{n-1}$ there exist numbers $a_1^X(x),\dots,a_N^X(x)$ such that*

1) $\sum_{j=1}^{N}a_j^X(x)Y(x_j)=Y(x)$ *for all* $Y\in\pi_\ell(S^{n-1})$,

2) $a_j^X(x)=0$ *if* $d(x,x_j)>C_1h_{X,S^{n-1}}$,

3) $\sum_{j=1}^{N}|a_j^X(x)|\le C_2$.

Proof: This is Theorem 1.4 from [9]. □

Theorem 3. *Suppose $X = \{x_1, \dots, x_N\} \subseteq S^{n-1}$ is quasi-uniform. Suppose further that the constants C_1 and C_3 from Lemma 2 and (6), respectively, satisfy $C_3 \geq 2C_1$. Denote the moving least squares approximant to $f \in C(S^{n-1})$ based on X by $s_{f,X}$. Then there exist constants $h_0, C > 0$ such that for every X with $h_{X,S^{n-1}} \leq h_0$ and every $x \in S^{n-1}$ the error between f and $s_{f,X}$ can be bounded by*

$$|f(x) - s_{f,X}(x)| \leq C \inf_{Y \in \pi_\ell(S^{n-1})} \|f - Y\|_{\infty, B(x,\delta)},$$

where $B(x,\delta) = \{y \in S^{n-1} d(x,y) \leq \delta\}$ and the constants C, h_0 are independent of f and X.

Proof: For any $Y \in \pi_\ell(S^{n-1})$ we can bound the error using standard arguments by

$$\begin{aligned} |f(x) - s_f(x)| &\leq |f(x) - Y(x)| + \sum_{j \in I(x)} |a_j^*(x)||f(x_j) - Y(x_j)| \\ &\leq \left(1 + \sum_{j \in I(x)} |a_j^*(x)|\right) \|f - Y\|_\infty, B(x,\delta). \end{aligned}$$

Thus it suffices to prove that the local Lebesgue constants $\sum_{j \in I(x)} |a_j^*(x)|$ are uniformly bounded. This will be done by starting with

$$\sum_{j \in I(x)} |a_j^*(x)| \leq \left(\sum_{j \in I(x)} |a_j^*(x)|^2 \theta_\delta(x, x_j)\right)^{1/2} \left(\sum_{j \in I(x)} \phi(\frac{d(x,x_j)}{\delta})\right)^{1/2}, \quad (7)$$

and estimating both factors on the right hand side separately.

To handle the first term, we assume that $h \equiv h_{X,S^{n-1}} \leq h_0$ with h_0 from Lemma 2. Then we find $a_j(x)$ that reproduce spherical harmonics and vanish if $d(x,x_j) > \delta/2$. Thus if we set $\tilde{I}(x) = \{j : d(x,x_j) \leq \delta/2\}$ and use the minimal property of the moving least squares coefficients, we have

$$\begin{aligned} \sum_{j \in I(x)} |a_j^*(x)|^2 \theta_\delta(x, x_j) &\leq \sum_{j \in \tilde{I}(x)} |a_j(x)|^2 \theta_\delta(x, x_j) \\ &\leq \frac{1}{\min_{j \in \tilde{I}(x)} \phi(d(x,x_j)/\delta)} \sum_{j \in \tilde{I}(x)} |a_j(x)|^2 \\ &\leq c_\phi^{-1} \left(\sum_{j=1}^N |a_j(x)|\right)^2 \\ &\leq C_2^2 c_\phi^{-1}, \end{aligned}$$

where we used

$$\min_{j\in\tilde{I}(x)} \phi(d(x,x_j)/\delta) \geq \min_{y\in B(x,\delta/2)} \phi(d(x,y)/\delta) = \min_{r\in[0,1/2]} \phi(r) =: c_\phi.$$

For the second factor in (7) we remark

$$\sum_{j\in I(x)} \phi(\frac{d(x,x_j)}{\delta}) \leq |I(x)| \|\phi\|_{\infty,[0,1]}$$

and use a packing argument from [12] to bound $|I(x)|$. Since any cap centered at x_j with colatitude q_X is essentially disjoint from a cap centered at any other x_k, $j \neq k$, and is contained in a cap with colatitude $q_X + \delta$ centered at x, we see

$$|I(x)| \leq \frac{S(q_X+\delta)}{S(q_X)},$$

where $S(q)$ denotes the surface area of a cap with colatitude q. From [12] we know

$$S(q) = \frac{2\pi^{\frac{n-1}{2}}}{\Gamma(\frac{n-1}{2})} \int_0^q \sin^{n-2}\vartheta d\vartheta$$

satisfies

$$\left(\frac{2}{\pi}\right)^{n-2} \frac{2\pi^{\frac{n-1}{2}}}{(n-1)\Gamma(\frac{n-1}{2})} q^{n-1} \leq S(q) \leq \frac{2\pi^{\frac{n-1}{2}}}{(n-1)\Gamma(\frac{n-1}{2})} q^{n-1}.$$

Thus we derive

$$|I(x)| \leq \left(\frac{q_X+\delta}{q_X}\right)^{n-1} \left(\frac{\pi}{2}\right)^{n-2} \leq (1+c_1C_3)^{n-1} \left(\frac{\pi}{2}\right)^{n-2}.$$

Collecting all results leads to $\sum_{j\in I(x)} |a_j^*(x)| \leq C$. □

Note that the condition $C_3 \geq 2C_2$ that describes the choice of the support of the weight function is problematic if C_2 is unknown, which is the case so far. Explicit constants might help here.

The condition that X is quasi-uniform is, as said before, not as restrictive as it seems at first sight. The reason for its assumption is based on the same choice of δ everywhere. Since the method is local, a local, continuously varying δ helps in the situation of clustered data.

From the last result we can read off convergence orders in case of smooth functions. Remember that the smoothness of a function f on the sphere (or any manifold) is defined by the smoothness of $f \circ T$ with a convenient map $T: U \subseteq \mathbb{R}^{n-1} \to \mathbb{R}^n$, $U \subseteq \mathbb{R}^{n-1}$ open.

Theorem 4. *Under the assumptions of Theorem 3, the error can be bounded for $f \in C^{\ell+1}(S^{n-1})$ and $x \in S^{n-1}$ by*

$$|f(x) - s_{f,X}(x)| \leq c_f h_X^{\ell+1}.$$

Proof: Without restriction we can assume that $x = (0, \ldots, 0, 1)^T$. Then

$$B(x, \delta) = \{y \in S^{n-1} : d(x, y) < \delta\} = \{y \in S^{n-1} : y_n > \cos\delta\}.$$

Define the bijective map $T : U \to B(x, \delta)$ by $\tilde{y} \mapsto (\tilde{y}, \sqrt{1 - \|\tilde{y}\|_2^2})^T$, where $U = \{\tilde{y} \in \mathbb{R}^{n-1} : \|\tilde{y}\|_2^2 < 1 - \cos^2\delta\}$. Its inverse is obviously given by $T^{-1}(y) = \tilde{y} = (y_1, \ldots, y_{n-1})^T$. Now $f \in C^{\ell+1}(B(x, \delta))$ means $g = f \circ T \in C^{\ell+1}(U)$. The Taylor expansion around $\tilde{x} = 0$ is given by

$$g(\tilde{y}) = \sum_{\substack{\alpha \in \mathbb{N}_0^{n-1} \\ |\alpha| \leq \ell}} \frac{g^{(\alpha)}(0)}{\alpha!} \tilde{y}^\alpha + \sum_{\substack{\alpha \in \mathbb{N}_0^{n-1} \\ |\alpha| = \ell+1}} \frac{g^{(\alpha)}(\xi)}{\alpha!} \tilde{y}^\alpha.$$

Hence

$$f(y) = g \circ T^{-1}(y) = \sum_{\substack{\beta \in \mathbb{N}_0^{n} \\ |\beta| \leq \ell}} c_\beta y^\beta + \sum_{\substack{\alpha \in \mathbb{N}_0^{n-1} \\ |\alpha| = \ell+1}} \frac{g^{(\alpha)}(\xi)}{\alpha!} \tilde{y}^\alpha.$$

The polynomial $Y(y) = \sum c_\alpha y^\alpha$ is a polynomial of degree at most ℓ which can be expressed by spherical harmonics of degree at most ℓ. Moreover, we find

$$\begin{aligned} |f(y) - Y(y)| &\leq c_f \|\tilde{y}\|_2^{\ell+1} \\ &= c_f (1 - y_n^2)^{(\ell+1)/2} \\ &\leq c_f (1 - \cos^2\delta)^{\ell+1/2} \\ &= c_f (\sin\delta)^{\ell+1} \\ &\leq c_f \delta^{\ell+1} \end{aligned}$$

for every $y \in B(x, \delta)$. Finally, Theorem 3 and the definition of δ lead to

$$|f(x) - s_{f,X}(x)| \leq C\|f - Y\|_{\infty, B(x,\delta)} \leq C c_f C_3^{\ell+1} h_X^{\ell+1}. \quad \square$$

§4. Numerical Examples

In this section we apply the moving least squares approach, and show that it can deal with large data sets.

To this end we have to construct centers on the sphere which are quasi-uniform. This, in itself, is a difficult problem. Possible solutions are to minimize a functional like

$$\sum_{i \neq j} \frac{1}{d(x_i, x_j)}$$

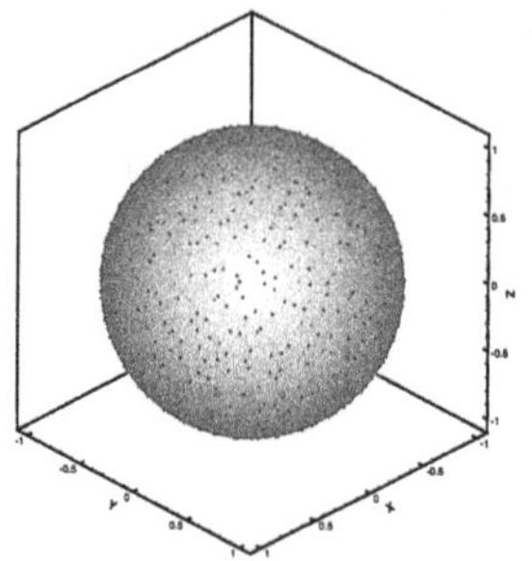

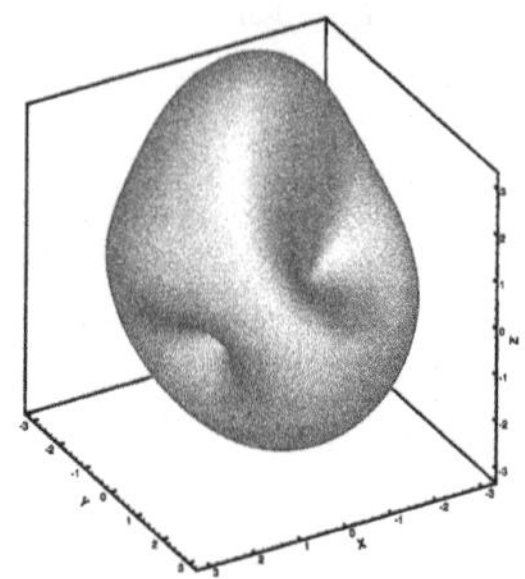

Fig. 1. The first 1000 Halton points on the sphere (left) and $(f(x)+3)x$.

(cf. [10]). But since this functional has an infinite number of minimum points, this method seems not to be applicable for the construction of large data sets. Instead, we use a method proposed by [17], which is based on so-called Halton points. The construction goes like this: If p is a prime, every nonnegative integer n has a p-adic expansion of the form

$$n = a_0 + a_1 p + a_2 p^2 + \cdots + a_r p^r.$$

With this expansion we can form the number

$$\Phi_p(n) = \frac{a_0}{p} + \frac{a_1}{p^2} + \cdots + \frac{a_r}{p^{r+1}}.$$

The first N 2-dimensional Halton points with respect to the primes p_1 and p_2 are simply given by $(\Phi_{p_1}(n), \Phi_{p_2}(n))$, $1 \le n \le N$. The transformation

$$\begin{aligned}(\Phi_{p_1}(n), \Phi_{p_2}(n)) &\mapsto (t, \theta) = (2\Phi_{p_1}(n) - 1, 2\pi\Phi_{p_2}(n)) \\ &\mapsto \left(\sqrt{1-t^2}\cos\phi, \sqrt{1-t^2}\sin\phi, t\right)\end{aligned}$$

leads to points on the sphere. Figure 1 (left) shows the first 1000 Halton points on the sphere using the primes $p_1 = 2$ and $p_2 = 3$.

Our test function f comes from [6]. It is the sum of five exponentials f_i of the form $f_i(x) = c_i e^{-\alpha_i (1 - y_i \cdot x)^{n_i}}$. The exact parameters can be found in [6]. Figure 1 (right) shows $(3 + f(x))x$.

Figure 2 reflects the influence of the support radius on the approximation in case of 100,000 points and $\ell = 1$. The discrete ℓ_2- and ℓ_∞-error are plotted. The picture shows that a support radius too small leads to a bigger error, also causing a bumpy surface. On the other hand, as seen in the proof of Theorem 4, a larger constant C_3 in the definition of $\delta = C_3 h_{X,S^{n-1}}$ also causes a bigger error. Table 1 contains the ℓ_∞- and ℓ_2-errors measured on a rectangular grid of angles of size 600×600.

Note, that in case of such a smooth test function the same approximation results can be achieved with less centers using a smooth "radial" basis function (cf. [6]). But since we deal here with only local problems, the computational effort is not much higher in our situation. Furthermore, if the test function is less smooth, radial basis functions have to deal with the problem of large non-sparse matrices, something we can completely avoid here.

		1000	10000	100000
	δ	0.27	0.16	0.08
$m = 0$	ℓ_∞	2.191078326e-1	5.281097764e-2	1.290082316e-2
	ℓ_2	4.239320187e-2	1.105376863e-2	2.64121179e-3
	δ	0.38	0.14	0.06
$m = 1$	ℓ_∞	4.986797196e-2	5.910919042e-3	5.7271218e-4
	ℓ_2	7.303700189e-3	5.76678963e-4	5.423775504e-5
	δ	0.51	0.14	0.08
$m = 2$	ℓ_∞	1.738234312e-2	1.119974013e-3	3.293247082e-5
	ℓ_2	2.74449414e-3	6.241901336e-5	3.383783815e-6

Tab. 1. Errors for MLS approximation.

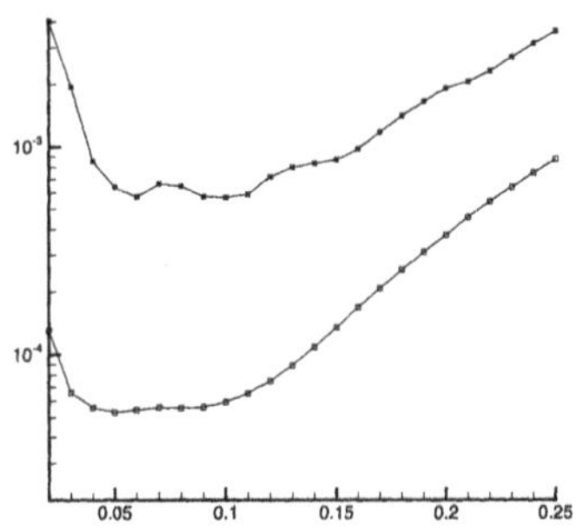

Fig. 2. ℓ_∞- (above) and ℓ_2-error as a function of δ.

References

1. Alfeld, P., M. Neamtu, and L. L. Schumaker, Fitting scattered data on sphere-like surfaces using spherical splines, J. Comput. Appl. Math. **73** (1996), 5–43.

2. Belytschko, T., Y. Krongauz, D. Organ, M. Fleming, and P. Krysl, Meshless methods, an overview and recent developments, Computer Methods in Applied Mechanics and Engineering, special issue on Meshless methods, **139** (1996), 3–47.

3. Farwig, R., Multivariate interpolation of arbitrarily spaced data by moving least square methods, J. of Comp. and App. Math. **16** (1986), 79–93.

4. Fasshauer, G., and L. L. Schumaker, Scattered data fitting on the sphere, in *Mathematical Methods for Curves and Surfaces II*, M. Dæhlen, T. Lyche, and L. L. Schumaker (eds.), Vanderbilt University Press, Nashville, 1998, 117-166.

5. Freeden, W., T. Gervens, and M. Schreiner, *Constructive Approximation on the Sphere with Applications to Geomathematics*, Clarendon Press, Oxford, 1998.
6. Jetter, K., J. Stöckler, and J. D. Ward, Error estimates for scattered data interpolation on spheres, Math. Comp. **68** (1999), 733–747.
7. Lancaster, P. and K. Salkauskas, Surfaces generated by moving least squares methods, Math. Comp. **37** (1981), 141–159.
8. Levin, D., The approximation power of moving least-squares, Math. Comp. **67** (1998), 1517–1531.
9. Light, W. and M. von Golitschek, Interpolation by polynomials and radial basis functions on spheres, Preprint 1997, Constr. Approx., to appear.
10. Maier, U. and J. Fliege, Charge distribution of points on the sphere and corresponding curbature formula, in *Multivariate Approximation: Recent Trends and Results*, W. Haußmann, K. Jetter, and M. Reimer (eds.), Mathematical Research, vol. 101, Akademie-Verlag, Berlin, 1999, 147–159.
11. Müller, C., *Spherical Harmonics*, Lecture Notes in Mathematics, Springer, Berlin, 1966.
12. Narcowich, F., N. Sivakumar, and J. Ward, Stability results for scattered-data interpolation on Euclidean spheres, Adv. in Comput. Math. **8** (1998), 137–163.
13. Shepard, D., A two-dimensional interpolation function for irregularly spaced points, Proc. 1968 A.C.M. Nat. Conf., 517–524.
14. Sloan, I. H., and S. Womersley, Constructive polynomial approximation on the sphere, J. Approx. Theory **103** (2000), 91–118.
15. Wahba, G., Spline interpolation and smoothing on the sphere, SIAM J. Sci. Stat. Comput. **2** (1981), 5–16.
16. Wendland, H., Local polynomial reproduction and moving least squares approximation, IMA J. of Numerical Analysis, to appear.
17. Wong, T., W. Luk, and P. Heng, Sampling with Hammersley and Halton Points, preprint.

Holger Wendland
Institut für Numerische und Angewandte Mathematik
Universität Göttingen
Lotzestr. 16-18
37083 Göttingen
Germany
`wendland@math.uni-goettingen.de`

NURPS for Special Effects and Quadrics

Joris Windmolders and Paul Dierckx

Abstract. NURPS are the rational extension of Powell–Sabin splines in their normalized B–spline representation. The weights associated with the control points allow to represent (parts of) some particular quadric surfaces. By choosing coincident control points, special effects on the surface can be generated.

§1. Preliminary Concepts

Given a simply connected subset $\Omega \subset \mathbb{R}^2$ with polygonal boundary $\delta\Omega$, and a conforming triangulation Δ of Ω having n vertices V_i with coordinates (u_i, v_i), $i = 1, \ldots, n$, a Powell–Sabin refinement Δ^* which divides each triangle $\rho \in \Delta$ up into 6 nondegenerate subtriangles can easily be found (see, e.g., [8]). A Powell–Sabin (PS) spline is a piecewise quadratic polynomial with C^1 continuity on Ω, and has a quadratic Bézier representation on each PS–subtriangle $\tau \in \Delta^*$:

$$\boldsymbol{b}(t) = \sum_{|i|=2} \boldsymbol{b}_i B_i^2(t), \quad t = (t_1, t_2, t_3) \in \tau, \quad t_1 + t_2 + t_3 = 1.$$

Definition 1. *A PS–spline surface has a normalized B–spline representation*

$$s(u,v) = \sum_{i=1}^{n} \sum_{j=1}^{3} \boldsymbol{c_{i,j}} B_i^j(u,v), \qquad (u,v) \in \Omega, \tag{1}$$

where $\boldsymbol{c_{i,j}} = (c_{i,j}^x, c_{i,j}^y, c_{i,j}^z)$ *are the B–spline control points and* $B_i^j(u,v)$ *are the normalized B–splines.*

The remainder of this section summarizes the relevant properties of this representation. For details we refer to [3]. The normalized B–splines are locally nonzero and constitute a convex partition of unity on Ω, hence this representation is affine invariant and has the local control and convex hull properties. Furthermore, the PS–spline surface is C^1–continuous regardless

Mathematical Methods for Curves and Surfaces: Oslo 2000
Tom Lyche and Larry L. Schumaker (eds.), pp. 527–534.

ISBN 0-8265-1378-6

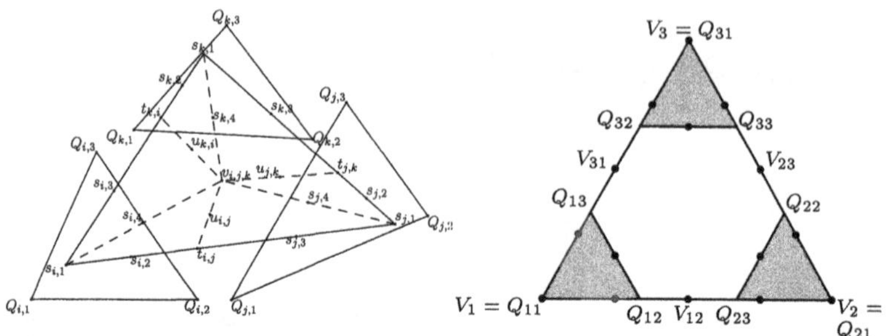

Fig. 1. A domain triangle.

of the choice of the control points. A control triangle $T_l(\mathbf{c}_{l,1}, \mathbf{c}_{l,2}, \mathbf{c}_{l,3})$ can be associated with each vertex V_l, $l = 1, \ldots, n$ which is tangent to the surface at $\mathbf{s}(V_l)$. The projection of each T_l onto the domain plane yields the PS–triangles $t_l(Q_{l,1}, Q_{l,2}, Q_{l,3})$, $l = 1, \ldots, n$, having the B–spline ordinates $Q_{l,j}(U_{l,j}, V_{l,j})$, $j = 1, 2, 3$, as its vertices.

Figure 1 (left) shows a domain triangle with its PS–subdivision and PS–triangles t_l. The Bézier ordinates are denoted $s_{v,l}$, $v = i, j, k$, and $l = 1, 2, 3, 4$; $t_{l,m}, u_{l,m}$, $(l, m) \in \{(i,j), (j,k), (k,i)\}$ and $v_{i,j,k}$. Given a PS–subdivision, any PS–triangle t_l must contain the PS–points $s_{l,j}, j = 1, 2, 3, 4$, in order for the B–splines to constitute a convex partition of unity. In that case, the Bézier ordinates can be written as unique convex barycentric combinations of the B–spline ordinates:

$$
\begin{aligned}
s_{v,l} &= \alpha_{v,l}\ Q_{v,1} + \beta_{v,l}\ Q_{v,2} + \gamma_{v,l}\ Q_{v,3} \\
t_{l,m} &= \delta_{l,m}\ s_{l,2} + \epsilon_{l,m}\ s_{m,3} \\
u_{l,m} &= \delta_{l,m}\ s_{l,4} + \epsilon_{l,m}\ s_{m,4} \\
v_{i,j,k} &= \lambda_{i,j,k}\ s_{i,4} + \mu_{i,j,k}\ s_{j,4} + \nu_{i,j,k}\ s_{k,4}.
\end{aligned}
\tag{2}
$$

The same combinations are valid for the Bézier control points, which can in turn be found as convex barycentric combinations of the PS control points.

Definition 2. *A* Non Uniform Rational Powell–Sabin (NURPS) spline *surface has the form*

$$
s(u,v) = \frac{\sum_{i=1}^{n} \sum_{j=1}^{3} \mathbf{c}_{i,j} w_{i,j} B_i^j(u,v)}{\sum_{i=1}^{n} \sum_{j=1}^{3} w_{i,j} B_i^j(u,v)}, \qquad (u,v) \in \Omega, \tag{3}
$$

where $w_{i,j} > 0$ in order for $s(u,v)$ to be defined everywhere on Ω.

For the evaluation of this rational extension, calculating the corresponding rational Bézier representation, the geometric interpretation of the weights, and the modelling of planar effects using relatively large weights, we refer to [10].

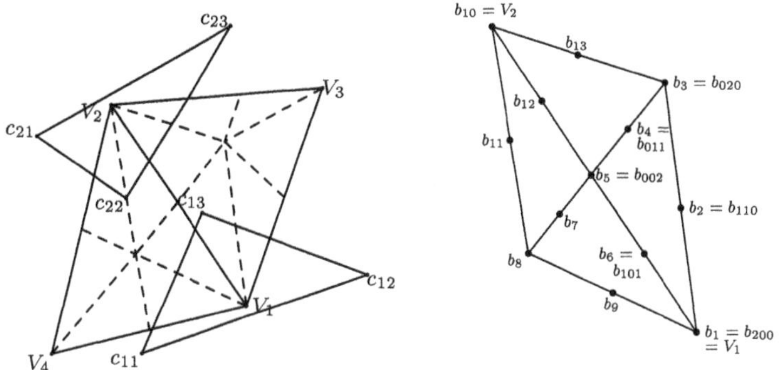

Fig. 2. Degenerate control triangles.

§2. Degenerate Control Triangles

In spite of the global C^1–continuity of (1), modeling in practice often requires the representation of C^0–continuous effects. Here we show how corners can be represented by choosing coincident control points.

Let $u = (u_1, u_2, u_3)$ and $v = (v_1, v_2, v_3)$ be the barycentric coordinates of two distinct domain points with respect to a triangle ρ. The difference $d = u - v = (d_1, d_2, d_3)$ defines a vector $(d_1 + d_2 + d_3 = 0)$. Suppose we are given a rational Bézier surface on ρ:

$$\boldsymbol{b}(t) = \frac{\sum_{|i|=2} w_i \boldsymbol{b}_i B_i^2(t)}{\sum_{|i|=2} w_i B_i^2(t)}, \quad t = (t_1, t_2, t_3) \in \tau. \tag{4}$$

Definition 3. *The directional derivative of a surface (4) at $\boldsymbol{b}(u)$ with respect to d is given by*

$$D_d \boldsymbol{b}(u) = d_1 \boldsymbol{b}_{t_1}(u) + d_2 \boldsymbol{b}_{t_2}(u) + d_3 \boldsymbol{b}_{t_3}(u), \tag{5}$$

where $\boldsymbol{b}_{t_i}(u)$ is the partial derivative of $\boldsymbol{b}(t)$ with respect to $t_i, i = 1, 2, 3$, evaluated at u. Note that these are not tangent vectors to the surface at $\boldsymbol{b}(u)$.

Consider the domain triangles $\rho_1, \rho_2 \in \Delta$ with common boundary V_1V_2. The PS–refinement and PS–triangles are shown in Figure 2 (left). The Bézier subtriangles along $V_1 - V_2$ are shown in Figure 2 (right). Suppose we are given the barycentric coordinates of a direction d and a point $u = (u_1, 0, u_3)$ with respect to $\tau(b_{200}, b_{110}, b_{101})$. The directional derivative of $\boldsymbol{b}(u)$ with respect to d is given by

$$\begin{aligned} D_d \boldsymbol{b}(u) = 2(&d_1(u_1 \boldsymbol{b}_{200} + u_3 \boldsymbol{b}_{101}) \\ &+ d_2(u_1 \boldsymbol{b}_{110} + u_3 \boldsymbol{b}_{011}) \\ &+ d_3(u_1 \boldsymbol{b}_{101} + u_3 \boldsymbol{b}_{002})) \end{aligned} \tag{6}$$

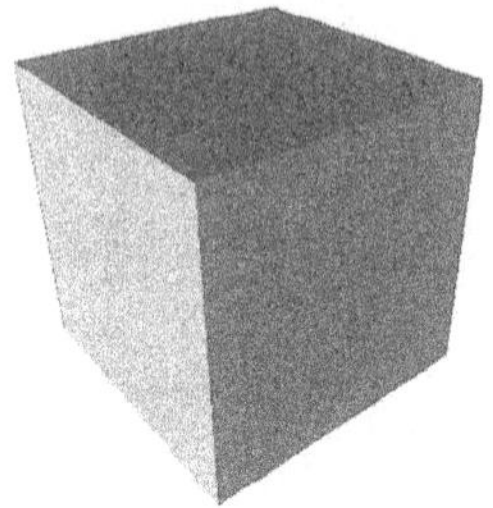

Fig. 3. Degenerate control triangles.

First, consider the case where one of the control triangles is degenerate, e.g., $\boldsymbol{c}_{1,1} = \boldsymbol{c}_{1,2} = \boldsymbol{c}_{1,3} = \boldsymbol{C}$. It follows from (2) that

$$D_d\boldsymbol{b}(u) = 2u_3(d_1\boldsymbol{C} + d_2\boldsymbol{b}_{011} + d_3\boldsymbol{b}_{011}).$$

The directional derivate is seen to degenerate to the zero vector at $u_3 = 0$, and hence the surface has a singulartity at V_1. Figure 3 (left) shows an example.

Next, consider the case where both control triangles at V_1 and V_2 are degenerate: $\boldsymbol{c}_{1,1} = \boldsymbol{c}_{1,2} = \boldsymbol{c}_{1,3} = \boldsymbol{C}$ and $\boldsymbol{c}_{2,1} = \boldsymbol{c}_{2,2} = \boldsymbol{c}_{2,3} = \boldsymbol{D}$. We then have

$$D_d\boldsymbol{b}(u) = 2\alpha u_3 d_1(\boldsymbol{C} - \boldsymbol{D})$$

for some $\alpha \in (0,1)$. This vector degenerates to the zero vector for $u_3 = 0$ and $d_1 = 0$, and coincides with $\boldsymbol{d_c} = \frac{\boldsymbol{C}-\boldsymbol{D}}{||\boldsymbol{C}-\boldsymbol{D}||}$ elsewhere. As an application, a corner along $V_1 - V_2$ in the direction of $\boldsymbol{d_c}$ can be represented. Figure 3 (right) shows a cube represented as a NURPS surface.

§3. An Optimal Conversion

In [10] a general formula is provided for calculating the NURPS representation of a given quadratic rational Bézier surface on one triangle. Following Dierckx [3], this representation is not *optimal* because the PS–triangles are not the smallest triangles containing the PS–points. The advantage of optimal PS–triangles is that the control triangles are closer to the surface than in the nonoptimal case. Suppose we are given a quadratic Bézier control net. We are free to choose the domain triangle to be equilateral. The optimal PS–triangles for this case can be found by solving a quadratic programming problem (see, again, [3]). The results are depicted in Figure 1 (right). The dots represent the PS–points, the shaded triangles are optimal PS–triangles. It can be shown that these are calculated from the Bézier ordinates by

$$\begin{aligned} Q_{i,1} &= V_i \\ Q_{i,2} &= \frac{2}{3}V_{i,i_3+1} + \frac{1}{3}V_i \\ Q_{i,3} &= \frac{2}{3}V_{(i+1)_3+1,i} + \frac{1}{3}V_i, \end{aligned} \tag{7}$$

where $k_3 = k \bmod 3$ and $V_{i,j} = \frac{1}{2}(V_i + V_j)$. The B–spline control points can be found from these equations by replacing the Bézier ordinates with the corresponding control points.

§4. Quadrics as NURPS

Representing quadrics as rational Bézier patches has been investigated in [1,2,4,5,6,9]. Here we present a constructive approach to the problem of representing quadrics as NURPS. The purpose is to derive closed formulae for the control triangles of (patches on) a cylinder, cone and sphere of given dimension, which is usefull in CAGD systems. The calculations have been made using modern computer techniques, such as the symbolic mathematical manipulation package Maple. We restrict ourselves to the case of the cylinder, and mention the results for a cone and a sphere only briefly.

4.1. The Cylinder

Using symmetry operations, we break up the surface into isometrical triangular parts. Then we calculate the rational Bézier representation

$$\boldsymbol{b}(t) = \frac{1}{q(t)} \begin{pmatrix} x(t) \\ y(t) \\ z(t) \end{pmatrix} = \frac{\sum_{i=0}^{5} w_i \boldsymbol{P_i} B_i^2(t)}{\sum_{i=0}^{5} w_i B_i^2(t)}, \quad t = (t_1, t_2, t_3) \in \tau, \tag{8}$$

of one such triangle. Note that we use a different indexing here then in former equations, in order to simplify notation. Application of (7) immediately yields the NURPS representation.

Figure 4 shows the octant of the cylinder with radius r and height equal to h:

$$\left(\frac{x(t)}{q(t)}\right)^2 + \left(\frac{y(t)}{q(t)}\right)^2 = r^2. \tag{9}$$

The Bézier control points follow from the fact that the control net is tangent to the surface at the corners of the patch:

$$\begin{array}{lll} \boldsymbol{P_0} = (r, 0, 0) & \boldsymbol{P_1} = (r, r, 0) & \boldsymbol{P_2} = (0, r, 0) \\ \boldsymbol{P_3} = (r, r, \frac{h}{2}) & \boldsymbol{P_4} = (r, 0, h) & \boldsymbol{P_5} = (r, 0, \frac{h}{2}). \end{array}$$

Setting $w_0 = w_2 = w_4 = w_5 = 1$ since the corresponding control points are on the surface, we can find the weights w_1 and w_3 by imposing that the edge curves satisfy (9), e.g.,

$$\begin{aligned} R_1 &\leftrightarrow x^2(t_1, 1-t_1, 0) + y^2(t_1, 1-t_1, 0) - q^2(t_1, 1-t_1, 0) = 0 \\ &\leftrightarrow t_1^4(4w_1^2 - 2) + t_1^3(4 - 8w_1^2) + t_1^2(4w_1^2 - 2) = 0 \\ &\leftrightarrow w_1 = \frac{\sqrt{2}}{2}. \end{aligned}$$

Analogously we find $w_3 = \frac{\sqrt{2}}{2}$. Substituting these weights in (8), one can verify that (9) is satisfied for all $t \in \tau$. The corresponding B–spline control

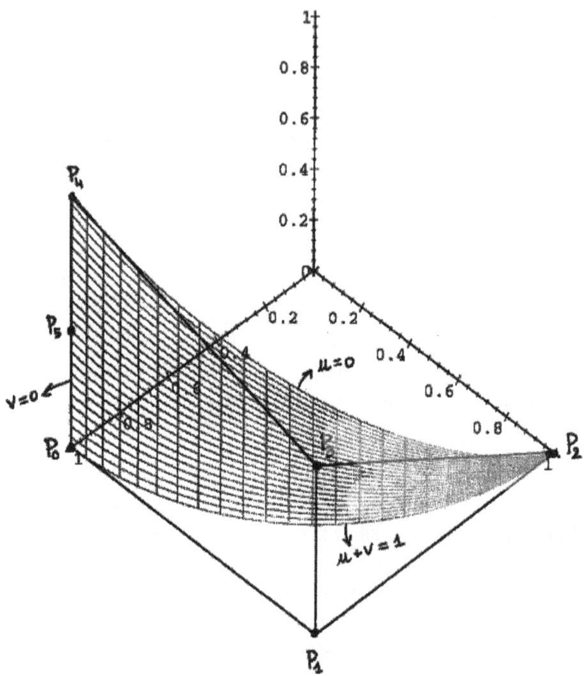

Fig. 4. Quadric segment of a cylinder.

points for a cylinder of radius r and heigth h can be found from (7), referring to Figure 1 (right):

i	$c_{i,1}$	$w_{i,1}$	$c_{i,2}$	$w_{i,2}$	$c_{i,3}$	$w_{i,3}$
1	$(r,0,0)$	1	$(r,\frac{2}{3}r,0)$	$\frac{\sqrt{2}+1}{3}$	$(r,0,\frac{1}{3}h)$	1
2	$(0,r,0)$	1	$(\frac{2}{3}r,r,\frac{1}{3}h)$	$\frac{\sqrt{2}+1}{3}$	$(\frac{2}{3}r,r,0)$	$\frac{\sqrt{2}+1}{3}$
3	$(r,0,h)$	1	$(r,0,\frac{2}{3}h)$	1	$(r,\frac{2}{3}r,\frac{2}{3}h)$	$\frac{\sqrt{2}+1}{3}$

By combining 8 isometrical patches, we have a complete NURPS cylinder, see Figure 5 (left).

4.2. The Cone

The NURPS control points for the 1/6 part of a cone with radius r and height h are:

i	$c_{i,1}$	$w_{i,1}$	$c_{i,2}$	$w_{i,2}$	$c_{i,3}$	$w_{i,3}$
1	$(r,0,0)$	1	$(r,\frac{2\sqrt{3}}{9}r,0)$	$\frac{\sqrt{3}+1}{3}$	$(\frac{1}{3}r,0,\frac{2}{3}h)$	1
2	$(r,\frac{\sqrt{3}}{2}r,0)$	1	$(\frac{1}{6}r,\frac{\sqrt{3}}{6}r,\frac{2}{3}h)$	1	$(\frac{5}{6}r,\frac{7\sqrt{3}}{18}r,0)$	$\frac{\sqrt{3}+1}{3}$
3	$(0,0,h)$	1	$(0,0,h)$	1	$(0,0,h)$	1

The complete cone is shown in Figure 5 (middle).

Fig. 5. Quadric NURPS.

4.3. The Sphere

It is not possible to represent a complete sphere by collecting non-degenerate quadratic rational Bézier triangles (see [1,2,9]). The NURPS patch given in the table below however allows to approximate the sphere, leaving small gaps around the equator plane, see Figure 5 (right).

i	$\mathbf{c}_{i,1}$	$w_{i,1}$	$\mathbf{c}_{i,2}$	$w_{i,2}$	$\mathbf{c}_{i,3}$	$w_{i,3}$
1	$(\frac{\sqrt{3}}{2}r, -\frac{r}{2}, 0)$	1	$(\frac{11\sqrt{3}}{18}r, -\frac{1}{6}r, \frac{2}{9}r)$	$\frac{5}{6}$	$(\frac{\sqrt{3}}{2}r, -\frac{1}{2}r, \frac{2}{3}r)$	$\frac{\sqrt{2}+1}{3}$
2	$(\frac{\sqrt{3}}{2}r, \frac{1}{2}r, 0)$	1	$(\frac{\sqrt{3}}{2}r, \frac{1}{2}r, \frac{2}{3}r)$	$\frac{\sqrt{2}+1}{3}$	$(\frac{11\sqrt{3}}{18}r, \frac{1}{6}r, \frac{2}{9}r)$	$\frac{5}{6}$
3	$(0, 0, r)$	1	$(\frac{\sqrt{3}}{3}r, -\frac{1}{3}r, r)$	$\frac{\sqrt{2}+1}{3}$	$(\frac{\sqrt{3}}{3}r, \frac{1}{3}r, r)$	$\frac{\sqrt{2}+1}{3}$

4.4. Parametric Continuity

In spite of the visual smoothness of the surfaces obtained by combining isometric NURPS patches, , the $q(t)$ component from (8) in the global representation as a piecewise rational Bézier surface, or as a piecewise 4D Bézier surface in homogeneous space, is not C^1–continuous. Hence, there does not exist a global NURPS representation with respect to the given parameter space.

References

1. Boehm, W. and D. Hansford, Bézier patches on quadrics, in *NURBS for Curve and Surface Design*, G. Farin (ed.), SIAM, Philadelphia, PA, 1991, 1–14.
2. Boehm, W., Some remarks on quadrics, Comput. Aided Geom. Design **10** (1993), 231–236.
3. Dierckx, P., On calculating normalized Powell–Sabin B–splines, Comput. Aided Geom. Design **15** (1997), 61–78.

4. Dietz, R., J. Hoschek, and B. Jüttler, An algebraic approach to curves and surfaces on the sphere and on other quadrics, Comput. Aided Geom. Design **10** (1993), 211–229.
5. Dietz, R., J. Hoschek, and B. Jüttler, Rational patches on quadric surfaces, Computer-Aided Design **27** (1995), 27–40.
6. Farin, G., B. Piper, and A. J. Worsey, The octant of a sphere as a nondegenerate triangular Bézier patch, Comput. Aided Geom. Design **4** (1987), 329–332.
7. Farin, G., *Curves And Surfaces For Computer Aided Geometric Design: A Practical Guide*, 4th edition, Academic Press, Boston, 1997.
8. Powell, M. J. D. and M. A. Sabin, Piecewise quadratic approximations on triangles, ACM Trans. Math. Software **3** (1977), 316–325.
9. Sederberg, T. W. and D. C. Anderson, Steiner surface patches, IEEE Comp. Graphics Appl. **5** (1985), 23–36.
10. Windmolders, J., and P. Dierckx, From PS–splines to NURPS, *Curve and Surface Fitting: Saint-Malo 1999*, Albert Cohen, Christophe Rabut, and Larry L. Schumaker (eds.), Vanderbilt University Press, Nashville, 2000, 45–54.

Joris Windmolders and Paul Dierckx
Department of Computer Sciences
Celestijnenlaan 200A
B-3001 Herverlee, Belgium
`joris.windmolders@cs.kuleuven.ac.be`
`paul.dierckx@cs.kuleuven.ac.be`

Computational Experiments with Resultants for Scaled Bernstein Polynomials

Joab R. Winkler

Abstract. The Bernstein basis is used extensively in geometric modelling because of its elegant geometric properties and simple algorithms that are available for processing it. Although resultants are used for several important operations in geometric modelling and computer graphics, it is necessary to perform a polynomial basis transformation because the established theory of resultants assumes that the polynomials are expressed in the power (monomial) basis. In this paper, the numerical behaviour of a resultant matrix for a scaled Bernstein polynomial (a polynomial of degree n whose basis functions are $(1-x)^{n-i}x^i, i = 0, \ldots, n$) is investigated. In particular, a companion matrix M for a scaled Bernstein polynomial $r(x)$ is developed and this is used to form a resultant matrix $s(M)$, where $s(x)$ is a scaled Bernstein polynomial. Computational evidence is presented that suggests that this method of computing the resultant of two Bernstein basis polynomials is superior to the established method of using a simple parameter substitution to perform a change from the Bernstein basis to the power basis.

§1. Introduction

There exist several different types of resultants, for example, the Sylvester, Bézout and Dixon resultants, and they may be considered theoretically equivalent because they all yield necessary and sufficient conditions for two polynomials to have a common root. It was realised in the 1980's that resultants can be applied to many problems in computer–aided geometric design, for example, the computation of the points of intersection of parametric curves, the processing of the curves of intersection of parametric surfaces, and ray tracing parametric patches [3,6,7].

Resultants were originally developed for power basis polynomials, and hence the direct use of the established theory of resultants for applications in computer–aided geometric design requires a polynomial basis transformation. However this transformation may be ill–conditioned [8,9,10], and regularisation may therefore be required to obtain a computationally reliable solution.

Mathematical Methods for Curves and Surfaces: Oslo 2000
Tom Lyche and Larry L. Schumaker (eds.), pp. 535–544.

ISBN 0-8265-1378-6

An alternative method that is widely used involves a parameter substitution [3], and this enables the entire theory of the resultant of power basis polynomials to be reproduced for Bernstein basis polynomials. This substitution is adequate for theoretical analysis or symbolic computations, but it cannot be used in a floating point environment because all computations are performed in the power basis, which is numerically inferior to the Bernstein basis. It is therefore desirable to develop a resultant matrix for two Bernstein basis polynomials directly, without reference to the power basis. This desire is relaxed slightly because the scaled Bernstein basis, whose basis functions for polynomials of degree n are $(1-x)^{n-i}x^i, i = 0, \ldots, n$, is used. However the numerical results in Section 4 show that the answers that are obtained with this basis are numerically superior to those that are obtained with the parameter substitution.

A review of previous work is considered in Section 2, and a measure of the numerical condition of a resultant matrix is developed in Section 3. This is used in Section 4 to compare the numerical condition of the resultant matrix of several pairs of scaled Bernstein polynomials with the numerical condition of the resultant matrix after the parameter substitution is used to transform the polynomials to their power basis forms. Section 5 contains a discussion of the results.

§2. Previous Work

The parameter substitution

$$t = \frac{x}{1-x}, \quad x \neq 1, \tag{1}$$

transforms the Bernstein polynomial

$$p(x) = \sum_{i=0}^{n} a_i \binom{n}{i} (1-x)^{n-i} x^i, \tag{2}$$

to

$$q(t) = (1+t)^n p\left(\frac{t}{1+t}\right) = \sum_{i=0}^{n} c_i t^i, \quad c_i = a_i \binom{n}{i}, \quad t \neq -1,$$

which is a power basis polynomial, and thus the established resultant matrices for power basis polynomials can be employed. The disadvantages of this method were noted in Section 1, and a numerically superior method of constructing a resultant matrix for scaled Bernstein polynomials is now considered.

The scaled Bernstein form of the polynomial (2) is

$$p(x) = \sum_{i=0}^{n} b_i (1-x)^{n-i} x^i, \quad b_i = a_i \binom{n}{i}, \tag{3}$$

and it is shown in [11] that if

$$A = \begin{bmatrix} 0 & 1 & 0 & 0 & . & 0 & 0 \\ 0 & 0 & 1 & 0 & . & 0 & 0 \\ . & . & . & . & . & . & . \\ 0 & 0 & 0 & 0 & . & 0 & 1 \\ -b_0 & -b_1 & -b_2 & -b_3 & . & -b_{n-2} & -b_{n-1} \end{bmatrix},$$

and $E = I + A$, then

$$\det\left(A - \lambda E\right) = (-1)^n \sum_{i=0}^{n} b_i \left(1-\lambda\right)^{n-i} \lambda^i, \quad b_n = 1.$$

It follows that

$$M = E^{-1}A = \left(I + A\right)^{-1} A, \quad \det E \neq 0, \tag{4}$$

is a companion matrix for the scaled Bernstein polynomial (3). A closed form expression for the elements $m_{ij}, i, j = 1, \ldots, n$, of M is developed in [11]. The general implication of the condition $b_n = 1$ is that $b_n \neq 0$, and thus it follows that $(1 - x)$ must not be a divisor of $p(x)$. If $b_n = 0$, then a polynomial of degree $n - 1$ is considered by removing the factor $(1 - x)$. It is noted that a similar situation arises when the parameter subsitution (1) is used because this substitution is not valid at $x = 1$.

It is shown in [11] that the companion matrix (4) allows a resultant matrix for two scaled Bernstein polynomials to be developed. In particular, let

$$r(x) = \sum_{j=0}^{n} r_j \left(1-x\right)^{n-j} x^j \quad \text{and} \quad s(x) = \sum_{j=0}^{m} s_j \left(1-x\right)^{m-j} x^j, \tag{5}$$

be two scaled Bernstein polynomials with coefficients $\left\{r_j\right\}_{j=0}^{n}, r_n = 1$, and $\left\{s_j\right\}_{j=0}^{m}$, respectively. Consider the matrix polynomial $s(M)$,

$$s(M) = \sum_{j=0}^{m} s_j \left(I - M\right)^{m-j} M^j,$$

where M is the companion matrix (4) for $r(x)$, and let $\left\{\lambda_i, x_i\right\}_{i=1}^{n}$ be the eigenpairs of M. It is shown in [11] that

$$\det\left(s(M)\right) = \prod_{i=1}^{n} s\left(\lambda_i\right),$$

and thus the determinant of $s(M)$ is equal to zero if and only if λ_i is a root of $s(x)$. Since the eigenvalues $\left\{\lambda_i\right\}_{i=1}^{n}$ are the roots of $r(x)$, it follows that $s(M)$ is a resultant matrix for the polynomials $r(x)$ and $s(x)$. The degree and coefficients of the greatest common divisor of $r(x)$ and $s(x)$ are obtained from $s(M)$ using the following theorem [2,3].

Theorem 2.1. *Let $w(x)$ be the greatest common divisor of $s(x)$ and $r(x)$. Then*

1) *The degree of $w(x)$ is equal to $n - \text{rank}\, s(M)$ where M is the companion matrix of $r(x)$.*
2) *The coefficients of $w(x)$ are proportional to the last row of $s(M)$ after it has been reduced to row echelon form.*

§3. The Numerical Condition of a Resultant Matrix

This section considers a measure to quantify the numerical stability of a resultant matrix. It will be used in Section 4 to compare the numerical condition of $s(M)$ and the resultant matrix that is obtained by using the parameter substitution (1).

It follows from Theorem 2.1 that the accuracy of the computation of the degree of the greatest common divisor of $r(x)$ and $s(x)$ is determined by the accuracy of the computation of the rank of $s(M)$, and thus the numerical condition of $s(M)$ should be proportional to the reciprocal of the distance to singularity (loss of unit rank). This is an intuitively appealing definition because as the distance from singularity increases, the magnitude of the perturbations in $s(M)$ that are required to cause a unit loss of rank increase, and thus the matrix becomes better conditioned. This measure of condition must be considered with care because it can be defined in both a componentwise and normwise manner, and although a componentwise measure yields a more refined analysis because it makes full use of all the entries of $s(M)$, but the normwise measure only makes use of $\|s(M)\|$, the normwise measure will be used because it is much easier to compute [5]. The following theorem [4] shows that the numerical rank deficiency of a matrix X is characterised by its singular value decomposition because the singular values define the distance between X and the nearest matrix of lower rank. Since $s(M)$ is always square, the theorem is restricted to square matrices.

Theorem 3.1. *Let $X \in \mathbb{R}^{n\times n}$ be of rank r, and let USV^T be its singular value decomposition. Let $X_k = US_kV^T, k < r$, where $S_k \in \mathbb{R}^{n\times n}$ is the diagonal matrix with elements*

$$[\sigma_1 \quad \sigma_2 \quad \cdots \quad \sigma_k \quad 0 \quad 0 \quad \cdots \quad 0], \quad \sigma_1 \geq \sigma_2 \geq \cdots \geq \sigma_k > 0.$$

Then the rank of X_k is k, and

$$\sigma_{k+1} = \|X - X_k\|_2 = \min_{\text{rank } Y = k} \|X - Y\|_2 \,.$$

This theorem states that of all matrices Y of rank k, X_k is the closest to X, measured in the normwise sense, and that the distance between them is σ_{k+1}.

The determination of the numerical condition of $s(M)$ requires that only a unit loss of rank be considered, and thus using the notation of Theorem 3.1,

$k = r - 1$. The minimum normwise perturbation in $s(M)$ that is required to cause this loss of rank is therefore

$$\sigma_r = \left\| s(M)^+ \right\|_2^{-1},$$

where

$$s(M)^+ = V\text{diag}\,[\sigma_1^{-1} \quad \sigma_2^{-1} \quad \cdots \quad \sigma_r^{-1} \quad 0 \quad 0 \quad \cdots \quad 0\,]\,U^T,$$

is the pseudo–inverse of $s(M)$.

Let $\tilde{s}(A)$ be the resultant matrix of the polynomials (5) when the parameter substitution (1) is used to transform them to the power basis,

$$\tilde{s}(A) = \sum_{i=0}^{m} s_i A^i,$$

where

$$A = \begin{bmatrix} 0 & 1 & 0 & 0 & . & 0 & 0 \\ 0 & 0 & 1 & 0 & . & 0 & 0 \\ . & . & . & . & . & . & . \\ 0 & 0 & 0 & 0 & . & 0 & 1 \\ -r_0 & -r_1 & -r_2 & -r_3 & . & -r_{n-2} & -r_{n-1} \end{bmatrix},$$

which is a companion matrix for the monic power basis polynomial [1]

$$\tilde{r}(x) = x^n + r_{n-1}x^{n-1} + \cdots + r_1 x + r_0.$$

The proof that $\tilde{s}(A)$ is a resultant matrix for two power basis polynomials follows identically the equivalent result for $s(M)$ in Section 2. It is important to emphasize the difference between $s(M)$ and $\tilde{s}(A)$ because both are resultant matrices. Specifically, $s(M)$ is a scaled Bernstein polynomial in the companion matrix M of a scaled Bernstein polynomial, and $\tilde{s}(A)$ is a power basis polynomial in the companion matrix A of a power basis polynomial.

The ratio of the smallest non–zero singular value of $s(M)$ to the smallest non–zero singular value of $\tilde{s}(A)$ is not an adequate measure of the ratio of the distances to singularity because each matrix can be scaled arbitrarily by scaling the coefficients of $s(x)$. It is recalled that $r(x)$ is a monic polynomial, and thus M and A are normalised. Effective comparison therefore requires that a normalisation be imposed, and this can be achieved by normalising $s(M)$ and $\tilde{s}(A)$ with respect to $\|s(M)\|_2$ and $\|\tilde{s}(A)\|_2$, respectively, in which case the ratio of the distance to singularity of $\tilde{s}(A)$ to the distance to singularity of $s(M)$ is

$$d(\tilde{s}(A), s(M)) = \frac{\frac{\|\tilde{s}(A)^+\|_2^{-1}}{\|\tilde{s}(A)\|_2}}{\frac{\|s(M)^+\|_2^{-1}}{\|s(M)\|_2}} = \frac{\|s(M)\|_2\,\|s(M)^+\|_2}{\|\tilde{s}(A)\|_2\,\|\tilde{s}(A)^+\|_2}. \tag{6}$$

If $r(x)$ and $s(x)$ are coprime, then $s(M)$ and $\tilde{s}(A)$ are non–singular and $d(\tilde{s}(A), s(M)) = \kappa_2(s(M))/\kappa_2(\tilde{s}(A))$, where $\kappa_2(R)$ is the 2–norm condition number of R, that is, the ratio of their distances to singularity is equal to the reciprocal of the ratio of their condition numbers.

Lower and upper bounds for the ratio $d\,(\tilde{s}(A), s(M))$ are easily obtained for the case in which $\tilde{s}(A)$ and $s(M)$ are non–singular. This restriction is required because $(XY)^+ = Y^+X^+$ only if X is of order $m \times n$, Y is of order $n \times k$ and rank X = rank $Y = n$. This condition implies that it is not easy to obtain lower and upper bounds of $d(\tilde{s}(A), s(M))$ for resultant matrices of arbitrary rank.

It is shown in [11] that

$$s(M) = (I - M)^m\, \tilde{s}(A),$$

and hence from (6),

$$d(\tilde{s}(A), s(M)) = \frac{\kappa_2(s(M))}{\kappa_2(\tilde{s}(A))} \leq \kappa_2\left((I - M)^m\right),$$

since it is assumed that $s(M)$ and $\tilde{s}(A)$ are non–singular. It follows from (4) that $I - M = (I + A)^{-1}$, and thus the upper bound of $d(\tilde{s}(A), s(M))$ is

$$d(\tilde{s}(A), s(M)) \leq \kappa_2\left((I - M)^m\right) = \kappa_2\left((I + A)^{-m}\right).$$

The lower bound of $d(\tilde{s}(A), s(M))$,

$$d(\tilde{s}(A), s(M)) \geq \frac{1}{\kappa_2\left((I - M)^m\right)} = \frac{1}{\kappa_2\left((I + A)^{-m}\right)},$$

is obtained in a similar manner.

§4. Examples

This section contains several examples that compare the numerical condition (and therefore distance to singularity) of $s(M)$ and $\tilde{s}(A)$.

Example 4.1. The resultant matrices $s(M)$ and $\tilde{s}(A)$ of several polynomials $s(x)$ with the truncated Wilkinson polynomial

$$r(x) = \prod_{i=1}^{19}\left(x - \frac{i}{20}\right) = \sum_{i=0}^{19} r_i\,(1 - x)^{19-i}\,x^i, \tag{7}$$

were computed and the ratio of the distances to singularity $d(\tilde{s}(A), s(M))$ was calculated. The upper indices in (7) are 19 and not their usual value of 20 because the root $x_0 = 1$ must be excluded, as noted in Section 2. The results of the computational experiments are shown in Table 1 and it is seen that

$s(M)$ is better conditioned than $\bar{s}(A)$ in 13 out of the 15 tests because it is further away from singularity. The ratio of the distances may be several orders of magnitude, and this substantial improvement may also be obtained when $s(x)$ has a multiple root. Experiments 1–4 show that as the degree k of the polynomial $(x - 0.50)^k$ increases, the ratio $d(\bar{s}(A), s(M))$ of the distances to singularity decreases and thus $s(M)$ displays significantly better performance than $\bar{s}(A)$ in this situation. The degree of the greatest common divisor of the polynomials in experiment 5 is 3, and the factor $(x - 0.05)$ is changed slightly to $(x - 0.0501)$ in experiment 6, thus reducing the degree of the greatest common divisor to 2. It is seen that $s(M)$ is better conditioned than $\bar{s}(A)$ by several orders of magnitude for both experiments.

Experiments 7–15 show that as a root of the polynomial $s(x)$ approaches one and increases in multiplicity, the improved numerical condition of $s(M)$ with respect to $\bar{s}(A)$ is less marked. Detailed examination of the results showed that if $d(\bar{s}(A), s(M)) \approx O(1)$, then both $\bar{s}(A)$ and $s(M)$ are approximately equally ill–conditioned,

$$\|s(M)\|_2 \left\|s(M)^+\right\|_2 \approx \|\bar{s}(A)\|_2 \left\|\bar{s}(A)^+\right\|_2 \approx 10^{12}.$$

It is interesting to note that

$$\|s(M)\|_2 \left\|s(M)^+\right\|_2 = 13.5 \qquad \text{and} \qquad \|\bar{s}(A)\|_2 \left\|\bar{s}(A)^+\right\|_2 = 9.39 \times 10^5,$$

in experiment 11, and thus $s(M)$ is well–conditioned (distant from singularity) but $\bar{s}(A)$ is ill–conditioned, even though the common divisor $(x-0.95)$ is 'near' $(x-1)$. Furthermore, experiments 1 and 11 suggest that $s(M)$ is significantly better conditioned than $\bar{s}(A)$, by several orders of magnitude, if the greatest common divisor has degree one.

Expt.	Polynomial $s(x)$	$d(\bar{s}(A), s(M))$
1	$(x - 0.50)$	1.42×10^{-6}
2	$(x - 0.50)^2$	4.53×10^{-7}
3	$(x - 0.50)^3$	1.43×10^{-7}
4	$(x - 0.50)^4$	5.62×10^{-8}
5	$(x - 0.05)(x - 0.10)(x - 0.95)$	1.19×10^{-6}
6	$(x - 0.0501)(x - 0.10)(x - 0.95)$	1.75×10^{-7}
7	$(x - 0.85)^2(x - 0.90)^2$	1.09×10^{-2}
8	$(x - 0.90)^4$	1.10×10^{-1}
9	$(x - 0.95)^2(x - 0.99)^2$	3.62×10^{1}
10	$(x - 0.05)^3(x - 0.95)$	3.25×10^{-8}
11	$(x - 0.95)$	1.44×10^{-5}
12	$(x - 0.95)^2$	1.85×10^{-2}
13	$(x - 0.95)^3$	5.34×10^{-1}
14	$(x - 0.95)^4$	2.60×10^{1}
15	$(x - 0.80)^2(x - 0.85)^2(x - 0.90)^2$	3.52×10^{-2}

Tab. 1. The ratio $d(\bar{s}(A), s(M))$ for Example 4.1.

Example 4.2. The resultant matrices $s(M)$ and $\tilde{s}(A)$ of the truncated Wilkinson polynomial (7) and the polynomial

$$s(x) = \prod_{i=2}^{10}\left(x - \frac{i}{20}\right) = \sum_{i=0}^{9} s_i \,(1-x)^{9-i}\, x^i,$$

were calculated. The polynomials have 9 common roots and thus both $\tilde{s}(A)$ and $s(M)$ are of rank 10. The singular values of $s(M)$ are, to within an order of magnitude,

$$\left(10^3\ 10^1\ 10^0\ 10^{-1}\ 10^{-2}\ 10^{-3}\ 10^{-3}\ 10^{-4}\ 10^{-5}\ 10^{-6}\ 10^{-15}\ 10^{-15}\ \cdots\right).$$

The singular values continue to decrease steadily and the numerical rank of $s(M)$ is equal to 10, the correct value. However the singular values of $\tilde{s}(A)$ are, also to within an order of magnitude,

$$\left(10^{17}\ 10^{11}\ 10^8\ 10^5\ 10^2\ 10^0\ 10^0\ 10^{-1}\ 10^{-2}\ 10^{-3}\ 10^{-3}\ 10^{-4}\ \cdots\right),$$

and they continue to decrease steadily. Clearly, $\tilde{s}(A)$ cannot be used to compute the greatest common divisor of the polynomials because the numerical rank is not defined.

Example 4.3. The ratio of the distances to singularity of the resultant matrices $s(M)$ and $\tilde{s}(A)$ for several pairs of polynomials $r(x)$ and $s(x)$ was calculated, and the results are shown in Table 2.

	Polynomial $r(x)$	Polynomial $s(x)$	$d(\tilde{s}(A), s(M))$
1	$(x-0.10)(x-0.20)^3$ $\times(x-0.21)^2$	$(x-0.10)(x-0.20)^3$	1.04×10^1
2	$(x-0.90)(x-0.97)$ $\times(x-0.98)(x-1.05)^2$	$(x-0.97)(x-0.98)$ $\times(x-1.05)$	1.80×10^{-2}
3	$(x-0.10)^2(x-0.15)^2$ $\times(x-0.20)^2$	$(x-0.10)(x-0.15)$ $\times(x-0.20)$	2.30×10^1
4	$(x-0.10)^3(x-0.15)^2$ $\times(x-0.20)$	$(x-0.10)(x-0.15)$ $\times(x-0.20)$	2.60×10^1
5	$(x-0.10)(x-0.20)$ $\times(x-0.30)(x-0.40)$ $\times(x-0.50)(x-0.60)$	$(x-0.105)(x-0.205)$ $\times(x-0.305)(x-0.405)$ $\times(x-0.505)(x-0.605)$	3.46×10^{-1}
6	$(x-0.80)(x-0.81)$ $\times(x-0.82)(x-0.83)$ $\times(x-0.84)(x-0.85)$	$(x-0.80)(x-0.82)$ $\times(x-0.84)$	2.64×10^{-3}

Tab. 2. The ratio $d(\tilde{s}(A), s(M))$ for Example 4.3.

The maximum degree of the polynomials is six, and it is seen that the numerical superiority of $s(M)$ over $\tilde{s}(A)$ is less marked than in Example 4.1. Although all the roots in experiment 2 in Table 2 lie near $x_0 = 1$,

$$\|s(M)\|_2 \left\|s(M)^+\right\|_2 = 10.7 \quad \text{and} \quad \|\tilde{s}(A)\|_2 \left\|\tilde{s}(A)^+\right\|_2 = 5.93\times 10^2,$$

and thus $s(M)$ is much better conditioned than $\tilde{s}(A)$. This result is consistent with experiment 6 because

$$\|s(M)\|_2 \left\|s(M)^+\right\|_2 = 54.3 \quad \text{and} \quad \|\tilde{s}(A)\|_2 \left\|\tilde{s}(A)^+\right\|_2 = 2.05 \times 10^4,$$

for this example. However in experiment 3, $\tilde{s}(A)$ is better conditioned than $s(M)$ because

$$\|s(M)\|_2 \left\|s(M)^+\right\|_2 = 50.3 \quad \text{and} \quad \|\tilde{s}(A)\|_2 \left\|\tilde{s}(A)^+\right\|_2 = 2.19,$$

but $s(M)$ is not ill–conditioned. Identical comments are appropriate for experiment 4.

Examples 4.1 and 4.3 show that $s(M)$ is, on average, better conditioned than $\tilde{s}(A)$. Although there exist situations for which $d(\tilde{s}(A), s(M)) = \alpha > 1$, Tables 1 and 2 show that $\alpha < 40$ for the specified polynomials $r(x)$ and $s(x)$. In this circumstance, either both $s(M)$ and $\tilde{s}(A)$ are well–conditioned, in which case there is little difference in their numerical behaviour, or they are both ill–conditioned, such that the determination of the greatest common divisor may be fundamentally ill–conditioned. However the results also show that there are many polynomials for which $d(\tilde{s}(A), s(M)) \ll 1$, and thus $s(M)$ is further away from singularity than $\tilde{s}(A)$ by several orders of magnitude. This improved numerical conditioning is associated with a greater accuracy in the computation of the coefficients of the greatest common divisor of $r(x)$ and $s(x)$. Although the companion matrix M is more complex than its power basis equivalent A, this computational cost is justified by the superior numerical performance of $s(M)$ with respect to $\tilde{s}(A)$.

The criterion for numerical stability is based on the minimum normwise perturbation that is required to achieve a unit loss of rank. This requires that there be a sharp distinction between the large and small singular values, so that the numerical rank of the resultant matrix is well–defined. This large decrease in magnitude was observed in the spectrum of the singular values of $s(M)$, and the distinction was in general clearer for $s(M)$ than for $\tilde{s}(A)$. This improved definition of the numerical rank is clearly demonstrated in Example 4.2, for which $s(M)$ has the correct rank but $\tilde{s}(A)$ fails to yield the correct result.

§5. Discussion

This paper has been motivated by the applications of resultants in geometric modelling, but the examples have been drawn from their original use in classical algebra. This approach was adopted in order to determine the simplicity (or complexity) of the use of resultants for scaled Bernstein basis polynomials for an application that is well understood. However, the close connection between Theorem 3.1 and its equivalent for the power basis suggests that other resultants (Sylvester, Bézout, etc.) should be considered with specific reference to applications in geometric modelling. In particular, it has been shown that the polynomial basis may have a significant effect on the numerical condition of a resultant matrix, and thus the extensions of these other resultants to their Bernstein forms, such that the parameter substitution (1) is not required, may yield further improvements in the computational implementation of resultants.

References

1. Barnett, S., *Polynomials and Linear Control Systems*, Marcel Dekker, New York, USA, 1983.
2. De Montaudouin, Y. and W. Tiller, The Cayley method in computer aided geometric design, Comput. Aided Geom. Design **1** (1984), 309–326.
3. Goldman, R. N., T. W. Sederberg, and D. C. Anderson, Vector elimination : A technique for the implicitization, inversion and intersection of planar parametric rational polynomial curves, Comput. Aided Geom. Design **1** (1984), 327–356.
4. Golub, G. H. and C. F. Van Loan, *Matrix Computations*, John Hopkins University Press, Baltimore, 1989.
5. Higham, N. J., *Accuracy and Stability of Numerical Algorithms*, SIAM, Philadelphia, 1996.
6. Kajiya, J. T., Ray tracing parametric patches, Computer Graphics **16** (1982), 245–254.
7. Sederberg, T. W., D. C. Anderson and R. N. Goldman, Implicit representation of parametric curves and surfaces, Computer Vision, Graphics and Image Processing **28** (1984), 72–84.
8. Winkler, J. R., Polynomial basis conversion made stable by truncated singular value decomposition, Appl. Math. Modelling **21** (1997), 557–568.
9. Winkler, J. R., Tikhonov regularisation in standard form for polynomial basis conversion, Appl. Math. Modelling **21** (1997), 651–662.
10. Winkler, J. R., An ill–conditioned problem in computer aided geometric design, Neural, Parallel and Scientific Computations **5** (1997), 179–200.
11. Winkler, J. R., A resultant matrix for scaled Bernstein polynomials, Linear Algebra and Its Applications **319** (2000), 179–191.

Joab R. Winkler
The University of Sheffield
Department of Computer Science
Regent Court
211 Portobello Street
Sheffield S1 4DP
UK
`j.winkler@dcs.shef.ac.uk`
`http://www.dcs.shef.ac.uk/~joab/`

Deformation With Hierarchical B-Splines

Zhiyong Xie and Gerald E. Farin

Abstract. Deformation has received wide attention in the field of Computer Graphics, Computer Vision, and Image Processing. This paper describes a new method which is based on the concept of hierarchical B-Splines. Compared with existing methods, our approach is adaptive in that it puts more effort in complex regions which need more attention. It also provides an elegent tool for curve or surface matching together with the Iterative Closest Point method.

§1. Introduction

The term "deformation" describes methods by which the shape of a 2D or 3D object is transformed into another shape. It has a variety of applications in animation, target recognition, image processing and medical image analysis [12,21,4,11]. In this research, deformation is used for image registration, which matches two image data sets (2D or 3D) taken at different time or from different sensors. For example, in brain studies, it is helpful to integrate MRI and PET data to display functional information and spatially detailed anatomical information together. It is also used to remove the position, orientation, and shape difference between two data sets such that a comparison can be carried out [6,16].

Over the last two decades, several deformation methods were developed. One class of these methods is point-based; such methods first find corresponding point features and construct the deformation by matching these discrete points. Another popular method is boundary-based. Rather than matching discrete correspondence points, it matches the boundaries of two objects.

In this paper, we propose a deformation method that is based on the concept of hierarchical B-Splines. It is adaptive in that it puts more effort in complex regions which need more attention. This method is computationally efficient and "well-behaved" when applied to point-based deformation. In addition, this coarse-to-fine strategy provides an elegant tool together with Iterative Closest Point (ICP), which is a popular method for curve or surface matching. By combing the hierarchical B-Spline with the ICP method, we

Mathematical Methods for Curves and Surfaces: Oslo 2000
Tom Lyche and Larry L. Schumaker (eds.), pp. 545–554.

ISBN 0-8265-1378-6

can match coarse shape features first, and then gradually refine them for more detail. This is difficult with conventional methods.

This paper is organized as follows. Section 2 focuses on the point-based deformation and evaluation of hierarchical B-Splines. An overview of previous work is given and our new method is proposed. In Section 3, we will discuss the basic idea of the ICP method, some problems and possible solutions using hierarchical B-Splines. Conclusions are given in Section 4.

§2. Point-Based Method

In studying diseases such as Alzheimer's, brains of patients and control groups are compared using MRI or PET imaging. If we are given two MRI brain "slices", a common comparison of anatomical features is carried out by subtracting the images (pixel intensities) and then studying the difference image. This method will produce false difference readings if the two brains were not aligned properly. But even after alignment (using linear or affine maps), corresponding major anatomical structures may have different shapes. Only after deforming these major structures onto each other, can a proper comparison of substructures such as the hippocampus be carried out.

The most straightforward deformation method is to identify points on the brain to be deformed (the source) and corresponding points on the second brain (the target). This problem can be solved by scattered data interpolation (or approximation). According to [18], this problem can be formulated as follows:

Point-Based Deformation Problem:

Input: L pairs of source points $\mathbf{p}_i$ and corresponding target points $\mathbf{q}_i$, $i = 1, \ldots, L$.

Output: A C^0 function $\boldsymbol{f} : \mathcal{R}^2 \to \mathcal{R}^2$ with $\boldsymbol{f}(\boldsymbol{p}_i) = \boldsymbol{q}_i$, $i = 1, \ldots, L$.

This problem naturally breaks down into two scattered data interpolation problems: one for the pairs $(\mathbf{p}_i, q_i^x)$ and one for the pairs $(\mathbf{p}_i, q_i^y)$, with q_i^x and q_i^y being the $x-$ and $y-$ components of the target points.

Many scattered data interpolation methods may be applied, such as Triangulation Based Methods, Distance Weighted Methods, or Radial Basis Functions, see [15,10,18,19,20].

2.1. Surface Fitting with B-Splines

If a large number of source and target points is given, then it seems natural to replace the above interpolation problem by an approximation problem. Our approach utilizes uniform bicubic B-splines, see [7] or [5].

A bicubic B-Spline deformation has the form:

$$\mathbf{f}(\mathbf{p}) = \sum_{i=0}^{m} \sum_{j=0}^{n} \mathbf{b}_{ij} N_i^3(p^x) N_j^3(p^y),$$

It is a map of $\mathcal{R}^2$ to $\mathcal{R}^2$ and may be viewed as the action of distorting a rubber sheet.

Our approximation problem may be stated as an overdetermined linear system of the form

$$\mathbf{q}_r = \sum_{i=0}^{m}\sum_{j=0}^{n} \mathbf{b}_{ij} N_i^3(p_r^x) N_j^3(p_r^y); \quad r = 1, \ldots, L.$$

The unknowns are the control points $\mathbf{b}_{ij}$; the system is overdetermined if $L > (m+1)(n+1)$. We set

$$\mathbf{Q} = (\mathbf{q}_0, \ldots, \mathbf{q}_L)^T$$

$$\mathbf{B} = (b_{0,0}, \ldots, b_{0,n}, \ldots, b_{m,0}, \ldots, b_{m,n})^T$$

$$N = \begin{pmatrix} N_0^3(x_0)N_0^3(y_0) & \ldots & N_m^3(x_0)N_n^3(y_0) \\ \vdots & \ddots & \vdots \\ N_0^3(x_v)N_0^3(y_v) & \ldots & N_m^3(x_v)N_n^3(y_v) \end{pmatrix},$$

and thus have to find the least squares solution of

$$N\mathbf{B} = \mathbf{Q}. \qquad (1).$$

We also use a shape constraint to obtain a "wiggle-free" transformation. The constraint minimizes $\frac{\partial^2 s}{\partial x \partial y}$ during the evaluation of each B-Spline function. We use

$$\Delta^{1,1}\mathbf{b}_{i,j} = (\mathbf{b}_{i+1,j+1} - \mathbf{b}_{i+1,j}) - (\mathbf{b}_{i,j+1} - \mathbf{b}_{i,j}) = min$$

as constraints. This constraint forces the control polygon to be close to a parallelogram mesh [7]. Another widely used fairness function in B-Spline surface fitting is minimizing the thin plate energy functional

$$\iint ((\frac{\partial^2 s}{\partial x^2})^2 + 2(\frac{\partial^2 s}{\partial x \partial y})^2 + (\frac{\partial^2 s}{\partial y^2})^2)\, dx\, dy.$$

Adding our shape constraints, we have a new linear system which is still overdetermined. Our constraint can be written as

$$\mathbf{CB} = \mathbf{O},$$

where $\mathbf{C}$ is an $mn \times mn$ matrix. Each row of $\mathbf{C}$ has four non-zero elements $(1, -1, -1, 1)$. $\mathbf{O}$ is a vector with mn zero elements. We shall find $\mathbf{B}$ such that

$$\mathbf{AB} = \mathbf{D}, \qquad (2)$$

where $\mathbf{A} = (\mathbf{N}, \mathbf{C})^T$, $\mathbf{D} = (\mathbf{Z}, \mathbf{O})^T$

$\mathbf{A}$ is an $(L + mn) \times mn$ matrix. The least squares solution of system (2) can be found by solving the linear system

$$\mathbf{A}^T\mathbf{AB} = \mathbf{A}^T\mathbf{D},$$

where $\mathbf{A}^T\mathbf{A}$ is a $mn \times mn$ symmetric matrix [13].

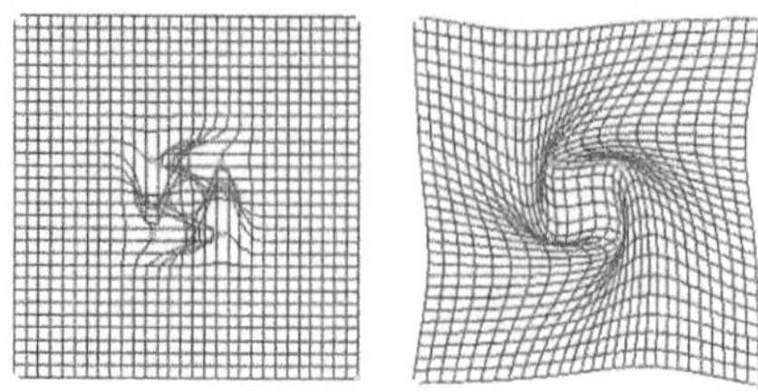

Fig. 1. Local deformation(left) and multilevel deformation(right).

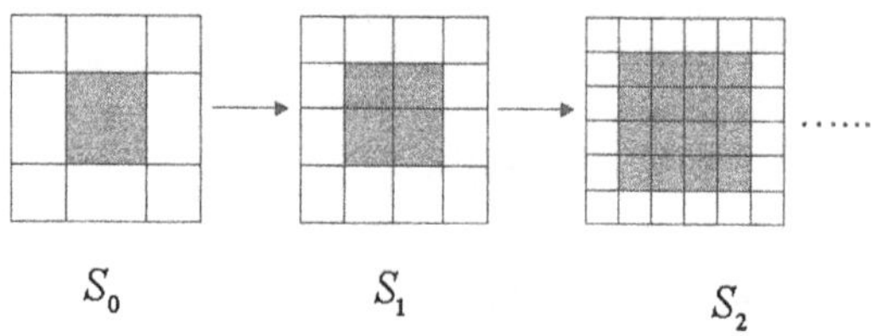

Fig. 2. Coarse to fine deformation. Gray regions represent the domain.

2.2. Surface Fitting with Hierarchical B-Splines

When applied to large point sets, the above methods can be expensive. Also, a tradeoff exists between the smoothness and precision of the deformation. If we use a small number of polynomial segments (small m, n), we may only achieve rough matching. If we use larger numbers, foldovers may occur.

Referring to the example shown in Figure 1, eight source points are mapped to eight target points. The source points are the vertices of the square containing the grid; they are left unchanged. The four vertices of a square in the center of the source are rotated by 90 degrees.

The left panel shows the result of the corresponding deformation using 11×11 spline segments. Notice the pronounced foldovers. This problem is avoided by the use of *multilevel B-Splines*, see Lee and Wolberg [14]. They presented an algorithm to fit a uniform bicubic B-Spline surface to scattered data. The basic idea of multilevel B-Splines is outlined in Figure 2. A 4×4 bicubic B-Spline is used for an initial deformation. Next, the differences between the target points and the deformed source points are computed. A 5×5 B-Spline is used to fit these differences. We can use it to refine the initial deformation and repeat this procedure until the deviation of any deformed source point from the target is small.

This method provides an acceptable result for deformation (see right panel of Figure 1). But it is computationally involved since at each level a uniform bicubic grid is used. This necessitates a change in source point coordinates, and causes the deformation to be influenced by the relative positioning between the data points and lattices [14].

Forsey and Bartels developed a surface fitting method which is adaptive based on hierarchical spline functions [9]. However, this method cannot be

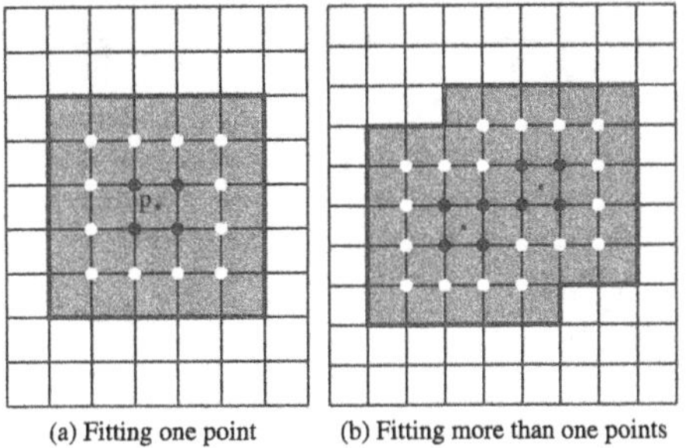

Fig. 3. Hierarchical B-Spline.

applied to our problem directly since it only deals with gridded data (our source and target point pairs do not come on grids).

For our approach, suppose that we have found an initial deformation to our source and data pairs, using a bicubic B-spline deformation and 4×4 spline segments. We now check for each source point $\mathbf{p}$ how closely it is mapped to its target $\mathbf{q}$. If the deviation is too large, the existing deformation is changed as follows:

1) The B-spline grid is locally refined by the process of knot insertion.
2) The four control points in the refined grid which are closest to $\mathbf{p}$ are selected and recomputed such that the transformed point $\mathbf{p}$ matches its target exactly.

With this method, only very few new control points have to be recomputed, see Figure 3. In Figure 3a, only one point $\mathbf{p}$ has to be changed, and hence only four control points have to be recomputed (shown in black). In Figure 3b, two points are recomputed, resulting in recomputation of eight control points (black). The white control points do not have to be changed.

§3. Boundary-based Method

In practice, it is difficult to produce a large number of source/target point pairs. By contrast, boundary features are much easier identified and paired. In 2D applications, boundaries typically consist of contour polygons (such as the outline of a skull). With a boundary-based method, we just need to know the correspondence between boundaries. We do not need to know the correspondence between individual points. Thus one important issue is how to match two boundaries correctly, i.e, how to match every part of the source boundary to the correct part of the target boundary.

3.1. Iterative Closest Point

In curve or surface matching, a popular method is Iterative Closest Point (ICP) [1,3,8,17]. With the ICP model, every point on the source curve will be

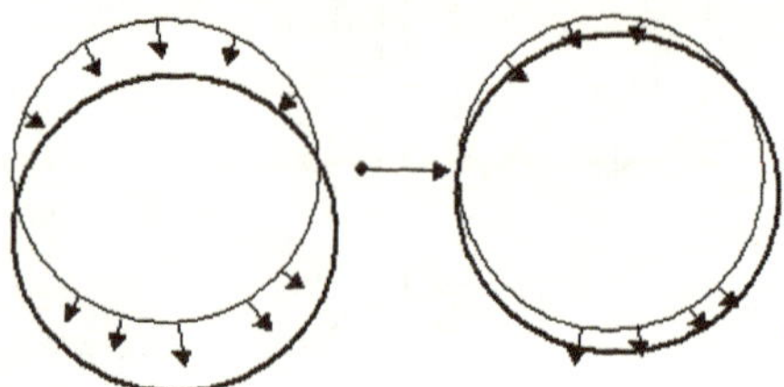

Fig. 4. ICP model.

pulled or pushed to the closest point on the target curve iteratively. Figure 4 gives a basic idea of the ICP method.

The distance transform provides an elegant method for computing the path from a source boundary point to the closest target boundary point. The distance transform assigns to each boundary pixel the value zero. Each non-boundary pixel will be given a value that is the distance to the nearest boundary pixel. Borgefors [3,2] gave an efficient method to get the distance image. With this pre-computed image, we can calculate the distance of any point from its closest point by bilinear interpolation, and find the direction toward the closest point by computing the gradient of that point in the distance image.

The ICP algorithm can be summarized as follows:

1) In iteration k, to each contour point (x_i^k, y_i^k), compute the displacement $(\Delta x_i^k, \Delta y_i^k)$.
2) Compute the deformation functions f, g.
3) Apply the deformation to each contour point (x_i^k, y_i^k) to get new a point set (x_i^{k+1}, y_i^{k+1}), where $(x_i^{k+1} = x_i^k + f(x_i^k, y_i^k), y_i^{k+1} = y_i^k + g(x_i^k, y_i^k))$.
4) Terminate the iteration according to a termination criterion.

   Several possible termination criteria were given in [8]:

   a) The distance between the two object contours is below a fixed threshold.
   b) The variation of the distance between the two data sets at two successive iterations is below a fixed threshold.
   c) A maximum number of iterations is reached.

3.2. Hierarchical B-Splines in ICP

A drawback of the ICP method is that it is hard to find a suitable deformation function. Another is the fact that a point on the source boundary may have several closest points on the target boundary. The use of hierarchical B-splines can help overcome these problems. We use a global affine transformation first, and then hierarchical B-Splines for local refinement. We can increase m, n gradually in hierarchical B-Splines. Referring to Figure 2, first we can use a global 4×4 control polygon for an initial deformation. Then we use successive B-Spline refinements until we get a satisfactory result.

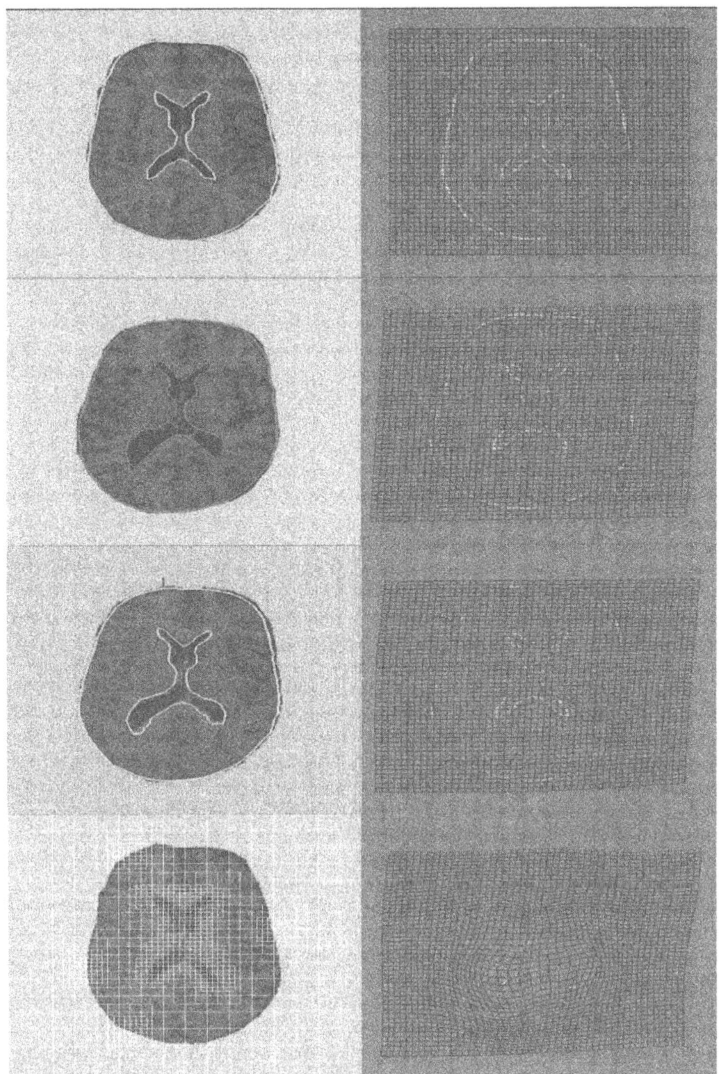

Fig. 5. An example of hierarchical brain deformation.

The basic algorithm can be outlined as follows:

1) Let $i = 0$, Each component of the deformation function $\boldsymbol{f}$ is a 4×4 B-Spline surface. Initial values of every control point are set to zero.
2) Let $m = 2^i + 3, n = 2^i + 3$, where i denotes the level in the hierarchy. Put an $m \times n$ knot lattice to overlay the image to be deformed.
3) Compute control points of new level and apply the deformation to the data set. Add these control point values to $\boldsymbol{f}$.

4) Checking termination criterion b from last section. If it is not satisfied, go to step 3 and repeat this procedure.
5) Check the termination criteria a) and c) from Subsection 3.1. If neither is satisfied, let $i = i+1$, refine the control polygons of $\boldsymbol{f}$ as $(2^i+3)\times(2^i+3)$ control polygons. Go to step 2 and repeat this procedure. Otherwise, stop.

With B-Spline functions, Criterion b) is equivalent to making every control point of the deformation function be almost zero. With rigid or affine transformations, Criterion b can also be checked by checking transformation parameters. So this criterion can be checked quickly in each iteration.

§4. Conclusion

Hierarchical B-Splines provide an elegant tool for adaptive deformation. When used together with the ICP method, it has both validity (each point is ultimately mapped to the correct target point by moving points slowly and smoothly), precision (deforming one shape into another shape exactly), and efficiency (only evaluating in regions with large deviation). The resulting deformation is smooth and "well-behaved." Figure 5 shows a result of our method. In the left column, we show from top to bottom: source brain, target brain, source brain after deformation, hierarchical control lattice. The right column contains the deformed contours with hierarchical deformations. It takes about 20 seconds (using an SGI Onyx) to get the shape match shown.

Acknowledgments. This work was funded in part by the Arizona Alzheimer's Research Center and Grant MH57899 from the National Institute of Mental Health and the National Institute on Aging.

References

1. Besl, P. J. and N. D. Mckay, A method for registration of 3-D shapes, IEEE Transactions on Pattern Analysis and Machine Intelligence **14**(2) (1992), 239–255.
2. Borgefors, G., Distance transformation in digital images, Computer Vision, Graphics, and Image Processing **34** (1986), 344–371.
3. Borgefors, G., Hierarchical chamfer matching: A parametric edge matching algorithm, IEEE Transactions on Pattern Analysis and Machine Intellignece **10**(6) (1988), 849–865.
4. Brown, L. G., A survey of image registration techniques, ACM Computing Surveys **24**(4) (1992), 325–376.
5. Dierckx, P., *Curve and surface fitting with splines*, Oxford, 1993.
6. Duncan, J. S. and N. Ayache, Medical image analysis: Progress over Two decades and the challenges ahead, IEEE Transactions on Pattern Analysis and Machine Intellignece **22**(1) (2000), 85–106.

7. Farin, G., *Curves and Surfaces for CAGD*, Academic Press, New York, 1996.

8. Feldmar, J. and N. Ayache, Rigid, affine and locally affine registration of free-form surfaces, Technical Report 2220, Institut National de Recherche en Informatique et en Automatique, 1994, Electronic version: http://www.inria.fr/RRRT/RR-2220.html.

9. Forsey, D. R. and R. H. Bartels, Surface fitting with hierarchical splines, ACM Trans. Graphics **14**(2) (1995), 134–161.

10. Franke, R. and G. M. Nielson, Scattered data interpolation and applications: A tutorial and survey, in *Geometric Modeling*, H. Hagen and D. Roller (eds.), Springer-Verlag, New York, 1991, 131–160.

11. Gomes, J. and L. Velho, *Image Processing for Computer Graphics*, Springer-Verlag, New York 1997.

12. Gomes, J., L. Darsa, B. Costa, and L. Velho, *Warping and Morphing of Graphical Objects*, Morgan Kaufmann Publishers, Inc., 1999.

13. Lawson, L. and R. J. Hanson, *Solving Least Squares Problems*, Prentice-Hall, 1974.

14. Lee, S., G. Wolberg, and S. Y. Shin, Scattered data interpolation with multilevel B-Spline, Visualization & Computer Graphics **3**(3) (1997), 228–244.

15. Lodha, S. K. and R. Franke, Scattered data techniques for surfaces, *Proceedings of the Dagstuhl'97 - Scientific Visualization Conference*, Electronic version: http://computer.org/proceedings/dagstuhl/0503/05030181abs.htm, 1997.

16. Maintz, J. B. A. and M. A. Viergever, A survey of medical image registration, Medical Image Analysis **2**(1) (1998), 1–16.

17. Malandain, G., S. Fernandez-Vidal, and J. M. Rocchisani, Improving registration of 3-D medical images using a mechanical based method, in *Proceedings of the Third European Conference on Computer Vision (ECCV'94)*, Springer Verlag, New York, 1994, 131–136.

18. Mueller, H. and D. Ruprecht, Spatial interpolants for warping, in *Brain Warping*, W. Toga eds.), Academic Press, New York, 1998, 199–219.

19. Ruprecht, D. and H. Muller, Free form deformation with scattered data interpolation methods, in *Geometric Modeling*, G. Farin, H. Hagen and H. Noltemeier (eds.), Springer Verlag, New York, 1993, 267–281.

20. Ruprecht, D. and H. Muller, Image warping with scattered data interpolation, IEEE Computer Graphics and Applications **15**(2) (1995), 37–43.

21. Wolberg, G., *Digital Image Warping*, IEEE Computer Society Press, 1990.

Zhiyong Xie
Computer Science and Engineering
Arizona State University
Tempe, AZ 85287-5406, USA
zxie@asu.edu

Gerald E. Farin
Computer Science and Engineering
Arizona State University
Tempe, AZ 85287-5406
farin@asu.edu